T0214674

Lecture Notes in Computer Science 11118

Commenced Publication in 1973
Founding and Former Series Editors:
Gerhard Goos, Juris Hartmanis, and Jan van Leeuwen

Olga Galinina · Sergey Andreev
Sergey Balandin · Yevgeni Koucheryavy (Eds.)

Internet of Things, Smart Spaces, and Next Generation Networks and Systems

18th International Conference, NEW2AN 2018
and 11th Conference, ruSMART 2018
St. Petersburg, Russia, August 27–29, 2018
Proceedings

 Springer

Editors
Olga Galinina (iD)
Tampere University of Technology
Tampere
Finland

Sergey Andreev (iD)
Tampere University of Technology
Tampere
Finland

Sergey Balandin (iD)
FRUCT Oy
Helsinki
Finland

Yevgeni Koucheryavy (iD)
Tampere University of Technology
Tampere
Finland

ISSN 0302-9743   ISSN 1611-3349   (electronic)
Lecture Notes in Computer Science
ISBN 978-3-030-01167-3   ISBN 978-3-030-01168-0   (eBook)
https://doi.org/10.1007/978-3-030-01168-0

Library of Congress Control Number: 2018955422

LNCS Sublibrary: SL5 – Computer Communication Networks and Telecommunications

This Springer imprint is published by the registered company Springer Nature Switzerland AG
The registered company address is: Gewerbestrasse 11, 6330 Cham, Switzerland

Preface

We welcome you to the joint proceedings of the 18th NEW2AN (Next Generation Teletraffic and Wired/Wireless Advanced Networks and Systems) and 11th Conference on Internet of Things and Smart Spaces ruSMART (Are You Smart) held in St. Petersburg, Russia, during August 27–29, 2018.

Originally, the NEW2AN conference was launched by ITC (International Teletraffic Congress) in St. Petersburg in June 1993 as an ITC-Sponsored Regional International Teletraffic Seminar. The first edition was entitled "Traffic Management and Routing in SDH Networks" and held by R&D LONIIS. In 2002, the event received its current name, NEW2AN. In 2008, NEW2AN acquired a new companion in Smart Spaces, ruSMART, hence boosting interaction between researchers, practitioners, and engineers across different areas of ICT. From 2012, the scope of ruSMART conference has been extended to cover the Internet of Things and related aspects.

Presently, NEW2AN and ruSMART are well-established conferences with a unique cross-disciplinary mixture of telecommunications-related research and science. NEW2AN/ruSMART is accompanied by outstanding keynotes from universities and companies across Europe, USA, and Russia.

The 18th NEW2AN technical program addressed various aspects of next-generation data networks. This year, special attention was given to advanced wireless networking and applications as well as to lower-layer communication enablers. In particular, the authors demonstrated novel and innovative approaches to performance and efficiency analysis of ad hoc and machine-type systems, employed game-theoretical formulations, Markov chain models, and advanced queuing theory. It is also worth mentioning the rich coverage of graphene and other emerging materials, photonics and optics, generation and processing of signals, as well as business aspects.

The 11th Conference on Internet of Things and Smart Spaces, ruSMART 2018, provided a forum for academic and industrial researchers to discuss new ideas and trends in the emerging areas of the Internet of Things and smart spaces that create new opportunities for fully customized applications and services. The conference brought together leading experts from top affiliations around the world. This year, we saw good participation from representatives of various players in the field, including academic teams and industrial world-leader companies, particularly representatives of Russian R&D centers, which have a good reputation for high-quality research and business in innovative service creation and applications development.

We would like to thank the Technical Program Committee members of both conferences, as well as the associated reviewers, for their hard work and important contribution to the conference. This year, the conference program met the highest quality criteria with an acceptance ratio of around 35%.

The current edition of the conferences was organized in cooperation with National Instruments, IEEE Communications Society Russia Northwest Chapter, YL-Verkot OY, Open Innovations Association FRUCT, Tampere University of Technology, Peter the Great St. Petersburg Polytechnic University, Peoples' Friendship University of Russia (RUDN University), The National Research University Higher School of Economics (HSE), St. Petersburg State University of Telecommunications, and Popov Society. The conference was held within the framework of the RUDN University Program 5-100.

We also wish to thank all those who contributed to the organization of the conferences. In particular, we are grateful to Nikita Tafintsev and Roman Kovalchukov for their substantial work on supporting the conference website and thier excellent job on the compilation of camera-ready papers and interaction with Springer.

We believe that the 18th NEW2AN and 11th ruSMART conferences delivered an informative, high-quality, and up-to-date scientific program. We also hope that participants enjoyed both technical and social conference components, the Russian hospitality, and the beautiful city of St. Petersburg.

August 2018

Olga Galinina
Sergey Andreev
Sergey Balandin
Yevgeni Koucheryavy

Organization

NEW2AN International Advisory Committee

Igor Faynberg	Stargazers Consulting, LLC; Stevens Institute of Technology, USA
Villy B. Iversen	Technical University of Denmark, Denmark
Andrey Koucheryavy	State University of Telecommunications, Russia
Kyu Ouk Lee	ETRI, South Korea
Sergey Makarov	Peter the Great St. Petersburg Polytechnic University, Russia
Svetlana V. Maltseva	National Research University Higher School of Economics, Russia
Mohammad S. Obaidat	Monmouth University, USA
Andrey I. Rudskoy	Peter the Great St. Petersburg Polytechnic University, Russia
Konstantin Samouylov	Peoples' Friendship University of Russia, Russia
Manfred Sneps-Sneppe	Ventspils University College, Latvia
Michael Smirnov	Fraunhofer FOKUS, Germany
Sergey Stepanov	MTUCI, Russia

NEW2AN and ruSMART Technical Program Committee

Naveed Abbasi	Koc University, Turkey
Bayram Akdeniz	Bogazici University, Turkey
Hassen Alsafi	IIUM, Malaysia
Baris Atakan	Izmir Institute of Technology, Turkey
Konstantin Avrachenkov	Inria Sophia Antipolis, France
Sergey Balandin	FRUCT Oy, Finland
Michael Barros	Waterford Institute of Technology, Ireland
Kalil Bispo	Federal University of Sergipe, Brazil
Jose Carrera	University of Bern, Switzerland
Paulo Carvalho	Centro Algoritmi, Universidade do Minho, Portugal
Oktay Cetinkaya	Koc University, Turkey
Youssef Chahibi	Georgia Institute of Technology, USA
Wei Koong Chai	Bournemouth University, UK
Ji-Woong Choi	DGIST, South Korea
Chrysostomos Chrysostomou	Frederick University, Cyprus
Meltem Civas	Koc University, Turkey
Gianpaolo Cugola	Politecnico di Milano, Italy
Bruno Dias	Universidade do Minho, Portugal

Contents

NEW2AN: Next Generation Wired/Wireless Advanced Networks and Systems

ruSMART: New Generation of Smart Services

Requirements for Energy Efficient Edge Computing: A Survey

Olli Väänänen[1($\boxtimes$)] and Timo Hämäläinen[2]

[1] Industrial Engineering, School of Technology,
JAMK University of Applied Sciences, Jyväskylä, Finland
olli.vaananen@jamk.fi
[2] Department of Mathematical Information Technology,
University of Jyväskylä, Jyväskylä, Finland
timo.t.hamalainen@jyu.fi

Abstract. Internet of Things is evolving heavily in these times. One of the major obstacle is energy consumption in the IoT devices (sensor nodes and wireless gateways). The IoT devices are often battery powered wireless devices and thus reducing the energy consumption in these devices is essential to lengthen the lifetime of the device without battery change. It is possible to lengthen battery lifetime by efficient but lightweight sensor data analysis in close proximity of the sensor. Performing part of the sensor data analysis in the end device can reduce the amount of data needed to transmit wirelessly. Transmitting data wirelessly is very energy consuming task. At the same time, the privacy and security should not be compromised. It requires effective but computationally lightweight encryption schemes. This survey goes thru many aspects to consider in edge and fog devices to minimize energy consumption and thus lengthen the device and the network lifetime.

Keywords: IoT · Edge computing · Fog computing · Sensor data compression

1 Introduction

The Internet of Things (IoT) has been in focus on recent years. There are already billions of devices connected to the Internet and the amount of the Internet connected things is estimated to grow exponentially in these years [1, 2]. There are forecasts that by 2020 there will be more than 50 billion devices connected to the Internet [3]. These connected devises and things are very heterogeneous and require very different and application specific solutions and approaches [1]. The IoT as a concept was first introduced in 1999 by Kevin Ashton and it was related to the devices connected to the Internet via RFID connection [1]. The term IoT was mainly forgotten for years after that but it was reinvented some years ago. The exact definition of the IoT is still not described clearly, [1] but the technologies, solutions and the use of the IoT is all the time emerging.

There are already solutions of the IoT in use but the real success of the IoT depends on the standardization, which allows the compatibility, interoperability, reliability and effectiveness of the IoT solutions. The IoT devices and things should be able to

© Springer Nature Switzerland AG 2018
O. Galinina et al. (Eds.): NEW2AN 2018/ruSMART 2018, LNCS 11118, pp. 3–15, 2018.
https://doi.org/10.1007/978-3-030-01168-0_1

autonomously communicate with other devices or things and connect data to the Cloud. The IoT describes the next generation of the Internet, where physical things are connected to the Internet and can be identified and accessed via Internet [1].

There are presented and used many solutions and techniques to save energy in the IoT devices. These methods are mainly based on reducing wireless broadcasting because it is more energy consuming to broadcast data than pre-analyze it in close proximity of the source (sensor) [4]. The IoT sensor data need to be compressed efficiently to reduce and minimize the cost of broadcast and storage [5]. At the same time, many IoT devices are battery powered wireless devices. Thus, these IoT devices can be located in places where changing the battery might be impossible or at least battery replacement cost is one of the most critical source of cost in this kind of devices [2]. These devices are often very limited in computing power. So often, it is the case that it is possible to perform only very light analysis of the collected data in locally. In addition, the IoT itself is very constrained in terms of bandwidth, energy and storage [5, 6].

The IoT systems and the whole IoT sector is very heterogeneous. The things vary a lot and may move geographically and they need to interact with other things and Cloud systems in real-time mode. When designing the IoT systems it should be taken account scalability and interoperability of the heterogeneous devices. Design of the IoT applications and systems require involvement of many factors like networking, communication, business models and processes, and security. The IoT architecture should be very adaptive to make IoT devices to interact with other devices and with the Internet [1].

2   Definition of Edge and Fog

The term Fog Computing was introduced by Flavio Bonomi in 2012 [7, 8]. It refers to dispersed Cloud computing which is vital in several applications where the IoT devices collect data in the local network and the actions required from analyzed data take place in the same local network [9]. In that kind of case, it is not efficient to send all the data to centralized Cloud to be analyzed. It is not even possible to send data to the Cloud for analysis in many latency critical applications. The term Edge Computing means that computing happens in close proximity of data sources in the edge of the network. In many cases the terms Edge Computing and Fog Computing are interchangeable. But it can be defined that Edge refers more to the device side very close to data sources and Fog refers more infrastructure side like gateways and routers [10].

Cloud service providers locate their data centers often in rural areas to minimize costs. This lead to high latencies because customers are often located far from data centers [11]. Many IoT applications require very short response times, some create a large amount of data that can be heavy for network and some applications are involved with sensitive private data. Cloud computing cannot reply all these requirements so the Edge Computing is one answer for these challenges [10]. Latency critical applications are for example many intelligent transportation and traffic systems, autonomous vehicles, virtual reality (VR) and augmented reality (AR) applications [7]. Also many safety critical applications cannot rely on the connection to the Cloud. For example, vehicle-to-vehicle connection or data from vehicles can be used to avoid collision, but that analysis need to be done locally or in very close proximity located Cloudlet [7]. The Cloudlet means smaller size local datacenter. Safety critical systems are also very common in industrial automations systems. These kind of applications cannot tolerate possible Cloud outages and they often need low and predictable latency [7, 11]. This kind of new Fog Computing paradigm is not a replacement of the centralized Cloud. These concepts are more complementary to each other [9, 11]. In some applications the Cloud is not even possible to be used; this kind of situation happens for example in the modern aircraft. The modern aircraft can generate nearly half a terabyte of data from its sensors in one flight [7]. This amount of data cannot be sent to the Cloud for real time analysis from the middle of the ocean. Only possibility is to analyze the data locally and then perhaps download the raw data after flight for further analysis that can be executed in the Cloud. Even in ground level, the current wireless networks will be challenged with the amount of data that the huge amount of devices will produce in the near future [10]. Most of the data produced by the IoT devices will be analyzed locally in the Edge devices and will never be transmitted to the Cloud [10].

In Fig. 1 is illustrated the basic architecture of the IoT infrastructure including Edge and Fog devices. The difference between the Edge and the Fog devices is not always as clear as presented in Fig. 1.

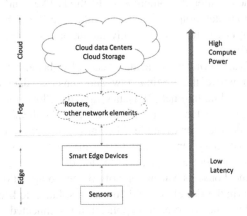

Fig. 1. Edge and Fog architecture in IoT [12].

Fog and Edge devices can be efficient data servers, routers, gateways, any kind of embedded systems or even end node like vehicles or sensors with some computational capability [11]. The Edge devices can be small-embedded devices with very energy efficient and limited micro controller or more capable single board Linux-computer like Raspberry PI. In Fig. 1. typically sensors are small wireless sensor tags and Smart Edge Devices are gateways for sensors. Smart Edge Device (gateway) is connected to the Internet via wireless or wired connection. Edge and Fog devices are very heterogeneous in nature with different hardware architectures and they run various different Operating Systems (OS). There are also available numerous different wireless access technologies and sensor network topologies [11]. This heterogeneous nature of Edge and Fog devices and systems avoid developing generic and easily adaptable solutions for Edge and Fog analytics. It is predicted that the Edge Computing could have as big impact in society as Cloud Computing has [10].

3 Benefits of the Edge Computing

While the Cloud Computing is very efficient method for data processing having a huge amount of computing power, [10] the Cloud Computing cannot meet and ensure the Quality of Service (QoS) in the IoT due to unstable latency and possible outages in the network connection and the Cloud servers. Fog or Edge Computing is an answer for the problem. In the Edge Computing the majority of the computing is carried out in close proximity of the data source. There are researches done that proof the Edge Computing reduction in response times and in energy consumption. By doing part of computation and analysis in the Edge reduce the needed wireless connection bandwidth. For example, photos can be compressed in the Edge before transmitting to the cloud [10]. Even if most of the data analysis is done in the Cloud, it is recommended to do some preprocessing for sensor data in the Edge before uploading it to the Cloud. In minimum this kind of preprocessing can be only filtering erroneous sensor data. More advanced preprocessing can mean different compression methods like sending only the information of the variation/alteration of the sensor values and not absolute values. This kind of preprocessing can reduce significantly the amount of data needed for upload data in the Cloud [10].

Security and privacy critical application can also benefit from the Edge/Fog Computing approach where the original raw and sensitive data is not sent to the centralized Cloud thru public Internet [7]. Data sent to the Cloud can be denatured data; for example, in images the faces can be blurred [7]. Applications producing very sensitive and private data are for example different healthcare applications.

Also home automation systems sending information to the Cloud could include some private sensitive data. For example, information of the water and electricity usage could easily tell if the house is vacant or not. If the computation is kept in close proximity of this data (in the Edge), it could be decent solution to keep sensitive data in private [10]. But if this home automation application is connected to the Internet, this sensitive data could be reachable for inappropriate quarters. So the cybersecurity is vital for all IoT applications whether the sensitive data is transferred to the Cloud or not.

4 Edge and Fog Computing Challenges

Fog and Edge devices are very heterogeneous [11]. It is difficult to design easily adaptable and generic solutions for the Edge Computing. Most applications are individual and cannot utilize generic computational, data aggregation and data analysis methods. There are different hardware platforms and different operational systems. Hardware platforms can vary from very simple micro-controller based platform with very limited memory to single board Linux-computer like Raspberry PI that is rather powerful platform. Virtualization is one way to handle multiplatform and multi-OS challenge.

One possibility towards generic solutions to be used in different and computationally restricted platforms is a container-based approach. Container-based virtualization can be considered as a lightweight virtualization solution. Because of lightweight nature, the containers can run in computationally limited IoT-platform like Raspberry PI [13, 14]. Containers could be used in the different platforms to perform same tasks. Anyway, these platforms could not be very limited basic embedded micro-controller based platforms, but require more computational power and generic operating system (OS) like Linux.

In [15], has been tested the ARM-based Single Board Computers with Docker containers and compared the overall efficiency in power consumption to the native executions. The performance evaluation showed almost negligible impact with container virtualization compared to native executions.

4.1 Methods for Reducing Energy Consumption in Wireless Sensor Networks

Several energy-efficient routing algorithms have been proposed for wireless sensor networks (WSN) but they are mostly not suitable for the IoT. Current IoT devices are mostly static and follow tree-based structure [16]. Dynamic routings developed for WSN architectures are not suitable for the IoT. The IoT network is often a complex large scale network and dynamic routing is difficult to be used effectively in this kind of network [17].

The Low-Energy Adaptive Clustering Hierarchy (LEACH) protocol utilizes several methods and techniques to reduce energy consumption in WSN [18]. LEACH is the most popular routing algorithms used in WSNs [19]. There are several variations and further developments of LEACH protocol like LEACH-C and ENHANCED LEACH for example [16, 20]. Weight energy efficient clustering (WEEC) is an extended version of LEACH. In WEEC the energy efficiency optimization is done by cluster head (CH) selection procedure. Every node in the sensor network can be elected as a cluster head. WEEC is a single-hop routing protocol [19].

In [16], the authors have presented a cluster head selection for energy optimization (CHSEO) algorithm to reduce the overall energy consumption in the IoT network. The CHSEO algorithm is based on selecting the optimal cluster head of the sensor nodes to reduce overall energy consumption. Hierarchical IoT sensor node framework is composed of different node types. Sensor node is sensing, aggregating and forwarding data, Relay node is receiving the data from sensor nodes and transmit it to the cluster head. Cluster head collects, aggregates and transmit the data to the base station. Base station collects, aggregates, analyses and process the data. The CHSEO algorithm was proved to have better performance than traditional WSN mechanism in energy consumption and network lifetime.

Other example of hierarchical network architecture to reduce IoT network energy consumption is presented in [17]. It is based on hierarchical relay node placement with energy efficient routing mechanism. Ad Hoc On-Demand Distance Vector (AODV) routing protocol has been used. This proposed network architecture gives balanced energy consumption and thus better network lifetime [17].

Modern long-range low-power IoT networks (NB-IoT, LoRa, SigFox) have star topology, so intelligent routing algorithms are out of the question [21]. In these technologies, the ultra-low energy consumption has been achieved by using very limited bandwidth and/or intelligent modulation.

4.2   Data Compression Methods in Edge Device: Lossy and Lossless Methods

In the IoT, huge amount of sensors are generating data and that data should be stored and processed with minimal loss of information. Sensor data compression is not a new discipline and several different compression algorithms are presented [5]. There are also very energy efficient contemporary compression methods for resource constrained IoT-nodes presented [6]. Data aggregation is also related to the data compression. Data aggregation here means for example to combine multiple sensor data and filter the redundant data. Data aggregation in wireless sensor network reduce the amount of data needed to transmit to the base station and thus reduce energy consumption [18]. Most of the compression methods presented for the IoT sensor data compression are lossy compression methods. Lossy methods are more efficient in compression compared to lossless methods. Lossy methods try to identify meaningful data points and discard redundant data. Different compression algorithms perform differently with different types of data sets. Also their computational complexity differs [5].

Lossy compression methods can be divided in two groups: Time domain and Transform domain. Time domain compression algorithms compress time series data directly without any transformation. Transform domain compression methods transform data into a different domain. Well-known transform domain methods are for example Discrete Fourier Transform (DFT) and Fast Fourier Transform (FFT) [5]. Different lossy compression algorithms are listed in Table 1.

Table 1. Lossy compression algorithms [5, 6].

Name of the algorithm	Type
Box-Car	Time domain
Backward slope	Time domain
OSIsoft PI software	Time domain
Compression extracting major extrema	Time domain
PLA, PCA	Time domain
Critical Aperture (CA)	Time domain
Fractal Resampling (FR)	Time domain
Lightweight Temporal Compression (LTC)	Time domain
Fast Fourier Transform (FFT)	Transform domain
Discrete Cosine Transform (DCT)	Transform domain
Chebyshev Transform (CH)	Transform domain
Wavelet Transform (CWT, DWT, WPT)	Transform domain

In ref. [5] the authors have selected four different lossy compression methods and compared their applicability to different signal characteristics. Compared methods were Critical Aperture (CA), Fractal Resampling (FR), Chebyshev Transform (CH) and Wavelet Packet Decomposition (WPD). Data used for comparison has been diverse publicly available sensor datasets. Comparison has been made by comparing the compression ratio with same Percentage Root mean square deviation (PRD). PRD level used in comparison has been 5%. Used datasets were different in composition. Some were quasi-periodic (QP), some non-stationary (NS) with sudden transient spikes and some non-stationary (NS) with periodic seasonal components [5].

As a result, the CH was the most effective method for QP data in terms of compression ratio. For NS with transient spikes data, the CA, FR and WPD were remarkably more effective than CH method. For NS with periodic seasonal data the WPD is the most effective method [5].

In [5], it is also shown that WPD requires considerably more computational time compared to the other methods. This means a higher energy consumption. In ref. [6] has been introduced lightweight compression algorithm for spatial data which is more energy efficient than wavelet compression. This lightweight compression algorithm can reduce energy consumption to half of the original consumption. This lightweight and energy-efficient compression algorithm is based on a lightweight temporal compression method named LTC [22]. LTC is tunable in accuracy and suitable for the datasets that are largely continuous and slowly changing. LTC is widely used method due to its good compression performance and low computational complexity [6]. LTC also requires very little storage compared to many other compression techniques. LTC is very effective for many environmental type data (temperature, humidity) which are approximately linear in small enough time window. Thus, LTC leverages temporal linearity of environmental data to compress that data [22].

5   Wireless Technologies for Energy Efficient IoT

For years the main wireless technology for transmitting sensor data with low energy consumption was IEEE 802.15.4 (mostly used protocol is called ZigBee). ZigBee was designed for ultra-low energy consumption and it has been popular in WSNs [21]. IEEE 802.11 (WiFi) has also been available for years but traditionally it has been used for high data rates and it has had rather high energy consumption. To address this energy consumption problem, there is available Power Saving Mode (PSM) in IEEE 802.11 [18]. This Power Saving Mode is developed for battery powered mobile devices. IEEE 802.11 was not designed for sensor applications but with PSM it has proofed to be potential alternative for other technologies used for WSNs. In some cases, the IEEE 802.11 PSM can outperform the IEEE 802.15.4 in energy consumption [23]. Bluetooth Low Energy (BLE) is very popular and widely used due to its availability. It is already available in most modern smartphones and it is widely used in wearable devices like heart rate monitors and other monitoring applications. ZigBee, BLE and WiFi uses the 2.4 GHz ISM frequency band while ZigBee is available also in sub-1 GHz band (868 and 915 MHz). IEEE 802.11ah version address for requirements of the IoT, like increased range, increased reliability and low energy consumption. IEEE 802.11ah is operated in sub-1 GHz range [21].

Using sub-1 GHz band increases the range and penetration thru obstacles (buildings, constructions). Sub-1 GHz band is also less crowded compared to popular 2.4 GHz band and thus these technologies are less vulnerable for interference [24].

ZigBee, BLE and WiFi all have rather short range, even if sub-1 GHz band is used (ZigBee and WiFi). As an answer for this limitation there are recent developments in long-range technologies like SigFox and LoRa. These are so called low-power wide-area-networks (LPWAN) [25]. SigFox is an ulta-narrow-band technology and it uses sub-1 GHz band (868 MHz in Europe). Its range is announced to be even up to 40 km. Direct competitor for SigFox is the LoRa. It uses the same frequency band as SigFox but its modulation is based on Chirp Spread Spectrum (CSS) [21]. CSS modulation was developed in the 1940's and it is very robust for interference and multipath fading. In CSS modulation the information in spread to different frequency channels and it has noise like properties [26].

Novel cellular based wireless technology for IoT solutions is Narrow Band-IoT (NB-IoT) which uses narrow bandwidth for lower power consumption [27]. The Third Generation Partnership Project (3GPP) introduced the NB-IoT in LTE Release 13. NB-IoT bandwidth for both uplink and downlink is set to 180 kHz. It is exactly size of one physical resource block (PRB) in LTE standard [28].

In Table 2 has been combined the main characteristics of the main WSN technologies used in the IoT. LPWAN technologies have long range and very limited data rate. ZigBee, BLE and WiFi have much higher data rate but the range is very limited.

Table 2. Wireless technologies summary for IoT [1, 23, 24, 26].

Technology	Band	Topology	Announced range	Data rate
802.15.4	2.4 GHz/0.9 GHz	Meshed	50 m	0.25 Mb/s
BLE	2.4 GHz	Scatternet	10 m	0.125–2 Mb/s
802.11 PSM	2.4 GHz	Star	100 m	11 Mb/s
802.11ah	0.9 GHz	Star	100 m–1 km	0.15–78 Mb/s
SigFox	0.9 GHz	Star	Up to 40–50 km	100 b/s or 1000 b/s
LoRa	0.9 GHz	Star	Up to 15 km (suburban), 45 km (rural)	0.25–50 kb/s
NB-IoT	700–900 MHz	Star	Up to 35 km	20–65 kb/s

As both SigFox and LoRa uses unlicenced ISM band, there is no guarantee for latency. For latency critical applications, the NB-IoT is better choice while SigFox and LoRa are suitable for low-cost projects with wide area coverage [26]. NB-IoT latency is maximum 10 s according to the standard, while SigFox and Lora can have latency of 10 s of seconds [27, 28]. Lora and SigFox are both very energy efficient technologies with very large range. BLE is also very energy efficient in its range [21].

6 Energy Efficient IoT Protocols

The most common IoT application protocols are MQTT, CoAP, XMPP and AMQP. MQTT (message queue telemetry transport) and CoAP (constrained application protocol) are designed especially for resource constrained devices like IoT end nodes and gateways [29, 30].

MQTT protocol is a publish-subscribe messaging protocol with minimal bandwidth requirements. It uses TCP (transmission control protocol) for transport. It is designed to be used in devices with restricted computational power and limited memory. MQTT is considered as a perfect messaging protocol for M2 M and IoT applications because of its ability to function within low power, low memory and cheap devices with low bandwidth networks [29].

CoAP protocol is a request-response protocol but it can function as a publish-subscribe mode too. CoAP uses UDP (user datagram protocol) for transport but it can be used for TCP too. CoAP has a wide acceptance for constrained devices [30].

In ref. [30] the authors have made comparison and experimental analysis between MQTT and CoAP. As a result they have found that MQTT consumes more bandwidth for transferring same payload than CoAP. But both protocols are efficient in terms of energy consumption.

In ref. [31] have been evaluated the performance, energy efficiency and resource usage of several IoT protocols (MQTT, CoAP, MQTT-SN, WebSocket and TCP). As a result, the authors found that MQTT and CoAP protocols are largely affected by the packet size. In generally CoAP is the most efficient in terms of energy consumption and bandwidth usage. But MQTT protocol is more reliable.

XMPP (extensible messaging and presence protocol) and AMQP (advanced message queuing protocol) are other popular protocols but they require more resources and they are not so suitable for resource constrained devices.

7 Security and Privacy Issues in the Edge

Privacy and security is a very big issue and concern in the IoT systems and applications. In the IoT systems, the end nodes (IoT devices) are connected to the Internet and thus these devices are reachable from all over the Internet. This kind of devices can be for example IP-cameras, health monitors and wearable devices or even WiFi connected toys. These devices can be connected by others if not protected properly.

Ownership of the collected data is other issue to take account. If the data is left on edge device for storage and analysis, then there are no ownership problems as the owner of the device can have all the rights for that data [10].

Battery powered IoT devices have very limited computational power, so complex encryption techniques require significant amount of computing and thus increase energy consumption. Lightweight encryption algorithms for the IoT devices have been developed.

Encryption scheme can be symmetric or asymmetric and both can be used in the IoT devices. In symmetric encryption scheme only one key is used to encrypt and decrypt the data. Both sender and receiver need to know the same key. In asymmetric encryption scheme two distinct keys are used. One for encrypting and other for decrypting. The advantage here is that the encrypting key can be public key and available to anyone. For asymmetric scheme the key need to be longer than in symmetric scheme to be secure. Thus calculations needed are longer than in symmetric scheme. Famous asymmetric encryption schemes are Rivest, Shamir, Adleman (RSA) scheme and Elliptic Curve Cryptography (ECC) [32].

Several researches have been done to compare ECC and RSA schemes to each other in regarding to encryption/decryption time and key length. The ECC has proved to be more efficient with shorten encryption/decryption time, smaller storage and in generally more energy efficient than RSA [33].

In ref. [34] the authors have presented lightweight asymmetric encryption scheme called AAβ and in ref. [35] the authors have made comparison in energy consumption between AAβ and RSA. The AAβ outperforms the RSA significantly in encryption and decryption.

8　Conclusions

In this study, a comprehensive study of the energy efficient Edge Computing has been carried out. There are a lot of research published from the different phases and aspects to reduce energy consumption in wireless end devices, but only few of them encompass the subject broadly. Minimizing energy consumption is one of the key aspects to carry out in the IoT device and system development. IoT end devices are often battery powered devices with wireless connection. Thus the computational resources are constrained but at the same time these devices should be able to do pre-processing and analysis for sensor data to reduce transferred data via wireless connection.

Most methods for reducing energy consumption in the IoT devices are concentrated to reduce wireless data transfer. Wireless data transfer is often the most energy consuming operation in the IoT device. In addition, many latency critical applications are pushing the development towards Edge Computing.

At the same time when more and more data analysis is carried out in close proximity of the sensors (in Edge and Fog); there are available several novel wireless technologies to transfer sensor data with low energy consumption. Considering energy consumption in every phase from the sensor to the Internet, it is possible to reduce energy consumption significantly. Many of these techniques are studied in this survey.

References

1. Li, S., Xu, L.D., Zhao, S.: The Internet of Things: a survey. Inf. Syst. Front. **17**(2), 243–259 (2015)
2. Montori, F., Contigiani, R., Bedogni, L.: Is WiFi suitable for energy efficient IoT deployments? a performance study. In: 2017 IEEE 3rd International Forum on Research and Technologies for Society and Industry (RTSI), Modena, pp. 1–5 (2017)
3. Jayakumar, H., Raha, A., Kim, Y., Sutar, S., Lee, W.S., Raghunathan, V.: Energy-efficient system design for IoT devices. In: 2016 21st Asia and South Pacific Design Automation Conference (ASP-DAC), Macau, pp. 298–301 (2016)
4. Stojkoska, B.R., Nikolovski, Z.: Data compression for energy efficient IoT solutions. In: 2017 25th Telecommunication Forum (TELFOR), Belgrade, pp. 1–4 (2017)
5. Bose, T., Bandyopadhyay, S., Kumar, S., Bhattacharyya, A., Pal, A.: Signal characteristics on sensor data compression in IoT - an investigation. In: 2016 IEEE International Conference on Sensing, Communication and Networking (SECON Workshops), London, pp. 1–6 (2016)
6. Ying, B.: An energy-efficient compression algorithm for spatial data in wireless sensor networks. In: ICACT (2016)
7. Satyanarayanan, M.: The emergence of edge computing. Computer **50**(1), 30–39 (2017)
8. Bonomi, F., Milito, R., Zhu, J., Addepalli, S.: Fog computing and its role in the internet of things. In: Proceedings of the First Edition of the MCC Workshop on Mobile Cloud Computing-MCC 2012, Helsinki, Finland, 17 August 2012, pp. 13–15 (2012)
9. Jalali, F., Khodadustan, S., Gray, C., Hinton, K., Suits, F.: Greening IoT with fog: a survey. In: 2017 IEEE International Conference on Edge Computing (EDGE), Honolulu, HI, pp. 25–31 (2017)
10. Shi, W., Cao, J., Zhang, Q., Li, Y., Xu, L.: Edge computing: vision and challenges. IEEE Int. Things J. **3**(5), 637–646 (2016)
11. Venkat Narayana Rao, T., Amer Khan, M.D., Maschendra, M., Kiran Kumar, M.: A paradigm shift from cloud to fog computing. IJCSET **5**(11), 385–389 (2015). http://www.ijcset.net
12. Yigitoglu, E., Mohamed, M., Liu, L., Ludwig, H.: Foggy: a framework for continuous automated IoT application deployment in fog computing. In: 2017 IEEE International Conference on AI and Mobile Services (AIMS), Honolulu, HI, pp. 38–45 (2017)
13. Morabito, R.: A performance evaluation of container technologies on Internet of Things devices. In: 2016 IEEE Conference on Computer Communications Workshops (INFOCOM WKSHPS), San Francisco, CA, pp. 999–1000 (2016)
14. Pahl, C., Helmer, S., Miori, L., Sanin, J., Lee, B.: A container-based edge cloud PaaS architecture based on raspberry Pi clusters. In: 2016 IEEE 4th International Conference on Future Internet of Things and Cloud Workshops (FiCloudW), Vienna, pp. 117–124 (2016)
15. Morabito, R.: Virtualization on Internet of Things edge devices with container technologies: a performance evaluation. IEEE Access **5**, 8835–8850 (2017)
16. Krishna, P.V., Obaidat, M.S., Nagaraju, D., Saritha, V.: CHSEO: an energy optimization approach for communication in the internet of things. In: GLOBECOM 2017 – 2017 IEEE Global Communications Conference, Singapore, pp. 1–6 (2017)
17. Cho, Y., Kim, M., Woo, S.: Energy efficient IoT based on wireless sensor networks. In: 2018 20th International Conference on Advanced Communication Technology (ICACT), Chuncheon-si, Gangwon-do, Korea (South), pp. 294–299 (2018)

18. Dargie, W., Poellabauer, C.: Fundamentals of Wireless Sensor Networks, Theory and Practise. Wiley, Hoboken (2010)
19. Bhushan, B., Sahoo, G.: A comprehensive survey of secure and energy efficient routing protocols and data collection approaches in wireless sensor networks. In: 2017 International Conference on Signal Processing and Communication (ICSPC), Coimbatore, pp. 294–299 (2017)
20. Kumar, S., Verma, U.K., Sinha, D.: Performance analysis of LEACH and ENHANCED LEACH in WSN. In: 2016 International Conference on Circuit, Power and Computing Technologies (ICCPCT), Nagercoil, pp. 1–7 (2016)
21. Morin, É., Maman, M., Guizzetti, R., Duda, A.: Comparison of the device lifetime in wireless networks for the internet of things. IEEE Access 5, 7097–7114 (2017)
22. Schoellhammer, T., Osterwein, E., Greenstein, B., et al.: Lightweight temporal compression of microclimate datasets. In: Proceedings of the 29th Annual IEEE International Conference on Local Computer Networks, pp. 516–524. IEEE Computer Society (2004)
23. Tozlu, S.: Feasibility of Wi-Fi enabled sensors for Internet of Things. In: 2011 7th International Wireless Communications and Mobile Computing Conference, Istanbul, pp. 291–296 (2011)
24. de Carvalho Silva, J., Rodrigues, J.J.P.C., Alberti, A.M., Solic, P., Aquino, A.L.L.: LoRaWAN — a low power WAN protocol for Internet of Things: a review and opportunities. In: 2017 2nd International Multidisciplinary Conference on Computer and Energy Science (SpliTech), Split, pp. 1–6 (2017)
25. Ayele, E.D., Hakkenberg, C., Meijers, J.P., Zhang, K., Meratnia, N., Havinga, P.J.M.: Performance analysis of LoRa radio for an indoor IoT applications. In: 2017 International Conference on Internet of Things for the Global Community (IoTGC), Funchal, pp. 1–8 (2017)
26. Poursafar, N., Alahi, M.E.E., Mukhopadhyay, S.: Long-range wireless technologies for IoT applications: a review. In: 2017 Eleventh International Conference on Sensing Technology (ICST), Sydney, NSW, pp. 1–6 (2017)
27. Wang, H., Fapojuwo, A.O.: A survey of enabling technologies of low power and long range machine-to-machine communications. IEEE Commun. Surv. Tutorials 19(4), 2621–2639 (2017)
28. Xu, J., Yao, J., Wang, L., Ming, Z., Wu, K., Chen, L.: Narrowband Internet of Things: evolutions, technologies and open issues. IEEE Int. Things J. 5(3), 1449–1462 (2018)
29. Yassein, M.B., Shatnawi, M.Q., Aljwarneh, S., Al-Hatmi, R.: Internet of Things: survey and open issues of MQTT protocol. In: 2017 International Conference on Engineering and MIS (ICEMIS), Monastir, pp. 1–6 (2017)
30. Bandyopadhyay, S., Bhattacharyya, A.: Lightweight Internet protocols for web enablement of sensors using constrained gateway devices. In: 2013 International Conference on Computing, Networking and Communications (ICNC), San Diego, CA, pp. 334–340 (2013)
31. Mun, D.H., Dinh, M.L., Kwon, Y.W.: An assessment of Internet of Things protocols for resource-constrained applications. In: 2016 IEEE 40th Annual Computer Software and Applications Conference (COMPSAC), Atlanta, GA, pp. 555–560 (2016)
32. Adnan, S.F.S., Isa, M.A.M., Hashim, H.: Analysis of asymmetric encryption scheme, AAβ performance on arm microcontroller. In: 2017 IEEE Symposium on Computer Applications and Industrial Electronics (ISCAIE), Langkawi, pp. 146–151 (2017)

33. Diro, A.A., Chilamkurti, N., Nam, Y.: Analysis of lightweight encryption scheme for fog-to-things communication. IEEE Access **6**, 26820–26830 (2018)
34. Ariffin, M.R.K., Asbullah, M.A., Abu, N.A., Mahad, Z.: A new efficient asymmetric cryptosystem based on the integer factorization problem of $N = P^{\wedge}2.q$. Malays. J. Math. Sci. **7**(S), 19–37 (2012). Special Issue 3rd International Conference on Cryptography and Computer Security 2012, vol. 7, pp. 1–6 (2012)
35. Adnan, S.F.S., Isa, M.A.M., Hashim, H.: Energy analysis of the AAβ lightweight asymmetric encryption scheme on an embedded device. In: 2016 IEEE Industrial Electronics and Applications Conference (IEACon), Kota Kinabalu, pp. 116–122 (2016)

Context-Based Cyclist Intelligent Support: An Approach to e-Bike Control Based on Smartphone Sensors

Alexey Kashevnik[1(✉)], Francesco Pilla[2], Giovanni Russo[3],
David Timoney[2], Shaun Sweeney[2], Robert Shorten[2],
and Rodrigo Ordonez-Hurtado[3]

[1] ITMO University, 49, Kronverksky Pr., 197101 St.Petersburg, Russia
alexey@iias.spb.su
[2] University College Dublin, Belfield, Dublin 4, Ireland
{francesco.pilla, david.timoney,
robert.shorten}@ucd.ie, shaun.sweeney@moixa.com
[3] Control and Optimization Group, IBM Research Ireland, Dublin 15, Ireland
grusso@ie.ibm.com, rodrigo.ordonez.hurtado@ibm.com

Abstract. Electrically assisted bicycles (e-bikes or pedelecs) have recently become popular as a means of personal transportation, particularly in cities. Pedelecs allow people to combine their muscular strength in varying proportions with the assistance of an electric engine. One of the challenges here is to determine the cyclist preferences, capabilities, and the context situation around the e-bike and, based on these, to make recommendations to the cyclist and also to control the degree of electrical assistance provided. The Smart Space concept is used here for context formation. The concept involves creation of a real-time model of the physical space that aids decision making about electrical engine utilization for the particular situation and generates a recommendation for the cyclist. An ontology-based publish/subscribe mechanism is used for information sharing in Smart Space.

Keywords: e-bike · Smart space · Publish/subscribe · Context
Ontologies

1 Introduction

Electric vehicles are currently experiencing growth in adoption, mainly due to their practicality and personal convenience benefits but also because of increased awareness around air-quality in densely populated areas. It is clear that e-bikes in particular can contribute to the solution of a variety of mobility problems in cities. These include problems associated with road or parking congestion, with greenhouse gas emission (depending on the fuels are used for electricity production) and with reduction of air pollution caused by internal combustion engine vehicles. It is in this city-context that electric bikes (known as 'e-bikes' or 'pedelecs') have the potential to be particularly beneficial.

© Springer Nature Switzerland AG 2018
O. Galinina et al. (Eds.): NEW2AN 2018/ruSMART 2018, LNCS 11118, pp. 16–22, 2018.
https://doi.org/10.1007/978-3-030-01168-0_2

The e-bike has an electrical motor that can assist the cyclist in completing journeys. The electric motor can be used to reject disturbances (such as wind and hills), and to provide new services to the cyclist. E-bikes are easily stored, do not require any infrastructure (the battery disconnects from the bike and can be charged from a regular wall socket), and their range (circa 40 km) makes them eminently suitable for use in urban environments. In addition, the on-demand electrical assist provided by the motor effectively removes many of the usual impediments to cycling (topology, wind, age of cyclist). Furthermore, the opportunity to develop services, for and from such bikes is very appealing, and it is in this context that we are interested in such vehicles. The opportunity to develop services to serve the cyclist is limitless. For example, we have already developed a system to help protect cyclists from the effect of air-born pollutants in [1]. Further applications include smart routing algorithms based on pollution levels, systems to optimize energy usage, as well as very basic services that could be developed to adjust the level of electrical assist based on topology, age, wind, etc.

The main contribution of this paper is an approach to provide intelligent support to e-bike cyclist based on information from both e-bike and smartphone-based sensors. It is proposed to mount the smartphone in front of the cyclist's face. The cyclist may use it as navigation application but at the same time the front-facing camera monitors the cyclist's face in order to identify dangerous situations and risk-prone behaviour. The DriveSafety application is used for this purpose [2, 3]. Based on the sensor information, the recommendations are generated for the cyclist and at the same time the bicycle engine is controlled to help the driver during his/her movement.

The rest of the paper is structured as follows: Sect. 2 describes the related work in the area of e-bike control and smartphone utilization for context determination. Section 3 describes the developed approach to cyclist intelligent support. Conclusions in Sect. 4 summarize the paper.

2   Related Work

Our e-bike work builds on several projects where context awareness has been used to protect cyclists. Most relevant is the work in [1] in which the breathing rate of cyclists are indirectly controlled in an effort to reduce pollution intake. Another related work is [4] in which a system was introduced to detect and alert the vehicle drivers of a nearby cyclist using passive RFID tags on the cyclist and an RFID antenna located on the vehicle. This work enables the detection of the cyclist to trigger the pollution mitigation system. Our work is also related to literature on engine management systems for hybrid electric vehicles. In [5, 6] engine management systems (EMS) were developed to minimize air pollution caused by the hybrid vehicle (HV), both directly and indirectly (whilst driving as well as vehicle/grid interaction). Furthermore, in [7] an emission trading framework was proposed whereby the problem of sharing an emission budget between HVs is formulated as a utility maximization problem. The authors in [5] used this framework to further implement a pollution mitigation system in a real HV. This system works by using a GPS-enabled Android smartphone to automatically switch the

HV into Electric Vehicle (EV) Mode when inside fixed geofences around residential or school areas. In [7] an environmental management system (EMS) was formulated with the aim to lower the emission output of the vehicle in areas of high pedestrian traffic, taking account of the uncertainty of routes the driver may travel to.

Other papers related to the smartphone utilization for context determination have been considered. Authors of the paper [8] are focused on the detection of affective states, their proper identification and interpretation with use of smartphones and wearable devices. They propose a method for personalization of emotion detection and present a series of experiments related to pulse and emotion determination for validation of the presented approach. The paper [9] presents a comprehensive state-of-the-art in the topic of context-aware recommendation systems that are operated in a mobile environment. The authors provide classification of these systems and comparative analysis by the following criteria: application field, context granularity, recommendation approach, recommendation technique, cold start problem, and evaluation. The work [10] discusses the role of smartphones as information sources in modern world. The authors note that more than half of the world's web traffic comes from smartphones. The paper presents a list of sensor information that a smartphone can provide based on operation of a system used and also discusses an approach to the driver behaviour determination in vehicle cabin. A computational system to assist wheelchair users, recording their trails, and providing context-aware assistance is presented in the paper [11]. The prototype is built on an Arduino device and allows assistance in indoor and outdoor environments, through the recommendation of contextualized accessibility resources. The paper [12] is aimed at exploration and evaluation of the smartphone's built-in sensors utilization and application of classification algorithms for context recognition. To implement this, the labelled sensor data has been collected at training and test datasets from volunteers' smartphones, the data being generated while performing daily activities. The authors present experiments with different algorithms, based on this data.

3 An Approach to Cyclist Intelligent Support

Presented here is an approach to the design of cyclist intelligent support system. Figure 1 describes the main system components and defines the relationships between them. In accordance with the reference model, an ontology is used to describe all concepts within a domain, using a shared vocabulary to denote the types, properties, and interrelationships of those concepts in the cyclist intelligent support system. The ontology formally represents all knowledge in the system. It includes the formal description of the e-bike (functions, capabilities, constraints, accessible built-in sensors), the cyclist (cyclist's profile, compatibility of the cyclist with e-bike, etc.), and the information about external sensors & services.

The cyclist's profile describes information about cyclist that is related to the cyclist intelligent support system. The profile contains cyclist preferences. Preferences are taken into account by the intelligent support system for personifying the interaction with the cyclist. Examples of preferences are: continuous cycling up to one hour then cycling in muscular power, angle of slope not more than twenty degrees when cycling under muscular power, maximum cycling speed of twenty five kilometres per hour, etc.

The electric engine is used to help the cyclist and provide him/her with possibilities of resting. The degree of electrical power assist is controlled by the engine control service. The e-bike can be moved by the engine and by combination of this with muscular power, in order to reduce the power consumption and save the battery.

The smartphone is used by the system to provide the sensor data (coordinates, speed, acceleration, images of the cyclist's face), to display status information, and to make recommendations for the cyclist. The smartphone navigation application suggests a route that is configured by the planning service. The DriveSafety application determines dangerous states while cycling by using the front-facing camera of the smartphone. In case of drowsy or distracted driving, the information about these dangerous states is used by the system to generate recommendations for the cyclist.

The context is defined as any information that can be used to characterize the situation of an entity. An entity is a person, place or object that is considered relevant to the interaction between a user and an application [13]. In accordance with the presented approach, the context is formed based on the ontology and information from accessible sensors and services (on-board e-bike sensors, smartphone sensors & DriveSafety application, cyclist profile, and Weather Service). The context is suggested to be modelled at two levels: Abstract and Operational. The Abstract Context is an ontology-based

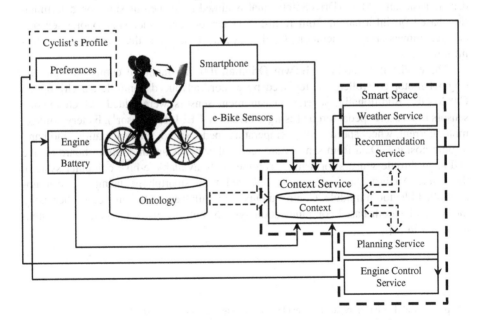

Fig. 1. An approach to the cyclist intelligent support system design.

model of a potential coalition participant related to the current task and is built by integration of information and knowledge relevant to the current problem situation. The Operational Context is an instantiation of the domain constituent of the Abstract Context with data provided by the contextual resources.

Smart Space [14] technology allows for provision of information sharing between different services of the system. This technology aims to achieve seamless integration of different devices by developing ubiquitous computing environments, where different services can share information with each other, make different computations and interact for joint tasks solving. There are following services participate in Smart Space: Context Service, Weather Service, Planning Service, and Recommendation Service. The Weather Service is responsible for determining and sharing with other services actual information about the weather situation at the region where the cyclist is located. The Planning Service is responsible for calculating the effective plan to use the e-bike engine based on the context information. The Planning Service provides a plan to the Engine Control Service that is responsible to control the engine power in accordance with the plan. Plan can be changed dynamically by the Planning Service and it should be adapted by the Engine Control Service. The Recommendation Service is responsible for generating the recommendations for the cyclist based on the context information. Some examples of the recommendations are: to reduce the speed, to increase the speed, to have a rest, to drink a cup of coffee, to concentrate on the road, etc.

The developed system involves the following main steps: (i) context generation based on the different accessible sensors and services; (ii) bespoke modification of a standard e-bike to enable context-aware behaviour; (iii) the development of low-level control and optimization algorithms to be deployed on the e-bike; (iv) the integration of these algorithms with an cyclist's smartphone that includes navigation system and drivers assistant system (DriveSafety) that is aimed at dangerous situation determination based on information from front-facing camera and other smartphone sensors; (v) recommendation generation to the cyclist based on the generated context information.

The e-bike is a modified BTwin Original 700. To facilitate control design, the original motor controller was replaced by a more advanced controller: a Grinfineon C4820-GR. Additionally, several measurement sensors were added, which include sensors to measure: pedal torque (using a THUN X-CELL RT sensor[1]), battery voltage, motor current, wheel speed, motor temperature, pedal speed, brake actuation, and hand throttle position. Data from sensors are read either by using an Arduino board (brake and hand throttle sensors) or using a commercially available e-bike computer system, the Cycle Analyst[2], and then communicated to a smartphone using an Arduino-controlled bluetooth module. Control inputs are finally sent to the bike controller using the same bluetooth based communication system. Full details of the hardware system are given in [1].

---

[1] http://www.ebikes.ca/shop/electric-bicycle-parts/torque-sensors/thun-120l.html.
[2] www.ebikes.ca.

4 Conclusion and Future Work

The paper presents an approach to context-based cyclist intelligent support system that uses information from different on-board bicycle sensors and a smartphone installed in front of the cyclist. Based on this information, the cyclist receives recommendations and the bicycle engine is controlled to reject disturbances (e.g., wind and hills), and to provide new services to the cyclist.

For the future work, authors are planning to conduct experiments based on the proposed approach. The experiments should show the usefulness of the DriveSafety system utilization for the cyclist. Also, the authors are planning to include a module in the DriveSafety system which assesses breathing rates, based on the detection of the cyclist's face skin colour through the use of the smartphone camera.

Acknowledgments. The presented results are part of the research carried out within the project funded by grants ## 16-29-04349, 16-07-00462 of the Russian Foundation for Basic Research. The work was partially supported by Government of Russian Federation, Grant 08-08 and by SFI grant 16/IA/4610.

References

1. Sweeney, S., Ordóñez-Hurtado, R., Pilla, F., Russo, G., Timoney, D., Shorten, R.: A context-aware e-Bike system to reduce pollution inhalation while cycling. IEEE Trans. Intell. Transp. Syst. **PP**(99), 1–12 (2018)
2. Smirnov, A., Kashevnik, A., Lashkov, I.: Human-smartphone interaction for dangerous situation detection and recommendation generation while driving. In: Ronzhin, A., Potapova, R., Németh, G. (eds.) Speech and Computer. SPECOM 2016. Lecture Notes in Computer Science, vol. 9811, pp. 346–353. Springer, Cham (2016). https://doi.org/10.1007/978-3-319-43958-7_41
3. Kashevnik, A., Lashkov, I., Parfenov, V., Mustafin, N., Baraniuc, O.: Context-based driver support system development: methodology and case study. In: Proceedings of the 21st Conference of Open Innovations Association FRUCT, Helsinki, Finland, 6–10 November 2017, ITMO University, St.Petersburg, pp. 162–171 (2017)
4. O'Faolain, C.: A Cyclist Collision Prevention System for In-Car Deployment, Master's thesis, University College Dublin, Ireland (2016)
5. Schlote, A., et al.: Cooperative regulation and trading of emissions using plug-in hybrid vehicles. IEEE Trans. Intell. Transp. Syst. **14**(4), 1572–1585 (2013)
6. Gu, Y., Liu, M., Naoum-Sawaya, J., Crisostomi, E., Russo, G., Shorten, R.: Pedestrian-aware engine management strategies for plug-in hybrid electric vehicles. IEEE Trans. Intell. Transp. Syst. (2017)
7. Crisostomi, E., Shorten, R., Studli, S., Wirth, F.: Electric and Plug-in Hybrid Vehicle Networks: Optimization and Control, CRC Press, Taylor & Francis Group (2017)
8. Nalepa, G., Kutt, K., Bobek, S.: Mobile platform for affective context-aware systems, Future Generation Computer Systems (2018). https://www.sciencedirect.com/science/article/pii/S0167739X17312207
9. Sassi, I., Mellouli, S., Yahia, S.: Context-aware recommender systems in mobile environment: on the road of future research. Inf. Syst. **72**, 27–61 (2017)

10. Kanarachosa, S., Christopoulosa, S., Chroneos, A.: Smartphones as an integrated platform for monitoring driver behaviour: The role of sensor fusion and connectivity, Transportation Research Part C (2018). https://www.sciencedirect.com/science/article/pii/S0968090X18303954

11. Barbosa, J., Tavares, J., Cardoso, I., Alves, B., Martini, B.: Trail care: an indoor and outdoor Context-aware system to assist wheel chair users. Int. J. Hum. Comput. Stud. **116**, 1–14 (2018)

12. Otebolaku, A., Andrade, M.: User context recognition using smartphone sensors and classification models. J. Netw. Comput. Appl. **66**, 33–51 (2016)

13. Dey, A., Abowd, G., Salber, D.: A conceptual framework and a toolkit for supporting the rapid prototyping of context-aware applications. Hum.-Comput. Interact. **16**, 97–166 (2001). https://doi.org/10.1207/S15327051HCI16234_02

14. Korzun, D., Balandin, S., Kashevnik, A., Smirnov, A., Gurtov, A.: Smart spaces-based application development: M3 architecture, design principles, use cases, and evaluation. Int. J. Embed. Real-Time Commun. Syst. **8**(2), 66–100 (2017)

Context-Driven Heterogeneous Interface Selection for Smart City Applications

Inna Sosunova[1(✉)], Arkady Zaslavsky[1,2], Alexey Matvienko[3],
Oleg Sadov[1], Petr Fedchenkov[1], and Theodoros Anagnostopoulos[1,4]

[1] Department of Infocommunication Technologies, ITMO University,
Kronverkskiy pr., 49, St.-Petersburg, Russia
{inna_sosunova, pvfedchenkov}@corp.ifmo.ru,
sadov@linux-ink.ru
[2] Digital Data61, CSIRO, Box 312, Clayton South, VIC 3169, Australia
Arkady.Zaslavsky@csiro.au
[3] Special Technology Center, Gzhatskaya Street, 21, St.-Petersburg, Russia
mail@t10000.ru
[4] Research and Education, Ordnance Survey, Southampton SO16 0AS, UK
Theodoros.Anagnostopoulos@os.uk

Abstract. With the diversity and variety of devices and interface modalities these devices offer, the choice of the right interface is still a significant research challenge. We propose a method of Context-driven Heterogeneous Interface Selection for Smart City Applications, which is based on context-driven and situation-aware modality selection mechanism. The method involves the use of a user model, a device model, and an environment model as an adaptation mechanism and a mechanism for selecting an appropriate modality or combination of modalities. Several scenarios of the functioning of the system are described. A series of tests was conducted for each scenario. Tests results are also given in the article. Benefits of the proposed approach are discussed and demonstrated.

Keywords: Heterogeneous interface · Knowledge representation
Context awareness · Smart city · Waste management

1 Introduction

Internet of Things (IoT) is enabling Smart Cites and supports data ingestion and digestion [1]. There is not a single universal IoT user interface that can support any possible use case or application scenario. For example, for the driver behind the wheel a voice and/or graphic interface is preferable, for the user in a public place a graphical or textual interface is more suitable. User interface, which provides simultaneously several means of communication: speech, gestures, text, etc., so that the user could choose from useful and convenient tools is called heterogeneous (multimodal) interface [2]. Heterogeneous interface for Smart City applications should also have mechanisms of automatic selection of an appropriate modality or combination of modalities.

O. Galinina et al. (Eds.): NEW2AN 2018/ruSMART 2018, LNCS 11118, pp. 23–32, 2018.
https://doi.org/10.1007/978-3-030-01168-0_3

In this paper we address the challenge of context-driven and situation-aware modality selection mechanism. We propose a method of Context-driven Heterogeneous Interface Selection for Smart City Applications. The proposed method is based on the concept that the intelligent system must be context-aware and situation-aware. Within the framework of the research, graphics, text, voice, audio, video modalities are conceded.

Validation and implementation of the proposed approach has been carried out as part of Solid Waste Management project, which in turn is a use case of bIoTope project [3]. Efficient waste management is an essential part of Smart City [4], which requires a complex multi-criteria approach [5]. The bIoTope project builds open IoT ecosystems on the basis of innovative IoT technologies and use cases. The project is funded under the EU Horizon 2020 Program. The system described in this paper is part of St. Petersburg pilot of bIoTope project on Smart Waste Management.

The rest of the paper is organized as follows. Section 2 presents Background and Related Work. Section 3 includes description of method of context-driven heterogeneous interface selection. Section 4 describes Implementation of proposed method. Section 5 describes Sample Scenarios and tests carried out at this stage. Section 6 concludes the paper.

2 Background and Related Work

2.1 Context and Situation

In this paper the context is understood as "any information that can be used to characterize situation of an entity" [6]. We define contextual information as ambient temperature, illumination, noise level, type and characteristics of the user's device, etc. A context-aware system is a system, which "uses context to provide relevant information and/or services to the user, where relevancy depends on the user's task" [6]. We consider a situation as context abstraction and generalization. The following situations are considered in the article: user is behind the wheel, user walks down the street, user is in a public place, etc.

2.2 Related Work

The article [7] describes a heterogeneous interface consisting of a graphical and a voice interface. The task of the interface is to simplify the work of doctors. This system is suitable only for medical specifics and uses only a mobile phone. The article [8] proposes the concept of an interactive system for people with disabilities, equipped with an adaptive context-sensitive user interface. The basic idea is to use a multi-level context processing. The main disadvantage is the absence of implication of the proposed idea. In [9], multi-level data processing is described. The algorithm is based on the use of contextual models and weights. The system can be easily adapted using three context models: user, device and environment. However, the proposed approach assumes that the developer must manually fill in the coefficients. Also, the system is suitable only for a small range of simple scenarios due to its architecture.

3 Method of Context-Driven Heterogeneous Interface Selection

Within the Solid Waste Management system, visual and multimedia modalities are necessary for users. Visual modalities include graphics (sending photo, graphical interface) and text; multimedia modalities include audio and video. Each modality has the following uses: input (receiving information from the user) and output (the user receives information from system). Heterogeneous interface can include a single modality or a combination of modalities (complex modality).

We developed a solid waste management (SWM) domain ontology, which is available online [10] for semantic manipulations over entities used in algorithms described further in this paper.

3.1 Appropriate Modality Selection Algorithm

To select a suitable modality or combination of modalities, a multi-step analysis of context is used. The main steps of the algorithm are shown in Fig. 1.

Algorithm contains 4 main parts. At the first stage, we need to personalize user interface. In our case, this means choosing a role. The following user roles were defined: citizen, fleet dispatcher of waste management company, driver, city administration, and janitor. Next step is detailing the situation. Situations determine the environment around the user, which can significantly affect the decision in the choice of interface. Within the framework of the Solid Waste Management use case, 6 main situations were identified:

1. user is in the building
2. user is at home
3. user is behind the wheel
4. user walks down the street
5. user is in a public place
6. user has his hands full.

For each situation, a set of context parameters defining this situation was proposed. To illustrate the algorithm, we use context matrixes. In total, we have three context matrices (T): the matrix of roles (R), the situation matrix (S) and the base matrix (B) that contains CR, CS and CB context rules respectively.

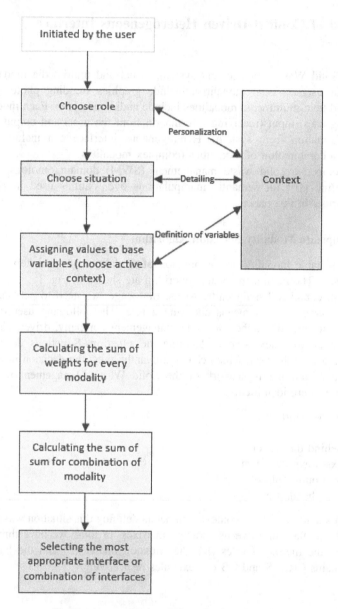

Fig. 1. Appropriate modality selection algorithm

For example, the roles matrix includes rules for define user level like advanced or beginner: *AdvancedIntelegence* and *NoobIntelegence*; situation matrix includes rules *NightEnviroment, NoiseEnviroment, LightnessEnviroment*. A base matrix is a matrix that contains all context rules that do not contain a role matrix or a situation matrix. For example, *SmarphoneType, CPUDevice*.

$$T_m = \begin{bmatrix} w_{11}^m & \cdots & w_{1i}^m \\ \vdots & \ddots & \vdots \\ w_{c1}^m & \cdots & w_{ci}^m \end{bmatrix}$$

Each row of the context matrix (T_m) describes how context rule (c) influences modality (i) for selection decision:

$$\text{influence} = \begin{cases} w > 0, & \textit{preferred for this context rule} \\ w < 0, & \textit{using is undesirable} \\ w = 0, & \textit{does not affect} \end{cases}$$

Active context rule is a rule that is activated by context. For example, we could make a rule that works only if the weather is rainy.

In the summation phase, we must select and summarize active context rules (c) for every modality (i) from the context matrices. After that we will calculate the sum of all coefficients for every modality $(MSUM^i)$.

$$ISUM^i = \sum_{c=0}^{C^R} R_i + \sum_{c=0}^{C^S} S_i + \sum_{c=0}^{C^B} B_i$$

The third stage represents the sum of coefficients of modalities $(ISUM^i)$ according to templates (t). A template is a predefined set of modalities (I) that could be useful together. For example, using text and graphics the best choice for PC in most types of information. Also, we have to separate the types in accordance with the direction (d) of interaction: input and output, because the interaction for reception and transmission can differ among themselves.

$$TSUM_d^t = \sum_{i=0}^{I_d} ISUM^i$$

At the 4th stage, we must select the template that has the maximum sum $(TSUM_d^t)$ and modalities (i) in this template (t) will be the best choice for interaction with user (W_d).

$$W_d = MAX(TSUM_d)$$

Pseudocode for appropriate modality selection algorithm:

```
function filterCoefficientsTable(table, activeContext)
    filtredTable = Table()
    foreach contextType in activeContext
       filtredTable.append(contextType, table[contextType])
    endfor
    return filtredTable
endfunction
function calcSumCoef(table, coefficientsUsed)
    fillzeroes(coefSum)
    foreach coefficient in table.keys():
        if (coefficient in coefficientsUsed):
            continue
        coefficientsUsed.add(coefficient)
        foreach modality in modalitySet
           coefSum[modality] += coefficientsTa-
ble[coefficient][modality]
  endforeach
    endforeach
    return coefSum
endfunction
function calcSumTemp(table):
    foreach templateTuple in table:
        foreach modality in templateTuple:
           templateSum[templateTuple] += sumAll[modality]
        endfor
      templateSum[templateTuple] /= length(templateTuple)
    endfor
    return templateSum
endfunction
-----------main application------------
Choose role
Choose situation
activeContextTypes = readContext()
roleTable, situationTable, baseTable = readCoefficients()
templateReceiveTable, templateSendTable = readTemplates()
roleTable = filterCoefficientsTable(roleTable, activeCon-
textTypes)
situationTable = filterCoefficientsTable(situationTable, active-
ContextTypes)
roleTable = filterCoefficientsTable(roleTable, activeCon-
textTypes)
fillZeroes(sumAll)
coefficientsUsed = List()
sumAll += calcSumCoef(roleTable[role], coefficientsUsed)
sumAll += calcSumCoef(situationTable[situation], coeffi-
cientsUsed)
sumAll += calcSumCoef(baseTable, coefficientsUsed)
templateReceiveSum = calcSumTemp(templateReceiveTable, sumAll)
templateSendSum = calcSumTemp(templateSendTable, sumAll)
templateReceiveWinner = max(templateReceiveSum)
templateSendWinner = max(templateSendSum)
```

4   Implementation

Implementation of the proposed method was carried out on the basis of ITMO SWM project [11]. ITMO SWM project is part of St. Petersburg pilot of bIoTope project, which aims to develop The Smart Waste Management System (SWMS). SWMS includes: Central Processing, Web Portal, Mobile applications, communication with IoT (means for collecting and processing data from SGB sensors and sensors installed on waste trucks), the model for IoT data sharing, dynamic route optimization in real time, security and privacy policies, and dynamic route optimization in real time. Web Portal supports three modes of operation: dispatchers of waste collection companies, citizens, and city administration; Mobile applications support two user roles: janitor and driver of waste truck.

The source language for this development is Python. CMS Plone and Zope built-in transactional object database (ZODB) are used for developing web applications. The development framework of Mobile applications is Kivy.

We extended and qualitatively enhanced the functionality of the ITMO SWM described in [11].

Web applications have the following functionality:

- City dashboard (visualization and presentation of current situation with the process of waste collection);
- Various reports on the quality of the collection of solid waste (tools for analyzing the volume and quality of work performed by drivers);
- Driving quality control;
- Traffic jams and road works;
- Complaints & Reviews (mechanism to leave or review a complaint about the quality of waste collection);
- Automated scheduling of vehicles.

Mobile applications have the following functionality:

- display the situation on the map, show the optimal route accordingly the cloud systems data;
- send messages about inability to pick up a garbage bin or other problems encountered, including text, photo, video and voice messages;
- display messages, including text, photo, video and voice messages.

Graphics modality (graphical interface) is represented as (1) City Dashboard, (2) various reports, (3) Traffic jams and road works visualization, (4) Complaints & Reviews. Graphics (sending photo), text and video modalities are implemented using the Messages Module of Mobile application. Audio modality (speech recognition and synthesis) is available for Mobile applications and Web Portal for all users. The implementation of audio modality is based on the algorithm described in [12].

Heterogeneous interface selection module is implemented in Python. The architecture of the developed web application is shown in Fig. 2. Sample scenarios and tests of this system are described in more details in the following section.

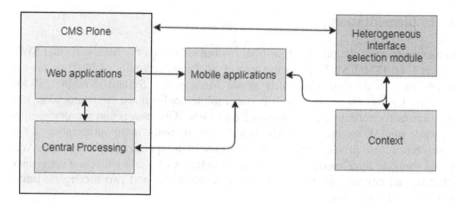

Fig. 2. Architecture

5 Sample Scenarios and Testing

Sample Scenario 1. The driver of the waste truck drives at night and requests a route to the garbage bin using smartphone or tablet.

```
-----role: Driver; situation: InCar
{'receive': (('AudioReceive',), {'Total': 3.0, 'Negative':
0.0}), 'send': (('AudioSend', 'GraphSend'), {'Total': 2.0,
'Negative': 0.0})}
```

Fig. 3. Admins interface

The admins interface is shown in Fig. 3.

Fig. 4. Driver's Mobile Application interface on devices Samsung Galaxy S7, Sony Xperia Z3 Tablet Compact и iPad Pro

The driver interface is shown in Fig. 4.

Sample Scenario 2. The citizen is in office (public place) and wants to complain about the quality of garbage collection using PC. The noise level is high.

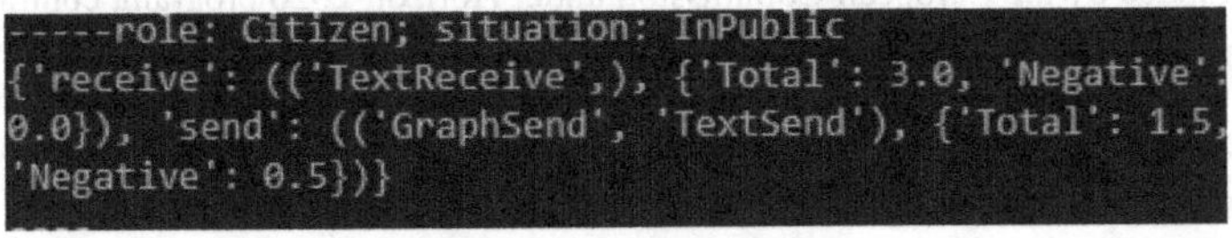

-----role: Citizen; situation: InPublic
{'receive': (('TextReceive',), {'Total': 3.0, 'Negative':
0.0}), 'send': (('GraphSend', 'TextSend'), {'Total': 1.5,
'Negative': 0.5})}

Fig. 5. Admins interface

The admins interface is shown in Fig. 5.

Fig. 6. Citizens interface

The citizens interface is shown in Fig. 6.

6   Conclusion

We developed and demonstrated method of Context-driven Heterogeneous Interface Selection for Smart City Applications. It implements automatic selection of an appropriate modality or combination of modalities based on context-driven and situation-aware modality selection mechanism. Several sample scenarios of the functioning of the system are described and tested. It is shown that heterogeneous interfaces allow IoT information system provide better service to its users.

In future work, we will refine the developed method by including larger number of situations and context rules.

Acknowledgements. Part of this work has been carried out in the scope of the project bIoTope which is co-funded by the European Commission under Horizon-2020 program, contract number H2020-ICT-2015/ 688203 – bIoTope. The research has been carried out with the financial support of the Ministry of Education and Science of the Russian Federation under grant agreement RFMEFI58716X0031.

References

1. Zanella, A., Bui, N., Castellani, A.V.L.: Internet of Things for smart cities. Internet Things J. **1**(2), 22–32 (2014)
2. Karpov, A.A.: Assistive information technologies based on audio-visual speech interfaces. In: SPIIRAS Proceedings, vol. 27, pp. 114–128 (2013)
3. The bIoTope Project Webpage [Electronic resource]. http://www.biotope-project.eu/
4. Sakai, S., Yoshida, H., Hirai, Y.: International comparative study of 3R and waste management policy developments. J. Mater. Cycles Waste Manag. **13**, 86–102 (2011)
5. Anagnostopoulos, T., et al.: Challenges and opportunities of waste management in IoT-enabled smart cities: a survey. IEEE Trans. Sustain. Comput. **2**(3), 275–289 (2017)
6. Dey, A.K., Abowd, G.D.: Towards a better understanding of context and context-awareness. Comput. Syst. **40**(3), 304–307 (1999)
7. Alnanih, R., Ormandjieva, O., Radhakrishnan, T.: Context-based and rule-based adaptation of mobile user interfaces in mHealth. Procedia Comput. Sci. **21**, 390–397 (2013)
8. Zouhaier, L., Yousra, B.H., Ben Ayed, L.J.: Building adaptive accessible context-aware for user interface tailored to disable users. In: Proceedings of International Computer Software and Applications Conference, pp. 157–162 (2013)
9. Macik, M.: Automatic User Interface Generation Doctoral Thesis, 173 p (2016)
10. SWM-DomainOntology [Electronic resource]. http://sdn.naulinux.ru:8128/SPB/swm-ontology.owl
11. Sosunova, I., Zaslavsky, A., Anagnostopoulos, T., Fedchenkov, P., Sadov, O., Medvedev, A.: SWM-PnR: ontology-based context-driven knowledge representation for IoT-enabled waste management. In: Galinina, O., Andreev, S., Balandin, S., Koucheryavy, Y. (eds.) NEW2AN/ruSMART/NsCC -2017. LNCS, vol. 10531, pp. 151–162. Springer, Cham (2017). https://doi.org/10.1007/978-3-319-67380-6_14
12. Sosunova, I., Zaslavsky, A., Anagnostopoulos, T., Medvedev, A., Khoruzhnikov, S., Grudinin, V.: Ontology-based voice annotation of data streams in vehicles. In: Balandin, S., Andreev, S., Koucheryavy, Y. (eds.) ruSMART 2015. LNCS, vol. 9247, pp. 152–162. Springer, Cham (2015). https://doi.org/10.1007/978-3-319-23126-6_14

An Artificial Intelligence Based Forecasting in Smart Parking with IoT

Petr Fedchenkov[1(⊠)], Theodoros Anagnostopoulos[1,3],
Arkady Zaslavsky[1,2], Klimis Ntalianis[3], Inna Sosunova[1],
and Oleg Sadov[1]

[1] Department of Infocommunication Technologies, ITMO University,
197101 Saint Petersburg, Russia
{pvfedchenkov, inna_sosunova}@corp.ifmo.ru,
thanag@teiath.gr, arkady.zaslavsky@csiro.au,
[2] CSIRO Computational Informatics, CSIRO,
Box 312, Perth, VIC 3169, Australia
[3] Department of Business and Marketing,
University of West Attica, Athens, Greece
{thanag, kntal}@teiath.gr

Abstract. Internet of Things (IoT) enables Smart Cities (SC) with novel services for the citizens' well-being. A Smart Parking (SP) system is an important part of the SC infrastructure, which enables the efficient handling of the demanding SC traffic congestion conditions. Such a system also protects the urban environment towards a green ecosystem. In this paper, we consider Artificial Intelligence (AI) algorithms towards processing of data produced by parking lots and disseminated by IoT technology in the SC of St. Petersburg in Russia. Such algorithms enhance the proposed SP system to predict the number of unoccupied parking lots within the SC parking places. In addition, the SP system uses vehicle navigation to decide the optimal parking place according the current vehicle location and the availability of parking lots in the SC.

Keywords: Internet of Things (IoT) · Smart cities · Genetic algorithms
Artificial Intelligence (AI) · Recurrent Neural Networks (RNN)

1 Introduction

IoT technology is actively expanding and have impact in various areas of SC infrastructure. With the rapid growth of vehicles, urban static traffic problems are becoming increasingly significant. This growth led to serious congestion due to excessive demand for road space at peak times. Providing drivers with detailed and more accurate information on the availability of parking lots leads to efficient and profitable parking solution for SC [1]. Drivers consider multiple factors when they are searching for a parking lot. Parking space fees and parking lot availability affect the driver's parking lot decision-making process [2, 3].

© Springer Nature Switzerland AG 2018
O. Galinina et al. (Eds.): NEW2AN 2018/ruSMART 2018, LNCS 11118, pp. 33–40, 2018.
https://doi.org/10.1007/978-3-030-01168-0_4

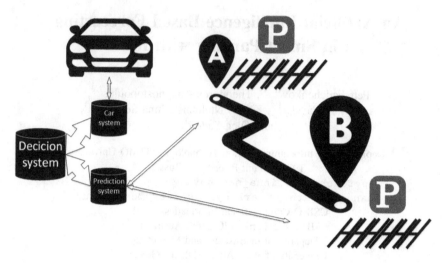

Fig. 1. Behavior of the driver

The proposed SP system studies AI algorithms for data processing enabling parking lot availability in SC. It is considered a use case scenario exploiting driver behavior starting at the current vehicle location in SC up to the parking lot recommendation inferred by the vehicle navigation system. Figure 1 presents the SP system architecture and use case scenario. Specifically, a vehicle that needs to park, asks the decision-making system to search the available destination parking lots of the nearest parking places from the current location within the SC. Concurrently, it is computed a Time-of-Arrival (ToA) estimation to reach each parking place from the current vehicle location in the Smart City. The proposed system predicts the availability of parking lots by analyzing parking places' IoT generated data. Information regarding unavailable parking lots through the nearest parking places enable the evaluation of the optimal parking lot choice based on AI algorithms.

The paper is structured as follows. Section 2 presents prior work in the research area. Section 3 describes the AI genetic algorithm for the SC generated parking data. Section 4 analyzes the AI RNN algorithm and the data forecasting. Section 5 presents the proposed decision system, while Sect. 6 concludes the paper and discusses further future research.

2   Related Work

Hisamitsu Kurogo, etc. [4] studied the role of the parking management system in Tokyo and the proposed parking system. The system can effectively simplify traffic management and improve the efficiency of transport. Ma Minghui [5] conducted a study on the development and implementation of an information forecasting subsystem, a proposed model of parking actions based on a neural network.

Ways to detect not only the congestion that has already been formed, but also to reveal the increased possibility of forming a congestion are considered in the article [6]. If we can identify the increased likelihood of traffic jams, we can take action to avoid this congestion: the closure of some exit roads, parking, informing drivers to avoid some suspicious places and thus avoid traffic jams.

Modern approaches Deep Learning in forecasting is a field of research in which the future state of an object is predicted based on a series of historical scenes of the same object. To find time series with long time intervals, neural networks with a long short memory (LNM) (NNs), which were proposed by Hochreiter and Schmidhuber [7]. In 1997, they were effectively applied in short-term motion prediction [8]. And they achieved remarkable results in ensuring the long-term time dependence of the traffic flow.

3   AI Genetic Algorithm for SC Generated Parking Data

3.1   IoT Generated Data at the SC Parking Places

IoT infrastructure generates source data from parking lots' conditions (i.e., whether a parking lot is available or unavailable) of the parking places within SC. The proposed SP system uses data from parking lots as well as spatiotemporal data relating to vehicle's distance from business centers and hypermarkets. Such a system also explores data regarding the near future availability of booked parking lots and discount programs offered by the parking places in the SC.

3.2   Incorporated Evaluation Real Datasets

In this paper, we incorporate certain evaluation real datasets with information about parking occupation (i.e. availability or unavailability) for a total duration of half a year [9]. The data format contains information about the moment of data acquisition, the number of available lots and the total number of parking lots. For modeling purposes, it is necessary to have information about specific available and unavailable parking lots within certain parking places. Such a SP system includes an array of parking places in which 0 value correspond to an unavailable parking lot, while value 1 corresponds to an available one.

3.3   Genetic Algorithm

To obtain IoT generated synthetic data from certain parking places, it is used an AI genetic algorithm which studies the genes that the current state of the parking lots is recorded for the time being examined [10]. The following criteria are used as the objective function:

- The total number of unavailable parking lots (1 in the genome) corresponds to the real value
- The change in the state of parking lots occurs as rarely as possible

The second requirement is introduced in order to achieve the corresponding values to the actual behavior of visitors to the parking place and the uniform distribution of parking lots.

3.4 AI Experimental Results

The AI based experimental results of the search for the best individual of the genetic algorithm is a set of parking conditions for the period under consideration. In this case, we are dealing with a sparse matrix, the main part of whose values 1 is concentrated in the central region of the rows. The time points at which measurements were taken and which are unevenly distributed in this matrix, interpolation of data is necessary to study measurements at equal intervals in order to predict the fullness.

4   AI Recurrent Neural Networks (RNN) and Data Forecasting

4.1   Incorporating AI RNN

RNN allow processing of a series of events in time or successive spatial chains. This is a great benefit in time series forecasting, where classical linear methods can be difficult to adapt to multivariate or multiple input forecasting problems. The Long Short-Term Memory (LSTM) network is well suited to learning the tasks of classifying, processing and forecasting time series in cases where important events are separated by time lags with uncertain duration and boundaries [11].

In this paper, we propose an LSTM model for multivariate time series forecasting of parking lots.

4.2   Data Preparation

The first step is to perform interpolation of data in order to uniformity of time stamps. As a step, the value of 10 min is selected, which will allow to predict the situation in the near future in the availability for a vehicle in a city environment at a distance of about 10 km.

The second step is to convert the data into a variant that is convenient for learning a recurrent neural network [12]. In this case, we add values at the instants of time (t–n, t–(n–1), …, t–1) to a string with the time t. At the same time, by varying the parameter n, it becomes possible to train the neural network for a set of values over a period of n steps. After the end of the neural network training, it becomes possible to obtain a prediction about the state in moment t, having information on n parking status values.

4.3   Learning on Datasets

Training was conducted on each of the received data samples. We used a network with a hidden layer of LSTM elements and an output layer with one output value (the percentage of parking filling). As a sample for training and for the test, the intervals of the matrix obtained during the preparation were used. We used the Keras library [13], which provides tools for building and learning neural networks, and the software product TensorFlow as a backend.

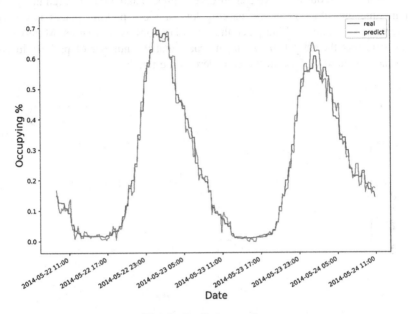

Fig. 2. Prediction results

The process of predicting the data of a trained neural network for 2 days is shown in Fig. 2. In this figure, lines of real and predicted values are constructed for one of their parking lots. The root-mean-square error is 0.035, i.e. 3.5%. Now we can use trained neural networks for the system of recommendations to drivers.

5 Decision System

5.1 Car System and Use Case

The proposed car system decides the vehicle navigation as well as the search and selection of the nearest parking place with available parking lots. The prosed system incorporates the Manhattan distance to compute the route. In addition, distance from the road map and traffic accounting is necessary for the prediction of the number of available parking lots in the SC. The system also takes into consideration information about a certain set of available lots, which are reserved for permanent tenants and VIP clients, as well as of parking lots for special services.

5.2   Finding the Best Parking Lot Through the Prediction System

Consider the situation when the car is located between two parking places, one near the SC center and the other in the Salling area. Also, consider that both parking places satisfy the driver's request. As we can observe in Figs. 3 and 4 (vertical red line), at this point in time the situation in the parking lots of these parking places is different. Therefore, in SC center parking place there is a large influx of visitors. Moreover, the parking data that the model gives out indicates that the number of parking lots will exceed the threshold of 50% in the considered time range.

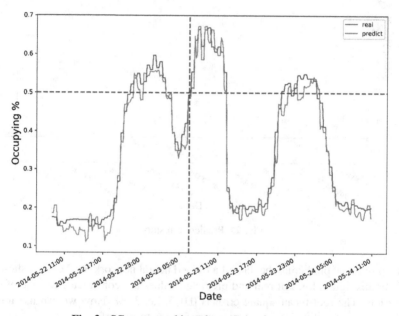

Fig. 3. SC center parking place (Color figure online)

In such situations, it is necessary to have additional information from the parking place, such as the estimated percentage of parking lots with pre-reservation. It is also necessary to consider the number of the SC drivers planning to visit the parking place. The proposed SP system can be used as a recommendatory driver system, which can ensure uniform distribution of parking lots within SC parking places.

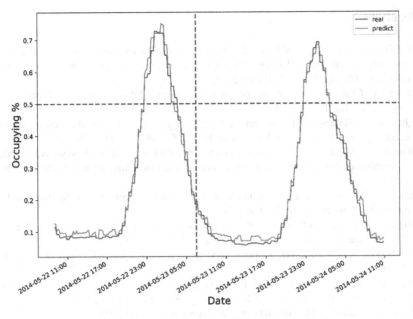

Fig. 4. Salling parking place (Color figure online)

6 Conclusions and Future Work

Current study is exploiting SC potentiality enabled by IoT. Specifically, it is proposed a SC parking system aiming to provide citizens' well-being in the SC of St. Petersburg in Russia. AI genetic algorithms and RNN we incorporated to the system. Such use of these techniques enabled the optimization of car navigation systems within the framework of the SC concept in the aspect of interaction with parking. We considered advanced techniques for data processing in order to provide accurate SC parking predictions based on current traffic conditions. In addition, data analyzed where further used by the proposed car navigation system to provide optimum solution for the specific problem.

In the future, it is necessary to consider the possibility of interaction of the navigation systems of a multitude of cars with the aim of making predictions based on the totality of vehicles that carry out the scenario of a particular parking.

Acknowledgments. Part of this work has been carried out in the scope of the project bIoTope which is co-funded by the European Commission under the Horizon-2020 program, contract number H2020-ICT-2015/688203 – bIoTope. The research has been carried out with the financial support of the Ministry of Education and Science of the Russian Federation under grant agreement RFMEFI58716X0031.

References

1. Yu, F., Guo, J., Zhu, X., Shi, G.: Real time prediction of unoccupied parking space using time series model. In: 2015 International Conference on Transportation Information and Safety (ICTIS), pp. 370–374. IEEE (2015)
2. Yong, S., Chunping, L., Yihuai, W.: A forecasting model for parking guidance system. Comput. Sci. Inf. Eng. 607–611 (2009)
3. Qian, Z.(S.), Rajagopal, R.: Optimal parking pricing in general networks with provision of occupancy information. Procedia – Social Behav. Sci. **80**, 779–805 (2013)
4. Kurogo, H., Takada, K., Akiyama, H.: Concept of a Parking Guidance System and Its Effects in the Shinjuku Area Configuration, Performance, and Future Improvement of System. IEEE (1995)
5. Ma, M.: Research and Implementation of Information Forecasting in Parking Guidance System. Suzhou University, Suzhou (2006)
6. Skszek, S.L.: "State-of-the-Art" Report on Non-Traditional Traffic Counting Methods. 505 N. Tanque Verde Loop Rd. Tucson, AZ 85748, 58 p (2001)
7. Hochreiter, S., Schmidhuber, J.: Long short-term memory. Neural Comput. **9**, 1735–1780 (1997)
8. Zhao, Z., Chen, W., Wu, X., Chen, P.C., Liu, J.: LSTM network: A deep learning approach for short-term traffic forecast. IET Intell. Transp. Syst. **11**, 68–75 (2017)
9. http://iot.ee.surrey.ac.uk:8080/datasets.html. Assessed 15 Mar 2018
10. Sargsyan, S., Brunton, S.L., Kutz, J.N.: Online interpolation point refinement for reduced-order models using a genetic algorithm. SIAM J. Sci. Comput. **40**(1), B283–B304 (2016)
11. Tian, Y., Pan, L.: Predicting short-term traffic flow by long short-term memory recurrent neural network. In: Proceedings of the IEEE International Conference on Smart City Socialcom Sustaincom, Chengdu, China, 19–21 December 2015, pp. 153–158 (2015)
12. https://machinelearningmastery.com/multivariate-time-series-forecasting-lstms-keras/. Assessed 20 Apr 2018
13. Vidnerová, P., Neruda, R.: Evolving KERAS architectures for sensor data analysis. Ann. Comput. Sci. Inf. Syst. **11**, 109–112 (2017)

On Data Stream Processing in IoT Applications

Dmitry Namiot[1]([✉]), Manfred Sneps-Sneppe[2], and Romass Pauliks[2]

[1] Faculty of Computational Mathematics and Cybernetics, Lomonosov Moscow
State University, Moscow, Russia
dnamiot@gmail.com
[2] Ventspils International Radio Astronomy Centre, Ventspils University
College, Ventspils, Latvia
manfreds.sneps@gmail.com, romass.pauliks@venta.lv

Abstract. This article is devoted to the issues of streaming data processing in
the Internet of Things applications. Stream processing is a natural fit for the
Internet of Things applications. Most of the data models on the Internet Things
are exactly the data streams. Accordingly, most applications (business applications) are oriented to processing real-time data streams (e.g., search for
anomalies, provide billing features, etc.). The paper considers the architecture of
data processing systems, classifies stream processing patterns. Much attention is
paid to time management in stream processing systems. The review is conducted
from the standpoint of the contents of the master's course on stream data processing in the Internet of Things and Industrial Internet of Things applications.
Also, the paper considers the specific application models and streaming data
architecture for the Internet of Things applications as well basic data analysis
algorithms that are used in such systems.

Keywords: Internet of Things · Stream · Data mining

1 Introduction

Stream processing could be defined as a computer science paradigm that deals with a
sequence of data (a stream) and a series of operations being applied to each element in
the stream [1].

The most obvious benefit of stream processing is its ability to treat data not as static
files (tables or other data structures) but as an infinite stream. In the most cases, stream
processing is associated with real-time processing. By this reason, stream processing is
a natural solution for Internet of Things (IoT) applications, where measurements and
their processing present majority of applications. The typical real-time stream processing, for example, should include the solutions for:

- integration with various data sources
- events detection (collecting, filtering, prediction, etc.)
- live data discovery and monitoring
- anomalies detection
- performance, scalability and real-time responsiveness

© Springer Nature Switzerland AG 2018
O. Galinina et al. (Eds.): NEW2AN 2018/ruSMART 2018, LNCS 11118, pp. 41–51, 2018.
https://doi.org/10.1007/978-3-030-01168-0_5

For example, Fig. 1 from EU FP7 CityPulse Project [2], illustrates an integrated approach to IoT and social media stream based smart city applications. It is a first key task in developing smart city (or smart home) frameworks – how to provide an integration across different application domains. For smart cities, for example, data streams should integrate data from traffic, smart transport cards, smart meter reading systems, street lighting, etc. [3]. For Home Automation data streams are responsible for collecting (and processing) data from various appliances. And in both cases, the ability to react in real time is a critical feature.

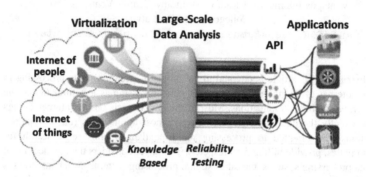

Fig. 1. An integrated approach to IoT and social media stream based smart city applications [3].

Also, data security is one of the major features for IoT. So, stream processing is also responsible for real-time security alerts for data.

Of course, stream processing and related questions should be a part of any IoT education [4]. In general, IoT study directions could be grouped into five topics [5]:

- Data measurement (sensing)
- Data transmission (networking)
- Data aggregation (middleware)
- Data analysis
- Data interaction (behavior)

The direction related to analytical processing includes descriptive analytics (with visualization), predictive analytics and recommendation systems. It is the place where we should highlight the high demand for stream analytics and real-time processing applications for IoT projects (e.g., CEP - complex event processing). Traditionally, this area (stream algorithms) has got a relatively poor reflection in educational programs. In our opinion, stream processing should be a key moment for IoT education [6].

The rest of this paper is organized as follows. In Sect. 2, we discuss stream processing software architectures. In Sect. 3, we present IoT related use cases for stream processing. And in Sect. 4, we discuss algorithms for stream processing.

2   On Streams-Based Architectures

At the first hand, in any educational program, we need to present the basics of data streaming. The key reasons for choosing stream processing architectures are:

- Achieving lower latency and adding more timely data to a processing
- Closing the gap between business models, oriented to never ended data volumes and computer architectures
- Stream processing lets proceed data as they arrive and create a more predictable workload

One of the first examples of streaming processing architectures is so-called lambda architecture (Fig. 2). Originally, the Lambda Architecture is an approach to building stream processing applications on top of MapReduce and Storm or similar systems. The main idea behind this model is the fact that an immutable sequence of records is captured and fed simultaneously (in parallel) into a batch system and a stream processing system [7].

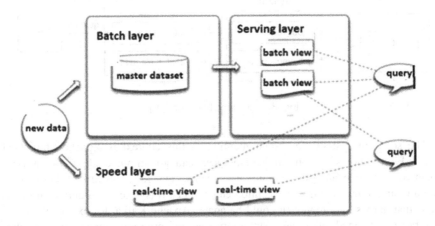

Fig. 2. Lambda architecture [8]

So, developers should implement business transformation logic twice, once in the batch system and once in the stream processing system. It is possible to combine the results from both systems at query time to produce a complete answer [8]. The biggest disadvantage for Lambda system is a need to build, provision, and maintain two independent versions of data pipeline, and then also somehow merge the results from the two pipelines at the end.

The next important moment is so-called Kappa architecture [9]. When the Lambda Architecture offers a dedicated real-time layer, in Kappa model everything is a stream. As per its definition, Kappa Architecture is a software architecture pattern. According to this pattern, rather than using a relational database or some NoSQL data store, the basic data store in a Kappa Architecture system is an append-only immutable log. From the log, data is streamed through a computational system and fed into auxiliary stores for serving. Actually, we can present it as a Lambda Architecture system with the removed batch processing sub-system. To replace batch processing, data is simply fed through the streaming system quickly. The basic proposal (elements) for Kappa pattern are [10]:

1. Use Kafka or some other system for retaining the full log of the data you want to be able to reprocess and that allows for multiple subscribers.
2. When you want to do the reprocessing, start a second instance of your stream processing job that starts processing from the beginning of the retained data, but directs this output data to a new output Table.
3. When the second job has caught up, switch the application to read from the new Table.
4. Stop the old version of the job, and delete the old output table.

In other words, it is a bunch of processes for retained data (Fig. 3).

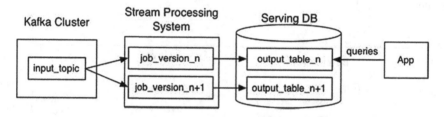

Fig. 3. Kappa architecture [10].

The software system we should mention in this context is Apache Flink [11]. Apache Flink follows a paradigm that embraces data-stream processing as the unifying model for real-time analysis, continuous streams, and batch processing both in the programming model and in the execution engine [11]. It is based on a durable message queue that allows data replay (it could be again, Apache Kafka, for example). As per the above-mentioned principles, stream processing programs make no distinction between processing the latest events in real-time, continuously aggregating data periodically in large windows or processing historical data. Actually, these different types of computations simply start their processing (job versions in Fig. 3) at different points in the durable stream and maintain different forms of state during the computation. Flink supports different notions of time: event-time, ingestion-time, processing-time in order to give programmers the flexibility in defining correlations for events (Fig. 4).

Time-management is one of the main features of stream processing systems. There are typically two domains of time associated with stream processing: event-time and processing time. Event-time is the time at which events actually occurred. And processing-time is the time at which events are observed in the system. Ingestion-time is the time at which events are selected for batch processing. Some of the papers (e.g., [12]) propose also tuple-based windowing as time domain. In this case, windows sizes are counted in numbers of elements. However, tuple-based windowing could be presented as a form of processing-time windowing where elements are assigned

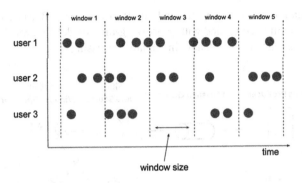

Fig. 4. Windows in Apache Flick [15]

monotonically increasing timestamps as they arrive at the system. Things that can affect the level of skew [13].

Time domain management for stream processing should be a significant part of an educational program. As per data processing patterns, we can highlight the following elements:

(1) Batch processing (a collection of bounded data sets appropriate for batch processing). For example, Spark Streaming is a really batch-processing system that appears as a streaming system. It simulates stream processing by creating so-called micro-batches across time which are processed individually [14].

(2) Windows-based processing. Fixed-sized windows are probably the most common way to process a data set. Fixed windows slice up time into segments with a fixed-size temporal length. The input data could be presented as a set of fixed-sized windows and each window then will be processed separately. And windowing could be implemented by processing-time or by event-time.

Sliding windows are defined by a fixed length and a fixed period. For the period equals the length, we will have fixed windows. For the period is less than the length, we will have windows overlapping. And if the period is greater than the length, we will have a sampling window (it shows only a subset of the data over time).

One of the biggest problems here are timestamps associated with the events (measurement). It is also a live scenario for IoT. Suppose we have some proxy, accumulated data from low-end sensors. It is a model that corresponds to oneM2M standard [16]. Each record will keep a real 'arrive time', but they could be uploaded to cloud, for example, sometimes later (e.g. by the scheduler on the proxy host). Also, the similar problem could be associated with the distributed systems and their delays in updating data.

Event-time windows let observe events in batches that reflect the times at which those events actually happened. As per Apache Flink manual, for example, an event-time model gives correct results even on out-of-order events, late events, or on replays of data from backups or persistent logs. In event-time programs must specify how to generate so-called Event Time Watermarks, which is the mechanism that signals progress in event time [15]. Event-time processing could face often some latency, due to waiting a certain time for late events and out-of-order events. Because of that, event-time programs are often combined with processing-time models. Figure 5 from Apache Flink manual illustrates time domains in stream processing.

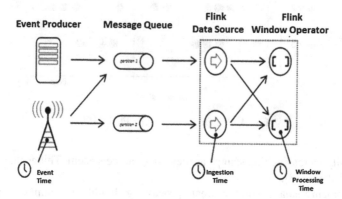

Fig. 5. Time domains in Apache Flink stream processing [15].

(3) Sessions are analogues of web sessions. Sessions are typically defined as periods of activity (e.g., for a specific process/user) terminated (divided) by some gap of inactivity. For example, it is a sequence of events terminated by a gap of inactivity greater than some predefined timeout [17]. Naturally, each session could be presented as a batch or dynamic window. In the most cases, their lengths cannot be defined a priori and they are dependent upon the actual data involved.

(4) Time-agnostic unbounded data stream. It is the case where a time of arrived data is irrelevant and all relevant logic is data-driven only. Batch systems could easily proceed time-agnostic streams. As a particular use case for a time-agnostic data stream, we can mention filtering [18]. It is collecting data that belongs to the given domain only. Another two elements are approximation [19] and aggregation [20].

As per practical software systems, the educational track, in our opinion, should include Apache Kafka [21]. At this moment it is some like industry standard. Apache Kafka is a distributed streaming platform. Originally, it was introduced as a messaging platform (Fig. 6). As a streaming platform, Apache Kafka provides low-latency, high-throughput, fault-tolerant publish and subscribe pipelines, durable storage, the ability to replay data and is able to process streams of events. Like many others publish-subscribe messaging systems, Kafka maintains feeds of messages in topics. There are Producers that can write data to topics and consumers that can read from topics. Kafka is a distributed system, so topics are partitioned and replicated across multiple distributed nodes. Messages are simply byte arrays and they can be used to store any object in any format.

Timestamps are automatically embedded into Kafka messages. Depending on Kafka's configuration these timestamps represent event-time or ingestion-time.

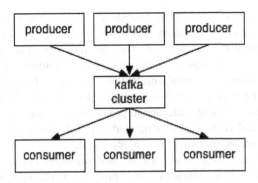

Fig. 6. Kafka architecture

3 IoT Applications and Streams

All futuristic scenarios converge on the fact that the number of devices in the Internet of Things systems will be very large. For example, Gartner [22] forecasts that the IoT will reach 26 billion units by 2020, up from 0.9 billion in 2009. IoT devices will generate enormous quantities of data that need to be aggregated and analyzed in real time to provide information regarding status, location, functionality, possible security alerts and environment of the devices.

All the devices work asynchronously and independently of each other, sometimes, perhaps, using some of their intermediate nodes (proxies). The number of models (types) of devices will also be large and, accordingly, there will be a large number of variations in the data structures to be collected. The model of data streams is the most suitable here, and often - the only possible model. Figure 7 illustrates oneM2M architecture (it is the only existing at this moment completed IoT standard). The IoT data flows in at a set of streams and their data needs to be analyzed in real time as it streams into the database.

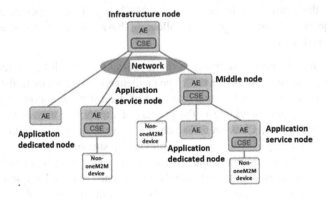

Fig. 7. oneM2M architecture [23]

Each node here is an independent data stream. And the whole system is just an aggregation of data streams. So, stream processing will analyze sensor and device data in real time and (probably) stream data through into persistent stores (e.g., some Hadoop platforms). It will provide continuous process automation from operational systems, as well as merge new data with historical datasets for analysis.

As the practical example, we could mention EU FP7 CityPulse project [2, 3]. CityPulse aims to provide large-scale stream processing solutions to interlink data from Internet of Things and relevant social networks. It will also extract real-time information for the sustainable and smart city applications.

The picture does not change if we look at cloud solutions for the Internet of Things. There is also a worldwide transition to stream processing. The main suppliers of cloud solutions offer exactly the stream processing for IoT. Here we can mention, for example, Cloud Publish/Subscribe messaging system that can act as an integrated model for incoming data streams. As per Google, Cloud Pub/Sub provides a globally durable messaging service. With a cloud, this service scales to handle data spikes that can occur when swarms of devices respond to events in the physical world. Another main feature is buffering these spikes from applications monitoring the data. And subscriptions for topics allow different functions of particular IoT application to subscribe to device-related streams of data without changing the primary data-import target (actually, without changing the data persistence scheme).

Another global cloud provider Amazon offers the similar approach with its Message Broker for AWS IoT. It is a publish/subscribe broker service that enables the sending and receiving of messages to and from AWS IoT. Also, Amazon supports such service as Device Shadow. It provides persistent representations of devices in the AWS Cloud.

4   On Streams-Based Algorithms

The next important question that should be covered in educational programs is mining for data streams and popular algorithms in this area.

Based on the analysis of literature (e.g., [24, 26]) and our experience in implementing projects, we can suggest the following basic elements for streams-based algorithms description:

(1) Sampling and shedding. Sampling is the process of probabilistic choice of a data item to be processed (not processed). The challenge with sampling in the context of data stream processing is the unknown size of data. It is an important process for checking anomalies. Shedding is the process of dropping some sequence of data streams. Load shedding has been. The challenge is the same – the unknown size of data. Load shedding is a technique employed by stream processing systems, for example, to handle unpredictable spikes in the input load whenever available computing resources are not adequately provisioned. So, a load shedder drops some data in order to keep the input load below a critical threshold and thus avoid system-wide problems [25].

(2) Sketching and synopsis. Sketching is the process of sample data streams. Sketching involves building a summary of a data stream using a small amount of memory, using which it is possible to estimate the answer to certain queries [26]. Creating synopsis of the data stream is the performing data reduction through synopsis data structures. The typical examples are histograms, frequencies, quantiles, wavelets.

(3) Aggregation. Aggregation of data streams is the process of computing statistical measurements (e.g., means, variance) that summarize data stream.

(4) Approximation. Approximation for data streams is the process of getting an approximate solution with error bounds. When our computations are limited to a bounded amount of memory, for example, it is not always possible to produce exact answers for data stream queries. However, high-quality approximate answers are often acceptable.

(5) Sliding. Sliding window refers to the process of the analysis of most recent data in source stream. For example, it is a typical solution for many time series analysis algorithms.

(6) One-time queries and continuous queries. Queries over continuous data streams have much in common with queries in traditional databases, but there are several distinctions. One-time queries are queries that are evaluated once over a point-in-time snapshot of the data set. And their answer will be returned to the user. Continuous queries are evaluated continuously as data streams continue to arrive. The answer to a continuous query is produced over time and it reflects the stream data seen so far. Continuous query answers may be stored and updated as any new data arrives, or they may be produced as data streams themselves.

(7) Predefined queries and ad-hoc queries. A predefined query is provided to the data stream management system before any relevant chunk of data has arrived. Usually (in the most cases), predefined queries are continuous queries. Ad-hoc queries are issued online after the data streams have already begun. The biggest challenge here is the fact that ad hoc query may require data elements that have already arrived on the data streams.

(8) Data mining algorithms for streams. Classically, it is about extracting knowledge from data streams. As the most often required and used approaches here, we could list here classification, clustering, and time series analysis. All of them are relevant to IoT applications. A perfect example of the latest development in this area is the paper [27]. Finding heavy frequent items is one of the most practical problems in the streaming algorithms. Time series analysis is, probably, one of the most interesting areas for IoT domain, because time series are the natural representation of the measurements. The typical model is presented in [28], for example.

On the practical level, we should mention Apache Spark streaming as a popular industry choice for IoT stream processing. Spark streaming divides the live stream of data into batches (called micro-batches) of a pre-defined interval (e.g. several seconds) and then treats each batch of data as processing block (in Spark it is so-called Resilient Distributed Datasets - RDDs). Then we can process these RDDs using the operations like map, reduce, join and window. The results of these RDD operations are returned in batches. We usually store these results in a data store for further analytics and to generate reports and dashboards or sending event-based alerts [29]. Also, as a significant result for the stream processing oriented systems, we should mention IBM Streams Processing Language [30]. It is the programming language for IBM Info-Sphere Streams, a platform for analyzing Big Data in motion.

5 Conclusion

In this article, we described the use of streaming data processing in the IoT projects. In our opinion, from a technical point of view, projects of the Internet of Things will be built mostly on the basis of stream processing. We analyzed the current practice of using streaming systems in the IoT projects. In this paper, we propose a classification of streaming systems. It can be noted that to date, some de facto standardized schemes of using stream processing systems in IoT projects have practically been formed. Of course, these decisions should be reflected in the curricula. Being employees of universities, we can note that currently, training courses (programs) related to databases (data processing) pay a very little attention (often not at all) to the processing of streaming data. Especially critical, in our opinion, is the situation with the explanation of streaming algorithms. It is possible that the beginning of the practical use of streaming SQL somehow will change this picture, by virtue of the acquaintance of the majority of developers (and teachers) with classical SQL.

Acknowledgement. We would like to thank the reviewers of the EUCNC conference for critical comments on the first versions of this work. Also we are grateful to the employees of the Laboratory of Open Information Technologies of the Lomonosov Moscow State University and Professor V.A. Sukhomlin for valuable discussions.

References

1. Garofalakis, M., Gehrke, J., Rastogi, R. (eds.): Data Stream Management: Processing High-Speed Data Streams. Springer, Heidelberg (2016)
2. EU FP7 CityPulse. http://www.ict-citypulse.eu. Accessed 24 May 2018
3. Tönjes, R., et al.: Real time iot stream processing and large-scale data analytics for smart city applications. Poster session, European Conference on Networks and Communications (2014)
4. Namiot, D., Ventspils, M.S.S., Daradkeh, Y.I.: On Internet of Things education. In: 2017 20th Conference of Open Innovations Association (FRUCT), pp. 309–315. IEEE, April 2017
5. Rose, D.: Enchanted Objects: Design, Human Desire, and the Internet of Things. Simon and Schuster, New York (2014)
6. Namiot, D., Sneps-Sneppe, M.: On Internet of Things and big data in university courses. Int. J. Embed. Real-Time Commun. Syst. (IJERTCS) **8**(1), 18–30 (2017)
7. Namiot, D., Sneps-Sneppe, M.: On data persistence models for mobile crowdsensing applications. In: Kalinichenko, L., Kuznetsov, S., Manolopoulos, Y. (eds.) Data Analytics and Management in Data Intensive Domains. DAMDID/RCDL 2016. Communications in Computer and Information Science, vol. 706, pp. 192–204. Springer, Cham (2017). https://doi.org/10.1007/978-3-319-57135-5_14
8. Lambda architecture. http://lambda-architecture.net. Accessed 24 May 2018

9. Kappa Architecture. http://milinda.pathirage.org/kappa-architecture.com. Accessed 24 May 2018
10. Questioning the Lambda Architecture. https://www.oreilly.com/ideas/questioning-the-lambda-architecture. Accessed 24 May 2018
11. Carbone, P., Katsifodimos, A., Ewen, S., Markl, V., Haridi, S., Tzoumas, K.: Apache flink: stream and batch processing in a single engine. Bull. IEEE Comput. Soc. Tech. Committee Data Eng. **36**(4), 28–38 (2015)
12. Akidau, T., et al.: The dataflow model: a practical approach to balancing correctness, latency, and cost in massive-scale, unbounded, out-of-order data processing. Proc. VLDB Endowment **8**(12), 1792–1803 (2015)
13. The world beyond batch: Streaming 101. https://www.oreilly.com/ideas/the-world-beyond-batch-streaming-101. Accessed 24 May 2018
14. Zaharia, M., et al.: Fast and interactive analytics over Hadoop data with Spark. USENIX Login **37**(4), 45–51 (2012)
15. Apache Flink Windows. https://ci.apache.org/projects/flink/flink-docs-release-1.1/apis/streaming/windows.html. Accessed May 2018
16. Sneps-Sneppe, M., Namiot, D.: About M2M standards and their possible extensions. In: 2012 2nd Baltic Congress on Future Internet Communications (BCFIC), pp. 187–193. IEEE, April 2012
17. Namiot, D.: On big data stream processing. Int. J. Open Inf. Technol. **3**(8), 48–51 (2015)
18. Golab, L., Özsu, M.T.: Issues in data stream management. ACM SIGMOD Rec. **32**(2), 5–14 (2003)
19. Motwani, R., et al.: Query processing, resource management, and approximation in a data stream management system. In: CIDR, January 2003
20. Gama, J., Gaber, M.M. (eds.): Learning from Data Streams: Processing Techniques in Sensor Networks. Springer, Heidelberg (2007)
21. Kreps, J., Narkhede, N., Rao, J.: Kafka: a distributed messaging system for log processing. In: Proceedings of the NetDB, pp. 1–7, June 2011
22. Gartner says the Internet of Things will transform the data center. http://www.gartner.com/newsroom/id/2684616. Accessed 24 May 2018
23. Standardization Activities of oneM2M. https://www.ntt-review.jp/archive/ntttechnical.php?contents=ntr201408gls.html. Accessed 24 May 2018
24. Gaber, M.M., Zaslavsky, A., Krishnaswamy, S.: Mining data streams: a review. ACM SIGMOD Rec. **34**(2), 18–26 (2005)
25. Rivetti, N., Busnel, Y., Querzoni, L.: Load-aware shedding in stream processing systems. In: Proceedings of the 10th ACM International Conference on Distributed and Event-Based Systems, pp. 61–68. ACM, June 2016
26. Babcock, B., Babu, S., Datar, M., Motwani, R., Widom, J.: Models and issues in data stream systems. In: Proceedings of the Twenty-First ACM SIGMOD-SIGACT-SIGART Symposium on Principles of Database Systems, pp. 1–16. ACM, June 2002
27. Larsen, K.G., Nelson, J., Nguyên, H.L., Thorup, M.: Heavy hitters via cluster-preserving clustering. In: 2016 IEEE 57th Annual Symposium on Foundations of Computer Science (FOCS), pp. 61–70. IEEE, October 2016
28. Papadimitriou, S., Sun, J., Faloutsos, C.: Streaming pattern discovery in multiple time-series. In: Proceedings of the 31st International Conference on Very Large Data Bases, pp. 697–708. VLDB Endowment, August 2005
29. Big Data Processing with Apache Spark - Part 3: Spark Streaming. https://www.infoq.com/articles/apache-spark-streaming. Accessed 24 May 2018
30. Hirzel, M., et al.: IBM streams processing language: analyzing big data in motion. IBM J. Res. Dev. **57**(3/4), 7:1–7:11 (2013)

Analysis of Assets for Threat Risk Model in Avatar-Oriented IoT Architecture

Ievgeniia Kuzminykh[(✉)] and Anders Carlsson

Blekinge Institute of Technology, Campus Grasvik, 371 41 Karlskrona, Sweden
{ievgeniia.kuzminykh,anders.carlsson}@bth.se

Abstract. This paper represents new functional architecture for the Internet of Things systems that use an avatar concept in displaying interaction between components of the architecture. Object-oriented representation of "thing" in the avatar concept allows simplify building and deployment of IoT systems over the web network and bind "things" to such application protocols as HTTP, CoAP, and WebSockets mechanism. The assets and stakeholders for ensuring security in IoT were specified. These assets are needed to isolate the risks associated with each of assets of IoT system. Example of Thing Instance's description and its functionality using JSON format is shown also in the paper.

Keywords: IoT · Avatar · Thing instance
Threat assessment · Security risk assessment

1 Introduction

The complexity of ensuring IoT security is that the system is heterogeneous, consists of many assets on each of the architecture layer. The experts from research organizations in the IT field, as well as equipment manufacturers, agree that providing IoT security on the vertical way is a complex task since security aspects will vary depending on the use case and scenario, the application domain and platform used [1].

Although, the threats in the IoT can be similar to those in the traditional IT network, the overall impact could be significantly different. That is why many experts in IoT security organizations focus on threat analysis [2–7] and risk assessments to estimate the impact if a security incident or a breach occurs.

Depending on the application domain of the IoT, a corresponding risk assessment is necessary to implement:

- highlight the specific threats inherent in this or that application and the assets on which it can affect;
- identify possible attack scenarios and distribute them in the context of a specific IoT service;
- determine what threats are critical and how they can be mitigated.

Supported by Swedish Institute Scholarship Programme.

O. Galinina et al. (Eds.): NEW2AN 2018/ruSMART 2018, LNCS 11118, pp. 52–63, 2018.
https://doi.org/10.1007/978-3-030-01168-0_6

Security threats for each of the application domain (such as Smart Cars, Smart Airports, Smart Hospitals, Smart Homes, Intelligent Public Transport, ICS/SCADA, etc.) are unique but there are also universal threats and attacks which most often appear in the applications that are based on the Internet.

In order to provide the general security requirements for the IoT system using threat risk modeling, the first thing to do is to identify the main security stakeholders, security assets, possible attacks, and, finally, threats for the IoT system. Using this general IoT threat model as a basis you can create a specific set of security objectives for a specific use case, IoT application domain.

In this work, we will try to highlight such assets that is necessary for further analysis of the treat risk model for the Internet of Things. We will also specify the stakeholders who are the connecting link between IoT devices, services and customers, as well as link between transfer and displaying the client commands onto smart things.

For describing the model of component interaction in IoT system we will use the avatar-oriented approach, since it allows us to merge objects into a system of objects, system of objects has more functionality than standalone object since the IoT application has complicate interface. If we assume Service as a key component of the IoT system then it displays only a single entity with a relatively simple interface that abstracts a significant amount of activity. Service is an atomic unit of functionality. Like a well-constructed object in object-oriented programming, from the Service Oriented Architecture (SOA), discipline begins, "the services are collections of capabilities". But the IoT Service has a more complex structure than a single entity. The application can use several services to display all information to the end user, can aggregate data from several devices. In simple words, opening the application on the smartphone, the customer wants to have access to the state of his home and all the devices in it, as well as to his car activities, monitor his health and nutrition, find free parking places and check traffic conditions in the city. At the same time, the customer, clearly, does not want to open the state of each physical sensor in his house but wants simply to see the general state, for example, "house is all right", and, also, user wants his position to be determined automatically for searching parking and route selection. Naturally, to perform such a functional, the service uses other services such as GPS and smart home control center in our cases, and after displays the data via the web interface to the user.

To manipulate the data objects the avatar representation approach is most appropriate, then you can easily connect or disconnect microservices, data from the things, or change the visual representation of data.

Avatars are designed to:

- expose objects as resources on the Web: the avatars can be invoked using semantic enabled service-oriented protocols;
- compose collaborative functionalities: they interact with the avatars of other objects to negotiate and fulfill requests requiring complex functionalities, thus enabling inter-object collaboration;

- manage context adaptation: they can adapt objects behavior according to their surrounding environmental changes;
- cope with pervasive setups: they allow network communication disruptions and support optimized communications with remote objects;
- deploy code on the objects: they either deploy application code modules onto the objects or execute them in a cloud infrastructure if objects do not have enough resources to do this [8].

2   Place of Assets in Threat Risk Model

First of all, to ensure security it is necessary to follow policies that are generally aimed at making the system more reliable and resistant to attacks. They should be adequate for a particular service or application platform and should contain well-documented information. When designing the IoT system it is necessary to take into account the features and context of the case of use itself, to determine the interfaces, communications and the instances that will be used during deployment. The IoT security system in the home environment will be different from the IoT in the critical infrastructure. Thus, the risk depends on the context, and regarding to this, security measures should be applied with this in mind.

To identify significant risks using a defense-in-depth approach, first of all, it is necessary to isolate the risks associated with each of assets of the IoT system. This should be done at the early stages of the life cycle of the program, at the design and testing stage. The Fig. 1 shows the interaction of the risk model components.

To compile a risk model we need to know:

- asset, A;
- application domain;
- list of potential threats to Asset, T;
- list of potential vulnerabilities, V;
- list of countermeasures and recommendations for risk mitigation.

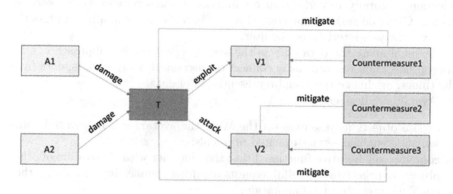

Fig. 1. Risk model components.

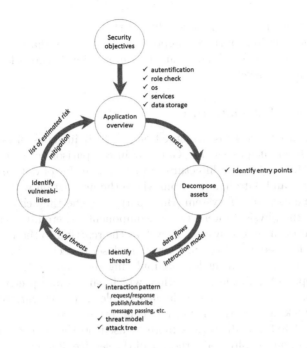

Fig. 2. Algorithm for threat assessment.

The list of threats will be based on a list of potential attacks, and the threat will be considered as the possibility of implementing an attack. The risk is estimated as the probability of the exploitability on Impact (Damage Potential). Regardless of which risk model will be chosen, for example, STRIDE, DREAD or Severity system, the threat algorithm for the risk model will look like this, as shown on Fig. 2.

After all operations to assess the risks of threats we obtain the estimated risk for each of the applications in this application domain and a list of recommendations for reducing these risks.

3   Architecture of IoT

Historically, each IoT solution was based on an application that required the interaction of things with each other and with the user, and therefore the system was developed for each application separately. Hence, there are such variety of architectures offered by different organizations in IoT field as NIST [9], ITU-T IOT [10], AIOTI, W3C WoT [11], and also architectures from manufactures as Microsoft Azure [12], Cisco [13], AWS, Google [14]. They suggest different layers in architecture, include different components, give different terminology, it is not difficult to get confused in this variety.

This paper will use terminology that is closer to the abstract representation of IoT but not to the physical representation. Such a decision was made because in the study we focus on the functionality of the IoT system components, and even on the combinations of these components. We can say that we consider the system as the interaction of functionals, the interaction between layers but not the elements on each level.

3.1   Classical IoT Architecture

A classical IoT architecture is presented on Fig. 3. It includes physical components on each layer such as Device, Gateway in comparison with the functional architecture. The classical architecture displays a transfer of data from the end device to storage and data handler from where the interaction with user begins. This is the time axis of IoT system. But apart from the time characteristics of the assets and the physical interaction of components, there are certain actions and events that occur in the system, as well as the reaction to these events. Such representation focused on actions and events is closer to security since it contains assets that are more convenient for manipulating the security language such as an action (property), an event, a reaction, which can be interpreted in the language of risk theory as: for action is damage, exploit, risk, threat, vulnerability; for event is attack; for reaction is logging, countermeasure.

The classical IoT architecture is more focused on the manufacturer of the device and cloud, on the physical structure of the service. For example, places for ensuring security are cloud, servers, routers, gateways, devices. But at the same time the transient processes are not considered when transferring an IoT object with its properties from the IoT device to the final user and vice versa. Data integrity can be lost when changing from one form of IoT objects to another.

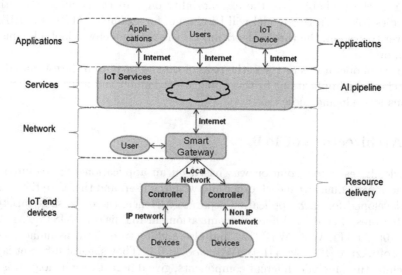

Fig. 3. Classical IoT architecture.

3.2   Avatar-Oriented IoT Architecture

The presentation of avatar-oriented model of interaction of the IoT components is shown on Fig. 4 and does not in any way eliminates the use of classical architecture. In addition, we take the assets which must be protected from potential attacks and threats from the classical model. It includes already not physical components but abstract components such as Thing Instance, Resource design, Service, Avatar.

In the classical model the lower level was a set of sensor devices that collect information and then forward it to the processing and control centers and to the storage with future visualization. In the functional model, all these functions of collecting, forwarding, processing are combined into one layer named *the resource delivery layer* the purpose of which is to deliver information to the service provider. Applications and services do not need primitive data from sensors, they need data of a higher level. At the level of the resource delivery a "thing instance" is created. *Thing instance* is an object representation of the merged data that contains the above-mentioned *data, metadata, interaction model* attributes, requirements for *communication* and *security*.

The resource delivery layer consists of the devices and the network instrumented so they can be addressed individually. The AI pipeline layer consists of platform that helps the resources to interconnect and of intelligence.

From the architecture on Fig. 4 it is not yet obvious which areas of IoT systems to protect. Also, it is not clear where the above-mentioned functionals of the model such as action (property), event, reaction. To move to language of the description of the system from the security point of view it is necessary to determine the so-called assets [8,15–19] which we will use in the threat risk modeling.

Over IoT system the data and meta data circulate, last one describes the type of data and the interaction models inherent for particular application platform or service [20]. The requirements for communication, security and privacy must be implemented for effective interoperability of platforms. The security means that the system must support its functionality even during an attack. The privacy means that the system should protect the confidentiality of personal identifiable information.

Data – information that thing provides to user.

Metadata – supporting information about Thing instance. It includes Protocols and ports, Data formats & encodings, Multiplexing and buffering of data, Efficient use of protocols, Devices specifications.

Interaction model – link from a Thing to the interaction patterns it provides.

Security – links a given Thing to the security information that indicates the access metadata information for securely transmitting information via all the resources of the Thing.

Link – provides Web links to arbitrary resources that relate to the specified Thing instance.

The *interaction model* should support multiple interaction patterns and messaging methods. By default, interaction patterns contain of such assets such as

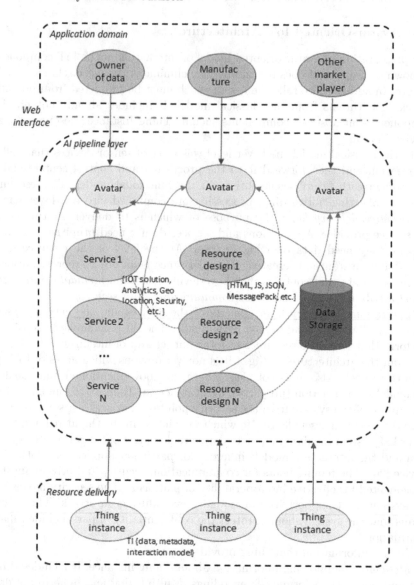

Fig. 4. Avatar-oriented architecture of IoT.

Property, Action, and *Event.* These assets were found to be able to cover the object model representation of any IoT Platforms.

Properties are abstract data points that can be read and often written, it displays the status of the object and stores a value, for example, boolean.

Actions are abstract invokable processes that may run for a certain time, they display object status changes, and are often a function.

Events are abstract interactions where the remote endpoint changes data asynchronously, most often the parameters of the function, threshold values and limited functions.

Simple example, for better understanding of "thing" representation via Thing instance, is shown on Listing 1. Listing 2 describes system of a smart room in a hotel where only a registered guest has access represented via JSON code and using avatar as object.

Door	Light switch	LCD Display

```
{                        {                        {
  "name": "door",          "name": "light",         "name": "screen",
  "description":           "description": "Light    "description": "A simple
  "A door that should       is switched on when       display that show
  be opened when valid      valid key is presented",  notification to client",
  key is presented",       properties: {            "events": {
                             on: {                    "write",
  events: {                   type: boolean,          "clear",
    bell: null,               writable: true          "blink",
    key: {                  }                         "color",
      valid: boolean       }                          "brightness"
    }                    }                          },
  },                                                "properties": {
  properties: {                                      "brightness": {
    is_open: boolean                                   "type": "integer",},
  },                                                 "content": {
  actions: {                                           "type": "string",},
    unlock: null                                     },
  }                                                 actions: {
}                                                     is_displayed: content
                                                    }
                                                  }
```

Listing 1. Simple example of Thing Instance representation.

Interactions between Things can be as simple as one Thing accessing another Thing's data to get or change representation of data such as metadata, status or mode. A Thing may also be interested in getting asynchronously notified of future changes in another Thing, or may want to initiate a process served in another Thing that may take some time to complete and monitor the progress (for example, in different IoT use cases where they need access to GPS location or weather server to provide their own functionality). Interactions between Things may involve exchanges of data between them. This data can be either given as input by the Thing User, returned as output by the Thing Provider or both who are the main stakeholders of Thing.

Each Thing instance can have one or more virtual representations of physical or abstract entities which are called avatars. The Things can also have a story, for example, a car has a story about previous owners. The Avatars have attributes such as a history, patterns of interaction, description, services, identifiers, access control policy, data processing policy, security policy. The Avatars have URIs and are accessible via the web interface. They allow us to simplify the collection of services and applications that can use information from different sources.

```
{
    {
        context: {
        link = http://hotel.de/room12 },
        dependencies: {
        door: door12,
        light: switch12,
        screen: lcd12
        }
    // invoked when service starts
    function start () {
        door.observe(key, unlock); }
        function unlock(key) {
        if (key.valid) {
        door.unlock();
        light.on = true;
        screen.display (Welcome!);
        } screen.display (Door is locked!)
    }
```

Listing 2. Simple example of avatar representation of smart room.

The avatar from the software developer point of view well presented in [11]. There are six functional modules that are responsible for interaction between avatar's attributes and with external components. In our avatar-oriented architecture from Fig. 4 on the top of AI pipeline layer that object of avatar described in [11] is partially formed. Partially is because some components are already inherited from previously formed objects, for example, the functions of the communication module that is responsible for selecting the right network interface and for selecting the right network was formed when Thing instance had been created.

4   Stakeholders and Assets

At each layer of the functional model it is possible to identify specific stakeholders. At the resource delivery layer when instance object is creating, the Manufacturer of the Device plays the role of the stakeholder. Stakeholder's function is to describe the characteristics of the model, properties, supported interaction models, all this information serves to create an instance.

The next stakeholder is the Thing Provider which uses the thing instance to build various specific solutions for different IoT domains. Thing Provider might define new instance or modify provided instance using the AI pipeline layer functionality. In addition, to maintain the integrity of the instance Thing Provider's privacy function increases at the AI pipeline layer. The thing instance can have several providers. In this case, the function of isolation of the Thing Provider increases with the subsequent division of the rights and preferences of users.

Stakeholder Thing User can be either a physical user or an abstract user, for example, if the instance is used by a business provider or company. Thing User must trust to two underlying layers his data and actions of physical objects (for example, video stream data from surveillance cameras or startup of the machine when certain thresholds are triggered). Thing User can differ in the functionality of using avatars and data itself, depending on the access rights. Some can change information, and some only read it. In this case, the function of ensuring the proper authentication and authorization of the user is increased.

Having information about the stakeholders and functional layers of IoT architecture, assets for security can be allocated. We can specify such assets as:

- Thing user data,
- Thing provider data,
- Thing instance itself,
- Interface (Administrative, Device Web Interface, Cloud Interface, Mobile Application) [21].

From the end user view the interaction with IoT system occurs in this way: the user through the IoT interface (for example, the browser on the smartphone, PC, smart TV) communicates through the IoT interfaces and IoT protocols with IoT network where the data about the object is stored in the form of Thing provider data and Thing instance, and user can perform certain actions on his own Thing user data.

Web of Things framework gives very limited approach for implementation security aspects in thing instance. Just couple of line presented in non-official draft of WoT standard [16]. For example, to provide simple security for Listing 2 we can add lines presented below:

```
"security": {
  "cat":"token:jwt",
  "alg":"HS256",
  "as":"https://authority-issuing.example.org"
}
```

Here as an example, JSON Web Token (JWT) type is assigned (cat), the corresponding hashing algorithm "HS256" (alg), and issuing authority of the security token (as).

5   Conclusions

Since IoT ecosystem is heterogeneous new vulnerabilities appear related to the interaction between the microservices on AI pipeline layer. More often the end user wants to have one application for many IoT systems where he can log in with the same credentials but do not download and open applications for each of the services. This multi-service also causes risks.

The security policy in the IoT must ensure the integrity of the thing instances and delivering it only to an authorized consumers, i.e. stakeholders, whether it is a service provider or an end user. According to the assets that were defined in the paper the attacks in IoT could be:

(1) Against Thing instances modifying
- property,
- action,
- event.
(2) Against Thing user data.
(3) Against Thing provider data.
(4) Against Interfaces.
(5) Against Communication.

For now, security metadata in Thing instance is defined as optional. That is why it is big challenge for researchers and software developers to implement security methods and mechanism for IoT that is avatar-oriented. Among the tasks under development are ensuring privacy and protecting Thing and related Assets against web attacks, DoS attacks, securing software and firmware updates.

Each IoT stakeholder should focuses on how devices and their resources must be secured so that they can only be accessed by authorized users and applications. Among the mechanisms that provides sharing Thing user data, Thing provider data, Thing instance itself in secure and flexible way there are well-known PKI, Encryption, TLS, OAuth, API tokens, JWT, delegated authentication, as well as specific to Web of Things concept mechanisms such as Social Wo Tot Social Networks authentication, WebSockets, Webhooks.

Once Things are connected to a public network, the most important problem to solve is how to ensure that only a specific set of users can access only a specific set of resources at a specific time and in a specific manner. In other words, if we back to smart room scenario, hotel guests (and only they) should have access to some services and devices in their room (and only there) during their stay (and only then).

For future work we are going to apply the threat risk model with the assets that were defined in the paper to the different IoT use cases, for example, to the hotel management system, health monitoring, SCADA network.

References

1. Baseline Security Recommendations for IoT. European Union Agency For Network And Information Security, ENISA (2017)
2. Ali, B., Awad, A.I.: Cyber and physical security vulnerability assessment for IoT-based smart homes. Sensors **18**(3), E817 (2018)
3. Hossain Md.M., Fotouhi, M., Hasan, R.: Towards an analysis of security issues, challenges, and open problems in the Internet of Things. In: 2015 IEEE World Congress on Services (SERVICES), pp. 21–28. IEEE (2015)

4. Macaulay T.: RIoT Control. Chapter 12 - Threats and Impacts to the IoT. Elsevier (2017)
5. Nurse, J.R.C., Creese, S., De Roure, D.: Security risk assessment in Internet of Things systems. IT Prof. **19**(5), 20–26 (2017)
6. Machine-to-Machine Communications (M2M). ETSI TR 103 167 V1.1.1 (2011)
7. Akatyev, N., James, J.I.: Evidence identification in IoT networks based on threat assessment. Future Gener. Comput. Syst. (2017). In press
8. Asset Avatars: Get a 360-Degree View of Your Assets - Whitepaper. Hitachi Vantara (2017)
9. Voas J.: Networks of 'Things'. NIST SP 800-183 (2016)
10. Reference architecture for IoT device capability exposure. Recommendation ITU-T Y.4115 (2017)
11. Mrissa, M., Mdini, L., Jamont, J.P., Le Sommer, N., Laplace, J.: An avatar architecture for the web of things. IEEE Int. Comput. **19**(2), 30–38 (2015)
12. Microsoft Azure IoT Reference Architecture. Version 2.0. Microsoft Inc. (2018)
13. Nivaggioli, P.: Cisco SP IoT Architecture. Cisco. https://www.cisco.com/c/dam/m/fr_fr/events/2015/cisco_day/pdf/7-ciscoday-10june2016-sp-iot.pdf. Accessed 29 May 2018
14. Guth, J., et al.: A detailed analysis of IoT platform architectures: concepts, similarities, and differences. In: Di Martino, B., Li, K.-C., Yang, L.T., Esposito, A. (eds.) Internet of Everything. IT, pp. 81–101. Springer, Singapore (2018). https://doi.org/10.1007/978-981-10-5861-5_4
15. Web of Things (WoT) Thing Description. W3C Draft. https://w3c.github.io/wot-thing-description/#introduction. Accessed 29 May 2018
16. Web of Things (WoT) Security and Privacy Considerations. W3C Draft. https://rawgit.com/w3c/wot-security/master/index.html#introduction. Accessed 29 May 2018
17. Zambonelli, F.: Towards a General Software Engineering Methodology for the Internet of Things. Cornell University Library, arXiv:1601 (2016)
18. ASAWoO project - Adaptive Supervision of Avatar/Object Links for the Web of Objects. https://projet.liris.cnrs.fr/asawoo/doku.php. Accessed 29 May 2018
19. Kuzminykh, I.: Avatar conception for "Thing" representation in Internet of Things. In: 14th Swedish National Computer Networking Workshop, Karlskrona, Sweden (2018)
20. BONSEYES - Artificial Intelligence Marketplace. https://www.bonseyes.com/. Accessed 29 May 2018
21. OWASP Internet of Things Project. https://www.owasp.org/index.php/OWASP_Internet_of_Things_Project. Accessed 29 May 2018

State of the Art Literature Review on Network Anomaly Detection with Deep Learning

Tero Bodström[✉] and Timo Hämäläinen

Faculty of Information Technology, University of Jyväskylä,
Agora, P.O. Box 35, 40014 Jyväskylä, Finland
`tero.bodstrom@gmail.com,timo.hamalainen@jyu.fi`

Abstract. As network attacks are evolving along with extreme growth in the amount of data that is present in networks, there is a significant need for faster and more effective anomaly detection methods. Even though current systems perform well when identifying known attacks, previously unknown attacks are still difficult to identify under occurrence. To emphasize, attacks that might have more than one ongoing attack vectors in one network at the same time, or also known as APT (Advanced Persistent Threat) attack, may be hardly notable since it masquerades itself as legitimate traffic. Furthermore, with the help of hiding functionality, this type of attack can even hide in a network for years. Additionally, the expected number of connected devices as well as the fast-paced development caused by the Internet of Things, raises huge risks in cyber security that must be dealt with accordingly. When considering all above-mentioned reasons, there is no doubt that there is plenty of room for more advanced methods in network anomaly detection hence Deep Learning based techniques have been proposed recently in detecting anomalies.

The papers reviewed showed that different Deep Learning methods vary in their performance to detect anomalies, but neural networks capability to adapt to rapidly changing network environments by self learning, to handle multi-dimensional data and to detect previously unknown attacks gives a huge advantage for detecting sophisticated attacks such as APT.

Keywords: Network attacks · Anomaly detection · Deep learning

1 Introduction

The purpose of this study is to highlight major challenges in network anomaly detecion with deep learning by focusing on recent research in the field. Among the growing number of data and network connected devices, the challenge is different attack types such as APT, DDoS and Zero-day. They each have a unique behavioural pattern and the difficulty is to come up with a solution that

© Springer Nature Switzerland AG 2018
O. Galinina et al. (Eds.): NEW2AN 2018/ruSMART 2018, LNCS 11118, pp. 64–76, 2018.
https://doi.org/10.1007/978-3-030-01168-0_7

has the capability to detect all of them efficiently in modern networks. In this study following aspects are considered: intended attack type detection, functional differences as well as the differences in detection accuracy.

By definition, anomalies are observations which differ from other observations enough to arise suspicion. Suspicious observations in network traffic can be caused by either legitimate events or non legitimate events and the purpose of anomaly detection is to divide normal and anomalous data with different techniques [1,2]. However, any suspicious event has to be treated as hostile, until it is verified and proved to be non-hostile.

Most of the presented studies in this paper are focused on DDoS, Zero-day and web attacks. There is less current research material on APT attacks for some unknown reason and one focus in this paper is to evaluate the possibility for using Deep Learning to detect APT attacks.

This paper unfolds as follows: in the second section research papers based on different anomaly detection technologies where deep learning methods were used to detect anomalies is discussed. The third section summarizes perceived improvements for the presented researches. The fourth section concludes this review discussing advantages and disadvantages of selected research presented along the paper.

2   Network Anomaly Detection with Deep Learning

In this section different methods that use deep learning for network anomaly detection are presented. These methods focus on one or more simultaneous attack types.

2.1   Deep Learning Approach for Intrusion Detection Using Recurrent Neural Networks

Yin et al. proposed a deep learning based intrusion detection (RNN-IDS) method in their paper. The purpose of the study was to improve intrusion detection systems with recurrent neural networks, system performance and review the possibility to solve two types of classification problems: (i) binary and (ii) multiclass classification. These classification problems were chosen since first, the data must be classified as anomalous or legitimate and then categorized for different attack types. Instead of more traditional machine learning methods, such as support vector machine (SVM), K-Nearest Neighbour (KNN), random forest (RF) and so on, the authors selected a recurrent neural network deep learning method as it surpasses traditional methods due to the ability of processing high-dimensional data. The proposed classification methods were tested with binary and multiclass classification with five categories as follows: (i) normal, (ii) DoS, (iii) User to Root (U2R), (iv) Probe (Probing) and (v) Root to Local (R2L) [3].

The authors used Python written Theano as a deep learning framework and selected RNN to be used because of high amount of dimensions in the data. They selected NSL-KDD dataset to be used during tests, as it resolves KDDCup99 dataset known problems, such as inherent redundant records [3–6]. NSL-KDD dataset includes 41 variables, of which 38 are numeric and 3 non-numeric. For training RNN and executing detection tests, they preprocessed the data which lead to an increase in the number of dimensions, from 41 to 122. This happens when a variable has multiple possible values and every value has to be presented as unique in a numeric matrix with zeros and ones [3].

As mentioned earlier, the authors executed comparison tests for both, binary and multiclass classification. They tested different hidden node numbers and learning rates for optimizing accuracy rate. For binary classification it was found that with 100 epochs, 80 hidden nodes and 0.1 learning rate the best accuracy was achieved. With multiclass classification it was found that 80 epochs, 80 hidden nodes as well and 0.5 learning rate presented the highest accuracy. Test results showed that RNN has higher accuracy and lower false positive rate than compared traditional machine learning methods [3].

For the future work Yin et al. will put focus on GPU acceleration to reduce training time and how to avoid exploding and vanishing gradients. Besides those, they will study how to improve classification performance of LSTM and Bidirectional RNNs algorithms for the intrusion detection purposes [3].

The implementation of basic RNN that the authors proposed handled 122 dimensional data with respectful results, thus outrunning traditional ML methods. Multidimensional handling is extremely important when detecting APT attacks since the difference to normal data can be almost non-existent, possibly only some bits at binary level.

2.2 DeepDefense: Identifying DDoS Attack via Deep Learning

Xiaoyong Yuan et al. proposed Bidirectional Recurrent Neural Network based DDoS detection in their paper. The purpose was to improve DDoS detection rate, as traditional machine learning methods are limited by small depth of representation models. They stated that DDoS attack traffic is hard to detect automatically since the traffic is very similar to normal traffic and attackers try to mimic normal high usage peaks by using Flash-crowd method. In addition, statistical based detection performs well with specific DDoS and needs preprocessing metrics for different attacks. They proposed a model called *DeepDefence* which they tested with the following deep learning methods for the model: (i) Recurrent Neural Network (RNN), (ii) Long Short-Term Memory Neural Network (LSTM) and (iii) Gated Recurrent Unit Neural Network (GRU), and executed comparison tests with more traditional Random Forest ML method [7].

In the proposed method, the authors designed Bidirectional Recurrent Neural Network, where input nodes has two separated parallel hidden layers: (i) Forward and (ii) Backward recurrent layer. Data passes through both hidden layer sets and results are connected in a latter layer before the prediction of an anomaly. RNN was selected as it handles historical information problem well (especially LTSM and GRU), that is, the method improves performance based on historical data patterns, whereas single-packet detection cannot perform [7].

For testing the proposed method, seven day recording of ISCX2012 dataset was selected and extracted to two separate datasets, *data14* and *data15*. First they tested different RNN methods with both datasets and compared each other to find the best performing method. All RNN's performed with high accuracy rate, but eventually LSTM showed highest accuracy rate with *data14* (97.996%) and 3LTSM with *data15* (98.410%). Secondly they tested random forest accuracy rate with same datasets. Tests showed lower accuracy rate, but also a slight gap between datasets: 97.117% accuracy with *data14* and 92.518% with *data15* [7].

Yuan et al. stated that for the future work they are increasing the variety of DDoS vectors and system settings in order to execute performance tests in different environments to verify robustness. They will build *Deep DDoS Defense system* based on proposed method and execute tests in real environments [7].

The proposed RNN showed interesting results. The accuracy rates were almost equal with 0.099% difference while with random forest the difference was 4.599%. Based on the results recurrent neural networks perform in a more stable manner than traditional machine learning methods. LTSM is also worthy for further study in APT attack detection as it can recall historical information which may be crucial when detecting APT attacks.

2.3 Network Anomaly Detection Using Artificial Neural Networks

Andropov et al. proposed multilayer perceptron Artificial Neural Network (ANN) with backspace propagation algorithm training in their paper. The purpose was to research the possibility to detect attacks that are unknown which is difficult if not even impossible with signature based detection methods [1,4]. Anomalies can be caused by network attacks, malware and hardware or software malfunctions, and any kind of anomaly should be treated as dangerous. Changes in network topology, such as new network device or software gets installed or new end-user device with previously unknown software is added, occurs from time to time and causes new patterns or behaviour in network. However, neural networks permit these changes and is able to adapt by adjusting its weights accordingly [1].

The proposed method has two functional states: (i) offline and (ii) online traffic analysis and it uses data aggregation to detect patterns in the elseways highly variable traffic data. The authors implemented Netflow protocol to gather information from network devices, which was then filtered and aggregated for anomaly detection process in ANN. Aggregation was executed with different criteria, for example: (i) number of packets per hour, (ii) average packet size and (iii) port usage, which were used as an input to neural network. They used three layer ANN: (i) input layer, (ii) hidden layer and (iii) output layer, where output layer has a neuron for every anomaly the method is able to detect and two extra neurons, one for normal and one for unknown. They selected six input neurons, 10 neurons in a hidden layer and seven output neurons, and used Sigmoid function for classification in both hidden and output layer [1].

For testing the proposed methods, a local ISP collected data for the authors from several hundred L2 nodes for a period of one month and this dataset was considered as a normal traffic. Authors created several small-scale anomalies for testing purposes, such as: (i) DoS and DDoS attacks, (ii) port scans, (iii) email spamming and (iv) routers turning off. They also created a custom anomaly to test if the neural network can detect an unknown anomaly class. These created anomalies were injected into normal traffic dataset before executing detection tests. The proposed ANN was able to detect both, known and unknown anomalies with high accuracy, however test results show that idle scan is the most difficult to detect. Custom anomaly had the lowest accuracy with 150000 classification iterations, ARP spoofing the second lowest and idle scan third lowest. After 300000 classification iterations all other anomalies had over 80% accuracy, while idle scan stayed below 80% [1].

Andropov et al. mentioned briefly that for the future work their focus is in *"reducing the false positive rates and optimizing the aggregation algorithms"* [1].

The proposed multi-output layer method gives more detailed information about the anomaly than the boolean alternative due to classification. This type of method can be useful in detecting APT attacks since it has multiple simultaneous attack vectors. It would be interesting to see how the idle scan detection accuracy could be increased, for example by optimizing the ANN, perhaps choosing different amount of hidden layers and neurons in a layer.

2.4   Network Anomaly Detection with Stochastically Improved Autoencoder Based Models

Aygun et al. proposed a deep learning based IDS for zero-day attack detection. The purpose was to improve zero-day attack detection with enhanced Autoencoder (AE) based models. Current IDS's are based on signature database from previous attacks and does not detect well unknown new attacks even if the database is kept up to date. They stated that, because of IDSs lack of ability to detect earlier unknown attacks, the research community is moving towards the machine learning based smart IDS, which can adopt new and constantly changing network attacks and reduces the existing problem from occurring [2,4-6].

The authors proposed AE with stochastic anomaly threshold determination algorithm. They tested and compared performance of stochastic and deterministic AE's with the proposed algorithm. AE has two basic functionalities, (i) encoding and (ii) decoding, whereas the number of nodes in input layer and output layer remain the same. The method uses encoding for trying to express input with smaller amount of units in a hidden layer and then reconstructs the encoded input in decoding phase as an output. However, AE's known problem is that it can become as an identity function due to training data and therefore may perform with low accuracy. To avoid that problem, the authors selected stochastic de-noising method. A semi supervised training method was used for training AE because of more satisfactory detection accuracy for zero-day attacks. They only used normal traffic data for the training process and afterwords the model was used for classifying instance. The data that was not recognized as normal traffic was labelled as anomalous. However, the detection results must be interpreted correctly with proper thresholds since the semi supervised training has a known problem called reconstruction error which will be high for anomalous and low for normal data [2].

NSL-KDD dataset was selected for testing the proposed smart IDS. The dataset includes different datasets for training and testing purposes, the test set also has completely different attack types than the training set and is therefore well suited for zero-day detection tests. The data was pre-processed and feature number increased from 41 to 121 and then dataset was scaled to range of [0, 1]. Evaluation test results showed that the deterministic detection method performed with 88.28% and the stochastic method with 88.65% accuracy respectfully [2].

The proposed method performed with decent detection rate for detecting zero-day attacks and therefore could be studied also for APT detection. However, automated interpretation for detection results is recommended.

2.5 An Anomaly-Based Network Intrusion Detection System Using Deep Learning

Van et al. proposed deep learning based Network Intrusion Detection System (NIDS) in their paper. The purpose was to implement DL method to NIDS to gain anomaly-based detection and to be able to detect known and unknown attacks [5]. Nowadays enterprise networks have a significant importance and the growing number of devices and vulnerabilities has lead to a situation were security solutions have to be more flexible to adapt to constantly changing environment variables, and deep learning methods are well suited to [2,4–6]. The authors stated that anomalies can be categorised in three ways: (i) point anomaly, (ii) contextual anomaly and (iii) collective anomaly, and these relate closely to network attacks such as Denial of Service (DOS), Probe, User to Root (U2R) and Remote to local (R2L). NIDS must have the ability to detect both known and unknown anomalies as well as to analyse and classify all attack types [5].

Intrusion traffic differs from normal and hence anomaly detection methods are suitable for intrusion detection, and the authors proposed a method where they used deep learning self-learning competence in an effective way. They used stacked method for constructing multi-layer neural networks. That is, they used three combined neural networks, where the output from the first neural network's hidden layer was used as an input data to the second neural network and finally the detection and classification was done based on the output from the third neural network. Restricted Boltzmann Machines (RBM) and Autoencoder (AE) were selected for neural network pre-training [5].

For detection tests, they selected KDDCup99 dataset and executed tests with both selected methods. Even though the authors stated that the widely used KDDCup99 dataset is not realistic since it is obsolete and lacks of modern network features, it was used in their tests [3–6]. Test results showed that both methods have high detection accuracy, but AE performs better than RBM when classifying data into normal and four different attack types. However AE has longer execution and pre-training time [5].

For future work Van et al. mentioned two things, first they will study how to implement system to parallel platforms for gaining more computational speed. Secondly they want restudy pre-training techniques to optimize and reduce oscillation in order to decrease training errors and increase detection accuracy [5].

Stacked neural network is an interesting idea and could be studied for APT detection. However, more relevant datasets should be used for modern detection system testing to verify if the test results get affected.

2.6 Unsupervised Labeling for Supervised Anomaly Detection in Enterprise and Cloud Networks

Baek et al. proposed Unsupervised Labeling for Supervised Anomaly Detection, that is, combination of functionality from both, supervised and unsupervised training methods to improve detection accuracy. This combined method was studied due to problems of individual trainings methods which they addressed as follows: (i) Supervised learning method uses labelled data which has high accuracy and it executes fast data point tests, however the downside is that all possible data is not available and thus cannot be labelled as detection expects, (ii) Instead of labelling, unsupervised learning method does data classification during the learning process and requires less data for detection, however it has low detection accuracy and high runtime complexity [6]. The purpose was to improve anomaly detection in an enterprise and ISP networks, where fast detection is critical for business [2,5,6].

In the proposed method, the authors were more interested in detecting anomalies than classifying attack types. By preliminary research they defined traditional anomaly detection as follows: *"Assumption: Normal data instances belong to large and dense clusters, while anomalies either belong to small or sparse clusters"* [6]. Based on the mentioned phrase, parameters were defined for size *(size := {small or not})* and density *(density := {sparse or not})*, and by defining this way mathematical definitions were created. The authors rephrased earlier assumptions for anomaly detection by proposing the following method: (i) extremely dense cluster is labelled as anomalous, (ii) small or sparse cluster is labelled as anomalous and (iii) otherwise cluster is labelled as normal. The authors redefined the mathematical format of clustering, based on their new definition, so that it suited the proposed method. They tested and optimized the method and finally labelled testing data with it. The authors then selected four supervised known and widely used methods: (i) Naive Bayes, (ii) Adaboosting, (iii) SVM and (iv) Random Forest and trained and tested all with the labelled data [6].

For testing purposes NSL-KDD dataset was selected due the know problems in KDDCup99 dataset [3–6]. First they tested performance of earlier mentioned traditional anomaly detection, which overall detected anomalies poorly. Tests were executed with five different cluster size (16, 32, 64, 128 and 256) and highest performance was found with cluster size 16, only 62% accuracy. However, the accuracy improved from 62% to 88% with the authors proposed rephrased method using the same cluster size 16. Finally they tested four supervised methods with labelled data and different cluster sizes. Even though the cluster size and dataset varied the test results showed that, by training supervised method with earlier labelled data by unsupervised method, the anomaly detection performs with high accuracy [6].

Baek et al. mentioned that for future work they will develop a method for minimizing randomness of K-Means clustering. They were thinking that one possible method could be to including results of multiple runs for gaining more coherent results, with the help of quorum method. Another possibility would be reducing data dimensions to gain improvement for classification [6].

Test results showed that by combining highly performing functionality from different methods, it is possible to improve overall performance. On the other hand, it would be interesting to see what is the performance accuracy for classifying different attack types with the proposed method. As in some cases only detecting anomalies does not fulfil the priorities of incident response - attack type classification is required so that most urgent and dangerous attacks can be attended and dealt with at first.

2.7   IEEE 802.11 Network Anomaly Detection and Attack Classification: A Deep Learning Approach

Thing proposed Network Anomaly Detection and Attack Classification method for IEEE 802.11 standard wireless networks. The purpose was to classify different type of attacks in wireless networks with deep learning. Besides legitimate traffic, attacks were classified for three types: (i) flooding, (ii) injection and (iii) impersonation type attacks. Wireless network anomaly detection was selected due to growing number of smart home, smart city and IoT solutions and devices, as their communication largely depends on wireless networks and IEEE 802.11 is de-facto standard. Wi-Fi protocol has various vulnerabilities and it has been intensively surrounded by attacks. This causes high level risks to end users and enterprises, including espionage and identity, credit card and money theft. According to authors, it is not enough to treat anomalies as a binary problem but classification is also needed for later analysis and possible recovery operations [4], including software vulnerability patching.

Stacked Auto-encoder (SAE) neural network was selected for implementing the proposed method. The authors proposed two SAE frameworks, with two and three hidden layer neural networks. In both frameworks, 256 neurons in the first and 128 in the second hidden layer were used. In addition, the latter framework had 64 neurons in the third layer. For the activation function selection they executed a test with following functions: (i) Sigmoid function, (ii) Rectified Linear Unit (ReLU), (iii) Leaky ReLU (LReLU) and (iv) Parametric Rectified Linear Unit (PReLU). After the activation function, the method executed Softmax regression for prediction, as it supports multi-class classification, while logistic regression supports only binary classification [4].

For testing purposes the authors created the dataset to be used, as they could not use raw TCP dumps due to the different approach. They created a lab environment (WEP encrypted access point (AP)) with desktop machine, two laptops, two smart phones, a tablet and a smart TV to generate wireless legitimate traffic and capture it directly from the air. Kali Linux was used to execute 15 type of attacks and in addition several penetration testing tools were used. They classified and defined attacks as follows: (i) injection, high number of correctly encrypted data frames, (ii) flooding, high volume of management frames per unit time and (iii) impersonation, introduce an access point to broadcast beacon frames to advertise a pre-existing network(victims network). Tests were executed with two SAE frameworks and all four action functions with the created data set, including 15 attack types, where seven were previously unknown. Results showed that two hidden layer framework with PReLU action function had highest prediction overall accuracy, respectful 98.6688%, and it was able to detect all categories with high and balanced manner. Based on test results, they were able to determine that impersonation attacks are most challenging to detect, but proposed method improved accuracy significantly compared to earlier studies [4].

The proposed method used standard protocol traffic for anomaly detection instead of raw TCP dump. It would be worth studying if this type of approach can be addressed with wireless protocols such as Bluetooth, LoRaWan, ZigBee and others used in sensor networks employed in IoT-systems. The research results showed also that it is truly important to optimize neural networks as varying the amount of hidden layers and different activation functions can give significantly different predictions with the same dataset.

2.8   Malware Traffic Classification Using Convolutional Neural Network for Representation Learning

Wang et al. proposed malware traffic classification with Convolutional Neural Network (CNN) by transforming traffic data to images. This selected and proposed method does not require any hand-designed features but instead it takes raw data as an input and classifies the raw data by transforming it to images. The author claims, this was the first attempt to use representation of raw data to classify malware traffic [8].

Classifying traffic, the authors used traffic granularity and packet layers, which enabled OSI or TCP/IP layer selection for each packet. In their proposed method, traffic granularity included: (i) TCP connection, (ii) flow, (iii) session, (iv) service, and v) host and besides those, they used flow and session information. Flow packets were defined as follows: (i) source IP, (ii) source port, (iii) destination IP, (iv) destination port and (v) transport-level protocol, while sessions were defined as bidirectional flows. Four types of data representations were defined to reduce image size and eliminate session problems with IP and MAC addresses. The defined presentations were: Flow + All, Flow + L7, Session + All, Session + L7 (L7 is OSI layer 7) [8].

For testing purposes the authors created USTC-TFC2016 dataset and developed data-preprocessing toolkit USTC-TK2016. They mentioned that KDD-Cup99 and NSL-KDD offers multiple features but did not meet the requirements for raw data detection. The specific dataset was created by collecting data from multiple sources, ten different sets of malware traffic data from public websites and normal traffic was collected with IXIA BreakingPoint simulator and included ten common office software traffic datasets. They executed tests with two separate datasets for all four representations, that is eight tests in total. The test data was preprocessed with following steps: (i) traffic split, (ii) traffic clear, (iii) image generation and (iv) IDX conversion, that is, conversion from raw traffic data to CNN input data. They selected a static image size, 784 bytes, and for smaller image sizes they added 0×00 padding to achieve the correct size and larger image sizes were trimmed down to correct size. With the toolkit they developed they generated 752,040 records for testing purposes. The authors took also visual test for images and mentioned that different classes had *"obvious discrimination degree and each class of traffic has high consistency"* [8]. Comparison test results showed that *Session + All* representation had highest accuracy and average accuracy was respectful 99.41% [8].

Wang et al. briefly stated that for the future work they are planning to improve the proposed method's capability for malware traffic data detection and identification [8].

An interesting different approach for anomaly detection with extremely high accuracy. One thing that paper did not mention though is what is the processing time for raw data to CNN input data, that is, real time detection performance. In case the transformation is time-consuming, the proposed method is not suitable for real time detection. The raw data approach could be used for APT detection without the traffic clear process.

2.9   Real-Time Detection of False Data Injection Attacks in Smart Grid: A Deep Learning-Based Intelligent Mechanism

He et al. proposed a real-time False data injection (FDI) detection from Smart Grids with deep learning. Smart power grid monitoring and communications are moving towards IP networks to gain quality and intelligent functionalities. This progress also increases the possibility to FDI attacks, and the main focus was to improve detecting electricity thefts executed by FDI. Smart power grids are complex interconnected systems including Phasor Measurement Units (PMUs), smart meter, Remote Terminal Units (RTUs) and Supervisory Control and Data Acquisition (SCADA) system, where the latter is the main target for FDI attacks. System state information is vital for stability and efficiency of power grids, which is commonly part of the state level critical infrastructure [9].

The authors proposed an extended Deep Belief Network (DBN) for anomaly detection, that is, *"Conditional Deep Belief Network (CDBN), that exploits Conditional Gaussian-Bernoulli RBM (CGBRBM) to extract high-dimensional temporal features"* [9]. The proposed CDBN method differs from earlier studied CDBNs by functions: while earlier has been studied with time-series model, the authors designed a method to act as a classifier. Other differences were, that the proposed method carried out CGBRBM only in the first hidden layer and Restricted Boltzmann Machine (RBM) for the rest of the hidden layers.

Different methods for hidden layers were selected to reduce training and execution time complexity. System performance depends heavily on sensitivity of the pattern detections, difference in patterns of normal data and FDI compromise data. The proposed method leans heavily to static physical topology of power system [9].

For testing purposes the authors trained CGBRBM and RBMs using unsupervised methods and final binary prediction was executed with Sigmoid activation, which indicated normal or compromised data. They collected data from real world sources to test simulation purposes, which contains small amount of compromised data and they generated artificially labelled compromised data with similar patterns as gathered real world compromised data had. Instead of TCP dump or raw data, the authors used IEEE 118-bus and IEEE 300-bus systems and their specific load profile data, that is used for tracking user power consumption. The proposed system was tested against various methods such as Artificial Neural Network (ANN) and Support Vector Machine (SVM) to compare their detection and scalability performances. Test results showed that the proposed method overall had higher detection accuracy with extremely low false positive and false negative rate [9].

For the future work He et al. mentioned two separate things, first they will expand research for modelling FDI attacks behaviour more practical way. Secondly, they will study what is the minimum number of required sensing units for their proposed method to still perform detection efficiently [9].

The authors research shows how versatile deep learning is for anomaly detection, it can be applied to multiple systems where data is processed. Moreover, it would be interesting to see what kind of changes the proposed method requires to make it more flexible and adaptive to changing power system topologies.

3   Summary of Further Improvements

In this section the perceived improvements for more precise anomaly detection is presented. These identified concerns vary from single to multiple papers.

Even though numerous papers mentioned that KDDCup99 is not realistic, data is obsolete and lack of modern network traffic, it is widely used for benchmarking new methods. The research community should go forward and look for present day datasets or create those to substitute KDDCup99, as in some papers researchers had already done.

In some of the papers only certain parts of IP and TCP headers were selected for the anomaly detection, such as source and destination IP's and ports. These types of solutions can limit the detection capability due to the vast amount of the possibly useful missing data.

Several research showed that Neural Networks have higher detection rate for the earlier unknown attacks. Due to the computational complexity and the hardware requirements, NN's have a negative impact for the detection systems cost structure. In a high velocity data centre hardware prices can increase significantly, in order to gain sufficient detection speed.

Another concern while using NN's is required amount of the training data. The detection rate highly depends on the amount of training data and cannot perform with a low detection rate without the sufficient amount of it.

Some of the papers focused on binary classification, normal traffic or anomalous. This type of solution do not give enough information of the actual attack vector, which can be crucial for incident investigations. Another important point with the binary classification is the false alarm rate, the false negative to be more precise. The false negative rate has to be extremely low or otherwise data will be incorrectly classified as a normal traffic.

4 Conclusion

The papers reviewed showed that different Deep Learning methods vary in their performance to detect anomalies. Every method had its advantages and disadvantages, but most of these methods detect previously unknown attacks extremely well. However, neural networks has a downside, they require more computational power that can cause also problems in real-time detection in high velocity networks, if hardware is not powerful enough. Another disadvantage is that it requires more training data to gain sufficient accuracy. On the other hand, neural networks capability to adapt to rapidly changing network environments by self learning, to handle multi-dimensional data and to detect previously unknown attacks gives a huge advantage for detecting sophisticated attacks such as APT since it tries to act undetected as long as possible and mimic normal traffic. They are known to be hiding in networks even for years and neural networks are well suited for detecting those by reducing detection time, as they are difficult to detect in a real-time anyway.

When developing Deep Learning anomaly detection systems and methods, these advantages and disadvantages should be further considered, as they can help to define what could be the actual focus of the work. With current practices a system or a method that could detect all types of attacks, not to mention in a real-time environment, requires enormous resources and might be still even impossible to implement.

References

1. Andropov, S., Guirik, A., Budko, M., Budko, M.: Network anomaly detection using artificial neural networks. In: 2017 20th Conference of Open Innovations Association (FRUCT) (2017). https://doi.org/10.23919/FRUCT.2017.8071288
2. Aygun, R.C., Yavuz, A.G.: Network anomaly detection with stochastically improved autoencoder based models. In: 2017 IEEE 4th International Conference on Cyber Security and Cloud Computing, pp. 193–198 (2017). https://doi.org/10.1109/CSCloud.2017.39
3. Yin, C., Zhu, Y., Fei, J., He, X.: A deep learning approach for intrusion detection using recurrent neural networks. IEEE Access 5, 21954–21961 (2017). https://doi.org/10.1109/ACCESS.2017.2762418

4. Thing, V.L.L.: IEEE 802.11 network anomaly detection and attack classification: a deep learning approach. In: 2017 IEEE Wireless Communications and Networking Conference (WCNC) (2017). https://doi.org/10.1109/WCNC.2017.7925567

5. Van, N., Thinh, T., Sach, L.: An anomaly-based network intrusion detection system using deep learning. In: 2017 International Conference on System Science and Engineering (ICSSE), pp. 210–214. https://doi.org/10.1109/ICSSE.2017.8030867

6. Baek, S., Kwon, D., Kim, J., Suh, S.C., Kim, H., Kim, I.: Unsupervised labeling for supervised anomaly detection in enterprise and cloud networks. In: 2017 IEEE 4th International Conference on Cyber Security and Cloud Computing, pp. 205–210 (2017). https://doi.org/10.1109/CSCloud.2017.26

7. Yuan, X., Li, C., Li, X.: DeepDefense: identifying DDoS attack via deep learning. In: 2017 IEEE International Conference on Smart Computing (SMARTCOMP) (2017). https://doi.org/10.1109/SMARTCOMP.2017.7946998

8. Wang, W., Zhu, M., Zeng, X., Ye, X., Sheng, Y.: Malware traffic classification using convolutional neural network for representation learning. In: 2017 International Conference on Information Networking (ICOIN), pp. 712–717 (2017). https://doi.org/10.1109/ICOIN.2017.7899588

9. He, Y., Mendis, G.J., Wei, J.: Real-time detection of false data injection attacks in smart grid: a deep learning-based intelligent mechanism. IEEE Trans. Smart Grid, 2505–2516 (2017). https://doi.org/10.1109/TSG.2017.2703842

Targeted Digital Signage: Technologies, Approaches and Experiences

Kurt Sandkuhl[1(✉)], Alexander Smirnov[1,2],
Nikolay Shilov[2], and Matthias Wißotzki[3]

[1] ITMO University, Kronverkskiy pr. 49, 197101 St. Petersburg, Russia
kurt.sandkuhl@uni-rostock.de
[2] SPIIRAS, 14 Line 39, 199178 St. Petersburg, Russia
{smir,nick}@iias.spb.su
[3] Wismar University, Philipp-Müller-Str. 14, 23966 Wismar, Germany
matthias.wissotzki@uni-rostock.de

Abstract. Information presentation to a wide audience on large screens (digital signage) is quite popular both in publicly accessible places (shopping malls, exhibitions) and in places accessible to limited groups of people (condominiums, company offices). It can be used for both advertisement and non-commercial information delivery. Though targeted information delivery to one person (e.g., advertisement banners on Web pages) is well developed so far, targeting of digital signage is not paid sufficient attention. The paper tackles this problem from three perspectives: new technologies of interactive digital signage at elevator doors are considered, an approach to provide for targeted digital signage is developed, and new business models taking advantage of the above technologies and the targeting approach are proposed.

Keywords: Digital signage · Targeting · Personalisation · Business model
Privacy

1 Introduction

Delivery of contextual personalized information, i.e. depending on the specific situation and taking into account the interests and preferences of specific users, is currently an essential function of modern decision support systems [1, 2]. Thus, the scientific community pays much attention today to improve the quality of information personalization. An illustrative example of providing contextual personalized information outside the area of decision support is targeted advertising, usually provided by such informational "giants" as Google and Microsoft. However, in targeted advertisement the information is provided and tailored to one particular user.

At the same time, information representation to a wide audience on large screens (Digital Signage [3, 4]) is quite popular both in publicly accessible places (shopping malls, exhibitions) and in places accessible to limited groups of people (condominiums, company offices). It can be used for both advertisement and non-commercial information delivery. However, in the case of providing targeted information not to one particular user, but to a group of users, a number of problems arise. For example, it is necessary to

© Springer Nature Switzerland AG 2018
O. Galinina et al. (Eds.): NEW2AN 2018/ruSMART 2018, LNCS 11118, pp. 77–88, 2018.
https://doi.org/10.1007/978-3-030-01168-0_8

identify what information would be of interest and useful not to one user, but to all users of the group, i.e., it is necessary to identify common preferences and interests of the user group. Another important aspect is to observe the confidentiality of users' preferences and interests, since providing information to a group of users may violate the confidentiality of preferences and interests of individual users in the group - who would not prefer to disclose these (for example, illustrations with promising developments of a company should not be broadcasted to a wide audience).

At the moment, there are no solutions that would provide for personalized contextual information to user groups. However, the potential of such "targeted" digital signage are very promising. For example, such information can be presented using electronic displays installed in places where people gather (for example, shops, elevators, transport) to provide information of reference or advertising nature. Information boards in companies, office centers, and transport can also work in this way, displaying information relating to employees, visitors or passengers near them. Such information support would be very relevant for the implementation of such programs as "smart city", the development of public transport, tourism.

The paper is motivated by an industrial project aiming at new application scenarios and potential business models for targeted digital signage solutions connected to elevators. The project is driven by a company producing elevators and aiming for extending its business model with interactive content on elevator doors or blackboards connected to the elevator. In the context of this project, we were able to study the process of both, technology development and digital business model development.

The research focus of the paper is not only on targeted digital signage but also on an approach for developing digital innovations consisting of the four dimensions service design, architecture integration, business model and user experience (cf. Sect. 3). While our previous work focused on the relevance of these four dimensions (see [5]), this paper concentrates on their mutual dependencies and the process of integrating them. The main contributions of the paper are (1) the concept of targeted digital signage motivated by an industrial case, (2) a technical framework for targeted digital signage, and (3) a methodical approach for digital business model innovation and experiences using it in targeted digital signage.

The paper is structured as follows: Sect. 2 discusses the research method used for the work. In Sect. 3, the two industrial cases and the approach for developing digital innovations are summarized. Section 4 describes the development of target digital signage and discusses lessons learned. Section 5 proposes a framework for targeted digital signage, which is independent of the industrial cases. Section 6 summarizes the paper.

2   Research Method

The research method guiding our research work is design science research (DSR) as proposed by Hevner et al. [6]. DSR is a problem-solving approach that is motivated and triggered by an identified business problem and attempts to solve the problem by creating and validating IT artifacts, like prototypes, models, methods or architectures. DSR projects commonly include several iterations in search for the artifact design to the identified problem.

(4) Implementation
Usage of interactive
blackboard in real-word
setting (Prototyping;
ongoing)

(1) Problem Refinement
Industrial demand of
elevator digital signage
in housing industry
(focus group)

DSR
cycle

(3) Validation
Demonstration in industrial
projects at elevator
manufacturer and housing
company (expert studies)

(2) Solution Design
Integration of individual
process support in digital
signage solution
(technical action research)

Fig. 1. Third design science research (DSR) cycle performed in our work.

In our previous work [5], we reported on the first two DSR cycles: cycle one focused on the development of an innovative digital signage solution for elevator doors (EDS); cycle two addressed the integration of gesture recognition into this solution (for a summary, see Sect. 3.1). In both cycles, the artefacts in focus of the DSR project were (a) the approach for digital innovation development, i.e. the dimensions to consider and the process to perform (see Sect. 3.2) and (b) the actual digital innovation, i.e. the EDS. Within these two cycles, it was possible to establish relevance and utility of an integrated development of business model, service design and architecture integration in order to achieve the desired user experience of the envisioned innovative service. The third DSR cycle is subject of this paper and targets the refinement of the approach for digital innovation development and the functionality to achieve the desired user experience of targeted digital signage.

Figure 1 illustrates the third DSR cycle: starting from the results of the first cycles, we made a refinement of the problem, in particular from the perspective of the housing industry (see Sect. 3.2). This resulted in design of a new EDS solution with the specific purpose to support individual processes of stakeholder groups in housing industry with the EDS (see Sect. 4). Validation of the solution design made clear that additional technological concepts were required which are discussed in Sect. 5. Implementation is the prototype is still ongoing. Within the DSR cycle, different research methods are applied: started with a focus group [7] in housing industry for refining the industrial demand we continued with technical action research [8] and expert studies.

3   Industrial Case and Previous Work

As indicated in Sect. 2, our work was embedded in an industrial case aiming at the development of innovative digital signage solutions for elevator doors. This section will briefly present the case study company and summarize the approach for digital business model development, which was developed in previous work and is subject of improvement (Sect. 3.2).

3.1   Industrial Case: Manufacturer of Elevators

The industrial case in our work is a company developing and producing elevators with global market presence and more than 100 years of tradition. In 2016, the company started a new "digital" business line outside the established manufacturing value chain. The first digital service is to offer targeted advertisements and information on elevator doors, i.e. digital signage on elevators (EDS). The case study company has detailed knowledge who operates the elevators and what kinds of users (i.e., target groups for the ads) are frequently using the elevator. This makes the elevator an interesting space to sell in marketing campaigns, which is the core of the business model.

In our first work for this company, we added gesture recognition to the elevator door, which made it an interactive EDS and improved user experience. This work was presented at Hanover industry fair 2017 and reported in [5]. In the latest work, which is subject of this paper, we developed the prototype for another kind of digital signage solution, the so-called blackboard, which is offered to customers as alternative to the elevator door solution. Housing industry is considered as one of the main target market for the blackboard.

Figure 2 shows on the left side the principal components of the elevator door signage (EDS) solution. Above the elevator door, (1) a data projector (2) is mounted or, alternatively, on one side of the door, a shockproof large touch-screen is fixed (3). The data projector (or the touch screen) uses the data communication facility of the elevator shaft to connect to a communication device for all elevators in the building (4). This communication device exchanges information with the back-office, including content and interactions with EDS and for the purpose of elevator maintenance (EOM). Figure 2 shows on the right side a picture of the EDS in operation.

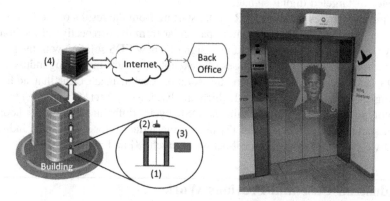

Fig. 2. Components of digital elevator signage (DES) solutions.

3.2 Previous Work on Approach for Digital Business Model Development

Research on digital transformation observed that digital product innovations have to adapt shorter innovation cycles than traditional physical products and that digital products are more dependent on actual user experience and implementing user needs. This observation was the basis for our proposal to perform digital product innovation and digital business development in a tight integration of several dimensions: the user experience (UX), which digital innovation aims to achieve, the product/service design required for the UX, the business model underlying the new product/service, and the integration into the enterprise architecture of the enterprise offering the new product/service.

The enterprise architecture defines how the digital business model is implemented and integrated into an existing enterprise by specifying the different architecture layers, such as defined in TOGAF: the technology used for implementing the business model, the applications required and how they interact, the data relevant for the business model, and the business processes and roles required. Service design includes the user interface, the content model, and specification for the actual service offered to the user (with functional and non-functional requirements). While the architecture dimension addresses the "macro" level of integration in the enterprise, the service design is more on the "micro" level of the service. The business model defines the value offering to the user, the "bundling" of the value proposition into actual services, the distribution, suppliers, and partners. All these dimensions are mutually dependent and aim at an optimal user experience.

4 Development of Targeted Digital Signage in the Industrial Case

4.1 Development Process and Results

The third cycle of our DSR work (see Sect. 2) aimed at the development of targeted digital signage solutions with focus on applications in housing industry and using the approach for digital business innovation.

In the housing industry we formed a focus group with managers of residential buildings and service processes to investigate digitization and innovation potential of tenant services. We developed a list of stakeholder groups of relevance for digitization, such as tenants (grouped according to different phases of their work life, e.g., students, works in shifts, retired, ...), the housing company (again in different groups, like facility managers, administration, ...), logistic service providers (mail, transport, delivery, ...), personal service providers (hairdresser, medical personnel, ...), relatives and friends to tenants. With the stakeholder groups defined, we created an interaction matrix showing all potential value exchange between the stakeholders. Example: the housing company might offer an additional service to elderly people to allow pharmacies to deliver medication to the tenants at a safe facility in the building. This service could be supported using the EDS by displaying relevant information to the tenant as soon as he/she returns to the building.

Based on the interaction matrix, a list of potentially interesting services, required information, and expected user functionality was provided to the design of the next version of EDS development in the industrial case. For the work in case 1, the prime results for developing the blackboard services were the target groups and the value proposition related to the personal assistance. From the perspective of our approach for the business model improvement, these two aspects related to the UX (stakeholder) and the business model (value proposition) dimensions. The service design in this context had to be massively based on the architecture integration, as personal assistance requires the presence of the right data what partly had to originate from the back-office systems of the housing company. Furthermore, required functionality for the blackboard to ensure UX became clear, for example to identify the user. This led to the technical work described in Sect. 5.

In case 1, we developed the prototype of a blackboard to be mounted adjacent to an elevator. The blackboard hardware consists of a shockproof touch-screen monitor with built-in camera, RFID, NFC, Bluetooth and WLAN reader. The blackboard is connected to the back-office via the communication link built into the elevator shaft. The blackboard software so far only supports two "roles" meant to represent stakeholder groups: tenant and facility manager. For supporting individual processes and demands of tenants, in the first step an import of the tenant's profile information is foreseen. The blackboard so far only was tested in the lab. Further tests are scheduled.

4.2 Discussion

The experiences from the blackboard development and the previous EDS and interactive EDS development can be used to derive lessons learned for future work from the perspective of the importance of the different dimensions of our approach for business model development for different categories of EDS solutions (Fig. 3). We define the categories of according to interactive and non-interactive solutions, and according to not personalized, personalized based on preferences and personalized based on individual processes. Based on this categorization, we define targeted digital signage solutions as preferably interactive but in any case personalized solutions. Figure 3 shows the categorization with examples for each category.

The importance of the different dimensions of our approach was determined based on the time consumed working on the dimension and based on the perceived importance from the perspective of the engineers involved. Table 1 shows the result of this investigation for the interactive elevator door solution and the blackboard solution. "1" indicates the highest and "4" the lowest time consumption and perceived importance. The table clearly illustrates the greater weight of user experiences in the interactive elevator door show, which is not providing personalized information, (i.e., no "targeted digital signage") as compared to the blackboard, which can be classified as targeted digital signage.

| | Not personalized | Targeted Digital Signage | |
		Personalized (preferences)	Personalized (individual process)
Non-interactive	Advertisements, campaign with fixed schedule, news	News for user based on preferences	Assistance in task by providing context-oriented information
Interactive	Ads, information and news with navigation possibilities	News for user based on preferences with navigation	Context-oriented assistance in task and dynamic features

Fig. 3. Categorization of digital elevator signage (DES) solutions.

Table 1. Comparison of time consumption and perceived importance

	Interactive elevator door solution		Blackboard	
	Time consumption	Perceived importance	Time consumption	Perceived importance
User experience	1	1	3	3
Business model	3	2	4	1
Architecture	4	4	1	2
Service design	2	3	2	4

5 Framework for Targeted Digital Signage

Based on the requirements and experiences form the industrial cases, we propose a conceptual framework for targeted digital signage which is currently independent from the actual elevator digital signage case. The framework exploits the possibility arising from presence of multiple people next to a digital signage screen to provide for personalized information based on the interests and preferences of these people. The complete developed framework can be split into three major components: viewer detection system, annotated content storage and content management system (Fig. 4). Viewer detection can be based on various techniques:

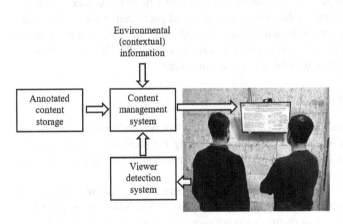

Fig. 4. The framework of the targeted digital signage system

Identification via Bluetooth. The Bluetooth range of several meters is the most appropriate for the task considered. Most of the portable and wearable devices (e.g., mobile phones, smartwatches, fitness bracelets, etc.) currently have Bluetooth and can be used for identification. However, the deeper research has showed that identification of Bluetooth-enabled devices is possible only in two cases: the Bluetooth is switched on in a discoverable mode (which is unlikely), or the personal device is paired with the scanning device (also not very likely due to potentially considerable amount of digital signage installations). Meeting these conditions is fine for experimenting but is difficult in real life.

Identification via Wi-Fi. Like the Bluetooth technology, Wi-Fi is now also integrated into most of portable electronics. The Wi-Fi signal is usually stronger and can span significantly further than that of Bluetooth, e.g., over several rooms in an office building. Technologies of in-door positioning via Wi-Fi exist [9–11], but the results are not precise enough without using additional complementing technologies. However, it is possible to identify if the signal source is really close (in the range of couple of meters) to the receiver what meets the requirements of the research.

Identification via Face Recognition. Another possible technology is identification of viewers via face through a camera photographing the space in front of the digital signage and application of face recognition techniques to the images taken. This is a complex task that requires additional research and it is not considered at the moment in the frame of the research.

The annotated content storage is a database storing the information pieces to be shown with tags that relate it to particular persons (e.g., Nikolay Shilov is asked to visit the HR department), positions at the organisation (e.g., an announcement to all PhD students), topics (e.g., architecture of the XVIII century), and environmental conditions (e.g., announcements related to upcoming weekend). The information about the relationships between particular persons and tags is either stored in the same database or potentially can be acquired from available sources. For example, in large scale the advertisement can be supplied by bigger provider (e.g., Google), which have its own database of users with their interests and preferences.

The content management system is responsible for integration of the environmental information (the context) including time, day of the week, weather, etc., the information about digital signage viewers and available content from the content storage. The content is selected based on the predefined rules and presented at the screen. For increasing the level of compatibility with different equipment, the content management system is located on an HTTP server, and the screen is controlled by a nettop or a single board computer with a web-browser as a thin client.

5.1 Semantic Graph for Domain Representation

In order to display information, which can be interested to a person, his/her preferences have to be analyzed. The research is based on describing the preferences and interests of users through profiling techniques and matching of those with the tags assigned to information pieces to be displayed.

However, when there are several people next to the screen, an intersection of preferences has to be defined so that the displayed information would be interesting for all of them (or at least to most of them). Intersecting preferences of various users is not an easy task. The closest area of research is the group recommendation systems based on the methods of collaborative filtering or similar. These methods are aimed at identifying groups of users with similar interests and preferences. In other words, they solve the problem opposite to the one set in the paper. Nevertheless, some techniques still can be applied within the frames of the research.

First of all, there are not so many works aimed at profile structuring so that it can be easily processed in an automatic way. Having carried out an analysis of the related works the authors of [12] state that their "work is one of the starting points for preference profiles which are "close" to being nicely structured". The authors of [13] use quantitative preference estimations for hierarchically organized preferences (with associative relationships like "the author of", "the genre of", etc.). This is an efficient preference organization when only one domain is considered (e.g., movies). However, when the domain is wide and not well defined such model will not work.

Application of ontologies for preference description can be considered as one of the most efficient solutions to this problem. Ontology is an "explicit specification of a conceptualization," which is, in turn, "the objects, concepts, and other entities that are presumed to exist in some area of interest and the relationships that hold among them" [14]. Ontologies have proven themselves as and efficient instrument for structuring knowledge about a problem domain [15]. The same applies to the organization and structuring of preferences. In [16] unique ontologies are used for every domain. The ontologies are described in OWL. The preferences are quantitative and assigned to various ontology nodes. The hierarchical organization of the ontologies makes it possible to generalize the preferences though this opportunity is not reflected in [16].

Another issue is collecting the preferences. It is not so easy to make people to set the preferences in an application or a web page manually. As a result, this process has to be automated. The authors of [17] track long-term user behaviour with the help of software agents to construct the ontology-based profiles of users. Another possibility is preference grokking [18] by crowd workers. Sometimes this can be an efficient approach, especially to solve the "cold start" problem of personalized systems. However, requirement to have a crowd available can put some significant constraints on its applications.

The organization and description of preferences with the use of ontology is considered to be efficient in the current research as well. As a result, preferences and interests are described in the form of quantitative estimates characterizing the degree of interest.

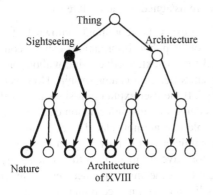

Fig. 5. Organization of preferences/interests and their intersection.

As already mentioned, one of the main problems of identifying common preferences is the absence of those for a relatively small number (3 to 10 people) of unrelated users due to the preference heterogeneity the problem. The existence of interconnections of different nature allows the generalization of preferences and interests (the relation "is a", for example, the interests of "architecture of the XVIII century" and "antique architecture" can be united by the interest of "architecture"), and the search for related interests or preferences (associative relations and relations "part of", for example, the interests of "Carlo di Giovanni Rossi" and "St. Basil's Cathedral" can also be united by the interest of "architecture"). In Fig. 5 it is possible to easily switch from detailed preferences or interests (e.g. Architecture of XVIII century) to more general ones (e.g., Architecture). The nodes do not necessary have only one parent but could have several ones (the relationship is "many to many"). The preferences are generalized and in this case, the match of preferences is higher. In Fig. 5, the preferences are indicated as bold circles, and the resulting node as black circle.

The processing of such structured preferences is carried out based on the graph theory methods (more precisely, semantic graphs). For this purpose the preferences on the weighted graph of ontology, whose vertices are the concepts of the subject domain, and the arcs are the relations between them, are "highlighted" and paths connecting them with supposedly common preferences are calculated. Based on the domain considered a system of weights for paths calculation on a weighted graph has to be identified. The rules for this are the subject of future research.

An important aspect in this research is the contextual dependence of the information provided. Under the context, in this case, we assume all the information that characterizes the environment of the entity and the entity itself at some point in time [19]. The application of methods and technologies of context management allow to analyze a concrete situation, and to determine which information is the most useful and interesting at the moment.

Description of the context is implemented based on the same ontology that is used to describe the preferences and interests of users. Such a description allows combining the context and information about the preferences of the users on the same semantic graph, which in turn makes it possible to better identify the information that corresponds to this situation and to the users.

5.2 Information Confidentiality

When information is presented to a group of people the confidentiality issues become important. In order to define which preferences/interests can be used in which context it was decided to use knowledge forms proposed by Wiig [20]: public, shared, and personal. The same levels of confidentiality were assigned to the preferences:

(1) Public preferences/interests are "known" to everyone (commonly known facts). Usually, the context (weather, date/time, other public information) can be considered as such an interest. It can be used for contextual advertising without any restrictions. E.g., advertising soccer souvenirs before a championship. However, this hardly can be referred to as "targeting". If the user explicitly assigns the level "public" to a certain preference, it is considered that this information can be used in a depersonalized way in groups of people that are not related to each other.

(2) Shared preferences/interests are limited within a community (e.g., a company, department, family). They can be used for advertisement targeting when only members of the same community can see it (e.g., meeting announcement within a department). Since shared information is limited within a community, it can be distributed within the same community. The most obvious example is a company's or department's building and the digital signage can be done through a

screen in an elevator, hall, etc. An example of such information can be an upcoming meetings with their times and locations, exhibitions, success stories or new software tools.

(3) Personal preferences/interests cannot be generally used for any public advertisements. However, it is still possible to use confidential preferences indirectly when searching for common preferences of a group of users, if this does not lead to a breach of confidentiality. This is a subject of future research.

6 Summary and Future Work

This paper focuses on the field of targeted digital signage motivated application needs from housing industry. We defined the category of targeted digital signage as interactive and personalized solutions.

The initial EDS solution developed in our previous work improved due to the new business requirements from the industrial partner from housing industry. The demonstration at the fair in Hanover created the impression that gesture recognition and display on elevator doors were more attractive for audiences seeking entertainment and leisure than for target groups with a certain personal plan or purpose on their minds. However, this impression was not seriously investigated at the fair. Thus, the third DSR cycle at the same time was supposed to address a differentiation of digital signage solution in door show, interactive door show, and classical blackboards.

From method perspective, the technical work described above will contribute to completing the ongoing third DSR cycle and probably motivate more cycles.

Acknowledgements. The research was supported partly by projects funded by grants# 18-07-01201 and 18-07-01272 of the Russian Foundation for Basic Research, by the State Research no. 0073-2018-0002, and by Government of Russian Federation, Grant 08-08.

References

1. Gallacher, S., Papadopoulou, E., Abu-Shaaban, Y., Taylor, N.K., Williams, M.H.: Dynamic context-aware personalisation in a pervasive environment. Pervasive Mob. Comput. **10**, 120–137 (2014)
2. Anagnostopoulos, A., Broder, A.Z., Gabrilovich, E., Josifovski, V., Riedel L.: Just-in-time contextual advertising. In: 16th ACM Conference on Information and Knowledge Management, pp. 331–340 (2007)
3. Schaeffler, J.: Digital Signage: Software, Networks, Advertising, and Displays: A Primer for Understanding the Business. CRC Press, Boca Raton (2012)
4. Want, R., Schilit, B.N.: Interactive digital signage. Computer **45**(5), 21–24 (2012)
5. Wißotzki, M., Sandkuhl, K., Smirnov, A., Kashevnik, A., Shilov, N.: Digital signage and targeted advertisement based on personal preferences and digital business models. In: 21st Conference of Open Innovations Association FRUCT, pp. 375–381 (2017)
6. Hevner, A.R., March, S.T., Park, J., Ram, S.: Design science in information systems research. MIS Q. **28**(1), 75–105 (2004)

7. Yin, R.K.: Case Study Research: Design and Methods. Applied Social Research Methods Series, Third Edition, vol. 5. Sage Publications, Inc., Thousand Oaks (2002)
8. Wieringa, R., Moralı, A.: Technical action research as a validation method in information systems design science. In: Peffers, K., Rothenberger, M., Kuechler, B. (eds.) DESRIST 2012. LNCS, vol. 7286, pp. 220–238. Springer, Heidelberg (2012). https://doi.org/10.1007/978-3-642-29863-9_17
9. Guo, J., Liu, X., Wang, Z.: Optimized indoor positioning based on WIFI in mobile classroom project. In: 11th International Conference on Natural Computation (ICNC), pp. 1208–1212. IEEE (2015)
10. He, S., Chan, S.H.G.: Wi-Fi fingerprint-based indoor positioning: recent advances and comparisons. IEEE Commun. Surv. Tutor. 18(1), 466–490 (2016)
11. Seshadri, V., Zaruba, G.V., Huber, M.: A Bayesian sampling approach to in-door localization of wireless devices using received signal strength indication. In: Third IEEE International Conference on Pervasive Computing and Communications (PerCom) 2005, pp. 75–84 (2015)
12. Bredereck, R., Jiehua, C., Woeginger, G.J.: Are there any nicely structured preference profiles nearby? Math. Soc. Sci. 79, 61–73 (2016)
13. Buvaneswari, N., Bose, S.: Quantitative preference model for dynamic query personalization. Asian J. Inf. Technol. 15(24), 5019–5027 (2016)
14. Gruber, T.R.: Toward principles for the design of ontologies used for knowledge sharing. Int. J. Hum. Comput. Stud. 43(5–6), 907–928 (1995)
15. Oroszi, A., Jung, T., Smirnov, A., Shilov, N., Kashevnik, A.: Ontology-driven codification for discrete and modular products. Int. J. Prod. Development 8(2), 162–177 (2009)
16. Chen, R.C., Hendry, C.Y.H., Huang, C.Y.: A domain ontology in social networks for identifying user interest for personalized recommendations. J. Univ. Comput. Sci. 22(3), 319–339 (2016)
17. Gao, Q., Xi, S.M., Cho, Y.I.: A multi-agent personalized ontology profile based user preference profile construction method. In: 44th IEEE International Symposium on Robotics (ISR), pp. 1–4 (2013)
18. Organisciak, P., Teevan, J., Dumais, S.T., Miller, R.C., Kalai, A.T.: Matching and grokking: approaches to personalized crowdsourcing. In: Twenty-Fourth International Joint Conference on Artificial Intelligence (IJCAI 2015), pp. 4296–4302 (2015)
19. Dey, A.K.: Understanding and using context. Pers. Ubiquit. Comput. 5(1), 4–7 (2001)
20. Wiig, K.M.: Knowledge Management Foundations: Thinking About Thinking – How People and Organizations Create, Represent, and Use Knowledge. Schema Press, Arlington (1993)

State of the Art Literature Review on Network Anomaly Detection

Tero Bodström[✉] and Timo Hämäläinen

Faculty of Information Technology, University of Jyväskylä, Agora, P.O. Box 35, 40014 Jyväskylä, Finland
tero.bodstrom@gmail.com,timo.hamalainen@jyu.fi

Abstract. As network attacks are evolving along with extreme growth in the amount of data that is present in networks, there is a significant need for faster and more effective anomaly detection methods. Even though current systems perform well when identifying known attacks, previously unknown attacks are still difficult to identify under occurrence. To emphasize, attacks that might have more than one ongoing attack vectors in one network at the same time, or also known as APT (Advanced Persistent Threat) attack, may be hardly notable since it masquerades itself as legitimate traffic. Furthermore, with the help of hiding functionality, this type of attack can even hide in a network for years. Additionally, the expected number of connected devices as well as the fast-paced development caused by the Internet of Things, raises huge risks in cyber security that must be dealt with accordingly. When considering all above-mentioned reasons, there is no doubt that there is plenty of room for more advanced methods in network anomaly detection hence more advanced statistical methods and machine learning based techniques have been proposed recently in detecting anomalies. The papers reviewed showed that different methods vary greatly in their performance to detect anomalies. Every method had its advantages and disadvantages, however most of the presented methods cannot detect previously unknown attacks but on the contrary, for example, detects DDoS attacks extremely well.

Keywords: Network attacks · Anomaly detection · Machine learning

1 Introduction

The purpose of this study is to highlight major challenges in network anomaly detection using statistics and machine learning, excluding deep learning, by focusing on recent research in the field. Among the growing number of data and network connected devices, the challenge is different attack types such as APT, DDoS and Zero-day. They each have a unique behavioural pattern and the difficulty is to come up with a solution that has the capability to detect all of them efficiently in modern networks. In this study following aspects are

© Springer Nature Switzerland AG 2018
O. Galinina et al. (Eds.): NEW2AN 2018/ruSMART 2018, LNCS 11118, pp. 89–101, 2018.
https://doi.org/10.1007/978-3-030-01168-0_9

considered: intended attack type detection, functional differences as well as the differences in detection accuracy.

By definition, anomalies are observations which differ from other observations enough to arise suspicion. Suspicious observations in network traffic can be caused by either legitimate events or non legitimate events and the purpose of anomaly detection is to divide normal and anomalous data with different techniques [1,2]. However, any suspicious event has to be treated as hostile, until it is verified and proved to be non-hostile. Most of the presented studies in this paper are focused on DDoS, Zero-day and web attacks. There is less current research material on APT attacks for some unknown reason and one focus in this paper is to evaluate the possibility for using presented methods to detect APT attacks.

This paper unfolds as follows: in the second section research papers based on different anomaly detection technologies are presented. The third section summarizes perceived improvements for the presented researches. The fourth section concludes this review discussing advantages and disadvantages of selected research presented along the paper.

2   Network Anomaly Detection

In this section different methods that use machine learning for network anomaly detection is presented. These methods focus on one or more simultaneous attack types.

The number of DDoS attacks are increasing due to the growing number of IoT devices with low security mechanisms and the fact that nowadays it is fairly easy to acquire attack tools. This has created a situation where large number of these devices can be used to perform distributed attacks, that is, to carry out more powerful attacks [3–7]. So far 24 different DDoS attack vectors have been found globally [3]. There exist researches that focuses on detecting these types of attacks specifically, to be used as a first line defence.

2.1   A Lightweight Network Anomaly Detection Technique

To serve as a frontier in a network anomaly detection system, Jinoh Kim et al. present in their research a new lightweight grid-based approximation technique. Their proposed method is based on a recursive algorithm that partitions data in D-dimensional space. In each recursion, the algorithm verifies if data belongs to a sub-block of a grid and if yes, the data is labelled and the execution of the algorithm stops. Otherwise, the algorithm continues to the next sub-block and execution continues until the data gets labelled. In case the algorithm ends without labelling, the data gets classified as *"Not Sure"*. In their experimental tests, they used only two variables to train the system and detect data: src and dst bytes. The former is the number of bytes from the source to the destination IP addresses, and the latter is the number of bytes from the destination to source hosts, including five continuous attributes (duration, src bytes, dst bytes, wrong segment and urgent) in the group. This method used six different classes for classification: *"Secure"*, *"Marginal Secure"*, *"Marginal Insecure"*, *"Insecure"*, *"Empty"*, and *"Not Sure"*. The authors also made comparison tests with more traditional methods such as decision tree and random forest. Tests for the proposed method showed the following results, (i) accuracy of 98.5% with the KDD data and (ii) 83% with NSL-KDD. In addition, the measured learning time for their method is significantly lower and approximately two orders of magnitude faster than decision tree and random forest [4].

The proposed method detected DDoS attacks successfully and the detection speed was fast. On the other hand, it would be interesting to see if these features will keep up when adding variables for more precise detection.

2.2 Distributed, Multi-level Network Anomaly Detection for Datacentre Networks

Mircea Iordache et al. presents in their research a distributed method for network anomaly detection. In the proposed method, all network devices are equipped with a detection algorithm and they contribute to detecting anomalies. The algorithm has consensus voting mechanism, where at first, each node independently detects if an anomaly exists and then peer nodes vote for the final decision. In addition, the devices are capable of creating attack path reconstruction for further investigation purposes. The actual algorithm has two detection layers to gain better accuracy. The first level performs a coarse-grain detection for fast analysis. In case the algorithm detects a potential anomaly, data is passed to the second, fine-grained detection level. The deeper sub level performs analysis and collects data metrics. The authors selected sketch-based algorithm for fine-grained detection. For in-depth analysis, they modified the Count-Min Algorithm. The purpose was to store source and destination IP addresses and transport layer ports to sketch but this approach enabled also storing the time variant state of the flow and transferring it to the sketch. Therefore, any changes in flow can be detected by comparing network data to stored metrics and used to detect presence of anomalies [5].

The authors tested the proposed method for various attack types, such as Brute Force access, 0-day attacks and Port Scans. Their test results concluded that the solution was able to offer complete path reconstruction at the onset of DDoS attacks that generally have high intensity. Also, partial path reconstruction was achieved for anomalies of lower intensity [5].

For future work M. Iordache et al. mentioned an improvement for synchronizing remotely invoked instances by advanced majority voting scheme. The purpose of the improvement is to extend the amount of classification information what will be used for anomaly detection. Secondly, they proposed further research in improving method deployment and management techniques by taking into account (i) data locality, (ii) system resource availability and demand and (iii) guaranteeing network topology changes seamlessly [5].

The proposed method was intended for Data centres, but it can be used also in a local networks. When comparing to grid-based approximation, this solution is more comprehensive as it also offers also full path reconstruction and can detect other type of attacks. However, the accuracy rate is lower and this type of solution does not fit well for APT attack detection as it has extremely low intensity when hiding in a network [8].

2.3 Distributed Network Anomaly Detection on an Event Processing Framework

Atanas Pamukchiev et al. research focused on distributed event process called Network Intrusion Detection System (NIDS) for data centres. Still today, approaches for NIDS are expensive to implement and they cover only segments of network focusing on data streams passing through fixed points. Another challenge in data centres internal networks is high speed, which can reach 40/100GbE. Distributed NIDS aims to reduce cost and also increase detection performance in these complex high speed networks that provides a large and varying numbers of services, such as cloud servers, data instances, data storage, image and facial recognition services [9].

Their proposed system architecture relies on topology of Apache Storm, Directed Acyclic Graph (DAG), which is used for distribution functionality. Apache Storm functionality is mapped directly to data center network topology. This is possible since Bolt nodes in Storm are similar to switches and Spout nodes are similar to core routers as well as to hosts in a data center. Similar topologies allows direct mapping regardless of the complexity of network. For bidirectional detection in a network, two identical topologies are needed due to one direction restriction in a Storm and DAG [9].

The authors proposed a lightweight Storm module directly integrated to fabric switch to facilitate network implementation. Their system detection modules are responsible for extracting data from packets and perform detection independently. Data extraction can be executed by fields or network layers and an aggregated method detects more complex anomalies. Though the proposed system is distributed, it requires a centralized management and configuration node. In all distributed nodes an API is required for runtime configuration. Controller node also monitor Bolts; it keeps track and check their state and behaviour. Controller also ensures quick recovery when problems occur [9].

Test results for the prototype of proposed system are as follows. According to the authors first tests were executed with various anomalies, such as invalid fields, blocked source and destination addresses, application layer packet contents marked as anomalous and even Denial of Service attacks. The first tests presented 20% contamination with 7,32% decrease in system throughput. The second test executed with 50% contamination and system throughput dropped 7,32%. All packets were processed in less than 7 ms [9].

Installing detection modules directly to network elements, for example in a switch, is one possible solution to resolve problems in a complex network. Tests demonstrated that integration decreases overall throughput when anomalies increase. In normal data center usage, these changes are not visible to end-users but when a solution requires almost real-time response time, even a latency of 7 ms starts to matter. The proposed system does not map directly to deep learning methods, as algorithm requires more calculation power than a fabric switch has. However, adding a lightweight deep learning module to the switch, might solve the problem.

2.4 Big Data Analytics for Network Anomaly Detection from Netflow Data

Duygu S. Terzi et al. proposed in their paper an anomaly detection method based on big data analytics from NetFlow data. Aim was to detect anomalies caused by UDP flood from specific IPs. Proposed method collects network, users, applications and routing traffic and uses clustering based unsupervised machine-learning for detection. They used six steps in their method, (i) Interval division in Netflows, (ii) Source IP Aggregation in Netflows, (iii) Standardization of data using z-score, (iv) aggregated NetFlows were clustered by K-mean algorithm as distributed, (v) Calculation of Euclidian distance between the cluster center and elements, (vi) Determination of normal and abnormal flow numbers from time intervals used in steps iv and v [6].

They implemented Netflow with Apache Spark Cluster and Azure HDInsight with python program. They reduced 6-dimensional data to 3-dimensions to help with accuracy and visualization using a statistical method called principal component analysis (PCA) [6].

The proposed method detected anomalies with 96% accuracy. The authors tested the proposed method with labelled botnet CTU-13 dataset. Data set includes 13 different scenarios and for the study, they selected the 10th due to size of dataset and number of attacks. "*The data set has 4.75 h records and 1309791 flows covering 106352 UDP DDoS flows*" [6].

Terzi et al. mentioned that most of the network traffic is normal, and due to this, anomalies and outliers based on network attacks are rare and this causes negative effect on anomaly detection research and development. For future work they proposed gathering more network data to gain better detection results as well as innovative algorithms and platforms [6].

The proposed method detected 96% of UDP DDoS attacks. Authors used PCA to reduce data dimensions from six to three, however, the above-mentioned reduction in dimensions caused some normal data to be detected as an anomaly. Using Support vector clustering (SVC) the possibility for false positive detection from occurring could be reduced. SVC maps data points from data space to high dimensional feature space with kernel function so dimension reduction is not needed, that is, all possible data is available for detection process.

2.5 Entropy-Based Network Anomaly Detection

Christian Callegari et al. proposed anomaly-based IDS system, where the detection is executed with a sketch algorithm for estimating the quantity of entropy in the data stream. The authors state that estimation of the entropy associated to the traffic descriptors has shown to be promising choice in anomaly detection. The proposed method is based on estimating different kinds of entropy. Their study includes following sections: (i) three-dimensional reversible sketch and (ii) combined sketches with entropy estimation, (iii) implementation of different kinds of entropy and (iv) impact of different traffic detection for system performance. For system and detection tests they used MAWILAB traffic recordings. They focused on finite discrete distribution and comparison of two empirical distributions. For entropy measures, the authors used "*some kind of entropy*". That is there exists two possible ways to do comparison: (i) comparing entropies of two distributions or (ii) checking relative entropy between distributions [10].

The Authors used NetFlow and Flow-Tools module Data Formatting for processing data, which was collected from router during fifteen minute time-bins. After the data was processed and formatted correctly, it was used as an input to sketch algorithm for constructing reversible sketch tables. Instead of two-dimensional array, sketch tables used three-dimensional data structures and the histograms were stored in third dimension. Random aggregation was added to the algorithm to avoid mimicry attack. That is, without randomization the attacker could try to mimic an actual attack and that would create a situation, where the histograms has exactly the same values. The constructed sketch tables was used for anomaly detection tests [10].

During the tests a scatter plot was created with two variables, Byte and Flow. Both variables were tested with five different entropy methods, (i) Shannon entropy, (ii) Tsallis entropy, (iii) Renyi entropy, (iv) Kullback-Leibler divergence and (v) Jensen-Shannon divergence. Test results show that different detection accuracies can be achieved by varying the entropy method used. Best detection accuracy, over 85%, resulted in a Flow test using Tsallis' method. Byte testing demonstrated good results with Shannon's and Thallis' methods as well as with Jensen-Shannon divergence. Test results were decent with known attacks, but unfortunately the proposed system was unable to detect unknown attacks so there was no improvement when comparing to the traditional IDS [10].

The proposed system has good detection performance, but tests were executed only with recorded data. It would be interesting to execute performance tests in a real-time environment and verify how the proposed system performs and what is the detection latency, if any.

2.6 Combination of R1-PCA and Median LDA for Anomaly Network Detection

Elkhadir et al. proposed a method with two combined dimension reduction algorithms for anomaly detection. Selected algorithms were Rotational Invariant L1-norm Principal Component Analysis (R1-PCA) and median Linear Discriminant Analysis (median LDA), and the focus was on detecting anomalies of Denial-Of-Service and Network Probe attacks [7].

The authors stated that the origin of PCA comes from minimizing the sum of squared errors and it is very sensitive for outliers. In their proposed method rotational invariance was used instead, which searches for the principal eigenvectors of a covariance matrix. Thus, R1-PCA has a significant drawback, as it may give more weight to features with higher variability disregarding their effectiveness. However, LDA searches first a projection matrix and then uses a class median vector to estimate a class mean vector. LDA has a known small sample size problem, that is, when the actual data has more dimensions than the training data, computing becomes impossible [7].

The authors used KDDcup99 dataset for testing the proposed method and they converted all discrete values of the dataset to continuous values. For accuracy testing two factors were used: (i) detection rate and (ii) false positive rate. Test results show that their proposed method was able to identify 95,5% of DoS attacks and 86,7% of Probe attacks, at highest. DoS attack detection was 94,7% and probe detection 71,6%, at lowest. Test result table shows that there was no linear dependency compared with training data and detection rate [7].

As future work Elkhadir et al. stated that they would like to test the proposed dimension reduction algorithms with multidimensional network with real multidimensional network data including images and text [7].

The proposed method is well suited for detecting DoS attacks, even with low number of training data. For detecting Probe attacks, the amount of training data has to be higher, as tests showed that lowest detection rate was 71,6% and that probably is not enough for good detection rate. Since tests were executed only with KDDcup99 dataset, it would be interesting to know what was the detection latency and also how well this method performs in a real environment. Moreover, due to the few drawbacks in the method that the authors presented, malware may be able to manipulate TCP packets in real environment which leads to non-standard data causing difficulties in detecting anomalies.

2.7   Integrating Short History for Improving Clustering Based Network Traffic Anomaly Detection

Juliette Dromard et al. proposed in their paper an unsupervised network anomaly detector. The proposed method's aim was to detect anomalies without prior knowledge or attack fingerprints caused by zero day attacks. The authors mentioned few most studied methods, such as K-means, SVM, DBSCAN which relies on clustering and PCA. Current detectors do not consider temporal information. Due to temporal existence of attack data and the mentioned lack of methods, they focused on studying H-ORUNDA(History Online Real-time Unsupervised Network Anomaly detection Algorithm), which is an improved version of ORUNDA algorithm. The algorithm was modified to keep temporal record of clustering results and it was implemented on Spark Streaming big data platform for reducing detection latency [11].

In the first phase, the authors define three rare incidents which may contain important and interesting historical data: (i) data flow that was similar to other data flows in the past, but has been modified since, (ii) data flow which statistic change suddenly and (iii) data flow which appear or disappear suddenly. Three new parameters were defined in the second phase: (i) length of history in seconds, (ii) threshold, if number of points of a cluster change, it can be considered as an anomaly and (iii) threshold d, if a point moves at least d distance, it is considered as a flow change. To improve the algorithm, 15 second-time slots were chosen since it gave good results in terms of true positive and false positive rates. Time-slots are then divided into micro-slots, which improves real-time detection. Collected data has to be aggregated to different flow levels and they chose seven aggregation levels. Every aggregated level has a unique flow matrix and every flow is a set of features and is stored in aggregation level matrix. The detector process every matrix independently. The authors also mentioned about the curse of dimensionality phenomena, what happens with high dimensions. *"In high dimensions, distance becomes meaningless and every point tends to become an outlier"*. To avoid this problem, they used subspace clustering and evidence assembly, which divided the entire space into subspaces and partitioned every subspace independently. To speed up the detection process, they also implemented incremental grid clustering algorithm (IGDCA) which can discover any shapes of cluster and it identifies noise too [11].

For the evaluation tests, SynthONTS dataset was selected. It is a real world sanitized data traffic gathered by a Spanish operator and it contains lot of different type real anomalies and artificially injected anomalies. For detection testing they made two different type of tests, (i) detection performance and (ii) detection time. For comparison methods PCA and DBSCAN were chosen. The tests results showed following: H-ORUNDA has high detection rate and low false positive rate and it performed better than methods compared in general. The performance test for detection time showed that reduction of micro-slot improved average runtime and proposed method can process incoming traffic faster than it arrives while the size of a micro-slot is at least 0,3 s. The authors also tested proposed method in Google Cloud Platform and its purpose was to test hardware requirements and scalability for real-time detection. Spark Streaming did not perform satisfactorily and they mentioned that *"a simple parallel implementation in C on a simple PC performs better"* [11].

The study had a few really interesting approaches, such as time-slotting flow and implementing IGDCA for any cluster shape. Time-slotting could help also in APT attack detection as those are really discreet and try to use slow traffic as a masquerade, among the other methods. Also multiple cluster shapes without restrictions can help to detect these type discreet attacks. Their proposed data flow aggregation may be worthy for further studies.

2.8 Network Anomaly Detection Based on Dynamic Hierarchical Clustering of Cross Domain Data

Yang Liu et al. proposed Dynamic Hierarchical Clustering of Cross Domain Data based anomaly detection in their paper. They focused on improving real-time detection and existing clustering methods, which are sensitive and easily fall to local optimal solution. The authors also mentioned that it is difficult to achieve real-time detection with machine learning and deep learning methods [12].

The proposed method uses cross domain hybrid data for anomaly detection. Hybrid data contains both, categorical and numerical data, and was added to unified framework instead of analysed separately. Euclidean distance was extended with frequency information to add dynamic clustering accuracy, that is, to measure similarity of cluster centres and samples. Their algorithm used the following execution sequence: (i) set dynamic clustering accuracy to evaluate accuracy of clustering process, (ii) execute new cluster analysis for classes that does meet accuracy requirements, (iii) repeat ii as long as is required and (iv) finally, tree clustering structure is trained by ongoing hierarchical clustering with training data. For defining clustering accuracy, disturbance in cluster class was used as an indicator. This accuracy determined if cluster needed K- means algorithm to execute second step, hierarchical cluster analysis [12].

KDDCup99 10% dataset was used for testing the proposed method and it contains different type of network attacks and intrusion behaviour. Also, the test set had 17 types of attacks which were not included in the training set. However, KDDCup99 10% dataset contains a lot of duplicates, which can cause that different types of attacks are added to same cluster leading to lower clustering performance. For comparison tests they selected following algorithms: (i) basic K-means, (ii) improved K-means, (iii) AGFCM and (iv) Naive Bayes. Test results show that the proposed method had highest detection rate (98.2%) and lowest false detection rate (5.72%] in a comparison test [12].

With the proposed method the authors were able to achieve good detection results, including low false detection rate. On the other hand, the method's performance in a real-time detection was not tested while it was mentioned that some other approaches are not suitable for real-time detection. It would be interesting to see how the proposed method performs in a real-time environment.

2.9 Probabilistic Transition-Based Approach for Detecting Application-Layer DDoS Attacks in Encrypted Software-Defined Networks

Elena Ivannikov et al. proposed a method for detecting Application-Layer DDoS Attacks in Encrypted Software-Defined Networks (SDNs), with Probabilistic Transition-Based Approach, that is, an algorithm which extracts statistics directly from data flows in SDN and compare behavioural patterns to normal traffic in order to detect significant anomalies. Research focus for cloud environments were defined because of the rapid growth in business use, thus it can be comprehended as a critical part of modern business and DDoS can cause huge problems for companies and their reputations by disabling services. Cloud environments can adapt quickly to fluctuating demands and SDN has eased network maintenance and configuration tasks [13]. However, it has created a situation where cloud networks and attacks are increasing in complexity every day [9,13]. The main difference between traditional and application layer DDoS attacks is that while traditional DDoS attacks are executed at the network layer, the application layer attacks are executed at the seventh OSI layer and it tries to mimic legitimate traffic to avoid detection. HTTP is commonly used in application layer DDoS attacks and it targets the vulnerabilities in operative systems and web applications [13].

The authors proposed a method where packet headers were extracted from data for detection purposes, which also enables using encrypted traffic. To start with, the authors built a model called *conversation* for normal user behaviour which was a collection of short time sequence data. They implemented four ways to characterize each *conversation*, (i) source IP, (ii) source port, (iii) destination IP or iv) destination port. Besides characterization the following information was extracted: (i) duration of *conversation*, (ii) number of packets sent in 1 second, iii) number of bytes sent in one second, (iv) average packet size and (v) presence of TPC flags such as URG, ACK, PSH, RST, SYN and FIN. For the detection method they selected two clustering algorithms, (i) k-means and (ii) Clustering Using REpresentatives (CURE), and then calculated probabilistic transition for clusters [13].

For executing tests, first they applied clustering algorithms to extracted features, which created representations of distinct groups and discovered hidden patterns in data traffic. In the second phase *conversations* were grouped together based on characterization and time interval. In the third phase each session was represented in every time window based on cluster sequence from the first step. Last phase was to estimate conditional and marginal probabilities for every sequence and calculate threshold. To solve a real time detection problem for previously unknown behavioural patterns of users, they used streaming k-means algorithm which allows updates to the behavioural data in the trained model. A so-called forgetting mechanism was also implemented to the update mechanism where after time the old models get less important for the actual updated model. Detection tests were executed with trained user behavioural models and by using normal net bank traffic and intermediate DDoS attack with several bots-attackers trying to mimic regular browsing behaviour. Test results showed that the proposed clustering algorithms with probabilistic transition-based approach *k-means+Prob* performed with the highest accuracy, respectful 99.58% and with false positive rate at zero. The authors also mentioned that the proposed method gained three improvements compared to earlier studies: (i) performance, that is, low false alarm rates and high detection accuracy (ii) reduced number of effective parameters in cluster, only one and (iii) significant reduce for required storing space [13].

For the future work Ivannikov et al. mentioned improving detection accuracy with bigger dataset and focusing on detecting more advanced DDoS attack with simulation [13].

Due to its ability to detect anomalies from the seventh OSI layer with encrypted data, the proposed approach could be used also for APT attacks. However, the forgetting algorithm would be needed to be revised due to the tendency for the APT attacks to hide unnoticed as long as possible to not forget old traces completely before the next *conversation*.

3   Summary of Further Improvements

In this section the perceived improvements for more precise anomaly detection is presented. These identified concerns vary from single to multiple papers.

Even though numerous papers mentioned that KDDCup99 is not realistic, data is obsolete and lack of modern network traffic, it is widely used for benchmarking new methods. The research community should go forward and look for present day datasets or create those to substitute KDDCup99, as in some papers researchers had already done.

In many of the proposed methods, the authors used dimension reduction to fit the multidimensional data to the two-dimensional space. This can cause losses of critical data and lead to lower accuracy rates and increases in false alarm rates.

In some of the papers only certain parts of IP and TCP headers were selected for the anomaly detection, such as source and destination IP's and ports. These types of solutions can limit the detection capability due to the vast amount of the possibly useful missing data.

In one paper, proposed method focus was in UDP DDoS attacks detection, hence the usage of method is highly limited.

In a few of the papers tests were executed only with recorded data and not a single test was executed in a real network environment. Results of these types of tests cannot be used for benchmarking anomaly detection performance in a real situation.

Some of the proposed methods required high intensity attacks to perform the anomaly detection properly and thus the usage is rather limited. In addition, almost every paper where the anomaly detection was executed by traditional machine learning or statistical approach suffered from the same performance shortage, lack of detecting the earlier unknown attacks. This is a common problem for the information security products which rely on signature based detection.

In one paper it was stated that some other approaches are not suitable for real-time detection but interestingly the authors did not perform any testing in real-time environment with their own proposed method.

Baddar et al. pointed out in their paper in 2014, that in several papers they investigated it was assumed that the majority of network traffic is normal. In reality that might be even the opposite, for example in attacks such as DDoS [14]. That assumption was also present in many of the papers which were presented in this review.

4   Conclusion

The papers reviewed showed that different methods vary greatly in their performance to detect anomalies. Every method had its advantages and disadvantages, however most of the presented methods cannot detect previously unknown attacks but on the contrary, for example, detects DDoS attacks extremely well. The methods mentioned were based on statistics and traditional machine learning, while those methods have slight disadvantages, they have advantages also, such as real-time detection and low memory consumption. Current network devices, such as firewall, IDS, switch and so forth, can be complemented with the proposed methods to gain more visibility to ongoing situation in networks. However, proposed methods does not fit for APT attacks detection due to its sophisticated behaviour and tendency to hide in networks even for years and mimic normal traffic.

When developing anomaly detection systems and methods, these advantages and disadvantages should be further considered, as they can help to define what could the actual focus of the work. With current practices a system or a method that could detect all types of attacks, not to mention in a real-time environment, requires enormous resources and might be still even impossible to implement.

References

1. Andropov, S., Guirik, A., Budko, M., Budko, M.: Network anomaly detection using artificial neural networks. In: 2017 20th Conference of Open Innovations Association (FRUCT) (2017). https://doi.org/10.23919/FRUCT.2017.8071288
2. Aygun, R.C., Yavuz, A.G.: Network anomaly detection with stochastically improved autoencoder based models. In: 2017 IEEE 4th International Conference on Cyber Security and Cloud Computing, pp. 193–198 (2017). https://doi.org/10.1109/CSCloud.2017.39
3. Yuan, X., Li, C., Li, X.: DeepDefense: identifying DDoS attack via deep learning. In: 2017 IEEE International Conference on Smart Computing (SMARTCOMP) (2017). https://doi.org/10.1109/SMARTCOMP.2017.7946998
4. Kim, J., Yoo, A., Sim, A., Suh, S., Kim, I.: A lightweight network anomaly detection technique. In: 2017 Workshop on Computing, Networking and Communications (CNC) (2017). https://doi.org/10.1109/ICCNC.2017.7876251
5. Iordache, M., Jouet, S., Marnerides, A.K., Dimitrios, P.P.: Distributed, multi-level network anomaly detection for datacentre networks. In: IEEE ICC 2017 Next Generation Networking and Internet Symposium (2017). https://doi.org/10.1109/ICC.2017.7996569
6. Terzi, D.S., Terzi, R., Sagiroglu, S.: Big data analytics for network anomaly detection from netflow data. In: 2017 International Conference on Computer Science and Engineering (UBMK), pp. 592–597 (2017). https://doi.org/10.1109/UBMK.2017.8093473
7. Elkhadir, Z., Chougdali, K., Benattou, M.: Combination of R1-PCA and median LDA for anomaly network detection. In: Intelligent Systems and Computer Vision (ISCV) (2017). https://doi.org/10.1109/ISACV.2017.8054985
8. Ussath, M. Jaeger, D., Cheng, F., Meinel, C.: Advanced persistent threats: behind the scenes. In: 2016 Annual Conference on Information Science and Systems (CISS) (2016). https://doi.org/10.1109/CISS.2016.7460498
9. Pamukchiev, A., Jouet, S., Pezaros, D.P.: Distributed network anomaly detection on an event processing framework. In: 2017 14th IEEE Annual Consumer Communications & Networking Conference (CCNC), pp. 659–664 (2017).https://doi.org/10.1109/CCNC.2017.7983209
10. Callegari, C., Giordano, S., Pagano, M.: Entropy-based network anomaly detection. In: 2017 International Conference on Computing, Networking and Communications (ICNC): Communications and Information Security Symposium (2017). https://doi.org/10.1109/ICCNC.2017.7876150

11. Dromard, J., Owezarski, P.: Integrating short history for improving clustering based network traffic anomaly detection. In: 2017 IEEE 2nd International Workshops on Foundations and Applications of Self* Systems (FAS*W), pp. 227–234 (2017). https://doi.org/10.1109/FAS-W.2017.152

12. Liu, Y., et al.: Network anomaly detection based on dynamic hierarchical clustering of cross domain data. In: 2017 IEEE International Conference on Software Quality, Reliability and Security (Companion Volume), pp. 200–204 (2017). https://doi.org/10.1109/FAS-W.2017.152

13. Ivannikova, E., Zolotukhin, M., Hämäläinen, T.: Probabilistic transition-based approach for detecting application-layer DDoS attacks in encrypted software-defined networks. In: Yan, Z., Molva, R., Mazurczyk, W., Kantola, R. (eds.) NSS 2017. LNCS, vol. 10394, pp. 531–543. Springer, Cham (2017). https://doi.org/10.1007/978-3-319-64701-2_40

14. Baddar, S., Merlo, A., Migliardi, M.: Anomaly detection in computer networks: a state-of-the-art review. J. Wirel. Mob. Netw. Ubiquitous Comput. Dependable Appl. (JoWUA) 5, 29–64 (2014). https://www.researchgate.net/publication/270274504_Anomaly_Detection_in_Computer_Networks_A_StateoftheArt_Review

Creating a Schedule for Parallel Execution of Tasks Based on the Adjacency Lists

Yulia Shichkina[✉] and Mikhail Kupriyanov

Department of Computer Science and Engineering,
Saint Petersburg Electrotechnical University "LETI", St. Petersburg, Russia
strange.y@mail.ru

Abstract. The article presents a method for transforming algorithm's information graph using adjacency lists. Algorithm's information graph always has a large number of vertices. For most algorithms, this graph contains more than 100 vertices. Manual analysis of this graph for the presence of internal parallelism is very difficult. The proposed method does not use conventional adjacency matrix for storing information about the connections between vertices and the adjacency lists. Adjacency lists allow to store information about the graph in a compressed form. As a result, the researcher gets a schedule of the algorithm on a computer, allowing parallel execution. The presented method can be successfully applied to queries in databases, to the distribution of tasks between nodes of a wireless network, to solving problems with large volumes of data in the field of the Internet of things.

Keywords: Parallel algorithm · Information graph · Graph width
Adjacency list · Algorithm schedule · Optimization

1 Introduction

In the development of parallel algorithms for solving complex scientific and technical problems the main point is to analyse the efficiency of parallelism. Usually it depends on evaluation of the resulting acceleration of calculation process (for how much time the problem solving decreases). Evaluation method of gained acceleration can be applied to the selected computational algorithms (evaluation of the effectiveness of a particular algorithm parallelization technique). Another important approach is to construct evaluation method of the maximum possible speed boost of solving the problem of a particular type (evaluation of parallel approach in solving that problem).

To this day a number of evaluation methods were developed to measure calculations boost, the effectiveness of high-performance computing systems, resources utilization, the real and peak performance.

But firstly, these evaluation methods were created for the abstract computing systems and algorithms. The user always has a specific task or a specific algorithm to be implemented on an existing computer system. And it needs to be assessed in advance how much computing resources he will need and what acceleration can be achieved.

O. Galinina et al. (Eds.): NEW2AN 2018/ruSMART 2018, LNCS 11118, pp. 102–115, 2018.
https://doi.org/10.1007/978-3-030-01168-0_10

In addition to the estimates of the rate and the amount of calculation, users often need to see at least a raw parallel implementation of their algorithm. Why raw? Because mathematical and software tools for automatic translation of sequential algorithms in their optimal parallel form do not exist today despite the fact that research in this area has been carried out for over 20 years.

The internal parallelism usage has an obvious advantage, so there is no need to spend extra effort to study properties of the newly created computational algorithms. The disadvantages are also obvious because of the need to identify and examine the algorithm's graphs. In those cases when the internal parallelism of an algorithm is insufficient for effective use of the particular computer's parallelism, it's necessary to replace it with another algorithm having the best properties of parallelism. Fortunately, for many tasks there was already developed a number of different algorithms. So it's almost always possible to select the appropriate algorithm. To make a choice between several algorithms will be easier on the basis of relevant information on the properties of parallel algorithms.

Getting the necessary information about the internal parallelism, its structure, the number of necessary processors, the execution time of the algorithm, density and uniformity of the processor load and other characteristics of the parallel algorithm can be performed by using the algorithm's information graph. In the initial state algorithm's information graph usually takes position between the extremes of fully sequential and fully parallel algorithms.

One of the problems to be solved is when parallelism depends on uniform loading of processors, because if the main computations will go to only a portion of them the advantage of parallelization will be reduced.

The best solution is to get parallel algorithm executed in the minimum time with minimum number of processors.

Usage of matrices in studies of information graph is a natural and successful solution due to existence of a large number of existing numerical methods of matrix algebra with internal parallelism in these methods. The result of these studies is the optimal parallel algorithm for converting the sequential algorithm for solving some problem to its parallel version.

Below we provide methods of parallelization of sequential algorithms which are not simply apply the matrix methods, they take into account the sparsity of information graph matrix. It's irrational to store and handle all elements of the information graph, the majority of which are zero. The methods proposed below are not based on adjacency matrix or consecution matrix, but on their analogy – adjacency lists which:

- save memory;
- reduce the conversion time of the graph;
- has large supply of internal parallelism.

In information graph construction it is assumed that the execution time of any computing operations is the same and equal to one conventional unit and the data transfer between computing devices is performed instantaneously without any time consuming.

2 Related Work

All the research conducted since the 1980s in the field of parallel computing software can be attributed to the following main areas:

- development of parallel computing systems;
- parallel computations efficiency analysis for estimating the resulting acceleration of computations and how effectively the computing equipment is used for parallel methods of solving different problems;
- formation of general principles for the development of parallel algorithms for solving complex computationally time-consuming tasks;
- construction and development of system software for parallel computing systems (e.g. MPI (Message Passing Interface), the implementation of which allows to develop parallel programs and, in addition, to significantly reduce the severity of the important problem of parallel programming – portability between different computing systems;
- construction and development of parallel algorithms for solving applied problems in different areas of practical applications.

Most of the research was conducted earlier and is being conducted today in the latter area. Here the number of works should be mentioned, for example, related to modeling the calculation of electromagnetic radiation based on multi-core processors and clusters of parallel architecture GPU using hybrid parallel algorithm with MPI-OpenMP and MPI-CUDA [1]. Studies are under way in the field of large data volumes in IoT (Internet of Things), where traditional methods of compression and encryption are neither competent nor effective. To solve such problems, a combined parallel algorithm was developed «CZ algorithm», which can effectively compress and encrypt large data [2]. Many studies of parallel algorithms are carried out in the field of differential equations [3] and systems of linear equations [4, 5]. Given the large amount of data and the need to solve real-time problems, a number of approaches to parallelize algorithms for finding the shortest path in a city road map is being developed [6].

Analysis of the scalability of parallel programs and computer systems began in 1967, when IBM employee Gene Amdahl (Gene Amdahl) published an article [7], which later became a classic. The article did not show any mathematical formula which describes the laws of development of parallel algorithms. Later, referring to Amdahl's ideas, such a formula was published and became known as Amdahl's law [8]. Further evaluation of the computing algorithms acceleration, resources utilization and computing systems performance have been obtained in many other studies [9–12].

The analysis of parallel computing efficiency, the effect of performance variability on the accuracy and efficiency of the optimization algorithm, and the strategy for minimizing the impact of this variability is given in [13].

In particular, the scalability of parallel algorithms is a vital problem, so different techniques have been developed to overcome it such as processors load balancing [14, 15] and reducing the number of synchronizations [16].

An article [14] presents a parallel algorithm of creating and deleting data copies, referred to as ghost copies, which localize neighborhood data for computation purposes while minimizing inter-process communication.

The latest supercomputers [17] have significantly increasing number of computing nodes/cores, but many practical applications cannot achieve better performance on more computing resources because enough parallelism in the applications has not been explored.

In the field of the formation of general principles for the development of parallel algorithms studies on planning the execution of operations of the algorithm should be mentioned.

Task Scheduling is one of the key elements in any distributed-memory machine, and an efficient algorithm can help reduce the inter-processor communication time [18, 19]. One of the flaws of these algorithms is their NP- completeness.

The problem of constructing a schedule for the parallel algorithm with optimal execution time was highlighted in various studies since the 80s. Among the algorithms there are some based on finding the maximum flow in the transport network; constructive (e.g. based on branch and bound, dynamic programming, greedy strategies), iterative (e.g. genetic algorithms, simulated annealing algorithm) [20, 21].

Most of these algorithms complete a timing diagram by placing each task in it and are used for a particular mathematical model of computing system. Using the same approximate models could lead to the fact that the resulting timing diagram can't be implemented in the real computing system. In addition, most algorithms that are based on finding the maximum flow in the transport network have a pseudopolynomial complexity [22, 23].

A paper [24] addresses the problem of scheduling parallel programs represented as directed acyclic task graphs for execution on distributed memory parallel architectures. Because of the high communication overhead in existing parallel machines, a crucial step in scheduling is task clustering, the process of coalescing fine grain tasks into single coarser ones so that the overall execution time is minimized. The task clustering problem is NP-hard, even when the number of processors is unbounded and task duplication is allowed. A simple greedy algorithm is presented for this problem which, for a task graph with arbitrary granularity, produces a schedule whose make span is at most twice optimal. Indeed, the quality of the schedule improves as the granularity of the task graph becomes larger.

The issues of parallelization of individual fragments of algorithms, usually the most difficult to parallelize, are discussed in [25, 26].

The authors of [27] developed a genetic algorithm (GA) approach to the problem of task scheduling for multiprocessor systems. This approach requires minimal problem specific information and no problem specific operators or repair mechanisms.

3 Schedule Construction Method Based on Adjacency Lists

In what follows we assume that:

- the response time is the same for all operations;
- the time required to transfer data between nodes equals zero;
- operation's execution time is equal among all nodes.

These conditions are the restrictions imposed on the algorithm and the computational system. In practice such algorithms and computer systems do not exist. But received schedule allows to evaluate the possibility of parallelization process of given algorithm and on the basis of the resulting schedules, you can get the new schedule without the restrictions imposed.

Adjacency list is a set of pairs of adjacent vertices, composed according to the rule: if the vertex v_i is following vertex v_j then the pair could be written as (v_i, v_j). We will call v_i - initial vertex v_j - final vertex.

Schedule construction method:

1. For a given information graph construct the corresponding adjacency list (V_1, V_2), where V_1 - the set of initial vertices of directed edges, V_2 - the set of final nodes.
2. Find the set $V = V_1 - V_2$. Vertices that are included in this set will form a group of vertices belonging to the same layer.
3. Remove from the adjacency list all pairs initial vertices of which coincide with vertices from the V set.
4. If the list is not empty, then return to step 2. If the list does not have a single pair, then the first part of the method - splitting set of vertices into groups - is over.

Example: let's check this method on information graph formed by a test program (Fig. 1).

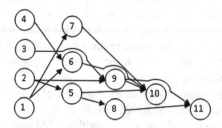

Fig. 1. Algorithm's information graph created by a test program

Adjacency list is shown below (Table 1):

Table 1. Corresponding adjacency list for information graph

№	V_1 V_2	№	V_1 V_2
1	1 6	8	5 10
2	1 7	9	6 10
3	2 5	10	7 10
4	2 9	11	8 11
5	3 9	12	9 10
6	4 6	13	10 0
7	5 8	14	11 0

In the last two pairs of adjacency list values of V_2 are zeros because there are no outgoing edges. Let's find the difference of initial and final vertices sets $V = V_1 - V_2 = \{1, 2, 3, 4\}$. This group of vertices will be the first layer of the information graph $M_1 = \{1, 2, 3, 4\}$.

Let's delete from initial list all pairs first elements of which are from M_1 (Table 2):

Table 2. Corresponding adjacency list for information graph after M_1 vertices removal

№	V_1	V_2
1	5	8
2	5	10
3	6	10
4	7	10
5	8	11
6	9	10
7	10	0
8	11	0

Again, let's find the difference of initial and final vertices $V = V_1 - V_2 = \{5, 6, 7, 9\}$. This group of vertices second layer of information graph $M_2 = \{5, 6, 7, 9\}$.

Let's remove all pairs with first elements from M_2 (Table 3):

Table 3. Corresponding adjacency list for information graph after M_2 vertices removal

№	V_1	V_2
1	8	11
2	10	0
3	11	0

After next difference of sets we get group $M_3 = \{8, 10\}$, after removing all pairs with first elements from M_3 we get list containing only one pair from set V_1 making the last group $M_4 = \{11\}$.

Results of the test program written for this paper are presented below (Table 4):

Table 4. Test program results

Parameter	Value	Groups
Max number of processors	4	{1, 2, 3, 4}
Max height	4	{6, 7, 5, 9}
Input vertices	4	{10, 8}
Max vertices	11	{11}

We should notice that on the last step of the algorithm there could be more than one row, but all pairs will have zeros in V_2.

Each group is a set of operations which could be executed separately. Schedule for computing nodes can look like this:

Node 1: 1, 6, 10, 11;
Node 2: 2, 7, 8;
Node 3: 3, 5;
Node 4: 4, 9;

Nodes can be loaded non-uniformly.

4   Schedule Optimization by the Number of Computing Nodes

By knowing the graph's height s it's possible to calculate the number of required processors – graph's width:

$$d = trunc\left(\frac{n}{s}\right) \tag{1}$$

where n - number of graph's vertices, function trunc(x) truncates number up to integer.

In this case calculations density should be $P = \frac{\sum_{i=1}^{s} N_i}{N \cdot s} = \frac{\sum_{i=1}^{s} d}{d \cdot s} = 1$. Acceleration coefficient is $K = \frac{T_s + T_p}{T_s + \frac{T_p}{N}} = \frac{0 + s}{0 + \frac{s}{N}} = N$, efficiency - $E = \frac{K}{N} = \frac{N}{N} = 1$.

But in real world these values are impossible to be achieved.

There is more: getting minimal values of s and d is desirable if doesn't make conflicts with capabilities of computing system. This nuance will be checked later.

Optimization method of obtained schedule by the number of computing nodes considers uniform distribution of vertices into groups starting the last one.

Optimization method of information graph by width:

1. Calculate value of theoretical minimal width d of information graph. Take the first group from the tail with the number of vertices lower than theoretical minimum width of graph d. Mark it as M_i.
2. Create adjacency list for each vertex from group $M_{i-1} : V_{i-1j}(V_1 - V_2)$, where $j = \overline{1,k}$, k - number of vertices in group M_{i-1}.
3. Find intersection of sets: $V = M_i \cap V_2$. If for some vertex from M_{i-1} the intersection is empty, then this vertex can be transferred in M_i. This process should be continued until M_i. won't reach theoretical minimum width of graph d.
4. If $i > 1$. then go to step 1, else – the method is over.

Example, let's take a look on the graph we used earlier (Fig. 2). Theoretical minimum width of this graph from formula (1) is $d = 3$.

Let's make an adjacency list for vertices from 3-rd group. For vertex № 8 (Table 5):

Table 5. Adjacency list for vertex № 8

№	V_1	V_2
1	8	11

Then find the intersection of sets $M_4 = \{11\}$ and $V_2 = \{11\}$: $V = M_4 \cap V_2 = \{11\}$. Because that intersection is not empty we can't move vertex № 8 from 3rd group to 4th. Let's make an adjacency list for vertex № 10 (Table 6):

Table 6. Adjacency list for vertex № 10

№	V_1	V_2
1	10	0

Then we find an intersection of $M_4 = \{11\}$ and $V_2 = \{10\}$: $V = M_4 \cap V_2 = \{10\}$. Intersection we found is empty and vertex № 10 can be transferred from 3rd group to 4th: $M_4 = \{11, 10\}$. In 3rd group there is one vertex left: $M_3 = \{8\}$.

After that let's find adjacency lists for 2nd group $M_2 = \{5, 6, 7, 9\}$ and intersections $V = M_3 \cap V_2$ (Tables 7, 8, 9 and 10):

Table 7. Adjacency list and intersection for vertex № 5

№	V_1	V_2	$V = M_3 \cap V_2$
1	5	8	$\{8, 10\} \times \{8\} = \{8\}$
2	5	10	

Table 8. Adjacency list and intersection for vertex № 6

№	V_1	V_2	$V = M_3 \cap V_2$
1	6	10	$\{10\} \times \{8\} = \{\}$

Table 9. Adjacency list and intersection for vertex № 7

№	V_1	V_2	$V = M_3 \cap V_2$
1	7	10	$\{10\} \times \{8\} = \{\}$

Table 10. Adjacency list and intersection for vertex № 9

№	V_1	V_2	$V = M_3 \cap V_2$
1	9	10	$\{10\} \times \{8\} = \{\}$

Intersection $V = M_3 \cap V_2$ is empty for vertices 6, 7 and 9. These vertices can be transferred in 3rd group: $M_3 = \{8, 7, 9\}$. We transfer only two of them because we need width $d = 3$. In 2nd group there are two vertices left: $M_2 = \{5, 6\}$.

Let's find adjacency lists for 1st group $M_1 = \{1, 2, 3, 4\}$ and intersections $V = M_2 \cap V_2$ (Tables 11, 12, 13 and 14):

Table 11. Adjacency list and intersection for vertex № 1

№	V_1	V_2	$V = M_2 \cap V_2$
1	1	6	$\{6, 7\} \times \{5, 6\} = \{6\}$
2	1	7	

Table 12. Adjacency list and intersection for vertex № 2

№	V_1	V_2	$V = M_2 \cap V_2$
1	2	5	$\{5, 9\} \times \{5, 6\} = \{5\}$
2	2	9	

Table 13. Adjacency list and intersection for vertex № 3

№	V_1	V_2	$V = M_2 \cap V_2$
1	3	9	$\{9\} \times \{5, 6\} = \{\}$

Table 14. Adjacency list and intersection for vertex № 4

№	V_1	V_2	$V = M_2 \cap V_2$
1	4	6	$\{6\} \times \{5, 6\} = \{6\}$

We found only one vertex with empty intersection $V = M_2 \cap V_2$. It's vertex № 3. It can be transferred from 1st group to 2nd: $M_2 = \{5, 6, 3\}$. There are three vertices left: $M_1 = \{1, 2, 4\}$.

As a result we aligned all groups by the number of vertices $d = 3$: $M_1 = \{1, 2, 4\}$, $M_2 = \{5, 6, 3\}$, $M_3 = \{8, 7, 9\}$, $M_4 = \{11, 10\}$.

It should be noticed that if we transferred on second step vertices 7 and 9 from 3rd group, then we could get another appropriate result like this (Table 15):

Table 15. Another achievable result when vertices 7 and 9 are transferred on step 2

Parameter	Value	Groups
Max number of processors	4	{4, 2, 3}
Max height	4	{9, 5, 1}
Input vertices	4	{8, 6, 7}
Max vertices	11	{11, 10}

As a result, we get only three active computing nodes instead of four.

5 Evaluation of Computing Resources Required for Maximum Acceleration

This method allows to decrease algorithms width dramatically. But the question about upper border of minimal graph width which could be easily achieved is still opened.

During optimization of graph by width moving vertices is allowed only from initial vertices to final, so after ending second part of the method we can calculate value of minimal width taking into account the number of groups:

$$d = \max_{1 \le k \le m} \left\lceil \frac{\sum_{i=i_k}^{n} d_i}{s - k + 1} \right\rceil \tag{2}$$

where m is number of groups, i_k is number of 1st vertex of group k.

Formula (2) is more practical method for evaluation of minimal graph's width. If outgoing vertex is single, then at least one group (the last one) will be consisted of only one vertex. So the density of other layers will be higher. Same thing could be said about initial vertices. So the calculation of upper border of minimal graph's width should be done with the number of outgoing vertices taken into account.

May $\Delta_{init} = d' - n_{init}$ be the difference of graph's width and the number of initial vertices, where n_{init}- number of initial vertices. $\Delta_{final} = d' - n_{final}$ where n_{final} will be the number of final vertices, so:

1. $\Delta_{init} > 0$, $\Delta_{final} > 0$. Density of other groups increases by:

$$\Delta = \left\lceil \frac{\Delta_{init} + \Delta_{final}}{g - 2} \right\rceil \tag{3}$$

Where g – number of groups we got after the first part of the algorithm, so $d' = d' + \Delta$.

2. $\Delta_{init} > 0, \Delta_{final} < 0$. The number of final vertices is bigger than minimal width. Density of the last group can't be reduced, so: $d' = n_{final}$.
3. $\Delta_{init} < 0, \Delta_{final} > 0$. The number of initial vertices bigger than minimal width

Notice that the number of final vertices has already been taken into account in formula (2), so we can refuse from the last three cases leaving the upper border of minimal width like this:

$$d' = d' + \Delta \tag{4}$$

where:

$$\Delta = \begin{cases} \frac{\Delta_{init} + \Delta_{final}}{g-2}, & (\Delta_{init} > 0, \Delta_{final} > 0) \\ 0, & otherwise \end{cases} \tag{5}$$

Let's use the number line to see the borders of minimal width (Fig. 2):

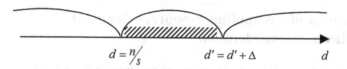

$d = {}^{n}\!/\!{}_{s}$ $d' = d' + \Delta$ d

Fig. 2. Interval of practical minimal width of information graph

So after the first part of the parallel algorithm's optimization method by width we can distinguish borders of graph's minimal width. After the second part we can measure the effectiveness of it. Also, in the second pa we can use practical value d' instead of theoretical d taking into account basic conditions of the task.

6   Comparison of the Method of Constructing a Schedule with the Help of Lists of Adjacencies with the Method Based on Matrices

One iteration of a cycle with a precondition in the algorithm for converting a sparse matrix to a block diagonal form (BDF) can be represented by the following information graph (Fig. 3).

The fragment of the adjacency matrix corresponding to this information graph is shown in Fig. 4

In the information box (Fig. 3) it is also possible to select a block that is repeated. When the matrix is converted to a block diagonal form, the size of 5×7 this fragment will contain 41 vertices. For one iteration of the algorithm for transforming the matrix to the BDF form, the adjacency matrix will be the size of 205×205. There are 290 non-zero elements in it. This is 6% of the total number of elements.

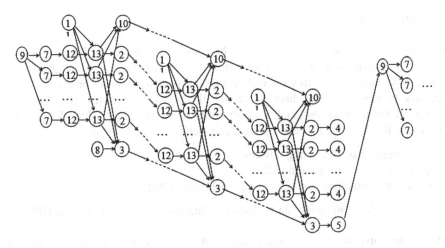

Fig. 3. Information graph of the algorithm

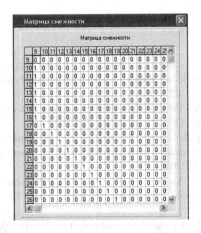

Fig. 4. The adjacency matrix corresponding to information graph

When applying the method based on the adjacency lists, as a result there will be a considerable saving of memory; a significant increase in performance due to the fact that iterative processing will not have 42025 elements, but only 580; The possibility of using other data structures besides arrays in programming, for example, sets or lists.

7 Conclusion

Complexity analysis of presented methods has shown that usage of adjacency lists is more effective than usage of adjacency matrices. Notice that after the second part of the algorithm the number of vertices in different groups sometimes won't be equal, so it's necessary to make one step back to change the transferring vertex. It's the main problem of the information graph's width optimization algorithm. Considered algorithms give the basis for the next optimization steps:

- by the number of computing devices;
- by the time of each operation [28];
- by the data throughput between computing nodes [29];

Also, considered methods can be used for database queries optimization [30].

Acknowledgments. The paper has been prepared within the scope of the state project "Initiative scientific project" of the main part of the state plan of the Ministry of Education and Science of Russian Federation (task № 2.6553.2017/8.9 BCH Basic Part).

References

1. He, B., Tang, L., Xie, J., Wang, X., Song, A.: Parallel numerical simulations of three-dimensional electromagnetic radiation with MPI-CUDA paradigms. Math. Probl. Eng. **2015**, 9 pages (2015). Article ID 823426
2. Qin, J., Lu, Y., Zhong, Y.: Parallel algorithm for wireless data compression and encryption. J. Sensors **2017**, 11 pages (2017). Article ID 4209397
3. Gong, C., Bao, W., Tang, G., Jiang, Y., Liu, J.: A parallel algorithm for the two-dimensional time fractional diffusion equation with implicit difference method. Sci. World J. **2014**, 8 pages (2014). Article ID 219580
4. Ma, X., Liu, S., Xiao, M., Xie, G.: Parallel algorithm with parameters based on alternating direction for solving banded linear systems. Math. Probl. Eng. **2014**, 8 pages (2014). Article ID 752651
5. Hou, J., Lv, Q., Xiao, M.: A parallel preconditioned modified conjugate gradient method for large sylvester matrix equation. Math. Probl. Eng. **2014**, 7 pages (2014). Article ID 598716
6. Yu, D.-X., Yang, Z.-S., Yu, Y., Jiang, X.-R.: Research on large-scale road network partition and route search method combined with traveler preferences. Math. Probl. Eng. **2013**, 8 pages (2013). Article ID 950876
7. Amdahl, G.M.: Validity of the single processor approach to achieving large scale computing capabilities. In: Processings AFIPS Spring Joint Computer Conference, Reston, pp. 483–485. AFIPS Press, VA (1967)
8. Ware, W.: The ultimate computer. IEEE Spectrum **9**, 84–91 (1972)
9. Grama, A., Gupta, A., Karypis, G., Kumar, V.: Introduction to Parallel Computing, Second Edition. Addison Wesley, Reading (2003)
10. Gergel, V.P., Strongin, R.G.: Parallel Computing for Multiprocessor Computers. NGU Publ, Nizhnij Novgorod (2003). (in Russian)
11. Quinn, M.J.: Parallel Programming in C with MPI and OpenMP, 1st edn. McGraw-Hill Education, New York (2003)
12. Wittwer, T.: An Introduction to Parallel Programming, VSSD uitgeverij (2006)

13. Tiwari, A., Tabatabaee, V., Hollingsworth, J.K.: Tuning parallel applications in parallel. Parallel Comput. **35**(8–9), 475–492 (2009)
14. Mubarak, M., Seol, S., Qiukai, L., Shephard, M.S.: A parallel ghosting algorithm for the flexible distributed mesh database. Sci. Program. **21**(1–2), 17–42 (2013)
15. Kruatrachue, B., Lewis, T.: Grain size determination for parallel processing. IEEE Softw. **5**(1), 23–32 (1988)
16. Lim, A.W., Lam, M.S.: Maximizing parallelism and minimizing synchronization with affine partitions. Parallel Comput. **24**(3–4), 445–475 (1998)
17. Meuer, H., Strohmaier, E., Dongarra, J., Simon, H.: Top500 supercomputing sites (2015)
18. Yang, T., Gerasoulis, A.: DSC: scheduling parallel tasks on an unbounded number of processors. IEEE Trans. Parallel Distrib. Syst. **5**(9), 951–967 (1994)
19. Darbha, S., Agrawal, D.P.: Optimal scheduling algorithm for distributed memory machines. IEEE Trans. Parallel Distrib. Syst. **9**(1), 87–95 (1998)
20. Liu, C.L., Layland, J.W.: Scheduling algorithms for multiprogramming in hard real-time environment. J. ACM **20**(1), 46–61 (1973)
21. Marte, B.: Preemptive scheduling with release times, deadlines and due times. J. ACM **29**(3), 812–829 (1982)
22. Burns, A.: Scheduling hard real-time systems: a review. Softw. Eng. J. **6**(3), 116–128 (1991)
23. Stankovic, J.A.: Implications of classical scheduling results for real-time systems. IEEE Computer Society Press (1995)
24. Darbha, S., Agrawal, D.P.: A task duplication based scalable scheduling algorithm for distributed memory systems. IEEE Trans. Parallel Distrib. Syst. **46**(1), 15–27 (1997)
25. Tzen, T.H., Ni, L.M.: Trapezoid self-scheduling: a practical scheduling scheme for parallel compilers. IEEE Trans. Parallel Distrib. Syst. **4**, 87–98 (1993)
26. Sinnen, O., Sousa, L.A.: Communication contention in task scheduling. IEEE Trans. Parallel Distrib. Syst. **16**, 503–515 (2005)
27. Wu, A.S., Yu, H., Jin, S., Lin, K.-C., Schiavone, G.: An incremental genetic algorithm approach to multiprocessor scheduling. IEEE Trans. Parallel Distrib. Syst. **15**(9), 824–834 (2004)
28. Kupriyanov, M.S., Shichkina, Y.A.: Applying the list method to the transformation of parallel algorithms into account temporal characteristics of operations. In: Proceedings of the 19th International Conference on Soft Computing and Measurements, SCM 2016, pp. 292–295. https://doi.org/10.1109/scm.2016.7519759, ISBN 978-146738919-8. 7519759
29. Shichkina, Y., Kupriyanov, M., Al-Mardi, M.: Optimization algorithm for an information graph for an amount of communications. In: Galinina, O., Balandin, S., Koucheryavy, Y. (eds.) NEW2AN/ruSMART -2016. LNCS, vol. 9870, pp. 50–62. Springer, Cham (2016). https://doi.org/10.1007/978-3-319-46301-8_5
30. Shichkina, Y., Degtyarev, A., Gushchanskiy, D., Iakushkin, O.: Application of optimization of parallel algorithms to queries in relational databases. In: Gervasi, O., et al. (eds.) ICCSA 2016. LNCS, vol. 9787, pp. 366–378. Springer, Cham (2016). https://doi.org/10.1007/978-3-319-42108-7_28

Measuring a LoRa Network: Performance, Possibilities and Limitations

Anders Carlsson[ID], Ievgeniia Kuzminykh[✉][ID], Robin Franksson, and Alexander Liljegren

Blekinge Institute of Technology, Campus Grasvik, 371 41 Karlskrona, Sweden
{anders.carlsson,ievgeniia.kuzminykh}@bth.se

Abstract. Low power wide area (LPWA) technologies becomes popular for IoT use cases because LPWA is enable the broad range communications and allows to transmit small amounts of information in a long distance. Among LPWA technologies there are LTE-M, SigFox, LoRa, Symphony Link, Ingenu RPMA, Weightless, and NB-IoT. Currently all these technologies suffer from lack of documentation about deployment recommendation, have non-investigated limitations that can affect implementations and products using such technologies. This paper is focused on the testing of LPWAN LoRa technology to learn how a LoRa network gets affected by different environmental attributes such as distance, height and surrounding area by measuring the signal strength, signal to noise ratio and any resulting packet loss. The series of experiments for various use cases are conducted using a fully deployed LoRa network made up of a gateway and sensor available through the public network. The results will show the LoRa network limitation for such use cases as forest, city, open space. These results allow to give the recommendation for companies during early analysis and design stages of network life circle, and help to choose properly technology for deployment an IoT application.

Keywords: IoT · LoRa · LPWAN · Transmission range · Arduino Raspberry Pi

1 Introduction

The interest and applications for Internet of Things has in recent years increased significantly. According to the forecast about IoT market from the global leaders such as Juniper Research, Research Nester, Cisco, Ericsson, Gartner, in just a couple of years tens of billions of different kind of "things" are estimated to be connected [1,2]. Most of these devices will be deployed in WAN (Wide Area Network) solutions [3]. This introduce a demand for new communication standards that targets the key features needed to deploy the technology in society such as low power consumption and long range coverage. This led to the creation of LPWANs (Low Power Wide Area Network) which is designed for low bit

© Springer Nature Switzerland AG 2018
O. Galinina et al. (Eds.): NEW2AN 2018/ruSMART 2018, LNCS 11118, pp. 116–128, 2018.
https://doi.org/10.1007/978-3-030-01168-0_11

rate long range communication [4]. The goal is to replace devices that today rely on cellular communications like GSM and 4G at a higher cost and power consumption with new ones that adapt this low bit rate communication style to provide battery lifetimes up towards 10 years while maintaining a communication range up to 30 Km under optimal circumstances.

Due to the LPWA technology being new and not thoroughly fleshed out together with being accessible and completely open to the public this paper will have a focus on measuring and evaluating one of these LPWAN technologies LoRa and its communication protocol LoRaWAN. Currently, the research on LoRa is very limited and mainly focused on how certain factors like temperature, humidity and precipitation affect the maximum range and battery life while the research on how a LoRa network actually performs and the research in this field is limited mostly by simulations and theories.

In this work we provide further insight over how performance of a LoRa network is affected by placement by showing signal quality, signal-to-noise ratios and packet loss in various environmental areas. By analyzing performance on the end-devices and gateways we hope to bring some light over how such kinds of systems should be set up to fit different use cases and also giving a hint to whether the technology is feasible given a specific use case.

2   Wireless Technologies for IoT

2.1   LPWAN Technologies

Currently, in the world a great variation of technologies is used for the IoT where each is made to fit a certain domain. For "smart homes" it fits Wi-Fi when possible and Bluetooth when not. A relatively new candidate, however, used for several home products is ZigBee which low-power and low-rang, it perfect fits for applications like home automation i.e. lighting, temperature, security and sensors. As distance increases the possibilities of using such technologies gets limited since both Wi-Fi and ZigBee are limited up to 100 m, so for products that require a higher range the cellular networks like GSM up to 5G are used, these however come at a heavy price in both licensing costs and battery lifetime [5]. The interest in adding this missing piece in radio communication technologies led to the creation of LPWANs. A comparison of the trade-offs in range and data rate for the various technologies can be seen on Fig. 1.

LPWAN or Low-Power Wide-Area Network is an umbrella term for technologies that focus on having very high power efficiency while maintaining a high transmission range to meet this goal a sacrifice in data rate is made. The goal of these technologies is to fill the niche of products that require long battery lifetime, have low duty cycles and require medium or long range [6]. Examples of such products can be agriculture and industrial sensors, "smart cities" applications such as traffic, trashcan and parking sensors. There are several variants of these LPWAN technologies such as LTE-M, SigFox, LoRa, Symphony Link, Ingenu RPMA, Weightless W, N and P, and NB-IoT.

Common features of LPWAN technologies are summarized in Table 1.

Table 1. High level comparison between different technologies competing in the LPWAN space.

Feature	LoRa	SigFox	LTE-M Rel 14	NB-IoT (NB-LTE) Rel 14	Ingenu RPMA	Weightless-W	Weightless-N	Weightless-P
Modulation	CSS	GFSK, BPSK	QPSK, 16 QAM	QPSK, BPSK	RPMA	OQPSK, QAM	DBPSK	PSK, GMSK
Maximum data rate, DL/UL	50 kbps / 250 bps	256 b/day / 100 bps or 600bps (depends on country regulations)	590 / 1100 kbps (HD CAT-M1) 2.35 / 1.5 Mbps (HD CAT-M2)	27 / 65 kbps (CAT-NB1) 85 / 150 kbps (CAT-NB2)	156 / 624 kbps	1 Mbps / 50 kbps	Only UL: 100 kbps	100 / 100 kbps
Frequency	Unlicensed ISM bands 869 MHz (Europe) and 915 MHz (North America)	Unlicensed ISM bands 869 MHz (Europe) and 915 MHz (North America)	20-1,4 MHz	Licensed LTE 200 MHz	2.4 GHz	470 - 790MHz	800-900 MHz	169 - 923MHz
Bandwidth	250kHz and 125 kHz	100Hz	5 MHz	180kHz	1 MHz	6-8 MHz	200 Hz	12,5 kHz
Range	5 (urban) 15 km (rural)	10 km (urban) 30 km (rural)	10 km	15 km	5 km (urban) 15 km (rural)	5 km (urban)	3 km (urban)	2 km (urban)
Coverage (MCL)	157 dB	150 dB	164 dB	164 dB	177 dB	148 dB	148 dB	134 dB
Transmit power	20dBm	20dBm	20dBm	23dBm	30 dBm	27 dBm (up to 30 dBm)	27 dBm	27 dBm
Standardization	LoRa Alliance	SigFox Inc.	3GPP	3GPP	Ingenu Inc.	Weightless SIG		

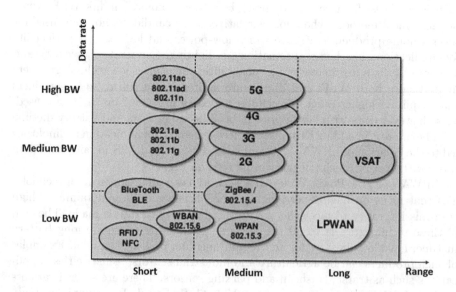

Fig. 1. Data rate vs. range capacity of radio communication technologies.

Lora is a technology based on open protocol but limited to Semtech's chips. It has low data rate that varies according to the SF and the bandwidth (27 kbps with SF = 12 and 500 kHz channel or 50 kbps with FSK). The LoRa uses Chirp Spread Spectrum (CSS) modulation when the signal is modulated by chirp pulses to increase resilience against interference, Doppler effect and multipath.

SigFox is not an open-protocol but one of the most developed. It is not limited to particular chips and the vendors can produce chips with SigFox compatible radios. SigFox is not cheap technology, it sells its network as a service. It is extremely narrow band and provide up to 30 km range. It has daily messages limit that depends on the contract you subscribed for. The maximum limit is then 140 messages in UL with a payload up to 12 bytes and 4 messages in DL with a payload of 8 bytes. These restrictions and business model shifted the interest to LoRaWAN that is considered more flexible and open.

LTE-M and NB-LTE (NB-IoT) from 3GPP group are part of LTE infrastructure. They use licensed spectrum and an LTE-based synchronous protocol and provide optimal QoS at the expense of cost.

Ingenu technology is open standard based on Random Phase Multiple Access (RPMA). It has high energy consumption that's why is not so popular.

The Weightless standards has three types of protocols: Weightless-W, Weight–less-N and Weightless-P. Weightless-W and Weightless-P have bidirectional communication. Weightless-N was developed to increase the transmission range of Weightless-W, decrease data rate and reduce the power consumption. Weightless-N operates in Ultra Narrow Band (UNB) 800–900 MHz and provides only uplink communication.

2.2   LoRa and LoRaWAN

LoRa is a long-range wireless communications technology promoted by the LoRa Alliance. It operates on the license free ISM bands and uses Chirp Spread Spectrum (CSS) radio modulation technique on physical layer, and a MAC layer protocol LoRaWAN. Depending on the frequency band the LoRaWAN duty cycle can be 0.1%, 1% or 10% but recommended is less than 1%. Value of duty cycle is considered as delay between the successive frames sent by the end node. If the value is 1% the device will have to wait 100 times of the duration of the last frame before sending again in the same channel. Last document [7] from LoRa Alliance specifies frequency and bandwidth parameters for different regions over the world. Also, document [8] describes all parameters required to the device communication on physical and MAC levels, the formats of messaging, explains how devices of different types should operate.

The payload of each transmission can range from 2 to 255 octets, and the data rate can reach up to 50 Kbps when channel aggregation is used. The maximum data rate varies according to the SF and the bandwidth. The spreading factor is a logarithm, in base 2, of the number of chirps per symbol, i.e. the number of bits encoded into each symbol. And a symbol is an instantaneous change in frequency. So, in an unique variation of frequency it is associated not only one bit, but more (e.g. in SF7 it is 7 bits). The chirp rate depends only on the bandwidth: the chirp rate is proportional to the bandwidth (one chirp per second per Hertz of bandwidth). Changing of spreading factor is a process in which we add more information, redundancy, to obtain a better receive sensitivity. The relationship between the data bit rate, sensitivity and chip rate for LoRa modulation can be observed in Table 2.

Table 2. Spread factor specification.

Mode	Equivalent bit rate (kbps)	Sensitivity (dBm)	Δ (dB)
FSK	1,2	−122	-
LoRa SF = 12	0,293	−137	+15
LoRa SF = 11	0,537	−134,5	+12,5
LoRa SF = 10	0,976	−132	+10
LoRa SF = 9	1757	−129	+7
LoRa SF = 8	3125	−126	+4
LoRa SF = 7	5468	−123	+1
LoRa SF = 6	9375	−118	−3

On the MAC layer LoRa uses the communication protocol LoRaWAN. So again, using the WiFi analogy, if LoRa is the WiFi connection, LoRaWAN is the IP protocol. It is intended primarily for wireless long battery time devices and targets key points of the Internet of Things such as mobility, localization and bi-directional communication. LoRaWAN [8] specifies three different types of end-devices to address the various needs of applications. Class A is bi-directional, Class B is bi-directional with scheduled receive slots and class C is bi-directional with maximal receive slots. Last type has the maximum power consumption because device is almost all time in open receiving mode.

LoRaWAN doesn't support device-to-device communications, data can be transmitted only in way device-to-gateway, or vice versa. Device-to-device communication could be set up through gateway communication. The architecture of a LoRaWAN network is typically laid out in a star-of-stars topology where gateways act like bridges relaying messages between the end-point devices and a central network server. Devices in a LoraWAN network are remote objects that can range from anything between a thermometer to a geolocation based tracking system.

3   Related Work

According to [9] many of LPWAN technologies will have a share on the IoT market in the future as each of the technologies are more fit for different use cases where some focus more on a higher battery lifetime while others on maximum range and throughput. A comparison is made for various possible implementations of LPWAN such as smart farming which heavily favors the LoRa and Sigfox as they do not rely on cellular coverage while applications such as a terminal for retailers sale require low latency due to do not limit the number of transactions, in such cases a Narrow Band implementation more attractive but the power consumption is more.

The research of Centenaro in [10] describes how in 2009 led by the company Sigfox there was a huge interest for a new wireless long range and lower bit-rate technology compared to the commonly used GSM and 3G.

LPWAN technologies can suffer greatly at various environmental variables. This is shown by Cattani in [11] in their studies of how variables such as weather conditions, temperature, humidity affect the performance of an LPWAN solution. For example, increasing the temperature by $10\,^{\circ}\mathrm{C}$ will reduce the RSSI by $1\,\mathrm{dBm}$. The decrease in signal strength caused by temperature change could in theory render a perfectly good LoRa link unusable.

This aspect is also mentioned by Wennerström [12] in their long-term study of meteorological affects where they show that in Uppsala, Sweden, the PRR can fluctuate more than 20% from day to night where a higher fluctuation is present during the dryer and hotter months June, July and August.

Another variable that can highly affect PRR is oversaturation of the network as there is a large risk for package collision when several devices transmit radio signals in the same time. This is shown by Ferre [13] in his mathematical theory and simulation of approximated packet loss. To avoid this congestion a specification was created with recommendation to limit duty cycles up to 1% for the EU bands that allows the device to transmit a signal for 1 time unit every 100 time units [14] .

Petäjäjärvi in the article [15] evaluates a LoRa network performance when end-node devices are mobile which could be a potential use case. Their results indicate that moving at speeds exceeding $40\,\mathrm{km/h}$ causes the communication performance to deteriorate while speeds around $25\,\mathrm{km/h}$ the communication is relatively reliable. They discuss that this behavior can be caused by the doppler effect and that lower SF values could be less affected by it.

As LoRa networks are open for anyone to use and neighboring networks introduce interference. Thiemo Voigh [16] investigates these interferences and how they can be avoided using directional antenna or multiple gateways. First method improves RSSI and at the same time reduces interference on the neighboring gateways. Deploying more gateways are economically irrational.

In the work [17] it was tested LoRa network but for particular gateway equipment of Cisco 910 industrial router. The authors evaluated the maximal throughput that a single device can obtain, network coverage and receiver sensitivity. Their testing environment was limited by suburban area of Paris with dense residential dwellings.

4   Materials and Methods

To provide testing of the LoRa network performance the preparatory step of literature study was conducted and represented in above section. Found discussions about technology limitations and the reasons behind them were figured out. Furthermore, the company supported us in investigation had their own purpose and were interested mostly on abilities of LoRaWAN in range up to 1 km. We will try to adapt our test based on these limitations and reasons.

4.1   Case Study Topology

The first stage of setting up any network is to ensure the necessary components are configured correctly, this, in turn, will ensure the accuracy of the data collected during the investigation and analysis of the network. But as you will note a LoRa network is not too complicated to setup and run as there are very few components to configure.

The topology of the LoRa network used in this study consists of the end-node device and gateway connected trough the public network to PC with analysing and visualization tool for representing measurements (Fig. 2).

The end-node device and gateway are powered by portable power supplies. The end-node device is configured to send a packet every 10 s on the 867.10–868.50 MHz frequencies. The gateway receives and decodes the packet and the forwarder software forwards the packet to LORIOT.io for presentation on a PC.

End-Node Device. In the real use case the end-node device collects data and then send it to the gateway. In our case and for the sake of the experiments this sensor just sends an array of chars.

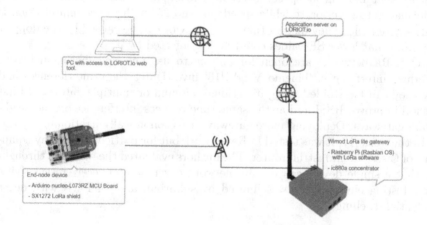

Fig. 2. The topology of the LoRa network.

The LoRaWAN node that we used is set up using the SX1272 LoRa Shield by Semtech and an STMicroelectronics NUCLEO-L073RZ MCU Board. This kit which can be seen in Fig. 2 uses the software I-CUBE-LRWAN provided by ST [18] and the app available called "End-Node". The software has support for SX1276, SX1276 and SX1272 LoRa shields. The IDE Atollic for embedded devices was used to edit the source-code to increase the duty cycle and mocking sensor data, Atollic was also used to flash the code onto the device.

Gateway. The gateway that we used is a "Wimod LoRa Lite Gateway" by IMST and is intended for development and evaluation purposes. It consists of a Raspberry Pi and a concentrator iC880A. The LoRa specific source code running on the gateway is provided by the open-source github project LoRa-net [19]. As the gateway is a commercial product it is pre-configured with the appropriate settings upon delivery. The Raspberry Pi is running Raspbian OS and is pre-installed with the repositories "lora_gateway" and "packet_forwarder" contained in the LoRa-net project.

All experiments were conducted using the following parameters for both end-device and gateway:

- Bandwidth: 867.1–868.5 Mhz
- Channel size: 125 kHz
- Spreading Factor (SF): variable = 7 to 12
- Coding Rate: fixed = 4/5
- Transmitting Power: 14 dBm

4.2 Types of Experiments

The measurements are conducted in similar weather and temperature conditions. All experiments are conducted during April in the timespan 11am–1pm. They are also performed in sunny weather and in a temperature range of 16 °C–20 °C.

In all of the experiments the end-node device transmits a package with a 2 byte payload every 10 s with a transmitting power of 14 dBm. The spreading factor can vary from SF7 to SF12. The SF value is chosen automatically based on the time it takes for the gateway to acknowledge the packet. For all experiments, except the gateway elevation, the experiment is done in two forms:

- with the antennas aligned that means both antennas are vertical,
- when the antennas are misaligned when one antenna is horizontal and pointing towards the other that is positioned vertically.

Open space. The equipment set up on a location where we have at least one kilometer with a free line of sight. The gateway is stationary and end-node device we move far and far away from the gateway, and on each 300 meters we do a measurement. As soon as we lose a single data-packages in the transmissions we decreased the step to 100 m. We then continue until either 0 packages are received or we run out of space that still maintain free line of sight.

Dense forest. In this test we find a location where the terrain was relatively flat and with a dense vegetation. We then do the measurements starting at 300 m. The distance is increased by 100 m each step since we can expect the signal to drop more drastically compared to open space experiment. Gateway is stationary, end device gets moved further away on a straight line. Urban. In this test we do our measurements inside of a city that has houses made of a variation of materials that are located on the same height. We start our measurements at 150 m and increasing the range by another 150 m for each measurement while maintaining the same increment of houses for each measurement.

Gateway elevation. For this test we place the gateway and end-node 400 m apart from one another. We then elevate the gateway in steps of 10 m starting at 0 while moving the gateway to maintain a 400 m distance between the two making measurements each step. The device is partly stationary while the device is continuously getting elevated.

5   Results

For the experiment in open space results are presented on Fig. 3. At 0 m distance we see a clear difference in RSSI between the aligned and misaligned antennas but no difference when it comes to SNR. As the distance increases this difference gets smaller for RSSI and larger for SNR. After 900 m the first package loss is observed for the misaligned antenna. Aligned antenna however led to 0 packets lost even at our maximum distance of 1100 m where we ran out of space.

These results show that even at a distance of +1100m we still have a strong signal that only fades at a tiny rate if at all which could potentially last for several more kilometers. In this scenario the LoRa technology is very suitable and the limitation is primarily given by the specification of LoRa itself namely the trade-off between range and bandwidth.

In the dense forest the SNR starts to decrease after only 300 m and reaches negative numbers on misaligned antennas at 400 m and for the aligned at 500 m Fig. 4. This would indicate that this environment has a big impact on SNR and by proxy on RSSI as well. SNR however doesn't follow a similar structure as a packet can seemingly be lost at an average of −5 dB while a packet with −13 dB can be read without problem. The dense forest environment did not affect the connectivity of LoRa.

The urban city experiment shows more drastic drop in both RSSI and SNR (Fig. 5) than in the previous experiments and it can be observed that the values become lower at 150 m than that of the previous experiments at 300 m. For the next distance measurement step to 300 m an extreme drop in signal quality can be observed where the PRR for having the antennas aligned gets as low as 32% while not a single packet can be read while having them misaligned.

While packets seemingly get lost due to poor signal quality the packets that actually get through maintain a mean SNR value of −7.1 dB while for the experiment in the forest the value could reach as low as −13.6 dB without the packet being lost. Gateway elevation improves the RSSI and gives a value of −75 dBm at 20 m height, after this RSSI value is basically unchanged up to 40 m, the SNR value however continues to rise although at a decaying rate.

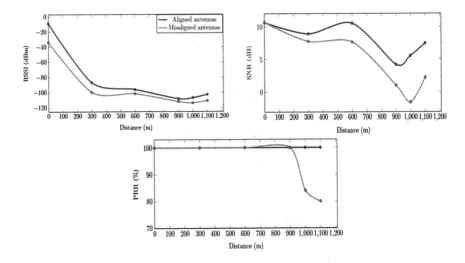

Fig. 3. Open space experiment results.

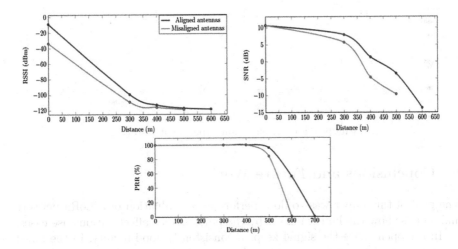

Fig. 4. Dense forest experiment results.

Height has as we expected a great impact on the performance to the point where free line of sight and the fresnel zone is clear of obstacles is achieved from that point on any further elevation might not be considered useful. Applying this elevation to the forest and city that were previously tested would most likely improve the performance greatly and allow for use cases that would otherwise be considered impossible, i.e. large forest agriculture and city based solution that span +1 km (Fig. 6).

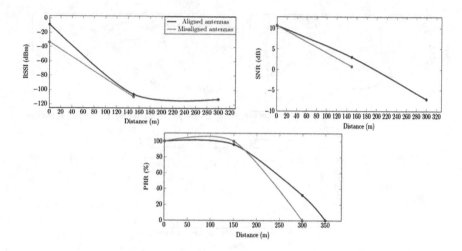

Fig. 5. Urban experiment results.

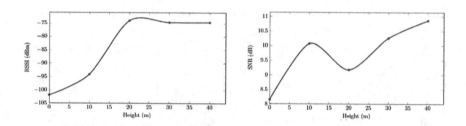

Fig. 6. Gateway elevation experiment results.

6   Conclusions and Future Work

The goal of this paper was to do a performance evaluation of a LoRa network and analyse how the limitation set by the performance affect various use cases.

In the open space the signal keeps a consistently good quality. In the forest and city tests the quality is significantly worse to the point that with our setup in a city environment reaching distances above 350 m is impossible due to the high amount of noise and quantity of dense materials. This problem can be circumvented by elevating the gateway and/or device to the point that free line of sight or close to it is achieved.

This means that large scale implementations deploying a gateway on a radio tower to prevent obstructions would have no problems reaching distances of +15 km. For implementations on a lower scale, problems can arise as barriers in the form of houses, terrain and/or vegetation quickly add up and in most cases limits the implementations range to a house or neighborhood. In some cases, elevating the gateway could be impossible then the solution is to use several gateways just like a normal telecommunication solution where a device can connect to any radio tower. Another factor is if the device is mobile as having misaligned antennas leaves a large mark on the performance and there can more than a 50% difference in PRR.

Regarding packet loss it appears that exceeding the threshold of $-120\,\mathrm{dBm}$ causes the packet to become corrupted, i.e. unreadable or lost completely. SNR, however, seems to have less of an effect as a packet is seemingly as readable at $-14\,\mathrm{dB}$ as it is at a positive value. This means that for most implementations it is very important to have packet loss in regard when planning placement as the RSSI can fluctuate and as in most cases there are no retransmissions in LoRa if the device is at the edge of the communication range the packet can be lost forever. This can be very painful for a lot of use cases as the duty cycles are very limited and the next transmission can be hours away.

For future investigation into this subject more case studies are needed, since this one is limited in long term tests and conducted under similar weather conditions for all of the test cases. Furthermore, an extensive comparison to theoretical and simulated results is needed to provide information about how the real-world performance compares to a simulated one as there is always external variables affecting that is not otherwise accounted for.

References

1. Sorrell, S.: The Internet of Things: Consumer, Industrial & Public Services 2016–2021. Juniper (2016)
2. IoT Market Forecasts. https://www.postscapes.com/internet-of-things-market-size/. Last Accessed 29 May 2018
3. Ericsson Mobility Report. Ericsson (2016)
4. A Comprehensive Look at Low Power, Wide Area Networks. Link Labs, Inc. (2016)
5. Lethaby, N.: Wireless connectivity for the Internet of Things: one size does not fit all. Texas Instrumentals (2017)
6. Adelantado, F.: Understanding the limits of LoRaWAN. IEEE Commun. Mag. **55**(9), 34–40 (2017)
7. LoRaWAN 1.1 Regional Parameters. LoRa Alliance, Inc. (2017)
8. LoRaWAN^{TM} 1.1 Specification. LoRa Alliance, Inc. (2017)
9. Mekki, K., et al.: A comparative study of LPWAN technologies for large-scale IoT deployment. ICT Express (2018). https://doi.org/10.1016/j.icte.2017.12.005
10. Centenaro, M.: Long-range communications in unlicensed bands: the rising stars in the IoT and smart city scenarios. IEEE Wirel. Commun. **23**(5), 60–67 (2016)
11. Cattani, M., Boano, C., Römer, K.: Experimental evaluation of the reliability of LoRa long-range low-power wireless communication. J. Sens. Actuator Netw. **6**(2), 7 (2017)
12. Wennerström, H. et al. : A long-term study of correlations between meteorological conditions and 802.15.4 link performance. In: IEEE International Conference on Sensing, Communications and Networking, pp. 221–229. IEEE, New Orleans, LA, USA (2013)
13. Ferre, G.: Collision and packet loss analysis in a LoRaWAN network. In: 25th European Signal Processing Conference, pp. 2586–2590. IEEE, Kos, Greece (2017)
14. European Commission ERC Recommendation 70–03. https://www.efis.dk/sitecontent.jsp?sitecontent=srd_regulations. Last Accessed 29 May 2018

15. Petäjäjärvi, J.: Performance of a low-power wide-area network based on LoRa technology: doppler robustness, scalability, and coverage. Int. J. Distrib. Sens. Netw. **13**(3), 1–16 (2017)
16. Thiemo, V., Utz, R., Bor, M., Juan, A.: Mitigating Inter-network Interference in LoRa Networks. In: Proceedings of the 2017 International Conference on Embedded Wireless Systems and Networks, pp. 323–328. ACM, Uppsala, Sweden (2016)
17. Aloÿs, A.: A study of LoRa: long range & low power networks for the internet of things. Sensors (Basel) **16**(9), 1466 (2016)
18. ST. I-CUBE-LRWAN. http://www.st.com/en/embedded-software/icube-lrwan.html. Last Accessed 29 May 2018
19. LoRa Gateway Project. https://github.com/Lora-net/lora_gateway. Last Accessed 29 May 2018

Testbed for Identify IoT-Devices Based on Digital Object Architecture

Mahmood Al-Bahri[1]([✉]), Anton Yankovsky[1], Alexey Borodin[2], and Ruslan Kirichek[1,3]

[1] The Bonch-Bruevich Saint-Petersburg State University of Telecommunications, St. Petersburg, Russia
`albahri.89@hotmail.com,ostnipx@gmail.com,kirichek@sut.ru`
[2] PJSC "Rostelecom", Moscow, Russia
`alexey.borodin@rt.ru`
[3] Peoples' Friendship University of Russia (RUDN University), Moscow, Russia

Abstract. Currently, the methods of identification of devices of the Internet of things are one of the main topics due to the increase in the number of devices connected to the Internet. The use of a large number of heterogeneous physical and logical identifiers led to the fact that there were instances of their falsification, which threatens many vital processes. This article describes the architectural network model of a software and hardware complex for IoT device identification. Using the digital object architecture (DOA), it is possible to implement any of the heterogeneous identifiers to create a unique IoT snapshot of the device and store it with a unique global identifier. The article presents a method and developed the software-hardware system for enabling and facilitating the use of DOA for Identify IoT-devices. It is also proposed to develop a software and hardware complex for testing the identification procedure of "smart" devices using.

Keywords: DOA · Identification · Internet of Things · Handle Counterfeit · Prefix · Suffix

1 Introduction

Today, the number of Things (devices) on the Internet is continuously growing where it is expected that the number of connected devices will be in the range of 50 billion with a market volume of a range of 7.5 trillion dollars by 2020 [5,6]. IoT is the communication paradigm that covers the connection and interaction among this growing number of devices in a smart way. The number of IoT connected devices around the world raise exponentially. Figure 1 illustrates the actual and the predicted number of IoT devices in billions from 2015 to 2025 with the market impact [7]. So the Internet of Things (IoT) is evolving faster. Because of the growing traffic, there is a need to improve communication networks to maintain their operability to ensure a given Quality of Service (QoS) [5].

© Springer Nature Switzerland AG 2018
O. Galinina et al. (Eds.): NEW2AN 2018/ruSMART 2018, LNCS 11118, pp. 129–137, 2018.
https://doi.org/10.1007/978-3-030-01168-0_12

This growing of devices entails some problems in the field of telecommunications [1]. However, the issues are not limited only to the physical capabilities of the networks, but they also affect the sphere of security. As an array of standards-setting organizations are strategically positioning themselves to address competition in the IoT space, a previously obscure alternative has emerged from the annals of 1990s-Internet Engineering Task Force (IETF) mailing lists [15]. One of the ways to improve the security of the Internet of Things is the implementation of DOA (Digital Object Architecture) systems, including identification - DOI (Digital Object Identifier). With the help of this system, the user can uniquely identify the object (IoT device) by receiving all the information in the network about the object by the identifier [6].

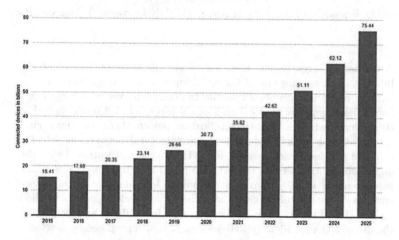

Fig. 1. Internet of Things (IoT) connected devices installed base worldwide from 2015 to 2025 (in billions) [9]

The article considers to develop and analyze the basic concept of identification for IoT devices using Digital Object Identifier from Digital Object Architecture depending on the network parameters [4]. The quality of services of the route can be judged on the basis of various parameters: time of data delivery, loss factor, throughput. Taking into account experience has proved that it is possible to use DOA for Identification objects in IoT, here in this work we will consider the software-hardware system for IoT-devices identification based on Digital Object Architecture [6].

2   Overview and Related Works

Currently, some studies describe hardware Identifiers of the Internet of Things by digital objects architecture as the basis for identifying IoT devices [3]. The principle of product's identification is based on allocation and attachment of the particular IoT unit/tag to the particular product. There are various types of IoT units/tags which can be used for combating counterfeiting, including passive tags (RFID, NFC, and SAW) and active tags such as MCU and MPU. Every product needs to be associated with IoT tag/unit which becomes a gate to the product's profile that contains the detailed information about the product itself. In this regards, a Universal Product's Identifier (UPI) needs to be assigned to every IoT tag/unit mentioned above [11].

In general, the Universal Identification System (UIS) should meet the following requirements:

- to be independent from the identified product/technology and should allow to identify services, processes and entities;
- provide access to the product's profile which contains different type of information about the product (e.g. shape/dimensions, picture, logo, SN, SW, etc.);
- to have a secure mechanism for preventing cloning and duplication of the used UPI.

In accordance with the above requirements, the IoT module which is used for this particular case needs to provide the suitable security mechanism which guarantees the high level of security. In general, the IoT module should be based on at least one of the following interfaces: IEEE 802.3, GPRS/EDG/3G,LTE, RFID, NFC (QR code), IEEE 802.1, IEEE 802.15.1, IEEE 802.15.4, IEEE 802.15.6, IEEE 802.16 and LPWAN (LoRa, Sigfox, NB-IoT).

Every technology indicated above has a particular technical feature (e.g. current deviation, power of transmitter, etc.) which is unique for this technology. These features will be used for creating identification mechanism/system for a particular technology. The relevant parameters of IoT unit are called as "hardware identifiers" and may be stored in IoT as a kind of "Electronic Digital Passport" of IoT [10].

3   Digital Object Architecture

The idea behind DOA was first originated by a research group released by National Research Initiatives (CNRI) in a project funded by the Defense Advanced Research Projects Agency (DARPA) in 1993 [12]. DOA is a system used for various purposes, generally for store and retrieve information for an Internet based system. DOA is a system considering information and digital material storing, accessing and managing. The digital object is information or digital material that contains two main components; the data and the metadata. The metadata consists mainly of the handle which is a global unique identifier for the digital object and may contain other fields that may come up for special purposes [1].

A general structure of the DOA system and the process involved is illustrated in Fig. 2. The first part of the system is the originator, which is the user who requests the service. The originator has a data and wants to form it in the shape of a digital object. For this purpose the originator communicate with handle generator asking for a handle to form a digital object for his data. The handle generator responds with a handle, which is a unique identifier that is independent of the logical or physical system [14]. The originator forms a digital object using the received handle with the digital data, and then forwards the data to a certain repository or a group of repositories [2,16].

A repository is a system used mainly for storing digital objects, services and management information. It works based on a repository access protocol which enables and manages the accessing and depositing mechanisms. The depositing mechanism is responsible for adding new digital objects and the accessing mechanism is used to control and manage the availability and accessing a certain repository [8]. Each repository has a unique name and an IP address which assigned by a local naming authority, which in contact with a global naming authority. Repository may define some services and properties to facilitate managing and controlling the stored data. An example of these properties is the record property, which allow collecting and gathering all data associated with the same digital object [13,14].

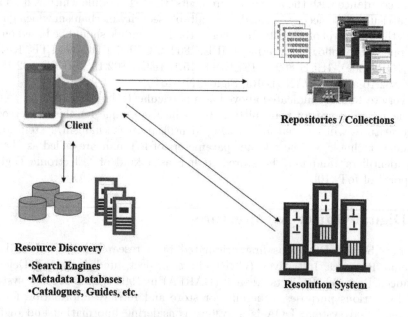

Fig. 2. General structure for the DOA system

Once a digital object is stored in a repository, the repository name or IP and the digital object's handle are transmitted to the handle server's system for registration purposes up on the user demand and the nature of the information, the originator can ask for the name of the repository and dedicated network or even the group of repositories in case of the digital object is stored in multi-repositories by sending the handle of the digital object to the handle servers system. The handle server's system responds to the request by the name of the repository where the user's data is available [9].

4   Testbed for Identify IoT-Devices

The main goal of previous studies was to develop and analyze basic concept of identification for IoT devices using Digital Object Identifier from Digital Object Architecture. Implemented system was based on direct interaction between the device that should be identified and handle-server via internet connection (Fig. 3). This particular study is aimed to review the scenario when intermediate verification device is used.

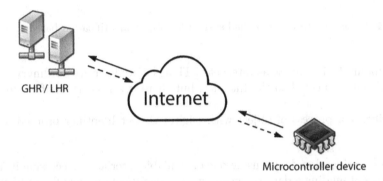

Fig. 3. Basic structure of original system for device identification based on DoA

Modified concept of identification system consists of several components (Fig. 4):

1. Handle-server, as the storage for information about device or object that needs to be identified.
2. Internet as the main infrastructure.
3. Endpoint device (IoT device or any particular object).
4. Object verification device for DoA system.

Considering differences between original and modified system, both Global Handle Register and Local Handle Register are merged into one object: for experimental purposes there is an access provided to the DoA testing zone using prefix "11.test" for all the participants of this study, which allowed us to manage personal identifiers in existing system. In perspective, it gives us an ability to properly understand many different characteristics of system on multiple layers.

Verification device represents as both software and hardware system with number of network interfaces, which allows to plug various devices in different ways: as using physical contact via NFC technology, as using any other network interface such as Wi-Fi or Bluetooth via application level.

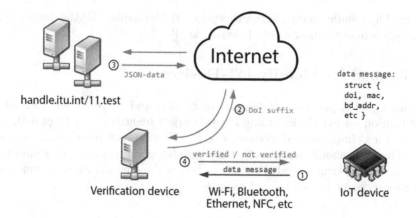

Fig. 4. Basic structure of modified system for object identification based on DoA

Endpoint device can be as Internet of Things module or even ordinary object, which is not connected to the Internet, but there is a need to verify it in some context.

Verification process of object with Digital Object Identifier proceed in four steps:

1. Access verification device using one of available interfaces. Accessing includes process of sending a data structure, which consists of Digital Object Identifier and, what is most important, additional data, which determines if target object is genuine. This data may be MAC address, Bluetooth address, unique product ID or realization date of product;
2. Verification device determines the right server to access and queries provided suffix;
3. Handle-server answers with JSON-structure;
4. Verification device compares required data fields from original object with data from DoA system, making decision if original object is genuine or not (also showing it on screen or sending result to original object);

These steps allows object to access only certain (genuine) DoA servers, which are protected from direct access for ordinary users, thus hiding important data about requested suffix. It also limits available ways to falsify objects with digital identifier and reduces load at endpoint device. Developed system in fixed design (as a stand) allows to properly demonstrate its speed and route of produced traffic in real time.

5   Results of Experiments

A simple experiment was conducted to identify an IoT device and provide a user friendly interface to read the device's identifier. For this purpose, we connect a

NOR flash memory (Atmel - AT26DF161 - SU) to the ESP32 board. The board is considered as the IoT device that is target to be identified. The flash memory is used as a register for the identifier. The flash memory is connected to the ESP32 board and the identifier is written to the memory using the operating system of Mongoose OS. A computer based interface is provided to enable reading the device identifier as illustrated in Fig. 5. Once the device identifier is extracted, it can go through the DOA system and provide the data associated with the identifier.

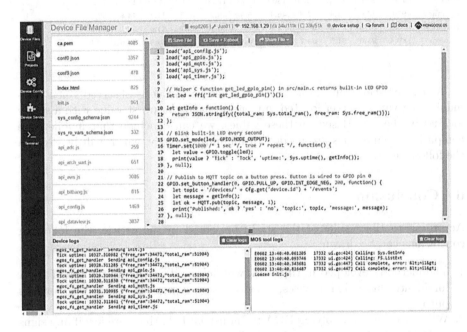

Fig. 5. Mongoose OS Web Interface

Using verification device as possible terminal for verification purposes leads us to make an experiment, which will define average time for system to access DoA-server and provide status of verification. Low-latency results could give us an ability to use this architecture in many different scenarios. Presented latency time only includes network delay. Accessing verification device using certain technologies such as NFC or Bluetooth does create additional delays, but it is not goal scenario for this particular study.

Delays were measured in two cases:

1. Using CNRI proxy server system, collection of web-servers that understand the handle protocol. The system consists of four different web servers that placed in three different geographic places.
2. Using main ITU server placed in Geneva.

Presented table contains latency time for each server and average latency based on ten experiments. Each request was performed using REST API, which enables to handle requests in JSON format. Time intervals were captured using Wireshark software.

Table 1. Results of measuring delay using different handle-servers

Delay (ms)	1	2	3	4	5	6	7	8	9	10	Average
USA	140,2	240,4	141,7	132,0	130,6	126,6	126,0	129,9	128,0	130,5	142,6
Germany	97,6	43,6	44,4	77,6	46,4	45,9	46,2	46,5	44,4	44,8	53,7
Ireland	53,8	57,3	54,6	54,6	54,7	54,8	55,7	55,7	54,6	56,2	55,2
Geneva	80,4	93,0	91,5	88,7	94,9	95,8	89,8	76,7	80,8	82,6	87,4

As can be seen from Table 1, the best latency value is observed when communicating with a server located in Germany and the worst value with a server located in the US. Based on these values, we can conclude that mobile edge computing technologies are needed to minimize latency, which will be done in the following articles.

6   Conclusions

According to the results of the study presented in the article, the following conclusions can be drawn:

1. The method developed in this work will help and accelerate the use of DOA for identification IoT devices, because there is not yet a safe way for this purpose.
2. When it comes to applying DOA to Industrial IoT, there are few if any signs of DOA adoption by the private sector outside the original usage context i.e. DOI. The IoT standardization process is extremely complex and diverse at the moment. It faces the same challenges as many other standards which combine aspects of both public and private goods.
3. According to the results of the experiment, the delay in requesting information from the IoT device to GRL server in Geneva shows low values.

References

1. Kahn, R., Lyons, P.: Representing value as digital objects. J. Telecommun. High Technol. Law **5**(1), 189 (2006)
2. Kahn, R., Wilensky, R.: A framework for distributed digital object services. Int. J. Digit. Libr. **6**(2), 115–123 (2006)
3. Kirichek, R., Koucheryavy, A.: Internet of Things laboratory test bed. In: Zeng, Q.-A. (ed.) Wireless Communications, Networking and Applications. LNEE, vol. 348, pp. 485–494. Springer, New Delhi (2016). https://doi.org/10.1007/978-81-322-2580-5_44

4. Muthanna, A., Paramonov, A., Koucheryavy, A., Prokopiev, A.: Comparison of protocols for ubiquitous wireless sensor network. In: 6th International Congress on Ultra Modern Telecommunications and Control Systems and Workshops, ICUMT 2014, pp. 334–337 (2014)

5. Paramonov, A., Koucheryavy, A.: M2M traffic models and flow types in case of mass event detection. In: Balandin, S., Andreev, S., Koucheryavy, Y. (eds.) NEW2AN 2014. LNCS, vol. 8638, pp. 294–300. Springer, Cham (2014). https://doi.org/10.1007/978-3-319-10353-2_25

6. Kirichek, R., Grishin, I., Okuneva, D., Falin, M.: Development of a node-positioning algorithm for wireless sensor networks in 3D space. In: 2016 18th International Conference on Advanced Communication Technology (ICACT), pp. 279–282 (2016)

7. Kirichek, R., Kulik, V., Koucheryavy, A.: False clouds for Internet of Things and methods of protection. In: 18th International Conference on Advanced Communication Technology, pp. 201–205 (2016)

8. Berber, F., Yahyapour, R.: A high-performance persistent identifier management protocol. In: 2017 International Conference on Networking, Architecture, and Storage (NAS), pp. 1–10 (2017)

9. Digital Object Architecture for IoT. http://www.wileyconnect.com/home/2016/11/8/what-governments-decided-on-digital-object-architecture-for-iot. Accessed March 2018

10. Ahson, S.A., Ilyas, M.: RFID Handbook: Applications, Technology, Security, and Privacy, 712 p. CRC Press, Boca Raton (2017)

11. Kirichek, R., Vladyko, A., Zakharov, M., Koucheryavy, A.: Model networks for Internet of Things and SDN. In: 18th International Conference on Advanced Communication Technology (ICACT), pp. 76–79 (2016)

12. Ateya, A.A., Muthanna, A., Gudkova, I., Vybornova, A., Koucheryavy, A.: Intelligent core network for Tactile Internet system. In: Proceedings of the International Conference on Future Networks and Distributed Systems, p. 15 (2017)

13. Mihovska, A., Sarkar, M.: Smart connectivity for Internet of Things (IoT) applications. In: Yager, R.R., Pascual Espada, J. (eds.) New Advances in the Internet of Things. SCI, vol. 715, pp. 105–118. Springer, Cham (2018). https://doi.org/10.1007/978-3-319-58190-3_7

14. Lund, D., MacGillivray, C., Turner, V., Morales, M.: Worldwide and regional Internet of Things (IoT) 2014–2020 forecast: a virtuous circle of proven value and demand. International Data Corporation (IDC), Technical report, vol. 1. 29 p. (2014)

15. Wang, J.: Copyright limitations and exceptions for education and research: unity in diversity. In: Wang, J. (ed.) Conceptualizing Copyright Exceptions in China and South Africa. CLS, vol. 6, pp. 49–89. Springer, Cham (2018). https://doi.org/10.1007/978-3-319-71831-6_3

16. The DONA Foundation. https://dona.net/. Accessed March 2018

The Application of Graph Theory and Adjacency Lists to Create Parallel Queries to Relational Databases

Yulia Shichkina[✉], Mikhail Kupriyanov, and Vladislav Shevsky

Saint Petersburg Electrotechnical University "LETI",
ul. Professora Popova 5, 197376 St. Petersburg, Russia
strange.y@mail.ru, mikhail.kupriyanov@gmail.com,
immortalghost@yandex.ru

Abstract. The increase in the volume of processed data and the requirements for accuracy and speed of their processing has been observed in the world. Therefore, the problem of finding effective methods for accelerating the execution of queries with the involvement of all possible software, mathematical and hardware tools is becoming increasingly important. This article presents the results of the authors' research in the field of creating parallel queries. These results can be used in practice to implement relational queries and in theory to improve the methods of parallelizing queries. In the article are considered various ways of performance of a complex queries both in sequential, and in a parallel type. It is proposed to use the theory of parallel computations for the transformation of queries. The results of numerical experiments confirming the authors' assumptions are presented at the end of the article.

Keywords: Database · Query · Parallel computing · Information graph
Adjacency lists

1 Introduction

A large number of software projects from a variety of organizations actively use database management systems (DBMS). The diversity of DBMS today is great, satisfying a wide range of data processing needs.

DB-Engines is an organization for collecting and presenting information about a database of different paradigms (SQL, NoSQL, NewSQL). It publishes monthly results of research on the popularity of DBMS. The research conducted by this organization shows that despite the increasing popularity of NoSQl and NewSQL databases, the majority of users in the world continue to use relational databases in practice [1].

Relational databases successfully work with their main tasks, namely:

- storage of new records;
- reading data;
- data search;
- protection of data from unauthorized use.

O. Galinina et al. (Eds.): NEW2AN 2018/ruSMART 2018, LNCS 11118, pp. 138–149, 2018.
https://doi.org/10.1007/978-3-030-01168-0_13

However, with the growth of data volume, many relational databases begin to run into a problem of low data reading speed.

The global solution was the change in the principle of storage and reading of data, which led to the creation of a new generation of DBMS - NoSql. These DBMSs offer storage of data as a collection of files, for example, in the JSON format. This approach allowed to speed up the processing of queries even under the condition of high server load. However, NoSql systems also have significant drawbacks. First, based on the speed of processing requests, these systems only partially make the ACID requirements. This means that NoSql databases can not provide the same reliability of work as a relational database. Secondly, they have their own query language. And although it is similar in syntax to sql, its introduction has several drawbacks, among which:

- variety of languages for each type of database, which inevitably forces the user of the current NoSql system to learn a new language;
- lack of support for the operator "join";
- in general, less capacity of languages than sql;
- lack of tools such as stored procedures.

In addition to SQL and NoSQL, there is also a NewSql database class. These are databases that took new approaches to distributed systems from NoSql and left the relational model of data representation and the language of SQL queries. These databases have emerged in the last decade, but are already intensively fighting for users, pushing away the good old SQL databases on the market, and slightly less old NoSql. The important thing is that when considering the NewSql class, we are still forced to consider relational models. So, the problem of optimizing queries and methods of speeding up their execution using computer systems that allow parallel calculation for NewSql are relevant to the same extent as for relational DBMSs.

This article presents the results of a query optimization study to a relational database by run-time by finding the optimal plan for their implementation on computational systems that allow parallel computations. As examples, there are considered various sql-queries and are presented the results of the analysis of their execution time at start-up in the form of sequential and parallelized plans.

2 Related Works

Due to the fact that the timeliness and quality of information processing is the key to the success of any organization, and the amount of information tend to continuously increase, it is natural that various aspects of data flow management have been given a lot of research since the appearance of the first computer. So, the authors of [2] propose a scalable method in the style of MapReduce, called ICBL, which reduces the execution time of queries in the neo4j database. The studies described in [3] are aimed at solving the problem of continuous processing of queries in distributed environments. The result of these studies are two algorithms that, through data analysis, allow to reduce the amount of data for processing and thereby reduce the cost of communication between database objects and the calculations. The emergence of mobile computing technology provides the ability to access information at anytime and anywhere.

However, since mobile computing environments have inherent characteristics such as power consumption, storage volumes, communication costs and throughput, for mobile computing efficient query processing and minimal query response time are of particular interest. The article [4] presents ways to optimize queries in mobile databases in two main categories: query processing strategy and cache management strategy. The authors of the article describe several methods for improving query performance by transferring data to local user repositories.

A model for converting SQL queries into a structure of a query tree intended for execution in a multiprocessor environment supporting a pipeline is proposed by the authors of [5]. At the last stage of the query presentation at no additional cost, they perform some elementary steps of query optimization and therefore the subsequent optimization can be focused on more complex query improvements. All these constructions since the nineties scientists have tried to represent with the help of graphs, for example, operator graphs [6], the graphs in which the nodes are the operators of relational algebra [7] and others [8, 9].

To represent subqueries using graphs are used different approaches. In System R [5], the query block is a tree node and is considered as the unit for which is selected the optimized execution plan. However, the inner block must be analyzed before analyzing the outer block containing the inner block as a subquery. It is shown that this leads to low query performance [5].

A widely recognized effective tool for large-scale data analysis is the MapReduce system. This system provides high performance through the use of parallelism between processing nodes, providing a simple interface for applications at the top level. Some vendors have improved their storage systems by integrating MapReduce into systems. However, existing query-processing systems based on MapReduce, such as Hive, do not give good results for traditional database systems. In [10] are described the query optimization scheme based on MapReduce. In particular, the authors implement the query optimizer in Hive, which is designed to create an effective query plan based on the cost model.

Recently, intensive research has been carried out to optimize queries to graphic databases [11, 12].

Thus, despite the variety and number of ongoing studies in the field of improving the process of obtaining data from various databases and creating query optimizers in most relational databases, methods and software for the equivalent conversion of queries to a form that is most effectively implemented on computational systems that allow parallel execution of queries there are very small today and they are far from perfect.

3 Formulation of the Problem

The input data is a query written in SQL and executed sequentially, call it the initial query. The output data is the plan for executing the initial query on a parallel computing system.

The task is to develop a method that would allow analyzing the initial query and creating a plan for its execution on a parallel computing system.

The current article discusses plans for executing queries that have been created manually. However, in the future it is supposed to develop and use the algorithm to automatically create a query execution plan.

Then after successful execution of the parallel query plan, there should be obtained disjoint sets of data S1–Sn (where n is the set of parallel branches of the plan). By combining these sets S1–Sn, we obtain the initial set: $S_0 = S_1 \cup \ldots \cup S_n$ (Fig. 1)

A subquery of each parallel part of the plan should read its database piece, thus eliminating the script of locking data for reading on systems that support this function.

4   Methods for Organizing Parallel Execution of Queries

The main goal of the research is to determine whether there will be an increase in performance when there will be a parallel execution of several subqueries of the initial query in comparison with the sequential execution of the initial query.

As a tested database was chosen MySql-database, which has the following architecture (Fig. 1):

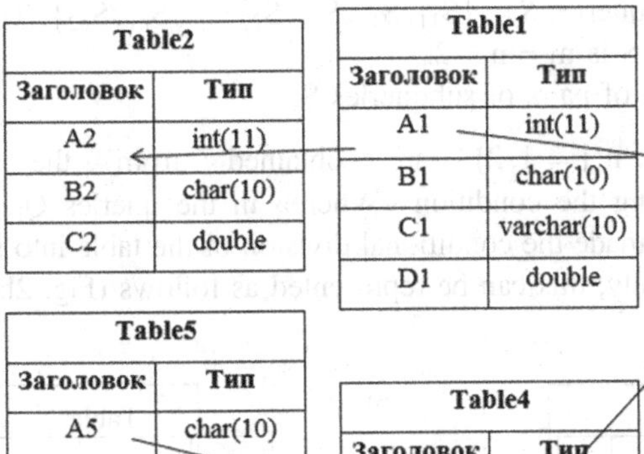

Fig. 1. Schema of the database being tested

Tables are created on the InnoDB low-level subsystem.
The volume of tables is the following:

- Table1: 565825;
- Table2: 565819;
- Table3: 318735;
- Table4: 565818;
- Table5: 318735.

For testing, it was written a program in the language java. The interaction with the database was organized by calling the program functions of the driver «jdbc». Parallelization of queries was done with the help of the basic threads represented in java as objects of the class «Thread», which is part of the java SE.

The testing was conducted in two directions:

- testing of queries consisting of one or more constructs «Select», but aimed at a significant amount of data selected from the database;
- testing large sql-queries composed of a number of interrelated subqueries;

The results of testing are given in this article.

4.1 Parallel Execution of Queries Consisting of One or More Constructs «Select»

Let there be given a set of SQL queries $Q = \{Q_1, Q_2, \ldots, Q_n\}$. All queries $Q_i, i = 1, 2, \ldots, n$ consist of a single construction «Select» and refer to one table. One of the schemes for the parallel execution of such requests is as follows:

- To each of queries Q_i to put in correspondence a pair of queries S_{i1}, S_{i2}. As a result, it will be obtained set of subqueries $S = \{S_{11}, S_{12}, S_{21}, S_{22}, \ldots, S_{n1}, S_{n2}\}$. The total number of queries in the set S is m = n * 2.
- Perform consistent execution of pairs of subqueries S.

Subqueries $S_{ij}(i = 1, 2, \ldots, n; j = 1, 2)$ are obtained from the query $Q_i (i = 1, 2, \ldots, n)$ by introducing the condition «Where» in the queries Q_i, which indicates by what principle it is made the conditional division of the table into several parts for processing. Schematically, this can be represented as follows (Fig. 2b)):

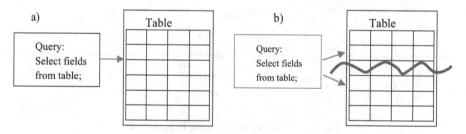

Fig. 2. (a) Scheme of the query execution to the whole table, (b) the execution scheme to the parts of the table.

In case (b) (Fig. 2), the condition «Where» specifies the principle that all records of a table can be uniquely divided into two roughly equal parts. For example: if the original query is «Select * from Table» and there is a primary key of type "counter" in the table, then this query can be divided into the following two subqueries:

$$Select * from Table where A <= Count/2;$$
$$Select * from Table where A > Count/2,$$

where Count – number of records in Table, A - key field-counter of Table.

For all methods in the program in the java language, each of the received subqueries in the code is initialized in a separate thread (the Thread object). After the threads are formed, they are executed in a loop.

4.2 Parallel Execution of Complex Queries Consisting of Set of Subqueries

Let there be given an SQL query Q that consists of a set of subqueries $S = \{S_1, S_2, \ldots, S_n\}$, n– the number of subqueries. All subqueries $S_i, i = 1, 2, \ldots, n$ consist of a single construction «Select». The number of tables to which subquery S is accessed in this context is not important. One of the schemes for the parallel execution of such queries is as follows:

- Divide the query Q into a set of subqueries $S' = \{S'_1, S'_2, \ldots, S'_m\}, m \le n$, m– number of subqueries S', each of which can be either equal to some query $S_i(i = 1, 2, \ldots, n)$ or contain $S_i(i = 1, 2, \ldots, n)$ as a subquery;
- Schedule the execution of subqueries S', on a given number of compute nodes.

When a complex query Q is executed in parallel, as in the case of a simple query, each subquery is executed in its thread in the Java application.

Notes:

- Both of the above options for parallel execution of simple and complex queries involve the presence of synchronization points and aggregation of data.
- Both of the above options for parallel execution of simple and complex queries can be combined to achieve the highest speed of query execution.

5 Testing

5.1 Testing Queries that Consist of One or More Constructions «Select»

To test the parallelization scheme for simple queries on the data it were selected the following queries taking into account the database schema from Fig. 2:

(1) select * from Table2 where C2 >= any (select B5 from Table5);
(2) select * from Table2 where B2 not in (select B3 from Table3);

(3) select Table2.A2, Table2.B2, Table2.C2 from Table2, Q5 where Table2. A2 = Q5.A2 and Table2.B2 = Q5.B2;

(4) select Table2.A2, Table2.B2, Table2.C2 from Table2, (select W1.A2, W1.B2 from (select A2, B2 from Table2 where C2 >= all (select B5 from Table5)) as W1, (select A2, B2 from Table2 where B2 not in (select B3 from Table3)) as W2 where W1.A2 = W2.A2 and W1.B2 = W2.B2) as W where Table2.A2 = W.A2 and Table2.B2 = W.B2;

(5) select * from Table2 where C2 >= all (select B5 from Table5) and B2 not in (select B3 from Table3);

(6) select * from Table2;

In the course of this test, it were created the following views:

(1) Q3 = select A2, B2 from Table2 where C2 >= all (select B5 from Table5);
(2) Q4 = select A2, B2 from Table2 where B2 not in (select B3 from Table3);
(3) Q5 = select Q3.A2, A3.B2 from Q3, Q4 where Q3.A2 = A4.A2 and Q3. B2 = Q4.B2.

Below is a diagram (Fig. 3) created by the results of this testing:

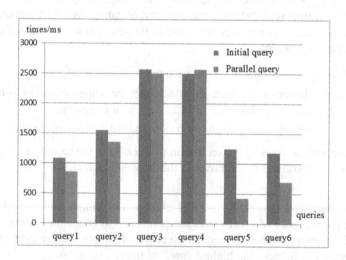

Fig. 3. The test results diagram, where the Y axis displays the execution time in milliseconds, the X axis - the query numbers.

The diagram in Fig. 3 shows that the parallelized form is more efficient for all queries except for the 4th query. The parallel query 4 is slower than the initial query 4 because the subqueries S41 and S42 for query 4, referring to shared memory locations, created conflicts, where:

S41 = select Table2.A2, Table2.B2, Table2.C2 from Table2, (select W1.A2, W1. B2 from (select A2, B2 from Table2 where C2 >= all (select B5 from Table5)) as W1, (select A2, B2 from Table2 where B2 not in (select B3 from Table3)) as W2 where W1. A2 = W2.A2 and W1.B2 = W2.B2) as W where Table2.A2 <= 159368 and Table2. A2 = W.A2 and Table2.B2 = W.B2.

S42 = select Table2.A2, Table2.B2, Table2.C2 from Table2, (select W1.A2, W1.B2 from (select A2, B2 from Table2 where C2 >= all (select B5 from Table5)) as W1, (select A2, B2 from Table2 where B2 not in (select B3 from Table3)) as W2 where W1.A2 = W2.A2 and W1.B2 = W2.B2) as W where Table2.A2 > 159368 and Table2.A2 = W.A2 and Table2.B2 = W.B2.

5.2   Testing Complex Queries Consisting of Set of Subqueries

Testing the scheme for parallelizing complex queries occurred on the basis of a query:

$$Q = S_1 \cup S_2 \cup \ldots S_n$$

where n = 29, $S = \{S_1, S_2, \ldots, S_n\}$ is the collection of subqueries:

S1: Select a5 From Table5 Where f5 >=value;
S2: Select a4 From Table4 Where a4 <> b4;
S3: Select * From Table4, Table5 where a4 = a5 and a5 in S1 and a4 in S2;
S4: Select a5 From Table5 Where value1 <=b5 and b5 <= value2;
S5: Select a4 From Table4 Where d4 >= (Select avg(f5) From Table5);
S5' Select avg(f5) From Table5;
S6: Select * From Table4, Table5 Where a4 = a5 and a4 in S5 and a5 in S4;
S7: Select * From Table1 Where c1 like '*a*';
S8: Select * From Table2, S7 Where a1 = a2 and c2 >= (Select min(d1) From Table1);
S9: Select min(d1) From Table1;
S10: Select * From Table3, S8 Where a1 = a3;
S11: Select * From Table1 Where c1 like '*v*';
S12: Select * From Table3 Where c3 like '*v*';
S13: Select a1 as a13 From S11, S12 Where a1 = a3;
S14: Select * From S13;
S15: Select * From S10 Where a1 not in S14;
S16: Select a3 as a17 From Table3, Table4 Where d3 = d4 and a3 <=d3;
S16': Select a17 From S16;
S17: Select a5 From S6;
S18: Select * From S3 Where a5 not in S17;
S19: S20 Union S21;
S20: Select avg(d1) as d11 From Table1;
S21: Select avg(c2) as d11 From Table2;
S22: Select * From S15 Where (d1 + c2) >=all(S23);
S23: Select max(d11) From S19;
S24: Select * From S15 Where a1 in S25 and a1 in S16';
S25: Select a17 From S10;
S26: Select * From S24, S18 Where d4 = d5;
S27: Select avg(a17)*2 as a18 From S17;
S28: Select * From S22 Where a3 <= S27
S29: Select a4, b4, d4, a5, b5, e5, f5, S26.a3, S26.b3, S26.c3, S26.d3, S26.a2, S26.b2, S26.c2, S26.a1, S26.b1, S26.c1, S26.d1 From S26, S28 Where S28.a3 = S26.a3;

To parallelize the Q query, it was selected the schedule:

1. Selection of the set of subqueries $S_1, S_2, \ldots, S_n$;
2. Run each subquery S_i in its thread;
3. Collecting the results.

The difficulty is that it is not possible to analyze such a number of queries manually.

5.3   Automatic Parallelization of Queries

The solution of this problem is the application of graph theory. Construct a dependency graph from the data between the subqueries $S_i(i = 1, 2, \ldots, n)$ (Fig. 4).

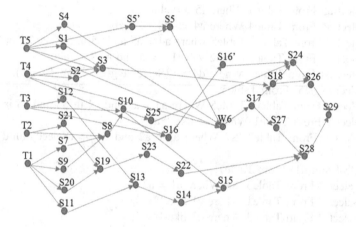

Fig. 4.   Graph of dependencies between subqueries $S_i(i = 1, 2, \ldots, n)$

Apply to this graph the method of parallelization based on the adjacency lists [13]. As a result, we get the following schedule for the subqueries $S_i(i = 1, 2, \ldots, n)$(Fig. 5):

$$t_0: S_{16}, S_1, S_2, S_4, S_{5'}, S_{20}, S_7, S_9, S_{21}, S_{16}, S_{12}, S_{11}$$
$$t_1: S_{16'}, S_8, S_{19}, S_{13}, S_3, S_5$$
$$t_2: S_{10}, S_{14}, S_{23}, S_6$$
$$t_3: S_{25}, S_{15}, S_{17}$$
$$t_4: S_{22}, S_{24}, S_{18}, S_{27}$$
$$t_5: S_{26}, S_{28}$$
$$t_6: S_{29}$$

Fig. 5.   Schedule of queries execution

After each stage of parallel execution $t_i, i = 0, 1, \ldots, 5$ except the last one, it is necessary to apply the functions for combining results.

Below is a diagram (Fig. 6), constructed from the results of testing a complex query Q:

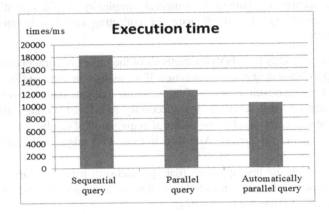

Fig. 6. The graph of test results, where the Y axis displays the execution time in milliseconds, the X axis - the type of query.

The diagram in Fig. 6 shows that the parallel query is more efficient than the sequential query. The fact that a query parallel to the application of graph theory and parallel computation methods is more efficient is for this experiment an accident. In the theory of parallel computing, there are other methods that allow to optimize parallel algorithms for various parameters (time, computational density, the amount of inter-processor data transfers, the amount of computing resources). And the application of these methods will give a more effective result on the speed of query execution. The purpose of our article was to demonstrate the necessity and possibility of applying such methods to queries in databases. As an experiment, it was selected the query that has a large number of subqueries, so that it can be seen the effectiveness of parallelization techniques.

6 Conclusion

Testing the methods of parallelizing queries to databases as a result of most tests showed a positive result. Analysis of the speed of query execution showed:

(1) Necessity of application for parallelization of queries to databases of two approaches known in the theory of parallel computations as data parallelism and task parallelism.

(2) Ability to apply methods to databases from the theory of parallel computing for parallelizing queries. Moreover, for the most part, these methods are much easier to apply to queries in databases than to algorithms in classical programming. This is due to the following factors:

- smaller number of SQL constructs than the number of constructs in the classical programming languages;
- smaller size of the code of the query;
- less complexity of query structure;
- more accurate estimate of the temporal complexity of SQL operations due to the availability of a mechanism for collecting statistics in most modern DBMSs

(3) The need to modify (or adapt) the methods of the theory of parallel computing for direct application in queries to databases. If we look at the last test (Fig. 6), we see that for fully automatic creation of the schedule and automatic conversion of the query code into this schedule, it is necessary to add additional vertices to the graph that will correspond to the points of aggregation of results after the series of previous parallel subqueries. And this, in turn, adds edges and complicates the graph.

(4) When applying the first approach (data parallelism), it is necessary to take into account that not all operations «Select» allow such an approach. So, obviously, if the operator «Select» looks like this:

$$Select \ * \ from \ table \ where \ field > value;$$

then the application of the first approach is possible in this way:

$$Create \ view \ V1 \ as \ select \ * \ from \ table \ where \ field > value \ and \ key <= count/2;$$
$$Create \ view \ V2 \ as \ select \ * \ from \ table \ where \ field > value \ and \ key > count/2;$$

Then, to combine the results, it is necessary to execute the command:

$$Select \ * \ from \ V1 \ union \ select \ * \ from \ V2;$$

It is possible to apply the approach to data parallelization and in the case when in the operation «Select» there are aggregate functions «count», «min», «sum», «max».

In order to select the values and combine the results in this case, it is necessary to execute the commands:

$$Create \ view \ V1 \ as \ select \ max(field) \ as \ M \ from \ table \ where \ field > value \ and \ key <= count/2;$$
$$Create \ view \ V2 \ as \ select \ max(field) \ as \ M \ from \ table \ where \ field > value \ and \ key > count/2;$$
$$Select \ max(M) \ from \ select \ M \ V1 \ union \ select \ M \ from \ V2;$$

But, the application of the approach to data parallelization is impossible if the construction «Select» has, for example, the function of aggregation «avg».

All these situations should be analyzed at the query parallelization. As a result, in the subquery dependency graph, one more parameter or vertex type can be added, indicating the divisibility or indivisibility of tables when performing operations corresponding to the vertices.

It must also be taken into account that the query graph will increase significantly in the number of vertices and edges when these two approaches are combined. But, this is justified, because it gives the researcher a mechanism that allows to evaluate existing queries and optimize them if necessary.

In the end, it should also be noted that the application of these methods is possible for queries of other types of databases, first of all this applies to NewSql databases, secondly to NoSql. For example, a query to a collection in MongoDB also often consists of subqueries. And the same graph can be constructed and the same methods can be applied to it.

Acknowledgments. The paper has been prepared within the scope of the state project "Initiative scientific project" of the main part of the state plan of the Ministry of Education and Science of Russian Federation (task № 2.6553.2017/8.9 BCH Basic Part).

References

1. https://db-engines.com/en/ranking
2. Biswas, R., et al.: A NASA perspective on quantum computing: opportunities and challenges. Parallel Comput. **64**, 81–98 (2017). https://doi.org/10.1016/j.parco.2016.11.002
3. Lu, W., Wang, Y., Juang, J., Liu, J., Shen, Y., Wei, B.: Hybrid storage architecture and efficient MapReduce processing for unstructured data. Parallel Comput. **69**, 63–77 (2017). https://doi.org/10.1016/j.parco.2017.08.008
4. Jin, P., Yang, P., Yue, L.: Optimizing B + -tree for hybrid storage systems. Distrib. Parallel Databases 33(3), 449–475 (2015). https://doi.org/10.1007/s10619-014-7157-7
5. Luo, Q., Teubner, J.: Special issue on data management on modern hardware. Distrib. Parallel Databases 33, 415–416 (2015). https://doi.org/10.1007/s10619-014-7168-4
6. Yasar, A., Gedik, B., Ferhatosmanoglu, H.: Distributed block formation and layout for disk-based management of large-scale graphs. Distrib. Parallel Databases 35(1), 23–53 (2017). https://doi.org/10.1007/s10619-017-7191-3
7. Amagata, D., Hara, T., Nishio, S.: Sliding window top-k dominating query processing over distributed data streams. Distrib. Parallel Databases 34(4), 535–566 (2016)
8. Waluyo, A.B., Srinivasan, B., Taniar, D.: Research in mobile database query optimization and processing. Mob. Inf. Syst. 1(4), 225–252 (2005)
9. Spiliopoulou, M., Hatzopoulos, M.: Translation of SQL queries into a graph structure: query transformations and pre-optimization issues in a pipeline multiprocessor environment. Inf. Syst. 17(2), 161–170 (1992)
10. Yao, S.B.: Optimization of query evaluation algorithms. ACM Trans. Database Syst. 4(2), 133–155 (1979)
11. Mikkilineni, K.P., Su, S.Y.W.: An evaluation of relational join algorithms in a pipelined query processing environment. IEEE Trans. Softw. Eng. 14(6), 838–848 (1988)
12. Jarke, M., Koch, J.: Query optimization in database systems. ACM Comput. Surv. 16(2), 111–152 (1984)
13. Shichkina, Y.A., Kupriyanov, M.S.: Applying the list method to the transformation of parallel algorithms into account temporal characteristics of operations. In: Proceedings of the 19th International Conference on Soft Computing and Measurements, SCM 2016, pp. 292–295 (2016). 7519759

NEW2AN: Next Generation Wired/Wireless Advanced Networks and Systems

On the Necessary Accuracy of Representation of Optimal Signals

Sergey V. Zavjalov(iD), Anna S. Ovsyannikova(iD),
and Sergey V. Volvenko(✉)(iD)

Peter the Great St. Petersburg Polytechnic University, St. Petersburg, Russia
zavyalov_sv@spbstu.ru, anny-ov97@mail.ru,
volk@cee.spbstu.ru

Abstract. The way to increase spectral efficiency without significant energy losses by using optimal signals is considered. SDR (software-defined radio) platform is proposed as transceiver prototype. Improvement of its performance may be achieved due to decreasing the number of digits after decimal point and the number of expansion coefficients, which define representation of signals. We used various forms of optimal signals obtained for different restrictions on out-of-band emissions, symbol rate and BER (bit error rate) performance. The influence of the number of digits after decimal point on spectral and energy efficiency of optimal signals is considered. The necessary accuracy of representation providing maximal spectral efficiency is found for different cases.

Keywords: Optimal signals · Accuracy · Energy efficiency
Spectral efficiency · Out-of-band emissions · Optimization problem
SDR-platform

1 Introduction

We can observe active development of next-generation wireless networks (5G). Numerous studies on 5G networks are actively conducted to improve efficiency [1–3]. Many scientific groups are concentrated at tendencies to increase the spectral efficiency in conditions of limited frequency bandwidth [4–7]. Spectral efficiency is calculated as $R/\Delta F$, where R is symbol rate, ΔF – occupied frequency bandwidth. So there are different ways to increase value $R/\Delta F$: increase R or reduce ΔF. Increasing symbol rate is known as Faster-than-Nyquist (FTN) signals [8, 9]. Reducing ΔF can be done by application of optimal signals [4, 6, 10] and by increasing duration of signals. Our approach consists of joint using of optimal signals with increased duration and increasing symbol rate [10–12].

The finite random sequence of N single optimal signals $s_{opt}(t)$ with duration $T_s = LT$ and energy $\xi^2 E_{opt}$ may be written as follows:

$$y(t) = \xi\sqrt{E_{opt}/T_s} \sum_{n=-N/2}^{N/2} c_n s_{opt}(t - n\xi T), \tag{1}$$

© Springer Nature Switzerland AG 2018
O. Galinina et al. (Eds.): NEW2AN 2018/ruSMART 2018, LNCS 11118, pp. 153–161, 2018.
https://doi.org/10.1007/978-3-030-01168-0_14

where coefficient ξ defines a symbol rate. This coefficient also is used to keep the average power of random sequence $y(t)$ constant. If we use the binary alphabet, the symbol rate is equal to a bit rate.

Forms of signals $s_{opt}(t)$ for different restrictions on out-of-band emissions, symbol rate, BER performance may be obtained by solving corresponding optimization problem. So we can control all time and spectral characteristics of signals. Application of optimal signals [10–12] allows increasing the symbol rate of data transmission without significant energy losses in BER performance (no more than 0.5 dB) [10, 12].

Next step is to develop the algorithm of formation and processing at the reception. We are planning to use SDR-platform HackRF One [13] to construct prototypes of transceivers. Performance of this SDR-platform is limited. Therefore, we must search ways to simplify algorithms and to increase performance.

Note, that we use limited Fourier series to present optimal signals. Initially the number of significant digits after the decimal point q is not limited and equal to 15. So we must use arithmetic of large numbers when going to integers. If we decrease the number of significant digits, resulting performance will improve.

In this article, it is proposed to consider the influence of the number of expansion coefficients and significant digits on optimal signals with increased duration.

2   Optimization Criteria of Signal Form

An optimality criterion of the signal form is based on the choice of fixed reduction rate of out-of-band emissions. The optimization task may be written in the form of linear functional J minimization [10] for signal $s_{opt}(t)$ with duration T_s and symbol rate $R = 1/\xi T$:

$$\arg\left\{ \min_{s_{opt}(t)} (J) \right\}, \; J = \int_{-\infty}^{+\infty} g(f) \left| \int_{-\infty}^{+\infty} s_{opt}(t) \exp(-j2\pi ft) dt \right|^2 df, \quad (2)$$

where $g(f) = f^{2p}$ $(p = 1, 2, \ldots)$ is a weighing function. Choosing $g(f)$ form determines reduction rate of out-of-band emissions.

Restriction on BER performance may be converted to restriction on correlation coefficient between two optimal signals on different time positions [10–12]:

$$\max_{n=1\ldots\lfloor L/\xi \rfloor} \left\{ \int_{-(L-2n)T/2}^{LT/2} s_{opt}(t) s_{opt}(t - n\xi T) dt \right\} < K_0. \quad (3)$$

We have not found analytical solutions of this optimization problem for arbitrary values of T_s, R and K_0. So we switched to numerical solutions. To solve this optimization task numerically we used presentation of $s_{opt}(t)$ in terms of limited Fourier series (m is a number of expansion coefficients).

$$s_{opt}(t) = \frac{s_{opt0}}{2} + \sum_{k=1}^{m-1} s_{optk} \cos\left(\frac{2\pi}{T}kt\right). \tag{4}$$

Then the original optimization task (2) can be transformed into the task of searching for expansion coefficients $\{s_k\}_{k=1}^m$, which minimize the function of several variables [10–12]:

$$\min_{\{s_k\}_{k=1}^m} J\left(\{s_k\}_{k=1}^m\right), \quad J\left(\{s_k\}_{k=1}^m\right) = T_s/2 \sum_{k=1}^m (2\pi k/T_s)^{2n} s_k^2. \tag{5}$$

The value of m is determined by the accuracy of representation $s_{opt}(t)$ and complexity of solution (5) caused by an ill-conditioned task. Target functional has a ravine-type shape, i.e. rises sharply along one direction and changes slightly along the other. It is taken into account in this work, therefore, the chosen values of m provide necessary accuracy of $s_{opt}(t)$.

When $s_{opt}(t)$ is obtained, energy spectrum $|S(f)|^2$ of random sequence of signals (1) may be calculated. For statistically independent modulation symbols it is defined in the area of positive frequencies by function $s_{opt}(t)$ and constant value Z [14, 15]:

$$|S(f)|^2 = \lim_{N\to\infty} \frac{1}{NT_s} m_1\left\{|S_j(f)|^2\right\} = (Z/T_s) \left| \int_{-T_s/2}^{T_s/2} s_{opt}(t) \exp(-j2\pi ft) dt \right|^2, \tag{6}$$

where value Z depends on signal constellation, $S_j(f)$ – spectrum of random sequence of signals (1), $m_1\{\ \}$ – mathematical expectation.

3 Results and Discussion

As $s_{opt}(t)$ we will use results, presented in [10–12] and obtained for next conditions: $p = 2$, $\xi = 0.5$, $K_0 = 0.01$. We can apply these solutions without loss of generality. Envelopes for $T_s = 6T$ and $T_s = 16T$ and corresponding energy spectra are presented on Figs. 1 and 2.

Our aim is to reduce the number of coefficients m. So let us consider the Euclid distance $d_{m,m-1}$ between envelopes formed with the use of m and $m - 1$ coefficients for different m. Firstly we should solve the optimization task for rather high value of m. Then array a is truncated by removing its last value, so we can obtain a new envelope. The Euclid distance between current envelope and previous one is calculated. Here we used $m = 27$ as the initial value and got some interesting results (Fig. 3). We can accept that $d_{m,m-1}$ must be no more than 10^{-3}. The minimal values of m providing such $d_{m,m-1}$ are presented in Table 1.

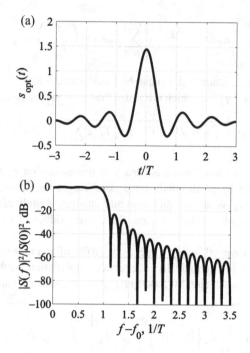

Fig. 1. Envelope (*a*) of optimal signal $T_s = 6T$ and corresponding energy spectrum (*b*).

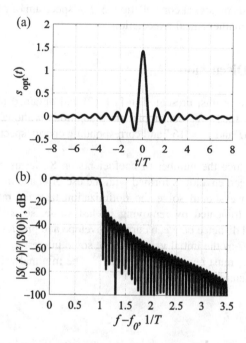

Fig. 2. Envelope (*a*) of optimal signal $T_s = 16T$ and corresponding energy spectrum (*b*).

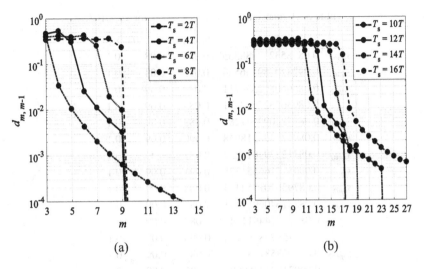

Fig. 3. $d_{m,m-1}$ vs m: (a) $T_s = 2T...8T$, (b) $T_s = 10T...16T$.

Table 1. The minimal values of m providing $d_{m,m-1} \leq 10^{-3}$.

T_s	m
2T	9
4T	10
6T	10
8T	10
10T	18
12T	18
14T	18
16T	24

Table 2. Expansion coefficients $\{s_k\}_{k=1}^{m}$ for $T_s = 6T$.

	$s_{opt}(t)$	$s_{opt(3)}(t)$	$s_{opt(2)}(t)$	$s_{opt(1)}(t)$
s_{opt0}	0.230754291462361	0.231	0.23	0.2
s_{opt1}	0.243020058615220	0.243	0.24	0.2
s_{opt2}	0.230277121912867	0.230	0.23	0.2
s_{opt3}	0.244036419329030	0.244	0.24	0.2
s_{opt4}	0.227953096881105	0.228	0.23	0.2
s_{opt5}	0.248713526443278	0.249	0.25	0.3
s_{opt6}	0.145222271533744	0.145	0.15	0.2
s_{opt7}	−0.011250298505672	−0.011	−0.01	0
s_{opt8}	0.005690067650762	0.006	0.01	0
s_{opt9}	−0.000000003319436	0	0	0

Table 3. Expansion coefficients $\{s_k\}_{k=1}^m$ for $T_s = 16T$.

	$s_{opt}(t)$	$s_{opt(3)}(t)$	$s_{opt(2)}(t)$	$s_{opt(1)}(t)$
s_{opt0}	0.083853277988335	0.084	0.08	0.1
s_{opt1}	0.092820789366586	0.093	0.09	0.1
s_{opt2}	0.086660852336211	0.087	0.09	0.1
s_{opt3}	0.089305327068671	0.089	0.09	0.1
s_{opt4}	0.088724099541282	0.089	0.09	0.1
s_{opt5}	0.089673438318048	0.090	0.09	0.1
s_{opt6}	0.087122794017252	0.087	0.09	0.1
s_{opt7}	0.090517166551378	0.091	0.09	0.1
s_{opt8}	0.087930707754474	0.088	0.09	0.1
s_{opt9}	0.088876406957491	0.089	0.09	0.1
s_{opt10}	0.088759244911348	0.089	0.09	0.1
s_{opt11}	0.089758708389713	0.090	0.09	0.1
s_{opt12}	0.087068587535874	0.087	0.09	0.1
s_{opt13}	0.090319412489984	0.090	0.09	0.1
s_{opt14}	0.088453068901215	0.088	0.09	0.1
s_{opt15}	0.088242284193880	0.088	0.09	0.1
s_{opt16}	0.052464281626184	0.052	0.05	0.1
s_{opt17}	−0.003487694515916	−0.003	0	0
s_{opt18}	0.001815455034236	0.002	0	0
s_{opt19}	−0.001160625678960	−0.001	0	0
s_{opt20}	0.000862815650940	0.001	0	0
s_{opt21}	−0.000661057279843	−0.001	0	0
s_{opt22}	0.000486939765318	0	0	0
s_{opt23}	−0.000380730842321	0	0	0

The next step is choosing the envelope with m defined in Table 1 and rounding the expansion coefficients upwards to fewer digits after decimal point. We decided to investigate the results with no more than three digits after decimal point. The expansion coefficients for $T_s = 6T$ and $T_s = 16T$ are presented in Tables 2 and 3 respectively.

To estimate energy and spectral efficiency simulation model was developed in Matlab (Fig. 4). The input data for this model are optimal envelope form $s_{opt}(t)$, its duration, transmission rate R and signal-to-noise ratio E/N_0. When simulation parameters are initialized, the model forms the random sequence of signals by generating random information bits and using BPSK modulation in the block "modulator". After this step, energy spectrum of random sequence of signals (1) may be calculated by averaging on various realizations. Here we used $N = 1000$ modulation symbols and 200 averages. As a result, we can compute spectral efficiency $R/\Delta F$ knowing ΔF for different level of energy spectra.

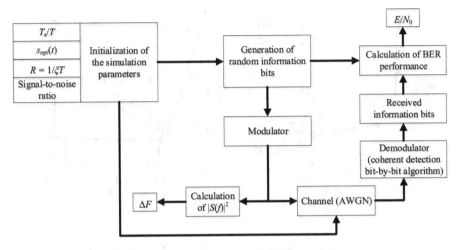

Fig. 4. Block diagram of simulation model.

Another branch of the model includes calculation of BER performance of the random sequence of signals gone through additive white Gaussian noise (AWGN) channel. The block "demodulator" is based on the coherent bit-by-bit detection algorithm, which is simple in realization and provides minimal delay for signal processing. At least 10^6 information bits were transmitted to check BER performance at each signal-to-noise ratio value. The output of this branch is energy efficiency defined as the value of signal-to-noise ratio E/N_0 providing error probability $p_{er} = 10^{-3}$.

Now we should take into account the dependency of energy efficiency on spectral efficiency. Figure 5 shows these relationships for different T_s. The results may be divided into four groups.

The first group includes $T_s = 10T$, $12T$ with maximal spectral efficiency provided by $q = 15$ (Fig. 5a). We decided to estimate spectral efficiency relatively to the results with $q = 15$ and energy losses relatively to the theoretical BER performance. Then the loss in spectral efficiency for $q = 3$ is about 12% for $T_s = 10T$ and 3.5% for $T_s = 12T$. Though it is possible in this case to reduce energy losses to the value 0.19 dB for $T_s = 10T$ and 0.26 dB for $T_s = 12T$.

The second group includes $T_s = 4T$, $14T$ with maximal spectral efficiency provided by $q = 3$ (Fig. 5b). The energy losses for $q = 3$ vary from 0.19 dB to 0.43 dB for $T_s = 4T$ and $14T$ correspondingly while the gain in spectral efficiency changes from 5.2% to 9.7%.

The third group is composed of $T_s = 2T$, $16T$ with maximal spectral efficiency provided by $q = 2$ (Fig. 5c). The energy losses for $q = 2$ are 1.33 dB for $T_s = 2T$ and 0.36 dB for $T_s = 16T$ while spectral efficiency increases by 36% for $T_s = 2T$ and by 34% for $T_s = 16T$.

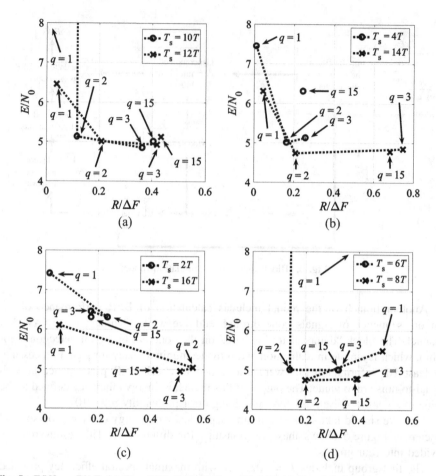

Fig. 5. E/N0 vs $R/\Delta F$: (a) $T_s = 10T$, 12T, (b) $T_s = 4T$, 14T, (c) $T_s = 2T$, 16T, (d) $T_s = 6T$, 8T.

The fourth group unites results with maximal spectral efficiency provided by $q = 1$ ($T_s = 6T$, 8T, Fig. 5d). Let us start with $T_s = 6T$. Increasing spectral efficiency by using $q = 1$ reaches 12.3% relatively to the spectral efficiency of $q = 15$, but energy losses relatively to the theoretical BER performance are huge (19.3 dB). Using $q = 3$ allows to reduce energy losses to the value 0.33 dB. However, in this case the increase in spectral efficiency is just 0.65%.

For $T_s = 8T$ the situation is almost the same. If we use $q = 1$, we can increase spectral efficiency by 25% comparing to $q = 15$ with energy losses relatively to the theoretical BER performance 0.73 dB. If we use $q = 3$, energy losses are reduced to the value 0.13 dB, but spectral efficiency increases just by 1.2%.

So we showed the possibility of reducing the number of coefficients and number of significant digits. These results will be applied in the next projects about realization of modem based on spectrally efficient signals.

The results of the work were obtained using computational resources of Peter the Great Saint-Petersburg Polytechnic University Supercomputing Center (http://www.scc.spbstu.ru).

References

1. Petrov, V., et al.: Achieving end-to-end reliability of mission-critical traffic in softwarized 5G networks. IEEE J. Sel. Areas Commun. **36**, 485–501 (2018)
2. Gelgor, A., Gorlov, A., Nguyen, V.P.: The design and performance of SEFDM with the Sinc-to-RRC modification of subcarriers spectrums. In: 2016 International Conference on Advanced Technologies for Communications (ATC), Hanoi, pp. 65–69 (2016)
3. Rashich, A., Kislitsyn, A., Fadeev, D., Nguyen, T.N.: FFT-based trellis receiver for SEFDM signals. In: 2016 IEEE Global Communications Conference (GLOBECOM), Washington, DC, pp. 1–6 (2016)
4. Gelgor, A., Gorlov, A.: A performance of coded modulation based on optimal Faster-than-Nyquist signals. In: 2017 IEEE International Black Sea Conference on Communications and Networking (BlackSeaCom), Istanbul, pp. 1–5 (2017)
5. Ghannam, H., Darwazeh, I.: SEFDM over satellite systems with advanced interference cancellation. IET Commun. **12**(1), 59–66 (2018)
6. Xu, T., Darwazeh, I.: Nyquist-SEFDM: pulse shaped multicarrier communication with subcarrier spacing below the symbol rate. In: 2016 10th International Symposium on Communication Systems, Networks and Digital Signal Processing (CSNDSP), Prague, pp. 1–6 (2016)
7. Jia, M., Yin, Z., Guo, Q., Gu, X.: Compensation of non-orthogonal ICI for SEFDM receivers. In: 2017 IEEE/CIC International Conference on Communications in China (ICCC), Qingdao, China, pp. 1–5 (2017)
8. Fan, J., Guo, S., Zhou, X., Ren, Y., Li, G.Y., Chen, X.: Faster-than-Nyquist signaling: an overview. IEEE Access **5**, 1925–1940 (2017)
9. Mazo, J.E.: Faster-than-Nyquist signaling. Bell Syst. Tech. J. **54**(8), 1451–1462 (1975)
10. Zavjalov, S.V., Volvenko, S.V., Makarov, S.B.: A method for increasing the spectral and energy efficiency SEFDM signals. IEEE Commun. Lett. **20**, 2382–2385 (2016)
11. Ovsyannikova, A.S., Zavjalov, S.V., Makarov, S.B., Volvenko, S.V., Quang, T.L.: Spectral and energy efficiency of optimal signals with increased duration, providing overcoming "Nyquist Barrier". In: Galinina, O., Andreev, S., Balandin, S., Koucheryavy, Y. (eds.) NEW2AN/ruSMART/NsCC -2017. LNCS, vol. 10531, pp. 607–618. Springer, Cham (2017). https://doi.org/10.1007/978-3-319-67380-6_57
12. Ovsyannikova, A.S., Zavjalov, S.V., Makarov, S.B., Volvenko, S.V.: Choosing parameters of optimal signals with restriction on correlation coefficient. In: Galinina, O., Andreev, S., Balandin, S., Koucheryavy, Y. (eds.) NEW2AN/ruSMART/NsCC -2017. LNCS, vol. 10531, pp. 619–628. Springer, Cham (2017). https://doi.org/10.1007/978-3-319-67380-6_58
13. https://greatscottgadgets.com/hackrf
14. Gelgor, A., Gorlov, A., Popov, E.: Multicomponent signals for bandwidth-efficient single-carrier modulation. In: 2015 IEEE International Black Sea Conference on Communications and Networking, BlackSeaCom 2015, Article no. 7185078, pp. 19–23 (2015)
15. Gelgor, A., Gorlov, A., Popov, E.: On the synthesis of optimal finite pulses for bandwidth and energy efficient single-carrier modulation. In: Balandin, S., Andreev, S., Koucheryavy, Y. (eds.) ruSMART 2015. LNCS, vol. 9247, pp. 655–668. Springer, Cham (2015). https://doi.org/10.1007/978-3-319-23126-6_59

On LDPC Code Based Massive Random-Access Scheme for the Gaussian Multiple Access Channel

Anton Glebov[1], Luiza Medova[2], Pavel Rybin[1,2] , and Alexey Frolov[1(✉)]

[1] Skolkovo Institute of Science and Technology, Moscow, Russia
anton.glebov@skolkovotech.ru, {p.rybin,al.frolov}@skoltech.ru
[2] Institute for Information Transmission Problems, Moscow, Russia
luiza.medova@phystech.edu

Abstract. This paper deals with the problem of massive random access for Gaussian multiple access channel (MAC). We continue to investigate the coding scheme for Gaussian MAC proposed by A. Vem et al. in 2017. The proposed scheme consists of four parts: (i) the data transmission is partitioned into time slots; (ii) the data, transmitted in each slot, is split into two parts, the first one set an interleaver of the low-density parity-check (LDPC) type code and is encoded by spreading sequence or codewords that are designed to be decoded by compressed sensing type decoding; (iii) the another part of transmitted data is encoded by LDPC type code and decoded using a joint message passing decoding algorithm designed for the T-user binary input Gaussian MAC; (iv) users repeat their codeword in multiple slots. In this paper we are concentrated on the third part of considered scheme. We generalized the PEXIT charts to optimize the protograph of LDPC code for Gaussian MAC. The simulation results, obtained at the end of the paper, were analyzed and compared with obtained theoretical bounds and thresholds. Obtained simulation results shows that proposed LDPC code constructions have better performance under joint decoding algorithm over Gaussian MAC than LDPC codes considered by A. Vem et al. in 2017, that leads to the better performance of overall transmission system.

Keywords: NOMA · Gaussian MAC · Massive random-access
LDPC code · PEXIT charts

1 Introduction

Current wireless networks are designed with the goal of servicing human users. Next generation of wireless networks is facing a new challenge in the form of machine-type communication: billions of new devices (dozens per person) with

The research was carried out at Skoltech and supported by the Russian Science Foundation (project no. 18-19-00673).

© Springer Nature Switzerland AG 2018
O. Galinina et al. (Eds.): NEW2AN 2018/ruSMART 2018, LNCS 11118, pp. 162–171, 2018.
https://doi.org/10.1007/978-3-030-01168-0_15

dramatically different traffic patterns are expected to go live in the next decade. The main challenges are associated with: (a) huge number of autonomous devices connected to one access point, (b) low energy consumption, (c) short data packets. This problem has attracted attention (3GPP and 5G-PPP) under the name of mMTC (massive machine-type communication).

There are $K \gg 1$ users, of which only T have data to send in each time instant. A base station (BS) sends periodic beacons, announcing frame boundaries, so that the uplink (user-to-BS) communication proceeds in a frame-synchronized fashion. Length of each frame is n. Each active user has k bits that it intends to transmit during a frame, where a typical value is $k \approx 100$ bit. The main goal is to minimize the energy-per-bit spent by each of the users. We are interested in grant-free access (5G terminology). That is, active users transmit their data, without any prior communication with the BS (without resource requests). We will focus on the Gaussian multiple-access channel (GMAC) with equal-power users, i.e.

$$y = \sum_{t=1}^{T} x^{(t)} + z,$$

where $z \sim \mathcal{N}(0, N_0/2 = \sigma^2)$ and $\mathbb{E}\left[|x^{(i)}|^2\right] \leq P$.

This paper deals with construction of low-complexity random coding schemes for GMAC (indeed we restrict our consideration to the case of binary input GMAC). Let us emphasize the main difference from the classical setting. Classical information theory provided the exact solutions for the case of all-active users, i.e. $T = K$. Almost all well-known low-complexity coding solutions for the traditional MAC channel (e.g. [10]) implicitly assume some form of coordination between the users. Due to the gigantic number users we assume them to be symmetric, i.e. the users use the same codes and equal powers. Here we continue the line of work started in [7,8,14]. In [8] the bounds on the performance of finite-length codes for GMAC are presented. In [7] Ordentlich and Polyanskiy describe the first low-complexity coding paradigm for GMAC. The improvement (in terms of required *energy-per-bit* E_b/N_0) was given in [14]. Recall, that E_b/N_0 is calculated as follows. Assume a user transmits k bits by means of n channel uses, then

$$E_b/N_0 = \frac{nP}{kN_0} = \frac{nP}{k2\sigma^2}.$$

In this paper we continue to investigate the coding scheme from [14]. The proposed scheme consists of four parts:

- the data transmission is partitioned into time slots;
- the data, transmitted in each slot, is split into two parts, the first one (preamble) allows to detect users that were active in the slot. It also set an interleaver of the low-density parity-check (LDPC) type code [2,12] and is encoded by spreading sequence or codewords that are designed to be decoded by compressed sensing type decoding;

- the second part of transmitted data is encoded by LDPC type code and decoded using a joint message passing decoding algorithm designed for the T-user binary input GMAC;
- users repeat their codeword in multiple slots and use successive interference cancellation.

The overall scheme can be called T-fold irregular repetition slotted ALOHA (IRSA, [4,6]) scheme for GMAC. The main difference of this scheme in comparison to IRSA is as follows: any collisions of order up to T can be resolved with some probability of error introduced by Gaussian noise.

In this paper we are concentrated on the third part of considered scheme. Our contribution is as follows. We generalized the protograph extrinsic information transfer charts (EXIT) to optimize the protograph of LDPC code for GMAC. The simulation results, obtained at the end of the paper, were analyzed and compared with obtained theoretical bounds and thresholds. Obtained simulation results shows that proposed LDPC code constructions have better performance under joint decoding algorithm over Gaussian MAC than LDPC codes considered in [14], that leads to the better performance of overall system.

2   Iterative Joint Decoding Algorithm

We consider T independent users, being sent to a single receiver. User t, $t \in \{1, \ldots, T\}$, is encoded by $\mathcal{C}^{(t)}$, where $\mathcal{C}^{(t)}$ is a irregular LDPC code with codeword length n and rate r. The codewords $\mathbf{c}^{(1)}, \mathbf{c}^{(2)}, \ldots, \mathbf{c}^{(T)}$ are BPSK modulated, and therefore the sequences $\mathbf{x}^{(1)}, \mathbf{x}^{(2)}, \ldots, \mathbf{x}^{(T)}$, $\mathbf{x}^{(i)} \in \{-1, +1\}^n$ are transmitted through a communication channel. The received signal $\mathbf{y}$ is an element-wise sum of these sequences affected by Gaussian noise. The joint multi-user decoder is expected to recover all the codewords based on that signal.

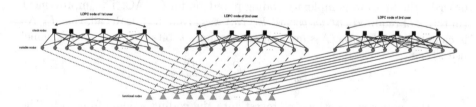

Fig. 1. Joint decoder graph representation for $T = 3$ (Color figure online)

The decoder employs a low-complexity iterative belief propagation (BP) decoder that deals with a received soft information presented in LLR (log likelihood ratio) form. The decoding system can be represented as a factor graph, which is shown in Fig. 1. The factor graph of the T-user LDPC-MAC is composed of the T LDPC graphs, which are connected through state nodes (marked with green color). These nodes correspond to the elements of the received sequence $\mathbf{y}$.

The belief propagation decoding algorithm proceeds as follows. The LLR values of variable nodes for each user are initialized with zero values assuming equal probability for 1 and -1 values and the joint decoder perform ℓ_O outer iterations, where each iteration includes the following steps:

- maximum likelihood decoding of state nodes;
- performing ℓ_I inner iterations of BP decoding for users' LDPC codes and updating LLR values of variable nodes (it's done in parallel);

The message update rules in the graph of each user follow from usual LDPC BP decoding algorithm but it is necessary to describe the update rule through state nodes. In accordance with principles of message-passing algorithms, the outgoing message from the ith variable node of user t to the connected state node is computed as

$$m^t_{vs,i} = \log \frac{p(x^t_i = 1)}{p(x^t_i = -1)}, \quad e^{m^t_{vs,i}} = \frac{p(x^t_i = 1)}{p(x^t_i = -1)},$$

where x^t_i denotes the ith transmitted code bit and y_i denotes the channel output.

Considering standard function node message-passing rules [9], we compute the message sent to i^{th} variable node of user t from the state node:

$$m^t_{sv,i} = \log \frac{p(x^t_i = 1|y)}{p(x^t_i = -1|y)} =$$

$$\log \left(\frac{\sum\limits_{\sim x^{(t)}_i} \prod\limits_{j \neq t} p(x^j_i = 1) p(y_i | x^{(1)}_i, ..., x^{(t)}_i = 1, ..., x^{(n)}_i)}{\sum\limits_{\sim x^{(t)}_i} \prod\limits_{j \neq t} p(x^j_i = -1) p(y_i | x^{(1)}_i, ..., x^{(t)}_i = -1, ..., x^{(n)}_i)} \right)$$

We can simplify it in the following way:

$$m^t_{sv,i} = \log \left(\frac{\sum\limits_{\sim x^{(t)}_i} \prod\limits_{j \neq t} e^{1_{x_j} X_j} p(y_i | x^{(1)}_i, ..., x^{(t)}_i = 1, ..., x^{(n)}_i)}{\sum\limits_{\sim x^{(k)}_i} \prod\limits_{j \neq t} e^{1_{x_j} X_j} p(y_i | x^{(1)}_i, ..., x^{(t)}_i = -1, ..., x^{(n)}_i)} \right), \quad (1)$$

where $1_{x_t} = \begin{cases} 1, & x^{(j)}_i = 1 \\ 0, & x^{(j)}_i = -1. \end{cases}$

The number of computations necessary to obtain the outgoing messages from state nodes grows exponentially with the number of users, nevertheless, this number of users usually remains small, and we will therefore not be concerned with this fact.

3 PEXIT Charts

Extrinsic Information Transfer (EXIT) charts [1] can be used for the accurate analysis of the behavior of LDPC decoders. But since the usual PEXIT analysis cannot be applied to the study of protograph-based [13] LDPC codes we will use a modified EXIT analysis for protograph-based LDPC codes (PEXIT) [5]. This method is similar to the standard EXIT analysis in that it tracks the mutual information between the message edge and the bit value corresponding to the variable node on which the edge is incident, while taking into account the structure of the protograph. In our work we use the notation from [5] to describe EXIT charts for protograph-based LDPC codes.

Let I_{Ev} denotes the extrinsic mutual information between a message at the output of a variable node and the codeword bit associated to the variable node:

$$I_{Ev} = I_{Ev}(I_{Av}, I_{Es}),$$

where I_{Av} is the mutual information between the codeword bits and the check-to-variable messages and I_{Es} is the mutual information between the codeword bits and the state-to-variable messages. Since the PEXIT tracks the mutual information on the edges of the protograph, we define $I_{Ev}(i, j)$ as the mutual information between the message sent by the jth variable node to the ith check node and the associated codeword bit:

$$I_{Ev}(i, j) = J\left(\sqrt{\sum_{s \neq i}[J^{-1}(I_{Av}(s, j))]^2 + [J^{-1}(I_{Es}(j))]^2}\right)$$

where $J(\sigma)$ is given by [1]:

$$J(\sigma) = 1 - \int_{-\infty}^{\infty} \frac{1}{\sqrt{2\pi\sigma^2}} \exp\left[-\frac{1}{2}\left(\frac{y - \frac{\sigma^2}{2}x}{\sigma}\right)^2\right] \log_2(1 + e^{-y})dy.$$

Similarly, we define I_{Ec}, the extrinsic mutual information between a message at the output of a check node and the codeword bit associated to the variable node receiving the message:

$$I_{Ec} = I_{Ec}(I_{Ac}),$$

where I_{Ac} is the mutual information between one input message and the associated codeword bit and $I_{Ac} = I_{Ev}$. Accordingly, the mutual information between the message sent by i^{th} check node to j^{th} variable node and the associated codeword bit is described as:

$$I_{ec}(i, j) = 1 - J\left(\sqrt{\sum_{s \neq j}[J^{-1}(1 - I_{ac}(i, s))]^2}\right).$$

The mutual information between the jth variable node and the message passed to the state node is denoted as $I_{Evs}(j)$ and is given by:

$$I_{Evs}(j) = J\left(\sqrt{\sum_{s}[J^{-1}(I_{av}(s, j))]^2}\right).$$

Next we need to compute the mutual information I_{Es}. In order to get an idea about the probability density function of (1) for user t, we generate samples of the outgoing LLRs through (1) based on the samples of the received LLRs from other users whose PDF is approximated with $\mathcal{N}(\mu_{Evs}, 2\mu_{Evs})$, where $\mu_{Evs} = \frac{J^{-1}(I_{Evs})}{2}$. To numerically estimate μ_{Es} and obtain the required mutual information as $I_{Es} = J(\mu_{Es})$, we refer to [11], where the following three approaches are proposed:

- Mean-matched Gaussian approximation : the mean μ is estimated from samples and we set $\mu_{Evs} = \mu$ and $\sigma^2_{Evs} = 2\mu$.
- Mode-matched Gaussian approximation : given a sufficiently large number of N samples generated through (1), the mode m is estimated from samples and we set $\mu_{Evs} = m$ and $\sigma^2_{Evs} = 2m$.
- Gaussian mixture approximation: mean values $\mu_1, ..., \mu_T$ and the weights $a_1, ..., a_T$ are estimated from samples and $I_{Es} = a_1 J(\mu_1) + ... + a_k J(\mu_T)$.

The rationale for using these approximations was shown in [11]. Furthermore, the authors compared the performance of these approaches. The mode-matched method was found to give the maximum output mutual information and the joint codes designed by using this approximation also yield the lowest decoding bit error probability compared to the other two approaches.

Each user calculate $I_{APP}(j)$, the mutual information between the posteriori probability likelihood ratio evaluated by the jth variable node and the associated codeword bit.

$$I_{APP}(j) = J\left(\sqrt{\sum_s [J^{-1}(I_{Av}(s,j))]^2 + [J^{-1}(I_{Es}(j))]^2}\right).$$

The convergence is declared if each of $I_{APP}(j)$ reaches 1 as the iteration number tends to infinity.

4   Numerical Results

In this section the simulation results, obtained for the cases $T = 2$ and $T = 4$, are represented. Let us first consider the simulation results for $T = 2$ (Fig. 2). For this case we compare the Frame Error Rate (FER) performance of rate-1/4 LDPC code (364, 91) from [14] obtained by repetition of each code bit of regular (3,6) LDPC code twice, rate-1/4 LDPC code (364, 91) optimized by PEXIT charts method described above and Polyanskiy's finite block length (FBL) bound for 2 user case.

As we can see in Fig. 2 proposed PEXIT-optimized LDPC code construction outperforms LDPC code construction from [14] by about 0.5 dB. In the same time the gap between Polyanskiy's FBL bound and PEXIT-optimized LDPC code is about 3 dB. But we would like to point out that used here Polyanskiy's FBL bound is for Gaussian signal and not for Binary Phase-Shift Keying (BPSK)

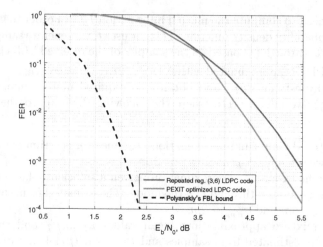

Fig. 2. Simulation results for T = 2 and LDPC code (364, 91)

modulation, used for simulation. So, we believe that this gap will be reduced is FBL bound for BPSK modulation is used.

Now let us consider simulation results for T = 4 (Fig. 3). For this case we obtain another PEXIT-optimized rate-1/4 LDPC code (364, 91) and compare FER performance of same LDPC code from [14] and Polyanskiy's FBL bound for 4 users.

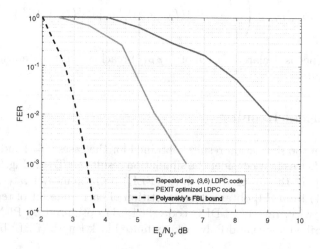

Fig. 3. Simulation results for T = 4 and LDPC code (364, 91)

As we can see in Fig. 3 proposed PEXIT-optimized LDPC code construction outperforms LDPC code construction from [14] by more than 3 dB. And again the gap between Polyanskiy's FBL bound and PEXIT-optimized LDPC code is a little bit less than $3\,\mathrm{dB}$.

5   Sparse Spreading of LDPC Codes

In this section we answer a very natural question: how to increase the order of collision, that can be decoded in a slot. E.g. consider the case from the previous section. Let the slot length $n' = 364$. We want to increase T up to 8. Here we face with two problems:

- The performance of LDPC joint decoder rapidly becomes bad with grows of T. We were not able to find $(364, 91)$ LDPC codes, that work well for $T = 8$.
- The number of computations necessary to obtain the outgoing messages from the functional node grows exponentially with the number of users T.

We address both these problems in a scheme, which is proposed below (see Fig. 4). The idea is to use sparse spreading signatures [3] for LDPC codes, such that the degree of functional node is reduced from T to d_c. The slot length is now n', $n' \neq n$.

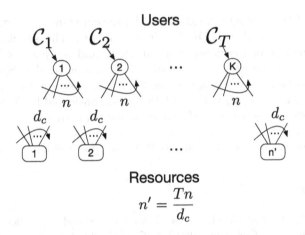

$$n' = \frac{Tn}{d_c}$$

Fig. 4. Sparse spreading of LDPC codes

In Fig. 5 we present the simulation results. As we were not able to find $(364, 91)$ LDPC codes, that work well for $T = 8$ we consider 2 times shorter LDPC codes and compare 2 strategies:

- split the slot into 2 parts and send 4 users in each part;
- use sparse spreading;

We see, that our approach is much better and works practically the same in comparison to the case of 2 times longer LDPC codes and 2 times smaller number of users (see the previous section).

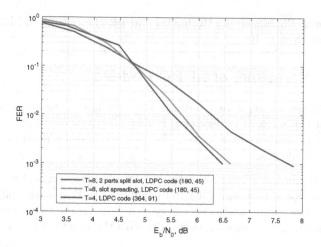

Fig. 5. Simulation results for spreading

6   Conclusion

We generalized the protograph extrinsic information transfer charts (EXIT) to optimize the protograph of LDPC code for GMAC. The simulation results, obtained at the end of the paper, were analyzed and compared with obtained theoretical bounds and thresholds. Obtained simulation results shows that proposed LDPC code constructions have better performance under joint decoding algorithm over Gaussian MAC than LDPC codes considered by A. Vem et al. in 2017, that leads to the better performance of overall system.

Acknowledgment. We want to thank Y. Polyanskiy for fruitful discussions.

References

1. ten Brink, S.: Convergence behavior of iteratively decoded parallel concatenated codes. IEEE Trans. Commun. **49**(10), 1727–1737 (2001). https://doi.org/10.1109/26.957394
2. Gallager, R.G.: Low-density parity-check codes (1963)
3. Hoshyar, R., Wathan, F.P., Tafazolli, R.: Novel low-density signature for synchronous CDMA systems over AWGN channel. IEEE Trans. Signal Process. **56**(4), 1616–1626 (2008). https://doi.org/10.1109/TSP.2007.909320
4. Liva, G.: Graph-based analysis and optimization of contention resolution diversity slotted aloha. IEEE Trans. Commun. **59**(2), 477–487 (2011). https://doi.org/10.1109/TCOMM.2010.120710.100054
5. Liva, G., Chiani, M.: Protograph LDPC codes design based on EXIT analysis. In: IEEE GLOBECOM 2007 - IEEE Global Telecommunications Conference, pp. 3250–3254, November 2007. https://doi.org/10.1109/GLOCOM.2007.616

6. Narayanan, K.R., Pfister, H.D.: Iterative collision resolution for slotted ALOHA: an optimal uncoordinated transmission policy. In: 2012 7th International Symposium on Turbo Codes and Iterative Information Processing (ISTC), pp. 136–139, August 2012. https://doi.org/10.1109/ISTC.2012.6325214

7. Ordentlich, O., Polyanskiy, Y.: Low complexity schemes for the random access Gaussian channel. In: 2017 IEEE International Symposium on Information Theory (ISIT), pp. 2528–2532, June 2017. https://doi.org/10.1109/ISIT.2017.8006985

8. Polyanskiy, Y.: A perspective on massive random-access. In: 2017 IEEE International Symposium on Information Theory (ISIT), pp. 2523–2527. IEEE (2017)

9. Richardson, T., Urbanke, R.: Modern Coding Theory. Cambridge University Press, New York (2008)

10. Rimoldi, B., Urbanke, R.: A rate-splitting approach to the Gaussian multiple-access channel. IEEE Trans. Inf. Theory $42(2)$, 364–375 (1996)

11. Shahid, I., Yahampath, P.: Distributed joint source-channel code design for GMAC using irregular LDPC codes. EURASIP J. Wirel. Commun. Netw. $2014(1)$, 3 (2014). https://doi.org/10.1186/1687-1499-2014-3

12. Tanner, R.: A recursive approach to low complexity codes. IEEE Trans. Inf. Theory $27(5)$, 533–547 (1981)

13. Thorpe, J.: Low-density parity-check (LDPC) codes constructed from protographs. Interplanet. Netw. Prog. Rep. 154, 1–7 (2003)

14. Vem, A., Narayanan, K.R., Cheng, J., Chamberland, J.F.: A user-independent serial interference cancellation based coding scheme for the unsourced random access Gaussian channel. In: Proceedings of the IEEE Information Theory Workshop (ITW), pp. 1–5 (2017)

Application of Optimal Finite-Length Signals for Overcoming "Nyquist Limit"

Sergey V. Zavjalov⬭, Anna S. Ovsyannikova⬭, Ilya I. Lavrenyuk,
Sergey V. Volvenko⁽⊠⁾⬭, and Sergey B. Makarov⬭

Peter the Great St. Petersburg Polytechnic University, St. Petersburg, Russia
zavyalov_sv@spbstu.ru, {anny-ov97, knaiser}@mail.ru,
{volk, makarov}@cee.spbstu.ru

Abstract. An opportunity of finite-length optimal signals application to excess Nyquist limit without energy losses and within defined occupied bandwidth is considered in this article. Finite-length optimal signals with different duration can be found by numerical solution of corresponding multistep optimization task. Optimality criterion of signal form is based on defined reduction rate of out-of-band emissions in condition of controlled intersymbol interference. Controlled intersymbol interference determines possibility of application of simple algorithm of coherent bit-by-bit detection. Solutions of optimization tasks for different conditions are compared with known signal forms in term of Euclidean distance between signal sequences according to Mazo method. It is shown that optimal signals provide higher symbol rate without significant losses in BER performance and may be applied in 5G standards instead of signals formed with the use of digital filtration.

Keywords: Optimal signals · Nyquist limit · Occupied frequency bandwidth Spectrum · Euclidean distance

1 Introduction

Nyquist limit overcoming problems with minimum energy losses were considered in many works [1–4]. In [1] the opportunity of data transmission at the symbol rate $R = 1/\xi T \, (0 < \xi < 1)$ with 25% Nyquist limit excess was shown. The increase in a symbol rate does not lead to the bit-error rate (BER) performance reduction. The additive white Gaussian noise channel (AWGN) with $N_0/2$ power spectral density is considered as a transmission channel. In this channel a random sequence of single signals $s(t)$ with length T_s and energy $\xi^2 E_s$ is transmitted:

$$y(t) = \xi\sqrt{E_s/T} \sum_{n=-M/2}^{M/2} c_n s(t - n\xi T); \quad s(t) = \sin(\pi t/T)/(\pi t/T);$$

$$c_n = \pm 1; T_s/T = \infty. \tag{1}$$

© Springer Nature Switzerland AG 2018
O. Galinina et al. (Eds.): NEW2AN 2018/ruSMART 2018, LNCS 11118, pp. 172–180, 2018.
https://doi.org/10.1007/978-3-030-01168-0_16

Symbols $c_n = \pm 1$ are independent and identically distributed. Coefficient ξ keeps the average power of the random sequence $y(t)$ transmission at a constant level with intersymbol interference (ISI). ISI value increases with ξ rising [1, 3].

Square of Euclidean distance $d^2(i, k)$ between two different implementations of random sequences $y_i(t)$ and $y_k(t)$ in (1), which defines BER, is calculated as follows:

$$d^2(i, k) = \int_{-\infty}^{-\infty} (y_i(t) - y_k(t))^2 dt \cdot \tag{2}$$

The normalized minimum Euclidean distance for each ξ value is defined as:

$$d_{min}^2 = \min_{i,k;\ i \neq k} \left\{ d^2(i, k) \right\} \Big/ 4\xi^2 E_s \cdot \tag{3}$$

It was shown that d_{min} does not decrease until $\xi \geq 0.802$ [1]. This is called the Mazo limit, and therefore, the data transmission with minimum error probability through the channel with a rectangular form amplitude-frequency characteristic in ΔF bandwidth is provided.

In [2] the signal $s(t)$ with the Raised Cosine shaping form outside the transmit filter is considered (expression (3) in [2]). Using the Mazo method, it was shown that applying such signals allows increasing a symbol rate by 42% over the Nyquist limit (including signals with finite length $40T$–$80T$) [5]. The main attention is paid to strong coding schemes, which help to reach BER close to potential possible values.

In Faster-Than-Nyquist (FTN) signaling discussion, infinite-length signals are considered, or a large enough length of transmitted $s(t)$ signal is specified. It requires very complex demodulation algorithms and summarized reception of signal sequence segments to obtain expected advantages in the bandwidth capacity.

There are some difficulties when FTN signals are length-limited. The first one is referred to determining occupied frequency bandwidth ΔF and the second one is related with comparison of obtained values $R/\Delta F$ with the Nyquist limit. On the other hand, some practical opportunities arise when length-limited FTN signals are used: transmission and receiver system simplification while maintaining the bandwidth capacity at a high level.

Optimal signals [6, 7] allow increasing the symbol rate of data transmission without affecting BER performance. Such signals' feature is ISI control. Optimal signal usage allows reaching at least a double symbol rate increase with energy losses no more than 0.5 dB [7, 8].

In this article, it is proposed to consider an opportunity to apply a finite time optimal signal in excess of the Nyquist limit without energy losses and within the defined occupied frequency bandwidth. We will try to compare the Euclidean distance of optimal signal sequences with the known one [2, 9] according to the Mazo method, and to understand how much R can be increased over the Nyquist limit.

2 Optimization Criteria of Signal Form

An optimality criterion of the signal form is based on the choice of extremely narrow spectrum $G(f)$. Note that the choice is determined by a reduction rate of out-of-band emissions. As a result, the spectrum in the occupied frequency bandwidth ΔF turns to the ideal rectangular form. This approximation may be used if ΔF is defined for a rather low level of energy spectrum, e.g. for -60 dB written as ΔF_{-60dB}.

The finite random sequence of N single optimal signals $s_{opt}(t)$ with duration $T_s = LT$ and energy $\xi^2 E_{opt}$ may be written as follows:

$$y(t) = \xi\sqrt{E_{opt}/T_s} \sum_{n=-N/2}^{N/2} c_n s_{opt}(t - n\xi T). \tag{4}$$

Coefficient ξ still keeps the average power of random sequence $y(t)$ constant in the presence of ISI and defines a symbol rate. If the binary alphabet used, the symbol rate is equal to a bit rate.

Energy spectrum $G(f)$ of the random sequence of signals (4) with complex spectrum $S(f)$ for statistically independent $c_n = \pm 1$ depends on the form of optimal signal $s_{opt}(t)$ with energy $\xi^2 E_{opt}$. Then for each ξ value the next expression takes place:

$$G(f) = \lim_{N\to\infty} \frac{1}{NT_s} m_1\{|S(f)|^2\} = \frac{\xi E_{opt}}{T_s T} \left| \int_{-T_s/2}^{T_s/2} s_{opt}(t) \exp(-j2\pi f t) dt \right|^2, \tag{5}$$

where expected value $m_1\{|S(f)|^2\}$ is determined by averaging over all possible finite combinations c_n. The optimization task of synthesis of signal form $s_{opt}(t)$ with duration T_s and symbol rate $R = 1/\xi T$ may be transformed into the task of linear functional J minimization [10]:

$$\arg\left\{ \min_{s_{opt}(t)} (J) \right\}, \quad J = \int_{-\infty}^{+\infty} g(f) \left| \int_{-\infty}^{+\infty} s_{opt}(t) \exp(-j2\pi f t) dt \right|^2 df, \tag{6}$$

where $g(f) = f^{2p}$ ($p = 1, 2, \ldots$) is a weighing function responsible for a reduction rate of out-of-band emissions. The choice of $g(f)$ allows adjusting the convergence of functional (6) and provides for $G(f)$ tendency to a rectangular form with increasing p.

A condition on the value of a correlation coefficient between two optimal signals on different time positions of the random sequence may be set as a restriction of the optimization task (6). In the latter case, the restriction on the correlation coefficient in the area of $t > 0$ is written as:

$$\max_{n=1\ldots\lfloor L/\xi \rfloor} \left\{ \int_{-(L-2n)T/2}^{LT/2} s_{opt}(t) s_{opt}(t - n\xi T) dt \right\} < K_0. \tag{7}$$

Choosing limits of integration in (7), we consider $s_{opt}(t)$ finite and symmetrical about $t = 0$. In this case, all the combinations should be sorted out just in the positive area of values n. Adding restriction (7) makes a task of searching for optimal signals much easier and determines a simple algorithm of coherent bit-by-bit detection on the interval $[-T_s/2, T_s/2]$.

The application of (7) is definitely different from (3). Firstly, a finite number of signals, which may be found within interval for analysis T_a of the sequence (4), limits it. Secondly, if K_0 in (7) is too low, the minimal Euclidean distance tends to $d_{min}^2 = 1$.

To solve this optimization task numerically, expansion of $s_{opt}(t)$ into the limited Fourier series was used (m is a number of expansion coefficients). Then the original optimization task (6) transforms into the task of searching for expansion coefficients $\{s_k\}_{k=1}^m$, which minimize the function of several variables [11]:

$$\min_{\{s_k\}_{k=1}^m} J\left(\{s_k\}_{k=1}^m\right), J\left(\{s_k\}_{k=1}^m\right) = T_s/2 \sum_{k=1}^m (2\pi k/T_s)^{2n} s_k^2. \tag{8}$$

The choice of m is associated with the accuracy of $s_{opt}(t)$ representation and complexity of solution (8) caused by an ill-conditioned task. Target functional has a ravine-type shape, i.e. rises sharply along one direction and changes slightly along the other. It is taken into account in this work, therefore, the chosen values of m provide necessary the accuracy of $s_{opt}(t)$.

For each set of ξ, R, p and K_0 the final form of $s_{opt}(t)$ will be different.

3 Solutions of Optimization Task

Let's consider forms of optimal signals and energy spectra of the random sequence of such signals for different values of symbol rate R. Examples of the obtained values s_k (8) after solving the optimization task for $p = 2$, $K_0 = 0.01$, $\xi = 0.5$ and $R = 2/T$ are presented in Table 1.

Table 1. Expansion coefficients $\{s_k\}_{k=1}^m$

T_s	$2T$	$6T$
s_0	0.63581	0.23075
s_1	0.78864	0.24302
s_2	0.41787	0.23028
s_3	−0.03426	0.24404
s_4	0.01056	0.22795
s_5	−0.00432	0.24871
s_6	0.00204	0.14522
s_7	−0.00109	−0.01125
s_8	0.00059	0.00569

Forms of the symmetrical function $s_{opt}(t)$ obtained as a result of solving the optimization task for $p = 2$, $K_0 = 0.01$ are shown in Fig. 1(a). Here the signal duration is fixed and equal to $T_s = 6T$, while R varies. In this figure each function $s_{opt}(t)$ (numbers 1 to 6) has its own value $\xi < 1$ (4) and symbol rate R. Doubling the symbol rate from $1/T$ to $2/T$ leads to decreasing the main petal's width. Besides, bipolar oscillations of the function become more significant due to the restriction on correlation coefficient K_0. It will be shown that this fact allows obtaining the value of the minimal Euclidean distance equal to $d_{min}^2 = 1$ for this range of symbol rates.

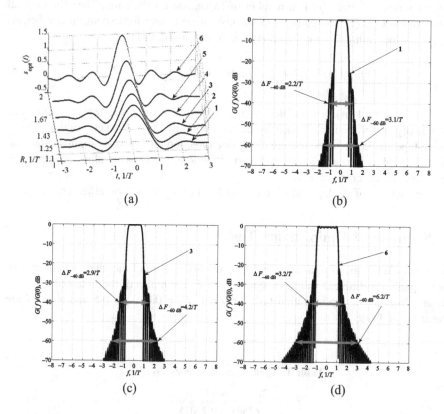

Fig. 1. Form of $s_{opt}(t)$ and $G(f)/G(0)$ for $R = 1/T$ (b), 1.43/T (c), 2/T (d).

The normalized energy spectra (5) of the random sequence of optimal signals (4) for different values of symbol rate R are given in Fig. 1(b–d). Numbers in these figures correspond to numbers in Fig. 1(a). As we can see, the shape of spectrum depends on R. While R increases, the form of $s_{opt}(t)$ changes (Fig. 1(a) as well as the occupied frequency bandwidth ΔF. Thus, ΔF_{-40dB} increases 1.5 times when R increases from $1/T$ up $2/T$. Gain in the occupied frequency bandwidth defined for the level of –60 dB reaches 1.8 times.

Let's consider how the form of $s_{opt}(t)$ depends on signal duration. Solutions to the optimization task for $p = 2$, $K_0 = 0.01$, fixed symbol rate $R = 2/T$ and different values of T_s are presented in Fig. 2(a). Numbers in this figure correspond to $s_{opt}(t)$ with signal duration T_s from $2T$ to $16T$. Figure 2(b–d) illustrate the normalized energy spectra of the random sequence of optimal signals. The occupied frequency bandwidth defined for the level of –40 dB decreases twice when T_s ranges from $2T$ to $16T$. The occupied frequency bandwidth defined for the level of –60 dB decreases 2.3 times.

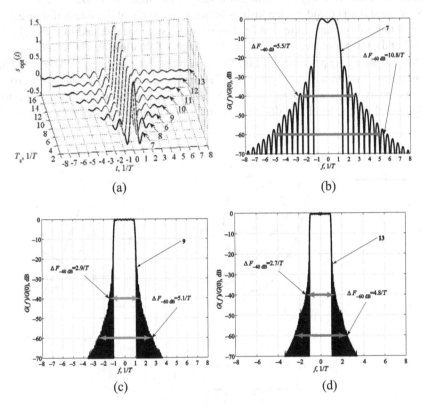

Fig. 2. Form of $s_{opt}(t)$ and $G(f)/G(0)$ for $T_s= 2T$ (b), $T_s= 8T$ (c), $T_s= 16T$ (d).

Dependencies in Fig. 2 are given for the optimal signal synthesized with symbol rate $R = 2/T$. The gain in ΔF will be obtained, if parameter p increases during solving the optimization task (6) and if p and T_s tend to infinity. In this case, energy spectrum will be of a rectangular shape.

4 Euclidean Distance

We will try to estimate Euclidean distances for sequences of optimal signals presented in Figs. 1 and 2 using the Mazo method. The minimal distances d_{min}^2 against R for optimal signals (Fig. 1) with symbol rate $R = 2/T$ are presented in Fig. 3. Additionally, there are dependencies of the Euclidean distance between different realizations of random sequences (4). Minimal Euclidean distance d_{min}^2 for definite R corresponds to a minimal value for a variety of Euclidean distances for different realizations.

Now we should consider relationships between the minimal Euclidean distance and a symbol rate for optimal finite signals. These signals were synthesized for symbol rate $R = 2/T$ and have parameters $p = 2$, $K_0 = 0.01$. Figure 3(b) illustrates how R influences d_{min}^2 for signals $s_{opt}(t)$ presented in Fig. 2(a) by graphs with numbers 7, 8, 6, 13. The maximum increase in a symbol rate (plot 7) equal to 2.8 times is provided by using optimal signals $s_{opt}(t)$ with duration $T_s = 2T$. However, the occupied frequency bandwidth defined for the level −60 dB is the broadest of all under consideration. It is equal

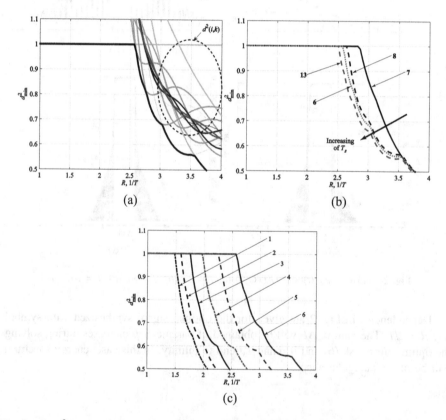

Fig. 3. (a) d_{min}^2 vs. R for optimal signals synthesized for duration $T_s = 6T$ and symbol rate $R = 2/T$; (b–c) d_{min}^2 vs. R for optimal signals with different duration synthesized for each specific symbol rate.

to $\Delta F_{-60 \text{ dB}} = 10.8/T$ (Fig. 2(b)). If optimal signals $s_{opt}(t)$ with duration $16T$ (plot 13 in Fig. 3(b) are used, the maximum increase in a symbol rate reaches 2.5 times, but bandwidth $\Delta F_{-60 \text{ dB}} = 4.8/T$ (Fig. 2(d)).

Dependencies of d^2_{min} from R for optimal signals with duration $T_s = 6T$ are given in Fig. 3(c). Note that forms of these signals were obtained because of solving the optimization task for each specific symbol rate (numbers of plots in Fig. 1 correspond to numbers in Fig. 3(c)). Values of $d^2_{min} = 1$ are provided for the symbol rate changing from $1.5/T$ to $2.6/T$ while the occupied frequency bandwidth $\Delta F_{-60 \text{ dB}}$ varies from $3.1/T$ to $6.2/T$.

5 Conclusions

Application of finite optimal signals under conditions of controlled ISI level allows achieving an increase in a symbol rate keeping the Euclidean distance equal to 1. This provides theoretical BER performance. The important thing is using a simple coherent bit-by-bit detection algorithm with the expanded interval for analysis equal to the wanted signal duration.

The increase in a symbol rate in relation to the Nyquist rate reaches from 2.5 to 2.8 times depending on signal duration. Besides, BER performance is close to the theoretical one. Nevertheless, this gain was obtained for a non-rectangular form of energy spectrum. The occupied frequency bandwidth defined for the level −60 dB (this level conforms to standards on measuring out-of-band emissions) ranges from $2.4/T$ to $5.4/T$ for low frequencies. The extra bandwidth gain may be achieved by increasing a controlled level of ISI (by increasing signal duration). It means we should use optimal signals synthesized for values of parameter $p = 4, 5, 6$ which influences $g(f)$ in (6). Adding coding (20% [2, 12]) for optimal signals allows increasing a symbol rate by about 2.3 times comparing to the obtained result.

Thus, since optimal signals provide a higher symbol rate without significant losses in BER performance, they may be applied in 5G standards instead of signals formed with the use of digital filtration.

The results of the work were obtained under the State contract № 8.2880.2017/ПЧ with Ministry of Education and Science of the Russian Federation and used computational resources of Peter the Great Saint-Petersburg Polytechnic University Supercomputing Center (http://www.scc.spbstu.ru).

References

1. Fan, J.C., et al.: Faster-Than-Nyquist signaling: an overview. IEEE. Access **5**, 1925–1940 (2017)
2. Liveris, A.D., Georghiades, C.N.: Exploiting Faster-Than-Nyquist signaling. IEEE Trans. Commun. **51**, 1502–1511 (2003)
3. Rusek, F., Anderson, J.B.: Constrained capacities for Faster-Than-Nyquist signaling. IEEE Trans. Inf. Theory **55**, 764–775 (2009)

4. Gorlov, A., Gelgor, A., Nguyen, V.P.: Root-raised cosine versus optimal finite pulses for Faster-Than-Nyquist generation. In: Galinina, O., Balandin, S., Koucheryavy, Y. (eds.) NEW2AN/ruSMART -2016. LNCS, vol. 9870, pp. 628–640. Springer, Cham (2016). https://doi.org/10.1007/978-3-319-46301-8_54

5. Anderson, J.B., Rusek, F., Owall, V.: Faster-Than-Nyquist signaling. Proc. IEEE **101**, 1817–1830 (2013)

6. Ovsyannikova, A.S., Zavjalov, S.V., Makarov, S.B., Volvenko, S.V., Quang, T.L.: Spectral and energy efficiency of optimal signals with increased duration, providing overcoming "Nyquist Barrier". In: Galinina, O., Andreev, S., Balandin, S., Koucheryavy, Y. (eds.) NEW2AN/ruSMART/NsCC -2017. LNCS, vol. 10531, pp. 607–618. Springer, Cham (2017). https://doi.org/10.1007/978-3-319-67380-6_57

7. Zavjalov, S.V., Volvenko, S.V., Makarov, S.B.: A method for increasing the spectral and energy efficiency SEFDM signals. IEEE Commun. Lett. **20**, 2382–2385 (2016)

8. Ovsyannikova, A.S., Zavjalov, S.V., Makarov, S.B., Volvenko, S.V.: Choosing parameters of optimal signals with restriction on correlation coefficient. In: Galinina, O., Andreev, S., Balandin, S., Koucheryavy, Y. (eds.) NEW2AN/ruSMART/NsCC -2017. LNCS, vol. 10531, pp. 619–628. Springer, Cham (2017). https://doi.org/10.1007/978-3-319-67380-6_58

9. Ishihara, T., Sugiura, S.: Faster-Than-Nyquist signaling with index modulation. IEEE Wirel. Commun. Lett. **6**, 630–633 (2017)

10. Xue, W., Ma, W., Chen, B.: Research on a realization method of the optimized efficient spectrum signals using Legendre series. In: 2010 IEEE International Conference on Wireless Communications, Networking and Information Security (WCNIS), vol. 1, pp. 155–159 (2010)

11. Zavjalov, S.V., Makarov, S.B., Volvenko, S.V., Xue, W.: Waveform optimization of SEFDM signals with constraints on bandwidth and an out-of-band emission level. In: Balandin, S., Andreev, S., Koucheryavy, Y. (eds.) ruSMART 2015. LNCS, vol. 9247, pp. 636–646. Springer, Cham (2015). https://doi.org/10.1007/978-3-319-23126-6_57

12. Vasilyev, D., Rashich, A., Fadeev, D.: Joint use of SEFDM-signals and FEC schemes. In: Galinina, O., Balandin, S., Koucheryavy, Y. (eds.) NEW2AN/ruSMART -2016. LNCS, vol. 9870, pp. 604–611. Springer, Cham (2016). https://doi.org/10.1007/978-3-319-46301-8_51

Influence of Amplitude Limitation for Random Sequence of Single-Frequency Optimal FTN Signals on the Occupied Frequency Bandwidth and BER Performance

Sergey B. Makarov[1] , Anna S. Ovsyannikova[1] , Sergey V. Zavjalov[1] ,
Sergey V. Volvenko[1(✉)] , and Lei Zhang[2]

[1] Peter the Great St. Petersburg Polytechnic University, St. Petersburg, Russia
{makarov,volk}@cee.spbstu.ru, anny-ov97@mail.ru,
zavyalov_sv@spbstu.ru
[2] School of Computer Science and Software Engineering, East China Normal University,
Shanghai, China
lzhang@ce.ecnu.edu.cn

Abstract. Application of optimal signals allows to solve the problem of improving spectral efficiency with minimal energy losses. However, the major disadvantage of such signals is high peak-to-average power ratio (PAPR) which leads to degradation of transmitter efficiency. This work considers the possibility of using rigid amplitude limitation for random sequence of optimal signals. PAPR values distribution is presented. It is shown that amplitude limiting does not cause widening of energy spectrum defined for the level higher than −20 dB. Decreasing PAPR value by 4 dB may provide energy gain up to 3.5 dB as a result of increasing average signal power.

Keywords: Optimal signals · Peak-to-average power ratio
Transmitter efficiency · Rigid limitation · BER performance

1 Introduction

The problem of transmitting binary messages through AWGN channel with the rate higher than "Nyquist Barrier" and Mazo limit with minimal energy losses is considered in numerous works [1–5]. There are two approaches to solve this problem. The first one supposes transformation of transmitted signals with duration T via linear filtration into the sequence of signals with uncontrolled intersymbol interference (ISI) [1, 2]. Another one implies application of optimal signals [5–10] with given time (signal duration $T_s \geq T$) and spectral parameters. Both approaches provide significant limitation of occupied frequency bandwidth ΔF for high transmission rates $R > 2/T$. The necessary decoding quality is achieved by means of different detection algorithms close to the summarized reception of blocks of signal sequence. Properties of random sequence of FTN signals (ΔF, BER) obtained with the use of linear filtration are fairly well understood.

© Springer Nature Switzerland AG 2018
O. Galinina et al. (Eds.): NEW2AN 2018/ruSMART 2018, LNCS 11118, pp. 181–190, 2018.
https://doi.org/10.1007/978-3-030-01168-0_17

The sequences of optimal FTN signals have high spectral and energy parameters (ΔF and BER) [9–11] and require less computer operations for their formation and reception. It allows to get digital modems with good technical specifications (less complexity of hardware, higher performance and reduced energy consumption). However, time parameters of random sequence of such signals need to be studied. In particular, when optimal FTN signals are used in telecommunication systems with the lack of radio link energy (e.g., satellite systems, mobile systems), the value of peak-to-average power ratio (PAPR) of random radiated oscillations becomes very important. High value of PAPR leads to reduction of radio transmitter's efficiency. An attempt to add rigid amplitude limitation causes immediate increasing of ΔF and BER degradation.

In this work, it is proposed to consider possibility of choosing acceptable level of amplitude limiting for random sequence of optimal FTN signals taking into account widening of occupied frequency bandwidth and energy losses at reception.

2   Signal Model

Let us take a look at optimal FTN signals with quadrature amplitude modulation. In general j-th realization of sequence containing N such signals with amplitude A_0, real envelope $a(t)$ and carrier frequency ω_0 may be written as follows:

$$y_j(t) = A_0 \sum_{k=0}^{N-1} (d_r^{(k)} a(t - kT) \cos(\omega_0 t) - b_r^{(k)} a(t - kT) \sin(\omega_0 t)), \tag{1}$$

where duration T_s of optimal FTN signal with real envelope $a(t)$ is equal to $T_s > T$. Symbol values of channel alphabet $d_y^{(k)}$ and $b_y^{(k)}$ depend on the index $r = 1, 2, 3 \dots m$. Here m is the volume of channel alphabet. For binary alphabet ($m = 2$) we have $d_y^{(k)} = 1$ for $r = 1$; $d_y^{(k)} = -1$ for $r = 2$; $b_y^{(k)} = 1$ for $r = 1$; $b_y^{(k)} = -1$ for $r = 2$. The total number of realizations of sequence containing N signals (1) is equal to m^N.

Function $a(t)$ may be found as a solution of optimization task stated in [9–12]. In this case $T_s = 2T, 3T, 4T$, etc. Parameters of optimization task are reduction rate of out-of-band emissions, signal energy, signal duration T_s and coefficient of mutual correlation K_0 [9–11]. Using K_0 allows to reach fixed level of BER for linear detection algorithms. It was shown that there is an ability to obtain forms of optimal FTN signals providing increasing of transmission rate up to 2.8 times relatively to the "Nyquist Barrier". Moreover, BER performance keeps being almost theoretical one, so we do not need to use error correcting coding in this case.

PAPR of random sequence of optimal FTN signals (1) with average power P_{average} and peak power $P_{\max}$ is calculated by next expression:

$$PAPR = P_{\max} / P_{\text{average}} = \max_j \left\{ \left| y_j(t) \right|^2 \right\} \bigg/ \frac{1}{m^N NT} \sum_{j=1}^{m^N} \int_0^{(N-1)T} \left| y_j(t) \right|^2 dt. \tag{2}$$

It ensues from (2) that for uniformly distributed symbols of channel alphabet $d_y^{(k)}$ and $b_y^{(k)}$ PAPR of random sequence of signals (1) is also a random value. Figure 1 shows forms of normalized instantaneous power $P_j(t)/P_{max}$ realizations depending on time for specific realization (1) of optimal binary ($m = 2$) FTN signals. Signal duration is equal to $T_s = 16T$ while coefficient of mutual correlation $K_0 = 0.01$. Data is transmitted with the rate $R = 2/T$.

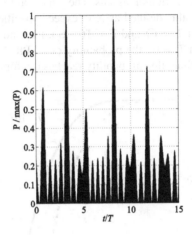

Fig. 1. Instantaneous power of sequence of optimal FTN signals.

Let us consider probability characteristics of PAPR distribution. On Fig. 2 histograms of PAPR distribution for $T_s = 6T$ (Fig. 2a) and $T_s = 16T$ (Fig. 2b) are given. Clearly, distribution of these random values of PAPR does not correspond to normal

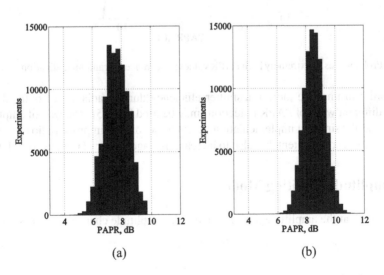

(a)   (b)

Fig. 2. Example of separate optimal signals (a), example of resulting truncated j-th realization of a random sequence of signals (b).

one, at least because PAPR values are bounded. Such distribution of PAPR is more likely to be close to truncated normal distribution. Indeed, when T_s increases, type of distribution tends to normal. Expected value of PAPR increases too. For example, it changes from 7.6 dB for $T_s = 6T$ to 8.6 dB for $T_s = 16T$. It correlates with the fact that longer duration of optimal FTN signals causes higher ISI level at transmission rates $R > 2/T$.

The issue of interest is to determine probability of extremely high PAPR values appearance for random realization of signals. The y-axis on Fig. 3 represents the probability of random PAPR value defined as X to exceed specified value of PAPR. This figure illustrates curves for $T_s = 6T$ and $T_s = 16T$. So, for random sequence of optimal FTN signals with duration $T_s = 6T$ the probability of PAPR to be greater than 8 dB is equal to 0.3. At the same time, this probability reaches 0.8 for $T_s = 16T$.

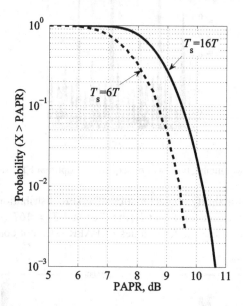

Fig. 3. Probabilities of extremely high PAPR values appearance for random realization of signals.

In order to improve radio transmitter efficiency during radiation of optimal FTN signals different ways of PAPR reductions may be used [13–15]. The way of amplitude limiting has the simplest implementation. It allows not only to improve radio transmitter efficiency to a certain extent, but also to increase average power of radiated oscillations.

3   Amplitude Limiting Model

Now we look at the amplitude limiting model. Its amplitude characteristic is shown on Fig. 4.

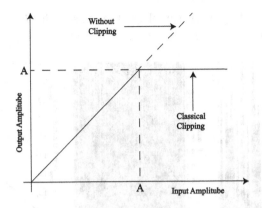

Fig. 4. Amplitude characteristic of limiter.

We will use piecewise linear approximation of amplitude characteristic of amplifying instruments. Suppose that frequency bandwidth of power amplifier, preliminary amplifier and amplitude limiter is wider than frequency bandwidth ΔF occupied by random sequence of signals (1). Besides, we will consider suppression of spectral components around frequencies multiple of central carrier frequency ω_0 to be rather strong. Then real signal $y_j^{(lim)}(t)$ on the output of amplitude limiter may be represented as follows:

$$y_j^{(lim)}(t) = \begin{cases} y_j(t), y_j(t) \le A \\ V_0, y_j(t) > A \end{cases}, \tag{3}$$

where A – limiting threshold of oscillation, V_0 – limiting level for random sequence of signals. A influences on PAPR (2) value on the output of the limiter (Fig. 4). Decreasing the limiting level A leads to reducing PAPR value of random sequence of signals (1) on the output of the limiter. But in this case value of $P_{average}$ on the output of amplitude limiter decreases too.

To obtain higher $P_{average}$ of random sequence (1) it is reasonable to use extra linear gain of oscillation power. Its level should reach P_{max} for which power amplifier is designed. Obviously, voltage gain coefficient of preliminary amplifier based on (3) is equal to:

$$K_u = \sqrt{P_{max}}/A = \sqrt{\max_j \left\{ \left| y_j(t) \right|^2 \right\}}/A. \tag{4}$$

Therefore, with due regard for (4) the sequences of signals on the output of power amplifier will always have constant value of peak power P_{max}, but different $P_{average}$ level.

Forms of realizations for normalized instantaneous power $P_j(t)/P_{max}$ vs. time for specific realization (1) of optimal binary ($m = 2$) FTN signals are given on Fig. 5(a, b). These signals have duration $T_s = 16T$ when coefficient of mutual correlation $K_0 = 0.01$. Transmission rate is $R = 2/T$. Figure 5(a, b) illustrates realizations of instantaneous power of signals (1) with amplitude limitation on 4 dB and 8 dB correspondingly. Here extra linear gain according to (4) takes place. It can be seen that P_{average} increases while P_{max} keeps constant.

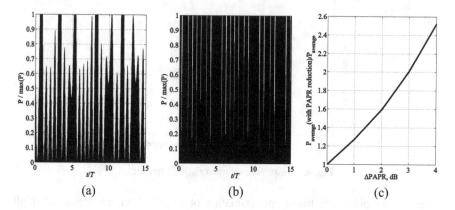

(a)   (b)   (c)

Fig. 5. Instantaneous power of sequence of optimal FTN signals for the case of amplitude limitation.

The y-axis on Fig. 5c represents averaged over ensemble average power with amplitude limitation P_{average} (with PAPR reduction) normalized to average power P_{average} without amplitude limitation. On the x-axis following values are displayed:

$$\Delta PAPR(dB) = PAPR_{\text{without limitation}}(dB) - PAPR_{\text{lim}}(dB), \tag{5}$$

where $PAPR_{\text{without limitation}}$ – PAPR value of sequence (1) without amplitude limitation, and $PAPR_{\text{lim}}$ – PAPR value after amplitude limitation. Figure 5c shows that limitation of oscillation amplitude leads to average power P_{average} increasing. So, for amplitude limitation on 4 dB the value of normalized average power increases 2.5 times.

4   Spectral Characteristics

Adding amplitude limitation causes distortion of energy spectrum $G(f)$ for random sequence of optimal FTN signals. To calculate spectral characteristics we used 10^4 realizations of information sequences. Figure 6(a, b, c) illustrates normalized energy spectra $G(f)/G(0)$ for random sequences of optimal FTN signals ($T_s = 16T$). There are $G(f)/G(0)$ for signal sequences without amplitude limitation (Fig. 6a) and for $\Delta PAPR = 2$ dB and $\Delta PAPR = 4$ dB (Fig. 6b–c correspondingly). As follows from Fig. 6a, occupied frequency bandwidth ΔF defined for the level of –30 dB reaches $2.2/T$. If ΔF is defined for the level of –40 dB, frequency bandwidth is $2.7/T$.

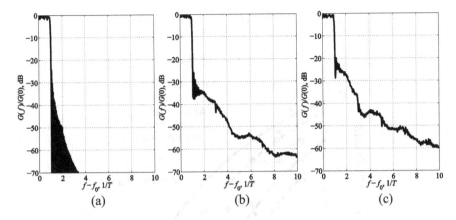

Fig. 6. Energy spectra of signal ($T_s = 16T$, $n = 2$, $K_0 = 0.01$, $R = 2/T$): (a) no limitation, (b) limitation on 2 dB, (c) limitation on 4 dB.

After amplitude limitation on $\Delta PAPR = 2$ dB ΔF increases up to $2.6/T$ for the level -30 dB and up to $6.2/T$ for the level -40 dB. Further change of $\Delta PAPR = 4$ dB leads to increasing ΔF to $4.6/T$ for the level -30 dB and to the value $6.3/T$ for the level -40 dB.

Analysis of spectral characteristics demonstrates widening of energy spectrum when amplitude limiting is used for random sequences of optimal FTN signals. However, it occurs for the levels of energy spectrum less than or equal to -20 dB.

5   BER Performance

Now we have to analyze BER performance of optimal FTN signal under presence of amplitude oscillation limiting. It was shown [8, 9] that signals synthesized with duration $16T$ and transmission rate $R = 2/T$ are characterized by theoretical BER performance. Coherent bit-by-bit detection algorithm with ideal synchronization system is used in AWGN channel. Dependence of bit-error rate (note that $m = 2$) on signal-to-noise ratio (SNR) is given on Fig. 7. On the one hand, adding amplitude limitation (Fig. 5c) leads to increase in average power of transmitted signal sequence. This results in error probability reducing. On the other hand, this causes an increase in the correlation of neighboring signals, so the BER performance is degraded. Figure 7 illustrates relationships between error probabilities and signal-to-noise ratio (SNR) for amplitude limiting of signal at reception on $\Delta PAPR = 1$ dB, 2 dB, 3 dB, 4 dB.

These curves allow estimating the possibility of choosing acceptable level of amplitude limitation for random sequence of optimal FTN signals. Further amplitude limitation ($\Delta PAPR > 4$ dB) will result in degradation of BER performance.

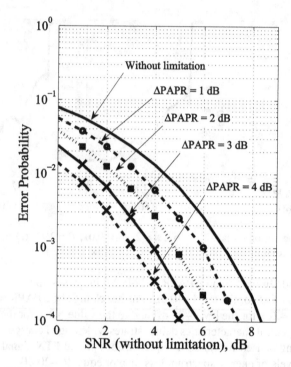

Fig. 7. BER performance of optimal FTN signals for different PAPR values.

6   Conclusions

1. Amplitude limitation for random sequences of optimal FTN signals leads to increasing the level of out-of-band emissions. Such widening of occupied frequency bandwidth takes place for the level of energy spectrum less than −20 dB.
2. Using amplitude limitation allows to improve energy efficiency of detection of optimal FTN signals, which provide transmission rate $R = 2/T$. It is shown that energy gain up to 3.5 dB may be achieved when PAPR decreases by 4 dB. It corresponds to results obtained in [16].
3. There is an optimal value of amplitude limitation level for random sequence of optimal FTN signals. Exceeding this value causes degradation of BER performance.

The results of the work were obtained using computational resources of Peter the Great Saint-Petersburg Polytechnic University Supercomputing Center (http://www.scc.spbstu.ru).

References

1. Fan, J., Guo, S., Zhou, X., Ren, Y., Li, G.Y., Chen, X.: Faster-Than-Nyquist signaling: an overview. IEEE Access **5**, 1925–1940 (2017)
2. Liveris, A.D., Georghiades, C.N.: Exploiting faster-than-Nyquist signaling. IEEE Trans. Commun. **51**(9), 1502–1511 (2003)
3. Rusek, F., Anderson, J.B.: Constrained capacities for faster-than-Nyquist signaling. IEEE Trans. Inf. Theory **55**(2), 764–775 (2009)
4. Gorlov, A., Gelgor, A., Nguyen, V.P.: Root-raised cosine versus optimal finite pulses for faster-than-Nyquist generation. In: Galinina, O., Balandin, S., Koucheryavy, Y. (eds.) NEW2AN/ruSMART 2016. LNCS, vol. 9870, pp. 628–640. Springer, Cham (2016). https://doi.org/10.1007/978-3-319-46301-8_54
5. Andreev, S., et al.: Exploring synergy between communications, caching, and computing in 5G-grade deployments. IEEE Commun. Mag. **54**(8), 60–69 (2016)
6. Waldman, D.G., Makarov, S.B.: Synthesis of spectral-effective modulation techniques for digital communication systems. In: 1st IEEE International Conference on Circuits and Systems for Communications, Proceedings, ICCSC 2002, pp. 432–435, June 2002
7. Xue, W., Ma, W., Chen, B.: Research on a realization method of the optimized efficient spectrum signals using Legendre series. In: 2010 IEEE International Conference on Wireless Communications, Networking and Information Security (WCNIS), 25–27 June 2010, pp. 155–159 (2010)
8. Xue, W., Ma, W., Chen, B.: A realization method of the optimized efficient spectrum signals using Fourier series. In: 2010 6th International Conference on Wireless Communications Networking and Mobile Computing (WiCOM), 23–25 September 2010
9. Ovsyannikova, A.S., Zavjalov, S.V., Makarov, S.B., Volvenko, S.V.: Choosing parameters of optimal signals with restriction on correlation coefficient. In: Galinina, O., Andreev, S., Balandin, S., Koucheryavy, Y. (eds.) NEW2AN/ruSMART/NsCC -2017. LNCS, vol. 10531, pp. 619–628. Springer, Cham (2017). https://doi.org/10.1007/978-3-319-67380-6_58
10. Zavjalov, S.V., Makarov, S.B., Volvenko, S.V., Xue, W.: Waveform optimization of SEFDM signals with constraints on bandwidth and an out-of-band emission level. In: Balandin, S., Andreev, S., Koucheryavy, Y. (eds.) ruSMART 2015. LNCS, vol. 9247, pp. 636–646. Springer, Cham (2015). https://doi.org/10.1007/978-3-319-23126-6_57
11. Zavjalov, S.V., Volvenko, S.V., Makarov, S.B.: A method for increasing the spectral and energy efficiency SEFDM signals. IEEE Commun. Lett. **20**(12), 2382–2385 (2016)
12. Zavjalov, S.V., Makarov, S.B., Volvenko, S.V.: Duration of nonorthogonal multifrequency signals in the presence of controlled intersymbol interference. In: 2015 7th International Congress on Ultra Modern Telecommunications and Control Systems and Workshops (ICUMT), Brno, pp. 49–52 (2015)
13. Baek, M.S., Yun, J., Lim, H., Kim, Y., Hur, N.: Joint masking and PAPR reduction for digital broadcasting system with faster-than-Nyquist signaling. In: 2017 IEEE International Symposium on Broadband Multimedia Systems and Broadcasting (BMSB), Cagliari, pp. 1–2 (2017)
14. Le, C., Schellmann, M., Fuhrwerk, M., Peissig, J.: On the practical benefits of faster-than-Nyquist signaling. In: 2014 International Conference on Advanced Technologies for Communications (ATC 2014), Hanoi, pp. 208–213 (2014)

15. Antonov, E.O., Rashich, A.V., Fadeev, D.K., Tan, N.: Reduced complexity tone reservation peak-to-average power ratio reduction algorithm for SEFDM signals. In: 2016 39th International Conference on Telecommunications and Signal Processing (TSP), Vienna, pp. 445–448 (2016)
16. Rave, W., Zillmann, P., Fettweis, G.: Iterative correction and decoding of OFDM signals affected by clipping. In: Proceedings of MC-SS 2005, pp. 443–452 (2005)

Spectral Efficiency Comparison Between FTN Signaling and Optimal PR Signaling for Low Complexity Detection Algorithm

Aleksei Plotnikov[✉] and Aleksandr Gelgor[ID]

Peter the Great St. Petersburg Polytechnic University, St. Petersburg, Russia
aplotnikov94@yandex.ru, a_gelgor@mail.ru

Abstract. This paper is devoted to detection of signals with intentionally introduced intersymbol interference (ISI). Faster-than-Nyquist (FTN) signaling and optimal partial response signaling (PRS) are considered. As the detection algorithm it is used a sub-optimal modification of BCJR algorithm, named Max-Log-M-BCJR. Signals are compared in the plane of spectral efficiency and energy consumptions for the fixed value of bit error rate (BER) and different grades of detection algorithm complexity. It is shown that using the sub-optimal BCJR algorithm provides a noticeable decrease in the computational complexity of the detection. For deep ISI optimal PR signaling provides higher spectral efficiency as compared with FTN signaling and vise versa.

Keywords: Spectral efficiency · Intersymbol interference
Faster-than-Nyquist signaling · Partial response signaling · Optimal pulse · BCJR
Viterbi algorithm

1 Introduction

Increasing the data rate is an actual problem in the field of radio engineering. Nowadays, in most single-carrier communication systems (e.g. satellite systems) the spectral efficiency is usually increased by enlarging the size of the signal constellation. However, the use of high order signal constellations inevitably leads to increase a peak to average power ratio of the emitted signal and energy consumptions.

Starting with Mazo's work [1], at the end of the 20th century another way to increase the spectral efficiency was evolved, that is the using signals with controlled intersymbol interference (ISI). Mazo suggested to symbol rate be greater than the threshold value, below which it is still possible to eliminate the ISI. Thereby the data rate is increased. Mazo called this approach as "modulation with a speed greater than the Nyquist limit" (Faster than Nyquist, FTN). Mazo used linear modulation with sinc-pulses. He accelerated the symbol rate by $1/\tau$ ($0 < \tau < 1$) times leaving pulse fixed. In [2] Liveris and Georghiades extended the idea from [1] by transition to root-raised cosine (RRC) pulses. RRC-pulses have parameter β which is named roll-off factor, where $0 \leq \beta \leq 1$. When β = 0, then RRC-pulse is equal to sinc-pulse.

© Springer Nature Switzerland AG 2018
O. Galinina et al. (Eds.): NEW2AN 2018/ruSMART 2018, LNCS 11118, pp. 191–199, 2018.
https://doi.org/10.1007/978-3-030-01168-0_18

Later, in 1998, Said and Anderson proposed partial response signaling (PRS) with optimal pulses [3] as another way to introduce ISI. They used the criterion of maximization the free Euclidean distance that is equal to the minimum Euclidean distance between a pair of signals corresponding to different information sequences. As the constraint they used the bandwidth value, which must contain no less than a given signal power concentration. Sought pulse is represented as convolution of optimal finite discrete pulse and infinite interpolation pulse. The proposed approach provides asymptotically the lowest bit error rate (BER) with increasing the signal-to-noise ratio (SNR). Note, that the optimization problem formulation and solving procedure are depend on signal constellation.

When increase the length of finite discrete pulse and the size of the signal constellation, then the computational complexity of optimization problem is increased exponentially. This is serious disadvantage of the technique from [3]. In such cases, the solution of optimization problem can be obtained by iterative solving a simplified problem, which contains fewer constraints. Setting of pulse energy to a fixed value is also used to ensure solution uniqueness.

Optimization technique in [3] has one lack, which is that the free Euclidean distance does not connected with the BER curves directly. Especially it is important at the BER values higher than 10–6. In more details, if the free Euclidean distance of signals with ISI is equal to the free Euclidean distance for signals without ISI, then it doesn't mean that BER curve of signals with ISI coincide with BER curve for signals without ISI. In region of relatively high BER values the first curve can lay noticeably higher than the second curve. Curves coincide only in the region of very low BER values.

As FTN and PRS both use an intentional ISI, it is impossible to use symbol-by-symbol detection. Instead of this, it is necessary to use the maximum likelihood sequence estimation (MLSE). The direct implementation of the MLSE is brute force as is computationally very complex. By considering the fact that the samples of signal with ISI depend on each other, the generator of such signals can be represented as a source with a memory. Thus, the forming of signals can be represented through the trellis. There are exist two popular algorithms for computationally effective implementation the MLSE in the trellis: the Viterbi algorithm [4] and BCJR algorithm [5]. The advantage of BCJR algorithm is the ability to calculate soft decisions at the de-modulator output to improve the performance of the FEC (forward error correction) decoder operation. The drawback of BCJR algorithm is the necessity to determine the noise dispersion and higher computational complexity compared to the Viterbi algorithm.

Although the Viterbi and BCJR algorithms are the computational effective implementations of the MLSE, nevertheless detection of signals with ISI using this algorithms is still have quite high computational complexity. In [6, 7] Anderson and others proposed an approach to the implementation of suboptimal versions of the Viterbi and BCJR algorithms – M-Viterbi and M-BCJR. Unfortunately, till now this suboptimal algorithms were mainly tested only for PRS.

In [8] it was performed the comparison between FTN and optimal PRS using original Viterbi algorithm. In [9] it was done for the fixed computational complexity of the Viterbi algorithm. To detect FTN signals demodulator assumed truncated pulses, thus a kind of suboptimal Viterbi algorithm was implemented.

The goal of this paper is to compare the spectral efficiency and energy consumptions of FTN signals with RRC-pulses and optimal PRS when using the computationally effective suboptimal implementation of the BCJR algorithm. One of the tasks is to identify the value of parameter of the suboptimal BCJR algorithm, which provide compromise between the computational complexity and proximity to the original BCJR (true-BCJR) algorithm.

2   BCJR Modification

Even now, the BCJR algorithm is not implemented in practice for detection of signals with ISI because of the high complexity of the algorithm. Therefore, it was decided to use sub-optimal versions of this algorithm, which, although they lose in the energy consumptions, but it is probable that the losses will be acceptable. Actually, we propose to couple different modifications of the BCJR algorithm, which reduce computational complexity in different ways, into one new modification.

One of the disadvantages of the BCJR algorithm is necessity to perform a large number of calculations of the exponent and multiplication operations. The first modification, Log-BCJR, performs transition to the logarithmic domain thus replacing the operation of multiplication with addition and reducing the complexity of the algorithm while remaining the same accuracy and performance. Nevertheless, this simplification still requires the calculation of exponents and logarithms. To further reduce the computational complexity of the Log-BCJR algorithm, in practice, an approximate method of calculating the metrics of states and paths is often used. For example:

$$\ln\left(\sum_i \exp(x_i)\right) = \max_i\{x_i\}. \tag{1}$$

A modification of Log-BCJR algorithm that uses rule (1) to calculate metrics is called Max-Log-BCJR [10].

In [7, 11] it was proposed one another way to reduce complexity of true-BCJR through calculating only the best M possible trellis paths. Thus, it is possible to maneuver between reducing the computational complexity of the algorithm and the depth of making decision. It is expected that the most probable paths will remain, while the less likely ones will not be considered at all. The algorithm assumes the calculation of the best metrics for the forward and backward passages independently of each other. This algorithm was named as M-BCJR.

Figure 1 demonstrates survived paths for M-BCJR. There are used only the best two of eight possible paths.

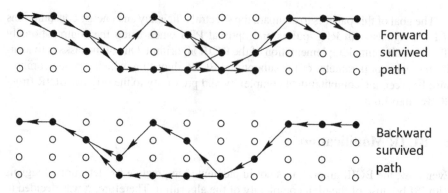

Fig. 1. M-BCJR survived path demonstration

We propose to use an algorithm that combines ideas of Max-Log-BCJR and M-BCJR, i.e. the calculation of the metric according to (1), and the choice of M best paths. Such a new algorithm we called Max-Log-M-BCJR.

3   Definition of Computational Complexity

Let us compare the computational complexity of algorithms observed above. By computational complexity, we mean the number of calculations of transition probabilities, metrics, and soft decisions. Let N to be a number of symbols entering the demodulator input; SL − 1 is the number of states, which is mainly determined by the depth of the interference L; S is the signal constellation size, e.g. the number of possible modulation symbols; M is the number of surviving paths for the M-BCJR algorithm.

It should be mentioned that in the Max-Log-BCJR and Max-Log-M-BCJR algorithms, all operations are additions, and in other algorithms, all operations are multiplications (and even exponents for true-BCJR). This is especially important when implementing in digital signal processors. From Table 1 it follows that the Max-Log-M-BCJR algorithm provides significant decrease of computational complexity, combining benefits of two suboptimal versions of true-BCJR. Using Max-Log-M-BCJR, it is possible to reduce the number of calculation of transition probabilities in SL − 1/M times, the number of calculation of metrics reduces in SL − 1/M times, and number of calculations of soft decisions reduces in SL − 1/2 times.

Table 1. Computational complexity of true-BCJR and its variations

Algorithm	Calculation of transition probabilities	Calculation of metrics	Calculation of soft decisions
True-BCJR	$2NS^{L-1}$	NS^{L-1}	$2NS^{L-1}$
Max-Log-BCJR	$2NS^{L-1}$	NS^{L-1}	$4N$
M-BCJR	$2NM$	NM	$2NM$
Max-Log-M-BCJR	$2NM$	NM	$4N$

4   Model Description

In practice, one of the common way to implement infinite pulses is weighting by finite window function in time domain. Thus, pulse is reduced to a finite form with minimal loss by some criterion. In this paper, we decided to use a rectangular window:

$$
\begin{cases} 1, & |t| \leq LT/2, \\ 0, & |t| > LT/2. \end{cases} \tag{2}
$$

The value L in (2) was selected such that the truncated pulse contains at least 99.9% of the energy of the initial infinite impulse. This is done accordingly to [9] and is provide negligible variation of bandwidth and BER. Figure 2 demonstrates infinite pulse and rectangular window function (2) for L = 8.

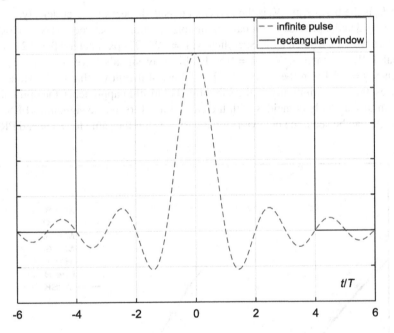

Fig. 2. Illustration of infinite pulse and rectangular window

In our simulations we used $N = 1000$ as the length of the information sequence and QPSK modulation. For the Max-Log-M-BCJR algorithm, the number of surviving paths M was chosen to be an integer power of 2, that is, 8, 16, 32, 64, 128.

5   Simulation Results

During the simulation, a large amount of statistical data were obtained. In this section, only some of them will be presented, which, if possible, will illustrate the general behavior of the remaining results.

To analyze the effectiveness of data transmission, a common analysis of energy and spectral efficiency or corresponding unit consumptions is usually carried out. By spectral efficiency, we mean the ratio of the bandwidth containing 99% of signal power $W_{99\%}$ to the data rate value R:

$$\gamma = \frac{R}{W_{99\%}}. \tag{3}$$

By energy consumptions, we mean the value of the signal-to-noise ratio (SNR) per bit, which corresponds to the BER $= 10^{-4}$:

$$h^2 = \frac{E_b}{N_0}, \tag{4}$$

where E_b is the bit energy, N_0 is the AWGN one-sided spectral power density.

Figures 3 and 4 provide a comparison of noise immunity between FTN and PRS for the pulses with the same length. Specifically, L = 12; FTN parameters: $\beta = 0.2$, $\tau = 0.8$; optimal PRS parameters: W99% = 0.8. In this way signals have the same spectral efficiency $\gamma = 2.5$ From the analysis of Figs. 3 and 4 it follows that as the value of M increases, the noise immunity of the suboptimal algorithm approaches the optimal one, which in turn actually coincides with the curve for full response signaling (FRS) with QPSK. As can be seen, the behavior of the curves is not the same for FTN and PRS. In

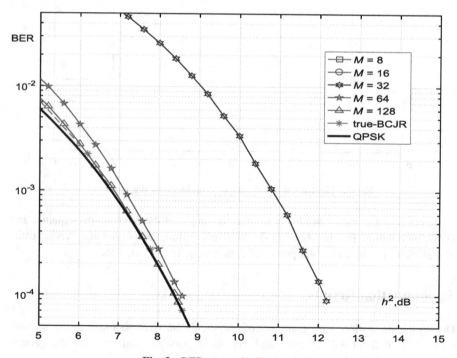

Fig. 3. BER curves for FTN signal

Fig. 3, as M increases, the curves converge more and more with the FRS curve for QPSK. Figure 4 shows that probably only in the region of low BER values (<10–4) the curves begin to converge to the FRS noise immunity curve. Thus, the optimal signals PRS lose to the FTN signals. That is, the difference in the signal-to-noise ratio at $M = 64$ is 1.7 dB.

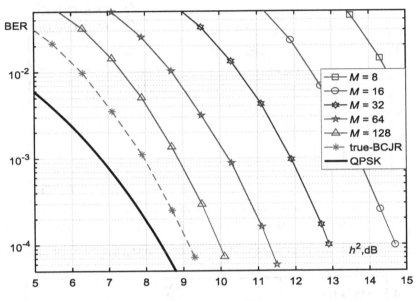

Fig. 4. BER curves for PRS signal

From Fig. 5 it follows that for the number of surviving paths M = 8, PRS signals show worse results comparing to FTN signals. For clarity, two curves for true-BCJR are added – one for FTN signals and one for optimal PRS signals. It is clear that without limitation on the number of surviving paths the optimal PRS signals provide better results than FTN signals. It is also worth noting that PRS signals have a lack consisting the curve is non monotonic. This is a consequence of the fact that the free Euclidean distance does not monotonically depends on bandwidth, therefore even though the spectral efficiency decrease, the energy consumptions can grow.

Figure 6 provides the results for the same signals set as in Fig. 5, but now the maximum number of paths $M = 64$ for Max-Log-M-BCJR algorithm. As can be seen from the Fig. 6 the behavior of the Max-Log-M-BCJR algorithm curves is similar to the true-BCJR algorithm and PRS signals provides higher spectral efficiency than FTN signals. Considering the fact that true-BCJR in this conditions has $S^{L-1} = 2^{11} = 2048$, it follows that Max-Log-M-BCJR algorithm provides 32 times lower computational complexity without noticeable lose in spectral and energy efficiency.

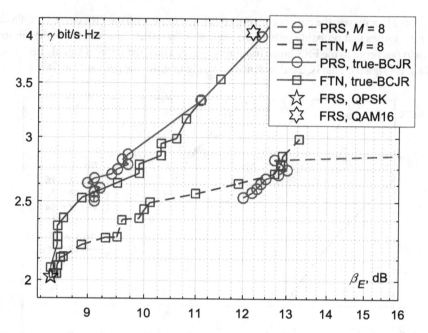

Fig. 5. Spectral efficiency by signal-to-noise, true-BCJR and Max-Log-M-BCJR with $M = 8$

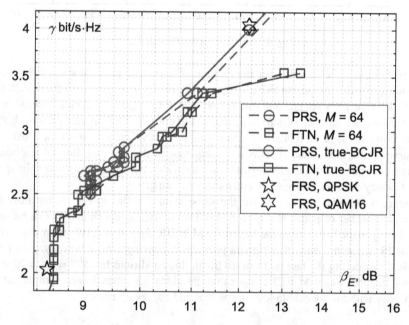

Fig. 6. Spectral efficiency by signal-to-noise, true-BCJR and Max-Log-M-BCJR with $M = 64$

Common conclusions are as follows. For the very low values of M the FTN signals provide better spectral efficiency than optimal PRS signals. This is due to the fact that with increase the value of M BER performance of Max-Log-M-BCJR for FTN is faster converges to BER performance of true-BCJR than for optimal PRS signals. There are exist such values of M that Max-Log-M-BCJR algorithm provides significantly lower computational complexity than true-BCJR without loss of energy and bandwidth.

6   Conclusions

In the work, we developed suboptimal Max-Log-M-BCJR demodulation algorithm. It is shown that it is possible to reduce the computational complexity of detection approximately in S^{L-1}/M times without significant energy losses. The comparison of FTN and PRS signals has shown that for high values of M the best spectral efficiency is achieved by PRS signals and vice versa. For pulse length $L = 12$ the Max-Log-M-BCJR detection with $M = 64$ provides the results close to true-BCJR detection, that is mean decrease of computational complexity in 32 times. Also, the lack of optimal PRS was founded, that is the spectral efficiency doesn't monotonically depends on energy consumptions.

References

1. Mazo, J.E.: Faster-than-Nyquist signaling. Bell Syst. Tech. J. **54**(8), 1451–1462 (1975)
2. Liveris, A.D., Georghiades, C.N.: Exploiting faster-than-Nyquist signaling. IEEE Trans. Commun. **51**(9), 1502–1511 (2003)
3. Said, A., Anderson, J.B.: Bandwidth-efficient coded modulation with optimized linear partial-response signals. IEEE Trans. Inf. Theory **44**(2), 701–713 (1998)
4. Forney, G.D.: The Viterbi Algorithm. Proc. IEEE **61**(3), 268–278 (1973)
5. Bahl, L.R., et al.: Optimal decoding of linear codes for minimizing symbol rate. IEEE Trans. Inform. Theory **IT-20**, 284–287 (1974)
6. Anderson, J.B.: Limited search trellis decoding of convolutional codes. IEEE Trans. Inf. Theory **35**, 944–955 (1989)
7. Franz, V., Anderson, J.: Concatenated decoding with a reduced search BCJR algorithm. IEEE J. Sel. Areas Commun. **16**(2), 186–195 (1998)
8. Rusek, F.: Partial response and faster-than-Nyquist signaling. Ph.D. dissertation. Lund University (2007)
9. Gorlov, A., Gelgor, A., Nguyen, V.P.: Root-Raised cosine versus optimal finite pulses for faster-than-Nyquist generation. In: Galinina, O., Balandin, S., Koucheryavy, Y. (eds.) NEW2AN/ruSMART 2016. LNCS, vol. 9870, pp. 628–640. Springer, Cham (2016). https://doi.org/10.1007/978-3-319-46301-8_54
10. Robertson, P., Villerbrun, E., Hoeher, P.: A comparison of optimal and sub-optimal MAP decoding algorithms operating in the log domain. In: Proceedings of International Conference on Communications, Seattle, WA, June 1995, pp. 1009–1013 (1995)
11. Fragouli, C., Seshadri, N., Turin, W.: On the rediced trellis equalization using the M-BCJR algorithm. Technical report TR 99.15.1, Florham Park, NJ, November 1999

A Method of Simultaneous Signals Spectrum Analysis for Instantaneous Frequency Measurement Receiver

Dmitrii Kondakov[✉], Alexey Kosmynin, and Alexander Lavrov

Peter the Great St. Petersburg Polytechnic University, St. Petersburg, Russia
{dmitrii.kondakov,lavrov_ap}@spbstu.ru, a.n.kosmynin@ya.ru

Abstract. In this paper the simplified problem of frequency determination for multiple simultaneously present harmonic oscillations through subsampling is considered. The proposed (used) subsampling is realized by Dirac function comb with principal frequency much less than input signal frequencies. So all signal frequencies are transformed to first Nyquist zone. We consider subsampling is implemented in three parallel channels when comb principal frequencies are differing but close one to another. Each channel includes also ADC, FFT unit and digital processing unit. Output channel information is a set of possible input frequencies, and these sets intersection is searched for finding input frequencies. Proposed system math model was developed for estimation of ambiguity of recovering values for original input frequencies. The subsampler with three parallel channels was realized as a small unit. It works as mixer with comb type heterodyne in superheterodyne receiver. The subsampler has analog bandwidth up to 5 GHz with 100 MHz principal frequencies. Its experimental characteristics are presented.

Keywords: Subsampling · Multicomponent signal
Frequency recovering · Nyquist zone · Dirac function comb

1 Introduction

Input signal spectrum analysis is an important task in many processing systems (receivers). Nowadays many types of complicated signals are used in various radioelectronics systems, such as phase and frequency modulation signal, chirp and so on. They occupied wide frequency range from RF to microwave, approx. 0.1–20 GHz, and may be present at the same time at a receiver input. One approach for such input signal spectral analysis is traditional superheterodyne receiver with tunable local oscillator, but there is a drawback – missing of pulse signal when local oscillator scans throw total frequency band. Another approach for spectral analysis is division of total frequency band to many subbands with parallel function of many receivers, but there is a drawback – an equipment will have large size and cost. One can use an instantaneous frequency measurement

© Springer Nature Switzerland AG 2018
O. Galinina et al. (Eds.): NEW2AN 2018/ruSMART 2018, LNCS 11118, pp. 200–209, 2018.
https://doi.org/10.1007/978-3-030-01168-0_19

(IFM) receiver. Conventional IFM receiver consists of a delay line, mixers and RF couplers. Built on this architecture receiver works in wide bandwidth and has high performance [1]. However, this approach has one serious disadvantage - well-known problem of frequency determination in case of multicomponent input signal. To overcome such problem Prony and Pisarenko method was applied in [2]. In addition, some other methods like MUSIC (Multiple Signal Classification) assist to distinguish the frequencies in multicomponent signal spectrum [3]. But these methods require digital processor with large processing power. Next approaches are based on input signal conversion in digital form and wide usage of digital techniques – one can say this is main way in modern processing systems realization. But it has become prohibitive to sample modern wideband signals because their Nyquist rates (for sampling) may exceed specifications of the best analog-to-digital converters (ADCs) by orders of magnitude. There are approaches with decreasing the sampling frequency – subsampling approaches with parallel work of some ADCs, but they to our opinion are not simple because of acquiring samples from a periodic but nonuniform grid or multi-coset sampling with specific strategy of this type [4–6].

The goal of this paper is to propose a more simple method of multicomponent input signal spectrum analysis which combines an approach of superheterodyne receiver and a concept of signal multi-rate subsampling in some parallel channels with constant but differing from channel to channel sampling rates. We consider channel synchronous subsampling as heterodyning with Dirac function comb.

2   Theory of Operation

For the beginning let's consider a frequency shift via heterodyning. Heterodyning is a radio signal processing technique that creates new two frequencies by combining two frequencies [7]. The down-conversion operation is illustrated in Fig. 1. The frequency of input radio signal $\mathbf{F_{RF}}$ and the image frequency $\mathbf{F_{RF}}$ also are transposed to the intermediate frequency $\mathbf{F_{IF}}$. The superheterodyne receiver operation over the wide frequency range is achieved by tunable local oscillator. Such tuning takes a certain time. In a conventional approach of heterodyning an image frequency signal must be filtered out before the frequency conversion.

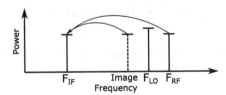

Fig. 1. Description of a conventional frequency down-conversion heterodyning.

The idea of parallel spectrum processing is based on using frequency Dirac comb generator instead of harmonic oscillator. Mathematically Dirac comb is a periodic tempered distribution constructed from Dirac delta functions [8]. The explanation of spectrum transform is depicted in Fig. 2.

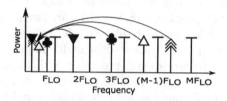

Fig. 2. Description of a parallel spectrum transform.

Here multiple harmonics from F_{LO} separate the spectrum into frequency zones. Every zone shifts to the beginning of spectrum. Thus with a knowledge of a law of spectrum transform it becomes possible to observe the wide frequency range in parallel mode. Performance of a receiver in this case depends only on Fourier transform speed and a time of frequency matrix calculation.

3   Basic Mathematical Relations

At first let's introduce key terms:
$s(t)$ is the incoming multicomponent RF signal; $\mathbf{s_{LO}(t)}$ is a signal frequency from comb generator; $\mathbf{F_{udf}}$ is a frequency range of unambiguous definition of frequency; $\mathbf{N}$ is a number of components of input signal; $\mathbf{M}$ is a number of comb generator's harmonics.

The expression for input signal is given by

$$\mathbf{s(t)} = \sum_{i=1}^{N} \mathbf{A_i cos(2\pi f_i t)} \tag{1}$$

The signal from frequency comb generator can be written as

$$\mathbf{s_{LO}(t)} = \sum_{k=1}^{M} \mathbf{B_k cos(2\pi k f_{LO} t)} \tag{2}$$

The signal on the mixer's output is equal to a multiplication of expressions for signal and local oscillator

$$\mathbf{s_{mixer}(t)} = \mathbf{s(t)} \cdot \mathbf{s_{LO}(t)} \tag{3}$$

A time domain representation of a signal on the mixer's output in general form is given by (5). Here for every input signal component like for conventional heterodyning we have two output components. The first term corresponds to intermediate frequency and the second is an image frequency.

$$\mathbf{s_{mixer}(t)} = \frac{1}{2}\left[\underbrace{\sum_{i=1}^{M} A_i \sum_{k=1}^{N} B_k cos\{(2\pi(f_i + kf_{LO}))t\} +}_{\textbf{Intermediate}}\right. \tag{4}$$

$$\left.\underbrace{+ \sum_{i=1}^{M} A_i \sum_{k=1}^{N} B_k cos\{2\pi(f_i - kf_{LO}))t\}}_{Image}\right] \tag{5}$$

We continue the discussion under the assumption of zero image frequency components. We note that there are some technical approaches to realize filtration of image frequencies in conventional approach of heterodyning.

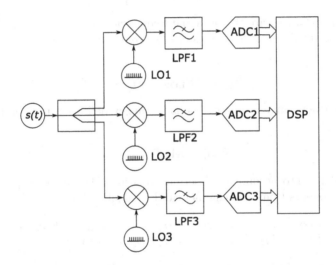

Fig. 3. Proposed IFM Receiver block diagram.

3.1 IFM Receiver operation algorithm

The block diagram of proposed instantaneous frequency measurement receiver is illustrated in Fig. 3.

The proposed IFM Receiver consists of three equivalent RF channels. The incoming signal is passed through in-phase power divider. On the next step the input signal is mixed with a signal from comb generator. Mixer output signals go to low-pass filter. The cutoff frequency is equal to $\mathbf{F_{LO}}$. Afterwards the response is sampled by analog-to-digital converter and is processed in digital form in DSP. First of all FFT is accomplished in DSP, so one can receive spectrum information: frequencies $\mathbf{F_{IF}}$ of all components and also its amplitude [9]. It should be marked that every channel gives the possibility of extracting additional information about incoming signal spectrum according to next assertion:

1. One channel provides frequency determination for one-tone signal for value not more than LO's principal frequency ($\mathbf{F_{LO1}}$).
2. Adding the second channel allows to uniquely identifying the frequency of one-tone signal in range up do $\mathbf{F_{udf}}$.
3. Three channels provide the frequency determination for N tones with values up to $\mathbf{F_{udf}}$.

where $\mathbf{M}$ is a maximum LO's harmonic number which is defined by (6) under condition (8).

$$\mathbf{M} = \frac{\mathbf{F_{LO1}}}{\mathbf{F_{LO2}} - \mathbf{F_{LO1}}} \tag{6}$$

where $\mathbf{F_{LO1}}$ and F_{LO2} are the principal frequencies of two comb generators
There is a requirement for values of LO's frequencies

$$\mathbf{F_{LO1}} < \mathbf{F_{LO2}} < \mathbf{F_{LO3}} \tag{7}$$

The frequency range of unambiguous frequency definition is given by

$$\mathbf{F_{udf}} = \mathbf{M} \cdot \mathbf{F_{LO1}} \tag{8}$$

With taking into account of described relations the values of possible frequency components from first channel can be calculated as

$$\begin{pmatrix} F_{LO1} - F_{IF_1} & 2F_{LO1} - F_{IF_1} & \cdots & MF_{LO1} - F_{IF_1} \\ F_{LO1} - F_{IF_2} & 2F_{LO1} - F_{IF_2} & \cdots & MF_{LO1} - F_{IF_2} \\ \cdots & \cdots & \cdots & \cdots \\ F_{LO1} - F_{IF_{N-1}} & 2F_{LO1} - F_{IF_{N-1}} & \cdots & MF_{LO1} - F_{IF_{N-1}} \\ F_{LO1} - F_{IF_N} & 2F_{LO1} - F_{IF_N} & \cdots & MF_{LO1} - F_{IF_N} \end{pmatrix} \tag{9}$$

where F_{LO1} is a first harmonic of comb generator and $F_{IF_1} \ldots F_{IF_N}$ are the intermediate frequencies. The values from the second and third channel can be obtained by the same way.
In order to get the initial frequencies of input signal the intersection between elements of matrices should be found.

4   Simulation

The simulation was performed with MATLAB. The graphical user interface (GUI) with main IFM receiver functional blocks according to above described functional diagram (see Fig. 3) was developed and it is shown in Fig. 4.

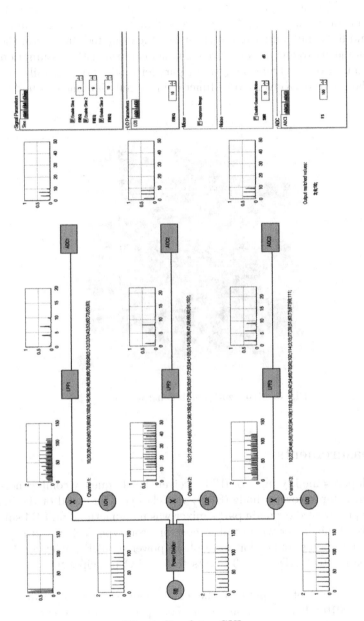

Fig. 4. Simulator GUI.

We choose 3 parallel working channels in IFM receiver model functioning. Moreover, the library of various simple signal types like sine, AM, chirp was also added. The spectrum after every block can be evaluated, see insertions - spectrum diagrams in Fig. 4 in different flow paths. The algorithm of peak search and frequency values intersection was encapsulated. In general the model of instantaneous frequency measurement has confirmed the assumption about the possibility of spectrum analysis in parallel mode. Some details about our approaches of IFM receiver model functioning simulation one can find in [9].

Fig. 5. Mixer with comb generator prototype.

5   Measurements

We developed some key units of IFM receiver. For frequency comb generation a circuit with step recovery diode (SRD) was chosen. The board of the instantaneous frequency receiver main part - mixer was manufactured on FR4 substrate. The photo of our mixer with comb generator prototype is shown in Fig. 5. One can see 3 inputs for connection to in-phase power divider (left side), 3 outputs for connection to 3 ADC and 3 inputs to feed LO principal frequencies (right side).

The spectrum of outputs of the comb generators were evaluated with Keysight N9010A EXA signal analyzer. The spectrogram of one channel comb is presented in Fig. 6, in this case frequency F_{LO} is 100 MHz. The unevenness of harmonics power was not more than 5 dB in frequency range up to 5 GHz.

On the input of the mixer prototype two-tone sine signal was applied with frequencies $f_1 = 2359$ MHz and $f_2 = 1261$ MHz. In the Fig. 7 the mixer one channel output spectrum is shown.

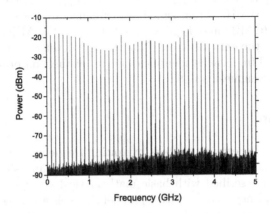

Fig. 6. The comb generator output spectrum.

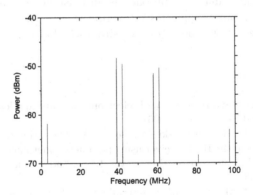

Fig. 7. The filter's output signal's spectrum for two-tone sine signal with frequencies 2359 and 1261 MHz.

It can be seen that there are four power peaks. Two of them are correspond to intermediate frequency and the two other are matched to image frequencies. We have to note that image signal frequency filtration not used here. From signal with frequency f_1 we received $f_{IF1} = f_1 - 230 \cdot F_{LO} = 59\,\text{MHz}$ and $f_{IM1} = f_1 - 231 \cdot F_{LO} = 41\,\text{MHz}$. From signal with frequency f_2 we received $f_{IF2} = f_2 - 120 \cdot F_{LO} = 61\,\text{MHz}$ and $f_{IM2} = f_2 - 121 \cdot F_{LO} = 39\,\text{MHz}$. This behavior is consistent with theoretical model. Now we are developing an algorithm of proper input signal frequencies identification (recovering) on processing data of 3 parallel receiver channels, see [9].

6   Conclusion

The idea of parallel spectrum analysis exploiting frequency comb generator as LO in receiver mixer was proposed. The algorithm for input signal frequencies determination was described with basic mathematical relations. The model of instantaneous frequency receiver with 3 parallel channels was developed. Simulation confirmed the assumption about the possibility of simultaneously presented signals (multicomponent signal) transformation into frequency baseband and its initial frequencies recovering. The receiver based on such approach has the following advantages. In comparison with the superheterodyne receiver it doesn't need any control of local oscillator. In addition, this approach eliminates the need of high-speed analog-to-digital converters. Compared to conventional approach of instantaneous frequency receiver the proposed receiver allows working with multicomponent signals. The developed mixer prototype has measured unevenness of spectral components power on the output of comb generator not more than 5 dB for frequency range up to 5 GHz. Nevertheless, the fundamental limitation of this parallel analysis approach is an existence of image channels. It leads to erroneous input frequencies recovering results. In the future work the image channel rejection will be made by using special algorithms.

References

1. Tsui, J.B.Y.: Microwave receivers with electronic warfare applications. Institution of Engineering and Technology (2005)
2. Gruchalla-Wesierski, H., Czyzewski, M., Slowik, A.: The estimation of simultaneous signals frequencies in the IFM receiver using parametric methods. In: MIKON 2008–17th International Conference on Microwaves, Radar and Wireless Communications, vol. 11, pp. 1–4 (2008)
3. Lioun, L.L., Lin, D.M., Tsui, J.B.: Determination of electronic warfare receiver's instantaneous dynamic range using music method. In: 2008 IEEE National Aerospace and Electronics Conference, pp. 59–67 (2008)
4. Tzou, N.: Low cost sparse multiband signal characterization using asynchronous multi-rate sampling: algorithms and hardware. J. Electr. Test. **31**, 85–98 (2015)
5. Mishali, M., Eldar, Y., Tropp, J.: Efficient sampling of sparse wideband analog signals. In: 2008 IEEE 25th Convention of Electrical and Electronics Engineers in Israel, pp. 290–294 (2008)

6. Venkataramani, R., Bresler, Y.: Perfect reconstruction formulas and bounds on aliasing error in sub-nyquist nonuniform sampling of multiband signals. IEEE Trans. Inform. Theory **46**(6), 2173–2183 (2000)
7. Graf, R.F.: Modern Dictionary of Electronics. In: Electronics & Electrical (1999)
8. Strichartz, R.: Guide to Distribution Theory and Fourier. World Scientific Publishing Company, Singapore (2003)
9. Kondakov, D.V., Kosmynin, A. N., Lavrov, A. P.: Multicomponent signal frequency estimation algorithm for digital receiver with subsampling. In: RLNC 2017–23th International Conference on Radio Location, Nagivation and Communication, vol. 2, pp. 481–486 (2017). (in Russian)

Analytical Models for Schedule-Based License Assisted Access (LAA) LTE Systems

Ekaterina Markova[1](✉), Dmitri Moltchanov[1,2], Anna Sinitsyna[1],
Daria Ivanova[1], Valeria Filipova[1], Irina Gudkova[1,3],
and Konstantin Samouylov[1,3]

[1] Department of Applied Probability and Informatics, Peoples' Friendship
University of Russia (RUDN University),
6 Miklukho-Maklaya Street, 117198 Moscow, Russia
{markova_ev, moltcanov_da, gudkova_ia,
samuylov_ke}@rudn.university, aa.sinitsyna@yandex.ru,
{daria.il996, valeryflp}@gmail.com
[2] Department of Electronics and Communications Engineering,
Tampere University of Technology,
10 Korkeakoulunkatu, 33720 Tampere, Finland
[3] Institute of Informatics Problems, Federal Research Center,
"Computer Science and Control" of the Russian Academy of Sciences,
44-2 Vavilova Street, 119333 Moscow, Russia

Abstract. The scarcity of resources available for commercial wireless access systems below 6 GHz coupled with constantly increasing traffic demands from the mobile users force network operators to seek additional spectrum. In addition to moving upper in the frequency band and occupying millimeter wave band with 3GPP New Radio access technology the set of solutions also includes implementing commercial LTE systems in unlicensed bands including 2.4 GHz and 5.1 GHz that are currently occupied by Wi-Fi. This technology, known as License Assisted Access (LAA), has recently received considerable attention within the 3GPP community. One of the solutions to provide fair division of air interface resources between competing technologies is to use schedule-based access, where LAA access point is in full control of shared medium and may dynamically schedule allocations to LTE and Wi-Fi traffic. The fine tuning of LAA technology requires careful understanding of various trade-offs and dependencies involved in Wi-Fi and LTE coexistence. In this paper, using the tools of the queuing theory we formulate and solve several analytical models targeting different implementation strategies of schedule-based LAA systems and traffic types of end users. We derive relevant performance characteristics including the session drop probabilities, probability that the session accepted to the system is drop before its service completion and average resource utilization of the system.

Keywords: LTE · 4G · LAA · License-assisted access · Analytical models

© Springer Nature Switzerland AG 2018
O. Galinina et al. (Eds.): NEW2AN 2018/ruSMART 2018, LNCS 11118, pp. 210–223, 2018.
https://doi.org/10.1007/978-3-030-01168-0_20

1 Introduction

While the subscriber base of commercial mobile systems no longer grows exponentially over the last few years, the amount of generated traffic preserves its exponential trend and is expected to reach unprecedented numbers by the year 2021 [1]. This trend is mainly explained by constantly increasing capabilities of handheld devices as well as appearance and further popularization of bandwidth greedy services [2]. This trend is expected to continue in the further with the prospective adoption of multimedia-rich wearable gadgets such as smart-glasses equipped with new services such as augmented and virtual reality [3].

The achievable data rate at the access interface is upper bounded by the famous Shannon result [4] relating achievable rate to the product of the available bandwidth, B, and spectral efficiency, $\log_2(1+S)$. With the wide adoption of advanced modulation and coding schemes and the use of nearly optimal medium sharing procedures the spectral efficiency of wireless channels is not expected to increase drastically in the coming years. This leaves the bandwidth B as the factor providing the decisive impact on the rate achievable at the air interface.

As the frequently resource below 6 GHz is tightly regulated and almost fully occupied, there are two principal approaches to improve the capacity of wireless systems. The straightforward way is to go upper in the frequency band to millimeter wave (mmWave) frequencies, where more bandwidth is available. The 3GPP New Radio (NR) systems operating at 28 GHz and higher frequencies is expected to provide extraordinary performance boost to bandwidth greedy applications. However, along with great promises, the use of mmWave band brings fundamental constraints. In particular, free spaces propagation losses are much more drastic significantly limiting the coverage area of a single AP [5, 6]. Furthermore, due to extremely small wavelength frequencies of this band are much more prone to scattering than to reflection leading to problems reaching receiver in non-line-of-sight (nLoS) conditions [7, 8]. As a result, mmWave links are expected to be characterized by intermittent connectivity while mmWave APs are likely to be deployed in crowded locations, where there is the need for extremely high access rates. Thus, LTE systems is expected to remain the integral part of mobile cellular systems in 5G providing access in areas where no mmWave APs are deployed and enabling data services for highly mobile subscribers.

One of the viable approaches to enhance capacity of LTE air interface is to deploy LTE micro/pico base stations (BS) in unlicensed band. Below 6 GHz there are two industrial scientific and medical bands available, including 150 MHz-wide 5.1 GHz band and 100 MHz-wide 2.4 GHz band. However, these frequencies are extremely crowded by non-commercial Wi-Fi systems and LTE systems operating in this band need to follow the set of rules ensuring seamless coexistence between competing systems.

The coexistence strategies have been studies by 3GPP in Release 13 under the term "license-assisted access" (LAA) and reported in TR.36.889 [9]. According to it, one may enforce listen-before-talk (LBT) functionality at the LTE interface such that fair competition between Wi-Fi and LAA is ensured. One of the inherent shortcomings of this approach is the lack of performance guarantees to LAA connections as LBT-based access cannot provide them by default. Preferable approach allowing provide performance guarantees to LTE applications is to fully delegate control functionality to LAA systems that dynamically schedule resources for Wi-Fi and LAA systems. Nevertheless, in both cases significant gains in terms of throughput is revealed due to the use of more robust and reliable physical and link-layer technologies.

The seamless implementation of schedule-based LAA system in unlicensed band requires dynamic fine tuning of resources between Wi-Fi and LAA systems. The latter heavily depends on the set of assumptions regarding LAA system operation as well as on traffic specifics. In this paper, we develop the set of models for performance assessment of scheduled-based LAA systems. Particularly, we consider systems with streaming and elastic traffic as well as systems, where upon arrival of Wi-Fi session the service process of LTE connections can be interrupted to ensure fairness between Wi-Fi and LAA systems. We provide numerical examples illustrating the crucial performance measures including the session drop probabilities, probability that the session accepted to the system is dropped and average resource utilization of the system.

The rest of the paper is organized as follows. In Sect. 2 we review the principles of schedule-based LAA systems and then formulate the system models encompassing the essentials of LTE and Wi-Fi coexistence in the unlicensed bands. In Sect. 3 we subsequently formalize and solve performance assessment models for various special cases of the system and different traffic types. The numerical examples are provided in Sect. 4. Conclusions are drawn in the last section.

2 System Model of LAA Schedule-Based System

In this section we first provide a brief account of recent developments in schedule-based LAA systems. Then we proceed formulating the system model and its special cases.

2.1 Schedule-Based LAA Systems

The major requirement imposed by 3GPP in TR.36.889 on the operation of LAA systems is that "LAA system operating in unlicensed band shall not interfere more than another Wi-Fi AP operating in this band" [9]. In practice this requirement implies that the presence of LAA shall divide the medium similarly to CSMA/CA protocol implemented in conventional Wi-Fi systems. Attempting to satisfy this requirement the first wave of proposals targeted the use of LBT-mechanisms by introducing a shim layer between physical and data-link layers in the LTE air interface protocol stack. Particular, solutions vary from simple ALOHA approach to very complex CSMA/CA-like access with multiple back-off stages of different size [10, 11].

Introducing random access procedure to LTE system, while achieving the major goal of LAA design, heavily affects performance provided to upper layers. Indeed, one of the most advantageous properties of commercial cellular system is well-defined quality-of-service (QoS) provisioning mechanisms allowing to satisfy performance requirements of applications. When LBT-based design is employed LTE scheduler operates in presence of uncertainty as no deterministic information about the channel access is available. This shortcoming was first discussed in [12], where the authors discussed shortcomings of LBT-based LAA-design and advocated the use of scheduled-based access. According to this design LAA system is provided the full control of the medium (via the use of Wi-Fi short-interframe spacing, SIFS) to allocate time shares for Wi-Fi and LAA. The major advantage of this schedule-based design is that the access schedule if fully known to LAA system. The concept was further elaborated in [13], where the use of duty cycle, a time period repeating itself and divided into Wi-Fi and LTE allocations, has been proposed. In [13], the authors first formulated the set requirements for LAA AP to ensure accurate scheduling of medium and then proceed adding adaptiveness to the duty-cycle-based scheduling. The latter is needed to ensure that applications with different delay requirements. They have shown that when perfect information about the session arrivals at Wi-Fi systems is available the adaptive duty-cycle-based LAA design may ensure fairness while still maintaining QoS guarantees. Deciding on the optimal values of Wi-Fi and LAA allocations in a duty cycle is a complex problem that depends on offered traffic load to both Wi-Fi and LAA.

2.2 System Model

The system we consider is similar to the one, proposed in [14], see Fig. 1. We assume that the duration of the duty cycle, T_D, is constant and set to some value that is dictated by the most delay critical LAA application. According to the system, LAA AP is in full control of the medium overtaking and releasing it using SIFS mechanism.

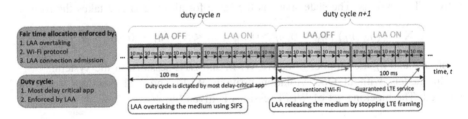

Fig. 1. Adaptive duty-cycle access strategy for LAA design (reproduced from [14]).

LAA system is designed to ensure (i) fair division of the medium between LAA and LTE sessions and (ii) providing throughput guaranteed to LAA sessions. Let $\Delta_{min,L}$ be the minimum time allocation required by LAA session. Thus, both requirements are satisfied by ensuring that all the sessions are provided at least the minimum required resources, i.e., $\Delta = \min(T_D/N, \Delta_{min,L})$, where N is the number of Wi-Fi and LAA sessions in the system. To enforce this requirement LAA implement the connection admission control (CAC) algorithm. Upon arrival of LAA session, CAC checks whether this requirement is satisfied when new session is admitted to the system. If yes, the session is accepted to the system. Otherwise, it is dropped.

The CAC algorithm does not have any control over Wi-Fi and all Wi-Fi sessions are admitted to the system. This implies that upon arrival of Wi-Fi session the requirement $\Delta = \min(T_D/N, \Delta_{min,L})$ can be violated. If this happens the system may either drop one or more LAA sessions or, alternatively, do nothing. In the latter case throughput requirements of LAA sessions might be violated. In what follows, we consider both cases.

As one may observe, Wi-Fi and LAA allocations of the duty cycle depends on the current offered traffic load. We assume that the Wi-Fi and LAA sessions arrive to the system according to Poisson process with intensities λ_W and λ_L, respectively. In what follows, we consider two cases of the service process affecting the resource require-ments and session holding times of sessions corresponding to elastic and streaming types of traffic. In the former case, resource requirements of Wi-Fi and LAA sessions are fully parametrized by constant demand sizes, θ_W and θ_L. For streaming traffic we assume that session holding times are μ_W and μ_L, respectively while the resource requests from the system are R_W and R_L, correspondingly.

In the next session we specify models for the following variants of the system: (i) streaming traffic without QoS violation, (ii) streaming traffic with QoS violation, (iii) elastic traffic without QoS violation, and (iv) elastic traffic with QoS violation.

3   Performance Evaluation Models

The behavior of all four models defined in Sect. 2 can be described by a two-dimensional continuous-time Markov chain (CTMC) $\mathbf{X}(t) = \{N_W(t), N_L(t), t \geq 0\}$, where $N_W(t)$ is the number of active Wi-Fi sessions, $N_L(t)$ is the number of active LTE sessions in the system at the moment t. Denote $N_L = \lfloor C/R_L \rfloor$ the maximum number of active LTE sessions. The state space is the same for all models and takes the form:

$$\mathbf{X} = \{(n_W, n_L) : n_W \geq 0, n_L \geq 0, n_W + n_L \leq N_L \cup n_L = 0, n_W > N_L\}, \qquad (1)$$

where n_W and n_L are number of active Wi-Fi and LTE sessions in the system.

In models without QoS violation arriving Wi-Fi and LTE sessions could be blocked. The blocking set of Wi-Fi sessions is given by:

$$B_W = \{(n_W, n_L) : n_W + n_L = N_L\}, \qquad (2)$$

The blocking set of LTE sessions takes the following form:

$$B_L = \{(n_W, n_L) : n_W + n_L = N_L \cup n_L = 0, n_W > N_L\}. \qquad (3)$$

In models with QoS violation only arriving LTE sessions could be blocked. The blocking set of the LTE sessions coincides with the set for the model without QoS violation and is provided by (3).

3.1   Streaming Traffic Without QoS Violation

Consider first the model without QoS violation with streaming Wi-Fi and LTE sessions. The form of the state transition diagram for this model is illustrated in Fig. 2, while the corresponding equilibrium are as follows:

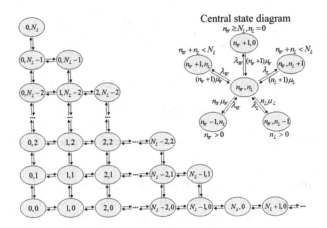

Fig. 2. The state diagram of the LAA system with streaming traffic without QoS violation.

$$
\begin{aligned}
p(n_W, n_L)[&\lambda_W.(1(n_W \geq N_L, n_L = 0) + 1(n_W + n_L < N_L)) \\
&+ \lambda_L.1(n_W + n_L < N_L) + n_L\mu_L.1(n_L > 0) + n_W\mu_W.1(n_W > 0)] \\
&= p(n_W + 1, 0)(n_W + 1)\mu_W.1(n_W \geq N_L, n_L = 0) \\
&+ p(n_W + 1, n_L)(n_W + 1)\mu_W.1(n_W + n_L < N_L)] \\
&+ p(n_W - 1, n_L)\lambda_W.1(n_W > 0) + p(n_W, n_L - 1)\lambda_L.1(n_L > 0) \\
&+ p(n_W, n_L + 1)(n_L + 1)\mu_L.1(n_W + n_L < N_L),
\end{aligned}
\tag{4}
$$

where $(p(n_W, n_L))_{(n_W, n_L) \in \mathbf{X}} = \mathbf{p}$ is the stationary state probability distribution.

Since CTMC $\mathbf{X}(t)$ is reversible the probability distribution $p(n_W, n_L)$, $(n_W, n_L) \in \mathbf{X}$ has the multiplicative form [14]:

$$
p(n_W, n_L) = \left(\sum_{i=1}^{\infty} \sum_{j=1}^{N_L - i} \frac{(\rho_W)^i}{i!} \frac{(\rho_L)^j}{j!} \right)^{-1} \cdot \frac{\rho_W^{n_W} \rho_L^{n_L}}{n_W! . n_L!},
\tag{5}
$$

where $\rho_W = \lambda_W / \mu_W$ and $\rho_L = \lambda_L / \mu_L$ are offered loads for the Wi-Fi and LTE sessions.

3.2   Streaming Traffic with QoS Violation

Consider now the second model with streaming traffic and QoS violation. The principal difference from the first model is absence of Wi-Fi session's blockage. As a result, the throughput guarantees of LTE sessions can be violated. The state transition diagram of this model is shown in Fig. 3. The system of equilibrium equations is as follows:

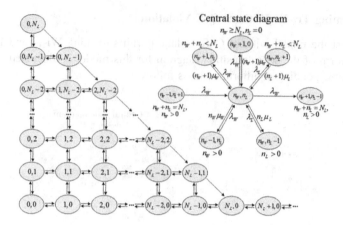

Fig. 3. The state diagram of the LAA system with streaming traffic with QoS violation.

$$p(n_W, n_L)[\lambda_W.(1(n_W \geq N_L, n_L = 0) + 1(n_W + n_L < N_L)$$
$$+ 1(n_W + n_L = N_L, n_L > 0) + \lambda_L \cdot 1(n_W + n_L < N_L)$$
$$+ n_L \mu_L \cdot 1(n_L > 0) + n_W \mu_W \cdot 1(n_W > 0)]$$
$$= p(n_W, n_L - 1)\lambda_L \cdot 1(n_L > 0)$$
$$+ p(n_W - 1, n_L + 1)\lambda_W \cdot 1(n_W + n_L = N_L, n_W > 0) \qquad (6)$$
$$+ p(n_W - 1, n_L)\lambda_W \cdot 1(n_W > 0)$$
$$+ p(n_W + 1, 0)(n_W + 1)\mu_W \cdot 1(n_W \geq N_L, n_L = 0)$$
$$+ p(n_W + 1, n_L)(n_W + 1)\mu_W \cdot 1(n_W + n_L < N_L)]$$
$$+ p(n_W, n_L + 1)(n_L + 1)\mu_L \cdot 1(n_W + n_L < N_L),$$

where $(p(n_W, n_L))_{(n_W, n_L) \in \mathbf{X}} = \mathbf{p}$ is the stationary state probability distribution.

As a result of service interruptions, the CTMC $\mathbf{X}(t)$ is no longer reversible and the probability distribution $p(n_W, n_L), (n_W, n_L) \in \mathbf{X}$ does not have the multiplicative form. However, one can determine it numerically. Rewrite the system (6) in the form of

$$\mathbf{p}.\mathbf{A} = 0, \mathbf{p}.\mathbf{1}^T = 1 \qquad (7)$$

where $\mathbf{A}$ is the infinitesimal generator whose elements $a((n_W, n_L)(n'_W, n'_L))$ are:

$$a = \left((n_W, n_L) \left(n'_W, n'_L \right) \right) = \begin{cases} \lambda_W, & n'_W = n_W + 1, n'_L = n_L - 1, n_L > 0, n_W + n_L = N_L \\ & \text{or } n'_W = n_W + 1, n'_L = n_L, n_W + n_L < N_L \\ & \text{or } n'_W = n_W + 1, n'_L = n_L = 0, n_W \geq N_L, \\ \lambda_L, & n'_W = n_W, n'_L = n_L + 1, n_W + n_L < N_L, \\ n_W \mu_W, & n'_W = n_W - 1, n'_L = n_L, n_W > 0, \\ n_L \mu_L, & n'_W = n_W, n'_L = n_L - 1, n_L > 0, \\ *, & n'_W = n_W, n'_L = n_L, \\ 0, & \text{otherwise,} \end{cases}$$

where $* = -[\lambda_W.1(n_W \geq N_L) + \lambda_W.1(n_W + n_L < N_L) + n_L \mu_L.(n_L > 0) + n_W \mu_W 1(n_W > 0) + \lambda_L 1(n_W + n_L < N_L) + \lambda_W 1(n_W + n_L = N_L, n_L > 0)]$.

3.3 Elastic Traffic Without QoS Violation

Similarly to the models with streaming traffic, the behavior of LAA system with elastic traffic is described by two-dimensional vector (n_W, n_L), where $n_W = \{0, 1, \ldots\}$, $n_L = \{0, 1, \ldots, N_L\}$. Wi-Fi sessions do not impose any requirements on the minimum throughput while LTE sessions are associated by the minimum bit rate, R_L. The state transition diagram of the system is shown in Fig. 4 while the system of equilibrium equations takes the following form

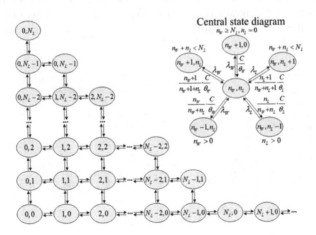

Fig. 4. The state diagram of LAA system with elastic traffic without QoS violation.

$$p(n_W, n_L)[\lambda_W.(1(n_W \geq N_L, n_L = 0) + 1(n_W + n_L < N_L))$$
$$+ \lambda_L.1(n_W + n_L < N_L) + \frac{Cn_L}{(n_W + n_L)\theta_L}.1(n_L > 0)$$
$$+ \frac{Cn_W}{(n_W + n_L)\theta_W}.1(n_W > 0)] = p(n_W, n_L - 1)\lambda_L.1(n_L > 0)$$
$$+ p(n_W - 1, n_L)\lambda_W.1(n_W > 0) + p(n_W + 1, 0)\frac{C}{\theta_W}.1(n_W \geq N_L, n_L = 0) \qquad (8)$$
$$+ p(n_W, n_L + 1)\frac{C(n_L + 1)}{(n_W + n_L + 1)\theta_L}.1(n_W + n_L < N_L)$$
$$+ p(n_W + 1, n_L)\frac{C(n_W + 1)}{(n_W + n_L + 1)\theta_W}.1(n_W + n_L < N_L),$$

where $(p(n_W, n_L))_{(n_W, n_L) \in X} = \mathbf{p}$ is the stationary state probability distribution.

Similarly, as for the model with streaming traffic, considered in the Sect. 3.1, the probability distribution $p(n_W, n_L), (n_W, n_L) \in \mathbf{X}$ for the model with elastic traffic without QoS violation can be represented in the multiplicative form [15]:

$$p(n_W, n_L) = \left(\sum_{i=1}^{\infty} \sum_{j=1}^{N_L - i} \frac{(\rho_W)^i}{i!} \frac{(\rho_L)^j}{j!} (i+j)! \right)^{-1} \frac{\rho_W^{n_W} \rho_L^{n_L}}{n_W!.n_L!} (n_W + n_L)!, \qquad (9)$$

where $\rho_W = \lambda_W \theta_W / C$ and $\rho_L = \lambda_L \theta_L / C$ are offered loads for the Wi-Fi and LTE sessions, and $(n_W + n_L)! = (n_W + n_L)(n_W - 1 + n_L) \cdot \ldots \cdot n_L(n_L - 1) \cdot \ldots \cdot 1$.

3.4   Elastic Traffic with QoS Violation

We finally address the model with elastic Wi-Fi and LTE sessions in which QoS violation may occur. Similar to the model discussed in Sect. 3.2 the particularity of the system consists in absence of Wi-Fi sessions blocking and presence of LTE sessions interruption in case of lack of free resources for the Wi-Fi session's servicing.

The state transition diagram is shown in Fig. 5 while the system of equilibrium equations takes the following form:

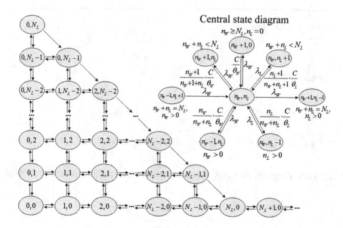

Fig. 5. The state diagram of the LAA system with elastic traffic with QoS violation.

$$p(n_W, n_L)[\lambda_W.(1(n_W \geq N_L, n_L = 0) + 1(n_W + n_L < N_L)$$
$$+ 1(n_W + n_L = N_L, n_L > 0) + \lambda_L.1(n_W + n_L < N_L)$$
$$+ \frac{Cn_L}{(n_W + n_L)\theta_L}.1(n_L > 0) + \frac{Cn_W}{(n_W + n_L)\theta_W}.1(n_W > 0)]$$
$$= p(n_W, n_L - 1)\lambda_L.1(n_L > 0) + + p(n_W - 1, n_L)\lambda_W.1(n_W > 0)$$
$$+ p(n_W - 1, n_L + 1)\lambda_W.1(n_W + n_L = N_L, n_W > 0) \qquad (10)$$
$$+ p(n_W + 1, 0)\frac{C}{\theta_W}.1(n_W \geq N_L, n_L = 0)$$
$$+ p(n_W, n_L + 1)\frac{C(n_L + 1)}{(n_W + n_L + 1)\theta_L}.1(n_W + n_L < N_L)$$
$$+ p(n_W + 1, n_L)\frac{C(n_W + 1)}{(n_W + n_L + 1)\theta_W}.1(n_W + n_L < N_L),$$

where $(p(n_W, n_L))_{(n_W, n_L) \in \mathbf{X}} = \mathbf{p}$ is the stationary state probability distribution.

Similarly to the model with streaming traffic, considered in the Sect. 3.2, the CTMC $\mathbf{X}(t)$ of the model with elastic traffic is not reversible implying that there is no multiplicative form for the stationary probability distribution $p(n_W, n_L), (n_W, n_L) \in \mathbf{X}$. However, one can numerically compute it as the solution of (10), written in the form (7), where the elements $a((n_W, n_L)(n'_W, n'_L))$ of the generator $\mathbf{A}$ are defined as:

$$a\left((n_W, n_L)\left(n'_W, n'_L\right)\right) = \begin{cases} \lambda_W, n'_W = n_W + 1, n'_L = n_L - 1, n_L > 0, n_W + n_L = N_L \\ \quad \text{or } n'_W = n_W + 1, n'_L = n_L, n_W + n_L < N_L \\ \quad \text{or } n'_W = n_W + 1, n'_L = n_L = 0, n_W \geq N_L, \\ \lambda_L, \qquad n'_W = n_W, n'_L = n_L + 1, n_W + n_L < N_L, \\ \frac{n_W}{n_W + n_L}.\frac{C}{\theta_W}, \quad n'_W = n_W - 1, n'_L = n_L, n_W > 0, \\ \frac{n_L}{n_W + n_L}.\frac{C}{\theta_L}, \quad n'_W = n_W - 1, n'_L = n_L, n_L > 0, \\ *, \qquad\qquad n'_W = n_W, n'_L = n_L, \\ 0. \qquad\qquad \text{otherwise,} \end{cases}$$

where $* = -[\lambda_W.1(n_W \geq N_L) + \lambda_W.1(n_W + n_L < N_L) + \lambda_L.1(n_W + n_L < N_L) + \lambda_W.1 (n_W + n_L = N_L, n_L > 0) + \frac{Cn_L}{(n_W + n_L)\theta_L} \cdot 1(n_L > 0) + \frac{Cn_W 1(n_W > 0)}{(n_W + n_L)\theta_W}]$.

3.5 Performance Measures

Having found the probability distribution $p(n_W, n_L), (n_W, n_L) \in \mathbf{X}$, one may compute performance measures of the considered LAA systems. General performance characteristics for both types of models, without and with QoS violation regardless of the type of generated traffic are the blocking probability of LTE sessions, as well as the average number of Wi-Fi and LTE sessions:

- LTE session blocking probability B_L is given by

$$B_L = \sum_{n_W=0}^{N_L-1} p(n_W, N_L - n_W) + \sum_{n_W=N_L}^{\infty} p(n_W, 0). \tag{11}$$

- The average number of Wi-Fi sessions in service is

$$\bar{N}_W = \sum_{n_W=1}^{N_L-1} n_W \sum_{n_L=0}^{N_L-n_W} p(n_W, n_L) + \sum_{n_W=N_L}^{\infty} n_W p(n_W, 0). \tag{12}$$

- The average number of LTE sessions in service is

$$\bar{N}_L = \sum_{n_L=1}^{N_L} n_L \sum_{n_W=0}^{N_L-n_L} p(n_W, n_L). \tag{13}$$

The additional characteristic of the model without QoS violation is the blocking probability of Wi-Fi sessions. It is computed as provided below

$$B_W = \sum_{n_W=0}^{N_L-1} p(n_W, N_L - n_W). \tag{14}$$

The additional characteristic of the model with QoS violation is the interruption probability of LTE sessions. We consider two types of interruption probability: I_{L_1} – the probability that one of the LTE sessions will be interrupted, and I_{L_2} – the probability that the particular LTE session will be interrupted. The calculation of these probabilities depends on the type of the traffic.

- The interruption probability I_{L_1}

  1. for the LAA system with streaming traffic

$$I_{L_1} = \sum_{n_W=0}^{N_L-1} \frac{\lambda_W}{\lambda_W + \lambda_L + n_W \mu_W .1(n_W > 0) + (N_L - n_W)\mu_L} p(n_W, N_L - n_W); \tag{15}$$

  2. for the LAA system with elastic traffic

$$I_{L_1} = \sum_{n_W=0}^{N_L-1} \frac{\lambda_W}{\lambda_W + \lambda_L + \frac{C n_W}{N_L \theta_W} .1(n_W > 0) + \frac{C(N_L - n_W)}{N_L \theta_L}} p(n_W, N_L - n_W). \tag{16}$$

- The interruption probability I_{L_2}

  1. for the LAA system with streaming traffic

$$I_{L_2} = \sum_{n_W=0}^{N_L-1} \frac{1}{N_L - n_W} \cdot \frac{\lambda_W}{\lambda_W + \lambda_L + n_W \mu_W .1(n_W > 0) + (N_L - n_W)\mu_L} p(n_W, N_L - n_W);$$

$$\tag{17}$$

2. for the LAA system with elastic traffic

$$I_{L_2} = \sum_{n_W=0}^{N_L-1} \frac{1}{N_L - n_W} \frac{\lambda_W}{\lambda_W + \lambda_L + \frac{Cn_W}{N_L\theta_W}.1(n_W > 0) + \frac{C(N_L-n_W)}{N_L\theta_L}} p(n_W, N_L - n_W). \quad (18)$$

4 Numerical Analysis

In this section we provide sample numerical results. We start with models having streaming traffic and then proceed to those characterized by the elastic traffic demands.

4.1 Models with Streaming Traffic

We consider a channel with the raw data rate $C = 40$ Mbps. Session holding time is 2 s for Wi-Fi and 3 s for LTE. The arrival rates of Wi-Fi sessions were chosen to be 1, 5 and 10 sessions per second. The bit rate R_L for LTE session is 1 Mbps. Fig. 6 illustrates LTE and Wi-Fi sessions blocking probabilities as a function of LTE sessions arrival rate for two considered systems with streaming traffic. As one may observe in Fig. 6a, for $\lambda_W = 1$ the LTE session blocking probability remains the same for the whole range of LTE arrival rates. However, as λ_W increases the system with guarantees is associated with lower blocking probability. This behavior is explained by the fact, that attempting to maintain throughput guarantees this system may also interrupt LTE sessions already accepted to the system. For higher values of λ_W these interruptions happens more frequently. The Wi-Fi session drop probability illustrated in Fig. 6b shows predictable behavior as LAA CAC does not have any control over Wi-Fi sessions.

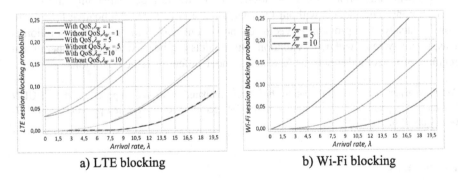

a) LTE blocking   b) Wi-Fi blocking

Fig. 6. LTE and Wi-Fi session blocking probability for LAA models with streaming traffic.

4.2 Models with Elastic Traffic

We illustrate metrics associated with models having elastic traffic demands using the same raw channel rate of 40 Mbps and arrival intensities of Wi-Fi sessions. The mean sizes of elastic sessions generated by Wi-Fi or LTE type sessions is set to 2 MB and 3 MB, respectively. The LTE and Wi-Fi session blocking probabilities are illustrated in Fig. 7. As one may observe in Fig. 7a, the difference between LTE session blocking probabilities for systems with and without throughput guarantees is negligible for all considered values of λ_W. However, this difference may become larger for other values of mean Wi-Fi and LTE session sizes. Wi-Fi session blocking probabilities, expectedly, shows exponential behavior.

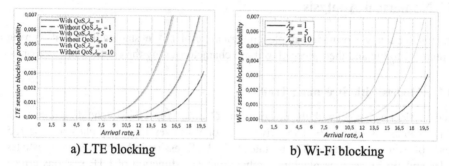

a) LTE blocking   b) Wi-Fi blocking

Fig. 7. LTE and Wi-Fi session blocking probability for LAA models with elastic traffic.

5   Conclusion

The fine tuning of LAA technology requires careful understanding of various trade-offs and dependencies involved in Wi-Fi and LTE coexistence. In this paper, we have developed analytical models for LAA system operating in unlicensed band and using scheduled-based access as a coexistence strategy. We addressed the cases of LAA systems with and without QoS violations for two traffic types, streaming and elastic data sessions. The models without QoS violation allows for multiplicative form solution for steady-state probability vector of the system while only numerical solution is feasible for systems with QoS violation. Our models can be used to identify operational regimes of scheduled-based LAA systems under different traffic conditions.

Acknowledgement. The publication has been prepared with the support of the "RUDN University Program 5-100" and funded by RFBR according to the research projects No. 18-37-00231 and No. 16-07-00766.

References

1. Cisco visual networking index: global mobile data traffic forecast, 2016–2021. Cisco white paper (2017)
2. Heuveldop, N.: Ericsson mobility report. Ericsson AB, Technol. Emerg. Business. – Technical report, EAB-17 5964 (2017)

3. Boulos, M.N.K., Lu, Z., Guerrero, P., Jennett, C., Steed, A.: From urban planning and emergency training to Pokémon Go: applications of virtual reality GIS (VRGIS) and augmented reality GIS (ARGIS) in personal, public and environmental health (2017)
4. Shannon, C.E.: The mathematical theory of communication, pp. 306–317 (1997)
5. GPP TR 38.912: Study on new radio (NR) access technology. Technical report, Release 14 (2017)
6. Rappaport, T.S., et al.: Millimeter wave mobile communications for 5G cellular: it will work! IEEE Access 1, 335–349 (2013)
7. GPP TR 38.901: Study on channel model for frequencies from 0.5 to 100 GHz. – Technical report, Release 14 (2017)
8. Gapeyenko, M., et al.: On the temporal effects of mobile blockers in urban millimeter-wave cellular scenarios. IEEE Trans. Veh. Technol. (2017)
9. GPP TR.36.889: Feasibility study on licensed-assisted access to unlicensed spectrum. Technical report, Release 13 (2016)
10. Kusashima, N., Nogami, T., Takahashi, H., Yokomakura, K., Imamura, K.: A listen before talk algorithm with frequency reuse for LTE based licensed assisted access in unlicensed spectrum. In: 83rd IEEE, Vehicular Technology Conference (VTC Spring), pp. 1–5 (2016)
11. Wang, H., Kuusela, M., Rosa, C., Sorri, A.: Enabling frequency reuse for licensed-assisted access with listen-before-talk in unlicensed bands. In: 83rd IEEE, Vehicular Technology Conference (VTC Spring), pp. 1–5 (2016)
12. Lien, S.-Y., Lee, J., Liang, Y.-C.: Random access or scheduling: optimum LTE licensed-assisted access to unlicensed spectrum. IEEE Commun. Lett. 20(3), 590–593 (2016)
13. Han, S., Liang, Y.-C., Chen, Q., Soong, B.-H.: Licensed-assisted access for LTE in unlicensed spectrum: a MAC protocol design. IEEE J. Sel. Areas Commun. 34(10), 2550–2561 (2016)
14. Maule, M., Moltchanov, D., Kustarev, P., Komarov, M., Andreev, S., Koucheryavy, Y.: Delivering fairness and QoS guarantees for LTE/Wi-Fi coexistence under LAA operation. IEEE Access 6, 7359–7373 (2018)
15. Gudkova, I., et al.: Service failure and interruption probability analysis for Licensed Shared Access regulatory framework. In: 7th International Congress on Ultra Modern Telecommunications and Control Systems (ICUMT), pp. 123–131 (2015)

Kinetic Approach to Elasticity Analysis of D2D Links Quality Indicators Under Non-stationary Random Walk Mobility Model

Andrey K. Samuylov[1,2(✉)], Anastasia Yu. Ivchenko[3],
Yu. N. Orlov[1,3], Dmitri A. Moltchanov[1,2], Ekaterina V. Bobrikova[1],
Yuliya V. Gaidamaka[1,4], and Vsevolod S. Shorgin[4]

[1] Department of Applied Probability and Informatics, Peoples' Friendship,
University of Russia (RUDN University), 6 Miklukho-Maklaya St., 117198
Moscow, Russia
{samuylov_ak, molchanov_da, bobrikova_ev, gaydamaka_yuv}
@rudn.university, yuno@kiam.ru
[2] Department of Electronics and Communications Engineering, Tampere
University of Technology, 10 Korkeakoulunkatu, 33720 Tampere, Finland
[3] Department of Kinetic Equations, Keldysh Institute of Applied Mathematics of
RAS, Miusskaya Sq. 4, Moscow 125047, Russian Federation
[4] Institute of Informatics Problems, Research Center "Computer Science and
Control" of the Russian Academy of Sciences, Vavilov St. 44-2, Moscow
119333, Russian Federation
vshorgin@ipiran.ru

Abstract. In device-to-device communications, the link quality indicators, such as signal-to-interference ratio (SIR) is heavily affected by mobility of users. Conventionally, the mobility model is assumed to be stationary. In this paper, we use kinetic theory to analyze evolution of probability distribution function parameters of SIR in D2D environment under non-stationary mobility of users. Particularly, we concentrate on elasticity of the SIR moments with respect to parameters of Fokker-Planck equation. The elasticity matrix for average SIR value, SIR variance and time periods, when SIR values is higher than a certain threshold are numerically constructed. Our numerical results demonstrate that the main kinetic parameter affecting SIR behavior is diffusion coefficient. The influence of the drift is approximately ten times less.

Keywords: Wireless communications · Device-to-device communications
Kinetic equation · Non-stationary random walk · Mathematical modeling
SIR distribution

1 Introduction

The use of direct device-to-device (D2D) communications is a prominent way to improve spatial frequency reuse in wireless communications system. Standardized by 3GPP in Release 13, D2D communications are expected to become integral part for future cellular systems.

© Springer Nature Switzerland AG 2018
O. Galinina et al. (Eds.): NEW2AN 2018/ruSMART 2018, LNCS 11118, pp. 224–235, 2018.
https://doi.org/10.1007/978-3-030-01168-0_21

The signal-to-interference ratio (SIR) is a crucial metric describing the quality of the link that depends on the propagation environment and distance between communicating entities. In D2D environment, the distance between communicating entities is affected by mobility of users involved in communications. Thus, when characterizing performance of D2D links one must explicitly track the mobility of users. Particularly, if the SIR fall below a certain threshold that depends on the sensitivity of a receiver as technology of interest two communicating entities experience outage conditions and service provisioning is no longer possible. Thus, the development of methods for the analysis of the duration of the connection availability and outage periods is of special interest. However, in most studies of D2D communications performed so far stationary mobility models have been assumed. However, in practice, mobility of users may be far from stationary assumption.

One of the way to capture various mobility patterns, including many well-known stationary models as well as non-stationary ones, is to use kinetic equation. This idea was originally proposed in [1], where the general methodology has been described and the dynamics of the SIR are shown for various special cases of users' mobility ranging from rectilinear motion to Brownian motion. Further, in [2] the kinetic based mobility model was extended to the general cases with a wide range of mobility characteristics including conventional stationary, fractal and even non-stationary ones. The analysis of time-dependent evolution of mean, variance and coefficient of variation of SIR metric demonstrated that under non-stationary motion of communicating entities the SIR may surprisingly exhibit stationary behavior. A non-trivial aspect of the theory is that the distribution function of SIR in the general case of a non-stationary random walk is not analytically deducible from the supposedly known distribution function of the user positions. This is due to the fact, that SIR is a complex nonlinear functional, which leads to significant technical complications of the evolution equations of the moments of the distribution function of this index. Thus, to model the trajectories of non-equidistant and non-stationary time series, when both the event stream and the set of event values are non-stationary, a software package was developed in [3].

The key point in the analysis of SIR behavior is the study of SIR over the ensemble of trajectories for given model parameters, for example the shape of drift, the diffusion coefficient, and the trajectory density. However, its behavior is heavily affected by parameters of Fokker-Planck equation – so-called drift and diffusion making sensitivity analysis a critical problem of interest. This problem is closely connected with the problem of stochastic control to ensure the reliability of wireless communication between users. The aim of the present article is to investigate these dependences theoretically and numerically.

In this paper, we perform sensitivity analysis of SIR under non-stationary mobility of users. Particularly, we use the kinetic theory to describe the evolution of various SIR metrics including the mean SIR of the ensemble of random trajectories, the variance of SIR and the normalized average. The letter is the ratio of the mean value to the standard deviation and can be considered as a stability factor of a D2D link in the case when the random walk of the users is non-stationary. The sensitivity is performed using the notion of elasticity which is defined as a logarithmic derivative of the investigated metric P with respect to the parameter Q. This value shows the relative variation of P (i.e. $\Delta P/P$) while the relative variation of Q is unit. We assume that users' random walk is a random process with independent differences and corresponding distribution function (DF) is a stable one. In this case, the movement of users can be modeled using the Fokker-Planck equation. In our numerical study we investigate the effect of the model's parameters on SIR identifying those providing the most important contribution. Together, this numerical analysis makes it possible to develop a methodology for obtaining optimal estimates for control parameters in the D2D model in direct connection networks.

The rest of the paper is organized as follows. In Sect. 2 we introduce the kinetic model for SIR analysis. Further in Sect. 3, we define the kinetic model is non-stationary random walk. Sensitivity analysis is performed in Sect. 4. Conclusions are drawn in the last section.

2 Kinetic Model for SIR Analysis

2.1 The Kinetic Approach

SIR is defined as a ratio of the useful signal power at a given spatial point to sum of signal powers from other sources. In open space environment the path loss coefficient is 2 (see e.g. [4, 5]) implying that the signal power is proportional to r^{-2}, where r is a non-zero distance between given transmitter and receiver. Thus, the SIR value is defined as $r^{-2}/\sum_k r_k^{-2}$, where r_k is a distance between given receiver of number k and other transmitters. As this metric is nonlinear with respect to users' location distribution function, the theoretical results can be obtained in some relatively simple cases, e.g. for a given spatial users distribution without any moving effects. However, in practice, user mobility patterns are often non-stationary, so that the appropriate method for modeling of statistical characteristics of corresponding ensemble of trajectories should be developed.

Our method of SIR analysis is based on simulation of devices trajectories with the use of Fokker-Planck kinetic equation for distribution function density (DFD) of independent coordinates differences. The main argument of this approach is that the SIR is non-linear and non-monotonic functional of DFD of devices positions. Thus, in general, the direct expression of SIR DFD does not exist. However, assuming that the spatial averaging over ensemble of device trajectories and averaging positions of an arbitrary device over sufficiently large time period are asymptotically the same, we can theoretically investigate the SIR DFD by means of kinetic equation for device coordinates DFD. Below we describe the proposed methodology.

When the independent coordinates differences have the same DFD as coordinates themselves, SIR average value can be expressed as a non-linear functional of coordinates DFD. Let

$$X(t) = X(0) + \sum_{k=1}^{t} x_k, \quad Y(t) = Y(0) + \sum_{k=1}^{t} y_k, \quad Z(t) = Z(0) + \sum_{k=1}^{t} z_k,$$

are trajectory coordinates for a certain device in a certain instant of time t, and $\mathbf{r}_k = \{x_k, y_k, z_k\}$ are corresponding coordinates differences for the time step k.

Let DFD $f(\mathbf{r}, t)$ obey to the Fokker-Planck equation, i.e.,

$$\frac{\partial f}{\partial t} + div(\mathbf{u}(\mathbf{r}, t)f) - \frac{\lambda(t)}{2}\Delta f = 0, \tag{1}$$

where Δ is Laplace differential operator. Boundary conditions are assumed to be zero.

The drift parameter $\mathbf{u}(\mathbf{r}, t)$ and non-stationary diffusion coefficient $\lambda(t) \geq 0$ are assumed to be known. In practice, they can be determined through the empirical sample mutual DFD $F(\mathbf{r}, \mathbf{v}, t)$ of values of $\mathbf{r}$ and its first differences $\mathbf{v}$, so that

$$f(\mathbf{r}, t) = \int F(\mathbf{x}, \mathbf{v}, t)d\mathbf{v}, \quad f(\mathbf{r}, t)\mathbf{u}(\mathbf{r}, t) = \int \mathbf{v}F(\mathbf{x}, \mathbf{v}, t)d\mathbf{v}, \quad \lambda(t) = \frac{d\sigma^2}{dt} - 2\mathrm{cov}_{r,v},$$

$$\sigma^2(t) = \int (\mathbf{r} - \mathbf{m}(t))^2 f(\mathbf{r}, t)d\mathbf{r}, \quad \mathbf{m}(t) = \int \mathbf{r}f(\mathbf{r}, t)d\mathbf{r}. \tag{2}$$

Observe that (2) directly follows from the evolution equation of DFD. If we differentiate the variance with respect to the time, we obtain

$$\begin{aligned}\frac{d\sigma^2}{dt} &= \frac{d}{dt}\int (\mathbf{r} - \mathbf{m}(t))^2 f(\mathbf{r}, t)d\mathbf{r} \\ &= -2\frac{d\mathbf{m}}{dt}\int (\mathbf{r} - \mathbf{m}(t))f(\mathbf{r}, t)d\mathbf{r} + \int (\mathbf{r} - \mathbf{m}(t))^2 \frac{\partial f(\mathbf{r}, t)}{\partial t}d\mathbf{r}.\end{aligned}$$

The first term in the right part of this equation is equal to zero, as $\mathbf{m}(t) = \int \mathbf{r}f(\mathbf{r}, t)d\mathbf{r}$, while the second term can be transformed with the use of kinetic Eq. (1). After integration by parts we have arrive at

$$\begin{aligned}\int (\mathbf{r} - \mathbf{m}(t))^2 \frac{\partial f(\mathbf{r}, t)}{\partial t}d\mathbf{r} &= -\int (\mathbf{r} - \mathbf{m}(t))^2 \left(div(\mathbf{u}(\mathbf{r}, t)f) - \frac{\lambda(t)}{2}\Delta f\right)d\mathbf{r} = \\ &= 2\int (\mathbf{r} - \mathbf{m}(t))\mathbf{u}(\mathbf{r}, t)f(\mathbf{r}, t)d\mathbf{r} + \lambda(t)\int f(\mathbf{r}, t)d\mathbf{r} = 2\mathrm{cov}_{r,v} + \lambda(t).\end{aligned}$$

2.2 Interference Assessment

Now we can use (1) for derivation of the evolution equation for interference. Let $\varphi(|\mathbf{r}|) = 1/|\mathbf{r}|^2$. Then, the field of interference at the point $\mathbf{r}$ is determined by

$$U(r, t) = \int \varphi(|\mathbf{r} - \mathbf{r}'|)f(\mathbf{r}', t)d\mathbf{r}' = \int \frac{f(\mathbf{r}', t)}{|\mathbf{r} - \mathbf{r}'|^2}d\mathbf{r}', \tag{3}$$

where $\mathbf{r}$ is a position of the first device (transmitter) with respect to second device (receiver), and $\mathbf{r}'$ denotes the respective positions of other devices. Similarly to the previous transformation we can show that

$$\frac{\partial U(r,t)}{\partial t} = \int \frac{1}{|\mathbf{r}-\mathbf{r}'|^2} \frac{\partial f(\mathbf{r}',t)}{\partial t} d\mathbf{r}',$$

where the term $\frac{\partial f}{\partial t}$ is expressed from the Eq. (1). Thus, after integration by parts we replace the term $\varphi(|\mathbf{r}-\mathbf{r}'|)$ by the derivative with respect to $\mathbf{r}$. We obtain the following

$$\frac{\partial U}{\partial t} = \frac{\lambda}{2}\Delta U - div\mathbf{J}, \quad \mathbf{J} = \int \varphi(|\mathbf{r}-\mathbf{r}'|)\mathbf{u}(\mathbf{r}',t)f(\mathbf{r}',t)d\mathbf{r}', \tag{4}$$

showing that the average interference field changes over time in the same way as DFD, i.e., according to the diffusion equation with the same parameters as in (1).

Let the total number of devices is N. Then the average over ensemble SIR value for two arbitrary devices is defined as

$$s(t) = \frac{1}{N} \int \frac{\varphi(r)}{U(r,t)} f(\mathbf{r},t)d\mathbf{r}. \tag{5}$$

Since $U(r,t)$ is defined as (3), the average SIR (5) is non-linear functional of coordinates DFD of users. The evolution equation for average SIR can still be obtained in the following form

$$N\frac{ds}{dt} = \int \frac{\varphi(r)}{U(r,t)} \frac{\partial f(\mathbf{r},t)}{\partial t} d\mathbf{r} - \int \frac{\varphi(r)}{U^2(r,t)} \frac{\partial U(r,t)}{\partial t} f(\mathbf{r},t)d\mathbf{r}.$$

The second term in this equation is transformed with the use of (4), and in the first term the partial derivative of DFD by time is expressed from (1):

$$\int \frac{\varphi(r)}{U(r,t)} \frac{\partial f(\mathbf{r},t)}{\partial t} d\mathbf{r} = - \int \frac{\varphi(r)}{U(r,t)} div(\mathbf{u}f)d\mathbf{r} + \frac{\lambda}{2}\int \frac{\varphi(r)}{U(r,t)} \Delta f d\mathbf{r}.$$

After integration by parts we obtain

$$\int \frac{\varphi(r)}{U(r,t)} \frac{\partial f(\mathbf{r},t)}{\partial t} d\mathbf{r} = \int f(\mathbf{r},t) \cdot \left(\mathbf{u}\nabla + \frac{\lambda}{2}\Delta\right)\left(\frac{\varphi}{U}\right)d\mathbf{r}.$$

As a result the evolution equation for average SIR has the form:

$$N\frac{ds}{dt} = \int \left(\left(\mathbf{u}\nabla + \frac{\lambda}{2}\Delta\right)\left(\frac{\varphi}{U}\right)\right)f(\mathbf{r},t)d\mathbf{r} - \int \frac{\varphi}{U^2}\left(\frac{\lambda}{2}\Delta U - div\mathbf{J}\right)f(\mathbf{r},t)d\mathbf{r}. \tag{6}$$

This equation is non-linear with respect to distribution function, i.e. with respect to the density of the ensemble of sample trajectories.

2.3 Sensitivity Coefficients

Let Q_i, $i = 1, \ldots, n$ be the input parameters and P_j, $j = 1, \ldots, m$ be the output parameters. The sensitivity coefficient of parameter P_j with respect to parameter Q_i is defined as logarithmic derivative, i.e.,

$$\alpha_{ji} = \frac{\partial \ln P_j}{\partial \ln Q_i}. \tag{7}$$

The set of these coefficients compose the sensitivity matrix $A_{m \times n}$. The norm $\|\Lambda\|$ of matrix $\Lambda_{n \times n} = A^T A$ presents the generalized sensitivity of the model. The determinant $\det \Lambda$ is considered as indicator of independence of input parameters.

For our problem there are five input parameters of the kinetic model,

$$Q_1 = N; \ Q_2 = \lambda; \ Q_3 = u_x; \ Q_4 = u_y; \ Q_5 = u_z. \tag{8}$$

We consider three output parameters. First parameter is average over ensemble SIR values, defined in (5), i.e. $P_1 = s$. The second input parameter is SIR variance, defined as

$$P_2 = \Sigma^2(t) = \frac{1}{N^2} \int\limits_V \left(\frac{\varphi(\mathbf{r})}{U(\mathbf{r}, t)} - \int\limits_V \frac{\varphi(\mathbf{r}')}{U(\mathbf{r}', t)} f(\mathbf{r}', t) d\mathbf{r}' \right)^2 f(\mathbf{r}, t) d\mathbf{r}. \tag{9}$$

The third parameter, $P_3 = T$, is an average time period SIR is above a certain threshold. In what follows, the outage is considered to happen whenever $s(t) < s^* = 0,01$. The value of T is obtained from numerical simulation. For the sake of simplicity, we assume that all coordinates are independent and have the same distribution function. Thus, in fact, we can restrict ourselves to three input parameter, letting $Q_3 = u = u_x$.

Further, from (5) and (9) it follows, that

$$\frac{\partial P_1}{\partial Q_1} = \frac{\partial s}{\partial N} = -\frac{s}{N}, \ \frac{\partial P_2}{\partial Q_1} = \frac{\partial \Sigma^2}{\partial N} = -\frac{2\Sigma}{N},$$

and hence the corresponding elasticity coefficients are known exactly: $\alpha_{11} = -1$, $\alpha_{21} = -2$.

The rest of the coefficients α_{ji} must be determined numerically.

3 The Kinetic Model of Non-stationary Random Walk

We propose the following method for computer simulation of non-stationary random walk of users. The base model is Fokker-Planck Eq. (1). The numerical solution of this equation for any given initial conditions (e.g., uniform spatial distribution) is constructed for the time horizon τ. The unit time step is considered. Thereafter, for each time step $k = 1, 2, \ldots, \tau$ a random coordinates difference x_k (and also similarly to y_k and z_k) is generated from DF, which is represented as

$$F(x,t) = \int\limits_0^x f(y,t)dy.$$

Let the solution of Eq. (1) is represented by a histogram $f_j(t)$, where j is a bin number. Then, the continuous strictly monotonic DF has the following form

$$F(x,t) = (nx - j) \cdot f_{j+1}(t) + \sum_{k=1}^{j} f_k(t), \quad x \in [(j-1)/n; \, j/n], \quad j = 1 \div n. \quad (10)$$

Further, we generate stationary series of numbers $\{\xi_k\}$ of size τ, uniformly distributed at $[0; 1]$. The corresponding series with distribution $F(x,t)$ from Eq. (10) is based on the quintile function moving in a sliding window of length τ, i.e.,

$$\xi_k = F_N(x_k, k) \tag{11}$$

We consider the case, when DFD in R^3 is factorized: $\tilde{f}(\mathbf{r},t) = f(x,t)f(y,t)f(z,t)$. We use tilde to distinguish single-coordinate DFD from three-dimensional DFD.

By generating a set of N uniformly distributed samples, denoted by the subscript $i = 1, 2, \ldots, N$, we obtain the corresponding set of trajectories, that can be considered as an ensemble of solutions of the kinetic Eq. (1). Such ensembles can be generated for various values of input parameters that can be further used to numerically determine the sensitivity matrix (7). The results of this modeling are provided in the next section.

4   System Model of LAA Schedule-Based System

A numerical sensitivity analysis for a randomly selected pair of users is carried out in two stages. At the first step, the effect of the model parameters is investigated, selecting from them the one whose influence is most significant. In the second step, we concentrate on parameters in isolation, revealing their effects on the metrics of interest.

Following the procedure described in Sect. 4 we construct the evolution of DFD according to Eq. (1) during the time period of 300 units from DFD $f_0(x)$ to DFD $f_n(x)$, $n = 1 \div 5$. The time series are presented in Fig. 1. The distances between DFD f_0 and f_n in L1-norm are equal respectively to 0,15; 0,30; 0,45; 0,60; 0,75. The corresponding drift parameters $u_1(x)$ and $u_5(x)$ for evolution from F_0 to F_1 and from F_0 to F_5 are presented in Fig. 2.

The relative variation of the output scalar parameter P with respect to vector input parameter $\mathbf{u}(x)$ is estimated in terms of distances between corresponding DFDs. The reson is that variations $\delta\mathbf{u}_n(x) = \mathbf{u}_{n+1}(x) - \mathbf{u}_n(x)$ and $du = \max\limits_x |\mathbf{u}_{n+1}(x) - \mathbf{u}_n(x)|$ are proportional to $\|f_{n+1}(x) - f_n(x)\|_{L1}$. Then for the point $\mathbf{u}_n$ the sensitivity coefficient can be calculated using finite difference formula, i.e.,

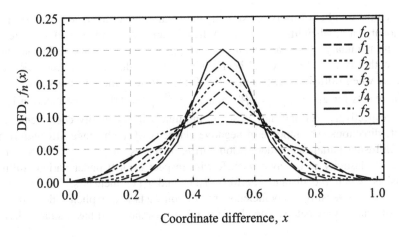

Fig. 1. The probability densities for coordinate differences.

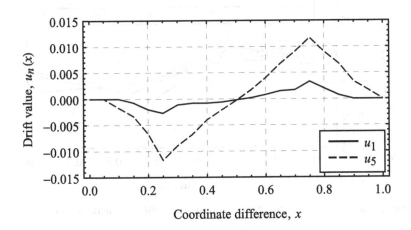

Fig. 2. Illustration of drift parameters.

$$\frac{\partial \ln P}{\partial \ln u_n} = \frac{1}{P(f_n)} \frac{P(f_{n+1}) - P(f_n)}{\|f_{n+1}(x) - f_n(x)\|}. \tag{12}$$

The diffusion parameter λ varies independently from 0.1 to 1, with the step 0.1.

Once (1) is solved for the horizon from $f(x, t = 0)$ to $f(x, t = 300)$ for all coordinates, we obtain DF in (10). We start from DFD $f_0(x)$ and after 300 time steps obtain one of the functions $f_1(x), \ldots, f_5(x)$. For N random trajectories we calculate SIR value by formulas (3) and (5). If the value of $s(k)$ is firstly below the level $s* = 0, 01$, our experiment is finished and corresponding time moment $T = k$ is stored. This experiment is repeated 10^5 times. The collected data are then used to derive time period T, when SIR remains above the defined threshold.

When the number of users is fixed, the maximal value of average SIR corresponds to stationary case, when $|\mathbf{u}| = 0$, $\lambda = 0$. When the area of a random walk is sufficiently large, the boundary effects are negligible. It appears, that drift and diffusion lead to decrease the average SIR. It has been shown in [3], that DF for low SIR values increases, when diffusion coefficient varies from zero to some positive value.

The results for $\alpha_{12} = \frac{\partial \ln P_1}{\partial \ln Q_2} = \frac{\partial \ln s}{\partial \ln \lambda}$ and $\alpha_{13} = \frac{\partial \ln P_1}{\partial \ln Q_3} = \frac{\partial \ln s}{\partial \ln u}$ are presented in Figs. 3 and 4, respectively. As one may observe, the trends in sensitivity coefficients are in opposite directions. The module of negative elasticity of the average SIR with respect to the diffusion parameter under various drifts increases when $\lambda \to 1$. However, the module of elasticity of the average SIR with respect to drift under various diffusion diminishes, when the distance between $f_0(x)$ and $f_n(x)$ increases. It is of special importance that sensitivity coefficients are not constants and explicitly depend on the values of kinetic parameters. The elasticity of SIR variance (9) has similar behavior.

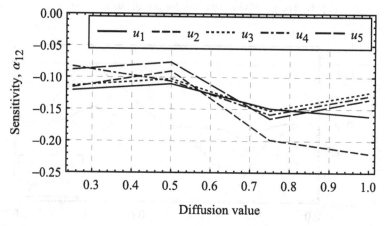

Fig. 3. Sensitivity of average SIR with respect to diffusion λ under various drift u

Consider now the sensitivity of average time periods of uninterrupted connectivity, T. It appears, that the distribution of the moment when SIR falls below a threshold is stationary under the fixed number of users. The corresponding DFD is shown in Fig. 5. Thus, the elasticity of T with respect to drift and diffusion are approximately zero. It should be noted that the asymptotic behavior of this empirical distribution has the same $T^{-3/2}$ type as a well-known theoretical result, which follows from the Andersen theorem [8, 9] for the processes with independent derivatives and stable distributions. Therefore, we can assume that the time-series of SIR values, generated by our method, has an identical property.

Now let us consider the value $\alpha_{31} = \frac{\partial \ln T}{\partial \ln N}$ under fixed drift $\mathbf{u}$ and diffusion λ. The value of T depends on average SIR and on the stability coefficient $\mu = \frac{s}{\Sigma}$. This coefficient represents a characteristic width of SIR variation. Denote $T = T(s, \mu)$ and observe that

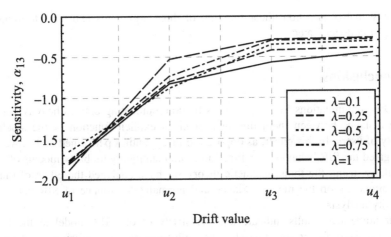

Fig. 4. Sensitivity of average SIR with respect to drift u under various diffusion λ

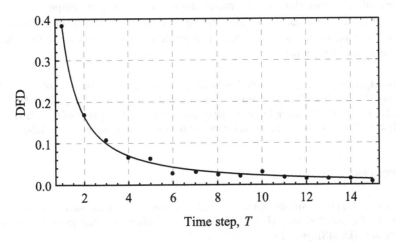

Fig. 5. Simulated DFD of the first instant of break of connection

$$\frac{\partial T}{\partial N} = \frac{\partial T}{\partial s}\frac{\partial s}{\partial N} + \frac{\partial T}{\partial \mu}\frac{\partial \mu}{\partial N}.$$

However, from (5) and (9) it follows, that μ does not depend on N. Then, we have

$$\alpha_{31} = \frac{\partial \ln T}{\partial \ln N} = \frac{N}{T}\frac{\partial T}{\partial s}\frac{\partial s}{\partial N} = \frac{s}{T}\frac{\partial \ln s}{\partial \ln N} = \frac{s}{T}\alpha_{11} = -\frac{s}{T}. \tag{13}$$

We may conclude that although the principal parameter is N, in the region of low drift velocity the sensitivity is still sufficiently large. Thus, non-stationary kinetic equation produces new effects compared to stationary situation. We also note that the sensitivity matrix is a non-linear function of three arguments and its determinant is strongly non-degenerated.

5 Conclusions

In D2D communications SIR experiences by communicating entities heavily depends on the distance between them that in turn in an explicit function of the stochastic mobility of users. In this paper, as opposed to many studies performed so far, we have investigated the SIR behavior as a function of non-stationary mobility models of users. Particularly, using the tools of kinetic theory we have analyzed the effect of random walk parameters on the mean, variance and normalized variance of SIR performing sensitivity analysis.

Our numerical results indicate that sensitivity of our SIR model to the kinetic coefficients is different. Particularly, the influence of the diffusion coefficient is approximately 10 times higher compared to the drift. Thus, one may conclude that the accuracy of drift modeling is of crucial importance, however, empirical diffusion coefficient must be estimated carefully. Additional, we emphasize, even in those cases when users motion has non-stationary distribution the distribution of the time till outage happens is stationary.

Acknowledgement. The publication has been prepared with the support of the "RUDN University Program 5-100" and funded by RFBR according to the research projects No. 17-07-00845, 18-37-00380. This work has been developed within the framework of the COST Action CA15104, Inclusive Radio Communication Networks for 5G and beyond (IRACON).

References

1. Fedorov, S.L., et al.: SIR distribution in D2D environment with non-stationary mobility of users. In: Paprika, Z.Z., et al. (Eds.) Proceedings 31st European Conference on Modelling and Simulation (ECMS), pp. 720–725 (2017)
2. Ivchenko, A., Orlov, Yu., Samouylov, A., Molchanov, D., Gaidamaka, Yu.: Characterizing time-dependent variance and coefficient of variation of SIR in D2D connectivity. In: Galinina, O., Andreev, S., Balandin, S., Koucheryavy, Y. (eds.) NEW2AN/ruSMART/NsCC -2017. LNCS, vol. 10531, pp. 526–535. Springer, Cham (2017). https://doi.org/10.1007/978-3-319-67380-6_49
3. Orlov, Yu., Pleshakov, R., Gaidamaka, Yu.: Software complex for modeling non-stationary event flows. In: Proceedings of the 9th International Congress on Ultra Modern Telecommunications and Control Systems (ICUMT), pp. 112–116 (2017)
4. Petrov, V., Moltchanov, D., Kustarev, P., Jornet, J.M., Koucheryavy, Y.: On the use of integral geometry for interference modeling and analysis in wireless networks. IEEE Commun. Lett. **20**, 2530–2533 (2016)
5. Samuylov, A., et al.: Analytical performance estimation of network-assisted D2D communications in urban scenarios with rectangular cells. In: Transactions on Emerging Telecommunications Technologies, vol. 28, no. 2 (2017)
6. Orlov, Yu.N, Fedorov, S.L.: Sample distribution function construction for non-stationary time-series forecasting. Matem. Mod. **29**(5), 61–72 (2017)

7. Orlov, Yu.N.: Kinetic Methods for Investigating Time-Dependent Time Series, p. 276. MIPT, Moscow (2014)
8. Andersen, S.E.: On sums of symmetrically dependent random variables. Scand. Aktuarietik-skr. **36**, 123–138 (1953)
9. Skorokhod, A.V.: Random Processes with Independent Derivatives, p. 280. Nauka, Moscow (1964)

The Phenomenon of Secondary Flow Explosion in Retrial Priority Queueing System with Randomized Push-Out Mechanism

Maria Korenevskaya, Oleg Zayats, Alexander Ilyashenko[✉],
and Vladimir Muliukha

Peter the Great St.Petersburg Polytechnic University, St.Petersburg, Russia
korenevskayamasha@gmail.com, zay.oleg@gmail.com,
ilyashenko.alex@gmail.com, vladimir@mail.neva.ru

Abstract. We consider a single-server queueing system with finite buffer size, Poisson arrivals and exponentially distributed service time. If the arriving customer finds the completely filled queue of the system, the customer joins a special retrial waiting group (called the orbit) and after a random period of time that has an exponential distribution tries to come to the system again. Primary customers take priority over secondary customers. We also introduce the so-called randomized push-out buffer management mechanism. It allows primary customers to push secondary ones out of the system to free up space. Such a queueing system can be reduced to a similar model without retrials, which had been studied by the authors earlier. Using generating functions approach, we obtain loss probabilities for both types of customers. Theoretical results allow to investigate the dependence of the loss probabilities on the main parameters of the model (such as the push-out and retrial probabilities). We considered in details the cases of preemptive and non-preemptive priorities and discovered an interesting phenomenon. When the intensity of the primary flow increases smoothly after it reaches a certain critical value, an avalanche-like increase in the intensity of the secondary flow occurs (up to tens of thousands of times). In other words, there is a kind of "explosion" of the flow of secondary customers. This article is a strictly quantitative study of this phenomenon, which is of great interest in the calculation of telematic devices.

Keywords: Priority queueing system · Retrial queueing system
Randomized push-out mechanism · Poisson arrivals · Exponential service time
Markov process · Steady-state distribution · Finite buffer · Preemptive priority
Non-preemptive priority · Explosion

1 Introduction

The main analytical approach of telematic devices investigation is based on their consideration as a special kind of queueing systems [1]. Telematic device means any device passing a large amount of information through itself. Such information is represented as a flow of network packets and is intended for some processing (for example, switching or filtering). Examples of such devices are firewalls or routers, with both backbones and

© Springer Nature Switzerland AG 2018
O. Galinina et al. (Eds.): NEW2AN 2018/ruSMART 2018, LNCS 11118, pp. 236–246, 2018.
https://doi.org/10.1007/978-3-030-01168-0_22

less-loaded ones. Models of queueing systems, which are necessary for a reasonable description of the network packets processing, prove to be rather complicated [2]. First, they must be multi-flow to reflect the real structure of the data flows. Second, they must take into account the priority of these or other flows in order to mathematically correct reproduce the real dispatch algorithms.

The priority of a particular type of customer (packet) means some of their preference in servicing. There are many possible types of priorities, but the main and most useful are preemptive and non-preemptive priorities [3]. With a preemptive priority, high-priority packets interrupt processing of low-priority ones and push them out from the service channel. Non-preemptive priority just gives the high-priority customers an opportunity to be the first in the queue, without interrupting the processing of low-priority customers. In order to provide the most important customers with the greatest possible advantages, the priority in servicing is reinforced by the push-out mechanism [4]. The push-out mechanism is a kind of priority in the queue. It allows the high-priority packets coming up to a system to push-out the low-priority ones from the buffer when it is completely filled. Such a mechanism generates a dilemma: if it is on, then high-priority packets will prevail in the buffer. If you do not enable it, the buffer will be filled mostly with low-priority packets. Meanwhile, it is well known that the effective functioning of a telematics device requires a certain balance of all types of customer.

To achieve such a balance, a randomized (probabilistic) push-out mechanism was proposed, when a high priority customer pushes-out a low priority one with a certain predetermined probability. The value of this probability can be treated as a control parameter and is used to adapt the telematics device to the network environment. Initially, the randomized push-out mechanism was considered for the case of non-preemptive priority [5, 6]. Subsequently, the authors of this article also studied the preemptive [7, 8], alternating [9] and randomized [10] priorities.

In practical cases it would be very interesting to take into account the presence in the system of retrial customers. The retrial customers can radically change the behavior of a queueing system. This can be explained by the appearance of a large number of secondary customers trying to get into the system again. The authors have previously considered the retrial queueing system with preemptive priority and randomized push-out mechanism in [11].

When considering retrial queueing systems, one new extremely interesting class of problems arises. These are the problems where the priority is granted to the primary packets only, but when the packet is reapplied it becomes a low-priority one. In many cases it is reasonable to assign a priority only to the customers who have been entered into the system for the first time. Recently we have studied such one-flow retrial queueing system with preemptive priority in [12]. Some additional publications on this problem by other researchers can be found in the same author's work. In the present paper, we will consider a combination of retrial queueing model, two main types of priority and randomized push-out mechanism. Randomized push-out mechanism allows to change characteristics of queueing system in a wide range. It makes the fine tuning of the primary and secondary customers balance possible. This balance is important in many practical applications, for example in the problems of robotic complexes controlling in space experiments describing in [11].

In [12], the authors focused on those effects that were previously discovered for two-flow systems without retrials. We are talking about the linear law of loss probabilities, as well as the phenomenon of locking the system relative to the low-priority flow. As it turned out, both of these effects remain valid also for single-flow retrial systems with a probabilistic push-out mechanism. In the present paper, we will describe a qualitatively new effect, which is possible only in single-flow retrial queueing systems. The effect is an avalanche-like increase of the secondary flow which can be observed in such systems under certain conditions. In the paper, the characteristics of this "explosion" of retrial customers are studied in detail and recommendations for its management are given.

2 Preemptive Priority System

Let us consider the queueing system with a single incoming elementary flow of intensity $\lambda_{1,0}$. We will denote the customers of this flow as the primary ones. These customers are granted with the highest priority. If a system has a limited buffer size, there is a chance for these primary customers to be lost due to a lack of available storage in the buffer. In retrial queueing systems the lost customers have a possibility to return to the system again. Such customers form one more, secondary flow with lower priority than first one. This flow actually turns the model into the one with two incoming flows. In more detail, the behavior of retrial customers in the system can be described as it was done in [12], where the lost primary customers returned to the system with the probability equal to one. In this paper such a model will be generalized.

The primary customers that are lost due to the lack of available storage in the buffer, fall into the orbit of the repeated customers with the probability q_1. To prevent the orbit from clogging with the pushed-out customers, let us introduce one more parameter into the model. It is a probability $q_2 \leq q_1$ to fall to the orbit for the pushed out secondary customers (which have already returned to the system from the orbit before). So, if a secondary customer is pushed out then it comes back to the orbit again with a probability q_2 and is irretrievably lost with the probability $(1 - q_2)$. The described queueing system is presented in Fig. 1.

The investigation of such a system can be reduced to the case of an ordinary two-flow system with randomized pushing-out and without retrial customers. However, for this purpose it is necessary to use the intensities of the primary and secondary incoming flows derived from the considered system. The process of obtaining these intensities was described in [12] in details and the system of equations that allows to determine the load factors ρ_1 and ρ_2 is as follows:

$$
\begin{cases}
\rho_1 = \rho_{1,0}; \\
\rho_2 = \frac{\rho_{1,0} q_1 \varphi_1(\rho_1, \rho_2, k, \alpha)}{1 - q_1 \varphi_1(\rho_1, \rho_2, k, \alpha)} \cdot \frac{1 - q_2 \varphi_2(\rho_1, \rho_2, k, \alpha)}{1 - 2 q_2 \varphi_2(\rho_1, \rho_2, k, \alpha)},
\end{cases}
\tag{1}
$$

where $\rho_1 = \frac{\lambda_{1,0}}{\mu}$, $\rho_2 = \frac{\lambda_2}{\mu}$, k means the total system capacity, α is the push-out parameter, μ is the service intensity and φ_1 and φ_2 mean the loss probabilities for primary and secondary customers, respectively, and can be evaluated as follows [4, 5]:

$$\begin{cases} \varphi_1(\rho_1, \rho_2, k, \alpha) = p_0 + (1 - \alpha) \sum_{i=1}^{k-1} p_i, \\ \varphi_2(\rho_1, \rho_2, k, \alpha) = \sum_{i=0}^{k} p_i + \frac{\rho_1}{\rho_2} \alpha \sum_{i=1}^{k-1} p_i + \frac{\rho_1}{\rho_2} p_k. \end{cases} \tag{2}$$

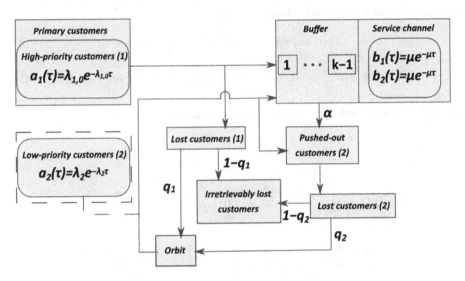

Fig. 1. The scheme of the retrial queueing system $\overrightarrow{M_2}/M/1/k/f_2^1$ with a single primary flow.

Using the above expressions for the load factors (1) and for the loss probabilities (2), it is easy to found a numerical value of the intensity of the secondary flow, and hence the load factor for this flow. After that one can find values of the probabilities of losing the customers, given all the system parameters and the probabilities q_1 and q_2 of repeated service.

3 Non-preemptive Priority System

Let us now consider a system with a single primary flow, randomized push-out mechanism, non-preemptive priority and retrial packets in a steady state. Some results on this system were obtained in 2003 by authors of [3]. But the generating functions approach gives a full and detailed solution, so let us consider it below.

The process in the described system is Markovian. It is also ergodic, which ensures that the final probabilities exist and do not depend on the initial state of the system. They satisfy the stationary system of Kolmogorov equations.

We first introduce the phase space and state probabilities of this model. The state of the system will be described by the number $N_j(t)$ of packets of type j in the queue. Let us denote

$$P(i,j;t) = P\{N_1(t) = i, N_2(t) = j\}. \tag{3}$$

Such an introduction of states for a system with non-preemptive priority causes the splitting of the case $\{N_1 = 0, N_2 = 0\}$ into two states. Further, let us denote the case of complete idle of the system by $\{\emptyset\}$, and the case of lack of queue by $\{0,0\}$. We introduce the final probability considering that

$$P_\emptyset(t) = P\{N_{oc}(t) = 0\}, P_\emptyset = \lim_{t\to\infty} P_\emptyset(t), \tag{4}$$

$$P_{0,0}(t) = P\{N_1(t) = 0, N_2(t) = 0, N_{oc}(t) = 1\}, P_{0,0} = \lim_{t\to\infty} P_{0,0}(t), \tag{5}$$

where $N_{oc}(t)$ is a number of occupied service channels.

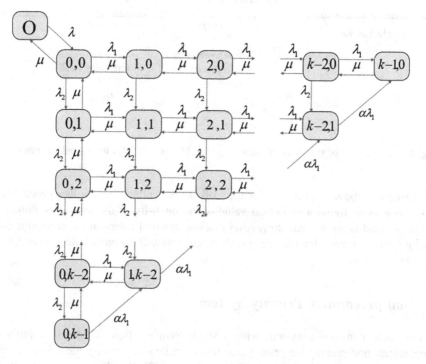

Fig. 2. The state graph of the retrial queueing system $\vec{M}_2/M/1/k/f_1^1$ with a single primary flow.

The state graph for this system is shown in Fig. 2. Let us call this graph the initial one. The probabilities at the initial graph must satisfy the normalization condition (6):

$$P_\emptyset + \sum_{i=0}^{k-1} \sum_{j=0}^{k-1-i} P_{i,j} = 1. \tag{6}$$

In the system of Kolmogorov equations, the state (4) will be connected only with the state (5), so the equation for it will be as follows:

$$-(\lambda_1 + \lambda_2)P_\emptyset + \mu P_{0,0} = 0. \tag{7}$$

Hence there is a simple linear relationship between the corresponding probabilities:

$$P_\emptyset = \frac{1}{\rho} P_{0,0}, \tag{8}$$

where $\lambda = \lambda_1 + \lambda_2$, $\rho_s = \frac{\lambda_s}{\mu_s}$, $(s = \overline{1,2})$, $\rho = \rho_1 + \rho_2 = \frac{\lambda}{\mu}$.

Now let us write equation of balance for the state $\{0,0\}$:

$$-(\lambda_1 + \lambda_2 + \mu)P_{0,0} + (\lambda_1 + \lambda_2)P_\emptyset + \mu(P_{1,0} + P_{0,1}) = 0. \tag{9}$$

Then, using (8), we obtain an equation of the following form:

$$-(\lambda_1 + \lambda_2)P_{0,0} + \mu(P_{1,0} + P_{0,1}) = 0. \tag{10}$$

Based on the above considerations, it is possible to construct a modified state graph, which no longer contains the "empty" state $\{\emptyset\}$. To use this graph, we should replace the normalization condition of the generating function by a new modified condition (11).

$$\sum_{i=0}^{k} \sum_{j=0}^{k-i} \left(1 + \frac{1}{\rho}\delta_{i,0}\delta_{j,0}\right) P_{i,j} = 1. \tag{11}$$

Then to calculate the generating function one can only keep on mind only the relation (8), which makes it possible to reconstruct the probability of this state using the data of the modified graph.

$$
\begin{aligned}
- &\left[\lambda_1\left(1 - \delta_{j,k-1-i}\right) + \alpha\lambda_1\left(1 - \delta_{i,k-1}\right)\delta_{j,k-1-i} + \lambda_2\left(1 - \delta_{j,k-1-i}\right) + \mu\left(1 - \delta_{i,0}\delta_{j,0}\right)\right]P_{i,j} \\
&+ \mu P_{i+1,j} + \mu\delta_{i,0}P_{i,j+1} + \lambda_2 P_{i,j-1} + \lambda_1 P_{i-1,j} \\
&+ \alpha\lambda_1\delta_{j,k-1-i}P_{i-1,j+1} \\
&= 0, \left(0 \leq i \leq k - 1; 0 \leq j \leq k - 1 - i\right).
\end{aligned}
\tag{12}
$$

The generating function of the final probabilities $P_{i,j}$ is defined as

$$G(u, v) = \sum_{i=0}^{k-1} \sum_{j=0}^{k-1-i} P_{i,j} u^i v^j. \tag{13}$$

Using the expression (12), we obtain following equation for generating function (13):

$$
\begin{aligned}
[\lambda_1 u(1 &- u) + \lambda_2 u(1 - v) + \mu(u - 1)]vG(u, v) \\
&= \mu(u - v)G(0, v) + \mu u(v - 1)G(0, 0) \\
&\quad + \alpha\lambda_1 u^k(v - u)P_{k-1,0} \\
&\quad + (1 - \alpha)\lambda_1 P_{0,k-1} v^{k-1} u(u - v) \\
&\quad + [\alpha\lambda_1(u - v) + \lambda_1(1 - u)v + \lambda_2(1 - v)v]u \sum_{i=0}^{k-1} P_{i,k-1-i} u^i v^{k-1-i}.
\end{aligned}
\tag{14}
$$

In a way as it was described in [7, 8] expand it into the series in powers of u and v. For this purpose let's use the expressions of the system characteristics well-known from the investigation of conventional single-flow system of the class $M/M/1/k$. Specifically, the distribution of the total number of customers in the system is given by

$$r_n = \sum_{i=0}^{n} P_{i,n-i} = \sum_{i=0}^{n} P_{n-i,i}. \tag{15}$$

Substituting the expression for the final probabilities from [7] into (15), we obtain the system of equations with respect to the "diagonal" probabilities $p_i = P_{k-1-i,i}$:

$$r_z = \sum_{j=0}^{z} P_{z-j,j} = p_0 \rho_1^{-1} \zeta_z - \alpha\phi_z p_{z+1} + \sum_{j=1}^{z} p_j \xi_{z,j}, \ (0 \le z \le k - 2), \tag{16}$$

where $\zeta_z, \phi_z, \beta, \xi_{z,j}$ can be evaluated as follows:

$$
\begin{cases}
\zeta_z = \sum_{j=0}^{z} \rho_1^{\frac{-k+z-j}{2}} \beta^j \left(C_{k-z-2}^{j+1} - \rho_1^{\frac{1}{2}} C_{k-z-3}^{j+1} \right) \\
\phi_z = \rho_1^{\frac{z-k}{2}} C_{k-2-z}^{1}, \\
\beta = -\frac{\rho_2}{\sqrt{\rho_1}}, \\
\xi_{z,j} = \sum_{s=j}^{z} \rho_1^{\frac{z-k+1-s+j}{2}} \left(\rho_1^{-\frac{1}{2}} C_{k-z-2}^{s-j+1} - C_{k-z-3}^{s-j+1} \right) \beta^{s-j} - \alpha\rho_1^{\frac{j-k+1}{2}} \beta^{z+1-j} C_{k-z-2}^{z-j+2}.
\end{cases}
\tag{17}
$$

Here C_n^v means $C_n^v(t)$ – the Gegenbauer polynomial of order n with index v, and $t = \frac{1+\rho}{2\sqrt{\rho_1}}$.

This system should be complemented by (18) to make it of full rank.

$$r_{k-1} = \sum_{i=0}^{k-1} P_{k-1-i,i}. \tag{18}$$

The difference between preemptive and non-preemptive priorities also appears in the equations for the loss probabilities for different type of customers. For non-preemptive priority the equations are as follows:

$$\begin{cases} \varphi_1(\rho_1, \rho_2, k, \alpha) = p_0 + (1 - \alpha) \sum_{i=1}^{k-1} p_i, \\ \varphi_2(\rho_1, \rho_2, k, \alpha) = r_{k-1} + \alpha \frac{\rho_1}{\rho_2} \sum_{i=1}^{k-1} p_i, \end{cases} \tag{19}$$

while for preemptive one the loss probabilities can be evaluated in the following way:

$$\begin{cases} \varphi_1(\rho_1, \rho_2, k, \alpha) = p_0 + (1 - \alpha) \sum_{i=1}^{k-1} p_i, \\ \varphi_2(\rho_1, \rho_2, k, \alpha) = r_k + \alpha \frac{\rho_1}{\rho_2} \sum_{i=1}^{k-1} p_i + \frac{\rho_1}{\rho_2} p_k. \end{cases} \tag{20}$$

4   Computational Results

The system described in Sect. 2 of this article has already been considered earlier in the work of the authors [12]. A study on a similar system with a non-preemptive priority, introduced in Sect. 3 of presented article, the authors expect to publish in the near future. Although the previous studies have been devoted to the same systems, the new statement of problem differs fundamentally from [12]. Earlier, the main attention was paid to the construction of loading areas, where the linear loss law was implemented, as well as the locking areas of the system for low-priority traffic.

Now we will focus on identifying the dependence of the intensity of the secondary flow of customers as a function of the intensity of the primary flow of customers. This dependence has a very interesting form and plays an important role in applications to telematics. As calculations have shown, there is such a critical value of the load factor for the primary flow, with a smooth excess of which the load factor on the secondary flow can increase sharply, and this increase can reach tens of thousands of times. It is well known that such sudden changes arising as a sudden response system to smoothly change external conditions, are called explosions or catastrophes [13]. It is important to note here that the randomized push-out mechanism makes it possible to effectively manage the process of "explosion" of secondary customers.

The graphs of how $\lg \rho_2$ depends on ρ_1 for a system with preemptive priority are given in Fig. 3, and for a system with non-preemptive priority – in Fig. 4.

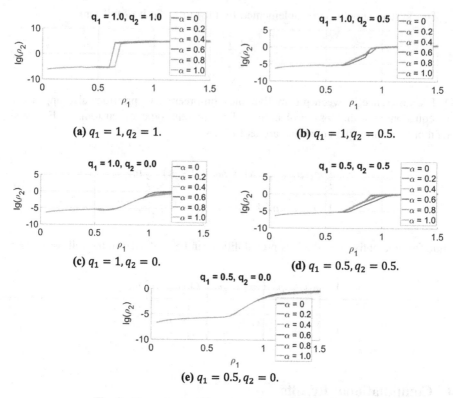

Fig. 3. Dependency of $\lg \rho_2$ on ρ_1: preemptive priority.

It can be seen that starting from a certain moment (when ρ_1 is approximately between 0.8 and 1, that is, when changing from a weak load of the system to a strong one), an "explosion" occurs – the intensity of the secondary flow increases, and for $q_1 = 1$ – the coefficient of growth is thousands. After this explosion, the growth in the intensity of the flow of secondary requirements slows down, but remains exponential.

For a system with a non-preemptive priority, the "explosion" for $\alpha = 0$ looks much more noticeable than for a system with a preemptive priority, although in both cases this effect takes a place.

There are several ways to control the "explosion" effect. First, one can choose small value load factor ρ_1 to avoid this effect. If it is no possibility to do this and the value of ρ_1 is large enough, it is still possible to reduce the effect. To do this, you should decrease the values of probabilities q_1 and q_2. Finally, the increasing of push-out parameter α also can significantly decrease the load factor ρ_2. Also it can be seen that the lower the probabilities q_1, q_2, the closer the critical point to 1.

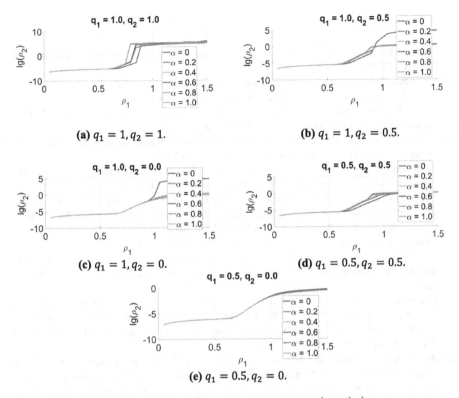

Fig. 4. Dependency of $\lg \rho_2$ on ρ_1: non-preemptive priority.

5   Conclusion

The single-flow queueing system with retrials and randomized push-out mechanism was investigated. Systems with preemptive and non-preemptive priorities were considered. The characteristics of the "explosion" of retrial customers are studied in detail and recommendations for its management are given. These results may be used to reduce the computational load on the telematic devices operating in the real time mode. The computations of such areas performed in advance make it possible to choose the necessary parameter values immediately instead of computing them in runtime. This will lead to increase in the systems performance and improvement of controlling them.

Acknowledgement. This research was supported by RFBR grant № 18-29-03250 mk.

Also this work related to the high performance computations and modelling was done using the infrastructure of the Shared-Use Center "Supercomputer Center Polytechnic" at Peter the Great St.Petersburg Polytechnic university registered at http://ckp-rf.ru/ckp/500675/ (shared-use center id 500676).

References

1. Vishnevsky, V.M.: Theoretical Bases if Designing Computer Networks. Technosphere, Moscow (2003)
2. Tanenbaum, A.S., Wetherall, D.J.: Network computers. Prentice Hall, Boston (2011)
3. Jaiswal, N.K.: Priority queues. Academic Press, New York (1968)
4. Basharin, G.P.: Some Results for Priority Systems. Queueing Theory in Data Transfer Analysis, pp. 39–53. Nauka, Moscow (1969). (in Russian)
5. Avrachenkov, K.E., Shevlyakov, G.L., Vilchevsky, N.O.: Randomized push-out disciplines in priority queueing. J. Math. Sci. **22**(4), 3336–3343 (2004)
6. Avrachenkov, K.E., Vilchevsky, N.O., Shevlyakov, G.L.: Priority queueing with finite buffer size and randomized push-out mechanism. Perform. Eval. **61**(1), 1–16 (2005)
7. Ilyashenko, A., Zayats, O., Muliukha, V., Laboshin, L.: Further investigations of the priority quiuing system with preemptive priority and randomized push-out mechanism. In: Balandin, S., Andreev, S., Koucheryavy, Y. (eds.) NEW2AN 2014. LNCS, vol. 8638, pp. 433–443. Springer, Cham (2014). https://doi.org/10.1007/978-3-319-10353-2_38
8. Muliukha, V., Ilyashenko, A., Zayats, O., Zaborovsky, V.: Preemptive queueing system with randomized push-out mechanism. Commun. Nonlinear Sci. Numer. Simul. **21**(1/3), 147–158 (2015)
9. Ilyashenko, A., Zayats, O., Muliukha, V., Lukashin, A.: Alternating priorities queueing system with randomized push-out mechanism. In: Balandin, S., Andreev, S., Koucheryavy, Y. (eds.) ruSMART 2015. LNCS, vol. 9247, pp. 436–445. Springer, Cham (2015). https://doi.org/10.1007/978-3-319-23126-6_38
10. Ilyashenko, A., Zayats, O., Muliukha, V.: Randomized priorities in queuing system with randomized push-out mechanism. In: Galinina, O., Balandin, S., Koucheryavy, Y. (eds.) NEW2AN/ruSMART -2016. LNCS, vol. 9870, pp. 230–237. Springer, Cham (2016). https://doi.org/10.1007/978-3-319-46301-8_19
11. Zayats, O., Korenevskaya, M., Ilyashenko, A., Lukashin, A.: Retrial queueing systems in series of space experiments "Kontur". Procedia Comput. Sci. **103**, 562–568 (2017)
12. Ilyashenko, A., Zayats, O., Korenevskaya, M., Muliukha, V.: a retrial queueing system with Preemptive priority and randomized push-out mechanism. In: Galinina, O., Andreev, S., Balandin, S., Koucheryavy, Y. (eds.) NEW2AN/ruSMART/NsCC -2017. LNCS, vol. 10531, pp. 432–440. Springer, Cham (2017). https://doi.org/10.1007/978-3-319-67380-6_39
13. Arnold, V.I.: Catastrophe Theory. Springer, Berlin (2004)

Comparison of LBOC and RBOC Mechanisms for SIP Server Overload Control

Oleg E. Pavlotsky[1], Ekaterina V. Bobrikova[2(✉)], and Konstantin E. Samouylov[2,3]

[1] Department of Communication Networks and Commutation Systems, Moscow Technical University of Communication and Informatics (MTUCI), 8 Aviamotornaya St., 11102 Moscow, Russia

[2] Department of Applied Probability and Informatics, Peoples' Friendship, University of Russia, (RUDN University), 6 Miklukho-Maklaya St., 117198 Moscow, Russia
{bobrikova_ev,samuylov_ke}@rudn.university

[3] Institute of Informatics Problems, Research Center "Computer Science and Control" of the Russian Academy of Sciences, Vavilov St. 44-2, Moscow 119333, Russian Federation

Abstract. The purpose of this article is to analyze mechanisms of Loss-based Overload Control, LBOC and Rate-based Overload Control, RBOC for server overload control of the Session Initiation Protocol (SIP). Overloading occurs when a server is unable to process an entire incoming message flow due to a lack of resources. Standards of IETF SOC recommend several overload control mechanisms, including most important LBOC and RBOC. This article proposes the mechanisms of LBOC and RBOC and describes the results of the comparative analysis of these mechanisms based on the hysteresis control over incoming stream of signaling messages. The system resides in one of three modes (normal, overload, discard) based on thresholds and a size of an input queue. A signal message source implements the Markov-modulated Poisson process, MMPP-2 model. The leaky bucket algorithm is applied to limit the number of incoming messages in the implementation of RBOC mechanism. The comparison of results showed that RBOC mechanism based on the hysteresis control over incoming stream of signaling messages demonstrates higher effectiveness of the congestion control, as a result of which the average time in the overload mode is less than for LBOC mechanism. However, LBOC mechanism has the ability to maintain its high efficiency for all RTT values on the same thresholds, while RBOC mechanism needs to have its own threshold dynamic control mechanism for different RTT values to supply maximum efficiency.

Keywords: SIP server · Mechanism of overload control
Hysteresis overload control · RBOC · LBOC

1 Introduction

The object of the research was the investigation of mechanisms of overload control for SIP server. The overload occurs when the SIP server is unable to process entire stream of received messages because of lack of resources. Two mechanisms of control are recommended to solve this issue – the mechanism LBOC (Loss-Based Overload Control) and the mechanism RBOC (Rate-Based Overload Control). We analyzed and

© Springer Nature Switzerland AG 2018
O. Galinina et al. (Eds.): NEW2AN 2018/ruSMART 2018, LNCS 11118, pp. 247–254, 2018.
https://doi.org/10.1007/978-3-030-01168-0_23

proposed a hysteresis control of overloads for these mechanisms, which realized with two thresholds [1]. The system under this hysteresis control operates in one of three modes: normal load mode, overload mode and blocking load mode. According to MMPP-2 model, the source of messages operated in one of two modes: generation mode or hollow stream mode. The paper investigated the dependence of the probability-time characteristics of the operation of the LBOC and RBOC mechanisms (such as the percentage of messages served, the average time in the overload mode, the average time of the control cycle) on the round-trip time [2, 4].

2   Simulation Model of System with Hysteretic Input Control

The processing of messages by the SIP server is defined by the queueing system $M|M|1|\langle L,H \rangle|R$ with a finite queue of size R and with hysteretic overload control with the thresholds L, $1 \leq L \leq R$, and H, $1 \leq H \leq R$. The system operates in one of three modes: normal load mode (s = 0), overload mode (s = 1) and blocking load mode (s = 2). In normal load mode, when the queue reaches the value H, the system switches to the overload mode (s = 1), in which new messages are accepted with intensity in accordance with LBOC or RBOC mechanisms [12]. In order to prevent oscillations the intensity of the input stream is restored to the normal value only when the buffer occupancy decreases to the overload abatement threshold L. In overload mode, when the queue reaches the value R, the system switches to the blocking load mode (s = 2), in which new messages are not accepted [10, 11]. The dependence of the intensity of receiving messages on the length of buffer is shown in Fig. 1.

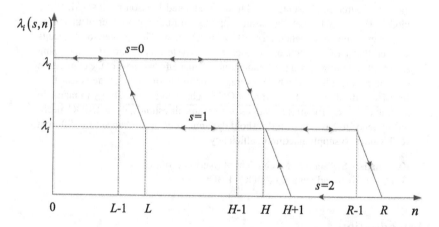

Fig. 1. Function of intensity of receiving messages.

According to the LBOC mechanism in case of overloading, the downstream SIP server demands the upstream SIP server to reduce traffic volume by N percent, and the upstream SIP server assigns a random number from the interval [0,100] to each upcoming message. If this random number exceeds N, the message is sent, otherwise it is redirected to another SIP server or reset [3, 8]. According to RBOC mechanism in case of overload, the downstream SIP server periodically estimates the load and reports the maximum number of messages per second it can process to the upstream SIP server, which filters messages sent to the downstream SIP server according to the leaky bucket algorithm [5, 9, 14] A simulation model has been developed in line with these mechanisms and its structural diagram is shown in Fig. 2.

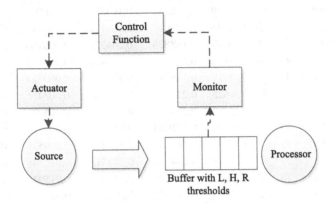

Fig. 2. Structural diagram of simulation model of LBOC and RBOC mechanisms.

3   Results

The simulation model has been used for a numerical experiment and a comparative analysis of the LBOC and RBOC mechanisms. It should be noted that the model allows performing the comparison under identical conditions using the same hysteresis load control algorithm, which provides the high quality of the experiment results and conclusions. The simulation model uses data with different thresholds and a buffer size of R = 100, chosen from another researches [13]. The total number of modeling iterations was total_step = 1000000. The minimum value of the average time in the overload mode (Mτ12) has been taken for each RTT for both of the LBOC and RBOC mechanisms, where Mt ≥ 0.45 ms (presented in Table 1). The table below has the following notation: J is the number of successfully processed messages, Bx1 is the probability of message loss in the overload mode (s = 1); Bx2 - probability of message loss in the blocking load mode (s = 2); Mt - average time of the control cycle (overload mode and no overload mode).

The results in the table above demonstrate that at RTT = 0 for the RBOC mechanism, the value of the probability of message loss in the blocking load mode (Bx2) is 0 due to the fact that the mechanism without delays activates the leaky bucket algorithm for messages sent to the downstream SIP server when the load is increased, thereby preventing the further growth of the loading and the transition of the system to the blocking load mode. An increase in RTT leads to an increase in the time interval between the time of the overload detection and the start of the leaky bucket algorithm on the upstream SIP server, which leads to increasing of blocking load mode time and the probability of message loss in this mode.

Table 1. Minimal value of average time in the overload mode.

Mechanism	RTT	L	H	J	Mτ12	Bx1	Bx2	Mt
RBOC	0	24	37	834153	0,134103	0,295259	0	0,453713
RBOC	0,01	79	90	833751	0,080224	0,172813	0,000059	0,464746
RBOC	0,02	78	87	831835	0,080417	0,170318	0,000098	0,472266
RBOC	0,05	78	82	834659	0,082582	0,164871	0,000505	0,498857
RBOC	0,1	69	72	834644	0,10361	0,218111	0,000147	0,474093
RBOC	0,2	52	55	826849	0,138825	0,264328	0,000164	0,525764
RBOC	0,3	43	46	784192	0,17583	0,294487	0,000152	0,595387
RBOC	0,4	32	35	749619	0,188214	0,775112	0,000794	0,485193
LBOC	0	75	83	831625	0,23902	0,155068	0,003456	0,452636
LBOC	0,01	78	86	833752	0,236122	0,151992	0,005271	0,450783
LBOC	0,02	79	87	831839	0,241984	0,151578	0,00722	0,457405
LBOC	0,05	79	86	831793	0,24545	0,150492	0,009529	0,459986
LBOC	0,1	79	85	833228	0,244596	0,133313	0,012346	0,50414
LBOC	0,2	78	82	835942	0,205288	0,117608	0,015946	0,46085
LBOC	0,3	79	82	832290	0,180382	0,090237	0,022737	0,478645
LBOC	0,4	79	82	830418	0,195105	0,074845	0,027071	0,573116

Consider the dependence of Mτ12 on RTT in the graph (Fig. 3) constructed from the data of Table 1.

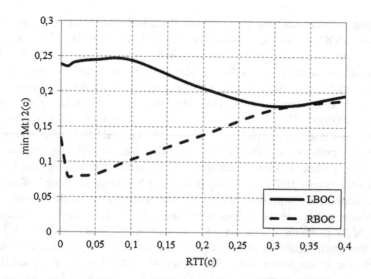

Fig. 3. The dependence of minimal average time in the overload mode on RTT.

Figure 3 demonstrates that for small RTT values RBOC mechanism results in significantly lower average time in the overload mode than LBOC mechanism. This is related to more efficient management (from the point of view of Mτ12) of upcoming message stream from the upstream SIP server, the speed limit of RBOC mechanism works more efficiently than the reset of a certain percentage of messages of LBOC mechanism. With the further growth of RTT for the RBOC mechanism, the time in modes s = 1 and s = 2 is increased.

The time behaviour of Mτ12 for LBOC mechanism differs from RBOC mechanism, including Bx2 ≠ 0 for RTT = 0 (Table 1). This means that dropping a portion of the load by the upstream SIP server does not prevent downstream SIP server's overloading and transition to blocking load mode. With the growth of RTT, the efficiency of the mechanism decreases, which leads to a more rapid transition to the blocking load mode (s = 2), i.e. the system is less in the state of overload mode (s = 1) and more in s = 2 mode. Since for s = 2 all incoming messages are discarded by the downstream SIP server, the downstream SIP server is unloaded faster in this mode and leaves the blocking load mode. Thereby the value of Mτ12 decreases due to a stay time increase of mode s = 2 comparing to the time of stay in the mode s = 1.

Figures 4 and 5 present the results of modeling with different values of γ at the source, where γ is the ratio of message generation time to the total time.

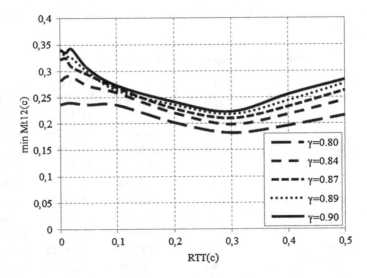

Fig. 4. The dependence of LBOC minimal average time in the overload mode on RTT for different value of γ.

However, if L and H thresholds in Table 1 are taken into account, it could be observed that the minimum value of Mτ12 for different RTT values by the LBOC mechanism is reached at thresholds closed to the values of L = 78 H = 86, while the achieving the minimum value of Mτ12 RBOC mechanism needs to set different thresholds based on RTT, which requires the development of a special mechanism.

Consider two fixed thresholds (L = 78 and H = 86) and how the test characteristics vary depending on RTT (Table 2).

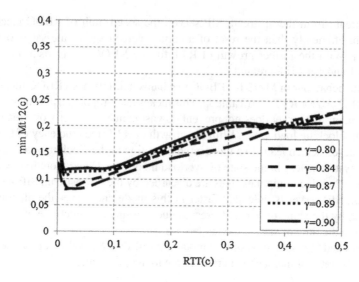

Fig. 5. The dependence of RBOC minimal average time in the overload mode on RTT for different value of γ.

Table 2. Modeling results for L = 78 and H = 86.

Mechanism	RTT	L	H	J	Mτ12	Mt
RBOC	0	78	86	832461	0,089201	0,295765
RBOC	0,01	78	86	834441	0,103366	0,342074
RBOC	0,02	78	86	834438	0,078399	0,442972
RBOC	0,05	78	86	833915	0,111336	0,661432
RBOC	0,1	78	86	832639	0,165227	0,997162
RBOC	0,2	78	86	830848	0,25527	1,56036
RBOC	0,3	78	86	828356	0,331232	2,04877
RBOC	0,4	78	86	810682	0,402214	2,28402
LBOC	0	78	86	834009	0,224218	0,43715
LBOC	0,01	78	86	833752	0,236122	0,450783
LBOC	0,02	78	86	832938	0,250116	0,466945
LBOC	0,05	78	86	830968	0,269561	0,507844
LBOC	0,1	78	86	831943	0,290611	0,598642
LBOC	0,2	78	86	834243	0,316989	0,800698
LBOC	0,3	78	86	832188	0,352398	1,03975
LBOC	0,4	78	86	829453	0,387572	1,25807

From Fig. 6 it could be observed that for small RTT values, the RBOC mechanism provides the best results, including the number of messages processed and the time of stay in the overload mode (Mτ12) than in the LBOC mechanism. As the RTT value increases, the advantage goes to the LBOC mechanism.

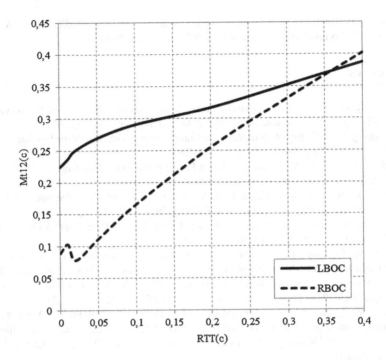

Fig. 6. Dependence of average time in the overload mode on RTT for L = 78 and H = 86.

4 Conclusion

The article suggests the mechanisms of LBOC and RBOC, based on the hysteresis load control and demonstrates the analysis of the mechanisms' performance characteristics. The comparison of these characteristics showed that RBOC mechanism provides better efficiency for the control of overloads, which leads to the decreased average time of the system stay in the overload mode comparing to the LBOC mechanism under other equal conditions and RTT ≤ 0.4 s. The advantage of LBOC mechanism is the ability to maintain its high efficiency for all RTT values on the same thresholds, while the RBOC mechanism needs to have its own dynamic control mechanism for the threshold values for different RTT values for maximum efficiency. But LBOC mechanism also requires threshold control in case of RTT < 0.05 because of lower value of average control cycle time (Mt) than 0.45, which means their inefficiency. Thus, with RTT ≥ 0.05, the LBOC mechanism is more efficient under a constant value of control thresholds, while for RTT < 0.05 for both mechanisms dynamic threshold control is required, and it is recommended to use the mechanism RBOC. The foregoing is a recommendation for the implementation of the realized LBOC and RBOC mechanisms for different RTT values.

Acknowledgement. The publication has been prepared with the support of the "RUDN University Pro-gram 5-100" and funded by RFBR according to the research project No.

16-07-00766. This work has been developed within the framework of the COST Action CA15104, Inclusive Radio Communication Networks for 5G and beyond (IRACON).

References

1. Pavlotsky, O., Talanova, M.: Analysis of overload control mechanisms on SIP servers for NGN/T-Comm. Telecommunications and Transport, no8, pp. 73–78 (2014)
2. Rosenberg, J.: RFC 5390: Requirements for Management of Overload in the Session Initiation Protocol (2008). http://tools.ietf.org/html/rfc5390
3. Gurbani, V., Hilt, Ed.V.: Session Initiation Protocol (SIP) Overload Control (2015). http://datatracker.ietf.org/doc/draft-ietf-soc-overload-control/
4. Hilt, Ed.V., Noel, E., Shen, C., Abdelal, A.: RFC 6357: Design Considerations for Session Initiation Protocol (SIP) Overload Control (2013). http://tools.ietf.org/html/rfc6357
5. Noel, E., Williams, P.M.: Session Initiation Protocol (SIP) Rate Control (2015). http://datatracker.ietf.org/doc/draft-ietf-soc-overload-rate-control/
6. Abaev, P., Gaidamaka, Y., Samouylov, Konstantin E.: Queuing model for loss-based overload control in a SIP server using a hysteretic technique. In: Andreev, S., Balandin, S., Koucheryavy, Y. (eds.) NEW2AN/ruSMART -2012. LNCS, vol. 7469, pp. 371–378. Springer, Heidelberg (2012). https://doi.org/10.1007/978-3-642-32686-8_34
7. Abaev, P., Gaidamaka, Yu., Pechinkin,A., Razumchik, R., Shorgin, S.: Simulation of overload control in SIP server networks. In: Proceedings of the 26th European Conference on Modelling and Simulation ECMS 2012, Koblenz, Germany, 29 May - 1 June, 2012, pp. 533–539 (2012)
8. Samouylov, K., Pavlotsky, O.: Rate-based overload control mechanism simulator for SIP server overload control. T-Comm Telecommun. Transp. **10**(3), 44–48 (2016)
9. Akbar, A., Basha, S.M., Sattar, S.A.: Overload Control in SIP Networks: A Heuristic Approach Based on Mathematical Optimization (2015). http://ieeexplore.ieee.org/document/7417081/
10. Montazerolghaem, A., Yaghmaee, M.H., Leon-Garcia, A., Naghibzadeh, M., Tashtarian, F.: A load-balanced call admission controller for IMS cloud computing. IEEE Trans. Network Service Manag. **13**(4), 806–822 (2016)
11. Samouylov, K.E., Abaev, P.O., Gaidamaka, Y.V., Pechinkin, A.V., Razumchik, R.V.: Analytical modelling and simulation for performance evaluation of SIP server with hysteretic overload control. In: Proceedings of the 28th European Conference on Modelling and Simulation (ECMS 2014), Brescia, Italy, 27–30 May 2014, pp. 603–609 (2014). https://doi.org/10.7148/2014-0603
12. Gaidamaka, Y., Pechinkin, A., Razumchik, R., Samouylov, K., Sopin, E.: Analysis of an MG1R queue with batch arrivals and two hysteretic overload control policies. Int. J. Appl. Math. Comput. Sci. **24**(3), 519–534 (2014). https://doi.org/10.2478/amcs-2014-0038
13. Abaev, P., Samouylov, K., Sinitsyn, I., Shorgin, S.: FSM based simulation for performance evaluation of SIP server loss-based overload control. In: Proceedings of the 29th European Conference on Modelling and Simulation (ECMS 2015), Albena, Bulgaria, May 26–29, 2015, pp. 675–681 (2015). https://doi.org/10.7148/2015-0675
14. Montazerolghaem, A., Moghaddam, M.: Design, implementation and performance evaluation of a proactive overload control mechanism for networks of SIP servers. Telecommun. Syst. **67**(2), 309–322 (2018)

Performance Analysis of Cognitive Femtocell Network with Ambient RF Energy Harvesting

Jerzy Martyna[✉]

Institute of Computer Science, Faculty of Mathematics and Computer Science,
Jagiellonian University, ul. Prof. S. Lojasiewicza 6, 30-348 Cracow, Poland
martyna@ii.uj.edu.pl

Abstract. Radio frequency (RF) energy harvesting is a promising technique to collect energy from the concurrent downlink transmissions. This energy after converting it into DC power can power up such devices as cell phones, Wi-Fi networks, etc. In this paper, a model of RF energy harvesting in the cognitive femtocell is presented. Additionally, an algorithm to maximise the average throughput of the secondary system over a given slot time is given. Increased throughput allows to improve the energy harvesting in the femtocell. Moreover, the effect of varying the different parameters such as the spatial density of BSs, significantly affects the values of energy harvesting in cognitive femtocell network. The obtained results of simulation tests confirm the obtained theoretical results of energy harvesting in cognitive femtocell networks.

Keywords: Cognitive radio · Femtocell network
RF energy harvesting · Spectrum sharing · Power control

1 Introduction

Recently, a radio frequency (RF) energy harvesting technique is emerging as an attractive solution to power low-energy wireless communicatiuon devices [4,11]. Such a technique allows to improve spectrum utilisation and convert electromagnetic waves from ambient RF sources (cellular base stations, PU base stations, etc.) into energy which can be used as to power up many devices such as cell phones, sensors, etc. This has been confirmed by numerous experiments and reports, a.o. results given by Ostaffe [10], which has been shown that with the transmit power of 0.5 W by a mobile phone, 0.4 mW of power can be harvested at the distance 10 m.

The idea of simultaneously transmitting both energy and information was first proposed by Varshney [19]. Author characterised the fundamental trade-off for capacity-energy function under the assumption of an ideal energy harvesting receiver. The basic relationships between the energy transferred by electromagnetic waves and the information contained in them is presented by Grover and

© Springer Nature Switzerland AG 2018
O. Galinina et al. (Eds.): NEW2AN 2018/ruSMART 2018, LNCS 11118, pp. 255–267, 2018.
https://doi.org/10.1007/978-3-030-01168-0_24

Sathai [3]. Two practical approaches for energy harvesting have been proposed in the paper by Zhou *et al.* [21]. The first approach is based on a power-splitting (PS) mechanism, where a PS receiver splits the received signal into two parts, based on the PS ratio. While the first part of the received signal is used for energy harvesting, the second one is used for information processing. The second approach is based on time-switching (TS) technique, where the total time is divided into two intervals: first for data harvesting and second for information processing. Some relaying protocols for wireless energy harvesting have been proposed by Nazir *et al.* [9] that can be implemented in an amplify-and-forward (AP) relay based one-way-communication networks.

Another approach to solving problems of energy harvesting is a concept of a cooperative network of simultaneously data relaying and energy harvesting [7]. An energy harvesting protocol and information processing in two-way multiplicative relay network using power and splitting-based relaying (PSR) protocol was proposed by Shah *et al.* [16]. The impact of the time switched-based relaing protocol at high transmission rates has been studied by Shah *et al.* [17]. On the other hand, a number of works investigated how the location of BS and the hierarchy in cellular networks affect energy harvesting. Among others, a performance evaluation of multi-tier uplink cellular network with RF energy harvesting and flexible cell association was developed by Sakr *et al.* [12]. An analysis of K-tier uplink cellular networks with ambient RF energy harvesting has been presented by the same author [15]. Nevertheless, none of these and other publications have analyzed the performance of harvesting in cognitive femtocell networks.

Cognitive radio network (CRN) is a technology that connects nodes in the form of cognitive radio (CR) systems [8] using network technologies. In general, these nodes are intelligent and have the ability to observe, learn, and optimise their performance. They can cooperate with others, but only then when cooperation can improve theiDow to develop cooperation among selfish nodes. Furthermore, the division of CRN equipments into two sub-networks: primary network (PN), using exclusive licensed band, and secondary network (SN), using both unlicensed bandwidth and unused at the moment, the licensed band, a system was created that allows to increase the efficiency of the use of spectrum resources. The PN network consists of all Primary Users (PUs) who use the licensed band, while the SN network includes secondar users (SUs).

Cognitive radio femtocell networks (CRFN) [13] are recent technology breakthroughs that aim to achieve throughput improvement by means of spectrum management and interference mitigation, respectively. Based on the CR technology, the access control scheme greatly improves the performance of cognitive users near to femtocells. The jointly designed distributed access and power control algorithm can be solved by game theory [14]. Downlink scheduling and power allocation in cognitive femtocell networks are studied a.o. by Elmaghrab [2]. According to the given results, the throughput maximisation of femtocell users allows to share spectrum resources with macrocell base station (MBS) while limiting interference between macrocell and femtocells. Distributed resource allocation with imperfect spectrum sensing information and channel uncertainty in cognitive femtocell networks has been studied by Huang *et al.* [6]. Nevertheless, none of the paper known to the author analyze RF energy harvesting by the cognitive femtocell network.

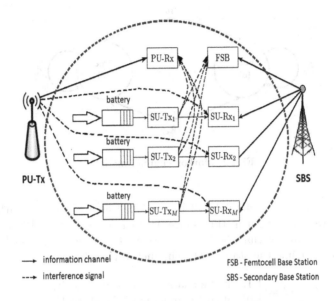

Fig. 1. System model of cognitive femtocell network.

The main purpose of the work is to create a model for acquiring energy transferred by electromagnetic waves in the cognitive femtocell network. Next, the second objective of the work is to formulate the basic dependencies allowing for the calculation of the obtained energy from the basic parameters of this network. Finally, the purpose of the work is to provide an algorithm that maximises bandwith for better performance SUs and guarantee QoS for PUs users.

The rest of the paper is organized as follows. The second section presents the system model. The third section presents the modelling of energy harvesting in the cognitive femtocell network. Section 4 presents the algorithm for obtaining energy from the femtocell network and the algorithm of bandwidth maximisation for better performance of SUs and QoS guarantee for PUs. Section 5 gives the results of simulation studies. The conclusion ends with this paper.

2   System Model

This section presents the model of the cognitive femtocell network.

Let the system model be a single PU receiver (PU-Rx) and M SUs, as shown in Fig. 1. It is assumed that each SU has an energy harvesting (EH) device. It is also assumed that all subcarriers are of the same band. Both PN and SN systems use the OFDMA scheme. In addition, it is assumed that one subcarrier can only be used by one SU at each time slot. The interference between individual SUs is ommitted. Each SU can use multiple subcarriers at each time slot.

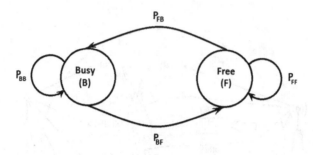

Fig. 2. Markov channel model.

The Rayleigh fading channel will be modelled as a two-state Markov chain (see Fig. 2) [18]. As shown in this figure, state B denotes that PU is inactive, while the state F means that the PU is inactive. For a time slot $k, k \in \{1, \ldots, K\}$ it is possible to define the state of the channel n, namely

$$x_k^n = \begin{cases} 1, & \text{if the channel is in the state } B \\ 0, & \text{if the channel is in the state } F \end{cases} \tag{1}$$

Let N be the number of subcarriers in the femtocell and L be the number of subcarriers occupied by PU receiver. Thus, the number of random subcarrier state can be expressed by

$$I = \binom{N-M}{L} \tag{2}$$

Let Q_k be the set of states of all channels at the k-th time slot. Then states of all channels available in cognitive femtocell at the time slot k can be given by

$$y_k = \{x_k^1, x_k^2, \ldots, x_k^N\}, \quad i \in \{1, 2, \ldots, K\} \tag{3}$$

For the transition matrix of PU receiver is defined the occupation state as $\mathbf{P}^o$. The state transition probability of the n-th channel can be given by

$$P_{BF}^n = \Pr\{x_{k+1}^n = 0 \mid x_k^n = 1\} \tag{4}$$

$$P_{FB}^n = \Pr\{x_{k+1}^n = 1 \mid x_k^n = 0\} \tag{5}$$

The transition probability of $\mathbf{P}^o$ can be described as follows

$$p_{ij}^o = \Pr\{O_{k+1} = y_j \mid O_k = y_i\} \tag{6}$$

After transformation

$$p_{ij}^o = \prod_{n=1}^{N} \Pr\{x_{k+1}^n \mid x_k^n\} \qquad (7)$$

3 RF Energy-Harvesting Model of Cognitive Femtocell Network

Regarding the RF energy as an energy-harvesting source, the following model is proposed here. First, by expanding the above-mentioned model it is assumed that the femtocell $\mathcal{F}$ is in the radio range of Ω primary transmitters (PUs-Tx). Next, each SU must be equipped with a power conversion circuit that can extract DC power from the received electromagnetic waves [10]. For each SU is defined the harvesting zone, which is a disk with radius r_h centered at each PU-Tx with the radius r_P. The radius r_h is determined by the energy harvesting circuit sensitivity for a given transmission power level of PU-Tx as P_P. It is assumed that given SU within a harvesting zone is entirely inside the femtocell and can receive power larger than the energy harvesting threshold, which is given by $P_P r_h^\alpha$, where $\alpha > 2$ is the path-loss exponent. The received power by a SU outside any harvesting zone is too small to activate the energy harvesting circuit, which means it can be omitted.

The probability that a SU lies in femtocell is equal to the probability that there is one PU-Tx inside the disk $SU(Y, r_h)$, if Y is a coordinate of SU and belongs to the area occupied by femtocell. Let the number of SUs inside $SU(Y, r_h)$ is denoted by Ω and is a Poisson random variable with mean $\pi r_h^2 \lambda_p$, where $\lambda_p = p\lambda_p'$, p is the probability of accessing PU-Tx at each time slot, λ_p' is the density of PU-Tx.

Thus, the probability mass function (PMF) inside the disk $SU(Y, r_h)$ is given by

$$\Pr\{\Omega = \omega\} = e^{-\pi r_h^2 \lambda_p} \frac{(\pi r_h^2 \lambda_p)^\omega}{\omega!}, \quad \omega = 0, 1, 2, \ldots \qquad (8)$$

The probability that the SU lies in femtocell within radio range of PUs-Tx, p_h, is given by

$$p_h = \Pr\{SU \in \mathcal{F}\} \qquad (9)$$

$$= \Pr\{\Omega \geq 1\} \qquad (10)$$

$$= \sum_{\omega=1}^{\infty} e^{-\pi r_h^2 \lambda_p} \frac{(\pi r_h^2 \lambda_p)^\omega}{\omega!} \qquad (11)$$

$$= 1 - e^{-\pi r_h^2 \lambda_p} \qquad (12)$$

In practice, values λ_p and r_h are both small. Thus, it is assumed that $\pi r_h^2 \lambda_p \ll 1$. This allows approximation of Eq. (12) by ignoring the higher-order terms with $\omega > 1$. It indicates that if SU is inside the harvesting zone of one single PU-Tx most probably most probably, which equivalently means that the harvesting zones of different PUs-Tx do not overlap at most time. Thus, the amount of average power harvested by SU in femtocell in a time slot can be lower-bounded by $\eta p_p R^{-\alpha}$, where $R \leq r_h$ indicates the distance between SU and its nearest PU-Tx, η is the harvesting efficiency.

It is necessary to define the remaining baterry energy for each m-th SU in the k-th time slot. It can be assumed using [18] that the battery energy is available at the next slot $k+1$ in the n-th channel can be defined

$$B_{k+1}^m = \min\{B_k^m - p_k^{m,n}T + E_k^m, B_{max}\}, \quad k \in \{1, \ldots, K\}, \ m \in \{1, \ldots, M\} \quad (13)$$

where $p_k^{m,n}$ is the transmission power allocated in the m-th SU in the n-th channel at time slot k, T denotes the duration of one time slot and B_{max} denotes the maximum energy battery capacity.

It is possible to define for a cognitive femtocell signal-to-interference-plus-noise ratio (SINR) $SINR_k^{m,n}$ of the m-th SU in the n-th channel at the time slot k, namely

$$SINR_k^{m,n} = \frac{h_k^{m,n} p_k^{m,n}}{\sum_{\omega=1}^{\Omega} h_k^{\omega,n} p_k^{\omega,n} + \sum_{j=1, j\neq m}^{M} h_k^{j,n} p_k^{j,n} + h_k^{F,n} p_k^{F,n} + \sigma^2},$$
$$k \in \{1, \ldots, K\}, \ j, m \in \{1, \ldots, M\}, \ n \in \{1, \ldots, N\} \quad (14)$$

where $p_k^{\omega,n}$ and $p_k^{F,n}$ denote the transmission power in the n-th channel of the ω PU-Tr and FSB, respectively, at the time slot k; $h_k^{m,n}$ is the channel coefficient at the m-th SU in the n-th channel, $h_k^{\omega,n}$, $h_k^{F,n}$ are the channel coefficient at the ω-PU-Tr and the FSB, σ^2 is the noise power, respectively.

Then it is possible to define

$$g_k^{m,n} = \frac{h_k^{m,n}}{\sum_{\omega=1}^{\Omega} h_k^{\omega,n} p_k^{\omega,n} + \sum_{j=1, j\neq m}^{M} h_k^{j,n} p_k^{j,n} + h_k^{F,n} p_k^{F,n} + \sigma^2},$$
$$k \in \{1, \ldots, K\}, \ j, m \in \{1, \ldots, M\}, \ n \in \{1, \ldots, N\} \quad (15)$$

The throughput of the m SU at the time slot k can be presented as follows:

$$R_k^m = \sum_{n=1}^{N} \log_2(1 + g_k^{m,n} \cdot p_k^{m,n}), \quad m \in \{1, \ldots, M\} \quad (16)$$

where $g_k^{m,n}$ means the channel gain distribution of the m-th SU in the subcarrier-occupied state of time slot k in the n-th channel.

The throughput of the m-th SU at the k-th time slot in the n-th channel can be maximised as follows

$$\max_{p_k^{m,n}} E\{\frac{1}{K} \sum_{k=1}^{K-1} \sum_{m=1}^{M} \sum_{n=1}^{N} \log_2(1 + g_k^{m,n} \cdot p_k^{m,n})\} \quad (17)$$

subject to

$$\sum_{k=1}^{K}\left(\sum_{m=1}^{M}\sum_{n=1}^{N}\sum_{\omega=1}^{\Omega}h_k^{\omega,n}p_k^{\omega,n} + \sum_{j=1,j\neq m}^{M}h_k^{j,n}p_k^{j,n} + h_k^{F,n}p_k^{F,n}\right) + \sigma^2 \leq I^{TH},$$
$$k \in \{0,\ldots,K-1\} \tag{18}$$

$$\sum_{n=1}^{N}p_k^{m,n} \leq \frac{B_k^m}{T}, \quad k \in \{0,\ldots,K-1\} \tag{19}$$

$$p_k^{m,n} \geq 0, \quad m \in \{1,\ldots,M\}, k \in \{0,\ldots,K-1\}, n \in \{1,\ldots,N\} \tag{20}$$

$$R_k^m \geq R_{min}^m, \quad k \in \{0,\ldots,K-1\} \tag{21}$$

The condition defined by Eq. (18) gives the interference power constraint to guarantee the interference to PUs. I^{TH} denotes the interference threshold acceptable for PUs. The condition given by Eq. (19) denotes the maximum transmission power constraint $\frac{B_k^m}{T}$ defines the total transmission power budget for the m-th SU at the time slot k. Equation (20) gives the minimum throughput requirement in cognitive femtocell network. The condition given by Eq. (21) guarantees the transmission power of each SU.

It remains to be defined how specific performance by measures are achieved by the RF-powered device as the expectation of RF energy harvesting rate, including average energy outage probability, and average transmission outage probability. The mathematical quantities of interest are then defined in the following. The expectation of the RF energy harvesting rate can be defined as:

$$E_{P_H} \triangleq E[P_H] \tag{22}$$

where the RF energy harvesting rate (in watts) by the device from the RF transmitter in a fading channel is given [21] by

$$P_H = \frac{\tau\beta P_S g_m}{d_\omega^\alpha} \tag{23}$$

where β is the RF-to-DC power conversion efficiency of the device, P_S is the transmit power of the PU transmitter, α is the path-loss exponent, h_ω is the channel power gain from the transmitter ω to the device, d_ω is the distance from the ω-th PU.

Energy outage occurs when the RF powered device cannot harvest sufficient RF energy from the ambiance to operate the circuit. The energy outage probability is defined as P_{CO}, P $(P_H < P_C)$, where P_H is the RF energy harvesting rate of the SU device, P_C is the circuit power consumption of the RF-powered device. QoS metric can be defined as a transmission outage probability.

Let $\kappa \geq 0$ denote the minimum information throughput requirement. If the RF-powered device fails to obtain enough throughput, it incurs a transmission outage. Thus, the transmission outage probability can be calculated as:

$$P_{TO} \triangleq P\left(P_H < P_C\right) + P(C < \kappa, P_H \geq P_C) \tag{24}$$

which indicates that the transmission outage occurs in two cases, namely when there is an energy outage, and when the decoded information throughput is less than the minimum requirement under the condition that there is enough harvested power.

4   An Algorithm for Energy Harvesting in Cognitive Femtocell Network

This part proposes an algorithm that can be used for RF energy harvesting in cognitive femtocell network.

Obtaining the highest possible amount of energy obtained by SU devices can be possible only when for all SUs are maximized their throughput, while maintaining guaranteed interference from the PU below a certain threshold. Therefore, an algorithm is proposed here that maximises the average throughput of SUs over a finite time interval. This algorithm uses a reward function, which is defined as the maximum of the sum of throughput at the current time slot and the expectected cumulative throughput at the future time slot from the current time system state.

The current reward function at time slot k is a function of the current energy budget B_k^m of each SU and the current system state S_k at time slot k and is given by

$$V_k(B_k^1, B_k^2, \ldots, B_k^M ; S_k) = \max_{p_k^{m,n}} E\{\sum_{v=k}^{K-1} \sum_{m=1}^{M} \sum_{n=1}^{N} log_2(1 + g_k^{m,n} p_k^{m,n})\}$$
$$k \in \{1, \ldots, K\} \tag{25}$$

The steps of throughput maximisation for energy harvesting in cognitive femtocell network are presented in Algorithm 1 (see Fig. 3). The presented algorithm checks the occupancy of all subcarriers available for each SU in cognitive femtocell. Then it maximises sum of the throughput at the current time slot. It uses Bellman's dynamic programming method.

5   Simulation Results

This section will present the results of simulation tests for cognitive femtocell network.

It was assumed that the system is composed of PU-Tx outside the femto-cell and a single PU-Rx inside fem-tocell. Inside the femtocell are located four SUs with energy harvesting devices. In addition, the OFDMA scheme is used OFDMA scheme, wherein the available spectrum is sharing into 12 subcarriers (N = 12). The number of subcarrier occupation state is equal to 70. It was assumed that the maximum battery capacity is equal to 6 J. Thus, the energy budget of femtocell is equal to 24 J. It is assumed that the constant depending on the energy budget at current time slot is equal to 0.001 W. In this case, the permissible interference at the PU-Rx is equal to 0.01 W.

Algorithm 1 Power allocation in CRFN system for energy harvesting

1: **procedure** PA IN CFN FOR ENERGY HARVESTING
2:    **for** $m \leftarrow 1, M$ **do**
3:       **if** $\sum_{n=1}^{N} \sum_{m=1}^{M} (a_k^{m,n} p_k^{P,n} + b_k^{m,n} p_k^{F,n}) \leq I^{TH}$ **then**
4:          **while** $\sum_{n=1}^{N} p_k^{m,n} \leq \frac{B_k^m}{T}$ **do**
5:             **for** $k \leftarrow 0, K-1$ **do**
6:                **for** $k \leftarrow K-1, 0$ **do**
7:                   $V_k(B_k^1, B_k^2, \ldots, B_k^M; S_k)$
8:                   $= \max_{p_k^{m,n}} E\{\sum_{v=k}^{K-1} \sum_{n=1}^{N} \sum_{m=1}^{M} \log_2(1 + g_k^{m,n} p_k^{m,n})\}$
9:                **end for**
10:            **end for**
11:          *Energy harvesting for the $m-th$ SU*
12:         **end while**
13:       **end if**
14:      *Energy harvesting for the $m-th$ SU*
15:    **end for**
16: **end procedure**

Fig. 3. Algorithm for energy harvesting in cognitive femtocell network.

Figure 4 shows the total throughput versus the number of slots. For comparison, the total throughput for the optimal area has been calculated by use the method proposed by [5]. It is evident that the proposed algorithm gives a minimally smaller values of total throuhput in comparison with the method described in [5].

Figure 5 shows the dependence of total energy budget depending on interference at PU-Rx. In this case, the proposed solution is slightly better than the method used in the paper by [5]. Nevertheless, this indicates the efficiency of the proposed algorithm for energy harvesting. This is due to the more accurate operation of the algorithm based on optimization than the proposed heuristic.

Figure 6 shows the total energy buget depending on the average bandwidth. It can be seen from the figure that the total energy budget is much higher for the optimal algorithm than for the used heuristic.

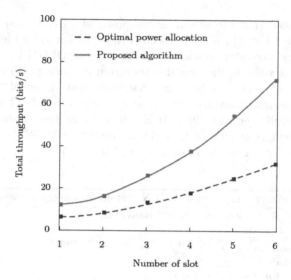

Fig. 4. Total throughput versus the number of slots.

Figure 7 shows P_{TO} as a function of an energy harvesting under different minimum information throughput requirement κ, which is associated with the specified QoS parameters. When the energy harvesting is small, transmission outage is mainly caused by insufficient harvested energy. Growth of energy harvesting causes decreasing value of transmission outage probability.

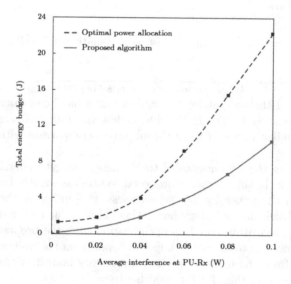

Fig. 5. Total energy budget versus average interference at PU-Rx.

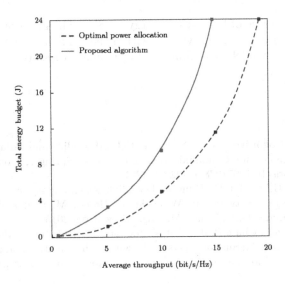

Fig. 6. Total energy budget versus the RF energy harvesting.

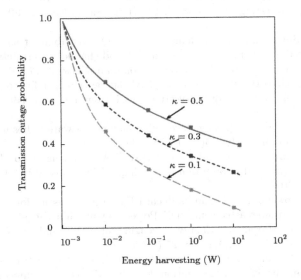

Fig. 7. Transmission outage probability versus RF energy harvesting.

6   Conclusion

This paper presents the model of energy harvesting in cognitive femtocell networks. This allow, among others SUs can generate energy from electromagnetic waves of PUs transmitters, and thus, the lifetime of these devices can be extended. The article presents the basic dependencies, combining SUs density with the value of harvesting energy by SUs. The given procedure shows the maximisation of energy harvesting within the femtocell. The compliance of the mathematical model was confirmed by simulation results, which were presented in the paper.

References

1. Arslan, H.: Cognitive Radio, Software Defined Radio, and Adaptive Wireless Systems. Signals and Communication Technology. Springer, Dordrecht (2007). https://doi.org/10.1007/978-1-4020-5542-3
2. Elmaghraby, H.M., Qin, D., Ding, Z.: Downlink scheduling and power allocation in cognitive femtocell networks. In: Weichold, M., Hamdi, M., Shakir, M.Z., Abdallah, M., Karagiannidis, G.K., Ismail, M. (eds.) CrownCom 2015. LNICST, vol. 156, pp. 92–105. Springer, Cham (2015). https://doi.org/10.1007/978-3-319-24540-9_8
3. Grover, P., Sahai, A.: Shanon meets Tesla: wireless information and power transfer. In: 1010 IEEE International Symposium on Information Theory, pp. 2363–2367 (2010)
4. Harba, A.: Energy harvesting: state-of-the-art. Renew. Energy 36(10), 2641–2654 (2011)
5. Ho, C.K., Zhang, R.: Optimal Energy allocation for wireless communications with energy harvesting constraints. IEEE Trans. Signal Process. 60(9), 4808–4818 (2012)
6. Huang, X., Shi, L., Zhang, C., Zhang, D., Chen, Q.: Distributed resource allocation with imperfect spectrum sensing information and channel uncertainty in cognitive femtocell Networks. EURASIP J. Wirel. Commun. Netw. 2017, 201 (2017)
7. Kirkidis, I., Timortheou, S., Sasaki, S.: RF energy transfer for cooperative networks: data relaying or energy harvesting? IEEE Commun. Lett. 16(11), 1772–1775 (2012)
8. Mitola, J., Maguire, G.Q.: Cognitive radio: making software radios more personal. IEEE Pers. Commun. Mag. 6(4), 13–18 (1999)
9. Nasir, A.A., Zhou, X., Durrani, S., Kennedy, R.A.: Relaying protocols for wireless energy harvesting and information processing. IEEE Trans. Wirel. Comm. 12(7), 3622–3636 (2013)
10. Ostaffe, H.: Power out of thin air: ambient RF energy harvesting for wireless sensors (2010). http://powercastco.com/PDF/Power-Out-of-Thin-Air.pdf
11. Paradiso, J.A., Starner, T.: Energy scavenging for mobile and wireless electronics. IEEE Pervasive Comput. 4(1), 18–27 (2005)
12. Sakr, A.H., Hossain, E.: Analysis of multi-tier uplink cellular networks with energy harvesting and flexible cell association. In: IEEE Global Communications Conference (2014)
13. Tariq, F., Dooley, L.S.: Cognitive femtocell networks. In: Grace, D., Zhang, H. (eds.) Cognitive Communications: Distributed Artificial Intelligence (DAI), Regulatory Policy and Economics, Implementation. Wiley (2012)
14. Li, Q., Feng, Z., Li, W., Liu, Y., Zhang, P.: Joint access and power control in cognitive femtocell networks. In: IEEE International Conference on Wireless Communicational and Signal Processing (2011)
15. Sakr, A.H., Hossain, E.: Analysis of K-Tier uplink cellular networks with ambient RF energy harvesting. IEEE J. Sel. Areas Commun. 33(10), 2226–2238 (2015)

16. Shah, S.T., Choi, K.W., Hasan, S.F., Chung, M.Y.: Energy harvesting and information processing in two-way multiplicative relay networks. Electron. Lett. **52**(9), 751–753 (2016)
17. Shah, S.T., Munir, D., Chung, M.Y., Choi, K.W.: Information processing and wireless energy harvesting in two-way amplify-and forward relay networks. In: 1016 IEEE 83rd Vehicular Technology Conference (VTC Spring), pp. 1–5 (2016)
18. Usman, M., Koo, I.: Access strategy for hybrid underlay-overlay cognitive radios with energy harvesting. IEEE Sens. J. **14**(9), 3164–3173 (2014)
19. Varshney, L.R.: Transporting information and energy simultaneously. In: 2008 IEEE International Symposium on Information Theory, pp. 1612–1616 (2009)
20. Zhang, Q., Kassam, S.A.: Finite-state Markov model for Rayleigh fading channels. IEEE Trans. Commun. **47**(11), 1688–1692 (1999)
21. Zhou, X., Zhang, R., Ho, C.K.: Wireless information and power transfer: architecture design and rate-energy tradeoff. IEEE Trans. Commun. **61**(11), 4754–4767 (2013)

Comparative Analysis of the Mechanisms for Energy Efficiency Improving in Cloud Computing Systems

A. V. Daraseliya[1]([✉])[iD], E. S. Sopin[1,2][iD], A. K. Samuylov[1][iD],
and S. Ya. Shorgin[1,2][iD]

[1] Peoples' Friendship University of Russia (RUDN University), Moscow, Russia
nastyadar6@gmail.com, {sopin_es,samuylov_ak}@rudn.university,
sshorgin@ipiran.ru
[2] Institute of Informatics Problems, FRC CSC RAS, Moscow, Russia

Abstract. We consider a cloud computing system with three different mechanisms for increasing the energy efficiency of cloud computing systems. We investigate how energy efficiency of a cloud system is
affected by a waiting time before a server goes to switch on/standby mode
and threshold-based switch. We developed four mathematical models of
Cloud computing system in terms of the queuing system and derived the
system of equilibrium equations, which makes it possible to obtain the
energy consumption indicators.

Keywords: Cloud computing · Energy efficiency · Queuing system

There are various mechanisms and approaches to improve the energy efficiency of
the cloud computing system. One way to improve energy efficiency is to upgrade
network equipment by implementing power-saving modes and adaptive transmission rates to achieve proportional power usage. Another example of an approach
to improve energy efficiency is scheduling and load balancing the servers, VMs,
and applications [4]. Speaking about VMs, several approaches can be taken for
improving energy efficiency. For example, VM self-adaption and hardware adjustment and choosing the most efficient physical machines for VM placements.
Finally, VM migrations allow dynamic consolidation of physical machines by
moving underutilized VMs onto fewer hosts and powering off the unused ones,
that permits reduced resource consumption at any given moment [3]. Similar
to virtual machines, it is possible to manage the power consumption of cloud
computing servers.

The servers can be put into standby state in order to improve the energy
efficiency of a cloud system in case of light load. On the one hand, the switching
to standby mode allows to reduce power consumption, and on the other hand, it

The publication has been prepared with the support of the "RUDN University Program 5–100" and funded by RFBR according to the research projects No. 16-07-
00766 and No. 18-07-00576.

O. Galinina et al. (Eds.): NEW2AN 2018/ruSMART 2018, LNCS 11118, pp. 268–276, 2018.
https://doi.org/10.1007/978-3-030-01168-0_25

leads to extra power usage to turn on/off the server. Therefore, it is important to understand under what conditions it will be advantageous to put the server in standby state, and under what conditions it is more profitable to leave it in the operating mode. In this regard, cloud providers employ various mechanisms that decrease server switching number. In this article we consider three mechanisms to improve the energy efficiency of the cloud computing server: the shutdown delay mechanism, the switch on delay mechanism and the threshold-based switch mechanism.

1 Analysis of Mathematical Models of Mechanisms for Increasing the Energy Efficiency of Cloud Systems

We proceed from the fact that arriving customers are distributed evenly on the virtual machine. Since we assumed that the computing resources are distributed evenly across all servers, we will consider a system consisting of one server with Processor Sharing policy.

We consider a base model [6] as a single-server queuing system with C virtual machines. Customers arrive according to the Poisson law with rate λ. Service times, switch on and switch off durations are exponentially distributed with the parameters μ, α and β, respectively. The system state is described by the vector (s, k), where k is the number of customers in the system, s is the server state. Here $s = 0$ means that the system is in the standby mode, $s = 1$ reflects switch-on mode and $s = 2$ and $s = 3$ represent operating and switch off modes, respectively. Arrival of a customer in an empty system cause change of the system state to the switch on mode. After exponentially distributed time with rate α, the system switches to the operating mode, in which serving of customers is started. When the system remains empty in the operating mode, it switches off immediately.

For the base model, the set of states S_1 is represented in the following form:
$S_1 = \{(s, k)|s = 1, 2,\ 1 \leq k \leq C\} \cup \{(s, k)|s = 3,\ 0 \leq k \leq C\} \cup (0, 0)$.

We derive the system of equilibrium equations, based on the transition intensity diagram [5, 6], which makes it possible to obtain stationary probabilities $p_{s,k}$ that the system is in (s,k) state:

$$p_{3,0} = \frac{\lambda}{\beta}p_{0,0}; \tag{1}$$

$$p_{1,1} = \frac{\lambda}{(\lambda + \alpha)}p_{0,0} + \frac{\beta}{(\lambda + \alpha)}p_{3,1}; \tag{2}$$

$$p_{1,k} = \frac{\lambda}{(\lambda + \alpha)}p_{1,k-1} + \frac{\beta}{(\lambda + \alpha)}p_{3,k}, \quad 2 \leq k \leq C - 1; \tag{3}$$

$$p_{1,C} = \frac{\lambda}{\alpha}p_{1,C-1} + \frac{\beta}{\alpha}p_{3,C}; \tag{4}$$

$$p_{2,k} = \frac{(\lambda + \mu)}{\mu}p_{2,k-1} - \frac{\lambda}{\mu}p_{2,k-2} - \frac{\alpha}{\mu}p_{1,k-1}, \quad 3 \leq k \leq C; \tag{5}$$

$$p_{2,C} = \frac{\alpha}{\mu}p_{1,C} + \frac{\lambda}{\mu}p_{2,C-1}; \tag{6}$$

$$p_{2,1} = \frac{(\lambda + \beta)}{\mu}p_{3,0}; \tag{7}$$

$$p_{2,2} = \frac{(\lambda + \mu)}{\mu}p_{2,1} - \frac{\alpha}{\mu}p_{1,1}; \tag{8}$$

$$p_{3,k} = \frac{\lambda}{(\lambda + \beta)}p_{3,k-1}, \quad 1 \le k \le C - 1; \tag{9}$$

$$p_{3,C} = \frac{\lambda}{\beta}p_{3,C-1}; \tag{10}$$

$$\sum_{i=0}^{3}\sum_{j=0}^{C} p_{i,j} = 1. \tag{11}$$

Due to the high energy consumption for shutting down the cloud server, in some cases it's more beneficial to leave it in operating mode pending the arrival of new customers. In [5] we consider the model with server shutdown delay mechanism. In contrast to the base model, where it was assumed that the server shuts down as soon as it remains empty, in this model the system does not switch off immediately, but waits exponentially distributed time with rate γ. If a customer arrives during that waiting period, then the system starts serving. Otherwise, the state is changed to the switch off mode. If a customer arrives during the switch off mode, then the system turns to the switch on mode immediately after the completion of the switch off. Otherwise, the system falls to the stand by mode. For this model, the set of states is represented in the following form: $S_2 = \{(s, k) | s = 1, 1 \le k \le C\} \cup \{(s, k) | s = 2, 3, 0 \le k \le C\} \cup (0, 0)$.

The system of equations for the model with shutdown delay mechanism differs from the system of equations for the base model by the following formulas: we added formula (12), and replaced formulas (7) and (8) by formulas (13) and (14).

$$p_{2,0} = \frac{\lambda + \beta}{\gamma}p_{3,0}; \tag{12}$$

$$p_{2,1} = \frac{\lambda + \gamma}{\mu}p_{2,0}; \tag{13}$$

$$p_{2,2} = \frac{(\lambda + \mu)}{\mu}p_{2,1} - \frac{\lambda}{\mu}p_{2,0} - \frac{\alpha}{\mu}p_{1,1}; \tag{14}$$

Also we consider the model with server switch on delay, as well as in base model, system passes in switch off mode at once after it remains empty. But it does not switch on immediately on arrival of a new customer, and waits exponentially distributed time with rate θ.

For this system, the set of states S_3 is represented in the following form: $S_3 = \{(s, k) | s = 0, 3, 0 \le k \le C\} \cup \{(s, k) | s = 1, 2, 1 \le k \le C\}$.

The system of equations for the model with switch on delay mechanism differs from the system of equations for the base model by the following formulas: we added formulas (18) and (19), and replaced formulas (2)–(4) by formulas (15)–(17).

$$p_{1,1} = \frac{\theta}{(\lambda + \alpha)} p_{0,1}; \tag{15}$$

$$p_{1,k} = \frac{\theta}{(\lambda + \alpha)} p_{0,k}, \quad 2 \le k \le C - 1; \tag{16}$$

$$p_{1,C} = \frac{\lambda}{\alpha} p_{1,C-1} + \frac{\theta}{\alpha} p_{0,C}; \tag{17}$$

$$p_{0,k} = \frac{\lambda}{\lambda + \theta} p_{0,k-1} + \frac{\beta}{\lambda + \theta} p_{3,k}, \quad 1 \le k \le C - 1; \tag{18}$$

$$p_{0,C} = \frac{\lambda}{\theta} p_{0,C-1} + \frac{\beta}{\theta} p_{3,C}; \tag{19}$$

Then we consider the mode with the threshold-based switch mechanism, in which system passes from standby mode in switch on mode only after arrived of a certain number κ of customers.

For this system, the set of states S_4 is represented in the following form: $S_4 = \{(s,k)|s = 0, \ 0 \le k \le \kappa - 1\} \cup \{(s,k)|s = 1, \ \kappa \le k \le C\} \cup \{(s,k)|s = 2, \ 1 \le k \le C\} \cup \{(s,k)|s = 3, \ 0 \le k \le C\}$.

The system of equations for this model differs from the system of equations for the base model by the following formulas: we added formula (25), and replaced formulas (2)–(4) by formulas (20) and (21) and formulas (5) and (6) by formulas (22)–(24).

$$p_{1,\kappa} = \frac{\lambda}{(\lambda + \alpha)} p_{1,\kappa-1} + \frac{\beta}{(\lambda + \alpha)} p_{3,\kappa}; \tag{20}$$

$$p_{1,k} = \frac{\lambda}{(\lambda + \alpha)} p_{1,k-1} + \frac{\beta}{(\lambda + \alpha)} p_{3,k}, \quad \kappa + 1 \le k \le C - 1; \tag{21}$$

$$p_{2,k+1} = \frac{(\lambda + \mu)}{\mu} p_{2,k} - \frac{\lambda}{\mu} p_{2,k-1}, \quad 2 \le k \le \kappa - 1; \tag{22}$$

$$p_{2,k+1} = \frac{(\lambda + \mu)}{\mu} p_{2,k} - \frac{\alpha}{\mu} p_{1,k} - \frac{\lambda}{\mu} p_{2,k-1}, \quad \kappa \le k \le C - 1; \tag{23}$$

$$p_{2,C} = \frac{\alpha}{\mu} p_{1,C} + \frac{\lambda}{\mu} p_{2,C-1}; \tag{24}$$

$$\lambda p_{0,k} = \lambda p_{0,k-1} + \beta p_{3,k}, \quad 1 \le k \le \kappa - 1; \tag{25}$$

We derived the system of equilibrium equations for each model, based on the transition intensity diagrams, which makes it possible to obtain stationary probability distribution of the system. Taking into account the normalization condition and using matrix methods, the system of equilibrium equations can be solved numerically, but we represent the analytical solution in [5].

2   Energy Consumption Indicators and the Performance Characteristics of Cloud Systems

With the system stationary distribution, we calculate the energy consumption indicators. We will assume that in the switch on/off mode, the power consumption is constant and equal to the average values P_1 and P_3, respectively. In the operating mode, the power consumption $P_{2,k}$ depends on the server occupancy. Through $P_{2,max}$ we denoted the maximum value of the server's power consumption in the operating mode, and through $P_{2,min}$ we denoted the power consumption in idle mode. The energy consumption in the standby mode will be calculated by P_0. By analogy with the formula given in [2], we derive the formula for the average server power consumption:

$$P = P_0 \sum_{i=0}^{C} p_{0,i} + P_1 \sum_{i=0}^{C} p_{1,i} + P_3 \sum_{i=0}^{C} p_{3,i} + \sum_{i=0}^{C} P_{2,i} p_{2,i} \tag{26}$$

where

$$P_{2,k} = P_{2,min} + \frac{P_{2,max} - P_{2,min}}{C} k \tag{27}$$

According to Little's law, the average number N of customers in the system is equal to the average effective arrival rate $\lambda(1 - \pi)$ multiplied by the average sojourn time T. Expressed algebraically the law is

$$N = \lambda(1 - \pi)T \tag{28}$$

where blocking probability π is

$$\pi = p_{0,C} + p_{1,C} + p_{2,C} + p_{3,C} \tag{29}$$

The average number N of customers is given by

$$N = \sum_{k=0}^{3} \sum_{i=1}^{C} i p_{k,i} \tag{30}$$

The average response time T follows directly from formulas (28) and (30):

$$T = \frac{\sum_{k=0}^{3} \sum_{i=1}^{C} i p_{k,i}}{\lambda(1 - \pi)} \tag{31}$$

3   Numeral analysis

On the energy profile of the cloud system installed at the University of Cardiff [1], it can be seen [1] that the inclusion of the server lasts 150 seconds, and the shutdown is 30 seconds. Further, for convenience, it was represented in minutes. The values of P_i were taken from [1], according to which $P_0 = 10$ W, $P_1 = 170$ W, $P_3 = 120$ W, $P_{2,min} = 105$ W and $P_{2,max} = 268$ W.

The results of numerical analysis for the values $C = 20$, $\mu = 20$, $\alpha = 1$, $\beta = 2$ are presented in Figs. 1, 2, 3, 4, 5 and 6.

The plots of the server's power consumption (Fig. 1) for base model show that the consumed power increases very fast for small values of the arrival flow intensity λ, also note that with the increase of waiting time, during which the system doesn't go into standby mode, the power consumption also increases.

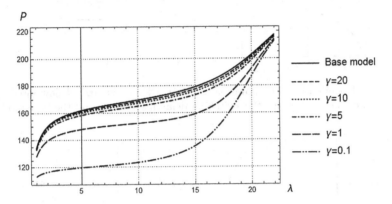

Fig. 1. The dependence of the power consumption P on the arrival flow intensity λ. Comparison of the model with the shutdown delay mechanism and the base model.

In Fig. 2 it is clearly seen that the greatest dependence of the average sojourn time T on the arrival flow intensity λ is observed at values of γ from 1 to 5.

The graph of the server power consumption P (Fig. 3) for the model with server shutdown delay increases most sharply at small values of the arrival flow intensity λ, also note that with an increase in the exponential time θ during which

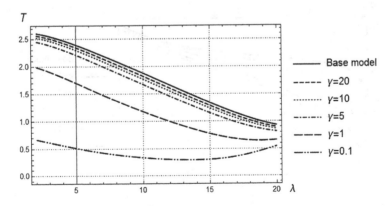

Fig. 2. The dependence of the average response time T on the arrival flow intensity λ. Comparison of the model with the shutdown delay mechanism and the base model.

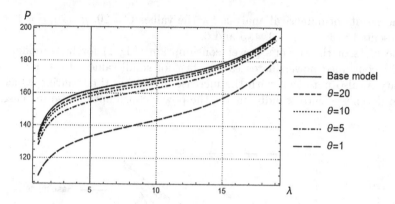

Fig. 3. The dependence of the power consumption P on the arrival flow intensity λ. Comparison of the model with the switch on delay mechanism and the base model.

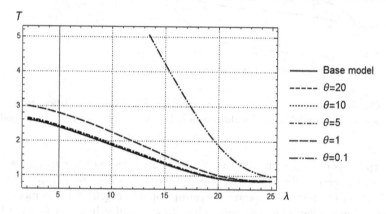

Fig. 4. The dependence of the average response time T on the arrival flow intensity λ. Comparison of the model with the switch on delay mechanism and the base model.

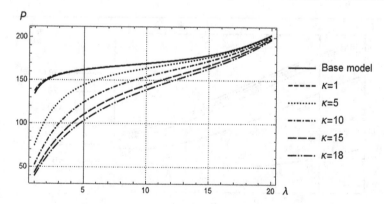

Fig. 5. The dependence of the power consumption P on the arrival flow intensity λ. Comparison of the model with the threshold-based switch mechanism and the base model.

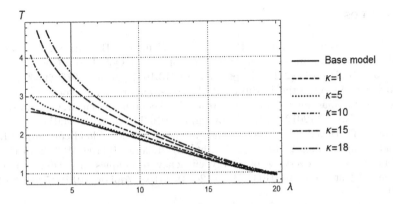

Fig. 6. The dependence of the average response time T on the arrival flow intensity λ. Comparison of the model with the threshold-based switch mechanism and the base model.

the system doesn't go into the operating mode after the first customer is received, the value of the power consumption increases accordingly. Note that for $\theta = 20$, the graph of the model with server shutdown delay is most closely approximated to the base model's graph, where the system was included immediately after the customer was received.

In Fig. 4 note that for small values of θ, the difference in the average sojourn time T is the greatest.

For each of the three models, it was noted that when $\gamma \to \infty$ (see Figs. 1 and 2), $\theta \to \infty$ (see Figs. 3 and 4) and $\kappa \to 1$ (see Figs. 5 and 6), respectively, each of the three systems tends to the initial baseline model.

4   Conclusion

We constructed mathematical models of cloud computing systems taking into account various mechanisms for increasing energy efficiency in terms of queuing theory, and analytical expressions for the main characteristics of energy consumption and server performance metrics were obtained. We conducted a comparative numerical analysis of the mechanisms for improving the energy efficiency of cloud computing based on the initial data close to the real ones. Numerical analysis showed that the mechanism with server shutdown delay gives an improvement in both power and response time, for a mechanism with server enable delay, there is an improvement at almost all intervals of power consumption, except service time at low loads. The mechanism with server thresholding gives an improvement for power, but deterioration in time.

References

1. Conejero, J., Rana, O., Burnap, P., Morgan, J., Caminero, B., Carrion, C.: Analysing Hadoop power consumption and impact on application QoS. Futur. Gener. Comput. Syst. **55**(C), 213–223 (2016). https://doi.org/10.1016/j.future.2015.03.009
2. Beloglazov, A., Abawajy, J., Buyya, R.: Energy-aware resource allocation heuristics for efficient management of data centers for cloud computing. Futur. Gener. Comput. Syst. **28**, 755–768 (2012). https://doi.org/10.1016/j.future.2011.04.017
3. Mastelic, T., Brandic, I.: Recent trends in energy-efficient cloud computing. IEEE Cloud Comput. Mag. **2**, 40–47 (2015). https://doi.org/10.1109/MCC.2015.15
4. Valentini, G.: An overview of energy efficiency techniques in cluster computing systems. Clust. Comput. **16**(1), 3–15 (2013). https://doi.org/10.1007/s10586-011-0171-x
5. Daraseliya A.V., Sopin E.S.: Analysis of an approach to increase energy efficiency of a cloud computing system. In: Selected Papers of the II International Scientific Conference "Convergent Cognitive Information Technologies" (Convergent 2017), vol. 2064, pp. 79–87. CEUR Workshop Proceedings, Moscow (2017). http://ceur-ws.org/Vol-2064/paper09.pdf
6. Daraseliya A.V., Sopin E.S.: Energy efficiency analysis of cloud computing system with setup and vacation perion of server. In.: Information and Telecommunication Technologies and Mathematical Modeling of High-Tech Systems (ITTMM 2017), pp. 119–121 (2017)

Blue Team Communication and Reporting for Enhancing Situational Awareness from White Team Perspective in Cyber Security Exercises

Tero Kokkonen$^{(\boxtimes)}$ and Samir Puuska$^{(\boxtimes)}$

Institute of Information Technology, JAMK University of Applied Sciences,
Jyväskylä, Finland
{tero.kokkonen,samir.puuska}@jamk.fi

Abstract. Cyber security exercises allow individuals and organisations to train and test their skills in complex cyber attack situations. In order to effectively organise and conduct such exercise, the exercise control team must have accurate situational awareness of the exercise teams. In this paper, the communication patterns collected during a large-scale cyber exercise, and their possible use in improving Situational awareness of exercise control team were analysed. Communication patterns were analysed using graph visualisation and time-series based methods. In addition, suitability of a new reporting tool was analysed. The reporting tool was developed for improving situational awareness and exercise control flow. The tool was used for real-time reporting and communication in various exercise related tasks. Based on the results, it can be stated that the communication patterns can be effectively used to infer performance of exercise teams and improve situational awareness of exercise control team in a complex large-scale cyber security exercise. In addition, the developed model and state-of-the-art reporting tool enable real-time analysis for achieving a better situational awareness for the exercise control of the cyber security exercise.

Keywords: Cyber security · Exercise · Training
Situational awareness · Communication

1 Introduction

Cyber security is an ongoing process where both organisations and individuals are training, working, and learning continually. Cyber security exercises are an excellent way to train and simultaneously test an organisation's or individual's capabilities under stressful cyber-attack situations. The exercise can be conducted in both public and private sectors. The cyber security strategy of the European Union notices the importance of national and international cyber security exercises [8]. Finland's security strategy for society states several times the

© Springer Nature Switzerland AG 2018
O. Galinina et al. (Eds.): NEW2AN 2018/ruSMART 2018, LNCS 11118, pp. 277–288, 2018.
https://doi.org/10.1007/978-3-030-01168-0_26

importance of regular exercises for improving the resilience against threats [23], whereas Finland's cyber security strategy states that cyber threats are evolving extremely rapidly, and therefore cyber security exercises should be conducted regularly for improving preparedness and cyber resilience [22]. Handbook for information technology and cyber security exercises [26] lists following exercise types: unannounced live exercises, initiation exercises, staff exercises, decision exercises, management exercises, cooperation exercises and Red Team - Blue Team exercises. The exercise type indicates the primary function of the exercise.

Cyber security exercises are usually organised using various teams with different tasks or missions. These teams are formed based on exercise type, training goals, and available resources and personnel. Blue Team (BT) is a group of people defending their information technology assets against cyber threats. They also report the observations to (simulated) management, create their own situational awareness and maintain their own security posture under cyber-attack. BT is very often modelled after a real organisation, team, or branch. There can be one or many BTs in the exercise that can represent different aspects of the real world. BTs often aim to role-play their normal organisational practices and procedures. Red Team (RT) is a group of people simulating the threat actors in the exercise by making real cyber-attacks against Blue Teams. White Team (WT) is responsible for controlling the exercise, making observations, collecting the data and handling the situational awareness of the exercise [5,13,25,26].

Sometimes the exercise control team is also called EXCON which has similar functions as WT. In that sense, the situational awareness of the WT is extremely important for controlling the exercise and for making the required decisions during the exercise. The communication patterns of the BTs are an important source for understanding what is happening in the exercises from the BT's perspective, and how they are communicating with the co-operation organisations under cyber-attack.

One of the most classical definitions of situational (or situation) awareness is as follows: *"Situation awareness is the perception of the elements in the environment within a volume of time and space, the comprehension of their meaning, and the projection of their status in the near future"* [7]. In this study, the term situational awareness (SA) is used. At the first level of SA there is the perception (observations and sensor information), the second level is the comprehension (understanding the current situation) and the third level is the projection (prediction of future events based on the information of earlier states and decision makers' pre-learned history). It is stated that with erroneous SA even the trained decision makers will make incorrect decisions [7]. In the cyber security the objective of SA is to know what is (and will be) the security level of organisation's assets in the networked systems [9].

Cyber security exercises enable a comprehensive platform for studying situational awareness in cyber security and behaviour or efficiency of individuals and teams under cyber-attack. In the study [6] a methodology is proposed for adjustment of situation awareness measurement experiments within the context of a cyber security exercise. The author of [10] states that cyber security exercises can be used as an empirical study of situation awareness in cyber security. Also, the paper [5] deploys cyber security exercise data for profiling the attacker. According to the authors of the studies [3,4], training and exercises have an important role for improving the competencies in the defence of the cyber security assets and for achieving the required level of preparedness especially in the resilience of critical infrastructure.

Situational awareness is important for all involved teams in the exercise. However, WT is required to have an understanding of the SA of the BTs in exercise in order to effectively adjust and steer the exercise towards fulfilling the desired learning and testing goals. Traditional monitoring of technical details of the exercise environment supplemented with the analysis of communication patterns provides an extensive view into Blue Team behaviour.

This study presents the study of Blue Team communication patterns and based on that the implementation of the state-of-the–art reporting tool for enhancing the SA of the White Team during the complex and hectic cyber security exercise. First the Finland's national cyber security exercise is introduced, the event timelines are studied, and analysis is made. In addition, the reporting tool is developed and studied to produce incident reports for enhancing the SA of the White Team. Finally, the conclusions are done, and future research ideas are found and introduced.

2   Finland's National Cyber Security Exercise

Finland's national cyber security exercise has been conducted annually since 2013 and every year, the Cyber Range of Finland's national cyber exercise has been Realistic Global Cyber Environment (RGCE) developed by JAMK University of Applied Sciences Institute of Information Technology [18].

Finland's national cyber security exercise of 2017 was executed from 8th of May to 11th of May and it was commanded by the Ministry of Defence with The Security Committee. The RGCE Cyber Range and the overall implementation was conducted by JAMK University of Applied Sciences. There were more than 100 individuals participating in the exercise forming several co-operating Blue Teams communicating with each other according to their operational tasks. The aim of the exercise was to practice co-operation between security organisations and security network organisations in Finland during cyber-attacks or incidents for verifying the performance of the participant organisations and ensuring their further development [18].

As described in the aim of the exercise, the Blue Teams of the exercise were formed from different security authorities of Finland. All of them were acting, communicating and co-operating according to their real operational tasks during the realistic cyber attacks of several simulated threat actors. Some of the Blue Teams mainly defend their own assets whereas some Blue Teams have highly co-operational role and act and communicate actively in accordance with that role.

2.1   RGCE Cyber Range

RGCE is a fully operational Cyber Range that mimics the structures, services and traffic of the real Internet. It allows the usage of real IP addresses and global GeoIP information with realistic end user traffic patterns automatically generated by botnet based special software. RGCE is a closed environment, which allows usage of real attacks or malware [12,14].

3   Event Timelines

Cyber security exercises consist of several components forming the core which the White Team uses to direct the overall flow. A typical exercise contains a background story that sets the general tone and mindset for the trainees. Several threat actors are created to portray real-world counterparts, such as hacktivist groups and more advanced organisations. Based on these actors and their modus operandi, various attack scenarios are prepared. The scenarios may include technical exploitations, denial-of-service attacks, social engineering, and advanced directed cyber operations.

3.1   Injects

Injects are pre-prepared actions in the Cyber Range. They are modelled after the threat actor's simulated campaigns. For example, a malicious group may want to use a denial-of-service (DoS) or a distributed denial-of-service (DDoS) attack to mask a more advanced exploit, targeted at one team. This could be achieved by two injects, one for each type of attack. The schedule for injects is drafted at the planning stage. However, due to the live nature of cyber exercises, White Team may choose to adjust their timing, targets or their potential execution, depending on the Blue Team response. Adjusting overlapping incidents and injects to support learning goals and desired stress levels is crucial for a successful exercise.

For the studied exercise, dozens of injects were prepared to simulate the cyber attack campaigns of threat actors. There were several realistic threat actors modelled and simulated in the exercise and the injects were prepared to simulate the behaviour of those threat actors. The attack campaigns varied from volumetric DoS/DDoS campaigns to targeted advanced persistent threat (APT) attacks including for example realistic behaviour of threat actors in social media.

Figure 1 illustrates the duration of the injects during the cyber security exercise. When WT decides to activate an inject, the actual time is recorded, as well as the moment when the inject in question is marked as 'executed', i.e. it does not require any further work from any of the teams. Figure 1 shows, that the approximate workload is relatively evenly distributed inside each exercise day, first and last being less intensive. This was the desired goal in the planning stage.

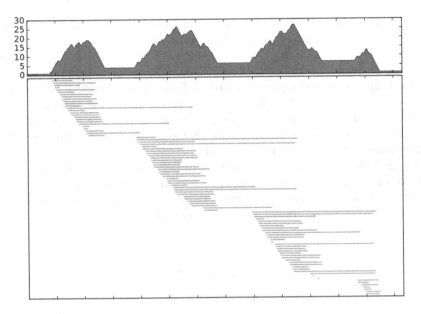

Fig. 1. Inject timing, durations (lower), and cumulative sum (upper) during the Exercise.

3.2 Communication Methods

Blue Teams were given various common methods for communicating between groups and internally. Each team had corporate email-accounts, two kinds of direct messaging options, and VOIP phones. Overall, the teams preferred e-mail over other forms of communication. Therefore, this study focuses on e-mails, and data fusion between other systems is considered as future work.

4 Analysis

Although Fig. 1 illustrates the approximate amount of desired work, it does not tell how the exercise teams actually react to the injects. In some cases the exercise teams may miss the inject entirely or fail to take appropriate measures. Direct monitoring or questionnaires disturb the flow of the exercise and require extra personnel.

E-mail patterns were analysed to see what communication patterns teams use during incidents. The mail headers were extracted from mail servers and analysed and visualised using Cytoscape software [24].

4.1 Team Communication Patterns

BTs in the exercise played several different roles. For example, one BT formed a common networking and service platform, which includes physical networks, as well as workstations and intranet services, and another BT was a cyber security service organisation offering services to all other teams.

During the exercise tens of thousands of emails were sent and received, also including an e-mail-based Denial of Service -attack, as well as general spam, and e-mails from automated reporting systems. BTs also forwarded information to each other using large mailing lists. Some teams included their own address into these lists, and therefore received many copies of their own mails. White Team also answered to requests and inquiries that were directed to higher levels of organisations not occupied in the exercise.

Figure 2 illustrates all used message paths between parties. Red nodes represent attacker-controlled domains, coloured ones are the Blue Teams. Edge colour indicates the sending party. The graph shows that Teams two and five never communicated directly, even though they should have.

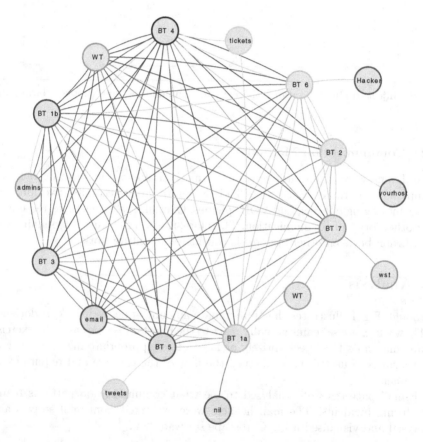

Fig. 2. The complete communication graph between domains.

The mailing patterns mostly reflect the nature and purpose of each team. Blue Team one, which was responsible for the core services, communicated with all other organisations actively. Their mails informed the organisations that were using their services about various disruptions, estimated repair times, and detected threats. Blue Team two was noticeably less active. They sent only a few notices of service disruptions, and mainly co-operated with Blue Team one, even though they were kept up to date by other teams. Blue Team three mostly co-operated with Blue Team five, which was expected. Blue Team six communicated actively with every other team, delivering threat intelligence and analysis services. Blue Teams four and one were also targeted by external Denial of Service and phishing campaigns. This may have affected their capability to send and receive mails.

In Fig. 3a, a typical set of service requests and responses is made. They indicate that the teams still have control over their infrastructure, and are able to take defensive measures. Figure 3b illustrates a phishing attempt, which later evolved into a spamming attack. Grey nodes represent mailboxes belonging to non-playing teams, while red nodes are controlled by threat actor (RT). In Fig. 3c Blue Team six has detected an unusually intensive port scanning originating from the Internet. The team informs others, and it can be seen that one team asks for more details.

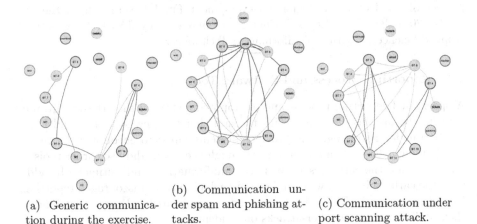

(a) Generic communication during the exercise.

(b) Communication under spam and phishing attacks.

(c) Communication under port scanning attack.

Fig. 3. Example of communication patterns.

Although the analysis of communication patterns revealed some omissions and errors that teams made, it does not have enough information for White Team to form a robust SA. Also, the analysis of communication pattern is not conducted in the real time and more real time reporting tool is required for improving the situational awareness of White Team. It can be concluded, that a special real time reporting system is required for obtaining data and understanding the Blue Team behaviour during the complex cyber security exercise.

5   Reporting Tool for Improving White Team SA

Situational awareness is required as a basis for decision making. OODA loop (Observation-Orientation-Decision-Action) is a classical model for decision making [15,21]. Another similar decision-making loop is introduced in four stages of an adaptive security architecture (Predict-Prevent-Detect-Respond) [17]. When reflected to both of those loops and earlier introduced definition of SA, SA is an extremely important element of decision making. When considering different data from different sources or sensors, there is a requirement for data fusion or multi-sensor data fusion, which is a process of synthesising overlapping and scattered data from the different sensors or sources to the user for achieving comprehensive SA of focused events [2,11].

In the cyber security exercises, the Blue Team reporting tool for gathering the SA is required in two functions. First the Blue Teams report (automatically from sensors or manually) their observations to the tool and forms their SA based on data fusion. Secondly, White Team is able to monitor what the Blue Teams are reporting and what mitigation actions they are executing [16].

The developed Reporting Tools was tested in the cyber security exercise in the industrial domain [20]. Industrial cyber security exercise is piloted in the project of the European Regional Development Fund/Leverage from the EU 2014-2020, called JYVSECTEC Center and managed by JAMK University of Applied Sciences Institute of Information Technology.

5.1   Reporting Process and Software Tool

A specialised reporting process and a supporting state-of-the-art software tool for Blue Teams was developed with the aim that the new system would lower the barrier for reporting. The previous systems failed to encourage the teams or reporting actionable information. Although the teams did use earlier tools to report events, the messages were short, uninformative, and untimely. In addition, the earlier platform was cumbersome, which further discouraged reporting. Reporting is seen in Blue Teams as an unnecessary artificial chore that hinders their ability under the cyber-attacks or incidents.

The goal of development was to construct a reporting tool and process that would be unobtrusive and quick to use. Comprehensive reporting was encouraged by providing a template which contained necessary headings and hints what to put under them. GUI with muted colours was opted to use instead of the console-based solutions.

5.2   Reporting Format

For helping the trainees during the complex exercise scenarios, the reporting format is kept relatively simple; it borrows elements from military-style situation report structure. Table 1 presents the main elements of the format. In addition to the presented elements, each report has a time-stamp and title.

Table 1. Report template fields, translations, and purpose.

Field (in Finnish)	Field (translated in English)	Purpose
Havainnon laatu	Type of observation	What is being reported? Error condition, support request, malicious program, etc.
Tapahtuma	Incident	What has happened?
Seuraukset	Consequences	What impact will this incident cause? What further measures will be likely taken to mitigate the impact?
Tarkennukset	Further information	Additional details about the incident or of the overall situation
Paikka	Place	Place, if relevant

A formal language was constructed for describing the reporting format in order to construct domain specific language (DSL). This domain specific language (DSL) allows the reports to be both human and machine readable. DSL is also expandable; other message types can be added in the future. The DSL was also equipped with syntax highlighting in the tool. As the DSL is verified using a formal language parser, the program can also notify user if values are missing or invalid.

The main view is illustrated in Fig. 4a By default, the user sees two windows, one of which lists all reports made by his/her team, the other window is for creating a new report. By clicking the reports, they can be opened into a new window and examined separately. The screen-shot shows one additional window that the user has opened.

Figure 4b is a screen-shot of the reporting screen. For keeping the tool simple during the complex and hectic exercise, there are only two buttons and one syntax indicator present in the editor. The button labelled "Tilanneilmoitus" (Situation report) will fill the editor with the report template. The indicator states if the document does not conform to our DSL specification. The reporting window is a text editor with additional syntax highlighting features.

The tool was implemented using Java programming language and JavaFX UI framework, making the tool cross-platform ready [19]. The program utilises a message bus for synchronising messages between team members and delivering a copy of each message to White Team. Our implementation used Apache ActiveMQ message bus for communication [1].

(a) Main view of the application. The list shows past reports, and the top window shows one of them in full detail. The report editor is in the background.

(b) Report editor with the report template loaded.

Fig. 4. Screen shots of reporting tool.

6   Conclusion

Monitoring Blue Team communication provides further insight into both exercise status and team behaviour. As the analysis suggests, communication monitoring can be a useful tool in measuring Blue Team performance during the cyber security exercise. The analysis revealed several omissions made by the Blue Teams. In addition, although the overall inject timing was successful, some teams might have benefited from intense workload.

When planning the injects, it could be useful to consider which teams are affected, and who is responsible for keeping them informed. By implementing real-time communication monitoring, the White Team can efficiently tell if the teams are acting correctly.

By using e-mail graphs in conjunction with other monitoring mechanisms, real-time mail visualisation aids White Team to build a more robust situational awareness over the exercise. This allows more fine tuned and accurate control, as well as more comprehensive results from the exercise.

However, the special reporting system is required to reliably monitor the Blue Team behaviour in real-time during the cyber security exercise. This requires additional timely reports from the Blue Teams, and a convenient, non-intrusive way for writing and delivering them. A specialised report format and state-of-the-art software tool was developed for achieving this goal. The tool was tested in the cyber security exercises within the industrial domain. It will also be used in the future exercises with improvements suggested in the initial tests.

Future work in the communication monitoring includes automating the message parsing and visualisation process so, that it is readily available to White Team during the exercise. This includes the development of a better visualisation system for monitoring purposes. In the future graphics will be designed to visualise multi-edged graphs efficiently for SA purposes. Future work with the reporting system will be more visualised SA of Blue Team behaviour for certain exercise inject and improvements of BT SA used for BTs' tactical leading and decision making.

Acknowledgment. This research is partially done in JYVSECTEC Center project funded by the Regional Council of Central Finland/Council of Tampere Region and European Regional Development Fund/Leverage from the EU 2014–2020.

References

1. The Apache Software Foundation: Apache ActiveMQ. http://activemq.apache.org/. Accessed 23 Apr 2018
2. Azimirad, E., Haddadnia, J.: The comprehensive review on JDL model in data fusion networks: techniques and methods. (IJCSIS) Int. J. Comput. Sci. Inf. Secur. **13**(1), 53–60 (2015)
3. Brilingaitė, A., Bukauskas, L., Krinickij, V., Kutka, E.: Environment for cybersecurity tabletop exercises. In: Pivec, M., Josef, G. (eds.) ECGBL 2017 11th European Conference on Game-Based Learning, pp. 47–55. Academic Conferences and Publishing Limited (2017)
4. Brilingaitė, A., Bukauskas, L., Kutka, E.: Development of an educational platform for cyber defense training. In: Scanlon, M., Nhien-An, L.K. (eds.) Proceedings of the 16th European Conference on Cyber Warfare and Security, pp. 73–81. Academic Conferences and Publishing Limited (2017)
5. Brynielsson, J., Franke, U., Tariq, M.A., Varga, S.: Using cyber defense exercises to obtain additional data for attacker profiling. In: 2016 IEEE Conference on Intelligence and Security Informatics (ISI), pp. 37–42, September 2016. https://doi.org/10.1109/ISI.2016.7745440
6. Brynielsson, J., Franke, U., Varga, S.: Cyber situational awareness testing. In: Akhgar, B., Brewster, B. (eds.) Combatting Cybercrime and Cyberterrorism. ASTSA, pp. 209–233. Springer, Cham (2016). https://doi.org/10.1007/978-3-319-38930-1_12
7. Endsley, M.: Toward a theory of situation awareness in dynamic systems. Hum. Factors **37**(1), 32–64 (1995). https://doi.org/10.1518/001872095779049543
8. European Comission: Cybersecurity Strategy of the European Union: An Open, Safe and Secure Cyberspace, February 2013
9. Evesti, A., Kanstrén, T., Frantti, T.: Cybersecurity situational awareness taxonomy. In: 2017 International Conference on Cyber Situational Awareness, Data Analytics And Assessment (Cyber SA), pp. 1–8, June 2017. https://doi.org/10.1109/CyberSA.2017.8073386
10. Franke, U., Brynielsson, J.: Cyber situational awareness - a systematic review of the literature. Comput. Secur. **46**, 18–31 (2014)
11. Han, X., Sheng, H.: A new method of multi-sensor data fusion. In: 2017 IEEE 3rd Information Technology and Mechatronics Engineering Conference (ITOEC), pp. 877–882, October 2017. https://doi.org/10.1109/ITOEC.2017.8122479
12. JAMK University of Applied Sciences, Institute of Information Technology, JYVSECTEC: RGCE cyber range. http://www.jyvsectec.fi/en/rgce/. Accessed 23 Apr 2018

13. Kick, J.: Cyber exercise playbook. The MITRE Corporation (2014). https://www. mitre.org/sites/default/files/publications/pr_14-3929-cyber-exercise-playbook. pdf. Accessed 23 Apr 2018

14. Kokkonen, T., Hämäläinen, T., Silokunnas, M., Siltanen, J., Zolotukhin, M., Neijonen, M.: Analysis of approaches to internet traffic generation for cyber security research and exercise. In: Balandin, S., Andreev, S., Koucheryavy, Y. (eds.) ruSMART 2015. LNCS, vol. 9247, pp. 254–267. Springer, Cham (2015). https://doi. org/10.1007/978-3-319-23126-6_23

15. Lenders, V., Tanner, A., Blarer, A.: Gaining an edge in cyberspace with advanced situational awareness. IEEE Secur. Priv. **13**(2), 65–74 (2015). https://doi.org/10. 1109/MSP.2015.30

16. Lötjönen, J.: Requirement specification for cyber security situational awareness, Defender's approach in cyber security exercises. Master's thesis, JAMK University of Applied Sciences, December 2017

17. van der Meulen, R.: Build adaptive security architecture into your organization, June 2017. https://www.gartner.com/smarterwithgartner/build-adaptive-security-architecture-into-your-organization/. Accessed 23 Apr 2018

18. Ministry of Defence Finland: The authorities of the state administration are trained in cyber-skills in Jyväskylä - Valtionhallinnon viranomaiset harjoittelevat kyberosaamista Jyväskylässä 8.-11.5.2017, official bulletin 3th of May 2017, May 2017. https://www.defmin.fi/ajankohtaista/tiedotteet/valtionhallinnon_viranomaiset_harjoittelevat_kyberosaamista_jyvaskylassa_8.-11.5.2017.8418.news. Accessed 23 Apr 2018

19. Oracle Corporation: Java programming language. http://www.oracle.com/technetwork/java/index.html. Accessed 23 Apr 2018

20. Pajunen, D.: Cyber security is ensured with genuine exercises, September 2017. https://www.fingridlehti.fi/en/cyber-security-ensured-genuine-exercises/. Accessed 23 Apr 2018

21. Révay, M., Líška, M.: Ooda loop in command control systems. In: 2017 Communication and Information Technologies (KIT), pp. 1–4, October 2017. https://doi. org/10.23919/KIT.2017.8109463

22. Secretariat of the Security Committee: Finland's Cyber security Strategy, Government Resolution 24.1.2013, January 2013

23. The Security Committee: Security Strategy for Society, Government Resolution 2.11.2017, November 2017

24. Shannon, P., et al.: Cytoscape: a software environment for integrated models of biomolecular interaction networks. Genome Res. **13**(11), 2498–2504 (2003). https://doi.org/10.1101/gr.1239303

25. Sommestad, T., Hallberg, J.: Cyber security exercises and competitions as a platform for cyber security experiments. In: Jøsang, A., Carlsson, B. (eds.) NordSec 2012. LNCS, vol. 7617, pp. 47–60. Springer, Heidelberg (2012). https://doi.org/10. 1007/978-3-642-34210-3_4

26. Wilhelmson, N., Svensson, T.: Handbook for Planning, Running and Evaluating Information Technology and Cyber Security Exercises. The Swedish National Defence College, Center for Asymmetric Threats Studies (CATS) (2014)

An Approach to Classification of the Information Security State of Elements of Cyber-Physical Systems Using Side Electromagnetic Radiation

Viktor Semenov[1,2], Mikhail Sukhoparov[3(✉)], and Ilya Lebedev[2]

[1] ITMO University, 49 Kronverksky pr., Saint Petersburg 197101, Russia
[2] SPIIRAS, 14-th Linia, VI, No. 39, Saint Petersburg 199178, Russia
[3] SPbF AO «NPK «TRISTAN», 47 Nepokorennykh pr., Saint Petersburg 195220, Russia
sukhoparovm@gmail.com

Abstract. The paper deals with problematic issues of information security in cyber-physical systems. Performance analysis of autonomous objects has been carried out. An information security monitoring system model based on the characteristics resulting from the analysis of electromagnetic radiation from electronic components in standalone devices of cyber-physical systems is presented. A typical scheme for determining the state of a system is shown. Due to the features of equipment sustaining the infrastructure, assessment of an information security state is aimed at analyzing normal system operation rather than searching for signatures and characteristics of anomalies during various types of information attacks. An experiment that provides statistical information on the operation of remote devices of cyber-physical systems has been disclosed, whereby data for decision-making are accumulated by comparing statistical information. The experimental results on information influence on a typical system are presented. The proposed approach for analyzing statistical data of standalone devices based on a naive Bayesian classifier can be used to determine information security states. A special feature of the approach is the ability to rapid adaptation and application of various mathematical tools and machine learning methods to achieve a desired quality of probabilistic evaluation.

Implementation of this type of monitoring does not require a development of complex system applications while allowing implementation of various architectures for system construction that are capable of processing on-board an autonomous object or of communicating data and calculating the state on external computer nodes of monitoring and control systems.

Keywords: Information security · Cyber-physical systems
Information security monitoring systems

1 Introduction

The present development stage of cyber-physical systems (CPSs) is characterized by the use of unmanned vehicles: flying objects, cars, trains. The absence of an operator or the presence of only a remote control afford ground for information influence on them

© Springer Nature Switzerland AG 2018
O. Galinina et al. (Eds.): NEW2AN 2018/ruSMART 2018, LNCS 11118, pp. 289–298, 2018.
https://doi.org/10.1007/978-3-030-01168-0_27

outside the controlled zone, making such means a very attractive target for attempts at various attacks [1–3]. Implementation of a large number of projects related to unmanned systems calls for ensuring the required level of security of the data circulating in them.

The introduction of unmanned vehicles is accompanied by the need to resolve additional challenging issues to ensure information security, such as [4, 5]:

- detection of unauthorized access to the major nodes at the software level;
- analysis and detection of anomalies in the technological cycles of an unmanned vehicle operation;
- detection of destructive information influence on programs and algorithms;
- monitoring software versions in order to detect undocumented features.

Identification of abnormal operating parameters, deviations of various characteristics, incorrect or inadequate instruction statements, a large number of repeated events is an important task for ensuring the information security of CPSs [6–8].

Additional means of protection consume information and energy resources and possess dimensions and weight characteristics, which is not always acceptable within the constraints of unmanned vehicles, and in case of unauthorized access and modification of the software code, they can become inoperative.

In this regard, a number of challenges are posed aimed at external monitoring of events of an object information security.

2   Problem Statement

Effective solutions in the field of information security are associated with the development of a research and methodical framework aimed at improving the quality indicators of security identification, which in turn necessitates the development of patterns and methods for monitoring information security of standalone computing facilities [9, 10] taking into account the CPS features.

One approach may be the use of electromagnetic radiation of various operating electronic components. To identify abnormal behavior, it is necessary to use characteristics that reflect the system states that can be used in statistical analysis [11].

In the case under consideration, the monitoring system D consists of a set of sensors $\{d_1, d_2, \ldots, d_n\}$ that pick up the emitted signals from device components. Each element of a system d_i processes a signal $s_{di}(t)$ from the component where it is mounted. As a result of transformation, a sequence of values $\{s_{di}(0), s_{di}(1), \ldots, s_{di}(m)\}$ appears at the time $t = 0, 1 \ldots m$ from the sensor d_i. By timing the process of reading the values of elements $\{d_1, d_2, \ldots, d_n\}$, one obtains tuples of characteristics $\{s_{d1}(0), s_{d2}(0), \ldots, s_{dn}(0)\}$, $\{s_{d1}(1), s_{d2}(1), \ldots, s_{dn}(1)\}, \ldots, \{s_{d1}(m), s_{d2}(m), \ldots, s_{dn}(m)\}$ at time intervals $t = 0, 1, \ldots m$.

Assuming that from sensors mounted on different components where software processes proceed, one can obtain a set of characteristic values at one time; hence, the task of determining the information security state can be reduced to a classification problem.

3   The Proposed Approach

The behavior of a system during runtime assumes that transitions from a state to a state can occur at any time. Data for a training sample are accumulated depending on the received amplitude-frequency radiation values of electronic components in various operation modes predetermined by the running software configuration. In course of data accumulation, there is a change in the statistical portrait of the electromagnetic radiation of operating devices.

The basis of the profile model of an object under investigation is $s_{di}(t)$ signals emitted from different circuit elements as a sequence of values $\{s_{di}(0), s_{di}(1), \dots, s_{di}(m)\}$. A lot of signals from the monitoring system D create a tuple of signal values from different sensors in time:

$$D(s_{di}(t), t) = \begin{vmatrix} s_{d1}(0) & s_{d1}(1) & \dots & s_{d1}(m) \\ s_{d2}(0) & s_{d2}(1) & \dots & s_{d2}(m) \\ \dots & \dots & \dots & \dots \\ s_{dn}(0) & s_{dn}(1) & \dots & s_{dn}(m) \end{vmatrix}. \tag{1}$$

The representation of data in the form (1) allows one to apply various methods of machine learning to implement a classifier. Against this background, it is possible to identify the state on the basis of a naive Bayesian classifier (NBC), a small amount of training data being its advantage:

$$C = \arg \max_{c} \; p(C = c) \prod_{i=1}^{n} p(D(s_{di}(t), i) = d_i | C = c), \tag{2}$$

Where C is a set describing possible states of a system; c is a certain state of the system ('normal' or 'abnormal'); i is a specific time interval; $s_{di}(i)$ is a signal value from sensor d_i at the time i; $D(s_{di}(t), i)$ is a tuple of signal values from different sensors in time.

By processing a tuple of features obtained through constant pre-assigned time intervals by the NBC, it is possible to determine abnormal conditions of a system, to which more attention should be paid.

4   Experiment

Taking into account the fact that external independent monitoring is required in a number of tasks [11], in this paper, side electromagnetic radiation and pick-ups (SEMRPs) from computer facilities arising from the operation of electronic devices are considered as a source of state information.

SEMRP measurements undoubtedly show better results in anechoic shielded chambers, but they are not always available to developers and are difficult to apply in the field environment. During the experiment, it was assumed that when a system is deployed,

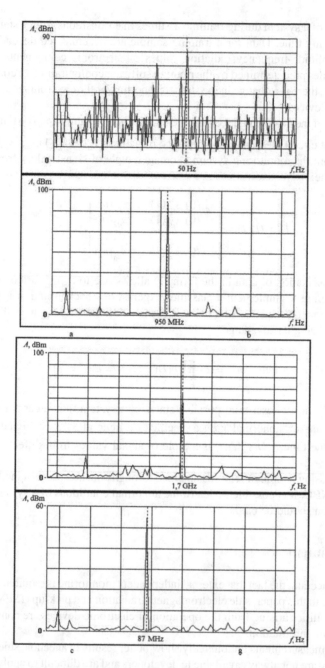

Fig. 1. Examined signal peaks for: sensors of the AP E3-50 type (a); near-field probes of the NFP-3-P1 type (b); near-field probes of the NFP-3-P2 type (c); proximity probes of the NFP-3-P3 type (g)

there is a certain time interval at which the current parameters of electromagnetic radiation can be taken prior to the destructive impact of a potential attacker.

The measurements were carried out at different frequencies by several sensors simultaneously from 50 Hz to 4 GHz for different elements. The selection of 'most informative' frequencies was made by an operator. At higher frequencies (f), priority was given to values with the largest feedback amplitude (A). Figure 1 shows the peaks under investigation.

Figure 2 shows the experiment design. Each sensor of the computational unit under investigation is connected to a spectrum analyzer. The information from the sensors was digitized, accumulated and timed; then, matrix tuples (1) were formed to be analyzed using a decision procedure based on the naive Bayesian classifier (2).

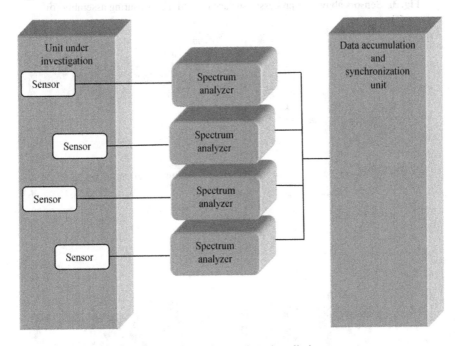

Fig. 2. The scheme of test installation

For analysis and processing, a profile was used where the parameters of external behavioral characteristics taken from the sensors were accumulated and stored for each set of processes.

In the experiment, electromagnetic radiation was taken above the nodal chips of the MSI G41M-P33 Combo motherboard (Fig. 3).

Fig. 3. Sensors above the processor surface (a) and the measuring assembly (b)

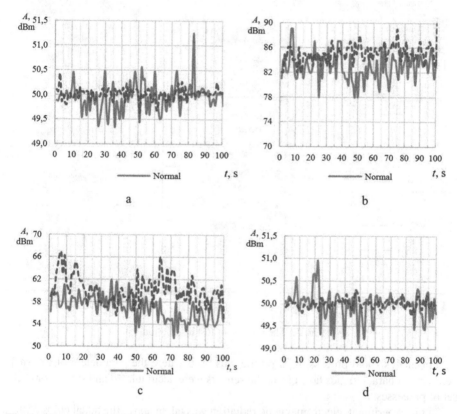

Fig. 4. View of graphs showing amplitude oscillations of the investigated signal peaks at normal and abnormal system operation in case of the sensors immediate proximity above the chip surface for: sensors of the AP E3-50 type (a); near-field probes of the NFP-3-P1 type (b); near-field probes of the NFP-3-P2type (c); near-field probes of the NFP-3-P3 type (d).

As the 'normal condition', an operating system-based process running and performing multiplication, division and result output operations to the console in the perpetual cycle was selected. As an 'abnormal condition' for analysis and comparison, characteristics were considered when, instead of the first process, a process was run that carried out only printing (output) of a symbol to the console in the perpetual cycle without additional calculations.

The obtained data were digitized and analyzed on the NBC basis. In the case under consideration, let us assume that the set of classes C takes values $C = \{c_0, c_1\}$, c_0 is a safe state with only pre-authorized processes running, c_1 is an unsafe condition where a 'modified' or unauthorized process has been started.

The graphs (Figs. 4 and 5) show values of signal amplitudes in 500 ms increments for the obtained data from different test programs from the sensors mounted next to the nodal chips.

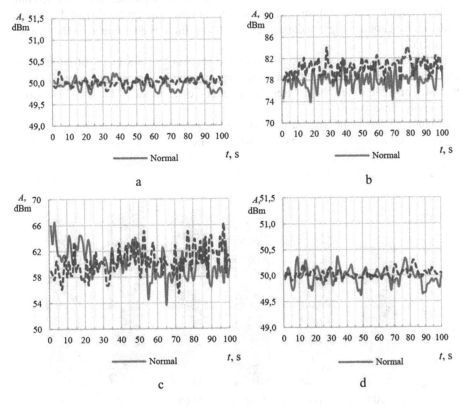

Fig. 5. View of graphs showing amplitude oscillations of the investigated signal peaks at normal and abnormal system operation in case of repositioning above the chip surface for: sensors of the AP E3-50 type (a); near-field probes of the NFP-3-P1 type (b); near-field probes of the NFP-3-P2 type (c); near-field probes of the NFP-3-P3 type (d).

During the experiment, some of the values were noisy, and some sensor reading obscured the condition determination, but digitized values from simultaneous information retrieval at the short-term stage make it possible to estimate the condition in the simplest binary classification.

Taking into account the specificity of products used for autonomous objects of cyber-physical systems, one of the problematic issues is arrangement of elements that serve to obtain initial data for analysis. In some cases, a monitoring system is unable to come into contact with the surface, so there is a need for a probabilistic assessment of the system condition determination at different sensor distances. By increasing or decreasing the distance and the tilt angles between the nodal chips and the sensor pickup surface, one obtains variations in signal portraits (Fig. 5).

The graphs shown in Figs. 4 and 5 for each experiment separately are converted into tuples of signal values D from different sensors in time. The probability of the system under investigation being in a dangerous or safe condition is determined on the basis of formula (2).

The experiment has shown that taking into account the 'noisiness' of information received from the sensors, after accumulating a training sample within one hour, it becomes possible to tell the differences in the processes running in the system with a probability close to 0.8.

Figure 6 shows the correctly identification probability for a running program depending on a training sample volume for the states shown in Figs. 4 and 5.

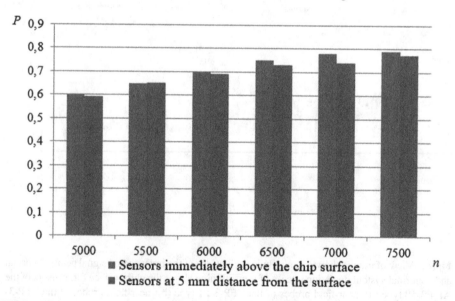

Fig. 6. Correct identification probability for running program (P) as a function of training sample volume (n) for the conditions in Figs. 4 and 5

The proposed solution does not require large computational effort; such a system can be quickly trained with various methods of machine learning and used as a solution aimed at detecting abnormal parameters of an autonomous object operating in conditions when, implementing measures to deploy the system in actual practice, it is possible to make a statistical data assessment at the initial operating stages with reference to which to later carry out an analysis, detection and identification of operation anomalies.

The statistics obtained based on the experiment shows the type of external response of the analyzed system sufficient for probabilistic determination of the information security state.

5 Conclusion

A typical solution to meeting the requirements for information protection in order to achieve confidentiality and integrity of information is the use of special software and hardware information protection tools, information security controls and information security management systems [12, 13].

However, introduction of implant tools and code modification can occur at the following life cycle stages: development, production, storage, transportation, commissioning, maintenance and upgrade of software and hardware, which enhances the applicability of external independent information security monitoring systems [14, 15].

An intruder can act at various stages of the life cycle of not only a cyber-physical system but also of built-in and integrated information protection tools.

The proposed method for monitoring autonomous objects based on statistical data of a system is aimed at analyzing external behavioral features of processes launched at a computing node; it does not affect the performance and does not require the computing and system resources of the node.

A special feature of the approach is the ability to rapid adaptation and application of various mathematical tools and machine learning methods to achieve a desired quality of probabilistic evaluation.

Implementation of this type of monitoring does not require a development of complex system applications while allowing implementation of various architectures for system construction that are capable of processing on-board an autonomous object or of communicating data and calculating the state on external computer nodes of monitoring and control systems.

References

1. Kotenko, I.V., Sayenko, I.B.: Creation of new systems for monitoring and managing cybersecurity. Bull. Russ. Acad. Sci. **84**(11), 993–1001 (2014)
2. Rogachev, G.N.: Production method of describing automated controllers in the analysis of continuous-discrete control systems. Autom. Control Comput. Sci. **48**(5), 249–256 (2014)
3. Leschev, S.V.: Electronic culture and virtual reality: the third digital wave of the NBIC paradigm. Bull. Hum. Fac. Ivanovo State Chem. Technol. Univ. **7**, 5–9 (2014)
4. Krivtsova, I., et al.: Implementing a broadcast storm attack on a mission-critical wireless sensor network. In: Mamatas, L., Matta, I., Papadimitriou, P., Koucheryavy, Y. (eds.) WWIC 2016. LNCS, vol. 9674, pp. 297–308. Springer, Cham (2016). https://doi.org/10.1007/978-3-319-33936-8_23

5. Sukhoparov, M.E., Lebedev, I.S.: Method for detecting abnormal behavior of personal networks. Prob. Inf. Secur. Comput. Syst. **1**, 9–15 (2017)
6. Bazhayev, N., Lebedev, I.S., Krivtsova, I.E.: Analysis of statistical data for monitoring the network infrastructure to detect abnormal behavior of the local segment of the system. Sci. Tech. Bull. Inf. Technol. Mech. Opt. **17**(1/107), 92–99 (2017)
7. Lebedev, I.S., Bazhayev, N., Sukhoparov, M.E., Petrov, V.I., Gurtov, A.V.: Analysis of the state of information security on the basis of surious emission electronic components. In: Proceedings of the 20th Conference of Open Innovations Association FRUCT, pp. 216–221 (2017)
8. Nikolaevskiy, I., Lukyanenko, A., Polishchuk, T., Polishchuk, V.M., Gurtov, A.V.: isBF: scalable in-packet bloom filter based multicast. Comput. Commun. **70**, 79–85 (2015)
9. Al-Naggar, Y., Koucheryavy, A.: Fuzzy logic and voronoi diagram using for cluster head selection in ubiquitous sensor networks. In: Balandin, S., Andreev, S., Koucheryavy, Y. (eds.) NEW2AN 2014. LNCS, vol. 8638, pp. 319–330. Springer, Cham (2014). https://doi.org/10.1007/978-3-319-10353-2_28
10. Chehri, A., Moutah, H.T.: Survivable and scalable wireless solution for E-health and emergency applications. In: Proceedings of the 1st International Workshop on Engineering Interactive Computing Systems for Medicine and Health Care (EICS4MED 2011), Pisa, Italy, pp. 25–29 (2011)
11. Sukhoparov, M.E., Lebedev, I.S.: Analysis of the state of information security based on spurious emission of electronic components. Inf. Secur. Prob. Comput. Syst. **2**, 92–98 (2017)
12. Royakkers, L., van Est, R.: A literature review on new robotics: automation from love to war. Int. J. Social Robot. **7**(5), 549–570 (2015)
13. Chernyak, L.: Cyberphysical systems at the start. Open Syst. **2**, 10–13 (2014)
14. Lee, E.A., Neuendorffer, S., Wirthlin, M.J.: Actor-oriented design of embedded hardware and software systems. J. Circ. Syst. Comput. **12**, 231–260 (2003)
15. Yusupov, R.M., Ronzhin, A.L.: From smart devices to the intellectual space. Bull. Russ. Acad. Sci. **80**(1), 45–51 (2010)

Signing Documents by Hand: Model for Multi-Factor Authentication

Sergey Bezzateev [ID], Natalia Voloshina [ID], Vadim Davydov [ID],
Tamara Minaeva [ID], and Nikolay Rudavin [✉] [ID]

ITMO University, St. Petersburg, Russia
bsv@aanet.ru, nataliv@ya.ru,
{vadim.davydov,tamara.minaeva,nikolay.rudavin}@niuitmo.ru

Abstract. The importance of verification of the user could not be neglected in the technological world of today. In this work, we focus on the Multi-Factor Authentication (MFA) system with a first factor as the user signature and with his fingerprint as a second biometrical factor. This authentication information about the user is further utilized to calculate a signature of the signed document. Obtained signature is embedded into the document as a digital watermark by using syndrome based steganographic algorithm for weighted container WF5.

1 Introduction

Today, the fundamental tool to enable data safety in any corporation is authentication. Many authentication frameworks utilize knowledge-based authentication strategies, i.e., passwords, PINs, etc., that has been out on a limb for years as one of the most insecure methods [1,2]. It is essential for a wide number of different attacks applicable to knowledge-based systems, which vary from very simple to complex, for instance, ranging from brute force and to peeping or John the Ripper attack [3].

Many authentication systems deployed today utilize biometric factors as a base as more secure data input method [4]. Therefore, users can be recognized by their physiological characteristics [5]. The most widely used method is fingerprint authentication, which has a relatively high degree of reliability and is one of the most comfortable and most convenient methods to use [6].

None the less, biometric authentication tools based on the behavior of the subject is to be considered particularly relevant because access to the subject does not require additional authentication threats and biometric data cannot be repeated as accurate. The collective use of several biometric and behavior methods in one authentication model is called a Multi-Factor Authentication (MFA) system being more reliable than standalone one [7].

In addition to the problem of authentication reliability, an equally significant task is the digital media management and security. Hiding information and its identification is also an important problem in the context of the rapidly developing infrastructures of network data exchange [8]. Due to its evolution, it

O. Galinina et al. (Eds.): NEW2AN 2018/ruSMART 2018, LNCS 11118, pp. 299–311, 2018.
https://doi.org/10.1007/978-3-030-01168-0_28

became possible to quickly and cost-effectively transfer digital media resources. As a result, a significant amount of transferred resources are often accompanied by illegal copying and distribution bringing us to the task of steganography or hiding the fact of secret actually being stored within the data. Based on the following set of tasks, in this manuscript, we propose a system model allowing signing confidential documents utilizing a pen with a number of embedded sensors that allows dynamic user authentication.

One of the most promising methods of dynamic authentication can be considered receiving biometric data of the subject from handwriting with the use of dynamic signature processing tools [9]. As a result of dynamic processing, data is more copy-protected than static data. This method is considered to replace obsolete and unreliable authentication methods such as a PIN or card.

In the case of dynamic biometric data, it is difficult for the intruder to repeat the signature procedure, since not only the positions of the signature device but also the time of the position change, as well as the effort applied by the subject during the movement of the hand, are to be considered during the authentication procedure.

In this manuscript, we focus on the MFA system with user signature as a first factor as behavior parameters and his fingerprint as a second biometrical factor. This authentication information about the user is further utilized to calculate a signature of the signed document. Obtained signature is embedded into the document as a digital watermark by using syndrome based steganographic algorithm for weighted container WF5 [10,11].

The rest of the paper is organized as follows. In Sect. 2 utilized hardware modules are discussed, and the authentication by handwriting is described, and the results are considered and discussed. Further, in Sects. 3 and 4, we show an estimate more two methods which are used in our model. In the last Sect. 5, we draw the succinct conclusion from our work.

2   Prototype Description

This section elaborates on the description of the sensors utilized for our prototype. Further, the architecture and methodology are explained. Nowadays, the electronic signature is a critical element in security and privacy of the document signed. In this section, we will consider several existing models and compare them with our model. Many solutions perform the task of capturing handwriting, including when using the necessary software, to generate an electronic signature. For instance, the Topaz system [12] or similar ones (for example, Wacom [13]) use a graphic touchpad as a signature capture device, which allows to receive dynamic data and accurately reproduce the signature draw. The main difference between our system and other similar solutions is the ability for a subject to sign the paper document itself directly physically, and this signature also will be a behavioral authentication factor. Moreover, some devices can also capture a signature without a unique subscribed object. For example, the Wacom pen allows a person to write on any surface, and the force of pressing the surface is essential. However, at the same time, this pen can work only with the device positioner, which must be placed on one horizontal surface with a signed object. Our model does not require additional particular surfaces and positioners, it is not able to read the pressure of the pen, however, unlike analogs, it reads data in three-dimensional space, allowing to collect more behavioral data of the subject. There are also analogs which use, in addition to behavioral factor, tokens for electronic signature, for example, [14,15]. The unique feature of our system is not only in the prototype that reads the signature but also in the technology of document signing. The analogs put an electronic signature to a separate file, which can be accidentally or intentionally lost. We, in turn, use the file itself completely as the carrier of the electronic signature. This makes it difficult for the attacker to distort or destroy the owner's data for the attacker. In our model we use fingerprinting and behavioural factors for authentication. Our choice is based on convenience and simplicity for the user. Fingerprinting allows not to waste time – the user can authenticate while signing the document. Also, the time of signing, the size and weight of the device and the number of elements must be taken into account.

3   Utilized Hardware Modules

The main goal of the developed 'smart-pen' is to analyze the behavior of the subject during the handwriting phase. The system is composed of a components set: (i) a microcontroller; (ii) the digital biometric fingerprint scanner; and (iii) the analog micro-accelerometer. A microcontroller is a unit of primary data processing. His task is to collect biometric data and prepare them for transmission of block secondary processing via a communication channel via a USB cable (in the future, data can be transmitted over a wireless channel). As a result of the initial processing, the behavior scanner signal is digitized, as well as pre-compression and signal encryption for transmission of information through the communication channel to the territory of the monitored zone (see Fig. 1).

For the processing unit, we utilize Arduino nano v3 microcontroller due to its sufficient processing power and the number of analogs and digital inputs to connect the accelerometer and the fingerprint scanner at a time [16].

One of the most popular compatible microcontrollers, which has proved itself to be compatible with the Arduino board, is the R308 biometric scanner (see Fig. 2), which also has small enough dimensions to accommodate it in the handle device. The fingerprint scanner R308 under consideration is optical. This method is based on the effect of the violated total internal reflection of Frustrated Total Internal Reflection [17].

To record fingerprint, we utilize software provided for Windows (the most comfortable and most convenient option, since you see a picture that is made) or a sketch for Arduino. To process biometric data in a sketch, special functions are applied from libraries provided by the manufacturer of the biometric scanner. As it is shown earlier, the scanner should fit the overall dimensions of the device-handle, and have an interface compatible with the platform Arduino nano v3.

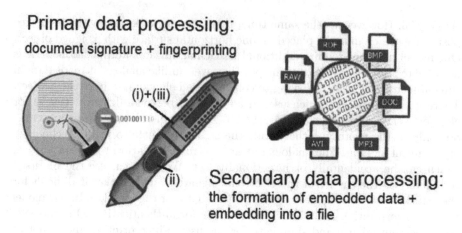

Fig. 1. Elements of the system.

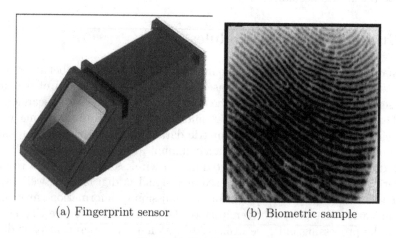

(a) Fingerprint sensor   (b) Biometric sample

Fig. 2. Utilized module.

Since the main serial port is already reserved for the fingerprint scanner, and the computing power of the microcontroller allows you to take on the initial processing (digitization) of the behavioural properties of the subject – the analog accelerometer ADXL335 was selected for our needs, which, also importantly, has smaller dimensions, high accuracy and low energy consumption [17]. ADXL335 is full-function triaxial accelerometer with voltage output signals and analog signal conversion circuits, as shown in Fig. 3. The accelerometer is tangled from 3.3 V to the PPO-1 controller, additionally occupying three analog inputs, the resulting voltage differences are converted to numbers from 0 to 1024 and fixed at a frequency of 10 ms.

Fig. 3. Selected accelerometer – ADXL335.

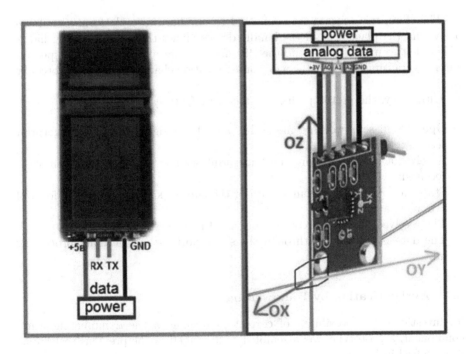

Fig. 4. Interfaces for connecting the fingerprint scanner and accelerometer.

Next, the fingerprint scanner is connected to a 5-volt power supply and sends data via a serial interface (RX/TX pins) with a supported speed of 57600 bytes/sec, as it is presented in Fig. 4.

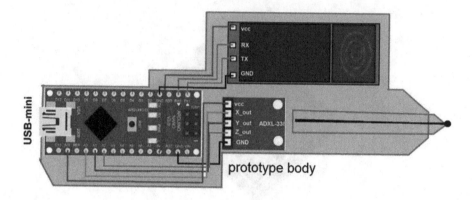

Fig. 5. Physical model of PPO-1.

Therefore, two sensors are connected to one hardware platform simultaneously, and the placement in the handle device allows the preliminary exchange of keys between the PPO-1 and the PRS for the encrypted data (in perspective wireless) transmission over the wire communication channel inside the protective case.

Ultimately, the prototype has the following features:

- Operates in semi-working mode, listening for a signal from the fingerprint scanner;
- Receives a command from the fingerprint scanner that the data sample is received;
- Starts recording the dynamic data of the subject's signature in the form of overload values in time;
- The received data is stored and encrypted by the system;
- The data is transferred through the serial port for on the main processing unit.

3.1 Authentication by Handwriting

To investigate the possibility of correct processing of the dynamic signature component, a prototype was assembled as described in the previous section, as it is drafted in Fig. 6.

The handwriting movements can be split divided into the horizontal plane $(O : X, O : Y)$ – directly the signature and the vertical axis $O : Z$ – setting the handle to the starting position and after the completion of the signature procedure.

Information from the controller is delivered in three axes in the form of different voltage on the wires on the analog outputs $A0$, $A1$ and $A3$ After processing, the signal is digitally transmitted to the monitored zone, where the extreme values of the signals and their time intervals are extracted and validated against the previously stored samples in the database.

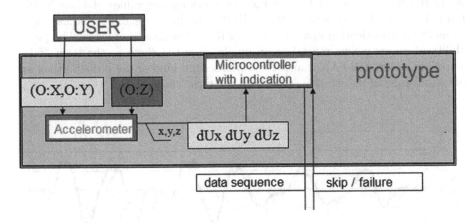

Fig. 6. Block diagram of the test system.

The primary signal processing is performed by a prototype - the pen on which the accelerometer and the microcontroller are located, which processes the signal and transmits it for secondary processing via the USB port.

As a result, it is possible to collect the signature data on two axes and transfer to a second signal processing program, whose primary task was to analyze a correspondence between the reference signal and the provided again. To determine the correspondence between the signal and the reference one, the value of the Pearson correlation coefficient is utilized [17]. This criterion allows determining whether there is a linear relationship between changes in the values of the two variables.

Fig. 7. A photo of a test prototype.

Figure 8 shows the values of the correlation coefficient (r) of the digitized signal of the hand movement along one of the axes. For the decision to admit the subject, the threshold values for the horizontal and vertical axis are set separately, and the access is allowed only if the threshold is exceeded by the values of the coefficients of both axes. It is evident that additional time synchronization is required for the signal matching, thus obtaining a high value of the linear dependence of the signal to the initial one. In the second signal processing phase, additional synchronization has been added, and tests have been conducted to verify the functionality of the method for checking the values of False Reject Rate (FRR) and False Accept Rate (FAR), the main reliability indicators in biometric authentication systems, where FRR is the first-kind error. The probability of an erroneous refusal to an employee is the same as the term "false alarm" in radar, and FAR is an error of the second kind of the probability of an erroneous miss of an attacker, similar to the term "false alarm" (Fig. 5).

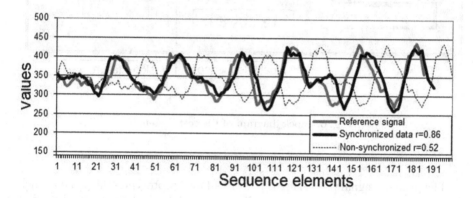

Fig. 8. Reference signature.

3.2   Numerical Results

During our applicability evaluation, the subject was asked to input 100 attempts of his signature. All subjects had the access to the picture (the graphical part of the signature) and everybody saw how the others signed the document. Next, a reference correlation coefficient was selected, with FRR being equal to one percent. To validate our system, two intruders attempted to replicate the signature of the subject for 50 times each. Eventually, from 100 attempts to forge the signature, it was possible to authenticate there times, i.e., FAR of the developed system is 3% (Fig. 7).

As can be seen from the results, the system is slightly inferior to analogs, which can be explained by the lack of signal filtering, which in turn simplifies the processing procedure, thereby speeding up the system and can be used in low-performance devices. Also, it is possible to increase the allowable threshold by the correlation coefficient to reduce the FAR, and the use of additional authentication factors can compensate the growth of the FRR value.

4   Benefits of Simultaneous Fingerprint Scanner Utilization

In this section, we elaborate on how simultaneous utilization of accelerometer and fingerprint scanner can improve the overall system reliability. Any identification in such a system could be divided into two phases – (i) registration and (ii) verification.

During the registration phase, a fingerprint image is obtained supplemented by its special features, such as control points and minuses highlighted, and the biometric template is formed. Then, the information is processed using a specific algorithm (which will be discussed later) and is written to the system. During the verification stage, the biometric sample is obtained from the sensor in the similar form, form a template, perform the steps of a specific algorithm, and, finally, the decision is made whether the presented and saved samples are related or not.

After receipt of biometric information on the scanner, three main stages could be selected from the fingerprinting processing perspective [18]:

1. Obtaining a biometric template – In modern systems, the most often used method is the selection of control points and minuses in the image;
2. Quantization – At this stage, the template obtained in step 1 is converted to a binary view. It should be noted that third-party information is also transferred to the database, which contains some data on randomness in the performance of quantization. This information is used in the verification phase to quantize the same way as when registering;
3. Key generation for the biometric template in the binary form obtained during stage two – At this stage, a fuzzy extractor is used. It is a biometric tool for authenticating a user with his biometric template as a key [19].

We obtain the following algorithm for recording biometric data:

1. Biometric information is sent to a biometric scanner, and the image is delivered for the initial processing;
2. The chosen formation algorithm forms the biometric template f;
3. Information is converted to a binary form z;
4. A random R line is generated from the resulting z by the logic of the fuzzy extractor. The database receives the received third-party string P;
5. It is considered a hash function from R and is written to the database.

The last step of the algorithm could be modified and utilized instead of counting the hash from the received string protocols with zero disclosure. As an example, consider the Schnorr protocol [20]. Now we estimate our system from information storage. We next consider two possible options: (i) the hashing is used, and (ii) when zero-knowledge protocols are utilized. Consider the case when the system stores a hash from the received string. The size of the hash depends on the selected algorithm. Suppose that the value is 256 bits. Also, a template f is stored in the database. The size of the template depends on the image. Usually, the image has dimensions of 25×14 mm, which corresponds to 500×280 pixels in our case. The point has 8 bits allocated. Therefore, about 140 KB of memory is required to the entire image. Since the registration process is the formation of a biometric template, this template consumes less memory. Usually, in print, there are about 30–40 min, according to which the template is formed. Note, the probability that all minuses coincide when applying fingers of different people is practically zero. Therefore, we can reduce the amount of memory required – assuming, the scanner has allocated 40 min each described by 4 bytes. Then the total volume will be equal to only 160 bytes, which is almost 900 times less than the image volume. Adding to the resulting value the size of the hash value is 32 bytes, it results in 192 bytes per fingerprint.

5   Applicable Stenographic Methods

The system uses stegano-paste as another factor of authentication. To sign an electronic document, you need to insert secret information into it. In order for the built-in information to be visually not detectable, you can use the $WF5$ method [10,11]. It is much less distorting the document when embedding information than the usual LSB replacement method [21]. The 24-bit bitmap of BMP format in RGB color system is considered as an electronic document for research. Each pixel is a combination of the brightness values of the three color components which occupy 1 byte each. The most common and least resistant steganographic methods of embedding information in a spatial area BMP image is the LSB replacement method [22]. This method of embedding does not require additional calculations and allows to 'hide' the vast amount of classified information in relatively small files [23], but at the slightest distortion of the image will result in the loss of built-in information. To solve the problem of increasing the volume of embedding concerning the standard LSB with a decrease in the visibility of the introduced distortions, a method of a multi-level embedding $WF5$ was used, which uses error correcting codes perfect in weighted Hamming metric [24] for embedding. In this method for forming the working area of the container, various combinations of the least significant bits of the color components of the pixels of the BMP image, namely several lower bit planes, can also be used.

The study revealed that the information could be embedded not only in the first least significant bits of the image but also in the next three. The analysis of visual quality indicators (distortion level) such as PSNR and SSIM show that there is no strong distortion of the container and such embedding method is not visually detectable [11].

By examining the different ways in which the workspace is formed, as shown in Fig. 9, it has been found that visual distortion practically does not occur, and the visual distortion index of PSNR does not fall below 60 dB, although for visual distortion detection the PSNR value should fall below 30–40 dB.

The implemented method allows combining the least significant bits for better distribution of embedded information in the container, thereby reducing visual distortion of the image. The study was conducted on various test images

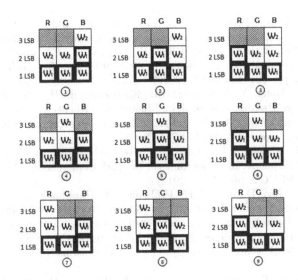

Fig. 9. Process of forming the pixel workspace (W1 – the first 4 bits of the image workspace vector, W2 – the last 3 bits of the image workspace vector).

of different types: portrait, landscape, text, etc. In the same way, one can extract the quilted insert from the image, which allows checking the signed image. The embedded information can be any: picture, key hash, encrypted message, etc. The limit for embedded information can only be in length, and the information should not be more than the amount of workspace obtained from the image used for embedding.

6   Conclusions

A new biometric/behavior two-factor authentication system is considered. The main advantage of it is using a pen-like device for collecting the authentication information about the user. Such solution makes this system more convenient for the user than previous one. By using an idea to embedding obtained user signature directly into the signed document as a watermark, the security of the final document is increased in the part of its integrity and authenticity.

Acknowledgment. This work was partly financially supported by Russian Foundation for Basic Research in 2017 (grant 17-07-00849-A).

References

1. Khan, S.H., Akbar, M.A., Shahzad, F., Farooq, M., Khan, Z.: Secure biometric template generation for multi-factor authentication. Pattern Recognit. **48**(2), 458–472 (2015)

2. Ometov, A., Bezzateev, S., Mäkitalo, N., Andreev, S., Mikkonen, T., Koucheryavy, Y.: Multi-factor authentication: a survey. Cryptography **2**(1), 1 (2018)
3. Weir, M., Aggarwal, S., De Medeiros, B., Glodek, B.: Password cracking using probabilistic context-free grammars. In: 2009 30th IEEE Symposium on Security and Privacy, pp. 391–405. IEEE (2009)
4. De Luca, A., Hang, A., Von Zezschwitz, E., Hussmann, H.: I feel like I'm taking selfies all day! Towards understanding biometric authentication on smartphones. In: Proceedings of the 33rd Annual ACM Conference on Human Factors in Computing Systems, pp. 1411–1414. ACM (2015)
5. Grankin, M., Khavkina, E., Ometov, A.: Research of MEMS accelerometers features in mobile phone. In: Proceedings of the 12th Conference of Open Innovations Association FRUCT, pp. 31–36 (2012)
6. Maltoni, D., Maio, D., Jain, A.K., Prabhakar, S.: Handbook of fingerprint recognition. Springer, London (2009). https://doi.org/10.1007/978-1-84882-254-2
7. Ometov, A., Bezzateev, S.: Multi-factor authentication: a survey and challenges in V2X applications. In: Proceedings of 9th International Congress on Ultra Modern Telecommunications and Control Systems and Workshops (ICUMT), pp. 129–136. IEEE (2017)
8. Bas, P., Furon, T., Cayre, F., Doërr, G., Mathon, B.: Watermarking Security: Fundamentals. Springer, Secure Designs and Attacks. Springer, Singapore (2016). https://doi.org/10.1007/978-981-10-0506-0
9. Xu, H., Zhou, Y., Lyu, M.R.: Towards continuous and passive authentication via touch biometrics: an experimental study on smartphones. In: Proceedings of Symposium on Usable Privacy and Security, SOUPS. vol. 14, pp. 187–198 (2014)
10. Bezzateev, S., Voloshina, N., Zhidanov, K.: Multi-level significant bit (MLSB) embedding based on weighted container model and weighted F5 concept. In: Abraham, A., Wegrzyn-Wolska, K., Hassanien, A.E., Snasel, V., Alimi, A.M. (eds.) Proceedings of the Second International Afro-European Conference for Industrial Advancement AECIA 2015. AISC, vol. 427, pp. 293–303. Springer, Cham (2016). https://doi.org/10.1007/978-3-319-29504-6_29
11. Voloshina, N., Zhidanov, K., Bezzateev, S.: Optimal weighted watermarking for still images. In: Proceedings of XIV International Symposium on Problems of Redundancy in Information and Control Systems (REDUNDANCY), pp. 98–102. IEEE (2014)
12. Topaz Systems INC.: [company website]. https://www.topazsystems.com/. Accessed 25 May 2018
13. Wacom Business Solutions: [company website]. https://www.wacom.com/ru-ru/enterprise/business-solutions/hardware/signature-pads. Accessed 25 May 2018
14. Comsigntrust: [company website]. https://www.comsigntrust.com/products/secure-portable-e-signature-token/. Accessed 25 May 2018
15. eSign: [company website]. https://www.esign.bg/. Accessed 25 May 2018
16. Gandra, M., Seabra, R., Lima, F.P.: A low-cost, versatile data logging system for ecological applications. Limnol. Ocean. Methods **13**(3), 115–126 (2015)
17. Grunthaner, M.P., Richards, P.W., Hotelling, S.P.: Frustrated total internal reflection and capacitive sensing (2018), uS Patent 9,891,759
18. Kevenaar, T.: Protection of biometric information. In: Tuyls, P., Skoric, B., Kevenaar, T. (eds.) Security with Noisy Data. Springer, London (2007). https://doi.org/10.1007/978-1-84628-984-2_11
19. Dodis, Y., Ostrovsky, R., Reyzin, L., Smith, A.: Fuzzy extractors: how to generate strong keys from biometrics and other noisy data. SIAM J. Comput. **38**(1), 97–139 (2008)

20. Schnorr, C.P.: Efficient signature generation by smart cards. J. Cryptol. **4**(3), 161–174 (1991)
21. Bezzateev, S., Voloshina, N., Zhidanov, K.: Weighted digital watermarking approaches comparison. In: Proceedings of International Symposium on Problems of Redundancy in Information and Control Systems (RED), pp. 172–174. IEEE (2016)
22. Fridrich, J., Long, M.: Steganalysis of LSB encoding in color images. In: 2000 IEEE International Conference on Multimedia and Expo, ICME 2000, vol. 3, pp. 1279–1282. IEEE (2000)
23. Anand, K., Sharma, E.R.: Comparison of LSB and MSB based image steganography. Int. J. Adv. Res. Comput. Sci. Softw. Eng. **4**(8), 906–909 (2014)
24. Bezzateev, S., Shekhunova, N.: Class of generalized Goppa codes perfect in weighted Hamming metric. Des. Codes Cryptography **1**(66), 391–399 (2013)

System for Secure Computing Based on Homomorphism with Reduced Polynomial Power

Viacheslav Davydov[(✉)] [iD]

National Research University Higher School of Economics, Moscow, Russia
v.davydov@hse.ru

Abstract. A significant interest recently emerged in the field of secure computations. Many systems were developed aiming at executing the summation and multiplication operation in a hidden way. Importantly, the cryptosystems enabling the execution of all four arithmetic operations are not yet present. This paper proposes a system for achieving this goal. The main benefit of its utilization is the possibility to continuous computation with no need for repetitive encryption of data.

1 Introduction

The high interest has recently emerged in developing a system allowing for secure computation [1,2]. One of the primary drivers for this requirement is the global integration and development of *Cloud computing* paradigm [3,4]. It allows reducing the cost of maintaining the network infrastructure significantly. It also provides an opportunity to quickly adapt to the requirements of the external environment, changing the size of the computations required for the specified task and performing more efficient system operation [5].

Indeed, the information systems of today are developing rapidly [6,7]. At the moment there is a vast number of commercial organizations providing cloud services [8]. In recent years, most major companies, for example, Microsoft, Google, Amazon, Citrix, offer similar services. However, there are several significant unresolved challenges in the cloud computing paradigm, and one of the central ones is the problem of ensuring data confidentiality [9]. Mainly, the user data is not protected while stored in the cloud in many cases. The cloud services provider, thus, obtains unlimited access to is in one form or another. As a consequence, there is a need to build a cryptosystem to protect the user sensitive information in cloud computing.

One of the solutions to avoid unauthorized access to private data is by utilizing homomorphic encryption for secure computation which does not require decryption of data while operating with it. In this paper, we provide an overview of homomorphic encryption in Sect. 2. Next, we propose a modified homomorphic system allowing summation, multiplication, and substantiation of the corresponding system elements in Sect. 3. The last section concludes the paper.

© Springer Nature Switzerland AG 2018
O. Galinina et al. (Eds.): NEW2AN 2018/ruSMART 2018, LNCS 11118, pp. 312–317, 2018.
https://doi.org/10.1007/978-3-030-01168-0_29

2   Secure Computation Background

The task of secure computation firstly appeared in late 70s in work [10]. Overall, the proposed framework consists of the algebraic system with set S. Let us define such algebraic system as $< S; f_1, f_2, \ldots; p_1, p_2, \cdots >$, where f_i is a function, p_i is a predicate s_i is a constant. The system could be thus defined as $< Z; +, -, *, /, \leq ; 0, 1 >$. This system describes the users U and there is a need for C system to be used by the computer system:

$$U = \quad < S; f_1, \ldots, f_k; p_1, \ldots, p_i; s_1, \ldots, s_m >, \tag{1}$$

and

$$C = \quad < S'; f'_1, \ldots, f'_k; p'_1, \ldots, p'_i; s'_1, \ldots, s'_m > . \tag{2}$$

Encoding and decoding shall then mean mapping elements from U to C and back: as $\phi^{-1} : S' \to S$ and $\phi : S \to S'$ as it's inverse.

The user has a set of data $d_1, d_2, \cdots \in S$ that he is willing to operate based on U. He is also willing to protect the operations from misuse and thus applies a certain transformation ϕ to those. Therefore, he received a set of $\Phi = \phi(d_1), \phi(d_2), \ldots$. In order for the system to execute such operations over the encoded date without decoding, ϕ should have a property of homomorphism for U and C:

$$\forall i(a, b, c, \ldots)[f_i(a, b, \ldots) = c \Rightarrow f'_i(\phi(a)\phi(b), \ldots) = \phi(c)], \tag{3}$$

$$\forall i(a, b, \ldots)[f_i(a, \ldots) = c \Rightarrow f'_i(\phi(a), \ldots) = \phi(b)], \tag{4}$$

$$\forall i s'_i = \phi(s_i), s_i = \phi^{-1}(s'_i). \tag{5}$$

If user is willing to calculate the result of $f_1(d_1, d_2)$ – it is necessary to calculate $f'_1(\phi(d_1), \phi(d_2))$. Due to the homomorphic property of ϕ:

$$\phi - 1(f'_1(\phi(d_1), \phi(d_2))) = f_1(d_1, d_2). \tag{6}$$

Next, authors of [11] proposed the homomorphic system enabling to execute the infinite number of summations and one multiplication. Further on, Gentry et al. [12] proved the possibility of developing the fully homomorphic system for the basis of OR and NOT. The proposed solution was not very effective regarding computational complexity, but an initial push towards the research in this direction was already made.

The main idea behind their method was to add the numerical "noise" to the data thus the decryption complex if the secret key is not known. The system was developed with three primary functions in mind: encryption, decryption, and calculation over data. The main drawback of the calculation is the increase of the numerical noise level. Therefore, the number of potential operations to be executed was limited.

Further development of this research topic was illustrated in [13,14] proposing the way to simplify the least common divisor. Later on, work [15] shown the cloud system based on homomorphism and polynomial ring. Here, it became feasible to securely sum, deduct and multiply the numbers from Z_n.

For each $a_0 \in Z_n$ there is a polynomial $a(x) = a_0 + a_1 x + \cdots + a_k x^k$, where k and $\{a_1, \ldots, a_k\}$ are selected randomly. Note, that for polynomial representation of a_0 and b_0, constant terms $a(x) + b(x) = a(0) + b(0)$ and $a(x)b(x) = a(0)b(0)$.

Therefore, $\phi : Z_n[x] \rightarrow Z_n[y], x = c_0 + c_1 y + \cdots + c_t y^t = \phi(y)$ is homomorphic and keeps the functionality of summation and multiplication. Authors propose to utilize the Gorner schema for division on $\phi(y)$.

The main drawbacks of this system are the uncontrollable growth of the polynomial power that can theoretically lead to the limitation of the computations number, and thus the system could become computationally not effective. Moreover, the division is not implemented thus leading to the computation of polynomials with increasing powers.

This paper proposes a homomorphic system with effective data representation compared to method in [15].

3   Proposed Homomorphic System

Let's assume a field $GF(Q) = \{q_1, q_2, \ldots, q_Q\}$, $m < Q$ and field extension $GF(Q^m) = u_1, u_2, \ldots, u_{Q^m}$. The secret of such homomorphic system is one of the primitive field elements $k\ in GF(Q^m)$. If k – is the primitive element of $GF(Q^m)$ than any other element could be derived as k^l, where l – is a relatively prime integer to $Q^m - 1$. Therefore, total number of elements in $GF(Q^m)$ could be obtained with an Euler function and equals to $\phi(Q^m - 1)$. In order to evaluate this number, we will further use [16]

$$\phi(Q^m - 1) > \frac{log2}{2} \frac{Q^m - 1}{log(Q^m - 1)}. \tag{7}$$

According to Theorem 1.86 in [10], the set of elements $\{1, k^1, k^2, \ldots, k^{m-1}\}$ is a basis over $GF(Q^m)$ is $GF(Q)$ and thus every $GF(Q^m)$ element could be represented based on $g(x) = \sum m - 1_{i=0} g_i x^i$ with coefficients from $GF(Q)$ when $x = k$.

The encryption process is primarily an association of a random message u_j $GF(Q^m)$ with a polynomial from x with a power $d > m$ with coefficients from $GF(Q)$

$$f(x) = \sum_{i=0}^{d} f_i x^i. \tag{8}$$

Here, $f(x)$ coefficients are formed according to:

1. $d - m + 1$ leading term coefficients are selected randomly for $f(x)$.
2. Next, the $GF(Q^m)$ element is calculated for the remaining m coefficients:

$$u_j^* = u_j - \sum_{i=m}^{d} f_i k^i. \tag{9}$$

Next, the linear system of equations should be solved to obtain the coefficients:

$$u_j^* = u_j - \sum_{i=0}^{m-1} f_i k^i. \tag{10}$$

Thus, we arrive at

$$f(k) = u_j. \tag{11}$$

We further define the encryption procedure as $\Im(u_j) = f_{u_j}(x)$ and decryption is given by 11, that in turn could be executed if k is known.

Next, we define α and β as two other elements of the field $GF(Q^m)$ as $f_\alpha(x) = \Im(\alpha)$ and $g_\beta(x) = \Im(\beta)$. Assume the defined polynomial $Z(x) = x^{Q^m} - x = \prod_{i=1}^{Q^m}(x - u_i)$ with a power of Q^m.

Thus, we define the basic operations (i.e., summation, multiplication, and subtraction) those but my modulus $Z(x)$ and prove that the homomorphism criteria is fulfilled.

Theorem 1.

$$\begin{cases} R(x) = f_\alpha \pm g_\beta(x) \\ R(x) = r(x) \mod Z(x) \end{cases} \Longleftrightarrow r(k) = \alpha \pm \beta . \tag{12}$$

Proof. Due to the fact that $deg(Z(x)) > deg(f_\alpha(x))$ and $deg(Z(x)) > deg(g_\beta(x))$, $R(x) = r(x)$. Considering $f_\alpha(k) = \alpha$ and $g_\beta(k) = \beta$ we arrive at $r(k) = \alpha \pm \beta$.

Theorem 2.

$$\begin{cases} R(x) = f_\alpha(x) \cdot g_\beta(x) \\ R(x) = r(x) \mod Z(x) \end{cases} \Longleftrightarrow r(k) = \alpha \cdot \beta . \tag{13}$$

Proof. According to the definition of $R(x)$, $R(x) = p(x) \cdot Z(x) + r(x)$ in addition to $deg(Z(x)) > deg(r(x))$. Thus, polynomial $Z(x) = x^{Q^m} - x = \prod_{i=1}^{Q^m}(x - u_i)$ has all elements of $GF(Q^m)$ as solutions. Therefore, $Z(k) = 0$ and $R(k) = r(k)$. Based on $R(x) = f_\alpha(x) \cdot g_\beta(x)$ we arrive at $R(k) = f_\alpha(k) \cdot g_\beta(k)$ and thus $r(k) = f_\alpha \cdot g_\beta(k)$.

Next we need to denote the division of $g_\beta(k)$ by $f_\alpha(k)$. We first define $f_\alpha(x)$ and $f_\alpha^{-1}(x)$ according to

$$f_\alpha^{-1}(x) = \begin{cases} \forall \lambda \in GF(Q^m), f_\alpha(\lambda) \neq 0, f_\alpha^{-1}(\lambda) = \frac{1}{f_\alpha(\lambda)} \\ \forall \lambda \in GF(Q^m), f_\alpha(\lambda) = 0, f_\alpha^{-1}(\lambda) = 0. \end{cases} \tag{14}$$

Based on the definition of $f_\alpha^{-1}(x)$, it could be explicitly obtained from $f_\alpha(x)$ if

$$f_\alpha^{-1}(x) = \begin{cases} \forall \lambda \in GF(Q^m), f_\alpha(\lambda) \neq 0, f_\alpha^{-1}(\lambda) \cdot f_\alpha(\lambda) = 1 \\ \forall \lambda \in GF(Q^m), f_\alpha(\lambda) = 0, f_\alpha^{-1}(\lambda) \cdot f_\alpha(\lambda) = 0 \end{cases} \tag{15}$$

Therefore, the division of $g_\beta(k)$ by $f_\alpha(k)$ is defined as multiplication of $g_\beta(k)$ and $f_\alpha^{-1}(k)$. Next, we prove that this operation has a homomorphic property.

Theorem 3.

$$\begin{cases} R(x) = g_\beta(x) \cdot = f_\alpha^{-1}(x) \\ R(x) = r(x) \quad mod \quad Z(x) \end{cases} \iff \begin{cases} \alpha \neq 0, r(k) = \frac{\beta}{\alpha} \\ \alpha = 0, r(k) = 0 \end{cases} . \tag{16}$$

Proof. According to the definition of $R(x)$, $R(x) = p(x) \cdot Z(x) + r(x)$ in addition to $deg(Z(x)) > deg(r(x))$. Thus, polynomial $z(x) = x^{Q^m} - x = \prod_{i=1}^{Q^m}(x - q_i)$ has all elements of $GF(Q^m)$ as solutions. Therefore, $z(k) = 0$ and $R(k) = r(k)$. Thus, $r(k) = g_\beta(k) \cdot f_\alpha^{-1}(k)$.

According to the definition of $f_\alpha-1(x)$, $f_\alpha-1(k)\frac{1}{\alpha}$ for each $\alpha \neq 0$. Therefore, fro each $\alpha \neq 0, r(k) = \frac{\beta}{\alpha}$. Next, $f_a^{-1}lpha = 0$ for $\alpha = 0$ and thus $\alpha = 0, r(k) = 0$.

Theorems 2–3 prove that the proposed system has the property of homomorphism. Note, that in order the key to be selected most efficiently, the size of $\phi Q^m - 1$ should be high.

Each message $q_i \in GF(Q^m)$ could be associated with Q^{d-m-1} different $f_{u_i}(x)$ polynomials with a power of d. Nonetheless, during the division and multiplication, the resulting polynomial power is increasing until $Q^m - 1$ due to the operation modulus $Z(x) = x^{Q^m} - x$. Note, that all the resulting coefficients are kept within $GF(Q)$ which is since the coefficients of the initial polynomials are also from $GF(Q)$.

If $d = Q^m - 1$ than $\aleph(Q^m)$ is a set of polynomials with a power of $d = Q^m - 1$. The encryption is thus $\Im(u_j) = f_{u_j}(x)$ splitting the entire set $\Im(Q^m)$ from Q^{Q^m} polynomials into Q^m subsets. Each subset $\Im(u_j)$ is associated with the element $u_j \in GF(Q^m)$. Note, $\Im(Q^m) = \bigcup_j^{Q^m} = 1\Im(u_j)$.

For each polynomial $f_{u_i} \in \aleph(u_i)$ exists a single inverse polynomial $f_{u_i}^{-1}(x) \in \aleph(u_i^{-1})$ for a message u_i^{-1}. As it is derived in (10), $\aleph(\alpha) \equiv \aleph(\frac{1}{\alpha})$ in case $\forall \alpha \in GF(Q^m), \alpha \neq 0$.

4   Conclusions

This paper proposes a modification of the system [15] limiting the maximum power of the polynomial to $Z(x) = x^{Q^m} - x = \prod_{i=1}^{Q^m}(x - u_i)$. Moreover, this limit could be lowered to $\phi(Q^m - 1)$ by utilizing not the entire $GF(Q^m)$ as roots but a set of it.

References

1. Bilakanti, A., et al.: Secure computation over cloud using fully homomorphic encryption. In: Proceedings of 2nd International Conference on Applied and Theoretical Computing and Communication Technology (iCATccT), pp. 633–636. IEEE (2016)
2. Zhao, F., Li, C., Liu, C.F.: A cloud computing security solution based on fully homomorphic encryption. In: Proceedings of 16th International Conference on Advanced Communication Technology (ICACT), pp. 485–488. IEEE (2014)

3. Mäkitalo, N.: Safe, secure executions at the network edge. IEEE Softw. **35**, 30–37 (2018)
4. López-Alt, A., Tromer, E., Vaikuntanathan, V.: On-the-fly multiparty computation on the cloud via multikey fully homomorphic encryption. In: Proceedings of the Forty-Fourth Annual ACM Symposium on Theory of Computing, pp. 1219–1234. ACM (2012)
5. Olshannikova, E., et al.: Conceptualizing big social data. J. Big Data **4**(1), 3 (2017)
6. Ometov, A., et al.: Feasibility characterization of cryptographic primitives for constrained (wearable) IoT devices. In: Proceedings of International Conference on Pervasive Computing and Communication Workshops (PerCom Workshops), pp. 1–6. IEEE (2016)
7. Florea, R., et al.: Networking solutions for integrated heterogeneous wireless ecosystem. In: CLOUD COMPUTING, p. 103 (2017)
8. Armbrust, M., et al.: A view of cloud computing. Commun. ACM **53**(4), 50–58 (2010)
9. Hashem, I.A.T., Yaqoob, I., Anuar, N.B., Mokhtar, S., Gani, A., Khan, S.U.: The rise of "Big Data" on cloud computing: review and open research issues. Inf. Syst. **47**, 98–115 (2015)
10. Rivest, R.L., Adleman, L., Dertouzos, M.L.: On data banks and privacy homomorphisms. Found. Secure Comput. **4**(11), 169–180 (1978)
11. Boneh, D., Goh, E.-J., Nissim, K.: Evaluating 2-DNF formulas on ciphertexts. In: Kilian, J. (ed.) TCC 2005. LNCS, vol. 3378, pp. 325–341. Springer, Heidelberg (2005). https://doi.org/10.1007/978-3-540-30576-7_18
12. Gentry, C., Halevi, S.: Implementing gentry's fully-homomorphic encryption scheme. In: Paterson, K.G. (ed.) EUROCRYPT 2011. LNCS, vol. 6632, pp. 129–148. Springer, Heidelberg (2011). https://doi.org/10.1007/978-3-642-20465-4_9
13. Brakerski, Z., Vaikuntanathan, V.: Fully homomorphic encryption from ring-LWE and security for key dependent messages. In: Rogaway, P. (ed.) CRYPTO 2011. LNCS, vol. 6841, pp. 505–524. Springer, Heidelberg (2011). https://doi.org/10.1007/978-3-642-22792-9_29
14. Brakerski, Z., Gentry, C., Vaikuntanathan, V.: (Leveled) fully homomorphic encryption without bootstrapping. ACM Trans. Comput. Theory (TOCT) **6**(3), 13 (2014)
15. Krendelev, S.: Homomorphic encryption (secure cloud computation). In: Proceedings of RusCrypto (2011)
16. Rosser, J.B., Schoenfeld, L.: Approximate formulas for some functions of prime numbers. Ill. J. Math. **6**(1), 64–94 (1962)

An Approach to Selecting an Informative Feature in Software Identification

Kseniya Salakhutdinova[1(✉)], Irina Krivtsova[1], Ilya Lebedev[2], and Mikhail Sukhoparov[3]

[1] ITMO University, 49 Kronverkskypr., Saint-Petersburg 197101, Russia
[2] SPIIRAS, 14-th Linia, VI, no. 39, St. Petersburg 199178, Russia
[3] SPbF AO « NPK « TRISTAN», 47 Nepokorennykh pr., Saint-Petersburg 195220, Russia
kainagr@mail.ru

Abstract. Statement of Research. A need to reduce the increasing number of system vulnerabilities caused by unauthorized software installed on computer aids necessitates development of an approach to automate the data-storage media audit. The article describes an approach to identification of informative assembly instructions. Also, the influence of a chosen feature that is used to create a unified program signature on identification result is shown. Methods. Shannon method allowing a determination of feature informativeness for a random number of object classes and not depending on the sample volume of observed features is used to calculate informativeness. Identification of elf-files was based on applying statistical chi-squared test of homogeneity. Main Findings. Quantitative characteristics of informativeness for 118 assembly instructions have been obtained. The analysis of experimental results for executable files identification with 10 different features used to create program signatures compared by means of the chi-squared test of homogeneity at significance levels $p = 0.05$ and $p = 0.01$ has been carried out. Practical Relevance. The importance of using a particular feature in program signature creation has been discovered, as well as the capability of considering several executable file signatures together to provide a summative assessment on their belonging to a certain program.

Keywords: Identification of executable files · Elf-files
Informativeness of a feature · Chi-square test · Information security

1 Introduction

A comparatively easy access to various webpages, Internet websites, open software and its modifiability necessitate electronic media audit to detect unauthorized software (SW) [1, 2].

Behavior of automated system users directed against the established security policy in an organization can lead to an increase in the number of system vulnerabilities. The reason for this is possible software weaknesses, presence of undocumented features, illegal use of intellectual property, as well as the use of special programs aimed at over-riding the installed security means or illegal actions inside the Intranet or on the Internet. The latter is especially relevant in terms of crimes related to computer information [3].

© Springer Nature Switzerland AG 2018
O. Galinina et al. (Eds.): NEW2AN 2018/ruSMART 2018, LNCS 11118, pp. 318–327, 2018.
https://doi.org/10.1007/978-3-030-01168-0_30

It should be noted that the authors consider herein conventional OS Linux user software (in particular, elf-files) and do not consider malicious programs or detection methods [4–6] identification or recognition of which have been presented in many papers [7, 8]. Yet, there have been extremely few research papers in the area under consideration [9, 10].

The use of some standard software analysis methods, such as physical inspection of typical installation locations of programs, comparison with a deep copy of data, hash total comparison, CRC control, hash coding, message authentication code and digital signature is not necessarily a successful way to monitor the installed programs because of possible absence of a reference sample (hash value, unmodified file copy, digital signature, etc.) used in the listed methods.

The approach to software identification considered in the article, that is, identification of an executable file with a known program is aimed at recognizing a program not based on its integrity. The identification process compares two signatures: unified one created from a training sample, and a signature of the identified program created just before the comparison stage.

Such a flexible approach allows a successful identification of different versions of the same program, even those that were not previously involved in creating a unified signature.

The authors have already developed new approaches to the creation of program signatures [11, 12], as well as identification methods based on these signatures [13, 14]. At the same time, attention was not paid to choosing a feature involved in the formation of signatures.

In this paper, the research focuses on the influence of a selected feature on the identification result accuracy. It is planned to consider the informativeness of 118 assembly instructions and compare the identification results for 10 selected instructions.

2 Calculation of Informative Features

When considering an executable file in terms of its disassembled code, it might be particularly helpful to select those assembly instructions that would most effectively help identify programs.

To solve this problem, Shannon method has been chosen as it allows a determination of feature informativeness for an arbitrary number of object classes, whereby not depending on the selection scope of the observed features.

This method proposes to assess informativeness as a weighted average amount of information (the eliminated entropy value) inherent in a feature $x \in X$, where X is the feature space.

To assess the informativeness of a feature x, the following formula is used:

$$I(x) = 1 + \sum_{i=1}^{G} \left(P_i \cdot \sum_{u=1}^{U} P_{i,u} \cdot \log_U P_{i,u} \right) \tag{1}$$

where G is the number of feature gradations (in this paper, two cases are considered: one gradation stands for the occurrence of a selected assembly instructions, two gradations stand for the occurrence of a selected assembly instruction and the occurrence of another instruction different from this one); U is the number of classes (this is the number of frequency distributions of assembly program instructions involved in calculating the informativeness of a feature); Pi is the probability that the feature value hits the i -th gradation calculated by the formula:

$$P_i = \frac{\sum\limits_{u=1}^{U} m_{i,u}}{N} \tag{2}$$

where $m_{i,u}$ is the occurrence rate of the feature value in the i -th gradation in the u -th class; N is the total number of a feature observations; $P_{i,u}$ is the occurrence probability of the i -th gradation of a feature in the k -th class calculated by the formula:

$$P_{i,u} = \frac{m_{i,u}}{\sum\limits_{u=1}^{U} m_{i,u}} \tag{3}$$

It should be noted that Shannon method gives an informativeness estimate to I(x) in the form of a normalized variable that takes on values over the range from zero to one. It is believed that the closer the I(x) value is to one, the higher the informativeness of feature x is and, conversely, the closer I(x) is to zero, the lower the informativeness of feature x is.

However, a limitation of this method is the inability to assess feature informativeness for the entire scope of the existing programs.

For 118 selected assembly instruction statements, an analysis was carried out to identify the most informative one of them. The experiment involved 10 different programs and their various versions. In general, 52 feature frequency distributions were generated reflecting the occurrence rate of each of the 118 instructions in the disassembled code of a program. The results of informativeness calculation for G = 1 and G = 2 are presented in (Table 1) and (Table 2), respectively, where, in order to optimize the table sizes, the I(x) informativeness values were rounded to fourth decimal. All the assembly instructions (also those that share a cell in the table) are arranged in ascending order of their informativeness.

Table 1. Informativeness values for 118 assembly instructions according to Shannon method with pre-assigned number of feature gradations G = 1

Assembly instructions	Informativeness, I(x)
cmpsb, cmpsw, esc, jc, jcxz, jna, jnae, jnb, jnbe, jnc, jng, jnge, jnl, jnle, jnz, jpe, jpo, jz, lodsb, lodsw, loopnz, loopz, movsb, movsw, repe, repne, retn, sal, scasb, scasw, stosb, stosw, wait	0
rep, jle, dec, lea, std, cmp, shr, jmp, jg, shl, and, cld, rol, js, jl, add, nop, mul	0.1408–0.1990
jne, push, ret, sub, xor, jge, ror, not, test, mov, div, jbe, jb, loopne, clc, je, jae, xchg, ja, or, repz, cli, adc, pushf, hlt, sar, sti, sbb, neg, jns, idiv, in, call, imul, jp, loop, jno, pop, repnz, out, iret, loope, jnp, xlat, int, rcl, cmc	0.2001–0.2972
jo, stc, lahf, retf, rcr, sahf, inc, cwd, popf, cbw	0.3026–0.3778
lock	0.4228
daa, les, into, aaa, lds, aam, aas, das, aad	0.6144–0.6785

Table 2. Informativeness values for 118 assembly instructions according to Shannon method with pre-assigned number of feature gradations G = 2

Assembly instructions	Informativeness, I(x)
cmpsb, cmpsw, esc, jc, jcxz, jna, jnae, jnb, jnbe, jnc, jng, jnge, jnl, jnle, jnz, jpe, jpo, jz, lodsb, lodsw, aaa, les, daa, aas, aam, ja, mul, js, loop, jge, jp, hlt, jl, jns, loopne, iret, das, rep, jae, idiv, loopnz, loopz, movsb, movsw, repe, repne, retn, sal, scasb, scasw, stosb, stosw, wait, cwd, cbw, div, neg, into, lds, aad	0.2024
rcl, std, sbb, lahf, ror, ret, jne, cmc, popf, rcr, adc, jno, in, shl, jbe, jb, pushf, not, clc, rol, int, jle, jg, sub, loope, xlat, cld, sti, shr, jnp, repnz, cli, dec, repz	0.2025
imul, sar, sahf, jmp, xor, nop, and, jo, je, stc, test, push, retf	0.2026
out, cmp, inc	0.2027
or	0.2028
xchg	0.2029
add, pop	0.203
lea	0.2032
call	0.2036
mov	0.2044
lock	0.2095

Obviously, the order of informativeness of assembly instructions depends on a selected value of a feature gradation number. This discrepancy occurs as a result of a significant influence of an additional gradation (not the occurrence of the feature under consideration, but the occurrence of another, a different assembly instruction statement) on the calculated parameters P_i and $P_{i,u}$ in formula (1). Thus, the ratio between the occurrence rate of the assembly instruction in question and the occurrence rate of all the remaining 117 assembly instructions is taken into account, which makes it possible to calculate informativeness not only based on a discrepancy in the instruction occurrence rate in different classes, but also in terms of its share in relation to the other instructions.

A limitation of establishing the most informative instructions using the number of feature gradations $G = 1$ is the infeasibility of further creation of program signatures, since their occurrence rate in disassembled program codes is too small to generate signatures with a sufficient number of non-zero occurrence rate values for an assembly instruction. The second approach using the number of feature gradations $G = 2$ makes it possible to eliminate this shortcoming.

The names of assembly instruction statements in bold will be later used in a program identification experiment. Following on from the results in (Tables 1 and 2), let us assume the following order of informativeness of the selected ten assembly instructions: mov, call, pop, push, je, lea, add, cmp, and, jmp.

3 Identification of Programs

Using one assembly instruction as a feature leads to development of a new approach in the formation of signatures based on dividing a disassembled program code into equal intervals and calculating a feature occurrence rate in them.

The process of creating a program unified signature is based on formation of a single sequence of a feature distribution proceeding from the similarity of several signatures of different versions of executable files that are related to the same program.

Recall that to form an archive of unified software signatures, a certain number of executable files is analyzed, for which a training sample $TS = \{v_1, v_2, \ldots, v_m\}, i = 1 \div m$ is formed, where v_i is a selection of various programs; m is the number of various programs; $v_i = \{f_1, f_2, \ldots, f_n\}, f_j$ are different versions of the i-th program, n is the number of files in a sample.

Each f_j file is disassembled and divided into equal intervals, with a fixed coefficient for forming the step length introduced to correct the number of intervals obtained for files of different sizes. In this case, the length of an interval is taken to be the number of different assembly instructions in one step interval. In the experiment, the coefficient has been chosen in such a way that the number of intervals was equal to thirty.

The feature frequency distribution for a file f_j is written as $L(f_j) = (a_k)$, where a_k is the feature occurrence rate in the k-th interval, $j = 1 \div n$, k is the number of intervals obtained and depends on the introduced coefficient.

Further development of unified signatures and signatures of identified files was previously described by the authors in the papers mentioned above.

The actual stage of comparing two signatures represents a test of the statistical hypothesis about the homogeneity of two samples verified by chi-square homogeneity test. This test makes it possible to compare empirical frequency distributions, that is, signatures whose distribution functions are not known.

If, as a result of the experiment, two independent samples of volumes n_1 and n_2 are obtained, whereby the samples fall into k intervals with frequencies $m_1, m_2, \ldots, m_k$ and

$m'_1, m'_2, \ldots, m'_k$ by the feature under consideration, the empirical value of chi-square test is calculated by the formula:

$$\chi^2 = n_1 n_2 \sum_{i=1}^{k} \frac{1}{m_i + m'_i} \left(\frac{m_i}{n_1} - \frac{m'_i}{n_2} \right)^2 \tag{4}$$

where $m_1 + m_2 + \ldots + m_k = n_1$ and $m'_1 + m'_2 + \ldots + m'_k = n_2$.

It is proved that these statistics for large values of n_1 and n_2 are distributed according to the law $\chi^2 \, c \, k - 1$ degrees of freedom [15].

It is known that the chi-square homogeneity test has a right-sided critical domain, therefore, if the inequality $\chi^2 < \chi^2_p$ is satisfied at the significance level p, there is no reason to discard the hypothesis of homogeneity of distributions.

4 Experiment Set up

In the experiment, 443 executable files of a training sample of different Linux OS versions and bitness (32x and 64x) related to 63 different programs were involved. The test sample included 123 files belonging to the same 63 programs, all of them being different from the involved files used in the training sample and of 32x and 64xbitness.

The training sample was formed by downloading programs from the official Linux repositories for x86 and x86-64 processor architectures. Then, to generate a test sample, two executable files of different versions and bitness for each program were extracted therefrom (except for one available only with one bitness type).

It should be noted that the signature archive is not fixed and inhibiting new signatures addition; on the contrary, it must be regularly updated and have up-to-date data to fulfill the tasks assigned by a researcher.

The test for homogeneity in the framework of the problem was used to test the background hypothesis H_0 – the signatures of the identified file and a program from the archive are similar and refer to the same program with the competing hypothesis H_1 – the signatures of the identified file and a program from the archive differ significantly and refer to different programs.

With the help of the STATISTICA software complex, the frequency distributions obtained from disassembled program codes were broken down into classes where they were subsequently formed into a single unified signature and entered into the archive. Separately, the process of constructing signatures occurred for identified executable files.

As informative features, the following assembly instructions were chosen: mov, call, pop, push, je, lea, add, cmp, and, jmp. For each of them, a frequency distribution was generated.

Regardless of the identified file bitness, its signature was compared with both the unified signature constructed for a 32x program and for a 64x program, with the decision made whether it belonged to the program in question in case after the statistical test was used, a conclusion was drawn to accept the background hypothesis.

The purpose of the experiments was to identify the dependence of software identification results on selection of one or another informative feature.

5   Results

In the process of comparing two signatures, the following results can be obtained:

- H_0 hypothesis is correctly accepted;
- H_0 hypothesis is incorrectly rejected;
- H_1 hypothesis is incorrectly rejected;
- H_1 hypothesis is correctly accepted;

In the first case, for two compared signatures pertaining to the same program, the hypothesis of their similarity was *correctly* accepted. In the second case (an error of the first kind), for two compared signatures pertaining to the same program, the hypothesis of their similarity was *incorrectly* rejected. In the third case (an error of the second kind), for two compared signatures pertaining to different programs, the hypothesis of their similarity was *incorrectly* accepted. And in the fourth case, for two compared signatures pertaining to different programs, the hypothesis of their similarity was *correctly* rejected.

The ratio of correct identification results and errors of the first and second kind is presented in Table 3.

Table 3.   Identification results for different assembly instructions

Relationship between the program signature from the archive and the signature of the identified file		Signatures pertain to the same program				Signatures pertain to different programs			
Hypothesis about similarity of signatures		Accepted		Rejected		Accepted		Rejected	
Significance level, p		0.05	0.01	0.05	0.01	0.05	0.01	0.05	0.01
Experimental results, %	Assembly instruction								
	mov	0.22	0.24	1.37	1.35	0.69	0.69	97.71	97.71
	call	0.34	0.4	1.25	1.19	2.19	2.19	96.22	96.22
	pop	0.33	0.39	1.26	1.21	2.34	2.34	96.07	96.07
	push	0.26	0.31	1.33	1.28	0.74	0.74	97.67	97.67
	je	0.45	0.47	1.14	1.12	3.43	3.43	94.98	94.98
	lea	0.47	0.53	1.12	1.06	11.4	11.4	86.97	86.97
	add	0.16	0.18	1.44	1.42	0.01	0.01	98.4	98.4
	cmp	0.44	0.48	1.15	1.11	6.34	6.34	92.07	92.07
	and	0.3	0.34	1.29	1.25	1.37	1.37	97.04	97.04
	jmp	0.33	0.39	1.26	1.2	1.15	1.15	97.26	97.26

Analyzing the data from Table 3, one can conclude that for the first half of the most informative assembly instructions (mov, call, pop, push, je) on average, the indicator of correct identification results is higher (96.85 for $p = 0.05$), and the number of errors of the second kind is smaller (1.88 for $p = 0.05$) than for the second half of assembly instructions (lea, add, cmp, and, jmp) that are less informative (94.69 and 4.062 with p = 0.05, respectively).

Obviously, the choice of an informative feature significantly affects identification results. One can consider not only one feature distribution but several ones, with subsequent formation of a general result for all distributions. The results of this approach are presented in Table 3.

Figure 1 shows a surface constructed of ten identifiable signatures (for selected assembly instructions) of the file amarok_2.3.0-0ubuntu4_i386. Here, along the abscissa axis, signature intervals (from one to thirty) are marked; along the ordinate axis, assembly instructions are marked where the first instruction is mov, the second instruction is call, etc. according to the pre-assigned order of informativeness, with the same order also in (Table 3); along the applicate axis, the occurrence rate of the assembly instruction in the file is marked.

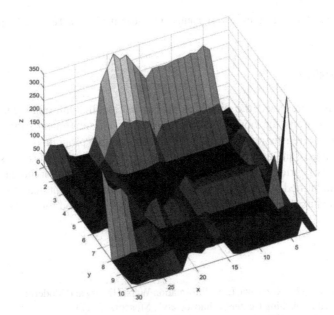

Fig. 1. Signatures of 10 assembly instructions for amarok_2.3.0-0ubuntu4_i386

Figure 2a and b show the surfaces constructed to the module of difference between ten distributions of the identified file and the ten distributions of the unified signatures. Obviously, for identical programs, the number of elevations is small and even minimal for half of the assembly instructions, while for different programs, the surface has a larger number of elevations, and their maximum value is twice as large as for identical programs.

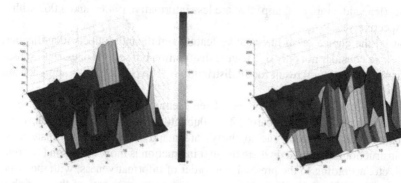

a) For amarok_2.3.0-0ubuntu4_i386 and   b) For amarok_2.3.0-0ubuntu4_i386 and ana-
                amarok   corn

Fig. 2. Absolute difference between a signature of the identified file and a unified signature from the archive

6   Conclusion

This paper has presented the approach to calculating the informativeness of 118 assembly instructions, the most informative of which were subsequently involved in the process of signature generation for training and test sample programs.

The outcome of the experiment on identification of executable files has shown that the percentage of correct identification results is on average higher for more informative instructions identified by Shannon method. It is becoming apparent that selection of an informative feature is an important part of the developed method for software identification.

References

1. Suleymanova, S.S., Nazarova, E.A.: Information Wars: History and Modernity: A Textbook.: International Publishing Center "Ethnosociety", Moscow (2017)
2. Lebedev, I., Korzhuk, V., Krivtsova, I., Salakhutdinova, K., Sukhoparov, M., Tikhonov, D.: Using preventive measures for the purpose of assuring information security of wireless communication channels. In: Proceedings of the 18th Conference of Open Innovations Association FRUCT, pp. 167–173 (2016)
3. Boukhtouta, A., Mouheb, D., Debbabi, M., Alfandi, O., Iqbal, F., El Barachi, M.: Graph-theoretic characterization of cyber-threat infrastructures. Dig. Invest. **14**(1), 3–15 (2015)
4. Alazab, M., Layton, R., Venkataraman, S., Watters, P.: Malware detection based on structural and behavioral features of API calls. In: Proceedings of the International Cyber Resilience Conference (ICR2010), pp. 1–10 (2010)
5. Shahzad, F., Farooq, M.: ELF-Miner: Using structural knowledge and data mining methods to detect new (linux) malicious executables. Knowl. Inf. Syst. **30**(3), 589–612 (2011)

6. Li, P., Liu, L., Gao, D., Reiter, M.K.: On challenges in evaluating malware clustering. In: Proceedings of the 13th International Symposium on Recent Advances in Intrusion Detection, pp. 238–255. Ottawa (2010)
7. Komashinsky, D.V., Kotenko, I.V.: Methods of data mining for the detection of malicious software objects: an overview of current research. Issues Inf. Prot. 4(102), 21–33 (2013)
8. Lai, Y., Liu, Z.: Unknown Malicious Identification. In: Ao, S.I., Gelman, L. (eds.) Advances in Electrical Engineering and Computational Science. LNEE, vol. 39, pp. 301–312. Springer, Dordrecht (2009). https://doi.org/10.1007/978-90-481-2311-7_26
9. Antonov, A.E., Fedulov, A.S.: Identification of the file type based on the structural analysis. Appl. Inf. 2(44), 068–077 (2013)
10. Kazarin, O.V.: Theory and practice of program protection. MGUL Press, Moscow (2004)
11. Krivtsova, I.E., Salakhutdinova, K.I., Kuzmich, P.A.: A method for constructing signatures of executable files for the purpose of identifying them. Bull. Pol. 5(3/5), 97–105 (2015)
12. Druzhinin, N.K., Salakhutdinova, K.I.: Identification of executable file by dint of individual feature. In: Proceedings of the International Conference on Information Security and Protection of Information Technology (ISPIT-2015). St. Petersburg, Russia, pp. 45–47 (2015)
13. Krivtsova, I.E., Salakhutdinova, K.I., Yurin, I.V.: The method of identifying executable files by their signatures. Bulletin of the State University of Marine and River Fleet named after Admiral S.O. Makarov. 1(35), 215–2242016
14. Krivtsova, I.E., Lebedev, I.S., Salakhutdinova, K.I.: Identification of executable files on the basis of statistical criteria. In: Proceedings of the 20th Conference of Open Innovations Association FRUCT, pp. 202–208 (2017)
15. Smirnov, N.V., Dunin-Barkovsky, I.V.: Course of Probability Theory and Mathematical Statistics. SNauka, Moscow (1969)

A-MSDU Frame Aggregation Mechanism Efficiency for IEEE 802.11ac Network. The Optimal Number of Frames in A-MSDU Block

Anton Vikulov[✉] and Alexander Paramonov

The Bonch-Bruevich State University of Telecommunications, 22 Pr. Bolshevikov, St. Petersburg, Russian Federation
asv012016@gmail.com

Abstract. This article discusses the effect of the A-MSDU frame aggregation mechanism on the efficiency of the IEEE 802.11 network. In order to define its main parameters as a function of the operation conditions the model of this mechanism is proposed. An analytical model is obtained for the probability of successful receipt of the code word as a function of bit error rate. An analytical model for the channel utilization efficiency as a function of bit error rate and number of frames in the A-MSDU block for the IEEE 802.11ac standard is also obtained. The proposed analytical models can be used to evaluate the network's performance. The method of determining the optimal number of frames in the A-MSDU block is proposed in terms of maximum efficiency of channel resource usage.

Keywords: Wireless access network · IEEE 802.11ac · Frame aggregation
A-MSDU · Channel efficiency · Modulation and coding scheme
Frame transmitting time

1 Introduction

One of the main objectives of the IEEE 802.11 group of standards is to increase data rates as well as the efficiency of the radio resources usage and thus to improve the quality of communication services. By the efficiency of the data transmission channel utilization, we will understand the ratio between the actual payload throughput and the nominal modulation and coding rate.

Due to high transmission overheads, frames are not transmitted one by one. Instead, various aggregation mechanisms are used in order to increase the efficiency of the radio resource usage. Here the aggregated MSDU (A-MSDU) frames aggregation mechanism for the IEEE 802.11-2016 standard [1] will be considered.

The MAC service data unit (MSDU) is a transmission unit used at the MAC layer (OSI model L2) when being received from a higher layer. As a result of the MSDU aggregation directly at the MAC layer, the A-MSDU blocks are constructed [5]. A-MSDUs are transmitted to the underlying physical layer (PHY-layer), where they are processed as MPDU. Each A-MSDU block contains a common header for multiple

© Springer Nature Switzerland AG 2018
O. Galinina et al. (Eds.): NEW2AN 2018/ruSMART 2018, LNCS 11118, pp. 328–339, 2018.
https://doi.org/10.1007/978-3-030-01168-0_31

MSDU frames that are assigned to one client and belong to the same IEEE 802.11e class of service. Each MSDU (except for the last one), when aggregated, is supplemented to be a multiple of 4 bytes.

In the IEEE 802.11ac amendment to the standard, the size of the A-MSDU is limited to a maximum MPDU size of 11454 bytes. The maximum number of MSDUs inside the A-MSDU can be 8, 16, 32, or have no restrictions depending on the corresponding value of the Extended Capabilities Element table [1].

The main purpose of this aggregation mode is that, since for the vast majority of clients, ethernet is the original frame format, it is convenient to combine multiple frames addressed to one client within a single transmission to form one A-MSDU. I.e. it is possible to optimize the overhead of L2 headers, which are identical for the large number of frames addressed to one client.

The complement to this mechanism in the standard since the IEEE 802.11n standard amendment is a lower layer A-MPDU aggregation process. Both mechanisms from the point of view of the channel layer are shown in Fig. 1.

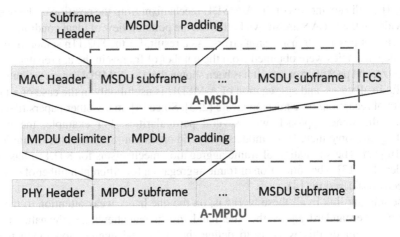

Fig. 1. IEEE 802.11n/ac frame aggregation mechanisms

MAC Protocol Data Unit (MPDU) is data block transmitted from the MAC level to the PHY level. In the 802.11ac standard amendment, the MPDU size can be limited to 3895, 7991 and 11454 bytes depending on the corresponding VHT Capability Information. Aggregated MPDUs (A-MPDUs) [1] are the MPDUs that are grouped together into one physical layer unit (physical protocol data unit (PPDU)). All frames have a common PLCP header and preamble. Each A-MPDU frame consists of several A-MPDU subframes, that include an MPDU delimiter frame and an optional MPDU frame.

The standard permits the usage of A-MPDU and A-MSDU mechanisms together or individually [1]. With the IEEE 802.11ac standard amendment A-MPDU is used, even if only one MPDU is to be transmitted. In other words, in VHT mode, the A-MSDU mechanism is optional and the A-MPDU is mandatory.

The main disadvantage of the A-MSDU mechanism is the fact that the whole frame sequence becomes one data element (PDU) and thus has only one CRC check. As the error rate increases with the increase of the frame size, and the retransmission of the part of an A-MSDU block is impossible, this will result in the retransmission of the entire aggregated block at lower speeds, which minimizes the advantages of aggregation. A-MPDU, in turn, consists of multiple PDUs, each with its own CRC. Therefore, in the case of an error, the PDUs can be retransmitted separately, thereby increasing overall efficiency. However, the performance gain carries an "overhead" in the form of an additional MAC header for each sub-frame.

2   Related Works

In the A-MPDU and A-MSDU mechanisms various overhead costs are reduced. A-MSDU - reduces the overhead on MAC headers in good quality channel conditions. A-MPDU, on the other hand, does not reduce the L2 overhead. Instead, it reduces the overhead of the CSMA/CA - IFS, ACKs, DIFS and backoff. In most real systems, the latter is more important, and therefore most of the actual IEEE 802.11n devices use A-MPDU [10].

In [4] the disadvantages of the A-MSDU mechanism from the position of the overall bandwidth of the WLAN are shown both from the point of view of ideal conditions, and in the case of possible transmission errors. As a result, for the 802.11n standard, bandwidth dependencies were obtained from the number of frames in the aggregated block, frame size, and packet error rate. It has been shown that the A-MPDU aggregation mode is useful in all cases, and aggregation of A-MSDU is useful only in the case of a fairly low rate of reception/ transmission errors [2, 3]. Since [4], many other papers like [15] consider the same approach with several particularities. For example, the works regarding this topic include the models of hybrid operation [8] and multicast traffic for 802.11n [12]. The simulation of frame aggregation mechanisms for 802.11n has been considered in [13]. The simulation of frame aggregation for different number of stations has been considered in [14].

The missing link in all these works is, on the one hand, weak attention to the last revision of the standard, and on the other hand, insufficient attention to the nature of the traffic. The aim of this work is to define the A-MSDU aggregation mode model depending on the channel parameters. We will also evaluate the limits of the applicability of this mechanism to supplement the 802.11ac standard and determine the optimal mode of its operation depending on the frame size and their number in the A-MSDU block.

3   The Model

We shall consider the efficiency of the channel resource utilization from the position of the frame aggregation modes. First, consider the ideal case for the mechanism of A-MSDU and UDP traffic. For the airtime being occupied, the transmission time of the data frame as a function of the frame size (L_{MSDU}) is defined [1] as:

$$T_{MSDU}(L_{MSDU}, SNR) = T_{PHY-PREAMBLES} + T_{SYML} \cdot \left\lceil \frac{T_{SYMS}}{T_{SYML}} \cdot \left\lceil \frac{8 \cdot L_{MSDU} + N_{Service}}{N_{DBPS}} \right\rceil \right\rceil, \tag{1}$$

where:

$T_{PHY-PREAMBLES}$ – is the total duration of all physical layer preambles.
N_{DBPS} – is the total number of data payload bits per OFDM symbol.
T_{SYMS} and T_{SYML} – is the duration of symbol transmission for short and long guard interval respectively.
$N_{Service}$ – is the size of « Service » field, equal to 16 bits.

A-MSDU block transmission time is:

$$T_{A-MSDU} = T_{PHY-PREAMBLES} + T_{MAC} + K \cdot (T_{A-MSDU-Header} + T_{MSDU}) + T_{SIFS} + T_{ACK}, \tag{2}$$

where:

K – is the number of frames with payload,
T_{SIFS} – is the duration of the short interframe spacing (16 µs for VHT channel).

$T_{A-MSDU-Header} = \dfrac{112bits}{PHYRate}$ – is the duration of A-MSDU header transmission (112 bits)

$T_{ACK} = \dfrac{112bits}{PHYRate}$ – is the duration of ACK frame transmission (112 bits)

$T_{MAC} = \dfrac{272bits}{PHYRate}$ – is the duration of MAC header transmission (272 bits),

where by *PHYRate* we understand the data transfer rate in Mbit/s, given by the modulation and coding sequence (MCS) index.
The transmission time of a single MSDU frame for one spatial stream is given by:

$$T_{MSDU} = T_{SYML} \cdot \left\lceil \frac{T_{SYMS}}{T_{SYML}} \cdot \left\lceil \frac{8 \cdot L_{MSDU} + N_{Service}}{N_{DBPS}} \right\rceil \right\rceil \tag{3}$$

While considering all frames to be of equal size, the total throughput shall be:

$$THPT = \frac{K \cdot L_{MSDU}}{T_{DIFS} + T_{BO} + T_{A-MSDU}}, \tag{4}$$

where:

T_{BO} – is the average back-off interval.
T_{DIFS} – is the duration of the long interframe spacing (34 µs for VHT channel).

The channel utilization efficiency in ideal conditions shall be:

$$ChUtil = \frac{THPT}{PHYRate} \tag{5}$$

Now consider the case for A-MSDU aggregation mechanism of the UDP traffic in a channel that has a certain probability of a bit error. Let's designate p_{BIT} to be the constant probability of erroneous receipt of one bit, or bit error rate (BER).

For a frame to be discarded by the receiver as received with an error, one corrupted LDPC code word with number of payload bits equal to CW_{size} is sufficient. The probability of the code word corruption is determined by the coding rate specified by the selected MCS index. All possible code word sizes for different coding rates are given in [1].

The standard provides three options for the code word size CW_{size} with a different payload size CW_{data}, depending on the selected coding rate. The number of code words N_{CW} required to transmit one frame payload of L_{MSDU} size (byte) is defined [1] as:

$$N_{CW} = \begin{cases} 1 & N_{avbits} \leq 1944 \\ 2 & 1944 < N_{avbits} \leq 2592 \\ \left\lceil \dfrac{8 \cdot L_{MSDU} + 16}{1944 \cdot Rate} \right\rceil & N_{avbits} > 2592 \end{cases}, \tag{6}$$

where the number of available bits in minimum number of OFDM symbols needed is:

$$N_{avbits} = N_{CBPS} \cdot N_{SYM} = N_{CBPS} \cdot \left\lceil \frac{8 \cdot L_{MSDU} + 16}{N_{CBPS} \cdot Rate} \right\rceil, \tag{7}$$

where N_{CBPS} – is the number of code bits per OFDM symbol. In this case, according to [1], the code word size CW_{size} is chosen as follows:

$$CW_{size} = \begin{cases} 648 & Cond1 \\ 1296 & Cond2 \\ 1944 & Cond3 \end{cases}, \tag{8}$$

where the conditions Cond1, Cond2 and Cond3 are defined by:

$$Cond1 = \left(N_{avbits} \leq 648\right) \wedge \left(N_{avbits} < 8 \cdot L_{MSDU} + 16 + 912 \cdot (1 - Rate)\right) \tag{9}$$

$$Cond2 = \begin{bmatrix} \left(N_{avbits} \leq 648\right) \wedge \left(N_{avbits} \geq 8 \cdot L_{MSDU} + 16 + 912 \cdot (1 - Rate)\right) \\ \left(648 < N_{avbits} \leq 1296\right) \wedge \left(N_{avbits} < 8 \cdot L_{MSDU} + 16 + 1464 \cdot (1 - Rate)\right) \\ \left(1944 < N_{avbits} \leq 2592\right) \wedge \left(N_{avbits} < 8 \cdot L_{MSDU} + 16 + 2916 \cdot (1 - Rate)\right) \end{bmatrix} \tag{10}$$

$$Cond3 = \begin{bmatrix} \left(648 < N_{avbits} \leq 1296\right) \wedge \left(N_{avbits} \geq 8 \cdot L_{MSDU} + 16 + 1464 \cdot (1 - Rate)\right) \\ 1296 < N_{avbits} \leq 1944 \\ \left(1944 < N_{avbits} \leq 2592\right) \wedge \left(N_{avbits} \geq 8 \cdot L_{MSDU} + 16 + 2916 \cdot (1 - Rate)\right) \\ N_{avbits} > 2592 \end{bmatrix} \tag{11}$$

Because of:

$$N_{CBPS} = \frac{N_{DBPS}}{Rate} \tag{12}$$

we can see that:

$$N_{avbits} = \frac{N_{DBPS}}{Rate} \cdot \left\lceil \frac{8 \cdot L_{MSDU} + 16}{N_{DBPS}} \right\rceil \tag{13}$$

Considering (6–13) together we can determine the number of code words and their size for the defined frame size and MCS index.

Because of the fact that the probability of the transmission error missing by the higher-layer MAC mechanism (CRC-32) is extremely small ($0.5^{32} = 2.3*10^{-10}$), and because of the absence of the error correction at the MAC layer, we assume that the receipt error probabilities of the individual frames on the channel and physical layers are equal. In other words, we will not consider the case of missing an error in the LDPC code word.

The successfully received code word must contain no more than t bit errors:

$$t = \left\lfloor \frac{d-1}{2} \right\rfloor, \tag{14}$$

where d – is the minimum code distance of the LDPC code implemented. The minimum code distances for different rates of the LDPC encoding are estimated in [9].

As we consider bit error rate p_{BIT} to be constant value, then the probability of successful bit transmission shall be:

$$q = 1 - p_{BIT} \tag{15}$$

Then, according to Bernoulli theorem, the probability of successful receipt of a code word P_{CW} of size CW_{size} is determined as sum of probabilities of successful code word receipt of the same size with number of corrupted bits k from 0 to t. Thus:

$$
\begin{aligned}
P_{CW} &= \sum_{k=0}^{t} P_{CW_{size}}(k) = \sum_{k=0}^{t} C_{CW_{size}}^{k} p_{BIT}^{k} (1 - p_{BIT})^{CW_{size}-k} \\
&= \sum_{k=0}^{t} \frac{CW_{size}!}{k! \cdot (CW_{size} - k)!} \cdot p_{BIT}^{k} (1 - p_{BIT})^{CW_{size}-k} \\
&= (1 - p_{BIT})^{CW_{size}} + \sum_{k=1}^{t} \frac{\prod_{i=0}^{k-1} (CW_{size} - i)}{k!} \cdot p_{BIT}^{k} (1 - p_{BIT})^{CW_{size}-k}
\end{aligned} \tag{16}
$$

Formula (16) gives the probability of successful reception of a code word depending on its size, coding rate and bit error rate. For example, Fig. 2 shows the probability of successful reception of the code word as function of the bit error rate for MCS5 and $L_{MSDU} = 700$ Bytes.

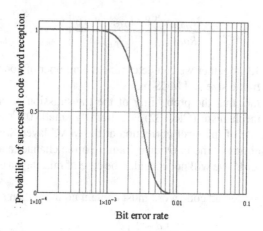

Fig. 2. Probability of successful reception of the code word

To have the whole frame corrupted, at least one code word out of total quantity N_{CW}, must contain an unrecoverable error. Thus the probability p_{MSDU} of the MSDU frame loss is given by:

$$p_{MSDU} = 1 - P_{CW}^{N_{CW}} \qquad (17)$$

The probability p_{A_MSDU} of A-MSDU block of frames loss is given by:

$$p_{A\_MSDU} = 1 - (1 - p_{MSDU})^K, \qquad (18)$$

where K – is the number of MSDU frames within A-MSDU block.

Combining (17) and (18), we can state:

$$p_{A\_MSDU} = 1 - (1 - (1 - P_{CW}^{N_{CW}}))^K = 1 - P_{CW}^{N_{CW} \cdot K} \qquad (19)$$

The probability that n-th consequent A-MSDU block transmission will be successful is:

$$(1 - p_{A\_MSDU}) \cdot p_{A-MSDU}^{n-1} \qquad (20)$$

Thus, the estimated quantity of retransmissions that will take place before a successful A-MSDU will be received is given by:

$$\sum_{i=1}^{\infty} \left((1 - p_{A\_MSDU}) \cdot p_{A-MSDU}^{i-1} \cdot i \right) = \frac{1}{1 - p_{A\_MSDU}} = \frac{1}{p_{CW}^{N_{CW} \cdot K}}. \qquad (21)$$

Thus, the total throughput of the channel with the fixed BER is:

$$THPT = \frac{K \cdot 8 \cdot L_{MSDU} \cdot p_{CW}^{N_{CW} \cdot K}}{DIFS + T_{BO} + T_{A-MSDU}} \qquad (22)$$

Combining (2), (5) and (22) we have:

$$ChUtil = \frac{THPT}{PHYRate} =$$
$$\frac{PHYRate^{-1} \cdot K \cdot 8 \cdot L_{MSDU} \cdot p_{CW}^{N_{CW} \cdot K}}{K \cdot (T_{A-MSDU-Header} + T_{MSDU}) + T_{MAC} + SIFS + T_{ACK} + T_{BO} + DIFS + T_{PHY-PREAMBLES}}, \qquad (23)$$

where T_{MSDU} is defined according to (3).

4   The Efficiency of the A-MSDU Frame Aggregation Mechanism

Let's now estimate the efficiency of the A-MSDU aggregation mechanism

As $ChUtil = f(K, L_{MSDU}, p_{BIT}, PHYRate)$, we will define the parameters in the following way:

- $L_{MSDU} = 700$ Bytes (average frame size, according to [6])
- $K = \{1; 2; 3; 5\}$. Here $K = 1$ means no frame aggregation.
- Both $PHYRate$ and N_{DBPS} are the characteristics of selected MCS index. Let's consider the operation in MCS8 mode with short guard interval of IEEE 802.11ac.

With the chosen mode $T_{BO} = 63$ μs [7, 11], $T_{PHY-PREAMBLES} = 40$ μs, $T_{SIFS} = 16$ μs and $T_{DIFS} = 34$ μs [1].

As the result of the assumptions above, we will calculate the channel utilization efficiency as a function of the bit error rate.

Now let's construct the dependencies $ChUtil (p_{BIT})$ for the noisy channel case for the traffic frames with different K and the characteristics above and determine the p_{BIT} value below which the A-MSDU mechanism is effective. The results for the chosen MCS8 are shown in Fig. 3.

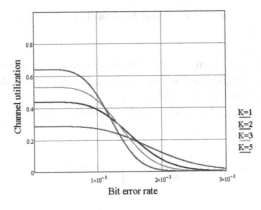

Fig. 3. $ChUtil(p_{BIT})$ for MCS8 for different K.

Note that there is a p_{BIT} value, which limits the application area for the A-MSDU mode. I.e. for large p_{BIT} values, the channel utilization efficiency decreases in relation to the non-aggregated ($K = 1$) transmission mode.

Let's now build the dependency $ChUtil(K)$ for fixed $pBIT$. Considering the operation in MCS8 see Fig. 4.

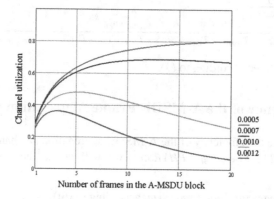

Fig. 4. $ChUtil(K)$ for MCS8 for different p_{BIT}.

It is clear that the graph has a maximum the more expressed, the higher is the p_{BIT} value.

5   Optimal Number of Frames Within A-MSDU Block

Now we will determine the optimal number of MSDU frames in A-MSDU block. The highest efficiency for the given bit error rate is achieved at a certain K value which we will consider to be the optimal number of frames K_{opt}.

It is given by:

$$K_{opt} = \arg\max_K (ChUtil(K)) \tag{24}$$

with $K \in N$, $K \le K_{MAX}$.

And K_{MAX} – is the maximum number of frames in a block for the selected operation mode. K_{MAX} – can be defined explicitly or otherwise be limited by A-MSDU maximum size, which is 11454 bytes for IEEE 802.11ac.

In order to solve the Eq. (24) we must determine the monotonic intervals for function (23). The maximum is given by:

$$K_{opt} = \arg \left\{ \frac{d(ChUtil)}{d(K)} = 0 \right\}, \text{ while } K > 0 \tag{25}$$

Let's denote:

$$T_1 = T_{A-MSDU-Header} + T_{MSDU}$$
$$T_2 = T_{MAC} + SIFS + T_{ACK} + T_{BO} + DIFS + T_{PHY-PREAMBLES}$$

The K-derivative:

$$\left(\frac{K \cdot 8 \cdot L_{MSDU} \cdot p_{CW}^{N_{CW} \cdot K}}{PHYRate \cdot K \cdot T_1 + PHYRate \cdot T_2} \right)'$$
$$= \frac{8 \cdot L_{MSDU} \cdot p_{CW}^{N_{CW} \cdot K}(T_2 + N_{CW} \cdot \ln p_{CW}(K^2 \cdot T_1 + K \cdot T_2))}{PHYRate \cdot (K \cdot T_1 + T_2)^2} \tag{26}$$

Then K_{opt} is given by:

$$K_{opt} = \arg \left\{ \frac{8 \cdot L_{MSDU} \cdot p_{CW}^{N_{CW} \cdot K}(T_2 + N_{CW} \cdot \ln p_{CW}(K^2 \cdot T_1 + K \cdot T_2))}{PHYRate \cdot (K \cdot T_1 + T_2)^2} = 0 \right\} \tag{27}$$

The equation has roots only when:

$$T_2 + N_{CW} \cdot \ln p_{CW}(K^2 \cdot T_1 + K \cdot T_2) = 0$$

And its positive root is given by:

$$K_{opt} = -\frac{T_2}{2T_1} - \frac{\sqrt{(T_2 N_{CW} \ln p_{CW})^2 - 4T_2 T_1 N \ln p_{CW}}}{2T_1 N \ln p_{CW}} \tag{28}$$

Formula (28) gives the analytical solution for optimal number of frames in A-MSDU block. Figure 5 shows the K_{opt} as function of the bit error for MCS8 mode, with L_{MSDU} = 700 bytes.

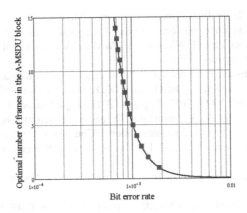

Fig. 5. $K_{opt}(p_{BIT})$ for MCS8

Now we have the quantitative practical assessments of the A-MSDU aggregation mechanism, which confirms the fact that this mode should be best used in the channel conditions close to ideal. The figures show that with the increase of the bit error rate, the range of values where the operation of this aggregation mode is useful is becoming narrower, and the absolute value of Channel Utilization is getting smaller.

Also we have obtained the method for calculating the optimal number of frames in the A-MSDU block for a given type of traffic in the specified channel conditions.

6 Conclusions

It has been proved that controlling the number of frames in the A-MSDU block permits to influence the IEEE 802.11ac channel efficiency. Without taking into account the bit error rate, the more is the number of transmitted frames the higher is the efficiency of channel utilization. However, in real-world conditions, when the bit error rate is nonzero, the overhead costs associated with retransmissions of frames increase, which brings the problem of finding the optimal number of frames.

An IEEE 802.11ac channel model has been developed to evaluate the efficiency of its utilization based on bit error rate. The proposed model takes into account the size of the frames (i.e., the type of traffic), the probability of a bit error, and the physical layer encoding mechanism for this standard amendment.

A method for selecting the optimal number of frames in the A-MSDU block has been proposed, which makes it possible to increase the efficiency of using the IEEE 802.11ac channel.

The results obtained in the work can be used for practical calculations and simulation of the transmission conditions and traffic characteristics of the channel.

References

1. IEEE Std 802.11 – 2016. IEEE Standard for Information technology — Telecommunications and information exchange between systems. Local and metropolitan area networks — Specific requirements. Part 11: Wireless LAN Medium Access Control (MAC) and Physical Layer (PHY) Specifications (2016)
2. Paramonov, A., Hussain, O., Samouylov, K., Koucheryavy, A., Kirichek, R., Koucheryavy, Y.: Clustering optimization for Out-Of-Band D2D communications. Wirel. Commun. Mob. Comput. **2017**, Article ID 6747052, 11 pages (2017). https://doi.org/10.1155/2017/6747052
3. Makolkina, M., Vikulov, A., A. Paramonov: The augmented reality service provision in D2D network. In: Proceedings of 20th International Conference. Distributed Computer and Communication Networks (DCCN 2017) Moscow, Russia, September 25–29, pp. 281–290 (2019)
4. Ginzburg, B, et al.: Performance Analysis of A-MPDU and A-MSDU Aggregation in IEEE 802.11n (2007)
5. Gautam Bhanage: AMSDU vs AMPDU: A Brief Tutorial on WiFi Aggregation Support. Report number: GDB2017-004. arXiv:1704.07015 [cs.NI], April 2017

6. Paramonov, A., Vikulov, A., Scherbakov, S.: Practical results of WLAN traffic analysis. In: Galinina, O., Andreev, S., Balandin, S., Koucheryavy, Y. (eds.) NEW2AN/ruSMART/NsCC -2017. LNCS, vol. 10531, pp. 721–733. Springer, Cham (2017). https://doi.org/10.1007/978-3-319-67380-6_68

7. Bianchi, G.: Performance analysis of the IEEE 802.11 DCF. IEEE J. Sel. Area Commun. **18**(3), 535–547 (2000)

8. Yazid, M., et al.: Performance Study of Frame Aggregation Mechanisms in the New Generation WiFi. VECoS (2016)

9. Butler, B.K.: Minimum distances of the QC-LDPC Codes in IEEE 802 Communication Standards. Arxiv.org https://arxiv.org/pdf/1602.02831.pdf. Accessed 05 May 2018

10. Westcott, D.A., Coleman, D.D., Mackenzie, P., Miller, B.: CWAP certified wireless professional official study guide (PW-270). Wiley Publishing, Chichester (2011)

11. Bianchi, G.: IEEE 802.11—saturation throughput analysis. IEEE Commun. Lett. **2**(12), 318–320 (1998)

12. Daldoul, Y., Ahmed, T., Meddour, D.: IEEE 802.11n Aggregation Performance Study for the Multicast. IFIP Wireless Days (WD 2011), October 2011, Canada, pp. 1–6 (2011)

13. Kolap, J.: Frame aggregation mechanism for high-throughput 802.11n wlans. Int. J. Wirel. Mob. Netw. **4**, 141–153 (2012)

14. García, M.A., Santos, M., Villalón, J.: IEEE 802.11n MAC mechanisms for high throughput: a performance evaluation. In: ICNS 2011. The Seventh International Conference on Networking and Services (2011)

15. Bourawy, A., Alokap, T.: Evaluation of frame aggregation in gigabit WLANs. Int. J. Eng. Appl. Sci. (IJEAS), **4**(4), April 2017. ISSN: 2394-3661

A Concise Review of 5G New Radio Capabilities for Directional Access at mmWave Frequencies

Giulia Sanfilippo[1,2], Olga Galinina[3,4](✉) ⓘ, Sergey Andreev[3] ⓘ, Sara Pizzi[1] ⓘ,
and Giuseppe Araniti[1] ⓘ

[1] Mediterranean University of Reggio Calabria, Reggio Calabria, Italy
`giuliasnfp@gmail.com`, {`sara.pizzi,araniti`}`@unirc.it`
[2] Vodafone Italy, Milan, Italy
[3] Tampere University of Technology, Tampere, Finland
{`olga.galinina,sergey.andreev`}`@tut.fi`
[4] Peoples' Friendship University of Russia (RUDN), Moscow, Russia

Abstract. In this work, we briefly outline the core 5G air interface improvements introduced by the latest New Radio (NR) specifications, as well as elaborate on the unique features of initial access in 5G NR with a particular emphasis on millimeter-wave (mmWave) frequency range. The highly directional nature of 5G mmWave cellular systems poses a variety of fundamental differences and research problem formulations, and a holistic understanding of the key system design principles behind the 5G NR is essential. Here, we condense the relevant information collected from a wide diversity of 5G NR standardization documents (based on 3GPP Release 15) to distill the essentials of directional access in 5G mmWave cellular, which becomes the foundation for any corresponding system-level analysis.

Keywords: mmWave · Beamforming · New radio · 5G NR
Numerology · Initial access · Random access

1 Introduction

In December 2017, the Third Generation Partnership Project (3GPP) released an early version of the first 5G specifications [14] – officially named 5G *non-standalone* (NSA) – to enable 5G New Radio (NR) deployments on top of the current 4G systems. In this case, a device fully relies on the existing LTE interface and protocols for control procedures while the data traffic can be split between the 5G NR and LTE, which corresponds to architecture option three: "LTE assisted, EPC[1] Connected" [17].

After this initial phase, further developments set the course for 5G *standalone* (SA) operation by incorporating a complete set of specifications for the new 5G

---

[1] Evolved Packet Core.

© Springer Nature Switzerland AG 2018
O. Galinina et al. (Eds.): NEW2AN 2018/ruSMART 2018, LNCS 11118, pp. 340–354, 2018.
https://doi.org/10.1007/978-3-030-01168-0_32

Core Network complementing the NSA version to enable operation not relying on the 4G infrastructure. Half a year later, in June 2018, the complete SA description has been "frozen" in Release 15. This signifies that its technical specifications[2] are considered sufficiently stable, i.e., all new features, along with the functionality required to implement them have been defined and addressed in the standardization documentation.

The completion of the SA 5G NR specifications not only opens the door to deploying 5G networks without relying on the existing infrastructure but also marks a decisive step into a new era of an interconnected society. Aiming at aggressive performance targets, the ongoing 3GPP efforts revolve around the following three emerging use-cases: (i) enhanced mobile broadband (eMBB) with the data rate requirements of up to 10 to 20 Gbps and support for high mobility (up to 500 km/h[3]); (ii) massive machine-type communications (mMTC) at high densities (up to one million connections per square km) calling for long battery life, broad range, and ultra-low cost; and (iii) ultra-reliable and low latency (URLLC) communications characterized by extremely reliable and available connectivity, high speeds, as well as 1 ms air and 5 ms end-to-end latencies [21].

Generally, although the 5G NR is defined with band-agnostic operations, which allows this technology to be deployed on any bands without restrictions, 3GPP specifies two major frequency ranges (FR) for Release 15 [3]:

- 450 MHz – 6 GHz (FR1, referred to as Sub-6 GHz) incorporating bands numbered from 1 to 255,
- 24.25 GHz – 52.6 GHz (FR2, commonly referred to as mmWave[4]) with the bands numbered from 257 to 511.

Albeit it is important *not to* misinterpret 5G as a strictly mmWave solution since the new standard provides high flexibility and supports a broad range of choices, the mmWave frequencies represent one of the most perspective capabilities of the 5G NR.

Naturally, mmWave communications exhibit certain undeniable advantages including the much wider – available and yet unoccupied – bandwidth as well as better spatial reuse and privacy aspects. The latter two are due to the utilization of highly directional transmissions that can be achieved with smaller wavelengths and hence, a higher number of antenna elements. At the same time, the defects of these qualities manifest in higher signal attenuation (including specific atmospheric effects) and implications of clustered multi-path signal structure, which may dramatically increase the bit error rate. Luckily, these negative effects can be mitigated by employing sophisticated beamforming and beam tracking mechanisms that become an indispensable part of NR research and implementation.

---

[2] 3GPP naming convention: TS = technical specifications, TR = technical report, CR = correction request.

[3] As of today, the numerology of Release 15 supports the speeds of up to 100 km/h [26], while higher values correspond to the eMBB use-case requirements and will be addressed in Release 16.

[4] Strictly speaking, mmWave starts at 30 GHz, but the community loosely assigns the slightly lower frequencies to mmWave as well.

In this paper, we provide a condensed vision of the 5G NR key features supported by Release 15, which should be taken into account by the engineers and theoreticians while searching for the fundamental trade-offs and evaluating the performance of mmWave-based NR systems, both analytically and through simulation studies. The remainder of this text is organized as follows. Section 2 outlines the main distinctive features of the NR technology according to Release 15, which boil down in this work to flexible NR numerology and 3D beamforming. Section 3 outlines the initial access procedure employed by the 5G NR, including a cell search mechanism and a random access procedure. Finally, we conclude with a discussion on open questions and new features expected in Release 16.

2 5G New Radio Features

The legacy LTE networks, which could easily be described as a "one-fits-all" solution, are unable to satisfy the increasingly stringent and highly diverse 5G requirements in terms of reliability, availability, latency, QoS, scalability, and throughput. To support a variety of vertical industries, the 5G NR – as a global standard for a new *OFDM-based air interface* – is specifically designed to support a tremendous variety of 5G services and use-cases, device types, and deployments. The officially completed Release 15 ratifies the 5G NR physical layer with an emphasis on constructing flexible *scalable numerology* and *scalable slot duration*.

In this section, we provide a brief outline of the key 5G NR features with a particular focus on cellular mmWave operation, which has received much attention in the standardization community recently. The most important technical specifications for the purposes of this review are listed in Table 1, whereas the complete list of documents may be accessed online [13]. A comprehensive interpretation of the standard may also be found in [1]. We structure the subsequent discussion as two dedicated subsections, one of which addresses the new numerology, while another one elaborates on the new beamforming features supported by the 5G NR.

2.1 New Scalable Numerology and Frame Structure

A *numerology* is defined as a set of parameters that specify the OFDM system design and includes Subcarrier Spacing (SCS), Cyclic Prefix (CP), symbol length, and Transmission Time Interval (TTI)[5]. The 5G NR numerology targets various deployments and performance requirements; therefore, it is designed to be configured flexibly to serve diverse purposes.

In particular, one significant difference between the LTE and 5G NR is that the latter defines several SCSs [6] as opposed to the only option of 15 kHz, which the current LTE standard specifies. Taking 15 kHz as a baseline, the NR numerology is based on the exponentially scalable SCS as defined by

---

[5] Also referred to as one slot, multiple slots, or one mini-slot (see below) [4].

Table 1. 5G NR specification map.

TS number	Title	Version, date*
38.101-1	UE Radio Transmission and Reception. Part 1	V15.2.0, 2018-07
38.201	Physical Layer General Description	V15.0.0, 2018-01
38.202	Services Provided by the Physical Layer	V15.2.0, 2018-06
38.211	Physical Channels and Modulation	V15.2.0, 2018-06
38.212	Multiplexing and Channel Coding	V15.2.0, 2018-06
38.213	Physical Layer Procedures for Control	V15.2.0, 2018-06
38.214	Physical Layer Procedures for Data	V15.2.0, 2018-06
38.215	Physical Layer Measurements	V15.2.0, 2018-06
38.300	Overall Description	V15.2.0, 2018-06
38.321	MAC Protocol Specification	V15.2.0, 2018-06

* Recent version as of July 2018.

$f[\text{kHz}] = 15 \cdot 2^{\mu}$ [4], where μ is referred to as the SCS configuration and takes the values of 0 (15 kHz), 1 (30 kHz), 2 (60 kHz), 3 (120 kHz), or 4 (240 kHz)[6].

In Table 2, we collect the range of SCSs that are advised by the current Release 15 as well as provide the respective slot durations and other parameters important for system-level evaluation. We intentionally highlight the *mmWave option*, since this direction remains the primary objective of our paper. Due to the impact of phase noise at higher frequencies, the carrier separation should be increased, which naturally divides our table: the left vertical part belongs to FR1 with narrower bands, while the right part corresponds to FR2, i.e., mmWave frequencies, as also indicated in Table 2.

Further, different SCS values are translated into a flexible frame structure. According to Release 15, downlink (DL) and uplink (UL) time is divided into *frames* of 10 ms duration, and each frame comprises ten *subframes* of 1 ms length (both values are constant). The basic transmission unit is a *slot* (TTI), which carries 14 OFDM symbols (or 12 with Extended CP) for SCS of up to 60 kHz and 14 symbols for higher SCSs [4]. In contrast to LTE, the slot duration can be flexibly modified from 1 ms to 0.0625 ms depending on the selected SCS option (i.e., the duration is calculated as a ratio $1/2^{\mu}$ ms, see Table 2). While shorter slot durations (larger SCSs) aim at supporting low latency and high reliability, longer values (lower SCSs) help increase spectral efficiency and may be suitable for larger cell sizes and thus for the lower frequency ranges as mentioned above (Fig. 1).

A slot may be used for only DL, only UL, or *mixed* UL and DL transmission (e.g., incorporate both control and data exchange within one slot interval, which may be managed dynamically via a slot format indication – valid for one or several consecutive slots). This flexibility makes it possible to exchange *TDD self-contained slots* [12,21] that incorporate UL/DL scheduling, data, and

---

[6] $\mu = -2$ (3.75 kHz), which corresponds to the LTE NB-IoT SCS, is also supported.

Table 2. Supported transmission numerologies in 5G NR.

μ	0	1	2	3	4
$\Delta f = 2^{\mu} \cdot 15\,[\text{kHz}]$	15	30	60	120	240*
Cyclic prefix**	Normal	Normal	Normal, Extended	Normal	Normal
For data***	+	+	+	+	−
For synchronization***	+	+	−	+	+
For data >6 GHz	−	−	+	+	−
For synch >6 GHz	−	−	−	+	+
Symbol duration, $1/\Delta f$ [μs]	66.67	33.33	16.67	8.33	4.17
Slot duration [μs]	1000	500	250	125	62.5
Number of slots per subframe	1	2	4	8	16
Number of slots per frame	10	20	40	80	160
Minimum bandwidth (MHz) [2]	4.32	8.64	17.28	34.56	69.12
Maximum bandwidth (MHz) [2]	49.5	99	198	396	397.44
Min. number of RBs, UL/DL [1]	24	24	24	24	24
Max. number of RBs, UL/DL [1]	275	275	275	275	138

*480 kHz is not adopted for Release 15 [5]
**Downlink: conventional OFDM with cyclic prefix (CP)
Uplink: conventional OFDM with CP with optional transform precoding
CP length is calculated based on slot and symbol length and number of symbols
per slot (14 for normal, 12 for extended CP [4])
***For either of two frequency ranges

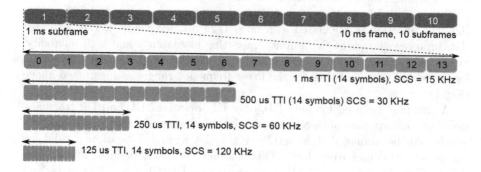

Fig. 1. Scalable NR TTI.

acknowledgment all at once and represent one of the key enablers for URLLC. Another new NR entity beneficial for URLLC is *a mini-slot* (minimal schedulable resource optimized for short data transmissions), which may occupy 2, 4, or 7 OFDM symbols and start at any time without waiting for the slot boundary [23]. Release 15 also supports *slot aggregation*, so that the data may be scheduled over multiple slots [3], even over slots having different formats [5].

2.2 Directivity and Beamforming at FR2

Another distinctive feature of 5G NR at mmWave frequencies is the possibility to rely on beam steering by highly directional antennas, which has become feasible for a wide range of use-cases due to smaller antenna elements and larger antenna arrays. Moreover, the use of highly directional antennas at the NR base station (gNB) and/or at the user equipment (UE) represents a natural solution to compensate for faster signal attenuation and improve the link budget. Importantly, beamforming is not an exclusive mmWave-specific feature – it can also be used at lower frequencies; however, when it comes to extremely high frequency range, beamforming becomes the only viable choice for most of the envisioned use-cases.

In general, beamforming techniques are responsible for controlling the properties of electromagnetic radiation patterns and the gain of an antenna array by aligning the amplitude and phase of transmit/receive signals. By doing so, a device is able to form an appropriate beam-pattern by increasing the antenna gains toward the desired direction and, at the same time, suppressing the radiation sideways and changing the interference footprint.

The most usable beamforming algorithms include exhaustive search (brute-force sequential beam searching over a predefined codebook, which is a set of beams to multiple directions covering the entire angular space) and iterative or hierarchical search (two-stage scanning, which transmits the signals over wide sectors and then refines within the best sector by steering narrower beams) [16, 18]. However, in case of narrow beams, these simple solutions may likely result in excessive delays and can be inefficient for certain use-cases. To properly align the beams within a limited delay budget, devices might need to exploit alternative intricate techniques for beam and mobility management that currently generate a particular interest within the research community.

Generally, beamforming may be categorized as two- or three-dimensional:

- 2D beamforming: controls the radiation pattern in one plane (the antenna elements are positioned as a linear array).
- 3D beamforming[7]: steers the antenna beams not only in the azimuth but also in the elevation plane (planar flat/volume arrays).

Compared to the conventional 2D techniques, 3D beamforming in 5G NR is built on up to 256 (32) antenna elements for gNB (UE) and supports vertical sectorization with an additional sector division in the radial direction. This extra

---

[7] also known as *elevation beamforming*.

sectorization allows reducing co-channel interference and creating a higher degree of freedom in optimizing the system performance without altering the existing physical architecture [25]. We note that 3D beamforming is supported not only by the NR but also by other mmWave technologies, such as IEEE 802.11ad/ay [19]. The beamforming architectures included in the 5G NR are [22]:

- *Analog beamforming*: exploits a single RF chain and multiple phase shifters, which results in simpler beam-search procedures (one beam at a time). Characterized by low power consumption and low complexity (used in, e.g., IEEE 802.11ad [20]).
- *Digital beamforming*: requires several RF chains, one for each antenna, and thus is able to support multi-stream operation (e.g., MU-MIMO). Characterized by higher flexibility in shaping the beams but also by increased costs, power consumption, and complexity (used in, e.g., LTE and supported by IEEE 802.11ay [19]).
- *Hybrid beamforming*: represents a compromise between the analog and digital options based on dividing the precoding between the analog and digital domains. Characterized by fewer supported streams than digital, lower complexity, and lower power consumption due to a decreased number of RF chains.

Hybrid beamforming constitutes a relatively recent solution to combine the strengths of both options above and promises nearly the same performance as achieved by the digital beamforming [15].

Finally, with respect to directional data transmission, the NR standard enables different design options for MIMO systems (up to 256 antenna elements). To increase the data rate and improve the spatial diversity, 5G NR supports eight streams for the single-user MIMO operation and twelve streams for the multi-user MIMO in DL, as well as four streams for the single-user MIMO operation in UL [22]. Since MIMO functionality requires continuous evaluation of the channel quality, it may also benefit from utilizing the self-contained subframe structure described above by transmitting the UL control information and sounding reference signals [24].

3   Initial Access in 5G NR

Generally, initial access in cellular systems comprises several consecutive steps, which we broadly divide into two stages (see Fig. 2):

- Stage I: cell search and synchronization (acquiring system information).
- Stage II: random access procedure.

In 5G NR, initial access resembles a standard procedure that the legacy LTE relies upon: in particular, it includes receiving synchronization signals, extracting system information, and establishing a connection via a random access procedure. However, regarding how the initial access is performed for FR2, 5G NR

Fig. 2. Steps of the initial access procedure.

differs from LTE operation significantly, which is primarily due to its highly directional nature at the frequencies above 6 GHz.

A key difference emerges already at Stage I during the cell search and synchronization. We remind that in LTE the synchronization signals are transmitted by using omnidirectional antennas (beamforming, if employed, applies only after synchronization, i.e., during the data transmission). In contrast, to extend communication distances, an NR gNB employs beam sweeping[8] already when broadcasting the synchronization signals. Hence, initial access for mmWave NR becomes much more challenging, since both the UE and the gNB have to detect the correct directions and align the beams before subsequent data transmission.

3.1 Stage I: Cell Search and Selection

At Stage I, the gNB periodically broadcasts *Synchronization Signal Blocks*[9] (SSBs), which contain (i) Primary Synchronization Signal (PSS), (ii) Secondary Synchronization Signal (SSS), and (iii) Physical Broadcast CHannel (PBCH) [4,5] as demonstrated in Fig. 3. In contrast to LTE, the synchronization signals and PBCH (carries system information) are inseparable in 5G NR. Each SSB is mapped onto a *different beam* and broadcasted by the gNB to its proximate UEs.

A cell search procedure is used by the UE to acquire time and frequency synchronization with the cell as well as to determine the Cell ID. The UE listening on a channel first detects the symbol timing and Physical Cell ID (PCI) in PSS over the time domain. Then, by utilizing SSS, the UEs obtain information regarding the frame timing in the frequency domain, CP length, as well as detect FDD/TDD and acquire the reference signals for demodulation [2].

Importantly, the periodicity of the SSB is configured by the network, while the default transmission periodicity, which is assumed by the UE before such notifications, is 20 ms (i.e., 2 NR frames). This interval is four times longer than that in LTE (5 ms) and aims at reducing the "always-on" transmission overheads. The frame and slot timings are defined by the identifiers of SBSs and acquired by the UE as described above.

More specifically, SSBs may be transmitted in a batch by forming an *SS Burst* (one SSB per beam) that may be used during beam sweeping; a collection of SS Bursts is referred to as an *SS Burst Set*. Both SS Burst and SS Burst Set

---

[8] A sequential transmission by using the entire codebook or its subsets.

[9] Also referred to as the synchronization signal and PBCH block in the specifications, but may be simply understood as a beacon.

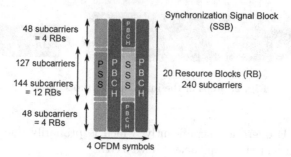

Fig. 3. Time-frequency structure of the synchronization signal and PBCH block (consists of PSS and SSS, which cover 1 OFDM symbol and 127 subcarriers (SCs) each, and PBCH that occupies 3 OFDM symbols and 240 SCs, respectively). One Resource Block (RB) = 12 SCs.

may contain one or more elements, while the maximum number of SSBs in an SS Burst is frequency-dependent and takes the values of 4 (below 3 GHz), 8 (3 to 6 GHz), or 64 (6 to 52.6 GHz) [27].

In Fig. 4, we illustrate the concept of SS bursts as well as outline the structure of one TTI and demonstrate the share of resources occupied by one SSB. SS Burst Set may occupy one half frame (5 ms), and the beam pattern repeats every 2 frames [7] (by default for the initial access). The overheads created by the occupied resources may be calculated based on the number of SCs (up to 3300), the size of the SSB (addressed in Fig. 3), and the number of SSBs per a time unit, which is defined by the required number of beams and their periodicity.

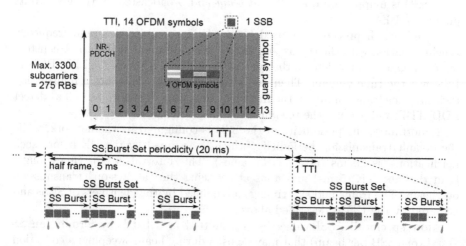

Fig. 4. SSB and SS burst composition [27]. One SSB corresponds to one beam.

The gNB may define multiple candidate positions for SSBs within a radio frame, and this number corresponds to the number of beams. Identification of which SSB is detected and thus acquisition of the frame and slot timings is facilitated by DeModulation Reference Signal (PBCH DMRS), which plays the role of a reference signal for decoding PBCH instead of Cell Specific Reference Signal (CRS) used in LTE.

3.2 Stage II: Random Access

A random access procedure in 5G NR may be triggered by, e.g., handover, initial access from idle/inactive modes, or beam failure recovery, and usually falls into either of the two categories: contention-free (CFRA) and contention-based random access (CBRA). Here, we focus on the latter option. As mentioned above, random access in mmWave cellular generally shares most of its functionality with LTE, which is built upon Physical Random Access CHannel (PRACH) preamble considerations; however, the nature of high directivity imposes new challenges and allows further options. Since single-beam operation (corresponding to omni-directional transmission/reception at the frequencies below 6 GHz) is similar to LTE RACH functionality, we concentrate on multi-beam operation that arises from using directional antennas [11].

At the beginning of the random access procedure, both the UE and the gNB are not aware of the appropriate beam directions; hence, the initial access signals are sent via multiple Tx beam sweeping. After detecting the initial synchronization signals, the UE is able to select the best gNB Tx beam for further DL data acquisition from the beams used to transmit the initial access signals. The gNB also utilizes multiple Rx beams to cover the entire angular space, since the position of this potentially attempting UE is unknown. The gNB provides multiple RACH resources (SSBs) to the UE and applies one Rx beam per each RACH resource that it announced previously. The number of RACH resources within one RACH occasion[10] and the number of contention-based preambles per one beam are also advised by the serving gNB (the maximum is 64 preambles [5]).

Importantly, the design of the random access procedure may vary depending on the presence of Rx/Tx reciprocity at the UE or the gNB (no reciprocity, partial, or full reciprocity) [9,10]. If the gNB relies on beam reciprocity, it maps the UL RACH resources onto the DL initial access signals before the UE starts the RACH procedure [7] and by that may significantly reduce the required beam training time.

In general, the NR RACH procedure includes the following four steps [11] (tailored here to the case of full reciprocity for simplicity [28]):

1. Based on the synchronization information from the gNB, the UE selects a RACH preamble sequence (MSG1) and sends it at the nearest RACH occasion (occurs every 10, 20, 40, 80, or 160 ms). Due to reciprocity, the UE may use the Tx beam corresponding to the best Rx beam determined during synchronization. If reciprocity is available at the gNB, the UE transmits only once; otherwise, it repeats the same preamble for all the gNB Tx beams.

---

[10] Time allocated for sending preambles.

2. The gNB responds to the detected preambles with a random access response (RAR) UL grant (MSG2) in PDSCH by using *one selected* beam. After that, the UE and the gNB establish coarse beam alignment that could be utilized at the subsequent steps.
3. Upon receiving MSG2, the UE responds by MSG3 over the resources scheduled by the gNB, which is thus aware where to detect the MSG3 and which gNB Rx beam should be used.
4. The gNB confirms the above by sending MSG4 in PDSCH using the gNB Tx beam determined at the previous step.

Without the beam reciprocity, the UE transmits identical MSG1 signals with the same UE TX beam during one RACH occasion, while the gNB receives the preamble by the gNB Rx beam sweeping. The UE changes its Tx beam at the next RACH occasion. Due to repeated transmission, preambles do not need a CP and a guard period; they hence shorten compared to the reciprocity case. Moreover, MSG2 also needs to be sent by sweeping the gNB Tx beams (unless the UE informs the gNB regarding the best gNB Tx beam). The time required for achieving beam alignment varies depending on the codebook length (i.e., the number of combinations of gNB beams and UE beams). We note that as full sweeping might significantly increase the delay, another potential solution for the UE is to act as if reciprocity holds; then in case of a failure the gNB may request to retransmit multiple preambles (Fig. 5).

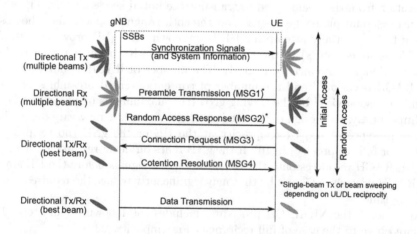

Fig. 5. An example of 5G NR CBRA procedure: no preamble collisions, reciprocity is not available at the UE (MSG1 is sent by multiple beams).

If two or more UEs select the same preamble, it may be decoded at the gNB as one preamble, and the gNB then transmits its RAR as for one UE. In this case, a preamble collision occurs at the third step above, when the UEs transmit their requests by using the same resources and perceive a preamble failure after the contention resolution timer expires (instead of receiving MSG4). After a collision,

Table 3. Antenna configuration options

Configuration	Strengths	Weaknesses
Omnidirectional Tx – Omnidirectional Rx (*Fully-Omnidirectional*)	Fast, has low complexity, lowest overhead	High signal attenuation, not suitable for most mmWave applications
Directional Tx – Directional Rx (*Fully-Directional*)	Compensated signal attenuation, wide coverage area, high throughput, reduced preamble collision probability	Challenging cell search, high complexity, high latency, accentuated deafness and blockage phenomena
Directional Tx – Omnidirectional Rx (*Semi-Directional*)	Reasonable compromise, low complexity, low latency	Subject to mmWave link instability

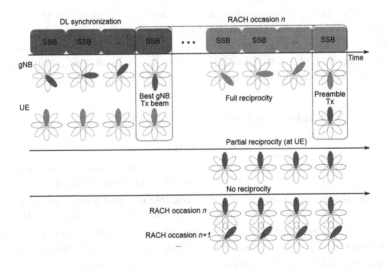

Fig. 6. Example of RACH operation with full/partial/no reciprocity.

the UE triggers a backoff time, which is selected randomly based on the backoff window (ranges from 5 to 1920 ms [8]), and restarts. The UE transmits with its default power or the power advised by the gNB. In case of an unsuccessful transmission, the UE follows a power ramping procedure [8] similar to LTE power ramping (while the power does not change during sweeping) (Fig. 6).

3.3 Antenna Configurations

Given the two alternatives – onmidirectional and directional transmission in 5G NR – we may differentiate between three possible antenna configurations that can be used during the initial access phase, namely, fully-omnidirectional (OD), fully-directional (FD), and semi-directional (SD) (see Table 3):

- Omnidirectional Transmissions – Omnidirectional Receptions (FO),
- Directional Transmissions – Directional Receptions (FD),
- Directional Transmissions – Omnidirectional Receptions (SD).

Here, the first configuration (FO) is the least preferable solution for most mmWave scenarios (except for short line-of-sight links), because a fully omnidirectional link suffers from uncompensated high path loss and thus the coverage area is reduced significantly; however, it allows for much faster and simpler beam search. The second configuration is the most beneficial in terms of the signal-to-noise ratio and throughput but requires complex beam search algorithms and causes delays as the UEs perform an entire angular-space scanning; however, when utilized during the RACH preamble transmission, it could significantly reduce preamble collision probabilities in dense scenarios.

Finally, the third configuration may become the desired compromise and successfully aid during the synchronization procedure, which requires minimizing delays at most. While directional transmission allows compensating the decreased link budget, omnidirectional reception reduces the complexity and delay of the beam search procedure; however, this option may also lead to the

Table 4. Selected LTE and NR differences*

LTE features	NR features
Microwave frequencies	Two available frequency ranges: below 6 GHz and above 6 GHz (mmWave)
Fixed physical layer parameters	Scalable physical layer parameters (e.g., flexible SCS, scalable TTI)
Synchronization signals are separately and omnidirectionally transmitted	Synchronization signals are grouped in SSBs and directionally transmitted
Cell search procedures are fast and have low complexity	Complex cell search procedures
Random access procedures have higher preamble collision probability	Directional preamble reception may reduce preamble collision probability during random access
90% bandwidth efficiency (100 RB cover 18 MHz of 20 MHz bandwidth carrier)	Higher bandwidth efficiency reaching 99%
Direct Current (DC, with no information) subcarrier helping to locate the frequency	No explicit DC subcarrier reserved for downlink nor uplink

*Detailed LTE/NR comparison w.r.t. numerology and channels may be found in [2].

coverage area mismatch. A choice between these three options can be based on the use-case requirements, which include, e.g., link length, UE density, target preamble collision probability, latency, and/or complexity of the device.

4   Conclusion

Our short review of the key differences between 5G NR and LTE is summarized in Table 4. More specific numbers in terms of the physical layer procedures are available in [2]. While Release 15 addresses the most essential features to deploy 5G networks, the following Release 16 – planned to be completed in December 2019 – targets to maintain the full-fledged 5G vision. The study items listed for Release 16 [13] as of today include the following developments beyond Release 15: NR-based access to unlicensed spectrum (to be studied for both licensed-assisted access and stand-alone deployments, similarly to LAA), Non Orthogonal Multiple Access (NOMA, allowing multiple UEs to access the channel over the same resources by relying on multi-user detection algorithms), evaluation of advanced V2X use-cases for NR and LTE, backhauling options for NR, industrial IoT scenarios, 5G-grade URLLC enhancements, solutions to support non-terrestrial networks, dual connectivity enhancements, and 5G for satellites.

Acknowledgment. The work is supported by the "RUDN University Program 5 − 100".

References

1. http://www.sharetechnote.com. Accessed June 2018
2. 5G Design Concepts Towards the Next Generation Networks. White Paper (2018). https://cdn-www.mediatek.com/page/5G_Design_Concepts_v12_MARCOM.pdf
3. 5G NR - A New Era for Enhanced Mobile Broadband. White Paper (2018). https://cdn-www.mediatek.com/page/MediaTek-5G-NR-White-Paper-PDF5GNRWP.pdf
4. 3GPP Technical Specifications Group RAN: TS 38.300: NR and NG-RAN Overall Description, Stage 2 (Release 15), December 2017
5. 3GPP Technical Specifications Group RAN: TS 38.211 (Release 15): Physical channel and modulation, June 2018
6. 3GPP Technical Specifications Group RAN: TS 38.213: NR Physical layer procedures for control (Release 15), June 2018
7. 3GPP Technical Specifications Group RAN: TS 38.213 (Release 15): Physical Layer Procedures for Control, June 2018
8. 3GPP Technical Specifications Group RAN: TS 38.321 (Release 15): Medium Access Control (MAC) protocol specification, June 2018
9. 3GPP Technical Specifications Group RAN: R1–1608966: Considerations on Sweeping Time Interval in NR, October 2016
10. 3GPP Technical Specifications Group RAN: R1–1609118: RACH design with and without beam reciprocity, October 2016
11. 3GPP Technical Specifications Group RAN: R1–R11609117: Discussion on RA procedure, October 2016

12. 3GPP Technical Specifications Group RAN: TS 38.802 (Release 14): Physical Layer Aspects, September 2017
13. 3GPP Technical Specifications Groups: 3GPP Features and Study Items (2018). http://www.3gpp.org/DynaReport/FeatureListFrameSet.htm. Accessed June 2018
14. 3GPP Technical Specifications Groups: About Release 15, June 2018. http://www.3gpp.org/release-15. Accessed June 2018
15. Alkhateeb, A., Mo, J., Gonzalez-Prelcic, N., Heath, R.W.: MIMO precoding and combining solutions for millimeter-wave systems. IEEE Commun. Mag. **52**(12), 122–131 (2014)
16. Desai, V., Krzymien, L., Sartori, P., Xiao, W., Soong, A., Alkhateeb, A.: Initial beamforming for mmWave communications. In: 2014 48th Asilomar Conference on Signals, Systems and Computers, pp. 1926–1930 (2014)
17. Deutsche Telekom AG: RP-161266: 5G architecture options - full set, June 2016
18. Giordani, M., Mezzavilla, M., Zorzi, M.: Initial access in 5G mmWave cellular networks. IEEE Commun. Mag. **54**(11), 40–47 (2016)
19. IEEE P802.11 Task Group ay: Status of Project IEEE 802.11ay. www.ieee802.org/11/Reports/tgay_update.htm. Accessed June 2018
20. IEEE Standard: 802.11ad-2012: Enhancements for Very High Throughput in the 60 GHz Band. https://ieeexplore.ieee.org/document/6392842/
21. Keysight Technologies: Understanding the 5G NR Rhysical Layer, November 2017
22. NTT Docomo: Status of Investigations on Physical-layer Elemental Technologies and High-frequency-band Utilization. NTT Docomo Tech. J. **19**(3), 24 (2018)
23. Qualcomm: Designing 5G NR: The 3GPP Release-15 global standard for a unified, more capable 5G air interface, April 2018. https://www.qualcomm.com/media/documents/files/the-3gpp-release-15-5g-nr-design.pdf
24. Qualcomm: Making 5G NR a reality. White paper, December 2016
25. Razavizadeh, S.M., Ahn, M., Lee, I.: Three-dimensional beamforming: a new enabling technology for 5G wireless networks. IEEE Signal Process. Mag. **31**(6), 94–101 (2014)
26. Rohde & Schwarz: Demystifying 5G - How mobile is 5G at mmWave frequencies? (2017). https://www.rohde-schwarz.com/ru/solutions/test-and-measurement/wireless-communication/5g/webinars-videos/demystifying-5g-enable-mobility-in-5g-systems_231144.html. Accessed June 2018
27. Rohde & Schwarz: Numerology and Initial Access Concept for 5G NR (2017). https://www.youtube.com/watch?v=eE_b7vWbkoI. Accessed June 2018
28. Yifei, Y., Xinhui, W.: 5G new radio: physical layer overview. ZTE Commun. **15**(S1), 11–19 (2017)

Energy - Aware Offloading Algorithm
for Multi-level Cloud Based 5G System

Abdelhamied A. Ateya[1,2(✉)], Ammar Muthanna[2,3], Anastasia Vybornova[2],
Pyatkina Darya[3], and Andrey Koucheryavy[2]

[1] Electronics and Communications Engineering, Zagazig University, Zagazig, Egypt
a_ashraf@zu.edu.eg
[2] St. Petersburg State University of Telecommunication, 22 Prospekt Bolshevikov,
St. Petersburg, Russia
ammarexpress@gmail.com, a.vybornova@gmail.com, akouch@mail.ru
[3] Peoples' Friendship, University of Russia (RUDN University), 6 Miklukho-Maklaya St.,
Moscow 117198, Russia
da@rudn.university

Abstract. Mobile edge computing (MEC) is a recent communication paradigm developed mainly for cellular networks. MEC is introduced to improve the whole network efficiency by offloading its operations to nearby clouds. Cellular networks are able to offer the cloud computing capabilities at the edge of the radio access network through MEC servers. Mobiles services and tasks can either be executed at the mobile device or offloaded to the edge server. In this work, we provide a latency aware and energy aware offloading algorithm for the 5G multi-level edge computing based cellular system. The algorithm enables the mobile device to request offloading or decide the local execution independently based on the available resources at the mobile device and edge server. The algorithm takes into consideration the energy consumption to handle the service and make the offloading decision that achieves higher energy performance. The system is simulated and numerical results are included for performance evaluation.

Keywords: Latency · Offloading · Mobile edge computing · Energy consumption
5G

1 Introduction

With the dramatic increase of wireless devices (e.g. Smart phones, Wearable devices and MTC devices), the fifth generation of cellular system (5G) must deploy new technologies to support these heterogeneous devices and achieve the user demands [1]. The main requirements of 5G include; high reliability, high throughput, ultra low latency, high mobility, high connectivity and high data rate up to ten Gbps [2, 3].

Mobile edge computing (MEC) technology is a powerful solution to achieve challenges and requirements associated with the design and realization of 5G cellular systems [4]. MEC improves the cellular network efficiency by offloading network operations to a cloud server deployed at the edge of the radio access network (RAN). MEC

© Springer Nature Switzerland AG 2018
O. Galinina et al. (Eds.): NEW2AN 2018/ruSMART 2018, LNCS 11118, pp. 355–370, 2018.
https://doi.org/10.1007/978-3-030-01168-0_33

offers computing capabilities and resources at the edge of RAN near to the end user (i.e. one hop away). Employing MEC servers achieves various benefits to the cellular networks includes the following [5, 6]:

1- Higher system bandwidth,
2- Providing an efficient way for data offloading,
3- Reduction of latency,
4- Reduction of network congestion, and
5- Introduction of new services and applications, as the MEC server access the network context information.

However, task offloading from mobile devices to edge servers consumes energy due to communication between MEC server and the mobile device and also increases latency of services [7]. In most cases, offloading is a necessary demand, due to limited hardware capabilities of mobile devices. Thus, offloading tasks to MEC server instead of core network or remote data centers is more efficient in these situations [8].

Researches in task offloading to MEC servers mainly focus on the latency aware and energy aware offloading, where mobile device only have the decision of offloading [9, 10]. Task offloading should be deployed in away to support the predefined quality of services (QoS), which puts constraints on the offloading scale [11]. In this work, we provide an offloading algorithm for multilevel edge computing system introduced in [6, 12]. The algorithm cares for both latency and energy consumptions. Mobile devices only have the decision of offloading based on available resources of devices and amount of consumed energy for the task. Multilevel edge computing system provides two levels of offloading for mobile devices; Micro-cloud units and Mini-cloud units, the system can be deployed for 5G cellular system to enable the requirements and overcome challenges.

2 Proposed Algorithm

In this part, the proposed energy aware selective offloading algorithm is introduced for 5G system with multilevel edge computing structure introduced in [6, 12]. We first provide the system structure and main parts included, and then different variables and parameters used in the offloading algorithm are defined. Finally we present the offloading algorithm.

A- System structure

The system provides two main levels of offloading for mobile tasks, besides the local handling of tasks at the mobile device when it is proper. Figure 1 indicates different levels of offloading and the corresponding place of task execution in each level. The mobile may handle the task without any offloading and this is indicated as the zero offloading level. This takes place, when the mobile device has available resources enough for handling task with the required QoS constraints and with energy consumption less than that consumes at the edge server and less than a threshold value that saves the mobile battery. The first offloading level is the offloading to Micro-cloud unit and the second offloading is the offloading to Mini-cloud unit. The proposed algorithm is implemented at each mobile device and is employed for each computing task, to decide where the task is executed (i.e. locally or offloaded).

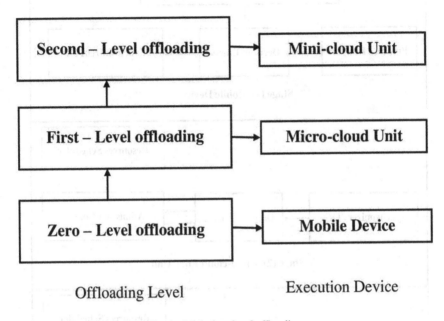

Fig. 1. Main levels of offloading.

The general structure of the three main parts of the system is presented in Fig. 2. The mobile device, as well as edge computing servers (i.e. Micro-cloud unit and Mini-cloud unit), employs a decision engine that decides whether to execute tasks locally or offload them to the appropriate edge server. The decision engine calculates all parameters required to take the decision.

B- Energy aware offloading algorithm

1- Annotation

Before introducing the offloading algorithm, the considered parameters are first defined. Notion of introduces variables and parameters are presented in Table 1. Different types of messages transferred between mobile device and heterogeneous edge servers are defined in Table 2. Various steps of offloading algorithm are indicated in Algorithm 1.

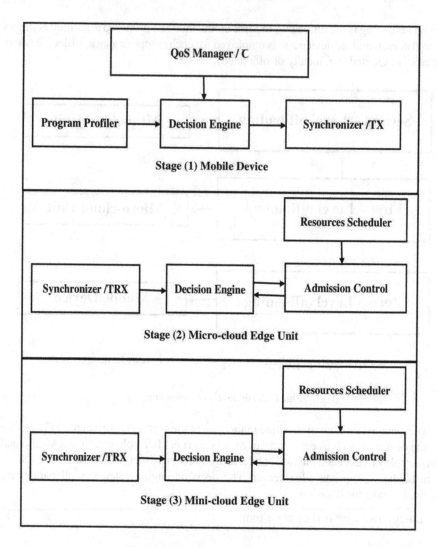

Fig. 2. Three stages system structure.

Table 1. Key Notations.

Notation	Description
L	Total length of input data of a task
S	Number of CPU cycles required to process one bit of task input data
N_{CYC}	Total number of CPU cycles required to process a task of length L
R_M	Mobile resources allocated for a task
	Maximum allowable latency of a task I determined by the task portioning scheme, in away to support QoS
δ_M	Energy consumption for one CPU cycle of mobile device
T_M	Total execution time of a task, when the task is executed locally at the mobile device
E_M	Energy consumption for a task executed mobile device
E_{th}	Threshold of energy consumption of mobile device
D_{T-M}	Binary time decision variable of mobile device
D_{E-M}	Binary energy decision variable of mobile device
D_{OFF}	Binary offloading decision variable of mobile device
I	Mode Indicator Variable
T_{tx}	Time spent on transmitting the input data of a task
T_{rx}	Feedback time of computation results
T_{eMicro}	Total execution time of a task, when the task is executed at the Micro-cloud unit
R_{Micro}	Micro-cloud resources allocated for an offloaded task from a mobile device
T_{Micro}	Total time to handle a task at Micro-cloud unit
R_b	Uplink achievable data transmission rate
ω	The system bandwidth
σ	Noise power at the receiver
h	Channel gain
P	Transmitting power of mobile devices
$D_{T-Micro}$	Binary time decision variable of Micro-cloud unit
E_{Micro}	Energy consumption for handling a task at Micro-cloud unit
δ_{Micro}	Energy consumption for one CPU cycle of Micro-cloud unit
η_c	Channel efficiency
T_{e-Mini}	Total execution time of a task, when the task is executed at the Mini-cloud unit
T_{Mini}	Total time to handle a task at Mini-cloud unit
D_{T-Mini}	Binary time decision variable of Mini-cloud unit
R_{Mini}	Mini-cloud resources allocated for an offloaded task from a Micro-cloud unit
$D_{Micro-Off}$	Binary offloading decision variable of Micro-cloud unit
f_M	Total mobile resources
f_{Micro}	Total Micro-cloud resources
f_{Mini}	Total Mini-cloud resources

Table 2. Exchanged messages.

Message	Type	T_X	R_X	Contents
Request message	I	Mobile device	Micro-cloud	L, S,
Respond message	I	Micro-cloud	Mobile device	$D_{T\text{-Micro}}$, E_{Micro}
Request message	II	Mobile device	Micro-cloud	L, S, , D_{Off}
Respond message	II	Micro-cloud	Mobile device	$D_{T\text{-Micro}}$, $D_{Micro\text{-Off}}$
Request message	III	Micro-cloud	Mini-cloud	L, S,
Respond message	III	Mini-cloud	Micro-cloud	$D_{T\text{-Mini}}$

Algorithm 1. Energy aware offloading algorithm

1:	Initialize $\square$, E_{thr}
2:	Mobile device calculates L, S
3:	Mobile device calculates T_M
4:	If ($T_M \leq \square$)
5:	$\quad$ $D_{T\text{-}M} = 0$
6:	$\quad$ Calculates E_M
7:	$\quad$ Sends request message of type I
8:	$\quad$ Receives respond message of type I
9:	$\quad$ If ($T_{Micro} > \square$)
10:	$\quad\quad$ $D_{Off} = 0$
11:	$\quad\quad$ Handle task locally
12:	$\quad$ else
13:	$\quad\quad\quad$ if (($E_M < E_{Micro}$)& ($E_M < E_{th}$)
14:	$\quad\quad\quad\quad$ $D_{Off} = 0$
15:	$\quad\quad\quad\quad$ Handle task locally
16:	$\quad\quad\quad$ else
17:	$\quad\quad\quad\quad$ $D_{Off} = 1$
18:	$\quad\quad$ end if
19:	$\quad$ else
20:	$\quad\quad$ $D_{Off} = 1$
21:	$\quad\quad$ Sends request message of type II
22:	$\quad$ end if
23:	If ($D_{off} == 1$)
24:	$\quad$ Micro-cloud calculates T_{Micro}
25:	$\quad$ If ($T_{Micro} > \square$)
26:	$\quad\quad$ $D_{Micro\text{-}off} = 1$
27:	$\quad\quad$ Send request message of type III
28:	$\quad$ else
29:	$\quad\quad$ $D_{Micro\text{-}off} = 0$
30:	$\quad\quad$ Handle task at Micro-cloud server
31:	$\quad$ end if
32:	end if

2- Offloading model

The offloading decision is first decided by the mobile device via a decision engine. The program profile calculates the total data length L and the total number of CPU cycles N_{CYC} needed to handle the pointed task. The devices should use this information to check if it can handle the task or take the decision of offloading. This depends on the resources available in the device that can be allocated for this task and the maximum latency of the task that supports the QoS.

The mobile device calculates the execution time T_M based on the available devices resources as in (1). Then, the device calculates the binary time decision value D_{T-M} by comparing T_M and as presented in (3). Each value of the binary time decision is considered as a case; Case (I) for the binary one decision and Case (II) for the binary zero decision.

$$T_M = \frac{N_{CYC}}{R_M} \quad , R_M \in f_M \tag{1}$$

$$N_{CYC} = L.S \tag{2}$$

$$D_{T-M} = I(T_M, \tau) = \begin{cases} 1 \; IF \; (T_M \le \tau) \\ 0 \; IF \; (T_M > \tau) \end{cases} \tag{3}$$

Case (I):

When the execution time is less than the maximum latency time, the task can be handled locally at the mobile device and the binary time decision in this case is one. In this case, the mobile device calculates the energy consumption to handle the task locally E_M using (4), and sends a request message of type I to the corresponding edge server (i.e. Micro-cloud or Mini-cloud). Based on the response message of the edge server, the mobile device takes the decision of offloading or processing the task locally. The edge server checks its availability to handle the task based on the procedures indicated later. The edge server sends a response message of type I to the associated mobile device, which decides whether to offload the task or process it locally based on the received information.

$$E_M = N_{CYC}\delta_M \tag{4}$$

Case (II):

However, if the time required to execute the task at the mobile device is larger than the maximum latency constraint , the task can't be handled locally and the decision of offloading should be took. In this case, the binary decision variable D_{off} is assigned one; and thus, the mobile devices send a request message of type II to the corresponding edge server (i.e. Micro-cloud unit or Mini-cloud unit).

The Micro-cloud unit receives the mobile device's request, and responds with appropriate message. If the Micro-cloud unit is prober to handle the task, it should send a response message of type II with the agreement of offloading and allocates the resources for task handling; otherwise, it sends a request message of type III to the corresponding Mini-cloud unit and waits for the response.

First, the Micro-cloud unit checks the latency constraints, once it receives a request message from a mobile device; either massage of type I or message of type II. The Micro-cloud unit calculates the execution time T_{eMicro} based on the available recourses at the edge server, as indicated in (6). Then, the total latency of task handling at Micro-cloud unit is calculated by summing both concerning delays; processing delay (i.e. execution time) and communication delay (i.e. uplink transmission time of input data and feedback time of computation result).

$$T_{Micro} = T_{eMicro} + T_{tx} + T_{rx} \tag{5}$$

$$T_{eMicro} = \frac{N_{CYC}}{R_{Micro}} \quad , R_{Micro} \in f_{Micro} \tag{6}$$

$$T_{tx} = \frac{L}{R_b} \tag{7}$$

The bit rate can be calculated as a function of the transmitted power and the channel power gain using Shannon-Hartley formula as the following [13]:

$$R_b = \omega \log_2 \left(1 + \frac{hp}{\sigma}\right) \tag{8}$$

The decision engine of the Micro-cloud unit calculates the binary time decision variable $D_{T-Micro}$, by comparing latency of task handling T_{Micro} with the maximum latency as illustrated in (9).

$$D_{T-Micro} = I(T_{Micro}, \tau) = \begin{cases} 1 \ IF \ (T_{Micro} \leq \tau) \\ 0 \ IF \ (T_{Micro} > \tau) \end{cases} \tag{9}$$

If the decision is zero, the Micro-cloud unit sends a request for the corresponding Mini-cloud unit. For the positive time decision (i.e. $D_{T-Micro} = 1$), the task can be handled at the Micro-cloud unit. If the received request is of type II, the task is then handled by the Micro-cloud unit, otherwise the energy constraints should be checked. The Micro-cloud unit calculates the energy consumption to handle the task at the Micro-cloud unit E_{Micro} as in (10).

$$E_{Micro} = N_{CYC}\delta_{Micro} + T_{tx}P\eta_C \tag{10}$$

If the received request is of type I, the Micro-cloud unit sends a respond message contains the E_{Micro} and the decision engine of the mobile device calculates the mobile binary energy decision value D_{E-M} by comparing both energies; E_M and E_{Micro}. If the decision is one the task is handled locally, otherwise the task is offloaded to the Micro-cloud unit.

$$D_{E-M} = I\left(E_M, E_{Micro}\right) = \begin{cases} 1 \ IF \ \left(E_M \leq E_{Micro}\right) \\ 0 \ IF \ \left(E_M > E_{Micro}\right) \end{cases} \tag{11}$$

For a negative time decision of Micro-cloud unit and mobile device, the Micro-cloud unit sends a request message of type III to the corresponding Mini-cloud unit. The Mini-cloud unit calculates the total latency of task handling T_{mini} based on the available resources as illustrated in (12).

$$T_{Mini} = T_{eMini} + T'_{tx} + T'_{rx} \tag{12}$$

$$T_{eMini} = \frac{N_{CYC}}{R_{Mini}} \quad , R_{Mini} \in f_{Mini} \tag{13}$$

Then, the Mini-cloud unit calculates the binary time decision D_{T-Mini} by comparing the total time required to handle the task at the Mini-cloud T_{Mini} with the maximum task latency .

$$D_{T-Mini} = I\left(T_{Mini}, \tau\right) = \begin{cases} 1 \ IF \ \left(T_{Mini} \leq \tau\right) \\ 0 \ IF \ \left(T_{Mini} > \tau\right) \end{cases} \tag{14}$$

The Mini-cloud unit responds with a respond message of type III, and the task is offloaded to mini-cloud unit for both condition; $D_{T-Mini} = 1$ and $D_{T-Micro}$ &$D_{T-M} = 0$.

3   Performance Evaluation

In this part, numerical results for the proposed algorithm are introduced to evaluate the performance. The system is simulated using Matlab and the introduced algorithm is implemented for a cellular system of two cells of radius 1 km and a base station at the cell center. The base station of cell (A) is assumed to be fed with a Micro-cloud edge server to enable the MEC facilities to cellular users. The neighbor cell (B) is assumed to have a Mini-cloud edge server that is connected to the Micro-cloud unit of cell (A).

Ten heterogeneous tasks are considered for the simulation, these tasks are randomly distributed for a ten mobile devices that are randomly distributed over the cell A. The considered tasks are corresponding to real workloads of processing web pages [14, 15]. Table 3 indicates the attributes of each task extracted by the program profile of each mobile device [16]. Devices are randomly distributed over the cell (A). Table 4 indicates simulation parameters. We consider two main scenarios, with three cases for each scenario; in each scenario a computational capabilities of mobile devices is assumed and in each case a certain value of QoS latency for each task is assumed. Table 5 presents the values of maximum latency in each case for each task and other considered simulation parameters [17, 18].

Table 3. Tasks parameters.

Task	Task(1)	Task(2)	Task(3)	Task(4)	Task(5)
L(kB)	1	1.2	1.35	1.65	1.75
S(Cyc/B)	6000	7200	8100	9900	10500
Task	Task(6)	Task(7)	Task(8)	Task(9)	Task(10)
L(kB)	1.8	1.97	2.2	2.25	2.48
S(Cyc/B)	10800	11820	13200	13500	14880

Table 4. QoS latency of tasks considered.

Task	Task(1)	Task(2)	Task(3)	Task(4)	Task(5)
$_1$(ms)	1	1.2	1.35	1.65	1.75
$_2$(ms)	2	2.4	2.7	3.3	3.5
$_3$(ms)	3	3.6	4.05	4.95	5.25
Task	Task(6)	Task(7)	Task(8)	Task(9)	Task(10)
$_1$(ms)	1.8	1.97	2.2	2.25	2.48
$_2$(ms)	3.6	3.94	4.4	4.5	4.96
$_3$(ms)	5.4	5.91	6.6	6.75	7.44

Table 5. Simulation parameters.

Parameter	Value
ω	1 MHz
σ	10-3 W
P	1 W
f_M (First Scenario)	ϵ [0.5,1.5] GHz
f_M (Second Scenario)	ϵ [1.5,3.0] GHz
$f_{Micro\text{-}cloud}$	ϵ [5.0,3.0] GHz
$f_{Mini\text{-}cloud}$	ϵ [1.5,3.0] GHz
δ_M	1 J/GHz
δ_{Micro}	1 J/GHz

Figure 3 illustrates the average latency of handling the ten considered tasks for the first conducted scenario. Some tasks are handled locally and others are offloaded to Micro-cloud units and the rest are offloaded at the Mini-cloud edge server. The offloading level of each task is indicated in Fig. 4.

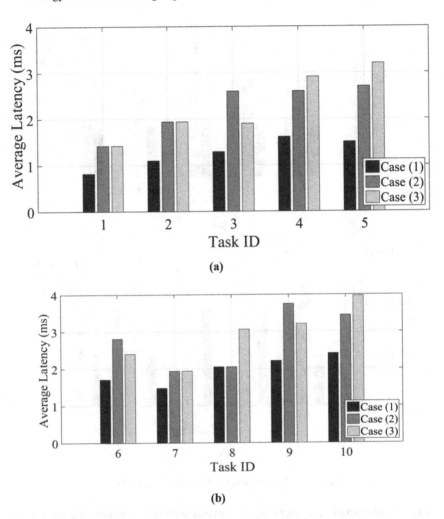

Fig. 3. Average latency of each task for scenario (1).

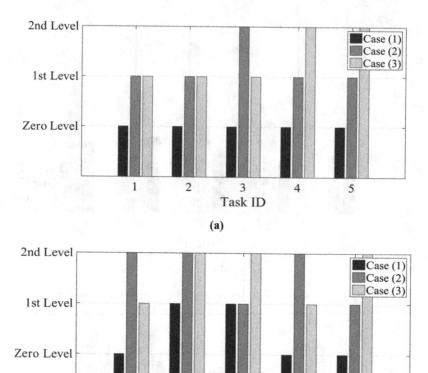

Fig. 4. Offloading level of tasks in scenario (1).

The second scenario considers mobile devises with higher computing capabilities. Thus, mobile devices can handle more complicated tasks and offer higher resources for tasks. Figure 5 illustrates the average latency of ten tasks in the second scenario. Mainly, all locally executed tasks of first scenario are executed in less time due to higher allocated resources offered by the mobile device in the second scenario. Moreover, some tasks from the offloaded tasks in the first scenario are handled locally in the second scenario, because of higher resources of mobile devices. Figure 6 illustrates the level of offloading of each task.

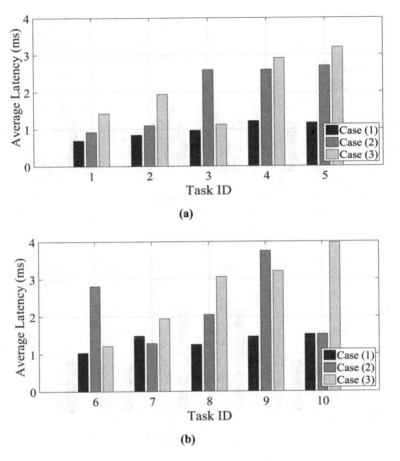

(a)

(b)

Fig. 5. Average latency of each task for scenario (2).

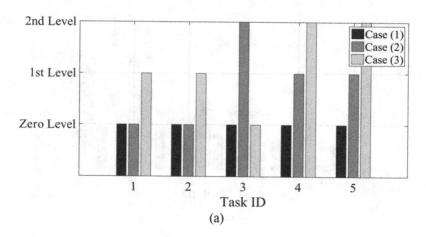

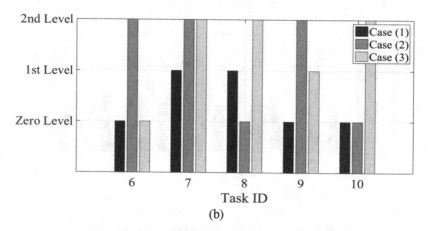

Fig. 6. Offloading level of tasks in scenario (2).

4 Conclusion and Future Work

MEC provides various benefits for cellular networks by introducing computing capabilities at the edge of RAN, one hop away from the end user. This work provides an offloading algorithm for the multilevel edge computing system introduced for the 5G cellular system. The algorithm allows mobile devices independently to take the decision of offloading or processing task locally. The offloading decision considers both task latency and energy conservation. The mobile offload tasks to MEC server if the mobile available resources are not enough to handle task in the required time to support QoS, or when the local execution executes much energy. The system provides two level of offloading; Micro-cloud edge units and Mini-cloud edge units. The system is evaluated for a cellular cell with ten mobile devices; each have a heterogeneous task and numerical results are included.

Our future vision is to deploy an intermediate level of offloading between mobile device and edge servers (i.e. cloudlet). The proposed algorithm should be deployed for the three level offloading system.

Acknowledgement. The publication has been prepared with the support of the "RUDN University Program 5-100".

References

1. Ateya, A., Muthanna, A., Koucheryavy, A.: 5G framework based on multi-level edge computing with D2D enabled communication. In: 2018 20th International Conference on Advanced Communication Technology (ICACT), pp. 507–512. IEEE, February 2018
2. Zheng, K., Zhao, L., Mei, J., Dohler, M., Xiang, W., Peng, Y.: 10 Gb/s hetsnets with millimeter-wave communications: access and networking-challenges and protocols. IEEE Commun. Mag. **53**(1), 222–231 (2017)
3. Tudzarov, A., Gelev, S.: Requirements for next generation business transformation and their implementation in 5G architecture. Int. J. Comput. Appl. **162**(2), 31–35 (2017)
4. Ateya, A.A., Muthanna, A., Gudkova, I., Abuarqoub, A., Vybornova, A., Koucheryavy, A.: Development of intelligent core network for tactile internet and future smart systems. J. Sens. Actuator Netw. **7**(1), 1 (2018)
5. Mobile Edge Computing A key technology towards 5G. ETSI White Paper, No. 11, September 2015
6. Ateya, A., Vybornova, A., Kirichek, R., Koucheryavy, A.: Multilevel cloud based tactile internet system. In: IEEE-ICACT2017 International Conference, Korea, Febuary 2017
7. Chen, M., Hao, Y.: Task offloading for mobile edge computing in software defined ultra-dense network. IEEE J. Sel. Areas Commun. **36**(3), 587–597 (2018)
8. Wang, F., Xu, J., Wang, X., Cui, S.: Joint offloading and computing optimization in wireless powered mobile-edge computing systems. IEEE Trans. Wirel. Commun. **17**(3), 1784–1797 (2018)
9. Zhang, K., et al.: Energy-efficient offloading for mobile edge computing in 5G heterogeneous networks. IEEE Access **4**, 5896–5907 (2016)
10. Cardellini, V., Personé, V.D.N., Valerio, V.D., Facchinei, F., Grassi, V., Presti, F.L., Piccialli, V.: A game-theoretic approach to computation offloading in mobile cloud computing. Math. Program. **157**(2), 421–449 (2016)
11. Lyu, X., et al.: Selective offloading in mobile edge computing for the green Internet of Things. IEEE Network **32**(1), 54–60 (2018)
12. Ateya, Abdelhamied A., Vybornova, A., Samouylov, K., Koucheryavy, A.: System Model for Multi-level Cloud Based Tactile Internet System. In: Koucheryavy, Y., Mamatas, L., Matta, I., Ometov, A., Papadimitriou, P. (eds.) WWIC 2017. LNCS, vol. 10372, pp. 77–86. Springer, Cham (2017). https://doi.org/10.1007/978-3-319-61382-6_7
13. Kartun-Giles, A., Jayaprakasam, S., Kim, S.: Euclidean matchings in ultra-dense networks. IEEE Commun. Lett. (2018)
14. Miettinen, A.P., Nurminen, J.K.: Energy efficiency of mobile clients in cloud computing. In: Proceedings of the 2010 USENIX Conference on Hot Topics in Cloud Computing. (HotCloud), pp. 1–7, June 2010
15. Chun, B., Ihm, S., Maniatis, P., Naik, M., Patti, A.: Clonecloud: elastic execution between mobile device and cloud. In: Proceedings of the Sixth cConference on Computer Systems, pp. 301–314. ACM (2011)

16. Huang, D., Wu, H.: Mobile Cloud Computing: Foundations and Service Models. Morgan Kaufmann, San Francisco (2017)
17. Habak, K., Ammar, M., Harras, K.A., Zegura, E.: Femto clouds: Leveraging mobile devices to provide cloud service at the edge. In: 2015 IEEE 8th International Conference on Cloud Computing (CLOUD), pp. 9–16. IEEE (2015)
18. Mao, Y., Zhang, J., Letaief, K.B.: Dynamic computation offloading for mobile-edge computing with energy harvesting devices. IEEE J. Sel. Areas Commun. 34(12), 3590–3605 (2016)

Performance Analysis for DM-RS Mapping in a High Speed Train System

Jihyung Kim[✉]⓪, Juho Park, Junghoon Lee, and JunHwan Lee

Electronics and Telecommunications Research Institute, Daejeon, Korea
savant21@etri.re.kr

Abstract. In this paper, we analyze performances for BLER and spectral efficiency in a high speed train (HST) system. The HST scenario is one of 5G mobile communication services. The performance analysis is evaluated in accordance with DM-RS mapping for channel estimation. DM-RS mapping is associated with high Doppler and frequency flat channel properties of the HST scenario. The performance results show that DM-RS in new radio (NR) satisfies the performance requirement of HST in ITU-R and a modified DM-RS mapping can be more efficient in HST channel properties.

Keywords: 5G NR · High speed train · DM-RS · Frequency selectivity

1 Introduction

1.1 A Subsection Sample

5G mobile communications should satisfy requirements to support various service scenarios, including high transmission speeds for enhanced mobile broadband (eMBB) service, short transfer delays with reliability for ultra reliable and low latency communication (URLLC) service, and large terminal connectivity for massive machine type communication (mMTC) service [1]. ITU-R working party (WP) 5D suggests specific technical performance requirements such as peak data rate, average spectral efficiency, reliability, and mobility [2]. In addition, it suggests guidelines for evaluation of radio interface technologies such as system and link level parameters, channel model, and network layout [3].

In order to address these requirements, 3rd Generation Partnership Project (3GPP) studied new wireless access technology called new radio (NR) [4]. Based on these studies, it is currently developing technical specifications [5–10]. NR does not have backward compatibility with long term evolution (LTE) and maintains forward compatibility for supporting various services in the future. One of

This work was supported by Institute for Information & communications Technology Promotion (IITP) grant funded by the Korea government (MSIT) (No. 2017-0-01973 (Korea-Japan) International collaboration of 5G mmWave based Wireless Channel Characteristic and Performance Evaluation in High Mobility Environments).

© Springer Nature Switzerland AG 2018
O. Galinina et al. (Eds.): NEW2AN 2018/ruSMART 2018, LNCS 11118, pp. 371–380, 2018.
https://doi.org/10.1007/978-3-030-01168-0_34

characteristics of NR utilizes a very wide spectral range from 1 GHz to 100 GHz. For this it basically considers a beamforming based system as well as scalable numerology. Beamforming can be applied to all signals and data transmitted to each user equipment (UE). In particular, unlike LTE, which is a single-beam-based system, NR is characterized by considering transmission using multiple beams. Accordingly, beam control procedures for setting control signals and beams used for data transmission are introduced. One of the other features is that it supports front-loaded demodulation reference signal (DM-RS) for fast decoding processing time. When control symbols occupy two or three symbols, the front-loaded DM-RS symbol can be located in the 3-rd or 4-th symbol in a slot. In addition, additional DM-RS symbols in a slot can be allocated for coping with the high Doppler effect.

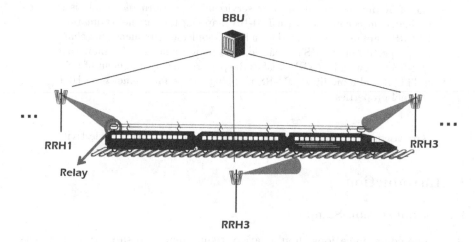

Fig. 1. Linear cell layout for a HST scenario

On the other hand, there are high-speed train (HST) scenarios as one of various service scenarios of 5G mobile communication [2–4,11]. [2–4] define two scenarios for HST. One is that transmission reception point (TRP) is directly linked with UE in below 6 GHz carrier frequency and the other is that TRP is linked with UE by the relay node in above 6 GHz carrier frequency. [11] defines a HST scenario connected in non-terrestrial networks (NTN) such as high altitude platforms (HAPs). In these scenarios the UE speed is considered up to 500 km/h. In addition, the line of sight (LOS) channel can be dominated. This is because the considered network layout for HST is generally a rural environment and the TRP is located very close to the HST. [2–4] take into account CDL/TDL-D and CDL/TDL-E channels for simulation evaluation parameters in above 6 GHz environment. These channels are constructed to represent the LOS property. In addition, the LOS probability is about 78–99.8% in NTN channel modeling [12].

When DM-RS mapping in NR is applied for channel estimation in HST scenarios, the Doppler effect for high speed may be ignored by using additional

DM-RS symbols with large subcarrier spacing. In the case of LOS channels, however, there may be unnecessary DM-RS resources for spectral efficiency. In other words, the low DM-RS density in the frequency domain can be considered. In this paper, we evaluate the performance using DM-RS of NR in a HST scenario and analyze whether it satisfies the requirement of HST in ITU-R. In addition, we take into account the low density DM-RS mapping design for LOS channel with performance comparison.

2   DM-RS Mapping in a HST Scenario

2.1   Requirements and Layout Configuration

Figure 1 shows linear cell layout configuration for high speed vehicular mobility at 500 km/h under rural-eMBB test environment [3]. In the figure a base station consists of 1 base-band unit (BBU) and 3 remote radio heads (RRHs). 3 RRHs in a cell are connected to 1 BBU. In this configuration one of requirements of HST in ITU-R is spectral efficiency. The spectral efficiency at 500 km/h mobility should satisfy 0.45 b/s/Hz [2]. Various parameters such as channel estimation and phase noise impact can be considered for performance evaluation. As mentioned above, we focus on DM-RS mapping for channel estimation. Phase noise is typically associated with phase tracking (PT)-RS in various reference signals of NR.

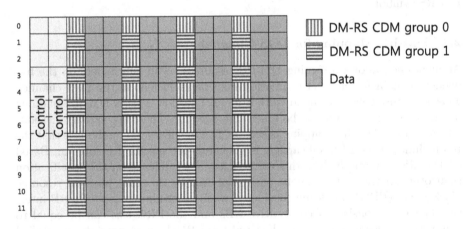

Fig. 2. DM-RS mapping for configuration type 1 in NR

2.2   DM-RS Mapping in NR

The DM-RS sequence at the p-th port for the k-th subcarrier and the l-th symbol for physical downlink shared channel (PDSCH) in NR can be written as follows [5]:

$$a_{k,l}^{(p)} = \beta \cdot w_f(k') \cdot w_t(l') \cdot r(2n + k') \tag{1}$$

where

$$k = \begin{cases} 4n + 2k' + \Delta & \text{Configuration type 1} \\ 6n + k' + \Delta & \text{Configuration type 2} \end{cases}$$
$$k' = 0, 1$$
$$l = \bar{l} + l'$$
$$n = 0, 1, \ldots$$

(2)

β is power constant and $r(n)$ is the pseudo-random sequence. $w_f(k')$, $w_t(l')$, and Δ are sequence values with cyclic shift of code division multiplexing (CDM) group for distinguishing channel per port. The values are listed in Tables 7.4.1.1.2-1 and 7.4.1.1.2-2 of [5]. The parameter l in (1) is associated with the position of DM-RS symbols in the time domain and the number of DM-RS symbols in a slot can be increased up to 4 in accordance with UE mobility. The values of $\bar{l}$ and l' for l in (2) are given in Tables 7.4.1.1.2-3, 4, and 5 of [5]. There are two kinds of mapping methods according to configuration types. The maximum number of ports in a DM-RS symbol is 4 for configuration type 1 and 6 for configuration type 2. Figure 2 shows an example of DM-RS mapping of configuration type 1 for a resource block (RB) in a slot. The number of subcarriers in the RB is 12. 1 front-loaded DM-RS symbol and 3 additional DM-RS symbols are allocated for coping with high Doppler effect. The parameter k is related to subcarrier mapping in the frequency domain. Based on (1) DM-RS resources in the frequency domain are mapped to all subcarriers allocated per port in each DM-RS symbol.

2.3   DM-RS Mapping for LOS Channel

As mentioned above, assigning DM-RS resources to all subcarriers on the frequency domain for channel estimation may be inefficient in the LOS channel. This is because coherence bandwidth is large. Thus low density DM-RS mapping in the frequency domain can be taken into account. There are various methods for low density DM-RS mapping based on (1). One of several methods is that the RB included in DM-RS subcarriers is allocated per specific RB interval. Based on DM-RS mapping in NR, this method has less influence on the specification than other methods and can be easily applied. In this case physical resource block group (PRG) should be considered when determining a specific RB interval. The PRG consists of consecutive RBs and the PRG size can be equal to one of the values among {2, 4, allocated total RBs} [8]. A precoder is applied per PRG. Since different precoders per PRG are applied to DM-RS, DM-RS in a particular PRG cannot be used for channel estimation for data demodulation in another PRG. Thus DM-RS with a specific RB interval should be included for each PRG. For applying this DM-RS mapping n of (2) can be substituted as follows:

$$k = \begin{cases} 4n + 2k' + \Delta & \text{Configuration type 1} \\ 6n + k' + \Delta & \text{Configuration type 2} \end{cases}$$
$$k' = 0, 1$$
$$l = \bar{l} + l'$$
$$n = S_C \cdot (D_I - 1) \cdot u + v \tag{3}$$
$$u = 0, 1, \ldots,$$
$$v = \begin{cases} 0, 1, 2 & \text{Configuration type 1} \\ 0, 1 & \text{Configuration type 2} \end{cases}$$

where S_C is the number of subcarriers in a RB. D_I is the interval of the RB included in DM-RS symbols. $D_I = i$ means that DM-RS subcarriers are included per i RBs. n is reset per PRG. When D_I is 1, the DM-RS mapping is same with (2). Figure 3 shows an example of DM-RS mapping of configuration type 1 with $D_I = 2$ for a PRG in a slot. As mentioned above, this method can be flexibly applied by adjusting D_I according to the channel environment while having little influence on the current NR standard. It can be applied by adding a radio resource control (RRC) parameter in [10] or associating with downlink control information (DCI) of PRG size in [6].

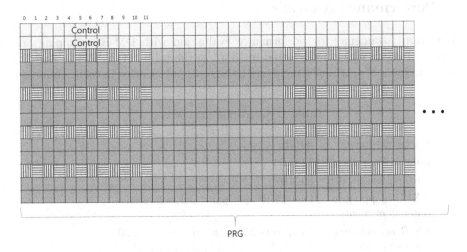

Fig. 3. DM-RS mapping for $D_I = 2$

On the other hand, as D_I becomes larger, channel estimation error may increase due to the decrease of DM-RS resources for channel estimation. On the contrary, one of two benefits for reduced DM-RS resources can be obtained. One can increase the data rate by sending more data to increased data resources. And the other can achieve a coding gain by reducing the coding rate with increased parity bits. For this DM-RS overhead and coding rate are analyzed. The ratio between DM-RS and data subcarriers for D_I can be derived as follows:

$$\text{OH}_{\text{DM-RS}} = \frac{R_S \cdot \left(D_F/D_I\right)}{D_S \cdot D_F + R_S \cdot D_F \cdot \left(1 - 1/D_I\right)} \tag{4}$$

where D_S, R_S, and D_F are the number of data symbols in a slot, the number of DM-RS symbols in a slot, and the number of allocated subcarriers in a symbol for data transmission, respectively. When D_I is 1, the number of allocated DM-RS subcarriers per symbol for R_S symbols is same as D_F. The effective coding rate can be written as follows:

$$C_{\text{eR}} = \frac{\text{TBS}}{Q_M \cdot \left\{D_S \cdot D_F + R_S \cdot D_F \cdot \left(1 - 1/D_I\right)\right\}} \tag{5}$$

where TBS is transport block size and Q_M denotes modulation order. TBS is determined by the number of allocated data symbols, control symbols, DM-RS symbols, modulation order, and so on [8].

3 Performance Analysis

Simulation parameters for performance evaluation with DM-RS mapping are listed in Table 1, which are referred by [3,4]. In the table, M, N, P, M_g, and N_g are the number of antenna elements with the same polarization in each column on

Table 1. Simulation parameters

Carrier frequency	30 GHz
Subcarrier spacing	120 kHz
Channel bandwidth	80 MHz
TXRU mapping to antenna elements	One TXRU per panel per polarization
BS/Relay antenna configurations	Unidirectional beam ($d_H = d_V = 0.5$) (M, N, P, M_g, N_g) = (8, 8, 2, 1, 1)
Codebook	Type 1 single panel
MCS	QPSK, 0.51
Channel coding	LDPC
Channel model	CDL-D with DS = 10 ns K-factor = 7 dB angle spread: 5 (ASD), 5 (ASA), 1 (ZSA), 1 (ZSD)
UE speed	500 km/h
Channel estimation, Data equalization	MMSE

each antenna panel, the number of columns on each antenna panel, the number of polarization on each antenna panel, the number of panels in a column, and the number of panels in a row, respectively. d_V and d_H denote the spacing in the vertical direction and the spacing of horizontal direction of antenna elements, respectively. BS and relay antenna element radiation patterns are same as Table A.2.1-10 in [4]. The applied beamforming scheme is the combination of analog beamforming and digital beamforming. The best beam pair among the limited set of DFT beams is selected for the decision of analog beam. The DFT beam candidate in this beam selection method is generated according to the uniform vertical and horizontal angular distribution shown as follows:

$$\theta_i = \frac{\pi}{rN_a}, \quad \text{for} \quad i = 1, \cdots, rN_a \tag{6}$$

where $r = 1$ (which is analogous to oversampling factor of 1), N_a denotes the number of vertical/horizontal antennas (M or N). Beam selection is based on the criteria of maximizing receive power after beamforming. In addition, we apply precoder cycling per PRG with type 1 single panel codebook for digital beamforming. The size of PRG is 4 RB. 8 RB is allocated for data transmission. In a slot, we assume that 8, 4, and 2 OFDM symbols are respectively allocated for data transmission of PDSCH, DM-RS for channel estimation, and control transmission of physical downlink control channel (PDCCH). The slot length is

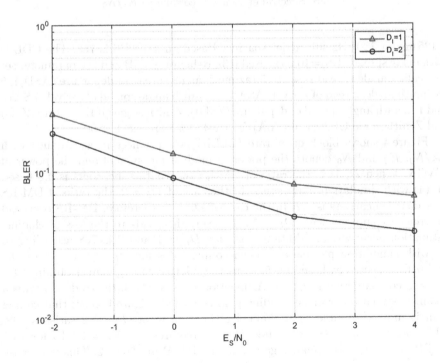

Fig. 4. BLER according to E_S/N_0

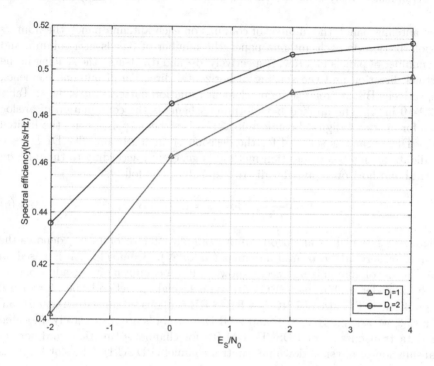

Fig. 5. Spectral efficiency according to E_S/N_0

0.125 ms for 120 kHz subcarrier spacing. Based on [8], TBS is 768. The CDL-D with delay spread (DS)=10 ns is used [13]. K-factor is 7 dB and the parameter set for scaling angle spread is set to 5 (azimuth angle spread of departure (ASD)), 5 (azimuth angle spread of arrival (ASA)), 1 (zenith angle spread of arrival (ZSA)), and 1 (zenith angle spread of departure (ZSD)). Zenith angle Of Departure (ZoD) and Zenith angle Of Arrival (ZoA) for cluster #1 are fixed at 90°.

Figure 4 shows block error rate (BLER) performances in accordance with E_S/N_0. E_S and N_0 denote the power of modulation symbol and the power of AWGN. When D_I is 1, the ratio of DM-RS over data subcarriers is 0.5 based on (4) with simulation parameters. In the case of $D_I = 2$, the ratio of DM-RS over data subcarriers is 0.2. In the case of $D_I = 2$, therefore, DM-RS overhead is reduced compared with $D_I = 1$. However, this leads to the loss of channel estimation accuracy. In addition, TBS for $D_I = 1$ and 2 is 768 according to [8] with simulation parameters. Then coding rates for $D_I = 1$ and 2 are 0.5 and 0.4, respectively, based on 5. Thus the relative coding gain is obtained for $D_I = 2$, compared with $D_I = 1$. As mentioned in Sect. 2.3, the tradeoff between channel estimation error and coding gain is existed. Figure 4 shows that coding gain is more profitable than channel estimation error. $D_I = 2$ satisfies 10% BLER at $E_S/N_0 = -0.3$ dB while $D_I = 1$ requires $E_S/N_0 = 1.1$ dB for 10% BLER. Thus the performance gain is about 1.4 dB for $D_I = 2$. This is because frequency selectivity of CDL-D channel is low and the channel estimation error in accordance with the increase of D_I can be ignored.

Figure 5 shows the spectral efficiency according to E_S/N_0. The spectral efficiency can be derived as follows:

$$SE = \frac{TBS \cdot (1 - BLER)}{D_F \cdot S_F \cdot S_L} \tag{7}$$

where S_F denotes the subcarrier spacing and S_L is the slot length. Since the requirement of spectral efficiency at $500\,km/h$ mobility is $0.45\,b/s/Hz$ in the HST scenario, the performances for $D_I = 1$ and 2 satisfy the requirement at $E_S/N_0 = -0.4\,dB$ and $-1.5\,dB$, respectively. The performance gain for $D_I = 2$ compared with $D_I = 1$ is about $1\,dB$ for $0.45\,b/s/Hz$. In addition, the average spectral efficiency is increased about 4% for $D_I = 2$.

4   Conclusions

In this paper, we analyze the performance of channel estimation for wireless channels with high mobility and frequency flat characteristics in HST scenarios. It is evaluated whether the application of DM-RS in NR meets the requirement of HST in ITU-R. In addition, we analyze performances for a efficient DM-RS mapping on the LOS channel. Performance analysis shows that the spectral efficiency of HST in ITU-R is satisfied by using DM-RS in NR. And, the efficient DM-RS mapping can obtain about $1\,dB$ E_S/N_0 gain and 4% spectral efficiency gain, compared with DM-RS in NR. The efficient DM-RS mapping can be considered in the NTN environment with a large flat fading characteristic.

Future work is performance analysis considering phase noise and retransmission in HST scenarios. It is also necessary to analyze performances for below $6\,GHz$ carrier frequency.

References

1. Recommendation ITU-R M.2083: IMT Vision - Framework and overall objectives of the future development of IMT for 2020 and beyond, September 2015
2. ITU-R WP5D: Minimum requirements related to technical performance for IMT-2020 radio interface(s), February 2017
3. ITU-R WP5D: Guidelines for evaluation of radio interface technologies for IMT-2020, June 2017
4. 3GPP TR 38.802 V14.2.0: Study on New Radio Access Technology Physical Layer Aspects (Release 14), September 2017
5. 3GPP TS 38.211 V15.1.0: NR; Physical channels and modulation (Release 15), March 2018
6. 3GPP TS 38.212 V15.1.1: NR; Multiplexing and channel coding (Release 15), April 2018
7. 3GPP TS 38.213 V15.1.0: NR; Physical layer procedures for control (Release 15), March 2018

8. 3GPP TS 38.214 V15.1.0: NR; Physical layer procedures for data (Release 15), March 2018
9. 3GPP TS 38.321 V15.1.0: NR; Medium Access Control (MAC) protocol specification (Release 15), March 2018
10. 3GPP TS 38.331 V15.1.0: NR; Radio Resource Control (RRC) protocol specification (Release 15), March 2018
11. 3GPP TR 38.811 V0.4.0: Study on New Radio (NR) to support non terrestrial networks (Release 15), March 2018
12. 3GPP TSG RAN WG1 Meeting #92bis: RAN1 Chairman's Notes, April 2018
13. 3GPP TR 38.900 V14.3.1: Study on channel model for frequency spectrum above 6 GHz (Release 14), July 2017

Characterizing mmWave Radio Propagation at 60 GHz in a Conference Room Scenario

Aleksei Ponomarenko-Timofeev[1]($\boxtimes$) (ID), Vasilii Semkin[1,2] (ID), Pavel Masek[3] (ID), and Olga Galinina[1] (ID)

[1] Tampere University of Technology, Tampere, Finland
`aleksei.ponomarenko-timofeev@tut.fi`
[2] Peoples' Friendship University of Russia (RUDN University),
Moscow, Russian Federation
[3] Brno University of Technology, Brno, Czech Republic

Abstract. In this paper, we provide a shooting and bouncing ray (SBR) based simulation study of mmWave radio propagation at 60 GHz in a typical conference room. The room geometry, material types, and other simulation settings are verified against the results of the measurement campaign at 83 GHz in [15]. Here, we extend the evaluation scenario by randomly scattering several human-sized blockers as well as study the effects of human body blockage models. We demonstrate that multiple knife-edge diffraction (KED) models are capable of providing meaningful results while keeping the simulation duration relatively short. Moreover, we address another important scenario, where transmitters and receivers are located at the same heights and are moving according to a predefined trajectory that corresponds, for example, to device-to-device interactions or inter-user interference.

Keywords: mmWave · 60 GHz · Radio propagation
Indoor propagation · Conference room

1 Introduction

The growing popularity of data-intensive sophisticated mobile and wearable devices is likely to cause a new surge of wireless data demand in the near future. However, the frequency spectrum below 6 GHz, highly populated today, will not be capable of accommodating the increased data traffic from various rate-hungry hi-tech devices that might flood the wireless market quite soon. To facilitate this growing demand the research community draws its attention to novel millimeter-wave systems, which due to the evolution of integrated circuits, progressively become available for multiple consumer and industrial use-cases in both indoor and outdoor scenarios [17].

In the light of this, in July 2016, the Federal Communications Commission (FCC) introduced a new guideline for the licensed operation of millimeter-wave

© Springer Nature Switzerland AG 2018
O. Galinina et al. (Eds.): NEW2AN 2018/ruSMART 2018, LNCS 11118, pp. 381–393, 2018.
https://doi.org/10.1007/978-3-030-01168-0_35

(mmWave) bands (centered at 28, 37, and 39 GHz[1]) as well as extended the regulations for the unlicensed 60 GHz frequency band (from 57–63 GHz to 71 GHz). In the aggregate, the nearly 11 GHz of frequency spectrum has been released, which opens up fundamentally new prospects for future wireless systems.

1.1 Motivation

In contrast to legacy frequencies below 6 GHz, the mmWave band offers channels with contiguous bandwidth orders of wider magnitude, hence accelerating the communications to multi-Gbps rates per link [23]. However, despite the high potential of mmWave communications, they come with many fundamental challenges. For example, while in an outdoor mesh network at 60 GHz high transmission directivity mitigates the interference between nonadjacent links [22], in indoor scenarios, due to the limited space, the assumption of "pseudowired" links may not be sufficiently accurate, and the signal propagation becomes less favorable.

To evaluate the system performance realistically, it is imperative to understand the behavior of a 60 GHz channel impulse response and to develop adequate channel models, which may be derived using the data obtained in the course of measurement campaigns or by running shooting and bouncing ray (SBR) simulation tools (see, for example, [16,21]). As human blockage is one of the most significant aspects of indoor mmWave communications, it attracts considerable attention from the research community [7,10,19], resulting in a wide variety of proposed models, including cylinders, knife-edge models, and cuboids [5,11,14].

In particular, most of the research efforts focus on the following three typical models for a human blocker: (i) circular cylinder – perfectly conducting cylinder (PEC) [18], dielectric cylinder [2]; dielectric-coated perfectly conducting cylinder [25]; (ii) dielectric elliptic cylinder [24]; (iii) and multiple knife-edge diffraction (MKED) model [10].

In this work, we focus on studying indoor signal propagation at 60 GHz in a conference room and in the presence of human blockers, which are simplified to MKED and elliptic cylinder models. After calibrating with openly available measurement data for 83 GHz, we perform a comparison of MKED and elliptic cylinder models in terms of performance and propagation paths. After that, we provide the results of SBR simulation for the said two types of human blocker models at 60 GHz.

1.2 Main Contributions

Although the properties of the mmWave channel have received considerable attention in the past, the assumptions under which the majority of channel models have been developed, however, limit their use to standard "a device to

---

[1] Report and Order and Further Notice of Proposed Rule-making, document FCC-16-89, Federal Communication Commission, July 2016.

an access point" links and cannot be directly applied to, for example, typical body-centric applications.

To fill this gap, we reconstruct an indoor scenario where multiple users carry wearable or hand-held mobile devices, potentially forming device-to-device (D2D) links [12, 26] or causing interference [26] from the neighboring devices. The main contributions of this paper are as follows:

- Analysis of the effects of the human body blockage at 60 GHz for D2D links in a conference/lecture room (see Fig. 1).
- Detailed discussion and comparison of different human body models, i.e., MKED and elliptic cylinder models within the studied scenario.

The paper is organized as follows. Section 2 briefly introduces our scenario of interest, as well as the process of calibration for 83 GHz. In Sect. 3, we present simulation results for 60 GHz frequency band in the presence of human blockers and compare the effect of different human body models. Section 4 refers to a D2D use-case and analyzes the simulation results for the case of the transmitter and receiver, moving along selected trajectories located at the same heights. Finally, Sect. 5 outlines the main conclusions of our study.

(a) 3D model

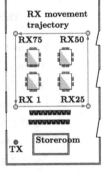
(b) Floor plan

Fig. 1. A 3D model of the conference room and a schematic top view of the room, with the receiver positions.

2   Our Scenario and Calibration with Measurement Data

In this work, we study radio channel propagation properties using the SBR method for a conference room 3D model (Fig. 1). To perform the simulations, we employ a commercial tool named Wireless Insite (Remcom), which is specifically developed for modeling signal propagation in various environments.

First, to calibrate our simulation tool settings and verify the correctness of our modeling approach, we study the scenario mentioned above at 83 GHz and compare the obtained results with the measurement data from [15]. In particular, we replicate the geometry of the conference room according to the available plan and specify the material properties so that our analysis corresponds to the measured results. We specify several materials, dominant in this scenario, such as wood, glass, and gypsum drywall. The electrical properties of these materials may be found in Table 1, while other simulation parameters are given in Table 2. For the better tractability of the results, omnidirectional antennas are assumed at the transmitter and the receiver side with 0 dBi gain, while the transmit power is set to 0 dBm.

Table 1. Material parameters

Material	Conductivity [S/m]	Relative permittivity	Thickness [mm]
Wood	0.000	5.00	30.0
Glass	0.567	6.27	3.0
Gypsum drywall	0.001	2.80	12.7

Table 2. Core simulation settings.

Parameter	Transmitter	Receiver
Antenna type	Isotropic	Isotropic
Gain (G)	0 dBi	0 dBi
Polarization	Vertical	Vertical
Input power	0 dBm	-
Transmission line loss	0 dB	-
Noise Figure	-	3 dB
Antenna heights	2.5 m	1.6 m
Location number	1	100

To verify the configuration of our scenario, we compare the simulation and measurement results as follows. Figure 2 shows the modeled angular delay profile for the line-of-sight (LOS) and multipath components (MPC) in comparison to the measured data [15]. The results for the LOS component are almost identical, while the first order MPCs obtained by ray-based modeling are reasonably close to the measurement data. Although the latter demonstrates a slight visual difference, this may be explained by the fact that the model used in the simulation does not include small details present in the real room.

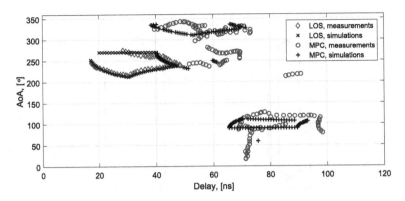

Fig. 2. Calibrating the SBR software. Comparison between the measurements from [15] and simulation results for LOS and MPC at 83 GHz.

3 Simulation Results at 60 GHz

3.1 Human Body Blockage Models

Wireless links at 60 GHz may provide up to 1.5 Gbit/s (assuming single carrier modulation and coding scheme MCS-6) [27], which is more than sufficient for HD video streaming. However, higher signal attenuation caused by human body presence causes certain limitations in terms of using the LOS link as a primary option at 60 GHz. Therefore understanding of multipath propagation characteristics is crucial for the further improvement of the network performance. The human blockage, as is well known, becomes a critical issue when it comes to the 60 GHz frequency range since mmWave signals are severely attenuated by the human body [3]. Therefore, analysis of human blockage effects in scenarios, which potentially may involve large numbers of human participants, opens one of the most important directions of mmWave research. In this work, we consider two human body blockage models: multiple knife-edge diffraction [8,9,13] and elliptic cylinder [6] models, illustrated in Fig. 3. The elliptic cylinder model is an extension of the cylindrical human blockage model, described in [4] and aims at replicating the proportions of the human body. In our simulations, the elliptic cylinder is represented by a cylinder polygon with ten faces on the sides of the cylinder (the purpose of presenting cylinder as a polygon is to simplify meshing in the SBR software). Importantly, the human blockage models in our simulations have two-layer surface: the first layer is equivalent to the human skin and the second layer is assumed to resemble muscle tissues. The parameters for the human tissue are taken from [1] and are summarized in the Table 3.

Here, the simulation is based on the same settings (including the environment and other parameters) as described in the previous section. The transmitter/receiver height is 2.5/1.6 m as the initial scenario advises. Figure 4 demonstrates the difference between MKED and elliptic cylinder models in terms of propagation. While these models show similar results for the LOS path, there

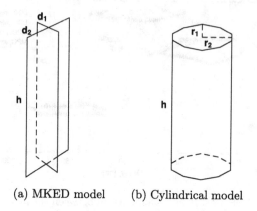

(a) MKED model   (b) Cylindrical model

Fig. 3. MKED and elliptic cylinder human body models used in the SBR simulations, where $h = 1.8\,\mathrm{m}$, $d_1 = 2r_1 = 0.28\,\mathrm{m}$, $d_2 = 2r_2 = 0.47\,\mathrm{m}$.

Table 3. Human body model parameters used in the SBR simulations.

Material	Conductivity [S/m]	Relative permittivity	Thickness [mm]
Skin	36.40	7.98	1.26
Muscle	52.83	12.86	5.00

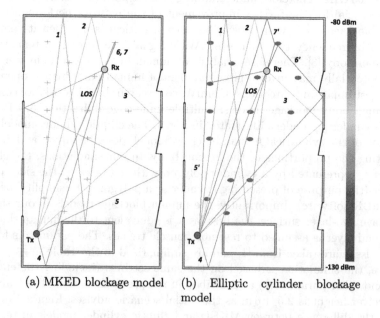

(a) MKED blockage model   (b) Elliptic cylinder blockage model

Fig. 4. Paths of the most dominant rays at 60 GHz (Tx/Rx height is 2.5/1.6 m).

are some differences between MPCs of higher orders. Figure 4 reveals that the LOS link exists for both models since the propagation path goes above all of the human blockers, which is due to the antenna heights. The rays from 1 to 4 are also retained since they are located far from the blockers (Fig. 4), and different blockage models do not affect these paths. However, as we may see, the rays from 5 to 7 are no longer present for the case, where the cylindrical model is used. If we examine ray №5 for the MKED case, we may observe that just before reaching the receiver, the ray passes very closely to the human body blockage model. Nevertheless, when the elliptic cylinder model is utilized, the ray №5 is blocked by one of the planes of the polygon cylinder (elliptic cylinder model) and cannot reach the receiver. Contrarily, the ray №5' appears when the signal reflects from the wall after traveling above all of the potential blockers and reaching the receiver after the second reflection. Furthermore, we notice that rays 6 and 7, which rely upon the reflection from the blocker, also disappears. In case of elliptic cylinder model, we observe that these rays are replaced by rays №6' and 7' (although they are reflected from the walls of the room, which results in the loss of the received power).

As one may see, there is a difference between the propagating rays of higher orders in the MKED and elliptic cylinder models. However, it is also necessary to take into account such significant factors as the simulation time and computational resources. Figure 5 shows the difference in the total simulation time that is required for equivalent experiments based on the models we study. For example, if we consider simulation with 5 people in the room the required time is 29 m in for MKED model and 48 for the elliptic cylinder human blockage model. Further, if we consider 15 people in the room, the simulation time increases linearly for the MKED model and exponentially for the elliptic cylinder model, i.e., resulting in 55 m in for MKED and 167 for the cylinder. Following computation times were measured on a computer with an Intel Core2Duo Q9500 CPU and Gigabyte nVidia GeForce GTX 550 Ti GPU with CUDA support. Given the total amount of the time required for modeling using elliptic cylinder model, we further focus only on the MKED human blockage model.

3.2   Channel Characterization

Importantly, we demonstrate the results that correspond to the case, when there are 15 human blockers in the room. We intentionally omit the cases with less than 15 blockers as they produce mostly low-power components. For 15 human blockers scattered in the room, we observe a large number of first and second order MPCs with considerable powers (Fig. 6). These components mainly correspond to the rays reaching the receivers at the beginning of the trajectory. Apart from that, for certain locations the LOS link is no longer present since the receiver is right behind the blocker and signal is received only via the MPCs. Additional first and second order components appear because the rays are reflected from the blockers.

Further, Fig. 7a shows a channel impulse response (CIR) at multiple receiver positions in the presence of 15 blockers as well as illustrates the dynamics of the

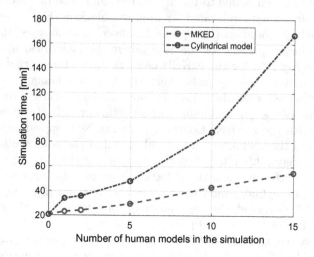

Fig. 5. Comparison of the simulation time for MKED and cylindrical models.

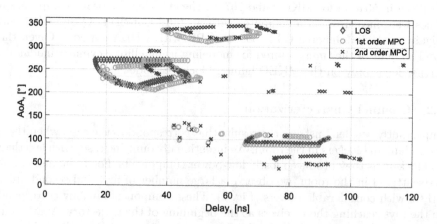

Fig. 6. Simulation results for LOS and dominant MPC at 60 GHz for the case of 15 people in the room.

delay when the receiver moves along the selected trajectory and blockage effects. For example, for the position №52 the LOS is blocked and only MPCs arrive at the receiver (Fig. 7b). In comparison, for the position №49 the LOS arrives at 46 ns. Since the distance difference between these positions is relatively small, the delay and power values for the LOS paths should be similar.

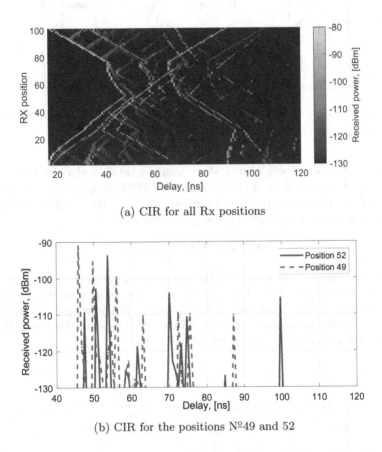

(a) CIR for all Rx positions

(b) CIR for the positions №49 and 52

Fig. 7. Channel impulse response at 60 GHz for the case of 15 people in the room.

4   Simulation Results for D2D Scenario at 60 GHz

In this section, we consider a slightly different scenario, which represents interactions between wearable devices of different owners (e.g., as in a D2D scenario). The transmitter and receiver heights are set to 1.6 m, while the number of blockers remains 15 and the MKED model is used for modeling blockers. The receiver moves along the same trajectory of 100 positions as before, capturing the snapshots of the CIR. At the same time, the transmitter travels around the area in

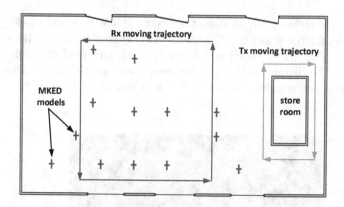

Fig. 8. A schematic view of the Rx and Tx positions in the conference room for the D2D scenario.

the conference room, where a small storeroom is located, and visits 18 positions (see Fig. 8). The snapshots of the CIR at the 60 GHz frequency are calculated for each of the transmitter/receiver positions. Based on the obtained results, we derive a logarithmic model [20] for the path loss:

$$L(d) = \alpha + 10\beta \log_{10}(d) + x, \tag{1}$$

where α and β are the parameters of the model and x is the random component, which reflects the effects of fading and is represented by a normally distributed variable with zero mean and standard deviation σ. Below we provide α, β, and σ for each of the components (Table 4). The path loss plots for the LOS case and first two orders MPCs are presented in Fig. 9.

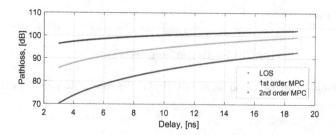

Fig. 9. Path loss model.

Table 4. Logarithmic model coefficients

Link type	α	β	σ
LOS	56.65	2.82	2.76
MPC-1	77.68	1.70	7.78
MPC-2	92.95	0.72	5.98

5   Conclusions and Discussion

In this paper, we study the mmWave indoor propagation in the conference room scenario, for which the channel measurements at 83 GHz are openly available. In particular, having reconstructed a 3D model that is reasonably close to the environment of the measurement campaign, we calibrate our simulation with the available data. The difference is negligible and indicates that the reconstructed geometry of the room and the selected materials (that is, their electrical properties) are correctly adjusted.

Further, we study the radio wave propagation at 60 GHz, extending the said scenario by placing human blockers into the room. We compare two types of human blockage models and show that the MKED model is preferable in terms of simulation time, since the time increases exponentially for the elliptic cylinder model and almost linearly for the MKED model.

Finally, we consider a D2D scenario, which represents communications between, for example, hi-end wearable devices, and may support prospective augmented/virtual/hyper-reality applications. In particular, we analyze the angular delay profile for the LOS and MPCs at 60 GHz in the presence of blockers, as well as derive a corresponding logarithmic path loss model, which may be instrumental in the further evaluation of the MAC protocols performance and high-level system-level analysis.

Acknowledgment. The publication has been prepared with the support of the "RUDN University Program 5 − 100". The described research is supported in part by the National Sustainability Program under the grant LO1401. For the research, the infrastructure of the SIX Center was used. The work of Dr. Olga Galinina is supported by Finnish Cultural Foundation and personal Jorma Ollila grant.

References

1. Andreuccetti, D., Fossi, R., Petrucci, C.: An internet resource for the calculation of the dielectric properties of body tissues in the frequency range 10 Hz - 100 GHz. Based on data published by C. Gabriel, et al. in 1996 (1997). http://niremf.ifac.cnr.it/tissprop/
2. Balanis, C.A.: Advanced Engineering Electromagnetics. Wiley, New York (1999)
3. Collonge, S., Zaharia, G., Zein, G.E.: Influence of the human activity on wide-band characteristics of the 60 GHz indoor radio channel. IEEE Trans. Wirel. Commun. **3**(6), 2396–2406 (2004). https://doi.org/10.1109/TWC.2004.837276

4. Ghaddar, M., Talbi, L., Denidni, T.A., Sebak, A.: A conducting cylinder for modeling human body presence in indoor propagation channel. IEEE Trans. Antennas Propag. **55**(11), 3099–3103 (2007). https://doi.org/10.1109/TAP.2007.908563
5. Ghaddar, M., Talbi, L., Denidni, T.A., Sebak, A.: A conducting cylinder for modeling human body presence in indoor propagation channel. IEEE Trans. Antennas Propag. **55**(11), 3099–3103 (2007)
6. Jacob, M.: The 60 GHz Indoor Radio Channel Overcoming the Challenges of Human Blockage. Ph.D. thesis, Technische Universität Braunschweig (2013)
7. Jacob, M., Mbianke, C., Kürner, T.: A dynamic 60 ghz radio channel model for system level simulations with MAC protocols for IEEE 802.11ad. In: IEEE International Symposium on Consumer Electronics (ISCE 2010), pp. 1–5, June 2010. https://doi.org/10.1109/ISCE.2010.5523241
8. Jacob, M., et al.: Fundamental analyses of 60 GHz human blockage. In: The 7th European Conference on Antennas and Propagation (EuCAP), pp. 117–121, April 2013
9. Jacob, M., Priebe, S., Maltsev, A., Lomayev, A., Erceg, V., Kürner, T.: A ray tracing based stochastic human blockage model for the IEEE 802.11ad 60 GHz channel model. In: The 5th European Conference on Antennas and Propagation (EuCAP 2011), Rome, Italy, pp. 3084–3088, April 2011
10. Jacob, M., Priebe, S., Maltsev, A., Lomayev, A., Erceg, V., Kürner, T.: A ray tracing based stochastic human blockage model for the IEEE 802.11ad 60 ghz channel model. In: Proceedings of the 5th European Conference on Antennas and Propagation (EUCAP), pp. 3084–3088, April 2011
11. Khafaji, A., Saadane, R., El Abbadi, J., Belkasmi, M.: Ray tracing technique based 60 GHz band propagation modelling and influence of people shadowing. Int. J. Electr. Comput. Syst. Eng. **2**(2), 102–108 (2008)
12. Kovalchukov, R., et al.: Modeling three-dimensional interference and SIR in highly directional mmWave communications. In: Proceedings of IEEE Global Communications Conference, pp. 1–7. IEEE (2017)
13. Kunisch, J., Pamp, J.: Ultra-wideband double vertical knife-edge model for obstruction of a ray by a person. In: 2008 IEEE International Conference on Ultra-Wideband, vol. 2, pp. 17–20, September 2008. https://doi.org/10.1109/ICUWB.2008.4653341
14. Kunisch, J., Pamp, J.: Ultra-wideband double vertical knife-edge model for obstruction of a ray by a person. In: IEEE International Conference on Ultra-Wideband, ICUWB 2008, vol. 2, pp. 17–20. IEEE (2008)
15. Lai, D., et al.: Multipath Component Tracking and Channel Model for Lecture Room. Standard, Document IEEE 802.11-16/846-ay, July 2016. https://mentor.ieee.org/802.11/dcn/16/11-16-0846-00-00ay-multipath-component-tracking-and-channel-model-for-lecture-room.pptx
16. Maltsev, A.: Channel models for 60 GHz WLAN systems. IEEE802.11-09/0334r8 (2010)
17. METIS: Wireless communications Enablers for the Twenty-twenty Information Society, EU 7th Framework Programme project. Technical report, ICT-317669-METIS
18. Pathak, P., Burnside, W., Marhefka, R.: A uniform GTD analysis of the diffraction of electromagnetic waves by a smooth convex surface. IEEE Trans. Antennas Propag. **28**(5), 631–642 (1980)

19. Peter, M., et al.: Analyzing human body shadowing at 60 ghz: systematic wideband MIMO measurements and modeling approaches. In: 2012 6th European Conference on Antennas and Propagation (EUCAP), pp. 468–472, March 2012. https://doi.org/10.1109/EuCAP.2012.6206013

20. Rappaport, T.S.: Wireless Communications: Principles and Practice, vol. 2. Prentice Hall PTR, New Jersey (1996)

21. Smulders, P.F.: Statistical characterization of 60-GHz indoor radio channels. IEEE Trans. Antennas Propag. 57(10), 2820–2829 (2009)

22. Solomitckii, D., Semkin, V., Naderpour, R., Ometov, A., Andreev, S.: Comparative evaluation of radio propagation properties at 15 ghz and 60 ghz frequencies. In: Proceedings of 9th International Congress on Ultra Modern Telecommunications and Control Systems and Workshops (ICUMT), pp. 91–95. IEEE (2017)

23. Sum, C.S., et al.: Virtual time-slot allocation scheme for throughput enhancement in a millimeter-wave multi-Gbps WPAN system. IEEE J. Sel. Areas Commun. 27(8) (2009)

24. Tai, C.T.: Dyadic Green functions in electromagnetic theory. Institute of Electrical & Electronics Engineers (IEEE) (1994)

25. Tang, C.C.: Backscattering from dielectric-coated infinite cylindrical obstacles. J. Appl. Phys. 28(5), 628–633 (1957)

26. Venugopal, K., Valenti, M.C., Heath, R.W.: Device-to-device millimeter wave communications: interference, coverage, rate, and finite topologies. IEEE Trans. Wirel. Commun. 15(9), 6175–6188 (2016)

27. Zhu, X., Doufexi, A., Kocak, T.: Throughput and coverage performance for IEEE 802.11ad millimeter-wave WPANS. In: 2011 IEEE 73rd Vehicular Technology Conference (VTC Spring), pp. 1–5, May 2011. https://doi.org/10.1109/VETECS.2011.5956194

Transmission of Augmented Reality Contents Based on BLE 5.0 Mesh Network

Maria Makolkina[1,2](✉), Van Dai Pham[1], Truong Duy Dinh[1],
Alexander Ryzhkov[1], and Ruslan Kirichek[1,3]

[1] The Bonch-Bruevich Saint-Petersburg State University of Telecommunications,
St. Petersburg, Russia
makolkina@list.ru, daipham93@gmail.com, din.cz@spbgut.ru,
aryjkov@mail.ru, kirichek@sut.ru
[2] St. Petersburg School of Economics and Management,
National Research University Higher School of Economics, St. Petersburg, Russia
[3] Peoples' Friendship University of Russia (RUDN University), Moscow, Russia

Abstract. In recent years, Augmented Reality (AR) applications have appeared on smart devices (smartphones, smart glasses, etc.) that have expanded the visual perception of users. With the appearance of new technologies, a huge number of new services are created and the quality of service provision for them also plays a really important role. Thus, the task of Quality of Experience (QoE) introduces interest to suppliers, service developers and users. This article discusses the issues of QoE in the implementation of AR services in big cities with a high density of users but not all users have the Internet access. To solve this problem, we propose to use Bluetooth 5.0 technology (BLE 5.0) to access the network. The interaction between the client and the server can be carried out by using D2D communication on BLE 5.0 through smartphones of other users, which are described by the queuing system model. The proposed model of providing augmented reality services was simulated. The simulation results show delays in the delivery of data, which depends on the number of BLE 5.0 nodes through which data is exchanged between the AR client and AR server, and also on the load factor.

Keywords: BLE 5.0 · Mesh network · Augmented reality
Delay · Quality of experience · Service system · Data transmission
Device-to-Device (D2D) communication

1 Introduction

Nowadays, the concepts of the Internet of Things and the Augmented Reality [1–3] are used very popularly, which represent the integration of a huge volume of data, information with real objects from the world around us. AR applications typically add information to an object that will be at first recognized and

© Springer Nature Switzerland AG 2018
O. Galinina et al. (Eds.): NEW2AN 2018/ruSMART 2018, LNCS 11118, pp. 394–404, 2018.
https://doi.org/10.1007/978-3-030-01168-0_36

then sent to the server. Depending on the application, the information coming in and out of the server may be different. For example, there can be a full-fledged video stream in one direction, and only text data in the reverse direction. Currently, this is not a problem due to the diverse range of modern devices' features. Smartphones, tablets and other smart devices are ideal elements for the implementation of AR services. Today smart devices not only support high-speed connections, but also have a powerful processor, camera, a graphical interface and an acceptable price that is accessible to any user.

Many AR applications can be implemented on these smart devices because they do not require complex processing of information. Typically, these AR applications are about the recognition of object and the display of information about it. However, there are other applications that require significant bandwidth, complex processing power, and a stable wireless communication channel. In big cities, with a high density of users and their constantly high activity to realize the representation of various AR services with a proper QoE can be quite problematic. One of the solutions to this problem is the use of D2D communications [4,5]. The basis for D2D communication can be Bluetooth 5.0 standard, which has a number of winning features [6]. Firstly, it provides a throughput of 2 Mbit/s to users, secondly, it operates at a distance of 70–100 m, and thirdly, it has the property of self-organization.

This article considers the possibility of using Bluetooth BLE 5.0 to provide AR services, in condition that a wireless channel is provided for the exchange of data between end-to-end devices so that the remote device can receive a response from the AR Server. In this case, the characteristics of the BLE mesh network and the queuing system model of the AR application in this network are considered. Simulation modeling in the AnyLogic program [7] was carried out. The results show delays in the delivery of data, which depends on the number of BLE 5.0 nodes and through these nodes data is exchanged between the AR client and the AR server, and also on the load factor.

2   Overview of Bluetooth 5.0 and BLE Mesh

Today, Bluetooth technology is one of the most popular wireless technologies in personal networks. Until now, different versions of the Bluetooth protocol are often used in many areas of life on various devices. Each new version of the protocol (1.2, 2.0, 2.1, 3.0, 4.0, 4.1, 4.2 and 5.0) introduces significant improvements in its operation. An important achievement is the maintenance of a low power mode, starting with version 4.0 (Bluetooth Low Energy - BLE). Recently a new version 5.0 is released, which improves many features such as speed, range, energy efficiency and the way in which the network is organized [9]. The following new features of Bluetooth 5.0 technology should be highlighted:

- 8x increased broadcast capacity.
- 2x data rate.
- 4x long range.
- Improving noise immunity.
- Improving energy efficiency.
- LE Advertising Extensions.

At the physical layer, Bluetooth 5.0 uses a variety of modulation schemes, encoding schemes, and data transfer rates. The physical layer data is presented in Table 1.

Table 1. The physical layer of Bluetooth 5.0

PHY	Modulation scheme	Encoding scheme		Data rate
		Access header	Payload	
LE 1M	1 Msym/s	Uncoded	Uncoded	1 Mbps
LE 2M	2 Msym/s	Uncoded	Uncoded	2 Mbps
LE Coded	1 Msym/s	S = 8	S = 8	125 Kbps
			S = 2	500 Kbps

Presently, the maximum data rate is doubled to 2 Mbit/s. Along with the achievements presented in Bluetooth 4.2, the bandwidth is increased 5 times compared to the original level of Bluetooth 4.0. The new version of Bluetooth 5.0 also supports BLE mesh, which creates more communication possibilities between several devices. BLE mesh was released independently and after the announcement of Bluetooth 5.0. This means that applications can use BLE mesh together with Bluetooth 4.x or 5.

In July 2017, the Bluetooth SIG (The Bluetooth Special Interest Group) published the first version of the mesh profile specification for Bluetooth [10]. It defines a mesh network based on a flooding-based solution that uses advertising channels to send messages so that the other nodes can receive and relay them. Any device on the network can send messages at any time if there is sufficient density of devices for listening and relaying the message. In order to limit the number of relaying messages, there are several methods presented in the specification [10]. The main methods are:

- Time To Live (TTL): Each message includes a TTL value that limits the number of times a message can be relayed.
- Network message cache (NMC): NMC is designed to prevent devices from relaying previously received messages by adding all messages to a cached list.
- Relay is optional: all nodes do not need to implement the relay function.

In the specification [10], the packet size is 33 bytes. The packet format is shown in Fig. 1. Each packet includes 12 or 16 payload bytes. In the case payloads exceed 12 or 16 bytes, there is a process of segmentation and reassembly. Therefore, in order to send a message in size of 120 bytes, it is required to segment the message into 10 packets.

Combining the capabilities of Bluetooth 5.0 and the mesh network has become a new solution for providing various services. In the next section, we consider using Bluetooth 5.0 Mesh for AR applications.

	1		1	3	2	2	12 or 16	4 or 8
IVI	Network ID	CTL	TTL	Sequence Number	Source Address	Dest Address	Packet Payload	NWK MIC

Fig. 1. Packet format of BLE mesh network

3   Implementation Model

With Bluetooth 5.0 support, mobile phones can exchange messages between a few devices. Taking into the discussed above features of Bluetooth 5.0, we propose a following model for providing augmented reality services, which is depicted in Fig. 2.

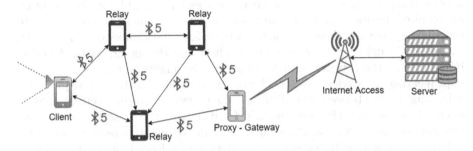

Fig. 2. Bluetooth 5.0 mesh network model for AR application

The Fig. 2 shows the following components:

- AR-Client: A node that uses its camera to recognize objects and sends requests to the server to find information about this object. In addition, with using the GPS sensor and the orientation sensor, the AR-Client can send requests to find information about the place where it is located. After receiving responses from the servers, information will be displayed on the device. Messages are transmitted by AR-Client over advertising channels.
- Relay: Nodes that receive the messages and then relay them over the advertising channels, if these messages did not arrive earlier. On these nodes, NMC and TTL method are used.
- Proxy-Gateway: A node that has access to the Internet. It provides data exchange between a AR-Client and a remote server.
- Internet Access: It provides Internet access service.
- Server: It receive requests from AR-clients and after request processing the answers will be sent back to AR-Client.

Thus, a waiting time between the client request time and the server response time is one of the main performances of service quality. Before sending a request, the subscriber terminal (smartphone) also spends time on recognizing the surrounding objects. Thus, this total time can affect user's quality of experience. Figure 2 shows that the waiting time depends on several components, specifically, subscriber terminals, communication network and cloud server. When implementing a particular service, the influence of each element of the service model is considered in more detail. The delay introduced by terminals depends on the characteristics of the devices and the functions of the AR application. The delay introduced by communication networks depends on the used technologies for data transmission, on the networks bandwidth and on the volume of transmitted data. In the BLE mesh network, there is a small bandwidth, as well as a small payload size. However, the new version of Bluetooth 5.0 announced an increase in payload size. And the delay introduced by the cloud server depends on its performance when data processing.

The Bluetooth mesh network allows to transmit small amounts of data, when sending a larger message segmentation and reassembly at the reception are required. In addition, in this network, packets are relayed through nodes to the source (gateway), therefore the network scale or the number of transit nodes will affect the delivery delay between the client and the server. Thus, the delay introduced by the network most of all affects the quality of transmission when implementing the service. We consider the delay, including the time required for delivering packets between the client and the gateway through a certain number of transit nodes, and the time required for delivering the message between the gateway and the server over the cellular network. We describe service process for AR-client by the a multiphase queuing system model. Each Relay node, gateway, and cloud server is a single-phase queuing system.

Assume that the packet flows arriving at each service phase can be described by the simplest flow model. The Bluetooth mesh network [6, 10] uses a flooding-based routing method. The packet flows can be transmitted through many nodes to other nodes. Each Relay node can receive the same message. With a sufficiently large number of users, the incoming packet flow will have properties close to the simplest flow.

A queuing system model is shown in Fig. 3.

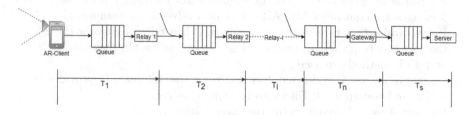

Fig. 3. Queuing system model of AR-Client

Assume that the AR client sends a message with a size of 120 bytes; consequently, this message is segmented into 10 packets. As follows from Fig. 3, data transmission process between the AR client and the server can be described in the following way:

- Each packet is sequentially transmitted to Relay-1 node with an average delivery time T_1.
- Then the Relay-1 node relays the packets to Relay-2 node. The average delivery time between the first and second nodes is T_2. To guarantee great reliability of delivery, the Relay-1 node can be configured to repeat retransmission of the same packet [10]. Generally, the Relay node repeats the packet retransmission 3 times with a minimum repetition delay, called the retransmission interval: $Interval_{retrans} = (Steps+1) \cdot 10 + (0 \rightarrow 10\,\text{ms})$, where $Steps = 1, 2, 3$.
- Similarly, the following Relay-i nodes have a packet delivery time T_i and retransmission intervals $Interval_{retrans}$.
- After the gateway received all of 10 packets, it collects the segmented packets into one message. Then the received message is sent to the server.

In such a queuing system, great interest is represented by a model, which describes the principles of the functioning of its elements. Obviously, when investigating in various models, it is required to determine the delay time for the augmented reality data delivery, i.e. it is necessary to determine probability that the data delivery time will not exceed a definite value T_0. Currently, models are known for typical queuing systems [11], for example, when the service time at each node, or at each phase, is random and has an exponential probability distribution. When describing the queuing system model M/M/1 at each phase, the average packet transit time at the i-th phase is represented by the following expression:

$$T_i = \frac{1}{\mu_i - \lambda_i} \tag{1}$$

In the case when the incoming packet flows at each phase of the queuing system are independent and have the same properties (the simplest flow), the average delivery time is equal to sum of the mean delivery times of each phase of queuing system. Also, when delivering a large message, it is necessary to consider the delivery time of packets that are received after segmentation. Thus, the average delivery time between the AR-client and the server is represented as:

$$T = m \cdot \sum_{i=1}^{n} T_i + T_S = m \cdot \sum_{i=1}^{n} \frac{1}{\mu_i - \lambda_i} + \frac{1}{\mu_S - \lambda_S} \tag{2}$$

Where:

- m is a number of packets after message segmentation
- n is a number of passing phases of queuing system to the server
- μ_i and μ_S are service rate at each phase of the Relays and Server
- λ_i and λ_S are arrival rate at each phase of the Relays and Server.

Let's assume that the arrival rate of incoming packets at each Relay node is the same. As follows from formula (2), the data delivery time between the

AR-client and the server is represented as:

$$T = \frac{m \cdot n}{\mu_i - \lambda_i} + \frac{1}{\mu_S - \lambda_S} \tag{3}$$

4   Simulation Modeling

In this article, simulation modeling is carried out with using the AnyLogic package. The model is constructed similarly to that shown in Fig. 3. In the process of modeling, the number of transit nodes Relay is changed, i.e. the number of passing phases of queuing system from the client to the server is changed. The following parameters were chosen as the experimental conditions: time of each experiment, $t_{exp} = 100$ s; data transfer speeds of Bluetooth 5.0, DR $= 1$ Mbps and 2 Mbps; message size (payload size), L $= 120$ bytes and 240 bytes; mean arrival rate at each Relay node, $\lambda_i = 0.7$; mean service rate at each node for types PHY 2 Mbps and 1 Mbps, $\mu_i = 1.1$ and 0.97; mean arrival rate on the server, $\lambda_S = 2.08$; and mean service rate on the server, $\lambda_S = 1$.

When changing the number of transit nodes from the client to the server, values of such parameters as the delivery time of all segmented packets between the client and the gateway and the delivery time of messages between the gateway and the server were measured. By the sum of these times, we can estimate the data delivery time between the client and the server. Measurements of these parameters were carried out when delivering messages in size of 120 and 240 bytes.

When modeling the delivery of a message with a size of 120 bytes, the load factor was changed, accordingly, the arrival rate of incoming packets at each phase of queuing system was changed, and the delivery time between the client and the server was measured.

As the simulation results, graphs of dependencies of packet delivery delay on the number of transit nodes Relay, on the size of the transmitted message, and on the load factor were obtained. Theoretically, according to formula (3), similar dependences for these parameters were also obtained. When transferring a packet with size of 240 bytes, the theoretical and simulated results are compared in Figs. 4 and 5, respectively, depending on the number of Relay nodes and depending on the load factor. Comparisons of simulation results with two sizes of the packet 120 and 240 bytes are shown in Figs. 6 and 7, respectively, depending on the number of Relay nodes and depending on the load factor.

As follows from graphs, in this case, the delivery time increases linearly with changing the number of transit nodes Relay. The results of theoretical calculations confirm the results obtained in the course of simulation modeling. Figures 4 and 6 show that the delivery time of message of 240 bytes is 1.8 s when using LE 1M mode (PHY 1 Mbps), and 0.9 s when using LE 2M mode (PHY 2 Mbps), in case where there are 25 transit nodes between the AR client and the server. In the latest version of the BLE mesh model specification, payload size of 16 bytes was determined; because message delivery occurs with data segmentation, the message delivery time will be more required. Thus, it is obvious that the

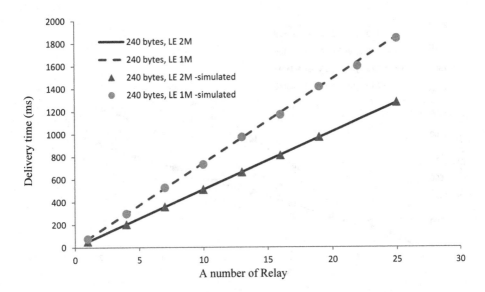

Fig. 4. Delivery time depends on number of Relay nodes

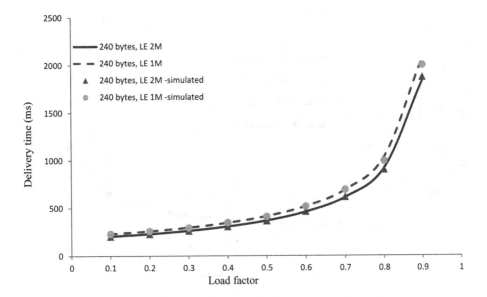

Fig. 5. Delivery time depends on the load factor

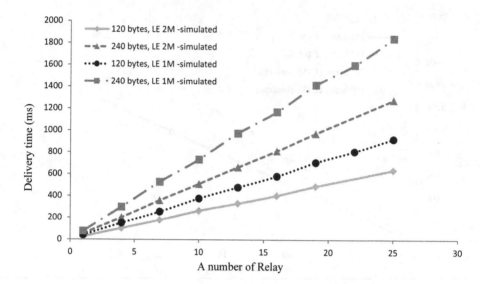

Fig. 6. Delivery time depends on number of Relay nodes (AnyLogic)

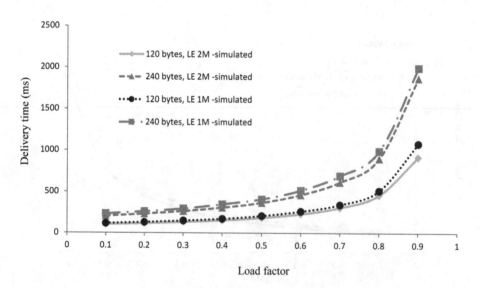

Fig. 7. Delivery time depends on the load factor (AnyLogic)

size of message affects the delivery time between the AR client and the cloud server. When comparing two messages with sizes of 120 and 240 bytes (Fig. 6), it is obvious that the delivery time of a message with size of 120 bytes is less than 2 times. Therefore, when implementing the augmented reality services, it is required to determine the amount of data to be delivered to AR clients.

The Figs. 5 and 7 show the effect of the load factor on the message delivery time between the AR client and the cloud server with a number of transit nodes equal to 10. The results in these figures show that the delivery time of message in size of 240 bytes was longer than 2 s with a load factor of 0.9, and the delivery time of message in size of 120 bytes was required more than 1 second. Thus, with a high user density, the data delivery time can reach up to several seconds. In this network, with a large number of transit nodes and a large load factor, the quality of service provision may not be provided in the proper way. Therefore, analyzing the results we can choose the appropriate delivery option for the implementation of a particular augmented reality application with considering the network scale, the number of transit nodes, and the amount of provided data.

5   Conclusion

In recent years, many new technologies have appeared which bring to the market a wide range of various services that can surprise the most demanding user. All sorts of information about the objects, from the world around us, are collected and stored. Today, every person has access to such information and there are a number of methods and technologies for displaying data to the users. One of methods is the augmented reality technology, which uses different ways of object recognition, data delivery of the object and the form of information presented to the user, which increases its mobility, convenience of searching and data perception.

The article considered the possibility of AR services based on the mesh network BLE 5.0 in big cities with a high density of users when not all the users have the Internet access. A service model as queuing system was proposed that describes the process of delivering a message between the AR client and the AR server. In this model, the AR data delivery over a mesh network BLE 5.0 was simulated in AnyLogic package. The simulation results showed a series of dependencies of the delivery time between the AR-client and the AR-server from various parameters.

Depending on the network scale, the delay in the delivery data between the AR client and the cloud server increases with the number of transit nodes Relay. It was also found that the size of message affects the packet delivery time over the network. With the increase of the load factor, the delivery time also increases. When considering the high density of users, i.e. at a great load factor, the delivery time can reach several seconds.

From these results, it can be concluded that the mesh network BLE 5.0 is capable of providing the required quality of information delivery for augmented reality services. Thus, when implementing an AR application in this network, it is necessary to consider the message size and the possible number of nodes in the network in order to minimize the delivery time between the client and the cloud server. In the near future, the specification of BLE mesh model is expected to be updated, which will make full use of the advantages of Bluetooth 5.0 for increasing the payload size.

References

1. Yuen, S.C.Y., Yaoyuneyong, G., Johnson, E.: Augmented reality: an overview and five directions for AR in education. J. Educ. Technol. Dev. Exch. (JETDE) **4**(1), 119–140 (2011)
2. Makolkina, M., Paramonov, A., Vladyko, A., Dunaytsev, R., Kirichek, R., Koucheryavy, A.: The use of UAVs, SDN, and augmented reality for VANET applications. In: DEStech Transactions on Computer Science and Engineering (AIIE), pp. 364–368 (2017)
3. Makolkina, M., Koucheryavy, A., Paramonov, A.: Investigation of traffic pattern for the augmented reality applications. In: Koucheryavy, Y., Mamatas, L., Matta, I., Ometov, A., Papadimitriou, P. (eds.) WWIC 2017. LNCS, vol. 10372, pp. 233–246. Springer, Cham (2017). https://doi.org/10.1007/978-3-319-61382-6_19
4. Paramonov, A., Hussain, O., Samouylov, K., Koucheryavy, A., Kirichek, R., Koucheryavy, Y.: Clustering optimization for out-of-band D2D communications. Wirel. Commun. Mob. Comput. **2017**, Article ID 6747052, 11 pages (2017). https://doi.org/10.1155/2017/6747052
5. Kirichek, R., Kulik, V.: Long-range data transmission on flying ubiquitous sensor networks (FUSN) by using LPWAN protocols. In: Vishnevskiy, V.M., Samouylov, K.E., Kozyrev, D.V. (eds.) DCCN 2016. CCIS, vol. 678, pp. 442–453. Springer, Cham (2016). https://doi.org/10.1007/978-3-319-51917-3_39
6. Woolley, M.: Bluetooth SIG. Bluetooth 5 Go Faster. Go Future. https://www.bluetooth.com/bluetooth-technology/bluetooth5/bluetooth5-paper. Accessed 14 Apr 2018
7. AnyLogic. https://www.anylogic.com/. Accessed 14 Apr 2018
8. Kirichek, R., Pham, V.-D., Kolechkin, A., Al-Bahri, M., Paramonov, A.: Transfer of multimedia data via LoRa. In: Galinina, O., Andreev, S., Balandin, S., Koucheryavy, Y. (eds.) NEW2AN/ruSMART/NsCC -2017. LNCS, vol. 10531, pp. 708–720. Springer, Cham (2017). https://doi.org/10.1007/978-3-319-67380-6_67
9. Bluetooth SIG. Bluetooth 5 Core Specification, Version 5, 06 December 2016. https://www.bluetooth.com/specifications/bluetooth-core-specification. Accessed 14 Apr 2018
10. Bluetooth SIG. Bluetooth Mesh Networking Specification.v.1.0. https://www.bluetooth.com/specifications/mesh-specifications. Accessed 14 Apr 2018
11. Kleinrock, L.: Queueing Systems, vol. 1, p. 448. Wiley, New York (1975)

Performance Limitations of Parsing Libraries: State-of-the-Art and Future Perspectives

Antonino Manlio D'Agostino[1], Aleksandr Ometov[2(✉)] (iD), Alexander Pyattaev[3],
Sergey Andreev[2] (iD), and Giuseppe Araniti[1] (iD)

[1] University Mediterranea of Reggio Calabria, Reggio Calabria, Italy
[2] Tampere University of Technology, Tampere, Finland
aleksandr.ometov@tut.fi
[3] Peoples' Friendship University of Russia (RUDN University), Moscow, Russia

Abstract. The acceleration of mobile data traffic and the shortage of available spectral resources create new challenges for the next-generation (5G) networks. One of the potential solutions is network offloading that opens a possibility for unlicensed spectrum utilization. Heterogeneous networking between cellular and WLAN systems allows mobile users to adaptively utilize the licensed (LTE) and unlicensed (IEEE 802.11) radio technologies simultaneously. At the same time, softwarized frameworks can be employed not only inside the network controllers but also at the end nodes. To operate with the corresponding policies and interpret them efficiently, a signaling processor has to be developed and equipped with a fast packet parsing mechanism. In this scenario, the reaction time becomes a crucial factor, and this paper provides an overview of the existing parsing libraries (Scapy and dpkt) as well as proposes a flexible parsing tool that is capable of reducing the latency incurred by analyzing packets in a softwarized network.

Keywords: SDN · Parsing · dpkt · Scapy · Performance evaluation

1 Introduction and Overview

Today, continuously growing numbers of interconnected devices push the telecommunication community towards developing new technologies for improved networking. Although several solutions have been proposed and implemented to address a steady increase in the mobile data consumption (e.g., the introduction of IPv6), they are still not ready for billions of new users/devices that are expected to join the network over a short period of time [1]. This projected acceleration suggests that the current and emerging (5G) mobile networks should evolve to become more "intelligent", efficient, secure, and, most importantly, scalable to enable future data communication that is incredibly diverse in nature [2,3].

© Springer Nature Switzerland AG 2018
O. Galinina et al. (Eds.): NEW2AN 2018/ruSMART 2018, LNCS 11118, pp. 405–418, 2018.
https://doi.org/10.1007/978-3-030-01168-0_37

The Open Networking Foundation (ONF) [4] is a nonprofit consortium dedicated to the development, standardization, and commercialization of one of the 5G enablers – Software Defined Networks (SDNs). The ONF provided the most explicit and well-received definition of SDN [5] as follows: "SDN is an emerging network architecture where network control is decoupled from forwarding and is directly programmable". Per this formulation, the SDN is shaped by main characteristics [6]: decoupling of control and data planes as well as programmability on the control plane. However, neither of these two SDN features is entirely new in the network architecture.

There is a pressing need to provide more capacity to the end users [7]. For this reason, LTE and WLAN integration may be attempted to improve the efficiency of mobile data offloading, which is a promising and low-cost solution to reduce the load on the cellular networks [8]. Further, the aggregate capacity of a heterogeneous network can be increased by utilizing short-range radio technologies [9] residing e.g., in unlicensed 2.4 GHz and 5 GHz spectrum [10]. This requires that two wireless interfaces are available on the smart phones [11].

Moreover, service operators have identified that offloading of bulky Internet traffic onto alternative access technologies constitutes a viable solution to relieve the high infrastructure costs [12]. Since 2000, there has been an extraordinary growth of research on SDN, initially in the area of wired networks and subsequently towards wireless technologies [13,14]. SDN can be utilized to configure not only the radio side of access points but also the end-user terminals. While not explicitly mentioned as SDN, injecting operator-specific offloading policies into the end-user terminals is also possible. Abstraction of such policies from the terminal side constitutes an important strategy for SDN deployment.

There are several technologies that aggregate LTE and WLAN, such as Access Network Discovery and Selection Function (ANDSF) [15], LTE WLAN integration with IPsec tunnel (LWIP), and LTE-WLAN Aggregation (LWA) [16]. Furthermore, network-assisted device-to-device (D2D) offloading enables user equipment to communicate directly with each other, without relying on the conventional infrastructure of APs or BSs [17,18]. 3GPP had invested considerable effort to ratify the IP traffic offloading solutions for the EPC: these approaches rely on tight cellular operator control and integration into the 3GPP network architecture.

Overall, rule-based policies like ANDSF are insufficient to represent the complex and/or stateful operation, such as in LWA and LWIP. For this reason, there is a need to introduce an appropriate finite state machine (FSM), which allows creating the desired stateful protocol operation via simple primitives (e.g., transmit a message, establish an IPsec tunnel, associate with an AP, etc.) that are pre-implemented in the device. Notably, the SDN technology evolved further due to the utilization of OpenFlow as a realistic and viable platform to the switch hardware [19]. The heart of OpenFlow is the "match/action" abstraction, which comprises {*rule, action*} pairs: if a rule is matched by the incoming packet, an action associated with this rule is executed.

There are three main operations that correspond to the said abstraction: (i) selection of the fields to be matched; (ii) query in the MAT (Match-Action Table) being efficiently supported in hardware by Ternary Content Addressable Memories (TCAMs); and (iii) execution of the corresponding action(s) selected among a fixed set of the standardized ones. Lately, the need for a more flexible OpenFlow emerged. It is thus vital to improve the programmability and the flexibility of the matching procedure, as well as the way we analyze the packets [20,21]. It is possible to develop a packet processor, and it is easy to understand that it can represent a bottleneck with respect to the delay of entry/exit of packets, as it is shown in Fig. 1.

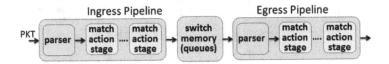

Fig. 1. A typical OpenFlow pipeline architecture

In this work, we evaluate the existing software libraries that allow for dynamic packet analysis and modification. We also elaborate on the development of flexible tools that enable fast and straightforward packet parsing, which may play a significant role in future softwarized networks. The rest of this paper is organized as follows. Section 2 outlines the design of the proposed parsing toolbox, which enables fast and reliable software oriented packet parsing. Further, the main functionality is detailed in Sect. 3. Section 4 provides a comparison of several existing parsers with the developed one. The main conclusions are drawn in the last section.

2   Design of Parsing Software for SDN

The best way to capture packets, analyze them, and understand which kinds of packets are to be processed is through a dynamic parser. Numerous packet parsers have been developed over the years, but it is still difficult to find *not a machine-oriented* one. For this reason, there is a need to develop a new more flexible parser. Our goal is to create a framework that is easily modifiable (including the source code), machine-oriented, and friendly to use. The developed parser should be compatible with any existing packet.

Our parser is written in Python, which contains all of the necessary functions required to analyze a packet (e.g., read data, compare data, convert data, etc.). It has a JSON "instruction file", which contains all the needed instructions and details to analyze the protocols as well as extract the requested conditions (see Fig. 2). Here, JSON was selected for its broad adoption. The purpose is to store primitive types as supported by JSON in a human-readable and straightforward format.

The bottleneck with parsing JSON and XML usually is not the parsing itself, but the interpretation/representation of the data. An event-based XML parser is typically very fast, but constructing a complex DOM tree with thousands of small objects is not. If it is necessary to parse XML to the nested native data structures, such as lists and dictionaries, the slow part will be the interpretation of the parsing results, not the actual string analysis. Since JSON parses directly into those primitive types rather than a complex object tree, it will likely be faster.

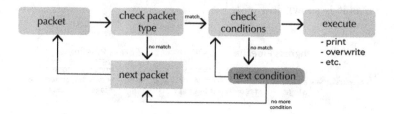

Fig. 2. Diagram of a custom parser

2.1   Configuration File Structure

Essentially, a JSON file that we need for retrieving the fields from the packets is divided into three blocks, and an example is given in Fig. 3.

```
1 - {
2      "enc_type": "1",
3 -    "objects": [{
4        "name": "ETH",
5 -      "properties": [{
6          "name": "Destination",
7          "read_from": 0,
8          "read_to": 6
9 -      },{
10         "name": "Source",
11         "read_from": 6,
12         "read_to": 12
13       }],
14       "read_from": 0
15     }],
16 -   "zones": [{
17       "name": "ALL",
18       "read_from": 0,
19 -     "find": {
20         "object": "ETH",
21         "multiple": false,
22         "print": ["Destination","Source"]
23       }
24     }]
25  }
```

Fig. 3. MAC address retrieval example

Explanation of "enc_type" is a field, which contains the encapsulation type of the entire capture. An integer within this field must be provided, which should correspond to the encapsulation type from the *pcap* file. In a logical order, it is actually the first parameter that the algorithm checks. Table 1 represents a part of the list of possible types of encapsulation.

Table 1. Examples of possible types of encapsulation

LINKTYPE_name	enc_type	DLT_ name	Description
LINKTYPE_ETHERNET	1	DLT_EN10MB	IEEE 802.3 Ethernet (10 Mb, 100 Mb, 1000 Mb, and higher)
LINKTYPE_IEEE802_11	105	DLT_IEEE802_11	IEEE 802.11 WLAN

Explanation of "objects" is an array containing the definitions and instructions of the objects. Inside this JSONObject, there are two mandatory fields and one optional field, as it is possible to observe in Table 2.

Table 2. List of fields inside "objects"

Name	Type	Mandatory
name	String	Yes
read_from	int/JSONArray	No
read_to	int/JSONArray	No
match	JSONArray	No
match_or	JSONArray	No
properties	JSONArray	Yes

Further, we describe the meaning of the following fields:

- *"name"* contains an easy to read string for the object to find;
- *"read_from"* and *"read_to"* indicate the position of bytes, which should be read from the packets;
- *"match"* and *"match_or"* check if the data extracted matches the chosen interval data;
- *"properties"* contains additional properties of the object.

Fields "read_from" and "read_to" represent the precise relative bytes where the parser can start (or stop) reading data (e.g., if TCP payload starts at the 50th byte, but it is only the 20th byte in an IP packet, *"read_from"* should be 20). For more complex protocols, there is no fixed position to start (or stop) reading, so we have to acquire this value from the packet itself. In this case, we can use a JSONObject instead of a simple *integer* number, which can contain all of the information to obtain the *read_from* value. The *"convert"* field can convert the read data (usually obtained in a byte format) into one of these formats: *int*; *int-DWORD* (it is an int value multiplied by four); *string*; or *binary*.

For more complex situations (e.g., IP header length), we have to use another parameter, *"edit_selection"*, which contains a JSON Object required to extract information from data. For example, an IP PDU starts after an IP header. The IP header length is stored inside the second half of the first IP byte, and the value is stored as an *int-DWORD*. Hence, one has to extract the first byte and convert it into a binary format. Then, one needs to acquire the last 4 bits, convert them into an integer, and multiply by four. The JSON code for this situation is represented in Fig. 4.

```
 1 ▾ {
 2 ▾   "read_from": {
 3       "read_from": 0,
 4       "read_to": 1,
 5       "convert": "binary",
 6 ▾     "edit_selection": {
 7         "convert_from": "binary",
 8         "read_from": 4,
 9         "read_to": 8,
10         "convert": "int-DWORD"
11       }
12     }
13 }
```

Fig. 4. Example of read_from and edit_selection fields

Field "properties" is a JSONArray containing all the snippets of information that one desires to extract from a packet. When a property has been extracted, it can be overwritten or printed (see below). It is composed of the following parameters: *"name"*; *"read_from"*; *"read_to"*; or *"convert"*.

Explanation of "zones". Further, we have to define the relative properties of an object required to be found inside a packet. We should also consider how to nest one object inside another. For example, if the parser is attempting to find an IP packet inside a TCP packet, there will be no output. Every zone must have the fields within Table 3.

While *"name"* is but a simple label for the zone to find, *"read_from"* behaves precisely as demonstrated in the previous text.

The field "find" is basically a JSONObject or a JSONArray of objects constructed with the parameters within Table 4. Here, *"object"* can set the name of the object to find inside this data interval. It must be one of the object names that have been declared previously; *"multiple"* is useful when multiple instances of the same object need to be found. For example, it can be used when multiple

Table 3. List of fields inside "zones"

Name	Type	Mandatory
name	String	Yes
read_from	int/JSONArray	Yes
find	int/JSONArray	Yes

tags do not have a fixed length and position; *"label"* is printed when the parser finds the required object. It is useful for debugging purposes; *"print"* prints the properties of an object. The array of strings must contain only the valid property names from the object zone, or a string "all" to print all of the packet sections.

Table 4. List of fields inside "find"

Name	Type	Mandatory
object	String	Yes
multiple	Boolean	Yes
label	String	No
print	JSONArray	No

3   Main Algorithm Functionality

Among the several developed functions, the most important one inside our main is *parseZones()*. This function, which is called from the main function, calls others two crucial functions, *readData()* and *findInData()*. These will be explained in the following text.

Function parseZones() divides a packet into one or more zones as well as performs operations on them by following the instructions in the JSON (Fig. 5).

```
1 ▾ def parseZones(data, json):
2       json_zones = json['zones']
3       newData = data
4 ▾     for json_zone in json_zones:
5           name = json_zone['name']
6           zone_data = readData(newData, json_zone)
7           newZoneData = findInData(zone_data, json_zone)
8           replace_from = getReadFrom(zone_data, json_zone)
9           replace_to = getReadTo(zone_data, json_zone)
10          newData = replace(newData, replace_from, replace_to, newZoneData)
11      return newData
```

Fig. 5. Representation of parseZones() function

The term "zone" refers to one or more parts of a packet separated from each other, where it is possible to perform operations defined by the JSON file. The partitions can be useful if there is a need to divide a packet and perform different operations for each of the corresponding zones. For example, a hypothetical 100-byte packet can be divided into four equal parts, but the payload may be located in a different position in each of them. Relying on the concept of zones, one can search for a specific payload in each partition – efficiently and timely.

Essentially, the said function performs the following two steps:

- reading the bytes obtained from the function *readData()*;
- processing the information contained by the JSON from the function *findIn-Data()*.

This procedure is performed cyclically for each zone described in the JSON file. At the end of a cycle, the modified package is returned if specified by the JSON file. Otherwise, a complete copy of the original package is obtained. Additionally, the script has the functionality to overwrite the original zone with that eventually modified according to the instructions.

Function readData() is used to read the bytes of a packet. Locating the zones from which it is possible to read the desired bytes is done by the *getReadFrom()* and *getReadTo()* methods. These two methods return an integer that represents the index where to start reading and the index where to finish reading the packet, respectively (Fig. 6).

```
1 ▾ def readData(data, json):
2        output = data
3        read_from = getReadFrom(data, json)
4        read_to = getReadTo(data, json)
5 ▾      if 'match' in json:
6            output = matchData(output, json)
7            if output == None: return None
8 ▾      if 'match_or' in json:
9            output = matchDataOr(output, json)
10           if output == None: return None
11 ▾     if read_from != None:
12 ▾         if read_to == None:
13               output = output[read_from:]
14 ▾         else:
15               output = output[read_from:read_to]
16       output = convertData(output, json)
17       output = editSelection(output, json)
18       return output
```

Fig. 6. Representation of readData() function

After that, any checks are performed by using the methods *matchData()* and *matchDataOr()*, which verify that a certain part of the packet is equal to a certain value present in the JSON file. At this point, the function extracts the real bytes from a packet and converts them, if necessary, into another format (e.g., byte ← string) via the *convertData()* function. After this conversion, it may be necessary to extract an even smaller part of the selection; in these cases, the *editSelection()* function is used.

Function findInData() allows to search for the objects specified in the JSON file within a zone. This search can be of either type: single or multiple (Fig. 7).

```
1 ▾ def findInData(data, json):
2 ▾     if 'find' not in json:
3           return data
4 ▾     else:
5           find = json['find']
6 ▾         if isinstance(find, list):
7               output = bytearray()
8 ▾             for my_find in json['find']:
9                   find_object = my_find['object']
10                  find_multiple = my_find['multiple']
11                  output = output + findObjectsInDataAndDoThings(data, find_object, my_find, find_multiple)
12              return output
13 ▾         else:
14              find_object = find['object']
15              find_multiple = find['multiple']
16              output = findObjectsInDataAndDoThings(data, find_object, find, find_multiple)
17              return output
```

Fig. 7. Representation of findInData() function

In the first case, the search stops when an object is found for the first time in the specified range of bytes, and then only a range is returned where to perform all of the following operations – as specified by the JSON file – using the *findObjectsInDataAndDoThings()* function. In the second case, the search stops when the last byte of the search interval has not arrived. In this way, if n intervals corresponding to the search terms are found, all of the subsequent procedures specified by the JSON for each of these n intervals are performed using *findObjectsInDataAndDoThings()*.

4   Performance Evaluation and Benchmarks

One of the goals of this work was to develop a fast parser to reduce the analysis time and overall delay as much as possible. A major challenge in this scenario is a large number of packets that can arrive simultaneously. The first version of the developed parser only includes analysis from a *pcap* file to evaluate the maximum reachable processing speed. Further versions will include the feature of directly scanning the interface (e.g., from a rooted Android smartphone one can access LTE or WLAN interfaces).

To evaluate the performance of the developed packet processor, simulations have been conducted with a *pcap* file composed of 1,000 DNS packets (UDP); 10,000 TCP packets; and 100 HTTP packets. To confirm the usability of our tool, we compared it with the well-known parsing libraries: dpkt[1] and Scapy[2].

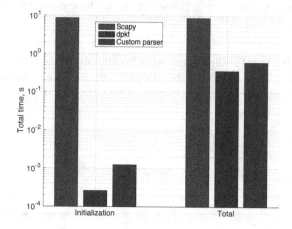

Fig. 8. Initialization and total time comparison

Initialization Phase. Before proceeding with the actual results, it is important to analyze the initialization time of the parsing software, see Fig. 8. Here, it is easy to see that the Scapy library is operating in the 'offline' mode, i.e., while working with a *pcap* file, it executes the actual parsing during the file read procedure. Our custom framework operates similarly to the dpkt tool, i.e., the actual parsing occurs when a standalone packet is analyzed. Hence, the initialization phase is high-speed. Another effect shown in this figure is a comparison of the total parsing time for the same set. The developed software demonstrates a relative gain even compared to dpkt.

Comparison of Parsers. Further, we analyze different packets per parser in the form of a cumulative distribution function (CDF). As it is displayed in Fig. 9(a), the parsing time for most of the packets is relatively similar. This is due to the effect of pre-parsing during the file read procedure. However, TCP parsing is consuming the most effort.

---

[1] "dpkt 1.9.1", 2018: https://pypi.python.org/pypi/dpkt.
[2] "Scapy library", 2018: http://www.secdev.org/projects/scapy/.

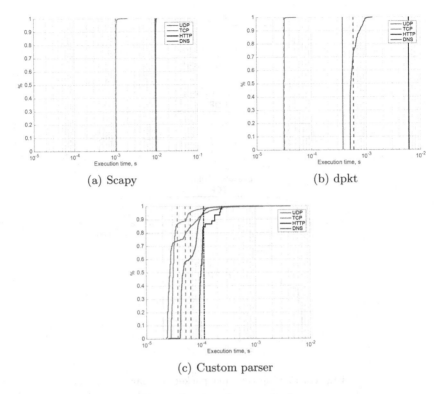

(a) Scapy

(b) dpkt

(c) Custom parser

Fig. 9. Parsing time comparison

Further, the dpkt framework is analyzed. As it is shown in Fig. 9(b), this parser operates under entirely different conditions. Since dpkt conducts parsing based on a pre-validation of the packet type, each of those provides completely different results. At the same time, the behavior in case of TCP remains the fastest.

Finally, we evaluate our developed tool, and the results are collected in Fig. 9(c). The fluctuating behavior of CDFs may be explained as a result of the different payloads of packets involved in the analysis. Overall, we conclude that the custom parser operates faster than Scapy or dpkt.

Comparison of Packet Types. Here, the focus is set on the packet parsing time comparison per parser. We show which one to select for the corresponding needs of a developer. As it is demonstrated in Fig. 10, UDP, DNS, and HTTP packets all confirm the benefits of utilizing our custom parser over the conventional alternatives. Only for TCP in case of dpkt, some difference in the execution speed is present. However, it can be considered negligible.

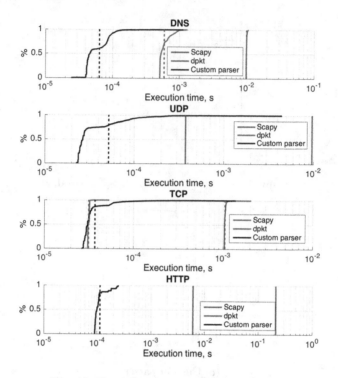

Fig. 10. Parsing time per packet comparison

5   Conclusions

Wireless networks are constantly evolving in the offered connectivity levels, thus strongly consolidating in our lives as a necessity. More and more devices are joining the networks to request continuous high-quality service and produce a vast amount of mobile data, which poses unprecedented challenges for the network design and implementation in the upcoming 5G era. Transformation of mobile user experience demands complex changes in both network infrastructure and device operation, where user experience is optimized by taking into account the surrounding network context.

Along these lines, Software Defined Networking can become essential to mitigate the network overload due to its programmable and centralized controller features, which decide – via the use of a finite state machine – how to manage the network offloading efficiently. The software libraries that exist today (e.g., Scapy and dpkt) may not be effective enough to support the requirements of emerging systems. In contrast, our proposed parser may be employed on any machine to help improve the SDN performance as well as introduce new features due to its universal compatibility with any packet. It demonstrates significant benefits over the counterpart parsing libraries with respect to the execution times.

Acknowledgment. This work is supported by the "RUDN University Program 5 – 100".

References

1. VNI Cisco: Global mobile data traffic forecast 2016–2021. White Paper (2018)
2. Mäkitalo, N., Ometov, A., Kannisto, J., Andreev, S., Koucheryavy, Y., Mikkonen, T.: Safe and secure execution at the network edge: a framework for coordinating cloud, fog, and edge. IEEE Softw. **35**(1), 30–37 (2018)
3. Karakus, M., Durresi, A.: A survey: control plane scalability issues and approaches in Software-Defined Networking (SDN). Comput. Netw. **112**, 279–293 (2017)
4. Florea, R., Ometov, A., Surak, A., Andreev, S., Koucheryavy, Y.: Networking solutions for integrated heterogeneous wireless ecosystem. In: Cloud Computing, p. 103 (2017)
5. Xia, W., Wen, Y., Foh, C.H., Niyato, D., Xie, H.: A survey on software-defined networking. IEEE Commun. Surv. Tutorials **17**(1), 27–51 (2015). https://doi.org/10.1109/COMST.2014.2330903
6. Ordonez-Lucena, J., Ameigeiras, P., Lopez, D., Ramos-Munoz, J.J., Lorca, J., Folgueira, J.: Network slicing for 5G with SDN/NFV: concepts, architectures, and challenges. IEEE Commun. Mag. **55**(5), 80–87 (2017)
7. Volkov, A., Khakimov, A., Muthanna, A., Kirichek, R., Vladyko, A., Koucheryavy, A.: Interaction of the IoT traffic generated by a smart city segment with SDN core network. In: Koucheryavy, Y., Mamatas, L., Matta, I., Ometov, A., Papadimitriou, P. (eds.) WWIC 2017. LNCS, vol. 10372, pp. 115–126. Springer, Cham (2017). https://doi.org/10.1007/978-3-319-61382-6_10
8. Laselva, D., Lopez-Perez, D., Rinne, M., Henttonen, T.: 3GPP LTE-WLAN aggregation technologies: functionalities and performance comparison. IEEE Commun. Mag. **56**(3), 195–203 (2018)
9. Ometov, A.: Short-range communications within emerging wireless networks and architectures: a survey. In: Proceedings of 14th Conference of Open Innovations Association (FRUCT), pp. 83–89. IEEE (2013)
10. Galinina, O., et al.: Capturing spatial randomness of heterogeneous cellular/WLAN deployments with dynamic traffic. IEEE J. Sel. Areas Commun. **32**(6), 1083–1099 (2014)
11. Ometov, A., Masek, P., Urama, J., Hosek, J., Andreev, S., Koucheryavy, Y.: Implementing secure network-assisted D2D framework in live 3GPP LTE deployment. In: Proceedings of International Conference on Communications Workshops (ICC), pp. 749–754. IEEE (2016)
12. Andreev, S., et al.: Exploring synergy between communications, caching, and computing in 5G-grade deployments. IEEE Commun. Mag. **54**(8), 60–69 (2016)
13. Feamster, N., Rexford, J., Zegura, E.: The road to SDN: an intellectual history of programmable networks. ACM SIGCOMM Comput. Commun. Rev. **44**(2), 87–98 (2014)
14. Volkov, A., Muhathanna, A., Pirmagomedov, R., Kirichek, R.: SDN approach to control internet of thing medical applications traffic. In: Vishnevskiy, V.M., Samouylov, K.E., Kozyrev, D.V. (eds.) DCCN 2017. CCIS, vol. 700, pp. 467–476. Springer, Cham (2017). https://doi.org/10.1007/978-3-319-66836-9_39

15. Grebeshkov, A., Gaidamaka, Y., Zaripova, E., Pshenichnikov, A.: Modeling of vertical handover from 3GPP LTE to cognitive wireless regional area network. In: Proceedings of 9th International Congress on Ultra Modern Telecommunications and Control Systems and Workshops (ICUMT), pp. 1–6. IEEE (2017)
16. Määttanen, H.L., Masini, G., Bergström, M., Ratilainen, A., Dudda, T.: LTE-WLAN aggregation (LWA) in 3GPP Release 13 & Release 14. In: Proceedings of Conference on Standards for Communications and Networking (CSCN), pp. 220–226. IEEE (2017)
17. Shen, X.: Device-to-device communication in 5G cellular networks. IEEE Netw. 29(2), 2–3 (2015). https://doi.org/10.1109/MNET.2015.7064895
18. Pyattaev, A., Johnsson, K., Surak, A., Florea, R., Andreev, S., Koucheryavy, Y.: Network-assisted D2D communications: implementing a technology prototype for cellular traffic offloading. In: Proceedings of Wireless Communications and Networking Conference (WCNC), pp. 3266–3271. IEEE (2014)
19. Gerasimenko, M., Moltchanov, D., Florea, R., Himayat, N., Andreev, S., Koucheryavy, Y.: Prioritized centrally-controlled resource allocation in integrated multi-RAT HetNets. In: Proceedings of 81st Vehicular Technology Conference (VTC Spring), pp. 1–7. IEEE (2015)
20. Pontarelli, S., Bruschi, V., Bonola, M., Bianchi, G.: On offloading programmable SDN controller tasks to the embedded microcontroller of stateful SDN dataplanes. In: Proceedings of IEEE Conference on Network Softwarization (NetSoft), pp. 1–4, July 2017. https://doi.org/10.1109/NETSOFT.2017.8004225
21. Pontarelli, S., Bonola, M., Bianchi, G.: Smashing SDN "built-in" actions: programmable data plane packet manipulation in hardware. In: Proceedings of IEEE Conference on Network Softwarization (NetSoft), pp. 1–9, July 2017. https://doi.org/10.1109/NETSOFT.2017.8004106

Optimization Algorithm for IPTV Video Service Delivery over SDN Using MEC Technology

Steve Manariyo[1], Abdukodir Khakimov[1], Darya Pyatkina[2], and Ammar Muthanna[1,2(✉)]

[1] St. Petersburg State University of Telecommunication, 22 Prospekt Bolshevikov, St. Petersburg 193232, Russia
mansteve06@mail.ru, khakimov.a@sdnlab.ru,
ammarexpress@gmail.com
[2] Peoples' Friendship University of Russia (RUDN University), 6 Miklukho-Maklaya Street, Moscow 117198, Russia
makolkina@list.ru

Abstract. Management of video content distribution through files allocation or caching in content delivery networks with some degree of reliable security measures is representing a big issue in video service delivery and user's requirements for quality of experience provision are constantly tightened. Operators are looking for new ways to efficiently deliver video content to specific customers or classes of customers, which allow the transfer of large amounts of traffic with the appropriate quality of experience. Mobile Edge Computing (MEC), initiated as an Industry Specification Group (ISG) within ETSI, is quickly gaining traction as a disruptive technology that promises to bring applications and content closer to the network edge, a move that will reduce latency and make new services optimization possible. The aim of this thesis is to provide optimization algorithms for accessing IPTV video services in managed way over Software-defined Networking (SDN) to meet the high Quality of service (QoS) reducing network latency and, ultimately, improving the end consumer's quality of experience (QoE). We also show the positive impact of SDN network using our algorithm noticeably reducing video delay.

Keywords: IPTV · Video traffic · Software-Defined network (SDN)
Mobile Edge Computing (MEC) · Quality of Experience (QoE)
Optimization algorithm

1 Introduction

Today's multimedia market has witnessed an increase in the popularity of video streaming over IP network (IPTV) [1]. Media delivery and streaming over public or private networks are becoming the highest rank of applications consuming Bandwidth and particularly sensitive to packet loss, latency/delay and jitter. Practically, they are rapidly increasing in network bandwidth utilization with the huge number of network users concerning video access. So, the performance optimization relevant to bandwidth

© Springer Nature Switzerland AG 2018
O. Galinina et al. (Eds.): NEW2AN 2018/ruSMART 2018, LNCS 11118, pp. 419–427, 2018.
https://doi.org/10.1007/978-3-030-01168-0_38

utilization and quality of Service improvement is a key factor for successful delivery and successful business based video. To satisfy the considerable amount of video content requests, operators have been pushing their content delivery infrastructure to edge networks–from regional CDN servers to peer CDN servers to cache content and serve users with storage and network resources nearby. The main objective from caching is to make the files very near to users so as to facilitate their accessing times.

Mobile Edge Computing (MEC) is technology that enables cloud computing capabilities and an IT service environment at the edge of the network and is a quickly ramping technology that brings applications and content closer to the network edge i.e., very close to the end consumer of that video content [2]. By putting content and applications at the edge, users can receive video content with minimum delay and the network owner or the enterprise can realize operational and cost efficiencies while reducing network latency and, ultimately, improving the end consumer's quality of experience [2, 3]. Technical standards for MEC are being developed by the European Telecommunications Standards Institute (ETSI), which created a new Industry Specification Group in 2014 for this purpose.

One of the great methods to simplify the main components and increase the capacity of any network (including IPTV) is Software Defined Network (SDN) [9, 11]. SDN concepts, in which data transmission and management are separated, are well suited for implementing a large set of complex video services. SDN is a promising solution that allows a more distributed, flexible, and scalable network. As a transport network, the technology SDN has been chosen, which today has a number of significant advantages when delivering content, by providing greater automation and orchestration of the network fabric, and by allowing dynamic, application-led configuration of networks and services. Our goal is to provide an optimization algorithm, which creates an environment characterized by ultra-low latency and high bandwidth as well as real- time access to network information that can be leveraged by video applications. The algorithm shows how software-defined networking (SDN) and network programmability can be used to reach high QoS using MEC technology.

The remainder of this article is organized as follows: Sect. 2 analyzes the other works in this field of research. Section 3 shows the experimental investigation structure and defines some of the elements. Section 4 describes the proposed algorithm. Section 5 presents testing results. Finally, Sect. 6 concludes the paper.

2   Background and Related Works

By now, a consortium of operators, manufacturers and vendors have been working together to develop an open architecture and API for delivering content and services from the edge. MEC is complementary to and supportive of both SDN and NFV [10]. Between them, these technologies are profoundly affecting the network. Thus, in [7] authors introduce an approach to offload video encoding efforts from mobile devices to external services in existing mobile network architecture and reduce the power consumption of mobile devices. The MEC is under standardization [4, 5], and surveyed [6] by ETSI. ETSI states the five important use-cases of the MEC that is targeted for future standardization [6]. Industry Specification Group (ISG) has been formed to foster the development of a broader MEC ecosystem based on open standards. Paper [8] proposes the Edge Cloud model by augmenting the common cloud data center with service nodes placed at the edges and it shows the advantages in two applications: an accurate and low-latency indoor localization and a scalable-bandwidth video monitoring stream. IBM, Huawei, Intel, Vodafone, NTT DoCoMo and Nokia Networks are founding members of the industry group supporting the MEC working group at the standards organization ETSI. This is important because the provision of a standardized yet open networking environment close to the access network edge will enable service providers to deliver content (especially optimized video content) and services to users in a much more timely manner because MEC allows time-critical content to be cached in local, proximate base stations thus greatly reducing congestion on the backhaul link to the network core.

Our proposed algorithm improves the network performance profiting from the way MEC jointly work with SDN.

3 Deployment Architecture of the Experimental System

This section gives an overview of the system and describes how MEC operate over software-defined network at the edge for an optimal video offloading from the core network. Figure 1 shows the deployment architecture. The system functions upon software-defined network consisting of multiple OpenFlow-enabled switches and SDN controller, whether it is physical or software, and fully separates the data plane from control functions.

In Fig. 1, the orchestrator is the control element that controls all the systems of the operator. The SDN controller has a global network topology view, so it realizes programming on the whole network, control the forwarding path, achieve flexible networking and get rid of the traditional network restriction. MEC controller is a mobile computing system that can rent its resources for a short period of time. The programmability of the Core Network provided by SDN is exactly where MEC can facilitate its programmability at the edge of the network and further delegate control decisions. SDN and MEC are complementary concepts and SDN has the same objectives as MEC in the way of applying specific rules to data plane. The IPTV user request video content through the network, then MEC is created in several switches or one of the OpenFlow switches. MEC assesses the probability of content requests from a group of users and in the case of high demand of the same content, the content is uploaded to the optimal node closer the users. This will ensure video content offloading from the core network, thereby providing high QoS.

4 The Proposed Algorithm

The considered scenario is illustrated in Fig. 2 and consists of 4 Openflow Switch lanner: Intel(R) Xeon(R) CPU E5-2650 v4 @2.20 GHz, core 12 Ram 40 GB, 1 orchestrator Brain Net service Platform and 3 virtual Openflow controller, 1 Video

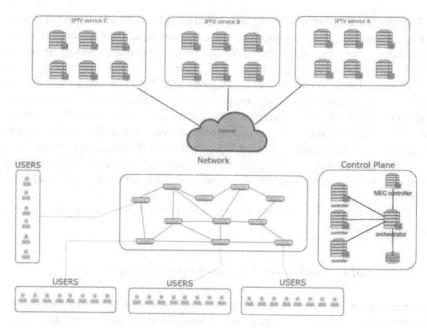

Fig. 1. Architecture of the experimental system

server Intel(R) Xeon(R) CPU E5-2620 v4 @2.10 GHz, core 32 Ram 48 GB, emulating an IPTV server using an RTP generator.

For each switch a random conditional probability of having the desired video content in that node was given: $P(A) = 0.25$ for switch 93(A), $P(B) = 0.64$ for switch 94(B), $P(C) = 0.71$ for switch 95(C), $P(D) = 0.25$ for switch 96(D).

Using the proposed algorithm in Fig. 3, one switch is selected for the MEC location. Then, based on probabilities of having the desired video content on a particular switch, the required content is downloaded closer to users.

In the proposed algorithm (Fig. 3), the main point of the sequence diagram starts from the authentication of the group of STB (Set top Box) by the IPTV server. After successful authentication, the server will initiate the procedure to transmit the information about the request sector of this group to the orchestrator. Then orchestrator determines the optimal allocation of MEC in one or several switches (decision making for resource allocation) based on the following criteria:

- Closest node to the STB group
- The ability to store content
- The ability to process node requests
- Possible downloads of the content from other MECs (the same content can be stored in other MECs at the current time)
- The evaluation of the optimality according to the previous parameters in comparison to video content delivery direct from the IPTV server.

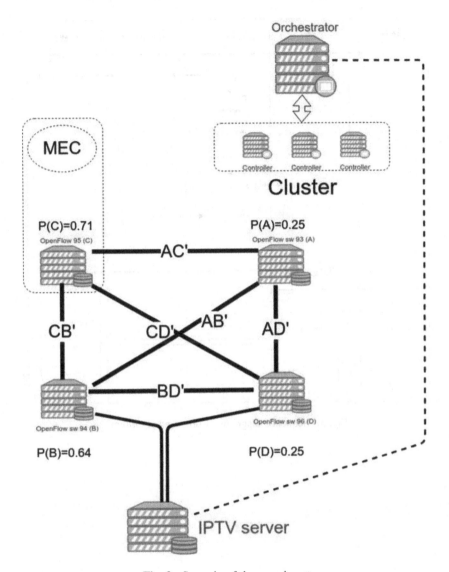

Fig. 2. Scenario of the experiment

In the case, when all criterias are met, a virtual MEC is created in a suitable node (allocate space). Then, user parameters (Billing, QoS,...) are sent from the IPTV server to the orchestrator (Options). Then video parameters and instructions (EPG) are sent to MEC, which farther upload content from other MECs or from the IPTV server and simultaneously broadcast to a group of users.

The algorithm helps offloading the core network and server interface, since all user requests go to the sector MEC, which provides a minimum delay, reduces network latency and, ultimately, improving the end consumer's quality of experience.

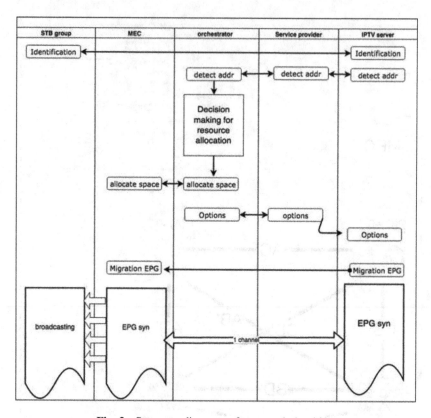

Fig. 3. Sequence diagrams of proposed algorithm

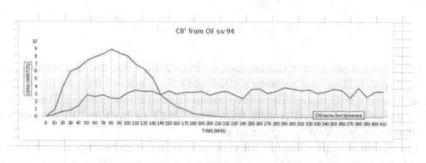

Fig. 4. Traffic load in CB' network segment in traditional systems and using our algorithm (Color figure online)

5   Results

As a result, in Fig. 4, shows that traffic load in the network in traditional system is uniform all time (blue line in Fig. 4). When using our algorithm, traffic load increases as the content is being downloaded and then there is a complete liberation of the channel.

When testing a traditional network, the average delay reaches 80 ms (Fig. 5).

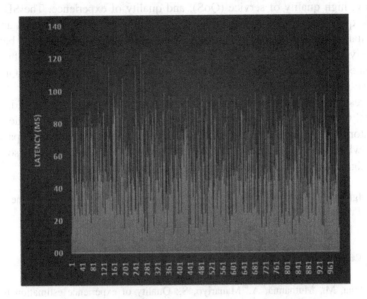

Fig. 5. Traffic delay in traditional system

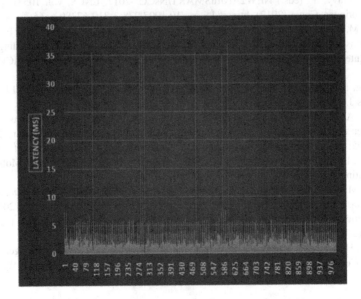

Fig. 6. Delay in a system using the proposed algorithm

When testing the network using the proposed algorithm, the average delay decreased by 20 times, improving QoS and ultimately, the end user's quality of experience (Fig. 6).

6 Conclusion

Software defined networking (SDN) approach provides security, network management, monitoring, high quality of service (QoS), and quality of experience. The SDN technology is quite universal, so the study of algorithms for SDN-networks and their implementation in the work of already existing systems for IPTV is becoming increasingly important. SDN allows load balancing, what makes possible the use of data channels much more efficiently. The use of MEC also improves user Quality of Experience in comparison to direct communication.

As a result, over Software Defined Networking (SDN)-based Mobile Edge Computing (MEC) platform, an algorithm for offloading the basic station was developed. So, operators will no longer need frequent equipment upgrades, network capacity will increase, which will significantly reduce operating costs. The algorithm solves the Mobile Edge management issues with respect to traffic management.

Acknowledgment. The publication has been prepared with the support of the "RUDN University Program 5-100".

References

1. Makolkina, M., Muthanna, A., Manariyo, S.: Quality of experience estimation for video service delivery based on SDN core network. In: Galinina, O., Andreev, S., Balandin, S., Koucheryavy, Y. (eds.) NEW2AN/ruSMART/NsCC -2017. LNCS, vol. 10531, pp. 683–692. Springer, Cham (2017). https://doi.org/10.1007/978-3-319-67380-6_65
2. Beck, M.T., Feld, S., Fichtner, A., Linnhoff-Popien, C., Schimper, T.: ME-VoLTE: network functions for energy efficient video transcoding at the mobile edge. In: Proceedings of the 18th International Conference on Intelligence in Next Generation Networks (ICIN 2015) (2015)
3. Chiang, M., Zhang, T.: Fog and IoT: an overview of research opportunities. IEEE Internet Things J. **3**, 854–864 (2016)
4. Ateya, A., Vybornova, A., Kirichek, R., Koucheryavy, A.: Multilevel cloud based Tactile Internet system. In: IEEE-ICACT2017 International Conference, Korea, February 2017
5. Mec005: Mobile edge computing (mec) poc framework. ETSI, Technical report (2015)
6. Patel, M., Joubert, J., Ramos, J.R., Sprecher, N., Abeta, S., Neal, A.: Mobile-edge computing. ETSI (2014)
7. Beck, M.T., Feld, S., Fichtner, A., Linnhoff-Popien, C., Schimper, T.: Mevolte: network functions for energy-efficient video transcoding at the mobile edge. In: 2015 18th International Conference on Intelligence in Next Generation Networks (ICIN), pp. 38–44. IEEE (2015)
8. Chang, H., Hari, A., Mukherjee, S., Lakshman, T.: Bringing the cloud to the edge. In: 2014 IEEE Conference on Computer Communications Workshops (INFOCOM WKSHPS), pp. 346–351. IEEE (2014)

9. Muhizi, S., Shamshin, G., Muthanna, A., Kirichek, R., Vladyko, A., Koucheryavy, A.: Analysis and performance evaluation of SDN queue model. In: Koucheryavy, Y., Mamatas, L., Matta, I., Ometov, A., Papadimitriou, P. (eds.) WWIC 2017. LNCS, vol. 10372, pp. 26–37. Springer, Cham (2017). https://doi.org/10.1007/978-3-319-61382-6_3

10. Ateya, A.A., Muthanna, A., Gudkova, I., Abuarqoub, A., Vybornova, A., Koucheryavy, A.: Development of intelligent core network for tactile internet and future smart systems. J. Sens. Actuator Netw. 7(1) (2018)

11. Vladyko, A., Muthanna, A., Kirichek, R.: Comprehensive SDN testing based on model network. In: Galinina, O., Balandin, S., Koucheryavy, Y. (eds.) NEW2AN/ruSMART -2016. LNCS, vol. 9870, pp. 539–549. Springer, Cham (2016). https://doi.org/10.1007/978-3-319-46301-8_45

Analytical Modeling of Development and Implementation of Telecommunication Technologies

Vladimir Gluhov, Valery Leventsov, Anton Radaev,
and Nikolay Nikolaevskiy[✉]

Peter the Great St. Petersburg Polytechnic University, Saint Petersburg, Russia
{vicerector.me,vleventsov}@spbstu.ru,
TW-inc@yandex.ru, triplenick@mail.ru

Abstract. The paper discusses the problem of defining the characteristics of processes related to development and implementation of telecommunication technologies. These characteristics must reflect an adequate assessment of expectations for the implementation of new technologies and be a basis for correct and relevant decision-making, which should lead to key competitive advantages, both for an individual enterprise and the economy of the country as a whole. The above circumstances have predetermined the topicality of the research study, whose aim at this stage is to assess the adequacy of the analytical model, which is developed on the basis of Gartner's hype cycle model and reflects the dependence of expectations for the technology on a time factor. The results of analytical modeling have been obtained on the example of the commonly used telecommunication technology Bluetooth. The source data included a number of articles on the relevant subject from Scopus database. Based on the obtained results, it has been concluded on the working capacity of the used model and recommendations have been given on its possible improvement.

Keywords: Analytical modeling · Gartner's hype cycle · Bluetooth
Technology development's completion indicator
Aggregate expectation indicator

1 Introduction

In today's conditions, where information and communication technologies are developing intensively and provide for prospective mass digitization of all sectors of manufacturing industry and economy, some issues are becoming really important, in particular, those related to the processes of implementation and development of telecommunication technologies, which, in turn, is an essential condition for key competitive advantages to form in the corresponding sectors of the economy both for an individual enterprise and for the entire country. A number of factors must be considered, which may include, for example, the expectations (and their adequate assessment) for new technologies and the characteristics of the approach to the implementation of the latter ones taking into account comparison of risks and investments. When making proper decisions about introducing innovations, it is necessary to

© Springer Nature Switzerland AG 2018
O. Galinina et al. (Eds.): NEW2AN 2018/ruSMART 2018, LNCS 11118, pp. 428–440, 2018.
https://doi.org/10.1007/978-3-030-01168-0_39

understand the general cycle of existence of the considered technologies with an aim of further comparative analysis in the historical aspect within retrospective research studies and determination of alternative scenarios of technology development in future. Today, for solving the above-mentioned problems related to the description of the process of development and implementation of innovation technologies in the context of life cycle, Gartner's hype cycle model has become widely spread [1].

This model describes the dependence of expectations as a characteristic of technology development on the time factor, which is determined by a representative curve of the Cartesian coordinate system in the corresponding graphic interpretation of the model. The reviewed literature on the relevant subject has shown that initially Gartner's hype cycle model had a conceptual character, i.e. it described the dependence of expectations on the time factor exceptionally in the qualitative aspect with the detection of the common sections (stages) of the technology development process [2–5]. In order to ensure an objective evaluation of the process of technology development based on Gartner's hype cycle model, quite a number of research studies were conducted, being dedicated to analytical description of the above model [6–8]. The authors of this paper, in turn, have also proposed a mathematical model [9], which is based on Gartner's hype cycle model and ensures a more detailed (in comparison to earlier models) assessment of the characteristics of the technology development process. This paper is aimed at verifying the working capacity of the developed analytical model on the example of a specific telecommunication technology, whose role is played by a commonly used technology for wireless data exchanging - Bluetooth. It is expected that analytical dependences of expectations for the technology on the time factor will be obtained and then analyzed on the basis of concrete values, which are presented in our case by a number of papers on the relevant subject.

The paper is organized in the following way. Section 2 contains the description of the used mathematical model. Section 3 describes the procedure of analytical modeling, which includes acquisition of the source data and determination of the design values for the parameters of the relevant functional dependences. Section 4 analyzes the obtained results.

2 Description of the Mathematical Model

The main principles of the considered mathematical model include the following:

- the model describes the process of technology development and implementation from the idea of its creation to its full-scale usage;
- the model reflects the dependence of the characteristic of expectation for the technology on the time factor, presented as analytical and graphic interpretation with both quantitative categories being expressed in dimensionless form - in percentage (correspond to the range of values from 0% to 100%);
- the time factor is identified with the Technology Development's Completion Indicator (TDCI);

– the main indicator of expectations for the technology is an Aggregate Expectation Indicator (AEI), presented as a sum of two components – Subjective Expectation Indicator (SEI) and – Objective Expectation Indicator (OEI).

It should be mentioned that the SEI characterizes hypothetical expectations, not confirmed with the objective results of the conducted research, while the OEI characterizes objective expectations, confirmed with the accumulated results of the conducted research. Below are presented analytical dependences of AEI, SEI and OEI values on the time factor values, while Fig. 1 shows a graphic interpretation:

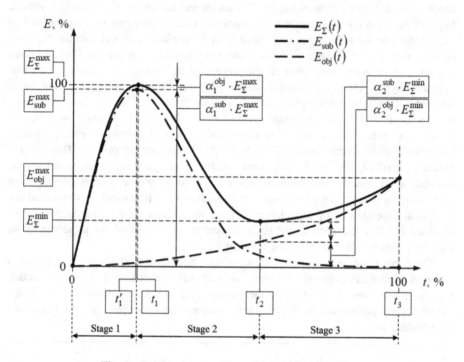

Fig. 1. Graphic interpretation of the used model [9]

$$E_{\sum} = E_{sub}(t) + E_{obj}(t), \qquad (1)$$

$$E_{sub}(t) = a_1 \cdot t \cdot b_1^{c_1 \cdot t}, \qquad (2)$$

$$E_{obj}(t) = a_2 \cdot t \cdot b_2^{c_2 \cdot t}, \qquad (3)$$

where $E_{\sum}$, E_{sub}, E_{obj} are correspondingly AEI, SEI and OEI, %; t is the time factor or TDCI, %; a1, b1, c1, a2, b2, c2 are the parameters of functional dependencies SEI and OEI on TDCI, whose description and purpose are given in paper [9].

3 Analytical Modeling

The research procedure included the following main stages:

1. Forming the source data acquired through browsing papers on the relevant subject in the scientometric base Scopus.
2. Determining the parameters of functional dependences of the SEI and OEI on the time factor (TDCI) in accordance with formulas (2) and (3) using Microsoft Excel.
3. Determining the parameters and constructing a functional dependency plot of the AEI on TDCI in accordance with formula (1).
4. Analyzing the obtained results, reflecting the characteristics of the process of development and implementation of the considered telecommunication technology in the context of individual stages, and, as a consequence, assessing the adequacy of the used model.

3.1 Forming the Source Data

The source data were formed on the basis of the specific values of the number of relevant articles in the scientometric base Scopus [10], with documents being searched under the following conditions:

- the inquired papers were selected by the key word "Bluetooth", and contained in the title or abstract the same word; The wording of the inquiry had the following form: «TITLE-ABS-KEY («*Bluetooth*») AND (LIMIT-TO (EXACTKEYWORD, "*Bluetooth*"))»;
- depending on the type of document, the papers in the obtained sample were distributed between the indicators of subjective and objective expectations, with the documents of "*review*" and "*conference review*" type being referred to the indicator of subjective expectation and those of the "*article*" type being referred to the objective expectation indicator.

It is important to notice that documents of the "conference paper" type were not considered since it is difficult to refer them to one or another category of expectations.

Because in the suggested model the AEI, SEI and OEI vary from 0 to 100%, the source data on the quantity of articles within the research study were rated, i.e. the source values were transformed in dimensionless ones, varying from 0 to 100%. For this purpose, the nominal values of the quantity of papers were designated for an elementary period. If division is done by them, transition is ensured from the initial values of the quantity of papers to the rated ones. Rating was carried out by formulas:

$$E_{sub}^{fact} = \frac{N_{sub}}{[N_{sub}]} \cdot 100\%, \tag{4}$$

$$E_{obj}^{fact} = \frac{N_{obj}}{[N_{obj}]} \cdot 100\%, \tag{5}$$

where E_{sub}^{fact}, E_{obj}^{fact} is the rated actual value of the indicator of subjective and objective expectations, %; N_{sub}, $[N_{sub}]$, N_{obj}, $[N_{obj}]$ are the initial and nominal quantity of documents for the indicator of subjective and objective expectations respectively, units.

The nominal values of the numbers of papers were designated in accordance with the following principles:

1. In point t_1' of the dependence extremum $E_{sub}(t)$ (in the suggested model, see Fig. 1) the share of indicator $E_{obj}(t_1')$ in the composite index $E_\Sigma(t_1')$ is rather small, accepted within this research study at a level of 0–5%. The composite index $E_\Sigma(t_1')$ in the stated extreme point has a value, close to 100% (accepted within this research study at a level of 95–100%), but not equal to it in the common case, since the extremum (maximum) of dependence $E_\Sigma(t_1)$ is a little further to the right than the extremum of dependence $E_{sub}(t_1')$, or $t_1 \geq t_1'$.
2. In the point of maximum value t_2 of dependence $E_{obj}(t)$ the share of the stated indicator $E_{obj}(t_2)$ in the composite index $E_\Sigma(t_2)$ is accepted as equal to 40–60% within this research study, whereas the value of indicator $E_{sub}(t_2)$ is very small.

Thus, given the above-mentioned, the source data are presented in Table 1.

Table 1. Source data

No.	Year	TDCI, t, %	Number of documents		E_{sub}^{fact}, %	E_{obj}^{fact}, %
			N_{sub}	N_{obj}		
0	–	0	0	0	0	0
1	2002	6.25	8	44	20.0	5.5
2	2003	12.50	7	80	17.5	10.0
3	2004	18.75	25	84	62.5	10.5
4	2005	25.00	38	89	95.0	11.1
5	2006	31.25	17	114	42.5	14.3
6	2007	37.50	11	123	27.5	15.4
7	2008	43.75	0	111	0.0	13.9
8	2009	50.00	3	107	7.5	13.4
9	2010	56.25	1	97	2.5	12.1
10	2011	62.50	0	89	0.0	11.1
11	2012	68.75	2	84	5.0	10.5
12	2013	75.00	0	99	0.0	12.4
13	2014	81.25	6	102	15.0	12.8
14	2015	87.50	1	153	2.5	19.1
15	2016	93.75	4	140	10.0	17.5
16	2017	100.00	4	161	10.0	20.1
Nominal number of documents, $[N_{sub}]$ and $[N_{obj}]$:			40	800		

3.2   Determining the Dependence Parameters of the SEI and OEI

It is important to note that in terms of the implemented analytical model, it is expected to describe the expectations for the technology through functional dependence on the time interval identified with the technology development's completion indicator. However, the source data cannot guarantee correctness in terms of the correspondence of the available actual time range to the real time interval, defining the technology development's completion. The above circumstances predetermined the reasonability of introducing additional parameters, characterizing the beginning and the end of the time interval (t_{begin} and t_{end}, %), which was directly identified with the TDCI and where the desired characteristics of functional dependences were sought for.

Thus, according to the information from the available sources [11], the idea of the considered Bluetooth technology dates back to the year 1998 and by today the fifth standard of the new generation is produced, but the data on the number of papers have been collected only starting from 2001, so parameters t_{begin} and t_{end} were designated constructively as 15 and 80% respectively. The time factor of the TDCI was transformed on the basis of the designated values by formula:

$$t' = \frac{t - t_{begin}}{t_{end} - t_{begin}} \cdot 100\%. \tag{6}$$

Taking into account formula (6) and the chosen limits of the initial and final values of time, it can be noticed that the transformed value t' can go beyond the limits of the interval from 0 to 100%. So, when concrete values of the SEI and OEI were calculated, an additional condition was introduced, which implies attributing a zero value of expectation for the transformed time indicators lying beyond the interval from 0 to 100%.

So, in order to identify the parameters of functional dependences of subjective and objective expectations in accordance with formulas (2) and (3) on the basis of the source data formed in Subsect. 3.1, Solver Add-in Microsoft Excel is used. We have to say that in both cases the result was obtained by the generalized reduced gradient (GRG) method for solving nonlinear problems. The target cells were set as the sums of squared differences of the actual values of expectation indicators ($E_{sub}^{fact}(t)$ and $E_{obj}^{fact}(t)$) and the design ones, which, in turn, were defined on the basis of variable coefficients a1, b1, c1, a2, b2, c2 given the corresponding limitations. On the one hand, the above limitations determined the search interval of the desired variables through admissible minimum and maximum values, whose numerical value was proposed in paper [9]. On the other hand, the limitations were necessary for the description of representative points, like, for example, the extremum of function.

In case the parameters of the subjective expectation indicator (SEI) were sought for, an additional desired time parameter t_{ex}, was introduced. It is a representative point of the plot where the function reaches the maximum value, i.e. has an extremum. The presence of the extremum point will be also considered in the limitations through the equality of the derived function in point t_{ex} to a zero value.

Thus, with the above assumptions being considered, some conditions were formulated so that a solution can be sought for in case of the SEI:

1. Target function:

– the sum of squared differences of the expected and actual values:

$$\sum_t \left(E_{sub}(t) - E_{sub}^{fact}(t) \right)^2 \rightarrow min$$

2. Unknown variables:

– parameters of functional dependence (2): a_1, b_1, c_1;
– time value, corresponding to the extremum of function: t_{ex}.

3. Limitations:

$$\begin{cases} a_1 \in (a_1^{min}; a_1^{max}), \ b_1 \in (b_1^{min}; b_1^{max}), \ c_1 \in (c_1^{min}; c_1^{max}), \\ \frac{dE_{sub}(t_{ex})}{dt} = 0, \quad t_{ex} \in [t_{ex}^{min}; t_{ex}^{max}], \\ E_{sub}(t_{ex}) \in [E_{sub}(t_{ex})_{min}; E_{sub}(t_{ex})_{max}], \\ E_{sub}(100\%) \in [E_{sub}(100\%)_{min}; E_{sub}(100\%)_{max}]. \end{cases}$$

Thus, on the basis of the above conditions, the parameter solution of the SEI was sought for by means of Microsoft Excel. The result is presented in Fig. 2. The determination coefficient R^2 was calculated additionally. Its value turned out to be equal to 0.91, which bears evidence of a high accuracy of the obtained result. The form of functional dependence of the SEI and the table of limitations with minimum and maximum values of the relevant parameters are also presented in Fig. 2.

Similarly to the principles used for determining the dependence parameters of the SEI on the TDCI, the problem was stated for seeking the characteristics of the OEI:

1. Target function:

– the sum of squared differences of the expected and actual values:

$$\sum_t \left(E_{obj}(t) - E_{obj}^{fact}(t) \right)^2 \rightarrow min$$

2. Unknown variables:

– parameters of functional dependence (3): a_2, b_2, c_2.

3. Limitations:

$$\begin{cases} a_2 \in (a_2^{min}; a_2^{max}), \ b_2 \in (b_2^{min}; b_2^{max}), \ c_2 \in (c_2^{min}; c_2^{max}), \\ E_{obj}(100\%) \in [E_{obj}(100\%)_{min}; E_{obj}(100\%)_{max}]. \end{cases}$$

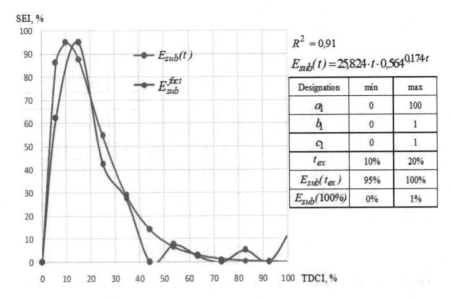

Fig. 2. Dependence plots of the actual and expected values of the SEI on the TDCI

The obtained result is presented in Fig. 3, which shows the general form of the OEI function and the table with limitations. In this case it should be noted that the value of the determination coefficient $R^2 = 0,41$ remains unsatisfactory. It is obvious that the cause of such a result lies in the divergence of the growth dynamic characteristics of objective expectations with the course of time. Namely, the used model suggests a monotonous growth of objective expectations for the technology with the course of time (see Fig. 1), while the actual curve (see Fig. 2) based on the available source data has a peak value, after which it decays. The obtained result allows us to conclude about the drawbacks of the used model, in particular, those related to source data formation. Detailed conclusions and comments are presented in Sect. 4.

3.3 Determining the Dependence Parameters of the AEI

In order to determine the functional dependence parameters of the AEI, coefficients a, b and c of the dependence of the SEI and OEI, defined in the previous section, were used. These values were taken as basic ones (a_1^b, a_2^b etc.), against which the optimum values of dependence coefficients of the AEI were sought for by the criterion of minimization of the sum of squared differences given the relevant limitations.

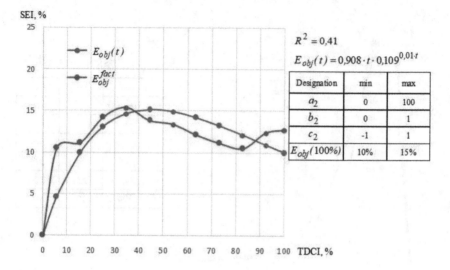

Fig. 3. Dependence plots of the actual and expected values of the OEI on the TDCI

In this case limitations were imposed on variable parameters and values $E_{\sum}(t)$ in representative points t_{ex1} and t_{ex2}, where the desired dependence takes on extreme values.

Thus, below are presented the conditions for seeking the solution to the AIE parameters by the GRG method:

1. Target function:

– the sum of squared differences of the basic and optimum values of the dependence coefficients of the SEI, OEI and AEI:

$$(a_1^b - a_1)^2 + (b_1^b - b_1)^2 + (c_1^b - c_1)^2 + (a_2^b - a_2)^2 + (b_2^b - b_2)^2 + (c_2^b - c_2)^2 \rightarrow min$$

2. Unknown variables:

– parameters of functional dependences (2) and (3): a_1, b_1, c_1, a_2, b_2, c_2;
– time value, corresponding to the extremum of functions: t_{ex1} and t_{ex2}.

3. Limitations:

$$\begin{cases} a_1 \in (a_1^{min}; a_1^{max}), \quad b_1 \in (b_1^{min}; b_1^{max}), \quad c_1 \in (c_1^{min}; c_1^{max}), \\ a \in (a_2^{min}; a_2^{max}), \quad b_2 \in (b_2^{min}; b_2^{max}), \quad c_2 \in (c_2^{min}; c_2^{max}), \\ \left. \dfrac{dE_{\sum}(t)}{dt} \right|_{t_{ex1}} = 0, \quad t_{ex1} \in [t_{ex1}^{min}; t_{ex1}^{max}], \\ E_{sub}(t_{ex1}) \in [E_{sub}(t_{ex1})_{min}; E_{sub}(t_{ex1})_{max}], \\ E_{obj}(t_{ex1}) \in [E_{obj}(t_{ex1})_{min}; E_{obj}(t_{ex1})_{max}], \\ E_{\sum}(t_{ex1}) = 100\%, \\ \left. \dfrac{dE_{\sum}(t)}{dt} \right|_{t_{ex2}} = 0, \quad t_{ex2} \in [t_{ex2}^{min}; t_{ex2}^{max}], \\ E_{sub}(100\%) \in [E_{sub}(100\%)_{min}; E_{sub}(100\%)_{max}], \\ E_{obj}(100\%) \in [E_{obj}(100\%)_{min}; E_{obj}(100\%)_{max}]. \end{cases}$$

Thus, with the implementation of the solution seeking procedure, the values of the functional dependence coefficients of the OEI and SEI were found (see Table 2). Figure 4 shows the final functional dependences of the SEI, OEI and AEI, their plots and a table with limitations. Representative points $t_{ex1} = 10,5\%$ and $t_{ex2} = 70,0\%$ are shown on the plot.

Table 2. Values of the functional dependence coefficients of the OEI and SEI

Designation		Value	
Basis	Optimal	Basis	Optimal
a_1^b	a_1	25.824	25.903
b_1^b	b_1	0.564	0.566
c_1^b	c_1	0.174	0.169
a_2^b	a_2	0.908	0.085
b_2^b	b_2	0.109	0.109
c_2^b	c_2	0.010	−0.003
Sum of squared deviations: **0.68**			

4 Analysis of Results and Conclusion

Based on the analyzed plot of dependence of the AEI on the TDCI time factor (see Fig. 4) in relation to representative points $t_{ex1} = 10,5\%$ and $t_{ex2} = 70,0\%$, it is possible to distinguish three consecutive sections, reflecting individual stages of the development and implementation process of the Bluetooth technology. Given the structure of the source data (Table 1) and the values of the representative points obtained with the studied period of technology development's completion being limited by the initial and final values, the stated stages correspond to the following time ranges:

- 1st stage from 1998 to 2004;
- 2nd stage from 2004 to 2010;
- 3rd stage from 2010 to 2017.

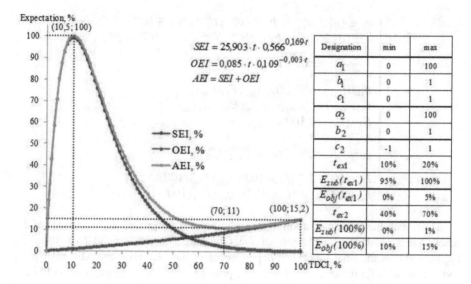

Fig. 4. Plots of functions of expectation on the time factor

The next step implied that the actual information about development of the Bluetooth technology, obtained from the available sources, was analyzed [11]. The obtained actual information in the context of time periods, characterizing the real level of expectations for the considered technology, was compared to the data of the implemented model, which is shown in Table 3.

The qualitative analysis of the information presented in the table above, having regard to the specifics of the used analytical model, allows us to make the following conclusions:

1. When comparing the characteristics of the stages of development and implementation of the Bluetooth technology to the actual events related to the formation of the considered technology, it can be concluded that the considered analytical model, on the one hand, adequately describes subjective expectations, which is proven by a high determination coefficient obtained in Subsect. 3.2. On the other hand, the indicator of objective expectations, composed on the basis of actual values and which, according to the structure of the model, should grow monotonously, has sections with a negative dynamics, which has predetermined the unsatisfactory value of the determination coefficient for the OEI curve.

2. The above circumstances, affecting the accuracy and adequacy of the obtained results, are caused, firstly, by the specifics of source data formation, characterized by a complexity of identifying objective expectations exceptionally on the basis of papers on the relevant subject. Moreover, a relatively low accuracy of the results is reflected in terms of defining the time intervals identified with the technology development's completion indicator, which predetermines the need to enhance the model in the context of substantiation of the initial and final values of the TDCI and

Table 3. Comparison of the analytical and actual information about technology development

Stage in accordance with the model	Characteristic of the stage in accordance with the model	Actual events in technology development
1st stage from 1998 to 2004	Characterizes by an intensive growth of subjective expectations. In other words, within this stage, ideas develop about the ways to study and apply the technology, which is preconditioned by the exceeding growth of the number of formulated hypotheses over the number of verified ones due to a relatively small quantity of the conducted research studies (low rate of growth of objective expectations)	**1998** The Bluetooth Special Interest Group (SIG) is formed and the name Bluetooth is officially adopted **1999** The Bluetooth 1.0 Specification is released **2001** First Bluetooth printer, laptop, hands-free car kit **2002** IEEE approves the 802.15.1 specification to conform with Bluetooth wireless technology **2003** The SIG adopts Bluetooth Core Specification Version 1.2 Bluetooth product shipments grow to 1 million per week
2nd stage from 2004 to 2010	Is characterized by an objective assessment of the capabilities provided by the technology. In other words, the number of formed hypotheses virtually do not change (the limit is reached). Moreover, the negative results (from the positions of technology development) prevail in the growing number of verified hypotheses	**2004** The SIG adopts Core Specification Version 2.0 Enhanced Data Rate (EDR) **2005** Bluetooth product shipments soar to 5 million chipsets per week **2006** Bluetooth wireless reaches an installed base of 1 billion devices **2009** The SIG adopts Core Specification Version 3.0 HS, making Bluetooth high speed technology a reality **2010** The Bluetooth SIG announces the formal adoption of Bluetooth Core Specification Version 4.0 with Low Energy technology
3rd stage from 2010 to 2017	Is characterized with the implementation of the technology in a full scale production and mass use. In other words, the number of verified hypotheses approaches the total number of formed hypotheses (and finally reaches it). The share of verified hypotheses grows faster that the share of hypothesis with a negative result	**2011** The SIG adopts the first new profile for Bluetooth v4.0 **2012** The first Bluetooth Low Energy tablets and music players hit the market **2013** The SIG rolls out Bluetooth 4.1, sets the stage for IoT **2014** The SIG introduces Bluetooth 4.2, adds features for IP connectivity, privacy and speed Bluetooth hits 90% penetration in all mobile phones **2016** The SIG announces Bluetooth 5 - increases data broadcasting capacity by 800%

in the context of source data formation (more parameters have to be assessed rather than just the number of papers on the relevant subject).

3. According to the results of the analysis of the development characteristics of the Bluetooth technology, it is important to note the specifics of the mentioned process, which lies, in the first turn, in the possibility of a cyclic repetition of growth stages of subjective and objective expectations. For example, according to the actual events, every launch of a new standard of the Bluetooth technology leads to growing expectations, part of which can remain unsatisfied. Moreover, the modern development of telecommunication technologies, including meeting subjective expectations, is characterized by a high degree of interrelation with the development of information and communication technologies as a whole. For example, some of the overstated subjective expectations, which have not been implemented due to data transfer security problems, can be put into practice if new coding technologies are developed.

Nevertheless, despite all the above circumstances, quite high values of determination coefficient R2, when data are approximated in terms of the development process of the Bluetooth technology for dependences of the SEI and OEI, bear evidence that the analyzed model can be used to describe development process of a wide range of technologies in the spheres of manufacturing, communication, transport, etc.

References

1. Gartner. Inc. Gartner Hype Cycle, 10 April 2018. https://www.gartner.com/technology/research/methodologies/hype-cycle.jsp
2. Nikolov, R., Shoikova, E., Krumova, M., et al.: Learning in a smart city environment. J. Commun. Comput. 13, 338–350 (2016)
3. Ukwuani, N., Bashir, E.: Emerging technologies: an exploration of novel interactive technologies. Int. J. Inf. Syst. Serv. Sect. 9, 30–43 (2017)
4. Palikaras, G., Kallos, E.: The gartner hype cycle for metamaterials. In: 8th International Congress on Advanced Electromagnetic Materials in Microwaves and Optics – Metamaterials, pp. 397–399 (2014)
5. Siow, K.S.: Graphite Exfoliation to Commercialize Graphene Technology. SainsMalaysiana 46(7), 1047–1059 (2017)
6. Steinert, M., Leifer, L.: Scrutinizing Gartner's hype cycle approach. In: IEEE Proceedings of Technology Management for Global Economic Growth, pp. 254–266 (2010)
7. Fenn, J., Raskino, M.: Mastering the Hype Cycle: How to Choose the Right Innovation at the Right Time. Harvard Business Press, Boston (2008)
8. Kim, J., Hwang, M., Jeong, D.H., Jung, H.: Technology trends analysis and forecasting application based on decision tree and statistical feature analysis. Expert Syst. Appl. 39(16), 12618–12625 (2012)
9. Leventsov, V., Radaev, A., Salkutsan, S.: Mathematic model of production technology transformation. In: SHS Web of Conferences (in press)
10. Scopus Database, 10 April 2018. https://www.scopus.com
11. The Bluetooth Special Interest Group (SIG), 10 April 2018. https://www.bluetooth.com

Resource Allocation for the Provision of Augmented Reality Service

Maria Makolkina[1,2(✉)], Alexander Paramonov[1], and Andrey Koucheryavy[1]

[1] The Bonch-Bruevich Saint-Petersburg State University of Telecommunications,
St. Petersburg, Russia
makolkina@list.ru, alex-in-spb@yandex.ru, akouch@mail.ru
[2] St. Petersburg School of Economics and Management, National Research
University Higher School of Economics, St. Petersburg, Russia

Abstract. This paper will look at the method for selecting the structural parameters of a service system when providing augmented reality service is proposed. The particular features of service provision, the quality of its perception are considered and the quality time indicators assessment model is proposed. As optimization goal, the formulation of the service system resources allocation task is given.

Keywords: Augmented Reality · Communication service
Data transmission · Quality of Service · Quality of Experience
Service system

1 Introduction

Augmented reality services are the next step in the development of mobile technology services. The combination of mobility properties of the terminal, its computing capabilities, the ability of interacting with the environment (discernment of video, sound and tactile images, coordinates calculation and orientation in space), as well as the potential of a modern communication network, allow a quality realization of a new level services with a high degree of interactivity [1–3]. In particular, they are augmented reality services (AR – Augmented Reality) today, such services as interactive maps of cities and settlements, the starry sky, various kinds of guidebooks, applications for ordering goods and services are already widely known and popular [4–6].

These services give the user the opportunity to receive necessary information in a timely manner, and the selection of the necessary information is performed automatically based on the user status data. For example, the position of the user in space (geographical coordinates), on the map and in location of the territory, based on the geolocation data, the position of the vehicle, etc. As shown in [5,15], the implementation of the service requires the organization of data exchange with the service server and/or directly with devices located in the communication zone of the user terminal, using D2D technologies [7,12]. Thus,

© Springer Nature Switzerland AG 2018
O. Galinina et al. (Eds.): NEW2AN 2018/ruSMART 2018, LNCS 11118, pp. 441–455, 2018.
https://doi.org/10.1007/978-3-030-01168-0_40

the time between the request and data delivery should not exceed a certain value at which the user does not feel yet a decrease in the quality of the service. That time is determined by a number of components: the request generating time (it depends on the service implementation), request delivery time from the terminal to the service server, request processing time, data delivery time from the service server to the terminal information presentation time to the user. They can be divided into three groups: time defined by data processing of user's terminal, data delivery time over the communication network and data processing time by the server. These components, in general are mutually dependent.

An essential role is played by the request generating process. The request is generated when the user's environment changes (or user's status) which can be determined by changes in some parameters. Such parameters can be sensor data, for example, geographic coordinates, terminal position in the area, speed, as well, the results of analysis of the image or sound obtained from the terminal cameras and microphones [8–10]. For example, if a data request is generated from the image recognition results (video filmed by the terminal camera), then image discernment functions can be implemented either in the terminal application or in the service server. In the first case, if computing performance of the terminal is low, the time will be spent on performing discernment functions by the terminal, in the second case, on video transfer through the communication network and the server video processing time. Clearly, the choice of the first or second option depends on the terminal's performance, network bandwidth, server loading and performance. Thus, the task is the choice of the optimal variant of service implementation. The model described in this example can be extended by introducing additional parameters. The dependence of the request processing time on data volume (database size) and request intensity can be taken into account. In this case, data clustering and organization of local service servers make sense.

Considering the fifth generation network perspective, D2D communication technologies, as well as SDN application, the service implementation structure shown in Fig. 1 is proposed.

Suppose that the communication network is built using SDN architecture with data centers of different levels [11, 13], what makes traffic possible and data localization "closer" to users. In the scheme, these data centers are depicted in micro, mini, and basic clouds. In a real network such levels can be as much as it will be necessary for the best services result. The network base station interacts with the AR terminal directly or through a mobile terminal, performing local cloud role that interacts with the AR terminal using D2D technologies, what increases the efficiency of using the radio frequency spectrum [12]. Here, the cloud is understood as a certain volume of computing and memory resources, which can be used to organize the server and database services. As shown below, service provision can be implemented at several levels of such servers and databases, what reduces network bandwidth requirements and improves the quality of service presentation due to data and traffic localization.

In this paper, the task of data clustering and localization of data processing is considered as resource allocation goal.

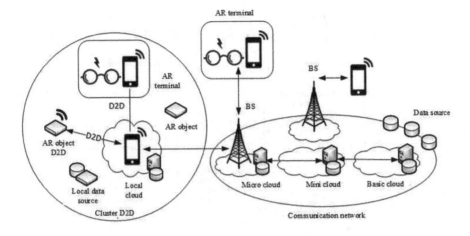

Fig. 1. Eventual service implementation structure

2   Service Model

For service model realization, it is necessary to associate the indicators (parameters) characterizing the quality of its provision with communication system parameters. As the basic indicator of quality, we will choose the response time to user's environment change τ. Let's assume that this time it includes all the components: changes in recognition time and the mobile terminal application preliminary processing t_r, data (request) transmission time to the service server through the communication network t_q, request processing time by the service server t_s, data delivery time through the communication network t_a and information presenting time to the user by the mobile terminal application t_d. The augmented reality services provision model is shown in Fig. 2.

Overall time can be represented by the arithmetic sum of all the components. We assume that each of the components is a random variable, then, assuming that they are self-sufficient, the average value of the response time will be determined as:

$$\overline{\tau} = \overline{\tau}_r + \overline{\tau}_q + \overline{\tau}_s + \overline{\tau}_a + \overline{\tau}_d \tag{1}$$

Let's consider each of the components separately.

Recognition time of the change in the user environment t_r, in it queue, it includes all the components associated with the detection of changes and information collection required for a request generation directed to the service server. Changes can be detected by analyzing data from various sensors and devices (geomagnetic field sensor, luminosity, acceleration, global positioning system signals receivers, the touch screen, etc.), as well as microphones and video cameras. This analysis can include, as relatively simple tasks of comparing several numerical values, and resource-intensive discernment tasks. Thus, the numerical value of t_r depends on the type of service, its implementation, and the computing resources of the mobile terminal. Thus, the mobile terminal resources affect the

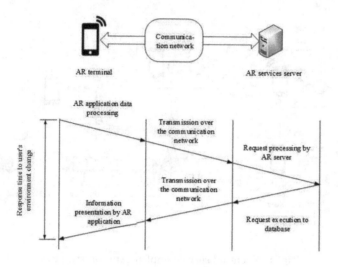

Fig. 2. AR services provision model

quality of the service through the t_r value. We will assume that there is a certain functional relationship between this time value and the mobile terminal computing resources:

$$\bar{t}_r = f_r(O) \tag{2}$$

where O – a parameter characterizing the mobile terminal performance, for example, the number of operations or commands performed per second, CPU clock speed, the capacity of memory, or some complex indicator.

Data transmission time to the service server t_q, is determined by the capacity of transmitted data and the throughput of the route between the mobile terminal and the service server C. Numerical value of this time, assuming that the time is only spent on data transmission, i.e. without taking into account the loss when waiting for transmission in the nodes on the route can be obtained as:

$$\bar{t}_q = f_q(C) = \frac{\bar{v}_q}{C} \tag{3}$$

where C – throughput (bit/s), $\bar{v}_q$ – the average capacity of data sent in the request (bit).

Average capacity of data in the request $\bar{v}_q$ depends on the type of service and the way it is implemented. For example, if an analysis of the images received from the device's video camera is required to identify the environmental change, then this analysis can be performed both by the mobile terminal application and by the server. In the first case, the request will contain relatively few data, which are only the identifiers of objects in the service database, information about which need to be provided. In the second case, it is necessary to transfer all the image data (or several images), required to be analyzed by the service server. An intermediate option is also possible, when only part of the video data is transmitted to the server.

Request processing time by the service server t_s, is the most complex characteristic as it depends on many parameters: analysis time of the request data τ_s, incoming data; intensity of requests from mobile user terminals λ_s; server performance μ_s, which, in turn, depends on database size n_s:

$$\overline{t}_s = f_s(\tau_s, \lambda_s, \mu_s(n_s)) \tag{4}$$

The server serves requests from multiple users; service time is determined by the request processing and the queue time. The server can be described by the Queuing system model, in which the request processing time is determined by the database size and the server performance. Technical implementation of the server can be different, so in this task it makes sense to consider it as a queuing system that have a general waiting distribution and there is a single server G/G/1. Assuming that the incoming requests flow can be described by the simple queue model (M/G/1), the average delay can be described by the Pollaczek-Khinchin formula [14], considering this:

$$\overline{t}_s = \frac{\rho_s}{\mu_s(v_s)2(1-\rho_s)}(1+V_s^2) + \frac{1}{\mu_s(v_s)} \tag{5}$$

where $\rho_s = \frac{\lambda_s}{\mu_s(v_s)}$ – server load, $V = \sigma_s\mu_s(v_s)$ is the coefficient of service time variation, σ_s^2 – standard deviation of service time.

Assuming the simple queue model is very useful, as it makes possible in many cases to obtain analytical expressions for the dependencies, especially when properties of the real flow are unknown. Server performance, which depend on database size $\mu_s(n_s)$, also presents a certain dependence, and is determined by the database implementation. In particular, the most common models describe this dependence as $\ln(n_s)$ or $n\ln(n_s)$ operations [15], where n_s the number of entry in the database. Let's take as an example the logarithmic dependence. Taking into account that the service time includes preliminary request processing, that have τ_s as the average execution time, we have:

$$\mu_s(v_s) = \frac{1}{\eta\ln(v_s) + \tau_s} \tag{6}$$

where η – execution time per entry.

Server response time t_a, as well as request transmission time, is determined by the volume of transmitted data and the throughput between the service server and the mobile terminal C. A numerical value of this time, under similar assumptions, can be obtained as:

$$\overline{t}_a = f_a(C) = \frac{\overline{v}_a}{C} \tag{7}$$

where C – throughput (bit/s), $\overline{v}_a$ – the average volume of data sent in the server's response (bit).

The average volume of data sent in the server's response $\overline{v}_a$ depends on the services type and its implementation. Transmitted data can be text, raster or vector graphics, sound, numerical values.

Message presentation time t_d includes all the components associated with data processing and presentation received by the mobile application. In general, a message can be presented visually: in text form, icon, video or other image; sound: speech or melody, tactile-vibration. Lets assume that there is a certain functional relationship between this time value and the mobile terminal computing resources:

$$\bar{t}_d = f_d(O) \tag{8}$$

As it can be seen from the models chosen above, the response time depends significantly on such parameters as, the mobile terminal performance, communication network bandwidth, and the request processing time by the server, determined by server load and performance.

In this paper, a method for selecting the equipment structure and parameters for ensuring the response time requirements for the Augmented Reality service was proposed.

3 Goal Formulation

Taking into account the above models, it can be seen that ensuring an acceptable response time is a task of choosing resources volume (bandwidth, performance and memory), as well as their distribution in the service system elements. Taking into account the above models, it's a task with several variables determined by the models. From a position of method construction for the service organization, all these variables are not accessible for modification. For example, if we assume that the mobile terminal performance and server hardware characteristics can be taken into account, however, they cannot be modified within this task, and then the expression for the response time can be as follow:

$$\bar{\tau} = \bar{t}_m + \frac{\bar{v}_q + \bar{v}_a}{C} + \frac{\rho_s}{\mu_s(n_s)2(1 - \rho_s)}(1 + V_s^2) + \frac{1}{\mu_s(v_s)} \tag{9}$$

where $\bar{t}_m = \bar{t}_r + \bar{t}_a$ – the total delay introduced by the mobile terminal application when processing input data and displaying information, we assume. This value depends on the terminal's performance and application features; we will assume that this value is constant.

$$\bar{\tau} = \bar{t}_m + \frac{\bar{v}_q + \bar{v}_a}{C} + \frac{\lambda_s}{\mu_s(n_s)2(\mu_s(n_s) - \lambda_s)}(1 + V_s^2) \tag{10}$$

Figure 3 shows the dependency obtained by (10) for a different number of entry in the database.

It can be seen from the dependency that the response time increases according to the law $1/(a - x)$. The constant time component is the time of request data transmission to the server, as well as the server processing time. According to the chosen model (6), the service time also depends on the number of entry in the database.

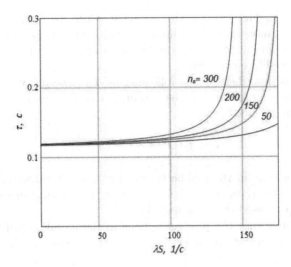

Fig. 3. The dependency of response time on the requests intensity and the size of the database

This, for a better service organization, it is necessary to provide the required response time. For this, based on the proposed models (9) and (10) the service delivery network structure is necessary to be chosen, taking into account the traffic, data delivery time, and the volume data in the database.

4   Processing of Environment Data

As noted above, the request generation is based on the results of environment change discernment. This discernment can be based on various data, both data from different sensors, or data obtained from video cameras and microphones. In the second case, the discernment task can have considerable computational complexity; therefore, the time spent can be quite significant. Thus, it makes sense to choose a solution to this problem: a mobile terminal or a service server. If the image is processed completely at the terminal, the request sent to the server contains only a relatively small data volume necessary for the object identification in the server database. In the case when the image is completely transmitted to the server for processing, a transmission of a large volume of data is required, determined by the camera resolution and the data presentation format. Intermediate solutions are also possible, for example, when the terminal application does not generate fully object identification, but select an object (useful data) in the image. In this case, only the selected part of the image is transmitted to the server for further processing. Thus, leaving the objects discernment task to the user terminal saves the data transmission time over the channel and the bandwidth resource. Leaving this task to the server allows saving the image processing time in the terminal but, leads to an increase in transmission delay and bandwidth consumption. Taking in account the use of network resources, the first option is more profitable, however, when implementing the service, the real terminal performance, the real network bandwidth and the response time requirements should be considered.

The above is described by the following model. The delay component conditioned by image processing in the terminal and in the server, and the data transmission time defined as:

$$\tau_P = f_r(O_r) + \frac{\bar{v}}{C} + f_S(O_S) \tag{11}$$

O_r – processing time in the mobile terminal at one bit calculation, O_S – processing time on the server at one bit calculation, C – data transfer rate (bit/s), $\bar{v}$ – volume of processed (transmitted) data.

We will assume that the processing time in the mobile terminal and in the server linearly dependent on the size of the processed data block (image or part of it).

$$f(O) = \bar{v}O \tag{12}$$

where O – image processing time at one bit calculation, typical for a mobile terminal or server (for a mobile terminal and a server, these values may vary significantly).

Then:

$$\tau_P = \eta\frac{\bar{v}}{O_r} + (1 - \eta_r)(\frac{\bar{v}}{C} + \frac{\bar{v}}{O_S}) \tag{13}$$

where η_r – part of data processed in the mobile terminal.

From (12) it's clear that the increase of η_r leads to the decrease of τ_P when the processing time in the mobile terminal is less than the sum of the transmission time and processing time in the server.

$$\frac{\bar{v}}{O_r} < (\frac{\bar{v}}{C} + \frac{\bar{v}}{O_S}) \tag{14}$$

Guided by (14) the service system can distribute the data processing functions between the server and the terminals, for example, depending on the server load.

5   Forming and Updating Data

The data used for the service may have a different provenance and location in the network. Data providers can be different people, organizations and technical systems that create or provide information support in areas that are accessible to the user of the service. Based on the request data generated by the subscriber terminal, the service server search for necessary data using the appropriate software. As was shown above, such searching systems (servers) can be several.

Each of these systems has its own database, which stores the most popular data, this reduces the time required to deliver data to the terminal. For example, a client-level system can be located on a user terminal, and its database can contain data about the current client environment. They are data about objects that can be identified based on the indications of the user terminal sensors. For example, data about objects located in close proximity to the user, what is assessed from information about the geographical location of the user (terminal).

The functionality of AR service searching systems is similar to the existing systems functionality, except specific service provision, which is determined by a composition of features that determine search realization and search results presentation. At this phase, the following requirements for such a searching system can be formulated:

- The ability of searching by such characteristics as geographical coordinates local coordinates (inside buildings), graphic and text objects identifiers (Bar and QR codes, text denotation), fixed and moving images of the objects (searching for graphics), speech and sounds (speech, music and other sounds discernment). For this realization, these characteristics need be sufficiently formalized, and their formation methods appropriate to mobile terminal or service server application need to be determined
- The ability to classify data about search objects by object function, application field, service type, geographic location, attachments, presentation type, data source type. For example, one object can be associated with several data blocks provided by different sources; in this case a method that selects the required data block, which corresponds to the parameters of the provided service, is required.
- Ability to select the data representation form, for example, text message, image or video, pictogram, sound or voice message, etc.

Data storage time in local databases should be determined based on their relevance and requirements to the quality of service (search time).

The main functionality of the AR service, as noted above, is the provision of the necessary information according to request data. Clearly, with this general formulation, the task is similar to that implemented by the search engines on the global Internet. However, in the case of AR service, it is necessary to take into account provided data and their requests particularities. The service organization principle itself determines, for example, such particularities like the correlation between the most requested contents and the user's environment information. Therefore, the probability of the server to receive a request to a specific information block depends on the object geographic location that matches with that information block and the user geographic location. It is fair to assume that the closer they are to each other, the higher this probability. This probability, in fact, determines the part of traffic generated by users on the server (service servers). From the observation above, it can be seen that, the AR information demand is different and depends on various factors, at least geographical.

This allows the AR data clusterization (localisation), Fig. 4.

At the same time, the data cluster, can be physically located in the server database, which in turn is geographically close to potential users. Of course, in this case, the communication network resource is saved. This approach can

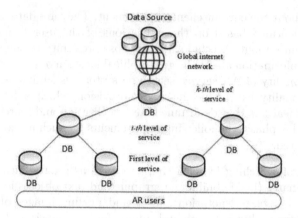

Fig. 4. The data allocation structure in the AR service system

be implemented by organizing several user requests service levels and data and traffic distribution.

6   Load and Data Distribution

A hierarchical structure that includes several levels of service can be organized for response time reduction, by reducing the load on the service server. Each level server is accessible to a different number of users. For example, first (lower) level server can be organized right in the mobile terminal and have a single user. An access to a higher-level server occurs in case when the required information is not found in the lower-level server database. The server database of each level contains information about each user environment for which this server is available, as well as the most demanded information by users, and the probability of this demand is p, Fig. 5.

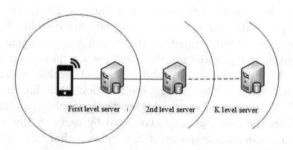

Fig. 5. Hierarchical structure of service provision

The service provision model in this case will be the following. If the environment change is detected, the user terminal transmits data (possibly an already formed request) to the first level service server. The server processes the data and the request. After a successful processing and data availability, the server sends a response to the user terminal. If for some reason the request is not executed by this level server, the request is sent to the next level server, and so on. The reason of the failure in this model is the lack of the necessary data at the current service level. The construction of the hierarchical service model allows, through traffic and data distribution between levels, to provide the required quality indicators for the services provision.

Request service time in a network with multiple servers can be described as:

$$\tau = \bar{t}_r + \sum_{j=1}^{k} p_j \sum_{i=1}^{j} (\bar{t}_q + \bar{t}_s) + \bar{t}_a + \bar{t}_d \tag{15}$$

where p_j – the probability that the requested data is in the j-th database server, k – number of levels.

Or taking into account the model (5):

$$\tau = \bar{t}_m + \sum_{j=1}^{k} p_j \sum_{i=1}^{j} (\frac{\bar{v}_q + \bar{v}_q}{C} + \frac{\rho_s}{\mu_s(n_s)2(1 - \rho_s)}(1 + V_s^2) + \frac{1}{\mu_s(v_s)}) \tag{16}$$

When organizing the service on an SDN network, its functionality can be used to dynamically manage the service by changing the number of service levels, i.e. increase or decrease k in the expressions (15) and (16).

The criterion for making a decision in this task is the response time value $\bar{\tau}$, truly its value in comparison with some normative value τ_0, which is the most acceptable value in the implementation of the service. Clearly, its normative value can be 0 (zero), However, it is also clear that this goal is unrealizable. Reducing data processing and delivery delays can be associated with significant financial costs, thus, the eligibility of τ_0 value is advisable to be considered as the maximum value, when the desired quality of service (QoS) and the quality of user perception of the service (QoE) are provided.

Thus, in this model, service management goal is to maintain a possible proximity between the real response time value and its normative value, i.e. ensuring $min\,|\bar{\tau} - \tau_0|$.

Then the objective function of this task can be written as a minimum of the difference between the response time value and the normative. The minimum of the difference can be expressed in terms of the minimum of the difference square, in this case let's use Ordinary Least Squares method.

Then the problem can be formulated as an optimization problem with the objective function:

$$\{k, p_i\} = \arg\max_{k, p_i} \left\{ \sum_{j=1}^{k} (\bar{\tau} - \tau_0)^2 \right\} \tag{17}$$

and constraints $k \in N$, $k \le k_{max}$, $0 \le p_i \le 1$, $\bar{\tau} > 0$, $\tau_0 > 0$ where $\bar{\tau}$ – is defined by expressions (15) or (16), response time normative value, k_{max} – maximum number of service levels allowed. It is worth noting that the expressions (15) and (16) are only possible models for describing the time parameters of the service. In their place, other analytical and simulation models that allow an adequate assess of needed parameters.

The above task (17) is formulated as the goal of finding the optimal number of service levels k and values p_i, which determine the composition of data in the database, and the traffic volume at each service levels.

In fact, data composition in the server database i level can be determined according to the rule, when the data block is stored in the database, If a part of requests exceeds the value p_i. In fact, it means that at this level of service the traffic generated from these requests will be closed.

Figure 6 shows the results obtained from a simulation of a three-level service model. The model included traffic source imitating a Simple Queuing Model and three queuing systems with waiting time simulating each of the service levels, and a database model was created for each of the service levels. When a request is received, a search of the corresponding entry is produced in the database, when the entry is found, the service time is equated to the composition of viewed number of entries and the time required for viewing one entry, and then request serving is considered complete. If there is no entry in this database level, the serving time is calculated analogically, but the request is transferred to the next service level, where the similar search process is performed, and then when the entry is found, the request is considered to be served, and the entry is also entered in the lower-level database. The entry storage time in the database is given by a certain constant time value, if there is no request made to the entry during this time value, the entry is deleted from the database. Deleting an entry from a higher-level database is not performed.

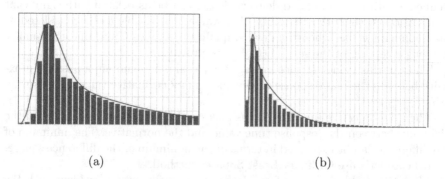

(a)   (b)

Fig. 6. The density of serving time probability of the first service level and all levels (based on simulation results)

The serving time at each level depends on the size of the database, which in turn is determined by the entry storage time. As this time increases, the database size and application serving time increase at this level, but decreases the probability of transferring the request to next levels. Figure 6 shows the empirical histograms of the density of serving time probability of the first service level (Fig. 6a) and for the whole system in general (Fig. 6b). The obtained histograms differ from the Erlang distribution and in this case are described by the sum of two gamma distributions.

7   Conclusion

1. The implementation of the augmented reality service involves a service system that performs data processing, transmitting, selection, storing functions and information presentation to the user. Performing each of these functions requires time resources, network bandwidth, server and memory capacity.
2. Information support of the augmented reality service can be formed in various ways, including the basis way of the search for information provided in the global Internet. Search and storage of data is performed by a service system that can have several processing levels, and the number of these levels influences the volume resources used (network bandwidth, server capacity and memory).
3. The main indicator of the quality of service provision is the response time, i.e. The time value, from the moment when the user's environment changes to the moment when the user receive the required message. This time depends on the distribution of the service delivery functions to the executive elements (user terminal, AR servers, data network link). The target value of this time should not exceed the response time of the user to the AR presented message.
4. To ensure the quality of the perception of the AR service, the resources of the mobile terminal can be used, what reduces data transmission time, as well as SDN resources that allow the implementation of a hierarchical service delivery model. The hierarchical model allows the localization a significant volume of data and traffic, what will save bandwidth resources of the communication network.
5. Parameters of the hierarchical service delivery model include a number of hierarchy levels and the probability of accessing each of these levels. The probability of accessing a certain level can be used as a criterion for the database organization.
6. Selecting the number of hierarchy levels in the service delivery model and the access probability to each of them is an optimization task whose goal is to provide the response time close to the given value. The solution of this problem allows obtaining the structural parameters of the service system based on the user traffic.

References

1. Makolkina, M., Paramonov, A., Vladyko, A., Dunaytsev, R., Kirichek, R., Koucheryavy, A.: SDN VANETs in 5G: an architecture for resilient security services. Adv. Intell. Syst. Res. **134**, 153–157 (2017)
2. Hussein, A., et al.: The use of UAVs, SDN, and augmented reality for VANET applications. In: Proceedings of the Fourth International Conference on Software Defined Systems (SDS), pp. 67–74 (2017)
3. Makolkina, M., Kirichek, R., Teltevskaya, V., Surodeeva, E.: Research of interaction between applications of augmented reality and control methods of UAVs. In: Koucheryavy, Y., Mamatas, L., Matta, I., Ometov, A., Papadimitriou, P. (eds.) WWIC 2017. LNCS, vol. 10372, pp. 186–193. Springer, Cham (2017). https://doi.org/10.1007/978-3-319-61382-6_15
4. Billinghurst, M., Clark, A., Lee, G.: A survey of augmented reality. Found. Trends Hum. Comput. Interact. **8**(2–3), 73–272 (2015)
5. Leppänen, T., Heikkinen, A., Karhu, A., Harjula, E., Riekki, J., Koskela, T.: Augmented reality web applications with mobile agents in the internet of things. In:2014 Eighth International Conference on Next Generation Mobile Apps, Services and Technologies (NGMAST), pp. 54–59. IEEE (2014)
6. Volkov, A., Khakimov, A., Muthanna, A., Kirichek, R., Vladyko, A., Koucheryavy, A.: Interaction of the IoT traffic generated by a smart city segment with SDN core network. In: Koucheryavy, Y., Mamatas, L., Matta, I., Ometov, A., Papadimitriou, P. (eds.) WWIC 2017. LNCS, vol. 10372, pp. 115–126. Springer, Cham (2017). https://doi.org/10.1007/978-3-319-61382-6_10
7. Makolkina, M., Vikulov, A., Paramonov, A.: The augmented reality service provision in D2D network. Commun. Comput. Inf. Sci. **700**, 281–290 (2017)
8. Makolkina, M., Koucheryavy, A., Paramonov, A.: The models of moving users and IoT devices density investigation for augmented reality applications. In: Galinina, O., Andreev, S., Balandin, S., Koucheryavy, Y. (eds.) NEW2AN/ruSMART/NsCC -2017. LNCS, vol. 10531, pp. 671–682. Springer, Cham (2017). https://doi.org/10.1007/978-3-319-67380-6_64
9. Makolkina, M., Koucheryavy, A., Paramonov, A.: Investigation of traffic pattern for the augmented reality applications. In: Koucheryavy, Y., Mamatas, L., Matta, I., Ometov, A., Papadimitriou, P. (eds.) WWIC 2017. LNCS, vol. 10372, pp. 233–246. Springer, Cham (2017). https://doi.org/10.1007/978-3-319-61382-6_19
10. Andreev, S., Galinina, O., Pyattaev, A., Johansson, K., Koucheryavy, Y.: Analyzing assisted offloading of cellular user sessions onto D2D links in unlicensed bands. IEEE J. Sel. Areas Commun. **33**(1), 67–80 (2014)
11. Makolkina, M., Paramonov, A., Vladyko, A., Dunaytsev, R., Kirichek, R., Koucheryavy, A.: The use of Uavs, SDN, and augmented reality for VANET applications. In: 3rd International Conference on Artificial Intelligence and Industrial Engineering (AIIE 2017), "DEStech Transactions on Computer Science and Engineering", pp. 364–368 (2017)
12. Paramonov, A., Hussain, O., Samouylov, K., Koucheryavy, A., Kirichek, R., Koucheryavy, Y.: Clustering optimization for out-of-band D2D communications. Wirel. Commun. Mob. Comput. **2017** (2017). Article ID 6747052, 11 p. https://doi.org/10.1155/2017/6747052

13. Vybornova, A., Koucheryavy, A.: Traffic analysis in target tracking ubiquitous sensor networks. In: Balandin, S., Andreev, S., Koucheryavy, Y. (eds.) NEW2AN 2014. LNCS, vol. 8638, pp. 389–398. Springer, Cham (2014). https://doi.org/10.1007/978-3-319-10353-2_34

14. Iversen, V.B.: Teletraffic Engineering and Network Planning. Technical University of Denmark (2010). http://www.osti.gov/eprints/topicpages/documents/record/982/1473132.html

15. Ateya, A., Muthanna, A., Gudkova, I., Abuarqoub, A., Vybornova, A., Koucheryavy, A.: Development of intelligent core network for tactile internet and future smart systems. J. Sens. Actuator Netw. **7**(1), 1 (2018)

Development of the Mechanism
of Risk-Adjusted Scheduling and Cost
Budgeting of R&D Projects
in Telecommunications

Sergei Grishunin[1,2]([⊠]), Svetlana Suloeva[1]([⊠]),
and Tatiana Nekrasova[1]

[1] St. Petersburg State Polytechnical University, St. Petersburg, Russia
sergei.v.grishunin@gmail.com, emm@spbstu.ru,
dean@fem.spbstu.ru
[2] Risk Advisory, Deloitte and Touche, Moscow, Russia

Abstract. We developed the mechanism of risk-adjusted scheduling and cost budgeting of research and development (R&D) projects in telecommunication. The relevance of this topic is explained by growing complexity and uncertainty of innovation activity; high importance of cost and time as key metrics of R&D projects' performance. The paper addresses decreasing efficiency of existing mechanisms which poorly incorporate risks and fills the gaps in the research in this area. Results included development of the concept of the mechanism, its block diagram, the specification of its comprising tools, the step-by-step description of its phases. Unlike the "conventional" peers, the developed mechanism allows timely identification of uncertainties; facilitates robust and transparent evaluation of risks; focuses management efforts on key threats; and ensures remediation of risks earlier in the process thus improving speed and quality of decision making. These advantages let us conclude that the suggested mechanism should have a significant practical use.

Keywords: Telecommunications · Investment controlling · Risk controlling
Project management · Project cost and schedule management
Risk-adjusted budgeting

1 Introduction

Telecommunication sector continues to stay a critical force for growth and innovation across multiple industries. On the other hand, it experiences the growing instability and uncertainties coupled with increasing cost and shortening time of innovations. The complexity of risks is increasing; the speed of their onset is growing while their impact on research and development (R&D) projects becomes more severe [1, 3].

In such conditions, executives of telecommunication companies (telcos) responsible for innovations demand the enhancements to R&D cost and schedule management systems. "Conventional" systems [5, 8, 12] fail to keep R&D costs and time within the limits of business plans with the acceptable precision. The solution is the development

© Springer Nature Switzerland AG 2018
O. Galinina et al. (Eds.): NEW2AN 2018/ruSMART 2018, LNCS 11118, pp. 456–470, 2018.
https://doi.org/10.1007/978-3-030-01168-0_41

and implementation of risk-adjusted project cost and schedule management systems. They are based on principles, methods and tools of investment controlling and risk controlling [2, 22] and ensure that critical project risks are timely identified and assessed while efficient risk mitigation decisions are made.

In the paper, we develop one of the critical modules of risk controlling-based complex cost and time management system - the mechanism of risk-adjusted scheduling and cost budgeting. The relevance of this topic is explained by: (1) high importance of cost and time as the key metrics of innovation process efficiency; (2) significant difficulties in prediction of project expenditures and time; and (3) growing role of budgeting module as a feedforward project control. In turn, cited in the literature attempts to integrate risk management into these mechanisms [10, 13, 15, 17, 18] are incomplete, discrete and do not provide the systemic view on risk management.

The novelty of the paper is driven by the advantages of developed mechanism over its "conventional" peers. It allows (1) timely identification of uncertainties (2) facilitating more robust and transparent evaluation of risks; (3) focusing management time and efforts on key risks and cost drivers; and (4) performing remediation of risks earlier in the process thus improving speed and quality of scheduling and budgeting.

In the Sects. 1 and 2 the paper presents the outlook for the global telecommunication industry and the literature review of risk-adjusted scheduling and cost budgeting. Section 3 explains the role of risk controlling in R&D project management and provides the reference model of complex R&D expenditures and time management system based on principles of investment controlling. In the Sects. 4 and 5 we develop the mechanism of risk-adjusted scheduling and cost budgeting of R&D project for telcos and formulate the conclusions. Theoretical and methodological basis was the research of Russian and foreign academics and business practitioners in R&D cost management.

2 Telecommunication Industry Outlook

The rapid development of technologies erases the boundaries between communication and information technologies. Nowadays, these two industries are merging into a single info-communication industry (ICT) [1]. In 2018–2023 telecommunication companies (telcos) will continue experiencing the headwinds and fierce competition from the disruptive technologies such as over the top content providers or artificial intelligence. Penetration of smartphones among adults will surpass 90% by 2023 from 85% currently. This growth will be fueled by the introduction of an array of innovations that are largely invisible for users but whose combined impact will be tangible in form of improved functionality and performance (the hardware and software), better entertainment ability (augmented and virtual reality) and deeper usage in a business context. The progress of smartphones will be also driven by penetration of 5G networks offering greater capacity and connectivity speed. Another new opportunities for telcos will come from (1) artificial neural networks, (2) machine learning and the associated hardware; (3) digital media; (4) in-flight connectivity; (5) biometrics; and (5) internet of things. To survive, telcos will continue to expand into these areas from traditional voice telephony and SMS business [7].

The industry outlook in 2018–2023 depends on geographies. In Europe, the Middle East and Africa (EMEA) telcos will sustain low revenue growth of around 1% annually; Central and Eastern Europe telcos is likely to post around 3% in this period on the back of decent macroeconomic growth. Russia's telcos revenue and EBITDA growth will turn positive; this growth will be in line with the growth of country's GDP of around 1.5%–2% annually. These companies, however, remain subject of geopolitical risks. In Asia-Pacific, year-on-year revenue growth is expected around 2%; the positive impact from GDP growth (of around 3%–4% annually) will be compensated by growing competition. Lastly, in the USA in this period, the revenue will stay flat while profit and cash flow may decline by single percentage digit due to price wars among players [6].

The challenges outlined above require telcos to sustain high level of capital expenditures (we project investments to remain between 15%–20% of revenue in the next 5 years) [3]. The second important target for investments after the core network infrastructure in ICT is the research and development (R&D) of new products and services. In the next chapter we will explain why budgeting of R&D cost has recently become a critical issue for innovation projects without exception of those from ICT industry.

3   Risk-Adjusted Cost and Schedule Management: A Literature Review

The development of efficient mechanisms of R&D cost and schedule management has been in focus of research both domestically and globally [5, 8]. This is underpinned by (1) growing complexity of innovation activity and increasing volatility of project environment; (2) short time-to-the market resulting to "faster-better-cheaper" philosophy; (3) the high costs of R&D projects coupled with low probability of R&D projects' success[1]. The latter is explained by various risks caused by (1) new technologies failures; (2) design errors; (3) non-adoption of products by customers; (4) vendors failures; (5) macroeconomic uncertainties and external shocks; (6) regulatory restrictions; (7) human resource issues; (8) actions of competitors; (9) operational issues; (10) coordination failures [1, 3].

However, in such business environment, "conventional" complex R&D cost and schedule management systems [2, 5, 8, 12, 14] fail to keep project costs/time within the forecasts/schedules with the suitable precision. These systems usually reported variances of actual project expenditures/time vs budgets/schedules of around 15%–40%. This magnitude of variances is above the acceptable levels nowadays [14].

The low efficiency of "conventional" systems is underpinned by that the scheduling and cost budgeting mechanisms in such systems do not serve as true feedforward controls. Their expenditures budgets and projections of duration are largely based on

---

[1] Research performed by Dos Santos [9] revealed that in the United States 80% of R&D projects failed before completion while among the survivors 49% exceeded expected costs; almost 63% of these projects ran late.

historical single-point estimates, standards and metrics. Their sensitivity analysis and stress testing approaches focused on single variance; their assumptions and cost standards are often based on poorly verifiable statistics/expert views as almost every R&D project is unique. Additionally, the feedback and concurrent controls in these systems often provide the late reaction due to rapid onset of the risks. Research shows: the efficiency of complex R&D cost and schedule management systems can be significantly improved if these systems integrate full-cycle risk management processes.

However, literature review shows that existing attempts of integration risk management into cost and schedule management systems are incomplete, discrete and isolated. They do not provide the systemic view on risk management. Researches are concentrated on developing of separate tools and techniques, for example: (1) analysis of internal and external environment and identification of risks [8]; (2) risk assessment techniques with various tools including project risk failure mode and effect analysis (RFMEA) [13, 15] or (3) risk-adjusted cost scheduling and budgeting [10, 17, 18].

We recommend building a complex approach which unites the above-mentioned tools and methods into the integrated risk-adjusted scheduling and cost budgeting mechanism. Such mechanism can be built on the principles and with the methodologies and tools of risk controlling [22].

4 Investment Controlling in Innovation Project Management

Investment controlling (IC) is an application of methods of controlling [2] to project management, a combination of processes, methods, skills and tools which ensuring the achievement of projects' goals in an uncertain and rapidly changing business environment [19]. In telcos, it is applied mainly for projects related to R&D of new technologies, products or services. Our own research and analysis of others' research has demonstrated that implementation of investment controlling, despite the cost of its implementation, allows to decrease the deviation of actual time spent and costs versus initial plans by around 50% [2, 19] and to range of 7%–20%. It helps to achieve earlier innovation project payback as well as gain competitive advantage to the firm.

Consequently, the risk controlling (RC) is an inherent part of all organizational processes of IC. It is a goals-oriented set of methods, processes and tools for organization of risk management in all processes of IC including planning, analysis, control and accounting, organization and regulation. RC solves the problems of development of architecture (infrastructure and processes) of risk management of complex innovation projects. Applying this architecture to particular risks, telcos' managers, as part of self-controlling process, perform risk-adjusted management of innovation project [22]. The functions of risk controlling are listed in Table 1.

Based on these principles, we developed a reference model of complex R&D expenditures and time management system (Fig. 1). Its advantages over "conventional" systems, such as presented in [5, 8, 11, 12], are (1) integration of risk management in all expenditure and time decision making; (2) ensuring integration and coordination of all stages of project cost/time management; (3) application of enhanced planning, control, reporting and decision-making tools with lower tolerance levels than their "conventional" peers. In the next chapter we will build the mechanism of risk-adjusted scheduling and cost budgeting which is the inherent part of this system.

Table 1. Functions of risk controlling in investment controlling system

Functions	Description of functions
Analysis	•Defines characteristics of external and internal environment which expose R&D projects to risks •Identifies and prioritized risks by elements of R&D process •Implements tools and methods of risks analysis •Identify non-compliances of telco's innovation policy with threats
Planning and budgeting	•Develops risk-adjusted planning and budgeting tools •Supplements system of key performance indicators of R&D activity with elements of risk limits and early warning signals
Product development and research	•Consults on management of exposures arising in any R&D project •Helps to integrate risk management practices into R&D process •Recommends remediation measures to reduce risks to acceptable level
Reproduction of project assets	•Develops tools of analysis of specific risks peculiar to R&D tasks •Supplements R&D portfolio management with risk management tools
Control	•Develops models of risk assessment and their impact on project goals •Projects and implement of early warning systems of risk prevention •Suggests "barriers" to stop risks from spreading •Consults about and develops control procedures
Monitoring and reporting	•Modifies and consults on creation of risk reporting system
Regulation	•Consults on issues related to decision making on risk management •Suggests options/consultations to manage risks

5 Developing of the Mechanism of Risk-Adjusted Scheduling and Cost Budgeting of R&D Project

5.1 Mechanism's Block Diagram

The developed block diagram of the mechanism is presented at Fig. 2.

It applies: (1) bowtie diagram for risk identification; (2) PERT technique for scheduling; (3) RFMEA method for risk analysis; (4) @Risk software for risk assessment and Monte-Carlo simulation; and (5) Hurwicz criteria for selection of optimal alternative [13, 16, 21, 23]. The prerequisite for applying the mechanism is the developed infrastructure of risk management in the project company. It specifies the roles, the responsibilities, risk communication and risk reporting structure [23].

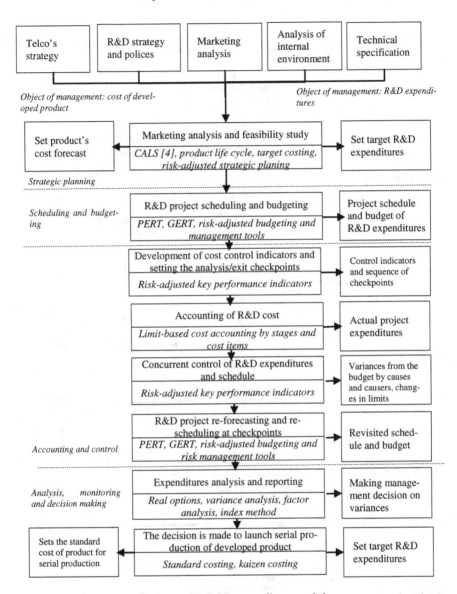

Fig. 1. Reference model of complex R&D expenditures and time management system

5.2 Phase 1: Generate Inputs of the Mechanism

For each option of fulfilling the R&D project (the option) the following key inputs should be formulated at the phase of strategic planning. (Table 2).

At this stage, the project budgeting and scheduling team (comprised from representatives of key diverse functions involved in the project) should be created and trained.

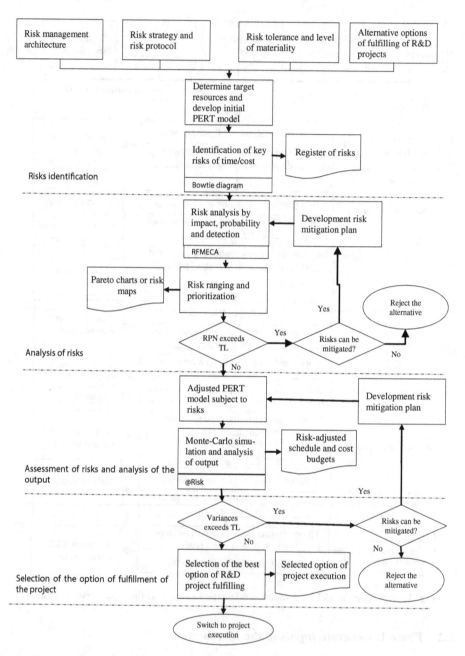

Fig. 2. The block diagram of risk-adjusted scheduling and budgeting mechanism of R&D project

Table 2. Key inputs to risk-adjusted scheduling and budgeting mechanism

Input name	Input definition	Notations/examples
Ultimate goals	Goals that defines total duration and total cost of the project	T – total duration C – total cost
Variances from ultimate goals	Duration (dT) and cost (dC) variances of actual duration (T_a) and costs (C_a) from the ultimate goals	$dV = T - T_a$ $dC = C - C_a$
Intermediate goals	Goals that are defined by splitting the ultimate goals by project stages and/or types of resources	T_i – duration of i-th stage; C_i – cost of i-th stage; c_{ij} – cost of j-th resource at i-th stage; q_{ij}, p_{ij} – consumption/price of j-th resource at i-th stage
Risk tolerance levels	The degree of variability in ultimate goals that project stakeholders willing to withstand	VT – tolerance level for duration VC – tolerance level for total cost
Risk appetite levels	The maximum level of variability in ultimate goals which is acceptable	RT – risk appetite for duration RC – risk appetite for cost
Risk materiality levels	The degree of variability in ultimate goals that project stakeholders consider as immaterial	MT – materiality level for duration MC – materiality level for total cost

5.3 Phase 2: Identification of Risks of Each Option

At the first step, the project team with application PERT techniques [24] and using inputs from phase one (1) identifies and calculates the critical path, critical jobs and time reserves of jobs; (2) models the relationship between the critical path and duration of jobs; (3) identifies the weak links and bottleneck as key risk areas; (4) models the total cost of the project depending on duration, quantity and prices of resources. The team can also perform the optimization of the network project to minimize the project cost at the minimal possible duration of the project [24]. Then, the team identifies key risks affecting the interim and ultimate goals with the application of bowtie diagram (Fig. 3) [23]. It provides structural analysis of risks and the visualization of the relationship between the risk, its causes, consequences; and helping to identify spots for the risk barriers, controls and contingency measures. Identified risks with attributes are recorded in the risk register.

5.4 Phase 3: Analysis of Risks with RFMEA

Each risk is analyzed by three dimensions: its impact on ultimate goals (I); the likelihood of risk event (P); and detection value (S), the ability to spot risk event before it occurs. The ratings are measured by the scale from 1 to 10 and are bundled into the risk priority number (RPN) which is the multiplication of the likelihood, impact, and detection values. The team shall use developed bowtie diagrams for the analysis. Tables 3 and 4 provide guidelines for the rating scores [13, 16].

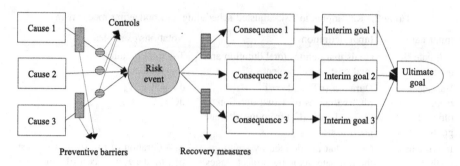

Fig. 3. Example of bowtie diagram

Table 3. Impact rating value guidelines

Impact value scale	Schedule (I_d)	Cost (I_c)
1. Insignificant		
2. Very minor		
3. Minor		
4. Very low		
5. Low	$I_d = \begin{cases} if\ dT < 0\ then \\ 1, if\ |dT| \le MT \\ \left\lceil \left(\frac{|dT|-MT}{VT-MT}\right) \times 7 \right\rceil \\ 8\ or\ 9 \\ if\ VT < |dT| < RT \\ 10, if\ |dT| \ge RT \end{cases}$ (1)	$I_c = \begin{cases} if\ dC < 0\ then \\ 1, if\ |dC| \le MT \\ \left\lceil \left(\frac{|dC|-MT}{VT-MT}\right) \times 7 \right\rceil \\ 8\ or\ 9 \\ if\ VC < |dC| < RC \\ 10, if\ |dC| \ge RC \end{cases}$ (2)
6. Moderate		
7. High	$if\ dT \ge 0\ then\ I_d = 1$	$if\ dC \ge 0\ then\ I_c = 1$
8. Very high		
9. Extremely high		
10. Dangerously high		

The project team may change these guidelines to ensure their fit the particular project (e.g. existence of key milestones which cannot be broken). The probabilities p_k c_{k1} and c_{k2} (k ∈ (D – duration, C – cost)) can be inferred statistically from the database of past R&Ds projects. For new and/or emerging risks (for which the data do not exist or unreliable), the expert methods can be used. Example of such methods include: additive - multiplication model or modifications of Elmery method [16].

Once the values for the individual ratings are entered, both the risk scores and the RPN values for individual risks are calculated and depicted at the Pareto chart (Fig. 4) [13]. If the team follows the guidelines from Tables 3 and 4 than RPNs for risk tolerance level, risk appetite and materiality levels equal 700, 1000 and 100 respectively [16], however, the team can establish the other boundaries. The team can also establish different decision levels for each cost or duration components. Additionally, the consolidated RPNs for the whole strategy should be calculated using the expert-based correlation coefficients (ρ) [16].

Table 4. The likelihood and detection value guidelines

Likelihood (P_d, P_c)	Detection (S_d, S_c)
1. Remote	1. Almost certain
2. Very low	2. Very high
3. Low	3. High
4. Mod. low	4. Moder. high
5. Moderate	5. Moderate
6. Moderately high	6. Low
7. High	7. Very low
8. Very high	8. Remote
9. Extremely high	9. Very remote
10. Certain	10. Uncertain

Likelihood column formula:

$$P_k = \begin{cases} 1, & if\ p_k \in [0, 0.1] \\ [10p_k], & p_k \in [0.1; 0.9] \\ 10, & p_k \in (0.9, 1] \end{cases} \quad (3)$$

-probability of risk occurrence, the ratio of frequency of failures to the total number of jobs

Detection column formula:

$$S_k = \begin{cases} 1, & if\ f_k \in [0, 0.05] \\ [10f_k], & p_k \in [0.05; 0.9] \\ 10, & f_k \in (0.9, 1] \end{cases} \quad (4)$$

$f_k = (1 - s_{1k} \times s_{2k})$

s_{1k} – the probability of detection of risk, the frequency of prevented risks to total number of occurred risks

s_{2k} – the probability of early detection of risk with controls

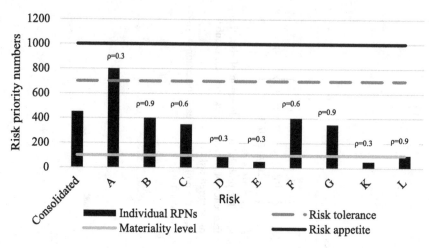

Fig. 4. Example of risk Pareto chart for duration

The Pareto chart allows the understanding, on the level of RPNs, which risk contributed the most to total exposure measured by consolidated RPN. If the individual risk's RPN is below materiality level, then this risk is excluded from further assessment and are not included into the risk management perimeter. For each option the following decision rules are suggested:

1. If consolidated RPNs of at least one component (duration or cost) exceed the risk tolerance, than the option is returned for reworking. The planning team develops and implement risk mitigation measures to return consolidated RPN within the risk tolerance boundaries. If risk mitigation is not possible then the option should be rejected. The team can consider immediate rejection of the alternative without reworking if at least one consolidated RPN materially exceeds risk appetite.
2. In rare cases, when both consolidated and individual RPNs for duration and cost components are below materiality level, such option is considered "risk free" and can be considered as "the optimal" without passing through phases 4 and 5. However, if RPNs from some individual risks exceed materiality level, the option may be sent to phases 4 and 5 for additional analysis.
3. If both consolidated RPNs for duration and cost of option is above materiality level but below risk tolerance level the option is considered acceptable and sent to Phase 4 for further assessment. The individual risks with RPNs above the materiality level are considered as material. However, if RPNs for some individual risks exceed tolerance level, the project team may consider sending the option for reworking if any of these risks are considered critical. The option should be sent for rework if RPN of any individual risk exceeds risk appetite.

Results of analysis of each risk as well as consolidated RPNs for each option must be recorded in the risk register for further quantitative assessment.

5.5 Phase 4: Quantitative Assessment of Risk and Analysis of Output

This assessment is performed for each option considered acceptable at phase 3. At first step, the PERT model of the option is adjusted taking into account the risk mitigation actions performed at phase 3. Then, the PERT model is exported into MS Excel with installed @risk modelling engine. For each job in PERT, the resources budget is modelled and the summary budget for the entire option is worked out.

At the second step, with the help of tools embedded into @risk, the project team determines (1) what variables of duration and cost (including prices or consumption volume or both) exposed to material risks are random variables: and (2) what kind of probability distribution theses variables follow. Given the restrictions of PERT [21], the only distribution option available for duration is the triangle distribution with parameters (a, m, b), where a is the best-case estimate, m is the most likely outcome and b is the worst-case estimate. For parameters of costs, however, @risk provides around 90 alternatives of distributions including normal, binominal, exponential, etc. The outcome of analysis of risk probabilities performed at phase 3 is applied in determining the parameters of these distribution. Once, the modelling of probability distributions has been completed, the team runs Monte Carlo simulation with pseudo-random number generator (the number of simulation varies from 1000 to 10,000 times). The simulation draws of duration/cost parameters values and making calculations according to the model constructed at step 2. The outcome is option's risk-adjusted total duration and cost.

The final step is the analysis of the outcome including: the mean, the median (the quantiles), the dispersion and standard deviation; the mode, the confidence intervals of duration and costs. Additional analysis includes understanding what jobs and/or resource components contributed the most to the deviation of outcome (Fig. 5). During this step the team determines: (1) the expected variance of option's outcome from the budget and the chances that option will be fulfilled in accordance to the budget; (2) the most probable duration/cost of the option; and (3) if the budget is tight and/or not realistic. If the expected variance of duration/cost exceeds the tolerance level (VT and/or VC); the project team will send the option for rework for development and implementation of risk mitigation actions. If variances exceed risk appetitive and/or risk mitigation/adjustments in the budget are not possible, the option is rejected.

Analysis of the output also helps to determine: (1) the risk-adjusted size of financing need to be attracted to fulfill the option; (2) what reserves should be maintained in case of realization of adverse scenarios; (3) what are the key risk areas to concentrate management attention while fulfilling the option; (4) what are the scenarios; and (5) what contingency plans need to be developed in case of realization of worst scenarios.

5.6 Phase 5: Selection of the Optimal Option of Fulfillment of R&D Project

The goal of the phase is to find the optimal option of fulfilling R&D projects among the alternatives (lets denote the entire set of options as Z). The inputs of the phase for each zth-option are: (1) consolidated RPNs for duration (RPN_{dz}) and cost (RPN_{cz}); and

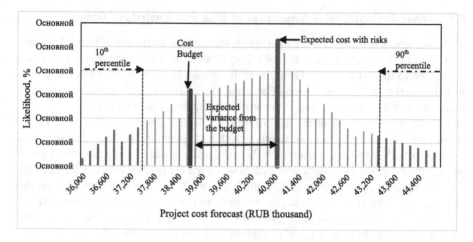

Fig. 5. Example of Monte-Carlo simulation output

(2) the outcome of duration and cost simulation. In the mechanism we apply Hurwicz's optimism – pessimism criterion which assumes that manager, while finding a suitable solution, is searching for a middle ground between the extremes posed by the optimistic and pessimistic cases [16]. At the start, the project team should define the importance of duration and cost in searching for the optimal option (the weights α and β). Then, the criterion (S_z) can be written in the following form:

$$\max\{S_z\}_{z\epsilon[1,Z]} = max\{\alpha GD_z + \beta GC_z\}_{z\epsilon[1,Z]} \qquad (5)$$

$$GD_z = \left\{\frac{RPN_{dz}}{1000} \times PD_z + \frac{(1 - RPN_{dz})}{1000} \times OD_z\right\} \qquad (6)$$

$$GC_z = \left\{\frac{RPN_{cz}}{1000} \times PC_z + \frac{(1 - RPN_{cz})}{1000} \times OC_z\right\} \qquad (7)$$

$$\alpha + \beta = 1; \alpha \in [0, 1], \beta \in [0, 1] \qquad (8)$$

GD_z, GD_c - Hurwicz's criteria of zth option for duration and cost respectively; PD_z, PC_z – pessimistic forecasts of duration and cost respectively of zth option; obtained from simulation outcome, given a specified degree of confidence. OD_z, OC_z – optimistic forecasts of duration and cost respectively of zth option; obtained from simulation outcome, given a specified degree of confidence (for example, 10%).

At this phase, project team calculates the value of S_z for each option and determines the option with the maximum value of S_Z. This option is considered by team as the optimal. This option should be presented by the team to the project managers for the discussion and the final approval.

6 Conclusion

We developed the mechanism of risk-adjusted scheduling and cost budgeting of research and development (R&D) project for companies in telecommunication; the inherent part of complex R&D cost and schedule management system. It outperforms the "conventional" approaches of scheduling and cost budgeting. It allows (1) timely identification of risks, (2) facilitating transparent evaluation of risks and uncertainties; (3) focusing management time and efforts on key risks; and (4) performing remediation of risks earlier in the cycle thus improving speed and quality of decision making. Unlike "conventional mechanism" which poorly addressed risks, it integrates risk management process with risk-adjusted budgeting making budgeting and scheduling process dynamic, iterative and responsive to changes in environment; enabling multi-risk scenario modelling thus enhancing confidence in project's schedules and cost budgets. We can, therefore, conclude that the suggested mechanism should have a significant practical use in performing R&D projects in telecommunications.

References

1. Glukhov, V., Balashova, E.: Economics and Management in Infocommunication: Tutorial. Piter, Saint Petersburg (2012)
2. Guseva, I., Covyrzina, K.: The algorithm of the R&D management and control involving controlling system. Controlling **53**, 50–56 (2014)
3. Nekrasova, T.P., Aksenova, E.E.: Economic Evaluation of Investments in Telecommunication Industry. St. Petersburg State Polytechnical University, Saint Petersburg (2011)
4. Sazonov, A., Dzhamay, V., Povekvechnykh, S.: Analysis of efficiency of implementation of CALS technologies on the example of domestic aviation. Organizator proizvodstva (Organ. Prod.) **26**(1), 84–92 (2018)
5. Suloeva, S., Gultceva, O.: Complex R&D cost management system. Organizator proizvodstva **25**(4), 57–66 (2017)
6. Moody's telecommunication industry outlook. https://www.moodys.com/Pages/rr003_0. aspx?bd=00300300B&ed=&rd=00300300B&tb=0&po=0&sb=&sd=&lang=en&cy= global&searchfrom=SearchWithin&kw=Telecom. Accessed 01 May 2018
7. Deloitte: Technology, Media and Telecommunications Predictions (2018). https://www2. deloitte.com/cn/en/pages/technology-media-and-telecommunications/articles/tmt-predictions-2018.html. Accessed 01 May 2018
8. Grekul, V., Korovkina, N., Kuprianov, Y.: Project Management in Information Technology Sphere, 2nd edn. BINOM Laboratory of Knowledge, Moscow (2015)
9. Dos Santos, F., Cabral, S.: FMEA and PMBoK applied to project risk management. J. Inf. Syst. Technol. Manag. **5**(2), 347–364 (2008)
10. Zhe, X., Jing, Y., Hogobo, L.: Analyzing integrated cost-schedule risk for complex R&D project. J. Appl. Math. **1**, 2–12 (2014)
11. Mikhailova, E., Remizova, N.: Budgeting in organizations which develop the tools for aviation complexes. Electron. J. Trydi Mai **69**, 24–29 (2013)
12. Sviridenko, V.: Complex cost management system in industrial enterprises. St. Petersburg State Polytechnical Univ. J. Econ. **4**(199), 59–65 (2014)
13. Carbone, T., Tippett, D.: Project risk management using the project risk FMEA. Eng. Manag. J. **16**(4), 28–35 (2004)

14. Bekleshov, V., Zavlin, P.: Norming in Research and Development Organizations. Economica, Moscow (1989)
15. Luppino, R., Hosseini, M., Rameezdeen, R.: Risk management in research and development projects. The case of South Australia. Asian Acad. Manag. J. **19**(2), 67–85 (2014)
16. Grichounine, S.: Developing the mechanism of qualitative risk assessment in strategic controlling. St. Petersburg State Polytechnical Univ. J. Econ. **10**(2), 64–74 (2017)
17. Batkovskiy, A., Konovalova, A., Semenova, E., Trofimets, V., Fomina, A.: Risks of development and implementation of innovative projects. Mediterr. J. Soc. Sci. **6**(4), 243–253 (2015)
18. Isidore, L., Back, W.: Probabilistic optimal-cost scheduling. J. Constr. Eng. Manag. **127**(6), 431–437 (2001)
19. Grishunin, S., Suloeva, S.: Strategic controlling and anti-crisis management. In: Strategy and Tactics of Anti-Crisis Management of a Firm. Specialnaya Literatura, Saint-Petersburg (1996)
20. Taylor, B.: Project management using GERT analysis. Proj. Manag. Q. **9**(3), 15–20 (1978)
21. Wyrozębski, P., Wyrozębska, A.: Challenges of project planning in the probabilistic approach using PERT, GERT and Monte Carlo. J. Manag. Mark. **1**(1), 101–107 (2003)
22. Grishunin, S., Mukhanova, N., Suloeva, S.: Development of concept of risk controlling for industrial enterprise. Organizator proizvodstva **26**(1), 45–56 (2018)
23. Sidorenko, A., Demidenko, E.: Guide to effective risk management 2.0, URL: https://www.risk-academy.ru/en/download/risk-management-book. Accessed 07 May 2018
24. Pleskunov, M.: The Tasks of Network Planning: Tutorial. The Publishing House of Urals University, Ekaterinburg (2014)

Towards Business Optimization and Development of Telecommunication Companies: Tools Analysis and Their Adaptation Opportunities

Vladimir V. Glukhov⑩, Igor V. Ilin⑩,
and Aleksandr A. Lepekhin⁽☒⁾ ⑩

Saint Petersburg Polytechnic University, Polytechnicheskaya Str. 29,
Saint Petersburg 195251, Russia
vicerector.me@spbstu.ru, ilyin@fem.spbstu.ru,
lepekhinalexander@gmail.com

Abstract. The telecommunication industry is complex and rapidly changing economy sector due to high vulnerability of new technologies, emerging on the market. The companies have to constantly adopt and develop their business and their services to these unstable conditions. The process of adaptation to market changes has to be supported by different tools and approaches, which are very different today. This study is aiming to analyze the possibilities of adaptation of different technologies for developing the business of telecommunication companies and make an attempt to reveal which technologies are core and baseline for company's development and growth.

Keywords: Telecommunication companies · Business development
Internet of Things · Business Intelligence · Enterprise architecture

1 Introduction

The digital economy is based on the telecommunications industry, which has transformed organizations and opened new opportunities for progress through technological advances. Important characteristics of this industry are high costs, rapid technological achievements, high level of obsolescence and intense competition in most segments.

The telecommunications industry was faced with the fact that many different types of Information Technologies began to penetrate telecommunication companies [1]. This phenomenon, on the one hand, has created a threat to the traditional telecommunications business, and on the other - opened new prospects for business both for existing operators and for owners of telecommunications infrastructure, as well as for new companies that received additional opportunities in the conditions of demonopolization of the telecommunications services market. Studies show that in the evolution of IT and telecommunication technologies convergence is expected - convergence of wireless and mobile communications, convergence in telecommunications, convergence of IT and telecommunications, convergent media, convergence in the evolution

© Springer Nature Switzerland AG 2018
O. Galinina et al. (Eds.): NEW2AN 2018/ruSMART 2018, LNCS 11118, pp. 471–482, 2018.
https://doi.org/10.1007/978-3-030-01168-0_42

of gadgets, convergence of IT with nano- and biotechnologies, and even convergence of virtual and social space.

Such changes in the industry led to a significant increase in competition. Cable operators, Internet providers and traditional telecommunications companies are increasingly coming into direct competition, providing a comprehensive service that includes fixed telephony, broadband Internet access, television and mobile telephony.

For example, the adoption of the Session Initiation Protocol (SIP) as a standard protocol for peer-to-peer communications allowed the provision of communication services and applications over the Internet. This protocol has become the main one for providing voice communication over the Internet (VoIP). The SIP protocol allows new players to provide voice and data services based on a broadband Internet connection, which leads to the emergence of new companies that compete with existing telecommunications and cable companies without having their own fiber-optic networks. The transition of telecommunications networks to digital standards led to the creation of a common electronic network infrastructure, which in turn helped to erase the differences between telephone networks and data networks, public networks and corporate networks. Television, which emerged as a convergence of radio technology and cinema, converged with IT.

The above factors of the industry development, modern trends of business digitalization and constantly developing technologies cause an urgent need for telecommunication companies to have a whole portfolio of business development tools. In this article, we will propose a set of methods and tools that allow a telecommunication company to function successfully in today's dynamically changing market.

2 Literature Review

In today's globalized and unlimited market, quality, productivity and customer satisfaction pose a challenge to the survival and growth of all firms. These requirements for growth and survival are further deepened due to the need to attract and retain loyal customers. Thus, the client is the main focus for any successful business.

One of the important services in the economy is telecommunications. We should not underestimate the role of the telecommunications industry in the economy. This is due to the fact that these are the means by which all daily transactions and activities are carried out. It helps to make decisions, organize, influence, activate, instruct, provide feedback, develop interpersonal and business relationships, and exchange information [2]. All social, economic, political, cultural, commercial and commercial activities are carried out using telecommunications. The nature of the country's telecommunications industry influences its pace of commercial and domestic activities. Telecommunication companies in Russia cover 14% of the total number of companies and have a significant impact on the economy (see Fig. 1).

In the telecommunications research, the main players of the industry were analyzed. Description of types of companies is presented below (see Fig. 2).

However, as already mentioned above, the telecommunications industry is undergoing significant changes due to the growth of the level of technology and the penetration of IT into the core of the business. This fact changes the telecommunications

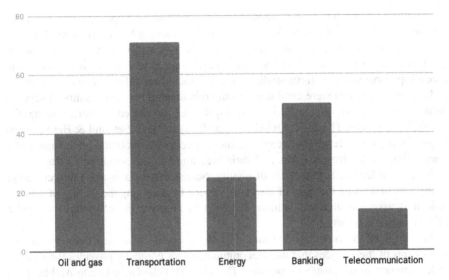

Fig. 1. Types of companies in Russia market, influencing economy.

Categories	Description
Network Providers Network Providers (landlines)	Companies that own and operate network based on physical links such as telephone and cable networks.
Network Providers Wireless and Satellite	Companies that own and operate network based on virtual links such as wireless and satellite networks
Tool Providers Hardware providers	Companies manufacturing hardware that is integral part of the communication network
Software Providers	Companies producing software that is integral part of the communication network
Transaction and Service Providers Service Providers	Companies providing diverse types of services such free Internet service, communication tools, Web hosting, VPN services, calling cards, etc.
Transaction Service Providers	Companies that enable buying and selling of goods and services through Internet
Internet/Content Providers Content Providers	Companies producing copyright material distributed through channels other than the Internet
Internet Content Providers	Companies producing copyright material distributed through the Internet

Fig. 2. Types of companies in telecommunications market.

business and can become a threat to many companies. At the same time, timely adaptation to this trend can open new prospects for business. However, it is important to note that due to the growing competition in the market and in the conditions of developing technologies, the core of the company's development is traditionally to remain customer loyalty.

Customer loyalty has been defined in the literature as a competitive tool for many companies. This is even much more pronounced in today's highly globalized, industrialized and competitive markets. The growth and survival of companies depend on how loyal their customers are, and telecommunications as an industry is no exception. In developed countries, various studies were conducted regarding customer loyalty.

In addition, studies were conducted on the relationship between quality of service, customer satisfaction, image and client loyalty in developed countries using the SERVQUAL model. The SERVQUAL model (Parasuraman, Ziethaml & Berry, 1988) suggests that the differences between customer expectations about the performance of a general class of service providers and their evaluation of the actual performance of a particular firm lead to a perception of quality. So the first step in achieving customer loyalty is to determine the level of customer service by assessing the quality of service, while it is assumed that competition will help to improve the efficiency of service delivery in the economy.

The level of customer service and the effectiveness of service delivery are the components of the services that the telecommunications company provides through the implementation of its business processes. The TOGAF (The Open Group Architectural Framework) defines a service as an element of behavior that provides certain functionality in response to requests from participants or other services. Examples of services include the analysis of loyalty programs for clients, the selection of relevant offers for the provision of services, and others. Services are implemented by the company's business processes, which in turn are supported by the entire application system and the company's infrastructure.

The development of the Applications layer and IT infrastructure of the telecommunications company is having a direct impact on the business processes of the organization and the implementation of services. This study examines the architecture of the telecommunications company in terms of the TOGAF standard and describes a list of technologies and tools for the development of a wide variety of aspects of the company.

3 Methodology

The TOGAF standard focuses on the business mission and uses the Architecture Development Method (ADM), which describes the process of transition from a basic architecture to a target in accordance with business objectives (Tang, Jun Han, & Pin Chen, n.d.). TOGAF framework is one of the main methods of enterprise architecture development due to availability of tools and applied software solutions. TOGAF represents the enterprise architecture in the form of 3 basic layers.

Business Architecture - system of business processes and functions of the organization, basic business units, organization locations, organizational structure.

Applications Architecture - application components, application functions, integration components of applications and IT services.

Technological Architecture - internal networks of the organization, server component of applications, databases, etc.

The basic structure of Enterprise Architecture in TOGAF standard is represented below (see Fig. 3).

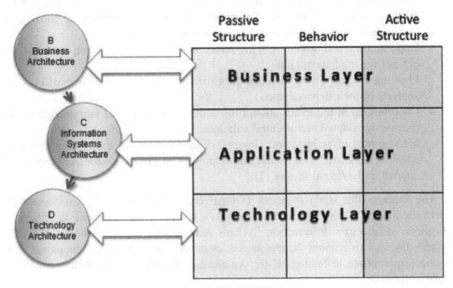

Figure 2 - Standard Concepts

Fig. 3. TOGAF baseline Enterprise Architecture representation.

TOGAF includes principles that support the decision-making process throughout the enterprise, form a guide to the management and development of IT resources and support architectural principles in the development and implementation of information technology support [3]. TOGAF makes it possible to formulate requirements based on an understanding of business organization, that is, the business architecture, using ADM [4].

ADM includes the following phases:

- Preliminary Framework and Principles – this phase describes the basis for the formation of the architecture through the definition of the basic principles of its formation.
- ADM Cycle
  - Vision architecture - a description of the basic principles of the architecture model "as is" and "as it should be" for all layers - business, applications and infrastructure;
  - Description of the Business Architecture and analysis of opportunities for improvement and target state.
  - Description of Information Systems Architecture and analysis of opportunities for improvement and target state.

- Description of the Technological Architecture and analysis of opportunities for improvement and target state. For this step (and the previous two), there are 8 sub-stages: the creation of a basic model, the use of different points of view, the formation of an architectural model, the choice of services, the confirmation of business objectives, the description of criteria, the definition of the target architecture and the analysis of discrepancies.
- Opportunities and solutions. This phase includes evaluation and selection of options for implementation.
- Planning for migration. Includes prioritization of implementation projects and analysis of their dependencies.
- Coordination of implementation. Project management by the introduction of all previously adopted architectural solutions.
- Management of architectural changes. A long process of monitoring changes, which is associated with changes in technology, business environment and other internal and external factors [5].

The focus of this study is points TOGAF 2b, 2c, 2d and 2e. It is necessary to determine how it is possible to develop 3 key layers of the enterprise architecture (Business Architecture, Information Systems Architecture and Technological Architecture), taking into account the characteristics of telecommunications companies and to select appropriate technological (or organizational) solutions for their subsequent implementation.

The analysis of the technology market showed that there are a large number of different types of technologies that have a serious potential for influencing the business of telecommunications companies.

Mobility-studies have shown that mobile devices and services based on the mobile Internet will be the fastest growing sector of the Digital economy [6].

Multi-screen - Already, most consumers use two or even three devices. Consumers want to have seamless and easy access to content - business and personal information - from any device. The concept of BYOD (Bring Your Own Device), which became popular among many companies, stimulating this trend. Ensuring the compatibility of multiple devices is both a challenge and new opportunities for the IT industry [7].

Mobile video content - With the development of mobile networks 3G and LTE, the consumption of video content will increase significantly. This trend affects the popularity of tablets and smartphones with a screen diagonal of 5 or more inches [8].

Electronics accessories - a promising trend, which was picked up by almost all the major manufacturers of consumer electronics: smart watches, bracelets, glasses and other accessories. While the popularity of wearable devices has been hampered by problems such as short battery life, lack of functionality and the controversial design of most of them.

Mobile, contactless payments - mobile payments are made through SMS and online banking applications, however, in the next few years, contactless payment systems based on NFC technology are expected to be distributed. The market is ripe for the mobile phone to become a full-fledged means of payment, and industry leaders have already started to create such devices [9].

Connected "smart" cars - today, cars connected to online systems are used mainly in the b2b market. In the field of transport and logistics - to monitor the location, route, speed, fuel consumption in the field of car insurance - to monitor the use of the car's functions and driving quality: measuring the amount of emergency braking, average driving time with increased speed, tracking faults, etc. The growth of mass demand for "smart" functions on the part of motorists is expected. The main advantages of a connected car lie in the field of security, economy, entertainment. We can expect new partnerships between auto brands and telecom operators that are already underway (Vodafone and Porsche, LG and Audi, AT & T and Cadillac) [10].

Smart House - one of the realizations of Internet of things. With the help of smart home technologies, consumers can remotely control almost everything in their home from the coffee machine and the contents of the refrigerator to lighting and door locks. At the recent technological exhibitions, a number of vendors have introduced special hubs - platforms that allow you to connect almost any home device to the Internet [11].

Big data - all Internet users during online shopping, using social networks, content consumption leave a "digital footprint." As a rule, these are huge arrays of unstructured information. One of the key problems that many companies try to solve that have access to such information is how to monetize and use it for the benefit of the consumer and business without violating the user's privacy. It is expected that in the coming years the income from using such information will amount to billions of dollars [12].

Augmented and virtual reality - adds a virtual layer to the image of physical objects, virtual - completely immerses the user in the virtual world. Both technologies open new applications in a variety of areas and can completely change the user experience of consuming content, from games and shopping to navigation, education and science [13].

All these technologies and trends will (or already have) a significant impact on the business of telecommunications companies and on its transformation. Their influence and the possibilities of introducing them into various layers of the architecture of the telecommunications company will be discussed in the next section.

4   Results

The first of the considered technologies is the Internet of Things. The paradigm of the Internet of Things is that a great many objects that surround us can interact with each other in a single network. Radio frequency identification (RFID) and a network of sensors can be used to solve this problem, and information and communication systems will be a supporting technology. This leads to the generation of huge amounts of data that must be stored, processed and provided in a seamless, efficient and easily interpretable form.

IoT will create the opportunity to unite a variety of telecommunications technologies and create new services. A good example is the use of GSM, NFC (Near Field Communication), Bluetooth with low power consumption, WLAN, multiprocessor networks, GPS and sensor networks together with SIM-card technology. In these types of applications, the reader is part of the mobile phone, and different applications use the SIM card. NFC provides a link between objects in a simple and safe way by simply associating them with each other. Therefore, the mobile phone can be used as an NFC-reader and transmit the read data to the central server. When used on a mobile phone, the SIM card plays an important role as a data repository and NFC authentication data (for example, ticket numbers, credit card accounts, identification information, etc.). Things can be networked and facilitate peer-to-peer communication for specialized purposes or improve the reliability of communication channels and networks. In disaster situations, peer-to-peer networks can be established to support the flow of vital information in the event of disruptions in the telecommunications infrastructure. The future M2M (Machine to Machine) ecosystems will be complex and cover many industries, including telecommunications and electronics. Unlike current M2M markets that are highly segmented and often depend on proprietary solutions, future M2M markets should be based on industry standards to achieve explosive growth. This process of standards will be much broader than writing a specification, since it includes not only interfaces, but also platforms and services. In addition to the above-described impact of technology, there is also the possibility of using IoT to monitor towers, their condition and the need for repair. Thus, the effect of IoT on the architecture of a telecommunications company can be summarized in the Table 1.

Table 1. IoT implementation potential.

Architecture layer	Impact	Comment
Business architecture	–	
Applications architecture	+	There is a need to integrate internal systems with new implemented services for customers, the structure of IT services is changing
Technological architecture	+	There is a possibility to monitor the company's infrastructure remotely using IoT tools

The next considered tool for business development is the mathematical optimization of the model of the telecommunications company.

Mathematical modeling and optimization is the most powerful device for business development in its various aspects. In the field of Telecommunications, the mathematical apparatus is essentially developed, and there are a large number of developed solutions for optimizing the activities of the organization [14].

Various studies include mathematical models to address the issue of ensuring high reliability and error tolerance, which is relevant in wireless communications networks. This helps to solve the problems associated with errors and failures at the physical level, with synchronization failures and random errors at the physical level [15].

A model is also developed to quickly and accurately assess the performance of telecommunications networks. This model can also be used to solve the following problems:

- Designing of telecommunication networks
- Capacity Assignments
- Development of protocols
- Addressing the congestion problems of telecommunications networks.

To develop the model, computer calculations and modeling were used using variables such as the speed of arrival of calls to routes, the structure of routing, the availability of alternative routes, the capacity of the channels and the means for selecting routes.

In addition, studies have shown that telecommunication networks are designed to serve with the required quality of the message flow entering its input. Service of the input message, continues for a certain time with the purpose of its delivery from the sender to the recipient. Servicing devices in this case are communication channels that provide information exchange between different subscribers. Communication channels have limited capabilities, which leads to the formation of queues. Thus, TCS is a typical queuing system (QMS). In this regard, the development of mathematical methods for determining the basic characteristics of queuing processes for quantifying the quality of the service system, including TSS, is important. To solve the described problems, the use of graph theory and network analysis is actual. The expediency of using the graph in solving connectedness problems is most obvious. So, in the general case, the task of delivering information from any point to any one can be of interest. With the help of graphs one can also solve synthesis problems. Suppose that a set of stations and the requirements for the magnitude of the maximum flow are given. We need to build an optimal system that meets these requirements. One of the possible criteria for optimality can be a minimum of cost [16].

Thus, the analysis showed that mathematical modeling and optimization is a serious and one of the most promising tools for the development of telecommunications companies, providing opportunities to influence all layers of the enterprise architecture (see Table 2).

Table 2. IoT implementation potential.

Architecture layer	Impact	Comment
Business architecture	+	The ability to provide high reliability and resistance to errors, improve the quality of services
Applications architecture	+	The ability to integrate network load estimation models into internal network monitoring applications
Technological architecture	+	Optimization of the structure of the telecommunications network, solution of the problem of constructing an optimal system

The third tool in question is Big Data. Currently, the adaptation of large data technology is just beginning. Therefore, the adaptation of technology in science, technology, medicine, medical care, finance, business, law enforcement, education, transport, retail and telecommunications - all this is an important and promising direction of research.

As in many other market sectors and industries, Big Data will also have many challenges and opportunities in the 5G wireless network. First of all, cellular networks must provide effective infrastructure support for this amount of data. For example, future M2M applications or Internet of Things (IoT) will generate a huge amount of data. Secondly, new network architectures can arise because of the need to run applications to process large data. There is a strong synergy between cloud computing, a network defined by software, and virtualization of network functions (NFV). The convergence of these technologies can be envisaged for the creation of highly reliable 5G platforms for large data. Third, making informed decisions and extracting information from large data is an extremely important and yet non-trivial task. For example, cellular operators can use different customer access data to reduce the outflow and look for new revenue opportunities. Methods of data mining and machine learning are necessary for efficient and optimized network operation.

Big Data technology and applications for working with it are the second step after the implementation of Business Intelligence (BI) class systems [17]. The analysis showed that few companies currently use these systems, although the most progressive in this direction are the telecommunications companies in comparison with all other business sectors. This is due to the fact that the telecommunications sector faces many challenges in terms of competition, rules, technology shifts and customer needs for new functions and services. One of the problems faced by telecommunications companies is the difficulty of obtaining information from all the data that they generate. Telecom generates a huge amount of data from the use of mobile phones, records of detailed information about calls, network equipment, information about billing, server logs and the growing ability to connect subscribers and users.

Analyzing these data with BI helps the telecoms companies gain a competitive advantage. They can improve overall network performance, optimize service levels, minimize overhead and maximize profitability, while ensuring customer loyalty. They can take data-driven investment decisions to achieve profit targets. The enterprise will have a single view of the data, even if it comes from several data sources and departments. Centralized BI solution is a serious tool for business development.

Such applications can analyze various data sources to determine fault-tolerant network points, analyze customer experience, find ways to reduce maintenance costs and solve other tasks. Users can also monitor basic business processes and KPIs.

BI can help organizations prioritize and make the right decisions when it comes to network expansion plans. Enterprises can analyze coverage data, such as the percentage of the population covered by services, the percentage of the population covered by services, the average area not available for services, using GEOanalysis.

In addition, telecommunications organizations can integrate and analyze data from several sources, such as the subscriber, network, traffic and location, to gain an understanding of how to improve network capacity planning. Using BI, organizations can analyze network traffic in real time to determine the periods when network usage is high, and take steps to ease congestion, prevent possible outages. They can then define a new expansion strategy for this area.

It is important to note that Big Data technology and BI type systems are impossible without a mathematical model, which was identified as a separate tool for business development. The joint use of the developed mathematical models and specialized tools is able to solve the tasks described above more effectively.

5 Conclusion

The analysis showed that technologies such as Internet of Things, Big Data and Business Intelligence systems are promising tools for business development. At the same time, the basis for using these technologies is a mathematical apparatus for modeling and optimizing the activity of telecommunication companies. Adaptation of various models is the basis for the development of business of telecommunication companies using various technologies.

References

1. Gluhov, V.V., Ilin, I.V.: Project portfolio structure in a telecommunications company. In: Balandin, S., Andreev, S., Koucheryavy, Y. (eds.) NEW2AN 2014. LNCS, vol. 8638, pp. 509–518. Springer, Cham (2014). https://doi.org/10.1007/978-3-319-10353-2_46
2. Glukhov, V.V., Ilin, I.V., Levina, A.I.: Project management team structure for internet providing companies. In: Balandin, S., Andreev, S., Koucheryavy, Y. (eds.) ruSMART 2015. LNCS, vol. 9247, pp. 543–553. Springer, Cham (2015). https://doi.org/10.1007/978-3-319-23126-6_47
3. Ilin, I., Kalinina, O., Iliashenko, O., Levina, A.: IT-architecture reengineering as a prerequisite for sustainable development in Saint Petersburg urban underground. Procedia Eng. **165**, 1683–1692 (2016)
4. Lankhorst, M.: Enterprise Architecture at Work: Modelling, Communication and Analysis. Springer, Heidelberg (2009). https://doi.org/10.1007/978-3-642-29651-2
5. Winter, K., Buckl, S., Matthes, F., Schweda, C.M.: Investigating the state-of-the-art in enterprise architecture management methods in literature and practice. MCIS **90** (2010)
6. Ghaleb, S.M., Subramaniam, S., Zukarnain, Z.A., Muhammed, A.: Mobility management for IoT: a survey. EURASIP J. Wirel. Commun. Netw. **2016**, 165 (2016)
7. Song, Y.: "Bring Your Own Device (BYOD)" for seamless science inquiry in a primary school. Comput. Educ. **74**, 50–60 (2014)
8. Oyman, O., Foerster, J., Tcha, Y., Lee, S.-C.: Toward enhanced mobile video services over WiMAX and LTE [WiMAX/LTE update]. IEEE Commun. Mag. **48** (2010)
9. Lacmanović, I., Radulović, B., Lacmanović, D.: Contactless payment systems based on RFID technology. In: MIPRO, 2010 Proceedings of the 33rd International Convention. pp. 1114–1119. IEEE (2010)
10. Sun, J., Wu, Z., Pan, G.: Context-aware smart car: from model to prototype. J. Zhejiang Univ. Sci. A. **10**, 1049–1059 (2009)
11. Hamed, B.: Design & implementation of smart house control using LabVIEW. Int. J. Soft Comput. Eng. IJSCE **1**, 98–106 (2012)
12. Chen, M., Mao, S., Liu, Y.: Big data: a survey. Mob. Netw. Appl. **19**, 171–209 (2014)
13. Sherman, W.R., Craig, A.B.: Understanding Virtual Reality: Interface, Application, and Design. Elsevier, New York (2002)

14. Kopeetsky, M.: Mathematical modelling of the wireless communication network. Math. Stat. **8**
15. Mathematical model developed to assess telecommunications network performance - UQ News - The University of Queensland, Australia. https://www.uq.edu.au/news/article/1999/01/mathematical-model-developed-assess-telecommunications-network-performance
16. Stavroulakis, P.: Chaos Applications in Telecommunications. CRC Press, Boca Raton (2005)
17. Telecom Business Intelligence (2017). https://www.panorama.com/telecom-business-intelligence/

A Prospect Theoretic Look at a Joint Radar and Communication System

Andrey Garnaev[1,2]([✉]), Wade Trappe[2], and Athina Petropulu[3]

[1] Saint Petersburg State University, Saint Petersburg, Russia
garnaev@yahoo.com
[2] WINLAB, Rutgers University, North Brunswick, USA
trappe@winlab.rutgers.edu
[3] ECE, Rutgers University, Piscataway, USA
athinap@rutgers.edu

Abstract. In this paper, we consider the problem of finding how a joint radar and communication system should divide its effort between supporting the radar and communication objectives when the system operates in an environment with hostile interference. Our model explores the uncertainty of the jammer's location by assuming the joint system knows only the a priori probabilities of jammer's positions. The underlying problem is formulated and solved as a Bayesian game involving the joint radar/communication system and a jammer. We then explore how irrational behavior by the rivals can affect the equilibrium strategies by using prospect theory (PT). It is shown that the PT system strategy is not sensitive to the jammer's probability weighting parameter, while jammer's strategy is sensitive to probability weighted parameters of both rivals.

Keywords: Communication · Radar · Bayesian game
Prospect theory

1 Introduction

Recently, there has been interest in enabling radar and communication systems to co-exist in the same frequency bands in order to allow spectrum to be utilized more efficiently [20]. This has given rise to a significant amount of research on methods for spectrum sharing between the two systems. One approach to achieve this is to formulate waveform design using OFDM signals and then optimally allocating the subcarriers [14,24]. Radar waveform design for controlled interference is considered in [2,3], while the cooperative design of the two systems was explored in [4,17].

In this paper, we consider a dual purpose communication-radar system that employs OFDM style waveforms and explore the *complementary aspect* of finding the optimal frequency of performance for radar and communication objectives when the joint system faces with hostile interference. Moreover, we consider

© Springer Nature Switzerland AG 2018
O. Galinina et al. (Eds.): NEW2AN 2018/ruSMART 2018, LNCS 11118, pp. 483–495, 2018.
https://doi.org/10.1007/978-3-030-01168-0_43

that the system might know only a priori probabilities about jammer's position. The problem is formulated and solved as a Bayesian game. To address the risk that the rivals' behaviour might be irrational, equilibrium strategies are found using prospect theory. We note that, although prospect theory originally was designed to take into account possibility of the risk of irrational rivals behaviour in economic problems [15,16], it has been applied in engineering applications [18,22,25].

The organization of this paper is as follows: in Sect. 2, we present the basic system model, and then we incorporate a jammer in the model in Sect. 3. In Sect. 4, a basic zero-sum game for selecting the mode of attack and transmission for a fixed jammer position is formulated, while, in Sect. 5, its PT solution is found. In Sect. 6, the basic game is generalized to a Bayesian game where the system knows only a priori probabilities associated with the jammer's possible location, while, in Sect. 7, the PT solution for this Bayesian game is found. Finally, in Sect. 8, discussion of the results and conclusions are given.

2   Basic Model

We begin our formulation by considering an operational scenario involving an RF transceiver that is attempting to support two different objectives: communication with a communication receiver that is distant and separate from the transmitter, while also supporting the tracking of a radar target through the reflections witnessed at the RF transmitter. In order to support these two different objectives, the transmitter uses a spectrum band that is modeled as consisting of n adjacent sub-channels, which may be associated with n different subcarriers. In this paper we employ a transmission scheme like OFDM, as considered in [12,13] for designing a bargaining strategy for a dual radar and communication system in the absence of hostile interference. With each of these n different subcarriers, two different (fading) channel gains are associated. Specifically, we let $h_{R,i}$ correspond to the i-th radar channel gain associated with the round-trip effect of the transmitted signal, reflected off the radar target, and received at the RF transceiver, while $h_{C,i}$ denotes the i-th channel gain associated with the i-th communication subcarrier between the transmitter and the communication recipient. Although there are two different objectives, there is nonetheless a single transmitter responsible for deciding how to allocate power across the different subcarriers. We consider a power-allocation strategy for the transmitter to be the power vector $\boldsymbol{P} = (P_1, \ldots, P_n)$ where P_i is the power assigned for transmitting on subcarrier i, and $\sum_{i=1}^{n} P_i = \overline{P}$ where $\overline{P}$ is the total power budget allocated for transmission. We assume that the system, to avoid mutual interference of the signals, can work in one of two modes: (i) *communication mode* for performing only the communication task, and (ii) *radar mode* for performing only the radar task. In order to unify the examination of radar and communication metrics, we note that radar detection/tracking and communication throughput are both closely related to the associated signal-to-interference-plus-noise ratio (SINR) as witnessed at the appropriate recipient. Let the radar SINR be given by $\sum_{i=1}^{n} h_{R,i} i P_i/\sigma_R^2$ and the communication SINR by $\sum_{i=1}^{n} h_{C,i} P_i/\sigma_C^2$, where σ_R^2 and σ_C^2 are corresponding background noises. For the communication objective, the SINR is used as the payoff function for two reasons: first, it is easily linearized; and, second, for a low SINR regime, SINR in an approximation to throughput. For the radar objective, SINR is used as the payoff function since it is closely related to the associated detection metrics [19].

3   The Jammer

Now, suppose there is an adversary present in the environment who seeks to introduce hostile interference to disrupt the functionality of either the radar or communication system. His strategy is a power vector $J = (J_1, \ldots, J_n)$ where J_i is the power assigned for jamming subcarrier i, and $\sum_{i=1}^{n} J_i = \bar{J}$ where $\bar{J}$ is the total power budget allocated for jamming. Under a jamming attack, the communication and radar SINRs are given by: $\text{SINR}_R(P, J) = \sum_{i=1}^{n} h_{R,i} P_i/(\sigma_R^2 + g_{R,i} J_i)$ and $\text{SINR}_C(P, J) = \sum_{i=1}^{n} h_{C,i} P_i/(\sigma_C^2 + g_{C,i} J_i)$, where $g_{R,i}$ are fading channels gains between the jammer and the radar, and $g_{C,i}$ are fading gains between the jammer and the communication receiver.

When the system works in a particular mode the corresponding SINR is the payoff to the system, i.e., $v_m(P, J) = \text{SINR}_m(P, J)$ for $m \in \{C, R\}$. Let us assume the jammer knows what mode the system is in, then the system payoff can be considered as the cost function for the jammer. Thus, here we deal with a zero-sum game-theoretical scenario [5]. Recall that a pair of strategies (P_m, J_m) is a (Nash) equilibrium for the game with payoff function v_m for $m \in \{C, R\}$ if and only if, for any strategies (P, J), the following inequalities hold:

$$v_m(P, J_m) \leq v_m(P_m, J_m) \leq v_m(P_m, J). \tag{1}$$

This game can be solved following [1]. See, also, [6,8,9,11,23,26], as examples of other jamming games.

4   Mode Selection Game

In this section, we consider the scenario where the jammer does not know which the mode the system is in, while the system does not know which mode the jammer's effort is optimized against. The system can choose its mode to use, while the jammer can choose the mode he wants to optimize his effort against. If a mode is selected, each of the rivals allocates power according to the optimal strategy for this mode. This problem can be described by the following 2×2 payoff matrix

$$M = \begin{array}{c} \\ R \\ C \end{array} \begin{array}{c} R \quad C \\ \left(\begin{array}{cc} A & a \\ b & B \end{array} \right), \end{array} \tag{2}$$

with $A = v_R(\boldsymbol{P_R}, \boldsymbol{J_R}), a = v_R(\boldsymbol{P_R}, \boldsymbol{J_C}), B = v_C(\boldsymbol{P_C}, \boldsymbol{J_C}), b = v_C(\boldsymbol{P_C}, \boldsymbol{J_R})$. By (1), we can assume that

$$A < a, \quad B < b. \tag{3}$$

In matrix M, the rows correspond to the system's strategies, i.e. signal transmission according to one of the two modes; and the columns correspond to the jammer's strategies, i.e. jamming against a particular system mode.

Let $\boldsymbol{x} = (x, 1 - x)^T$ and $\boldsymbol{y} = (y, 1 - y)^T$, be randomized (mixed) strategies [5] for the system and the jammer, i.e., x and $1 - x$ (y and $1 - y$) be the probabilities for employing (pure) strategies "R" and "C" by the system (respectively, the adversary). Then, the expected payoff to the system is given:

$$v(\boldsymbol{x}, \boldsymbol{y}) = \boldsymbol{x}^T M \boldsymbol{y} = Axy + ax(1 - y) + b(1 - x)y + B(1 - x)(1 - y). \tag{4}$$

while for the jammer v is cost function. We look for a (Nash) equilibrium [5], i.e., for a pair of strategies $(\boldsymbol{x}_*, \boldsymbol{y}_*)$ such that the following inequalities hold:

$$v(\boldsymbol{x}, \boldsymbol{y}_*) \leq v(\boldsymbol{x}_*, \boldsymbol{y}_*) \leq v(\boldsymbol{x}_*, \boldsymbol{y}) \text{ for all } (\boldsymbol{x}, \boldsymbol{y}) \tag{5}$$

This is a classical 2×2 matrix zero sum game [5], which has a closed form solution.

Theorem 1. *The game has the unique equilibrium* $(\boldsymbol{x}, \boldsymbol{y})$. *Namely,*

$$x = \begin{cases} \xi, & B < a\,\&\,A < b, \\ 1, & A \geq b, \\ 0, & B \geq a, \end{cases} \qquad y = \begin{cases} \Xi, & B < a\,\&\,A < b, \\ 1, & A \geq b, \\ 0, & B \geq a, \end{cases} \tag{6}$$

with $\xi := (b - B)/(a + b - A - B)$ *and* $\Xi := (a - B)/(a + b - A - B)$.

Note that two inequalities $A \geq b$ and $B \geq a$ cannot hold simultaneously since otherwise summing them up implies $A + B \geq a + b$, and this contradicts assumption (3). Thus, (6) defines $(\boldsymbol{x}, \boldsymbol{y})$ uniquely.

5   PT Solution for Mode Selection Game

In the previous section, we assumed that rivals' decisions were according to the expected utility theory. To describe decisions under the risk that the agents' behaviour might be irrational occur, prospect theory was developed [15]. According to [15], agents use their subjective probabilities w rather than objective probabilities p to weight the values of possible outcomes. Moreover, agents tend to over-weight low probability outcomes and under-weight moderate and high probability outcomes. This feature is captured by weighting the probability distribution by an S-shaped function, the so-called *weighting function*. The original example of weighting function [16] is given by

$$w_\gamma(p) = p^\gamma/(p^\gamma + (1-p)^\gamma)^{1/\gamma}, \tag{7}$$

with $\gamma \in [1/2, 1]$ the *probability weighting parameter,* where the lower bound on γ comes from [21]. In particular, this weighting function infinitely overweights infinitesimal probabilities and infinitely underweights near-one probabilities, i.e. $w_\gamma(p)/p$ tends to infinity for $p \downarrow 0$ and $(1 - w_\gamma(p))/(1-p)$ tends to infinity for $p \uparrow 1$.

Denote by α and β these parameters for the system and the jammer, respectively. Then, the PT-utilities for the rivals in the matrix game (2) are given as follows:

$$
\begin{aligned}
u_S^{PT}(x, y) &:= x\left(Aw_\beta(y) + aw_\beta(1-y)\right) + (1-x)\left(bw_\beta(y) + Bw_\beta(1-y)\right), \\
u_J^{PT}(x, y) &:= y\left(Aw_\alpha(x) + bw_\alpha(1-x)\right) + (1-y)\left(aw_\alpha(x) + Bw_\alpha(1-x)\right).
\end{aligned} \tag{8}
$$

Then, the PT-equilibrium is given as the solution of the best response equations:

$$x = \mathrm{BR}_S^{PT}(y) := \underset{x \in [0,1]}{\operatorname{argmax}}\, u_S^{PT}(x, y),\, y = \mathrm{BR}_J^{PT}(x) := \underset{y \in [0,1]}{\operatorname{argmin}}\, u_J^{PT}(x, y). \tag{9}$$

Theorem 2. *The game has the unique PT equilibrium* $(\boldsymbol{x}, \boldsymbol{y})$. *Namely,*

$$
x = \begin{cases} \xi(\alpha), & B < a\,\&\,A < b, \\ 1, & A \geq b, \\ 0, & B \geq a, \end{cases} \qquad
y = \begin{cases} \Xi(\beta), & B < a\,\&\,A < b, \\ 1, & A \geq b, \\ 0, & B \geq a, \end{cases} \tag{10}
$$

with $\xi(\alpha) := (b - B)^{1/\alpha}/((a - A)^{1/\alpha} + (b - B)^{1/\alpha})$ *and* $\Xi(\beta) := (a - B)^{1/\beta}/((a - B)^{1/\beta} + (b - A)^{1/\beta})$.

Proof: Since $u_S^{PT}(x, y)$ is linear on x while $u_J^{PT}(x, y)$ is linear on y solving the optimization problems (7) we find the best response strategies as follows:

$$
\mathrm{BR}_S^{PT}(y) = \begin{cases} 0, & (b - A)w_\beta(y) > (a - B)w_\beta(1 - y), \\ \text{any in } [0, 1], & (b - A)w_\beta(y) = (a - B)w_\beta(1 - y), \\ 1, & (b - A)w_\beta(y) < (a - B)w_\beta(1 - y), \end{cases} \tag{11}
$$

$$
\mathrm{BR}_J^{PT}(x) = \begin{cases} 1, & (a - A)w_\alpha(x) > (b - B)w_\alpha(1 - x), \\ \text{any in } [0, 1], & (a - A)w_\alpha(x) = (b - B)w_\alpha(1 - x), \\ 0, & (a - A)w_\alpha(x) < (b - B)w_\alpha(1 - x). \end{cases} \tag{12}
$$

Substituting (7) with $\gamma = \alpha$ into (12) implies

$$
\mathrm{BR}_J^{PT}(x) = \begin{cases} 1, & x > \xi(\alpha), \\ \text{any in } [0, 1], & x = \xi(\alpha), \\ 0, & x < \xi(\alpha). \end{cases} \tag{13}
$$

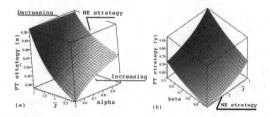

Fig. 1. (a) Probability x and (b) probability y for PT equilibrium.

Let $b \leq A$. Then, by (3), $a > B$. Thus, (11) implies that $\mathrm{BR}_S^{PT}(y) \equiv 1$. Then, by (13), $\mathrm{BR}_J^{PT}(x) \equiv 1$.

Let $a \leq B$. Then, by (3), $b > A$. Thus, (11) implies that $\mathrm{BR}_S^{PT}(y) \equiv 0$. Then, by (13), $\mathrm{BR}_J^{PT}(x) \equiv 0$.

Let $b > A$ and $b > A$. Then, substituting (7) with $\gamma = \beta$ into (11) yields that

$$
\mathrm{BR}_S^{PT}(y) = \begin{cases} 0, & y > \Xi(\beta), \\ \text{any in } [0,1], & y = \Xi(\beta), \\ 1, & y < \Xi(\beta). \end{cases} \tag{14}
$$

Thus, by (13) and (14), $x \in (0,1)$ and $y \in (0,1)$ is the PT equilibrium if and only if $x = \xi(\alpha)$ and $y = \Xi(\beta)$, and the result follows. ∎

It is clear that the PT equilibrium in mixed strategies (10) coincides with the Nash equilibrium (10) for $\alpha = \beta = 1$. The threshold condition for switching the PT equilibrium from mixed to pure strategies is stable with respect to the probability weighting parameters (i.e. it does not depend on them). Both mixed strategies $\xi(\alpha)$ and $\Xi(\beta)$ are monotonic with respect to the probability weighting parameters. Namely, if $a + B > b + A$ then $\xi(\alpha)$ is increasing and less than the NE strategy x. If $a + B < b + A$ then $\xi(\alpha)$ is decreasing and always greater than the NE strategy x. While if $b + B > a + A$ then $\Xi(\beta)$ is increasing and less than the NE strategy y. If $b + B < a + A$ then $\Xi(\beta)$ is decreasing and less than the NE strategy y. Fig. 1 illustrates these monotonic properties by an example $n = 5$, $\sigma_C^2 = 1, \sigma_R^2 = 1, h_C = (0.9, 0.7, 0.8, 1, 0.9), g_C = (1, 2, 3, 4, 5), h_R = (0.8, 0.9, 1, 0.7, 0.8)$, $g_R = (5, 4, 3, 2, 1)$ and $\overline{P} = 1$. Moreover, Fig. 1(a) illustrates that variation in network parameters (in this case, total jamming power) can change the type of monotonicity.

6   Uncertainty About the Jammer's Position

The channel gains depend on the positioning of the jammer and the receiver [27]. In this section we assume that the system does not know the jammer's position with certainty, but knows only a priori probabilities about jammer's position. Namely, let the jammer be at point (X_t, Y_t) with a priori probability γ_t, and let the corresponding channel gains be $\{h_{C,i,t}, g_{C,i,t}, h_{R,i,t}, g_{C,i,t}\}$. Let $\boldsymbol{P}_{m,t}$ and $\boldsymbol{J}_{m,t}$ be the equilibrium strategies when the system is in mode m and the jammer acts versus this mode. We assign the type, namely, *type t*, to the jammer associated with his location. Let M_t be matrix of payoffs for the system facing a jammer of type t designed based on strategies $\boldsymbol{P}_{m,t}$ and $\boldsymbol{J}_{m,t}$, while A_t, a_t, B_t, b_t be entries of this matrix. Following (3), we can assume that

$$A_t < a_t \text{ and } B_t < a_t. \tag{15}$$

Let $\boldsymbol{y}_t = (y_t, 1 - y_t)^T$ be the strategy of the jammer with type t, and $\mathcal{Y} = (\boldsymbol{y}_1, \ldots, \boldsymbol{y}_T)$. Let $\boldsymbol{x} = (x, 1 - x)^T$ be the strategy of the system. Then, the (expected) payoff to the system is given as follows $v(\boldsymbol{x}, \mathcal{Y}) = \sum_{t=1}^{T} \gamma_t \boldsymbol{x}^T M_t \boldsymbol{y}_t$, while the cost function for the jammer of type t is given by $v_t(\boldsymbol{x}, \boldsymbol{y}_t) = \boldsymbol{x}^T M_t \boldsymbol{y}_t$. The system wants to maximize its expected payoff while each of type of the jammer wants to minimize its cost function. Thus, here we deal with a Bayesian game. Recall that $(x_*, \mathcal{Y}_*)$ is the Bayesian equilibrium if and only if for any $(x, \mathcal{Y})$ the following inequalities hold:

$$v(\boldsymbol{x}, \mathcal{Y}_*) \le v(\boldsymbol{x}_*, \mathcal{Y}_*) \text{ and } v_t(\boldsymbol{x}_*, \boldsymbol{y}_{t*}) \le v_t(\boldsymbol{x}_*, \boldsymbol{y}_t) \text{ for } t = 1, \ldots, T. \tag{16}$$

Let us introduce the following axillary notations:

$$\Theta_t := \sum_{\tau=1}^{t} \gamma_\tau(a_\tau + b_\tau - A_\tau - B_\tau) \text{ for } t = 1, \ldots, T \text{ and } \Theta_0 = 0,$$

$$\xi_t := (b_t - B_t)/(a_t + b_t - A_t - B_t) \text{ for } t = 1, \ldots, T \text{ and } \xi_0 = 0, \tag{17}$$

$$\Theta := \sum_{t=1}^{T} \gamma_t(a_t - B_t).$$

By (15), $0 < \xi_t < 1$. To avoid bulkiness in formulas we assume that $\xi_t \ne \xi_\tau$ for any $t \ne \tau$. Then, without loss of generality we can assume that

$$0 = \xi_0 < \xi_1 < \xi_2 < \ldots < \xi_T < 1. \tag{18}$$

Theorem 3

(a) Let $\Theta \le 0$. Then the equilibrium $(\boldsymbol{x}, \mathcal{Y})$ is unique and $x = 0$, $y_t = 0$, $t = 1, \ldots, T$.
(b) Let $\Theta_T < \Theta$. Then the equilibrium is unique and $x = 1$, $y_t = 1$, $t = 1, \ldots, T$.
(c) Let $\Theta_{t_*-1} < \Theta < \Theta_{t_*}$ for a $t_* \in \{1, \ldots, T\}$. Then the equilibrium is unique and

$$x = \xi_{t_*} \text{ and } y_t = \begin{cases} 1, & t \le t_* - 1, \\ (\Theta - \Theta_{t_*-1})/(a_{t_*} + b_{t_*} - A_{t_*} - B_{t_*}), & t = t_*, \\ 0, & t \ge t_* + 1. \end{cases} \tag{19}$$

(d) Let $\Theta = \Theta_{t_*}$ for a $t_* \in \{1, \ldots, T\}$. Then each x such that $\xi_{t_*-1} < x < \xi_{t_*}$ is an equilibrium strategy for the system while the jammer equilibrium strategy is unique and given as:

$$y_t = \begin{cases} 1, & t \le t_* - 1, \\ 0, & t \ge t_*. \end{cases} \tag{20}$$

Note that although a continuum of equilibria might arise for the system they are equivalent since they return the same payoff.

Proof. By (16), $(\boldsymbol{x}, \mathcal{Y})$ is an equilibrium if they are best response strategy to each other, i.e., solution of the following equations:

$$x = \mathrm{BR}_S(\mathcal{Y}) := \underset{x}{\mathrm{argmax}}\, v(\boldsymbol{x}, \mathcal{Y}), y_t = \mathrm{BR}_{J,t}(x) := \underset{y_t}{\mathrm{argmin}}\, v_t(\boldsymbol{x}, \boldsymbol{y}_t), t = 1, \ldots, T. \tag{21}$$

Since $v(\boldsymbol{x}, \mathcal{Y})$ is linear on $\boldsymbol{x}$ while $v_t(\boldsymbol{x}, \boldsymbol{y}_t)$ is linear on $\boldsymbol{y}_t$ solving the optimization problems (21) we find the best response strategies as:

$$\mathrm{BR}_S(\mathcal{Y}) = \begin{cases} 1, & \sum\limits_{t=1}^{T} \gamma_t(a_t + b_t - A_t - B_t)y_t < \sum\limits_{t=1}^{T} \gamma_t(a_t - B_t), \\ \text{any in } [0,1], & \sum\limits_{t=1}^{T} \gamma_t(a_t + b_t - A_t - B_t)y_t = \sum\limits_{t=1}^{T} \gamma_t(a_t - B_t), \\ 0, & \sum\limits_{t=1}^{T} \gamma_t(a_t + b_t - A_t - B_t)y_t > \sum\limits_{t=1}^{T} \gamma_t(a_t - B_t), \end{cases} \tag{22}$$

$$\mathrm{BR}_{J,t}(\boldsymbol{x}) = \begin{cases} 0, & (a_t + b_t - A_t - B_t)x < b_t - B_t, \\ \text{any in } [0,1], & (a_t + b_t - A_t - B_t)x = b_t - B_t, \\ 1, & (a_t + b_t - A_t - B_t)x > b_t - B_t. \end{cases} \tag{23}$$

Then, (3), (17) and (23) imply

$$\mathrm{BR}_{J,t}(\boldsymbol{x}) = \begin{cases} 1, & x > \xi_t, \\ \text{any in } [0,1], & x = \xi_t, \\ 0, & x < \xi_t. \end{cases} \tag{24}$$

Let $x = 0$. Substituting into (24) implies that $y_t = \mathrm{BR}_{J,t}((0,1)) = 0, t = 1, \ldots, T$. Finally substituting $x = 0, y_t = 0.t = 1, \ldots, T$ into (22) implies that $\Theta \le 0$, and (a) follows.

Let $x = 1$. Substituting into (24) implies that $y_t = \mathrm{BR}_{J,t}((1,0)) = 1, t = 1, \ldots, T$. Finally substituting $x = 1, y_t = 1, t = 1, \ldots, T$ into (22) implies that $\Theta \le \Theta$, and (b) follows.

Thus, only the case $0 < x < 1$ is left to consider. Two subcases arise: (I) there is t_+ such $x = \xi_{t_+}$ and (II) there is t_+ such $\xi_{t_+-1} < x < \xi_{t_+}$.

(I) Let there exist t_+ such $x = \xi_{t_+}$. Then, by (18) and (24),

$$y_t = \mathrm{BR}_{J,t}(\boldsymbol{x}) \equiv \begin{cases} 1, & t < t_+, \\ \text{any in } [0,1], & t = t_+, \\ 0, & x > t_+. \end{cases} \tag{25}$$

Substituting these $\{y_t\}$ into (22) implies

$$x = BR_S(\mathcal{Y}) = \begin{cases} 1, & L_{t_+}(y_{t_+}) < 0, \\ \text{any in } [0,1], & L_{t_+}(y_{t_+}) = 0, \\ 0, & L_{t_+}(y_{t_+}) > 0, \end{cases} \tag{26}$$

where

$$L_{t_+}(y_{t_+}) := \sum_{t=1}^{t_+-1} \gamma_t(b_t - A_t) - \sum_{t=t_+}^{n} \gamma_t(a_t - B_t) + \gamma_{t_+}(a_{t_+} + b_{t_+} - A_{t_+} - B_{t_+})y_{t_+}. \tag{27}$$

By (27), $x = \xi_{t_+}$ if and only if $L_{t_+}(y_{t_+}) = 0$. Since $L_{t_+}(y_{t_+})$ is linear and increasing in y_+, this equation has a root $y_+ \in (0,1)$ (and it is unique) if and only if $L_{t_+}(y_{t_+}) < 0$ and $L_{t_+}(y_{t_+}) > 0$. This is equivalent to $t_+ = t_*$ where t_* is given by condition (a-iii). Then, solving $L_{t_*}(y_{t_*}) = 0$ by y_{t_*} implies (19), and (c) follows.

(II) Let there exist t_+ such $\xi_{t_+-1} < x < \xi_{t_+}$. Then, by (18) and (24),

$$y_t = BR_{J,t}(x) \equiv \begin{cases} 1, & t < t_+, \\ 0, & t \geq t_+. \end{cases} \tag{28}$$

Substituting these $\{y_t\}$ into (22) implies

$$x = BR_S(\mathcal{Y}) = \begin{cases} 1, & L_{t_+}(0) < 0, \\ \text{any in } [0,1], & L_{t_+}(0) = 0, \\ 0, & L_{t_+}(0) > 0. \end{cases} \tag{29}$$

By (29), $\xi_{t_+-1} < x < \xi_{t_+}$ if and only if $L_{t_+}(0) = 0$, i.e., $t_+ = t_*$ where t_* is given by condition of (b). This and (28) imply (d), and the result follows. ∎

7 PT Solution for Bayesion Game

Denote the probability weighting parameter for the system by α. We assume that the jammer's weighting parameter β does not depend on jammer's position. Then, the PT-utilities for the rivals in the Bayesian game are given as:

$$u_S^{PT}(x, \mathcal{Y}) := x \sum_{t=1}^{T} (A_t w_\beta(y_t) + a_t w_\beta(1 - y_t)) + (1 - x) \sum_{t=1}^{T} (b_t w_\beta(y) + B_t w_\beta(1 - y_t)),$$

$$u_{J,t}^{PT}(x, y_t) := y_t(A_t w_\alpha(x) + b_t w_\alpha(1 - x)) + (1 - y_t)(a_t w_\alpha(x) + B_t w_\alpha(1 - x)). \tag{30}$$

Then, the PT-equilibrium is given as the solution of the best response equations:

$$x = BR_S^{PT}(\mathcal{Y}) := \underset{x}{\operatorname{argmax}}\, u_S^{PT}(x, \mathcal{Y}), \quad y_t = BR_{J,t}^{PT}(x) := \underset{y_t}{\operatorname{argmin}}\, u_{J,t}^{PT}(x, y_t), t = 1, \ldots, T.$$

Let us introduce the following auxiliary notation: $\xi_t(\alpha) = (b_t - B_t)^{1/\alpha}/((a_t - A_t)^{1/\alpha} + (b_t - B_t)^{1/\alpha})$ for $t = 1,\ldots,T$ and $\xi_0(\alpha) = 0$. It is clear that $0 < \xi_t(\alpha) < 1$. To avoid bulkiness in formulas we assume that $\xi_t(\alpha) \neq \xi_\tau(\alpha)$ for any $t \neq \tau$. Then, without loss of generality we can assume that

$$0 = \xi_0(\alpha) < \xi_1(\alpha) < \xi_2(\alpha) < \ldots < \xi_T(\alpha) < 1. \tag{31}$$

Theorem 4. *The PT equilibrium coincides with the Bayesian equilibrium given by Theorem 3 except for the case where equilibrium strategies of both rivals are mixed, i.e., the case (c). In this case, i.e., if $\Theta_{t_*-1} < \Theta < \Theta_{t_*}$ then*

$$x = \xi_{t_*}(\alpha) \text{ and } y_t = \begin{cases} 1, & t \leq t_* - 1, \\ \Xi_{t_*}(\beta), & t = t_*, \\ 0, & t \geq t_* + 1, \end{cases} \tag{32}$$

where $\Xi = \Xi_{t_*}(\beta)$ is the root in $(0,1)$ of the equation:

$$
\begin{aligned}
L_{\beta,t_*}(\Xi) := &\sum_{t=1}^{t_*-1} \gamma_t(b_t - A_t) - \sum_{t_*+1}^{T} \gamma_t(a_t - B_t) \\
&+ \gamma_{t_*}(b_{t_*} - A_{t_*})w_\beta(\Xi) - \gamma_{t_*}(a_{t_*} - B_{t_*})w_\beta(1 - \Xi) = 0.
\end{aligned}
\tag{33}
$$

Proof: Since u_S^{PT} is linear on x while $u_{J,t}^{PT}$ is linear on y_t the best response strategies are given as follows:

$$
\mathrm{BR}_S^{PT}(\mathcal{Y}) \begin{cases} = 0, & \sum_{t=1}^{T} \gamma_t(b_t - A_t)w_\beta(y_t) > \sum_{t=1}^{T} \gamma_t(a_t - B_t)w_\beta(1 - y_t), \\ \in [0,1], & \sum_{t=1}^{T} \gamma_t(b_t - A_t)w_\beta(y_t) = \sum_{t=1}^{T} \gamma_t(a_t - B_t)w_\beta(1 - y_t), \\ = 1, & \sum_{t=1}^{T} \gamma_t(b_t - A_t)w_\beta(y_t) < \sum_{t=1}^{T} \gamma_t(a_t - B_t)w_\beta(1 - y_t), \end{cases} \tag{34}
$$

$$
\mathrm{BR}_{J,t}^{PT}(x) \begin{cases} = 1, & (a_t - A_t)w_\alpha(x) > (b_t - B_t)w_\alpha(1 - x), \\ \in [0,1], & (a_t - A_t)w_\alpha(x) = (b_t - B_t)w_\alpha(1 - x), \\ = 0, & (a_t - A_t)w_\alpha(x) < (b_t - B_t)w_\alpha(1 - x). \end{cases} \tag{35}
$$

Substituting (7) with $\gamma = \alpha$ into (35) implies

$$
\mathrm{BR}_{J,t}^{PT}(x) \begin{cases} = 1, & x > \xi_t(\alpha), \\ \in [0,1], & x = \xi_t(\alpha), \\ = 0, & x < \xi_t(\alpha). \end{cases} \tag{36}
$$

Then, by (7), (31), (34) and (36) all of the cases besides (c) can be proved similarly to Theorem 3. Thus, we have to consider only the cases where $x = \xi_{t_*}(\alpha)$ for a t_* such that $\Theta_{t_*-1} < \Theta < \Theta_{t_*}$. By (34), $x = \xi_{t_*}$ if and only if $L_{\beta,t_*}(y_{t_*}) = 0$. Note that $L_{\beta,t_*}(0) = \Theta_{t_*-1} - \Theta < 0$ and $L_{\beta,t_*}(1) = \Theta_{t_*} - \Theta > 0$. Thus, the root exists and the result follows. ∎

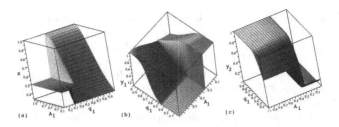

Fig. 2. (a) Probability x, (b) probability y_1 and (c) probability y_2 with $T = 2$, $q_1 \in [0, 1]$, $\alpha = \beta = 1/2$ $q_2 = 1 - q_1$, $A_1 \in (0, a_1)$ and $M_1 = ((A_1, 1), (1, 0.3))$, $M_2 = ((0.3, 0.9), (0.8, 0.4))$.

8   Discussion of the Results and Conclusions

By Theorems 3 and 4, the PT system strategy is not sensitive to the probability weighted parameter for the jammer, and it depends only on the probability weighted parameter of the system. On the other hand, the PT strategy for the jammer depends on both of these parameters. Although the system's strategy is any probability vector, i.e., the feasibility set consists of a continuum of elements, Theorems 3 and 4 shows that there is only a finite set $\Gamma := \{\xi_t(\alpha) : t = 0, \ldots, T\} \cup \{1\}$ of strategies given in closed form containing all of the equilibrium. While elements of this set depend on weighted parameter of the system, i.e., the PT system strategy is sensitive to this parameter, the selection rule to identify the equilibrium is stable with respect to this parameter. Figure 2 illustrates that the PT system strategy can be sensitive to a priori probabilities (namely, q_1) while the jammer's strategy is sensitive to the system's parameters (namely, A_1). This is similar to what can be observed in bandwidth protection games [7,10], where an agent sometimes has to react sharply to small variations in the environment parameters. Finally, we note that, in this paper, although we have employed the original probability weighting function (7) given in [16], the obtained result can be generalized for any such S-shaped weighting function. A goal for our future work is to develop solutions for the joint radar and communication system design using *cumulative prospect theory* [16], which is a variant of prospect theory that allows one to take into account risk aversion of the rivals.

References

1. Altman, E., Avrachenkov, K., Garnaev, A.: Jamming in wireless networks under uncertainty. Mob. Netw. Appl. **16**, 246–254 (2011)
2. Aubry, A., De Maio, A., Huang, Y., Piezzo, M., Farina, A.: A new radar waveform design algorithm with improved feasibility for spectral coexistence. IEEE Trans. Aerosp. Electr. Syst. **51**, 1029–1038 (2015)
3. Bica, M., Huang, K.W., Koivunen, V., Mitra, U.: Mutual information based radar waveform design for joint radar and cellular communication systems. In: IEEE International Conference on Acoustics, Speech and Signal Processing (ICASSP), pp. 3671–3675 (2016)

4. Bica, M., Koivunen, V.: Delay estimation method for coexisting radar and wireless communication systems. In: IEEE Radar Conference, pp. 1557–1561 (2017)
5. Fudenberg, D., Tirole, J.: Game Theory. MIT Press, Boston (1991)
6. Garnaev, A., Hayel, Y., Altman, E., Avrachenkov, K.: Jamming game in a dynamic slotted ALOHA network. In: Jain, R., Kannan, R. (eds.) GameNets 2011. LNIC-SSITE, vol. 75, pp. 429–443. Springer, Heidelberg (2012). https://doi.org/10.1007/978-3-642-30373-9_30
7. Garnaev, A., Trappe, W.: Stationary equilibrium strategies for bandwidth scanning. In: Jonsson, M., Vinel, A., Bellalta, B., Marina, N., Dimitrova, D., Fiems, D. (eds.) MACOM 2013. LNCS, vol. 8310, pp. 168–183. Springer, Cham (2013). https://doi.org/10.1007/978-3-319-03871-1_15
8. Garnaev, A., Trappe, W.: To eavesdrop or jam, that is the question. In: Mellouk, A., Sherif, M.H., Li, J., Bellavista, P. (eds.) ADHOCNETS 2013. LNICSSITE, vol. 129, pp. 146–161. Springer, Cham (2014). https://doi.org/10.1007/978-3-319-04105-6_10
9. Garnaev, A., Trappe, W.: Fair resource allocation under an unknown jamming attack: a Bayesian game. In: 2014 IEEE Workshop on Information Forensics and Security (WISP), pp. 227–232. IEEE, Atlanta (2015)
10. Garnaev, A., Trappe, W.: One-time spectrum coexistence in dynamic spectrum access when the secondary user may be malicious. IEEE Trans. Inf. Forensics Secur. 10, 1064–1075 (2015)
11. Garnaev, A., Trappe, W.: Stability of communication link connectivity against hostile interference. In: 2017 IEEE Global Conference on Signal and Information Processing (GlobalSIP), pp. 136–140. IEEE, Montreal (2018)
12. Garnaev, A., Trappe, W., Petropulu, A.: Bargaining over fair performing dual radar and communication task. In: 50th Asilomar Conference on Signals, Systems, and Computers, pp. 47–51. Pacific Grove, CA, November 2016
13. Garnaev, A., Trappe, W., Petropulu, A.: Bargaining in a dual radar and communication system using radar-prioritized OFDM waveforms. In: Galinina, O., Andreev, S., Balandin, S., Koucheryavy, Y. (eds.) NEW2AN/ruSMART/NsCC - 2017. LNCS, vol. 10531, pp. 382–394. Springer, Cham (2017). https://doi.org/10.1007/978-3-319-67380-6_35
14. Gogineni, S., Rangaswamy, M., Nehorai, A.: Multi-modal OFDM waveform design. In: IEEE Radar Conference, pp. 1–5 (2013)
15. Kahneman, D., Tversky, A.: Prospect theory: an analysis of decision under risk. Econometrica 47, 263–291 (1979)
16. Kahneman, D., Tversky, A.: Advances in prospect theory: cumulative representation of uncertainty. J. Risk Uncertainty 5, 297–323 (1992)
17. Li, B., Petropulu, A.P., Trappe, W.: Optimum co-design for spectrum sharing between matrix completion based MIMO radars and a MIMO communication system. IEEE Trans. Sig. Process. 64, 4562–4575 (2016)
18. Li, T., Mandayam, N.B.: When users interfere with protocols: prospect theory in wireless networks using random access and data pricing as an example. IEEE Trans. Wirel. Commun. 13, 1888–1907 (2014)
19. Poor, H.V.: An Introduction to Signal Detection and Estimation. Springer, New York (1994)
20. Federal Communications Commission (FCC): FCC proposes innovative small cell use in 3.5 GHz band, December 2012. https://apps.fcc.gov/edocs_public/attachmatch/DOC-317911A1.pdf
21. Rieger, M.O., Wang, M.: Cumulative prospect theory and the st. petersburg paradox. Econ. Theory 28, 665–679 (2006)

22. Sanjab, A., Saad, W., Basar, T.: Prospect theory for enhanced cyber-physical security of drone delivery systems: a network interdiction game. In: IEEE International Conference on Communications (ICC), Paris, France (2017)

23. Song, T., Stark, W.E., Li, T., Tugnait, J.K.: Optimal multiband transmission under hostile jamming. IEEE Trans. Commun. **64**, 4013–4027 (2016)

24. Turlapaty, A., Jin, Y.: A joint design of transmit waveforms for radar and communications systems in coexistence. In: IEEE Radar Conference, pp. 0315–0319 (2014)

25. Xu, D., Xiao, L., Mandayam, N.B., Poor, H.V.: Cumulative prospect theoretic study of a cloud storage defense game against advanced persistent threats. In: IEEE Conference on Computer Communications Workshops (INFOCOM WKSHPS), pp. 541–546 (2017)

26. Yang, D., Xue, G., Zhang, J., Richa, A., Fang, X.: Coping with a smart jammer in wireless networks: a stackelberg game approach. IEEE Trans. Wirel. Commun. **12**, 4038–4047 (2013)

27. Zhu, Q., Saad, W., Han, Z., Poor, H.V., Basar, T.: Eavesdropping and jamming in next-generation wireless networks: a game-theoretic approach. In: Military Communication Conference (MILCOM), pp. 119–124 (2011)

Algorithm for Positioning in Non-line-of-Sight Conditions Using Unmanned Aerial Vehicles

Grigoriy Fokin[(⊠)] and Al-odhari Abdulwahab Hussain Ali

The Bonch-Bruevich St. Petersburg State University of Telecommunications,
St. Petersburg, Russia
grihafokin@gmail.com, abdwru2011@yandex.ru

Abstract. The identification of Line of Sight (LOS) in the processing of navigational measurements is relevant for positioning in urban conditions, as well as in heterogeneous terrain such as mountains and hills, when there is no direct visibility between the radio source and the receiving stations. The purpose of this work is to develop and verify algorithm for positioning a transmitting radio source in Non-Line-of-Sight Conditions (NLOS) using Unmanned Aerial Vehicles (UAVs) in three dimensional space. Algorithm under consideration implements time difference of arrival (TDOA) measurements processing for identification of receivers with NLOS measurements. Algorithm operability is illustrated for the layout including terrestrial segment with ground receiver stations and flying segment with receiving sensors aboard UAVs. The method used for NLOS identification and mitigation exploits the comparison of variance for intermediate location estimates calculated for different TDOA measurements combinations among all possible sets of receivers with thresholds. Algorithm was realized in simulation model including system level, link level and visualization model subsystems. TDOA system level model represents positioning layout with distributed transmitter, receivers, obstacles and NLOS reflectors in three dimensional space. TDOA link level model represents radio links between transmitter, receivers, and NLOS reflectors taking into account actual pathloss, signal modulation, sampling rate, additive noise and cross-correlation calculation. Comparing with the case on the plane, TDOA measurements processing in three dimensional space case with flying receiver aboard UAVs reveals substantially higher thresholds of calculated variances to reliably identify and exclude NLOS source.

Keywords: TDOA · NLOS · UAV · Root mean square error
Measurement processing

1 Introduction

Positioning of transmitter stations or radio sources in wireless networks is an important trend in next-generation mobile communications applications. Time Difference of Arrival (TDOA) is most widely used technique for geolocation applications and was adopted by 3rd Generation Partnership Project in LTE [1] Positioning Protocol providing Observed Time Difference of Arrival (OTDoA) positioning scheme based on reference signal TDOA measurements which was investigated in [2] with simulation.

© Springer Nature Switzerland AG 2018
O. Galinina et al. (Eds.): NEW2AN 2018/ruSMART 2018, LNCS 11118, pp. 496–508, 2018.
https://doi.org/10.1007/978-3-030-01168-0_44

That investigation assumed configuration of receiver points or sensors on the plane, allocated around unknown radio source position to be determined and did not take into account Non-Line-of-Sight (NLOS) situations. In the case of NLOS several surrounding sensor TDOA measurements could be corrupted by reflection, diffraction and scattering leading to NLOS errors in range difference which result in erroneous location estimate and as a consequence it is necessary to exclude such sensors from measurement [3].

The complexity of positioning in hilly terrain and mountain areas is determined by the irregularity of the morphostructure of the terrain. To determine the location of radio source in such hard-to-reach hilly and mountain areas it is necessary to solve the problem of positioning in three-dimensional space, which is complicated, firstly, by the impossibility of proper arranging of ground receiver stations within terrestrial segment around radio source and at the same altitude level as radio source [3] and, secondly, by the probable absence of Line of Sight (LOS) between radio source and one and/or several receiver stations [4].

In LOS case when radiowave propagates through a straight line between the transmitter (Tx) and Receiver (Rx) the only source of error in TDOA measurements is measurement noise which can be modelled as zero-mean Gaussian random variable. In the absence of LOS between Tx and Rx radiowave suffers from reflection, diffraction and scattering leading to additional propagation distance and pathloss with respect to LOS case and result in NLOS propagation error ranging up to several hundreds of meters, which is much greater than the average Gaussian measurement noise [5].

However, in hilly terrain and mountains for certain applications, such as search and rescue operations, emergency medical services, law enforcement, personnel tracking and others, the task of radio source positioning has to be realized even in worse NLOS conditions, and one of the solutions for this problem is implementation of flying segment with receiving sensors aboard Unmanned Aerial Vehicles (UAVs) [6], which helps to improve the accuracy of the positioning of radio emission sources.

Analysis of publications about the area of time difference of arrival measurement processing for positioning in non-line-of-sight conditions has shown that this problem had already been studied before for the case of mobile networks on the plane in [7–10] and several NLOS error identification and mitigation techniques had already been proposed. However, these works lack of simulation model subsystems including radio links between transmitter, receivers, and NLOS reflectors in 2-dimensional lay-out taking into account actual pathloss, signal transmission and reception, with signal format, modulation, sampling rate, additive noise and signal cross-correlation calculation, peak detection and pairwise TDOA estimation. Such comprehensive simulation model was developed in [11], however it did not take into account effects of signal power loss due to propagation. Complex simulation model of radio sources positioning in NLOS conditions taking into account actual pathloss [12] was developed and realized in [13] for 2-dimensional layout on the plane. Three-stage TDOA measurements processing algorithm for positioning a radio source when there are up to two ground receivers with NLOS measurements was also verified in [13], however it was limited by terrestrial segment in 2-dimensional layout on the plane.

Recent research in flying ubiquitous sensor networks development [14, 15] demonstrated operability of joint cooperation of flying segment based on Unmanned Aerial Vehicles (UAVs) with terrestrial segment including ground based stations.

Implementation of flying segment based on UAVs was already investigated in [16, 17], however there was not taken into account NLOS problem.

The aim of this work is to develop and verify TDOA measurements processing algorithm introduced in [13] for positioning a radio source in NLOS conditions for three dimensional space layout including terrestrial segment with ground receiver stations and flying segment with receiving sensors aboard UAVs.

The material in the paper is organized in the following order. System level, link level and visualization model subsystems considering TDOA measurements are given in Sect. 2. TDOA NLOS measurements processing algorithm is formalized in Sect. 3. Simulation results for three dimensional space layout including terrestrial segment with ground receiver stations and flying segment with receiving sensors aboard UAVs are described in Sect. 4. Finally, we draw the conclusions in Sect. 5.

2 TDOA Measurement Processing Model

TDOA measurement processing system and link level models for terrestrial segment in 2-dimensional layout including ground based stations are detailed in [2, 3, 13]. Let's generalize it for accounting flying segment based on receiver stations aboard UAVs.

TDOA measurement processing system level model under consideration includes up to N receiving points termed receivers (Rx) located at the distances d_i from the radio source as the positioning target termed transmitter (Tx) and described in Fig. 1.

TDOA measurement processing model described further is based on the assumption of ideal synchronization between receiving points (stations) of the terrestrial and flying segments aboard UAV, which gather primary time-of-arrival (TOA) measurements. Primary TOA measurements are instantly sent (via radio communication channel) to the central processing unit with respective time stamps for further TDOA estimation by means of pair-wise signals cross correlation, presented in Fig. 2.

TDOA measurement processing link level model under consideration is detailed in [3, 11, 13].

TDOA measurement processing visualization model under consideration can be illustrated by an example scenario in Fig. 3: radio source Tx is located at the point (5, 4.3, 1.5); ground segment includes 5 stationary receiving points, which are located at the points (3, 8, 0.5), (7.5, 8, 0.9), (9.5, 5, 0.6), (7.5, 1.5, 1.1), (3, 1.5, 0.6), coordinates are measured in km; flying segment is represented by UAV, that flies circumferentially over the working area at a constant altitude z = 4 km.

The simulation described further is detailed in [13] and based on the assumption of ideal synchronization between receiving points (stations) of the terrestrial and flying segments aboard UAV.

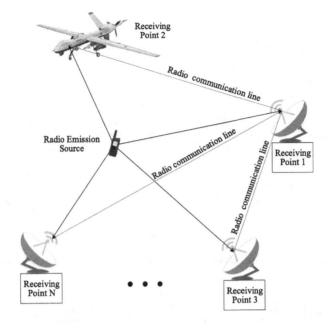

Fig. 1. TDOA measurement processing model in three dimensional space with UAV.

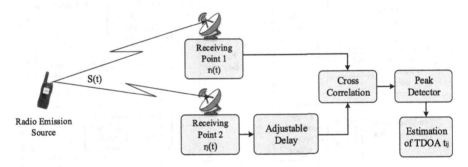

Fig. 2. TDOA estimation with pair-wise signals cross correlation.

3   TDOA NLOS Measurement Processing Algorithm

Consider TDOA NLOS measurement processing algorithm depicted in Fig. 4.

Realized algorithm for NLOS identification and mitigation exploits the comparison of variance for intermediate location estimates calculated for different TDOA measurements combinations among all possible sets of receivers with thresholds [13].

Let's define intermediate estimate $\widehat{\mathbf{x}}_i$, calculated with respect to reference receiver Rx_i, as an estimate for a given set i; the number of possible sets equals to the number of receivers N. Then the final transmitter location estimate $\widehat{\mathbf{x}}$ can be calculated as a mean from intermediate location estimates for N sets as

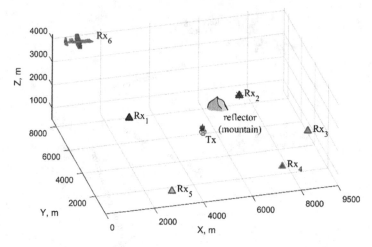

Fig. 3. Example layout with 5 ground stations and one UAV receiver; Rx_2 is NLOS source.

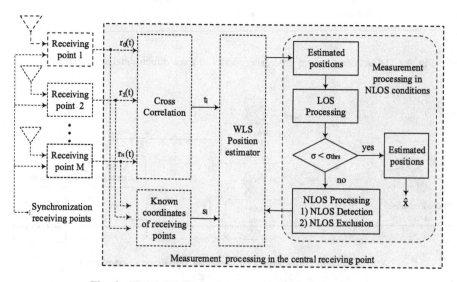

Fig. 4. TDOA NLOS measurement processing algorithm.

$$\widehat{\mathbf{x}} = \frac{1}{N}\sum_{i=1}^{N}\widehat{\mathbf{x}}_i. \qquad (1)$$

Variance σ of intermediate location estimates for N sets is

$$\sigma = D[\widehat{\mathbf{x}}_i] = \frac{1}{N}\sum_{i=1}^{N}(\widehat{\mathbf{x}}_i - \widehat{\mathbf{x}})^2. \qquad (2)$$

Variance σ of intermediate location estimates can reach the order of 10^3 m when one of the receivers has NLOS measurements for the case of positioning on the plane [11, 13]. Intermediate location estimates for a single set $\widehat{\mathbf{x}}_i$ in its turn are calculated as a mean from intermediate location estimates for different combinations $\widehat{\mathbf{x}}_{ic}$ in a given set

$$\widehat{\mathbf{x}}_i = \frac{1}{C}\sum_{c=1}^{C}\widehat{\mathbf{x}}_{ic}, \tag{3}$$

where the number of combinations in a given set is defined by

$$C = C_N^L = \binom{N}{L} = \frac{N!}{L!(N-L)!}, \tag{4}$$

where L is the minimum required number of receiver stations with LOS measurements; for positioning in three dimensional space L = 4. Variance σ_n of intermediate location estimates for c = 1,…,C combinations in each set is defined by

$$\sigma_n = D[\widehat{\mathbf{x}}_{ic}] = \frac{1}{C}\sum_{c=1}^{C}(\widehat{\mathbf{x}}_{ic} - \widehat{\mathbf{x}}_i)^2. \tag{5}$$

Let's make several novel assumptions for our TDOA measurement processing algorithm in NLOS conditions especially for the case of three dimensional space compared to [13]. At first, consider that receiver stations with NLOS measurements are identified and excluded from final location estimation calculation (1). Secondly, for reliable NLOS source identification the number of receivers with LOS measurements should be greater than minimum by one; that is, for reliable identification of a single NLOS Rx source in three dimensional space we need 6 stations with 5 LOS stations. Thirdly, we'll consider sets with the minimum required number of receiver stations in a single combination; that is L = 4 for three dimensional space. Now we can illustrate stages of TDOA NLOS measurement processing algorithm for example scenario in Fig. 3 with intermediate location estimates for sets and combinations in the next section.

4   Simulation Results

The stage of estimated positions in TDOA NLOS measurement processing algorithm for an example scenario in Fig. 3 with intermediate location estimates for sets (1) and variance of intermediate location estimates (2) is illustrated in Fig. 5.

Define intermediate estimate $\widehat{\mathbf{x}}_i$, calculated with respect to (wrt) reference receiver Rx_i, as an estimate for a given set i; the number of possible sets equals to the number of receivers N. Then the final transmitter location estimate $\widehat{\mathbf{x}}$ can be calculated as a mean from intermediate location estimates for N sets. Estimate 1 was obtained wrt reference receiver Rx_1, estimate 2 – wrt Rx_2 etc. Variance $\sigma = 1{,}12{\cdot}10^5$ m for 6 sets (2) and significant deviation of intermediate location estimates from true location is a sign of NLOS.

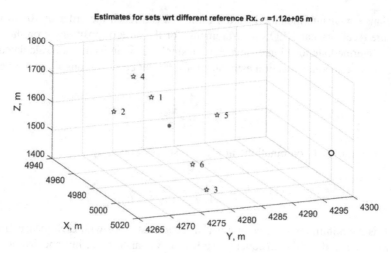

Fig. 5. Intermediate estimates for sets of receivers for example in Fig. 3.

The stage of LOS processing in TDOA NLOS measurement processing algorithm for an example scenario in Fig. 3 is illustrated by intermediate location estimates for combinations within sets in Fig. 6.

For example in Fig. 6 we have $C_{N-1}^{L-1} = C_5^3 = 10$ combinations in each set [13]. Intermediate location estimate $\widehat{\mathbf{x}}_1$ for the 1st set S_1 is calculated by (3) as a mean from 10 combinations within set $S_1 = \{Rx_{1234}, Rx_{1235}, Rx_{1236}, Rx_{1245}, Rx_{1246}, Rx_{1256}, Rx_{1345}, Rx_{1346}, Rx_{1356}, Rx_{1456}\}$ wrt Rx_1; similarly, $\widehat{\mathbf{x}}_2, \widehat{\mathbf{x}}_3, \widehat{\mathbf{x}}_4, \widehat{\mathbf{x}}_5, \widehat{\mathbf{x}}_6$ were calculated form sets S_2, S_3, S_4, S_5 and S_6 wrt to Rx_2, Rx_3, Rx_4, Rx_5 and Rx_6 respectively. Analysis of Fig. 6 shows also great variances of intermediate location estimates for combinations in each set, calculated by (5). For example, within set S_1 variance $\sigma_1 = 2{,}89 \cdot 10^5$m: six estimates for combinations $\{Rx_{1234}, Rx_{1235}, Rx_{1236}, Rx_{1245}, Rx_{1246}, Rx_{1256}\}$ are significantly scattered relative to the true location, while four estimates for combinations $\{Rx_{1345}, Rx_{1346}, Rx_{1356}, Rx_{1456}\}$, in which range measurement d_2 from the NLOS source Rx_2 is excluded, lead to correct estimates. Similar conclusion can be drawn for the other sets except for a set wrt NLOS source Rx_2: estimates for all combinations within set $S_2 = \{Rx_{2345}, Rx_{2346}, Rx_{2356}, Rx_{2341}, Rx_{2351}, Rx_{2361}, Rx_{2456}, Rx_{2451}, Rx_{2461}, Rx_{2561}\}$ contain NLOS erroneous distance measurements d_2.

The stage of NLOS processing in TDOA NLOS measurement processing algorithm for an example scenario in Fig. 3 is illustrated in Fig. 7. To reliably identify NLOS source we implement consecutive exclusion of a single receiver with subsequent calculation of intermediate estimates (3) and respective variances (5) for combinations in residual sets, illustrated by three-dimensional layout and NLOS processing log file in Fig. 7.

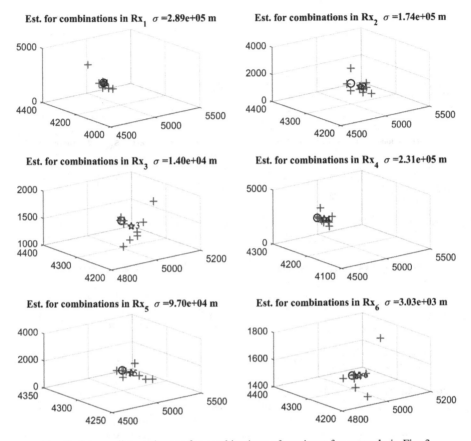

Fig. 6. Intermediate estimates for combinations of receivers for example in Fig. 3.

In Figs. 3, 5, 6 and 7 we have illustrated scenario when stationary ground receiver Rx_2 within terrestrial segment produced NLOS measurements. Let's now consider scenario, when moving receiver aboard UAV within flying segment produces NLOS measurements for a short time flight with three-dimensional layout and NLOS processing log file in Fig. 8.

Root Mean Square Error (RMSE) of current estimates in three axes according to UAV flight time for scenario in Fig. 8 is provided in Fig. 9.

From Figs. 8 and 9 it can be seen that the RMSE considerably increases in the interval from 42 s to 60 s, just when LOS between UAV and transmitting radio source is absent and NLOS radiowave comes after reflection from mountain during the UAV flight behind the obstacle; in Fig. 8 this flight interval is illustrated by reflected rays. Results in Fig. 9 also confirmed that RMSE along the z-axis increases substantially higher in comparison with RMSE along x-y axes on the plane [18, 19].

Resulting RMSE and variance of positioning accuracy for the case of L = 4 LOS receivers in three dimensional space wrt SNR is plotted in Fig. 10.

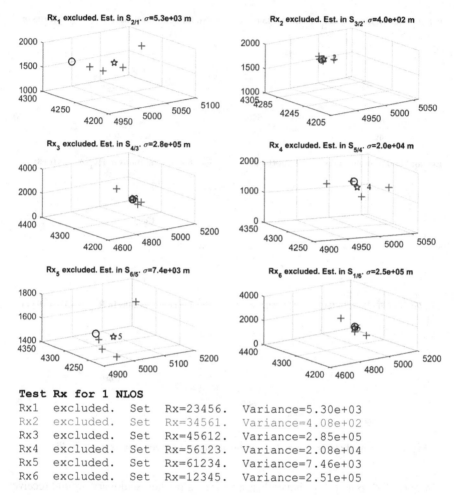

Fig. 7. Intermediate estimates for combinations of Rx's in residual sets for example in Fig. 3.

RMSE in Fig. 10 is approximately five times greater than for two dimensional case in [13], and can be explained with increased error along the z-axis in comparison with RMSE along x-y axes on the plane. Variance in Fig. 10 is by an order of magnitude greater than for two dimensional case in [3, 11, 13].

Value of NLOS threshold should be chosen through simulations according to the following considerations: it should be larger than LOS variance for worst SNR values and, at the same time, lower than variance of NLOS containing intermediate estimates in sets during calculations within TDOA NLOS measurement processing algorithm. This consideration, which held true with a margin for NLOS identification and mitigation in two-dimensional case on the plane [3, 11, 13], requires fine-tuning in three dimensional space using UAVs. For example, $\sigma_{thrs} = 10^4$, chosen as a threshold for instance, is slightly larger than LOS variance for low SNR values in Fig. 10, and, at the

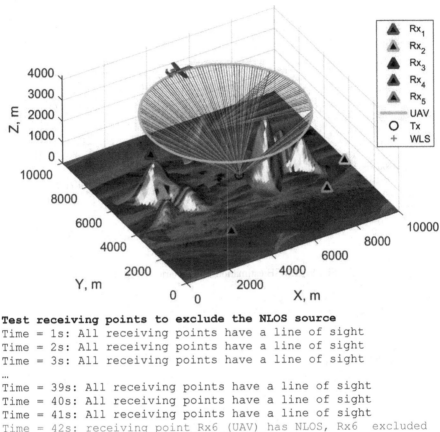

```
Test receiving points to exclude the NLOS source
Time = 1s: All receiving points have a line of sight
Time = 2s: All receiving points have a line of sight
Time = 3s: All receiving points have a line of sight
...
Time = 39s: All receiving points have a line of sight
Time = 40s: All receiving points have a line of sight
Time = 41s: All receiving points have a line of sight
Time = 42s: receiving point Rx6 (UAV) has NLOS, Rx6  excluded
Time = 43s: receiving point Rx6 (UAV) has NLOS, Rx6  excluded
...
Time = 59s: receiving point Rx6 (UAV) has NLOS, Rx6  excluded
Time = 60s: receiving point Rx6 (UAV) has NLOS, Rx6  excluded
Time = 61s: All receiving points have a line of sight
Time = 62s: All receiving points have a line of sight
Time = 63s: All receiving points have a line of sight
...
Time = 198s: All receiving points have a line of sight
Time = 199s: All receiving points have a line of sight
Time = 200s: All receiving points have a line of sight
```

Fig. 8. Example layout with 5 ground receivers and one UAV receiver with NLOS measurements.

same time, is not always lower than variance of intermediate estimates in sets and combinations in Figs. 5, 6 and 7.

Practical relevance of the obtained results lies in the choice of $\sigma_{thrs} = 10^3$ which is by an order of magnitude greater comparing with two dimensional case on the plane in [13] and is reasonable in three dimensional space case for positive SNR values.

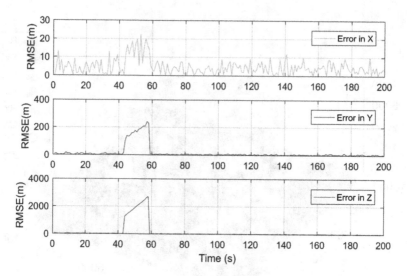

Fig. 9. RMSE estimation for scenario in Fig. 8.

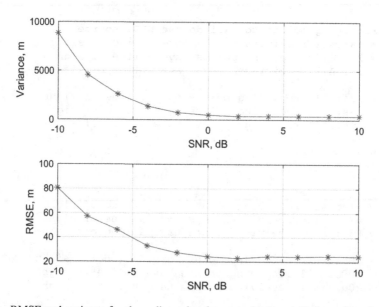

Fig. 10. RMSE and variance for three dimensional space with L = 4 (3 ground receivers and one UAV receiver).

5 Conclusion

Contribution of proposed TDOA measurement processing algorithm lies in accounting of calculated variances for different combinations within individual set of receivers and among all possible sets of receivers especially for the case of three dimensional space including terrestrial segment with stationary ground receiver stations and flying segment with moving receiving sensors aboard UAV. By means of computer simulation possibility to reliably identify and temporarily exclude single NLOS receiver aboard UAV from TDOA measurement processing was verified. Comparing with two dimensional case, threshold of calculated variance to reliably identify and exclude NLOS source in three dimensional space should be by an order of magnitude greater.

Acknowledgements. The reported study was supported by the Committee on Science and Higher School of the Government of St. Petersburg.

References

1. Gelgor, A., Pavlenko, I., Fokin, G., Gorlov, A., Popov, E., Lavrukhin, V.: LTE base stations localization. In: Balandin, S., Andreev, S., Koucheryavy, Y. (eds.) NEW2AN 2014. LNCS, vol. 8638, pp. 191–204. Springer, Cham (2014). https://doi.org/10.1007/978-3-319-10353-2_17
2. Sivers, M., Fokin, G.: LTE positioning accuracy performance evaluation. In: Balandin, S., Andreev, S., Koucheryavy, Y. (eds.) ruSMART 2015. LNCS, vol. 9247, pp. 393–406. Springer, Cham (2015). https://doi.org/10.1007/978-3-319-23126-6_35
3. Fokin, G., Kireev, A., Al-odhari, A.H.A.: TDOA positioning accuracy performance evaluation for arc sensor configuration. In: 2018 Systems of Signals Generating and Processing in the Field of on Board Communications, Moscow, pp. 1–5 (2018)
4. Kireev, A., Fokin, G., Al-odhari, A.H.A.: TOA measurement processing analysis for positioning in NLOS conditions. In: 2018 Systems of Signals Generating and Processing in the Field of on Board Communications, Moscow, Russia, pp. 1–4 (2018)
5. Zekavat, R., Buehrer, R.M.: Handbook of Position Location: Theory, Practice and Advances. Wiley, Hoboken (2011)
6. Al-odhari, A.H.A., Fokin, G., Kireev, A.: Positioning of the radio source based on time difference of arrival method using unmanned aerial vehicles. In: 2018 Systems of Signals Generating and Processing in the Field of on Board Communications, Moscow, pp. 1–5 (2018)
7. Cong, L., Zhuang, W.: Nonline-of-sight error mitigation in mobile location. IEEE Trans. Wirel. Commun. 4(2), 560–573 (2005)
8. Wylie, M.P., Holtzman, J.: The non-line of sight problem in mobile location estimation. In: Proceedings of ICUPC - 5th International Conference on Universal Personal Communications, Cambridge, MA, vol. 2, pp. 827–831 (1996)
9. Chen, P.C.: A non-line-of-sight error mitigation algorithm in location estimation. In: Proceedings of WCNC. IEEE Wireless Communications and Networking Conference (Cat. No. 99TH8466), New Orleans, LA, vol. 1, pp. 316–320 (1999)
10. Cong, L., Zhuang, W.: Non-line-of-sight error mitigation in TDOA mobile location. In: Proceedings of Global Telecommunications Conference. GLOBECOM 2001, vol. 1, pp. 680–684. IEEE, San Antonio (2001)
11. Montminy, M.B.: Passive geolocation of low-power emitters in urban environments using TDOA. Master's thesis, Air Force Institute of Technology (2007)
12. 3GPP TR 36.814 V9.2.0. Technical Specification Group Radio Access Network; Evolved Universal Terrestrial Radio Access (E-UTRA); Further advancements for E-UTRA physical layer aspects (2017)

13. Fokin, G.: Complex imitation model of radio emission sources positioning in the non-line-of-sight conditions. In: Proceedings of Telecommunication Universities, vol. 4(1), pp. 85–101 (2018)

14. Koucheryavy, A., Vladyko, A., Kirichek, R.: State of the art and research challenges for public flying ubiquitous sensor networks. In: Balandin, S., Andreev, S., Koucheryavy, Y. (eds.) ruSMART 2015. LNCS, vol. 9247, pp. 299–308. Springer, Cham (2015). https://doi.org/10.1007/978-3-319-23126-6_27

15. Kirichek, R., Paramonov, A., Vareldzhyan, K.: Optimization of the UAV-P's motion trajectory in public flying ubiquitous sensor networks (FUSN-P). In: Balandin, S., Andreev, S., Koucheryavy, Y. (eds.) ruSMART 2015. LNCS, vol. 9247, pp. 352–366. Springer, Cham (2015). https://doi.org/10.1007/978-3-319-23126-6_32

16. Kim, D.H., Lee, K., Park, M.Y., Lim, J.: UAV-based localization scheme for battlefield environments. In: MILCOM 2013 – 2013 IEEE Military Communications Conference, San Diego, CA, pp. 562–567 (2013)

17. Du, H.J., Lee J.P.Y.: Passive geolocation using TDOA method from UAVs and ship/land-based platforms for maritime and littoral area surveillance. Defence R & D Canada-Ottawa (2004)

18. Mashkov, G., Borisov, E., Fokin, G.: Experimental validation of multipoint joint processing of range measurements via software-defined radio testbed. In: 18th International Conference on Advanced Communication Technology (ICACT), Pyeongchang, pp. 268–273 (2016)

19. Mashkov, G., Borisov, E., Fokin, G.: Positioning accuracy experimental evaluation in SDR-based MLAT with joint processing of range measurements. In: 2016 International Conference on Radar, Antenna, Microwave, Electronics, and Telecommunications (ICRAMET), Jakarta, pp. 7–12 (2016)

Features of the Development of Transceivers for Information and Communication Systems Considering the Distribution of Radar Operating Frequencies in the Frequency Range

Alexey S. Podstrigaev[1,2], Andrey V. Smolyakov[2],
Vadim V. Davydov[3,4], Nikita S. Myazin[3(✉)],
and Maria G. Slobodyan[5]

[1] Scientific-Research Institute Vector OJSC, St. Petersburg, Russia
[2] Saint Petersburg Electrotechnical University "LETI", St. Petersburg, Russia
[3] Peter the Great Saint-Petersburg Polytechnic University, St. Petersburg, Russia
myazin.n@list.ru
[4] Department of Ecology, All-Russian Research Institute of Phytopathology,
Odintsovo District, B.Vyazyomy, 143050 Moscow Region, Russia
[5] Bryansk State Technical University, Bryansk, Russia

Abstract. The influence of out-of-band emitters on the performance of information and communication systems (ICS) is estimated. The broadband and narrowband ICS are analyzed. A significant influence of radars on the characteristics of ICS is shown. The distribution density histogram of more than 900 types of radars in frequency range is obtained and described. The prospects for the operation of ICS in the millimeter range are demonstrated.

Keywords: Information and communication systems
Information and communication systems · Radar · Frequency range
Radar operating frequencies · Microwave interference · Radio range load

1 Introduction

In time of designing systems for radio control, data transmission, communication, navigation and other information and communication systems (ICS), the choice of the transceiver is an important step.

At the design stage of the transceiver, the information about the load of various parts of the radio band allows predicting possible out-of-band emitters.

Thus, for the most common superheterodyne receiver to date, out-band radiation sources may interfere with the combination and intermodulation reception channels. Receiver mixers and amplifiers are nonlinear elements for high power signals and form an ensemble of harmonics. In the case of work in a narrowband channel, harmonics are formed in it, if there is an input broadband amplifier. Such case is typical for a multichannel system.

With the expansion of the channel band, it is highly likely that the harmonics will be also formed in the mixer. This is mainly due to the high power of out-of-band

© Springer Nature Switzerland AG 2018
O. Galinina et al. (Eds.): NEW2AN 2018/ruSMART 2018, LNCS 11118, pp. 509–515, 2018.
https://doi.org/10.1007/978-3-030-01168-0_45

radiation. In this case, the frequencies of parasitic harmonics practically cannot be predicted and, therefore, suppressed.

Thus, it is obvious that when using broadband path elements, the noise immunity of the info-communication system decreases under the influence of powerful out of band emitters.

However, even in a single-channel narrowband system, the influence of powerful interference is possible. It manifests itself in the overload of sensitive elements of the receiving equipment, and in some cases up to their failure. To improve the stability of the receiving equipment to the effects of high-power interference, it is possible to use a protective device. However, to protect the sensitive input amplifier, protective device must be installed between the antenna and the amplifier, which inevitably reduces sensitivity. Therefore, to increase the range of the ICS, protective devices are often not installed in them.

In addition, if the automatic gain control (AGC) is used, the occurrence of a strong interference leads to the loss of weak signals.

Another aspect of the influence of powerful interference on the quality of the ICS is the occurrence of intersymbol interference in the transmission of digital signals. Operation in the nonlinear mode, triggering of the AGC system and changing of any other parameters of the transceiver device in the continuous signal reception mode lead to a change in the phase-frequency characteristic (PFC) in time. As a result, not only the phase jitter increases, which in modern systems is minimized with the devices based on phase automatic frequency control (AFC) [1, 2], but also phase jumps can occur. In this case, the distinctiveness of the code sequence characters deteriorates even with the use of fiber-optic communication lines [3, 4], and the probability of erroneous reception increases.

Thus, in a single-channel narrowband system, the ingress of a powerful noise signal into the reception band (or into its surroundings in the case of low selectivity of the channel filter) also leads to disruption of the normal operation of the system, up to its failure.

In addition, when designing both narrowband and broadband microwave range systems, it is necessary to consider the periodic nature of the frequency-response function of band devices. In other words, there are side channels of interference penetration corresponding to resonance bands of filters and other microwave band devices. It is worth noting that the suppression of the signal in the parasitic trans-mission bands, as a rule, is insignificant. Therefore, one of the many phenomena described above may occur at a lower interference power.

Moreover, due to the widespread and ubiquitous use of radars (both for civilian and military use), their frequencies and powers in a given area are difficult to predict. Therefore, the development or modernization of ICS requires information about the distribution of operating frequencies of radar in the radio band.

Such parameters of the ICS as carrier frequency, channel bandwidth, radiated power, etc., are directly related to the workload of certain parts of the radio band. These parameters, in turn, determine the structure of the system and the quality of its work [5, 6].

In this regard, the quantitative assessment of the radio band load is an urgent task. Considering the constant increase in the number of sources of radio emission in the information space, the demand for results of this assessment in the future will only grow.

2   Frequency Ranges of Radars for Various Purposes

In this paper, the radio band load is indirectly estimated by the density of distribution of radar types in it. Information on operating frequencies of more than 900 radars was obtained from open sources, including the most representative [7]. The obtained results can be useful for a wide range of developers of electronic devices. In Fig. 1 shows graphs of the distribution density of civilian and military radar types in the frequency range up to 40 GHz.

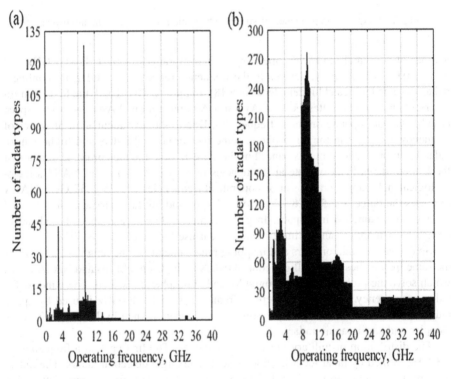

Fig. 1. Frequency distribution density of different radar types: (a) - civilian, (b) - military.

The choice of frequency range of radar primarily due to the non-uniform frequency dependence of attenuation of waves in the atmosphere. In Fig. 2 shows, as an example, one of such dependences, obtained at T = 293 K and normal atmospheric pressure 101325 Pa.

The vast majority of the civilian radars included in the sample are concentrated in the band up to 20 GHz. The two most noticeable peaks in the graph represent a large

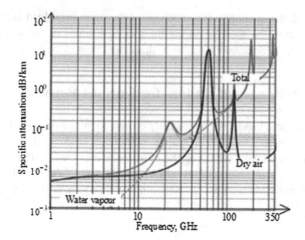

Fig. 2. Frequency dependence of specific attenuation of waves in the atmosphere

number of river and sea radars operating in narrow ranges of 9.41 ± 0.03 GHz and 9.41 ± 0.03 GHz. Mainly, devices in the frequency range up to 1 GHz are nonlinear radars designed to detect hidden electronics [8]. In the 1–2 GHz band, the flight control radars of the United States Federal Aviation Administration (FAA) and some space based radars are operate on those frequencies. Due to the low attenuation of waves of frequency 2–4 GHz in the atmosphere, some meteorological radars and flight control radar at airports operate in this band. For the same reason, the meteorological radars of the United States FAA operate at frequencies of 5.00–5.55 GHz, and NASA's meteorological and mapping radar stations operate at frequencies of 5.25–5.57 GHz. NASA also uses the range of 5.57–5.83 GHz to track civil and military missiles, satellites and other objects launched into space. In general, meteorological and space based radars designed for mapping and altitude measurement use the range of 4–8 GHz.

Space-based scatterometers and mapping radars operate in the range 8–18 GHz. At the same frequencies operate flight control radars in the United States and NASA radar, which solve the problems of oceanography and atmospheric research. The window for radio transparency of the atmosphere at frequencies of 30–40 GHz is used by space radio altimeters, cloud and precipitation research radars owned by NASA, as well as Doppler and cartographic radars.

The distribution of military radars is also uneven. The first peak at a frequency of 1.3 GHz formed by radar identification systems and target detection radars. The radars used by the military to detect and track missile and artillery fire operate within that range too. Frequencies 1.3–1.35 GHz are used by the military to monitor air traffic. The long-range airspace radar, as well as the long range air traffic control radar also use a range of 1–2 GHz.

The next peak refers to the range from 2.9 to 3.5 GHz. At frequencies 2.9–3.1 GHz a large number of radars (including mobile) for airspace surveillance, search and tracking targets, air defense radars and sea-based navigation radars operate. In the range 3.1–3.3 GHz, military use radars of various types of basing, designed to support the operation of weapon control systems, as well as to detect air and ground targets. In the range 3.3–3.5 GHz, multi functional radars (including sea-based systems) are used in weapons control systems, detection and tracking of ground and air targets.

The next small peak is observed in the 5–6 GHz band. The frequency band 5.03–5.091 GHz of this range is used in landing systems. In the band 5.255–5.925 GHz, there are radars of missile defense systems.

In the frequency range 8–12 GHz, attenuation of signals in the atmosphere is relatively low. Due to this, a balance is achieved between the resolution of the radar and its mass-size characteristics.

Meteorological and navigation stations, mobile stations for detection of artillery fire, weapon control radar, missile defense radar systems, airborne radar of interceptor aircrafts, fighter aircraft and ground attack planes, cartographic radars operate at these frequencies. At the current stage of the development of radio engineering, a long range of action cannot be ensured when operating at higher frequencies, so the radio frequency load decreases with increasing frequency.

At frequencies above 20 GHz, short range radars operate due to high attenuation of the signal in the atmosphere (Fig. 2). Also this range is operational for a target tracking.

3   Analysis of the Results

The total dependence of the distribution density of all types of radars is more convenient for use in the work and for the choice of operating frequencies (Fig. 3).

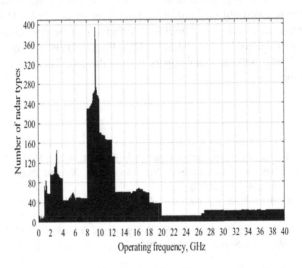

Fig. 3. Distribution density of radars for different purposes by frequency

Analysis of the data presented in Fig. 3 shows that most of the radars are concentrated in the X-band. The maximum density of distribution falls on a narrow band of frequencies about 9.1 GHz (main peak). In general, we can conclude that the frequency range from 0.5 to 20 GHz is extremely loaded.

Due to the short range of operation and technological difficulties [8, 9] in the implementation, in the frequency range above 40 GHz are currently working special radar for military purposes (for example, in the homing missiles, to direct special short-range weapons, etc.).

The millimeter range is used in the development of radar for unmanned aerial vehicles [10–12], as well by devices for checking the quality of the railway track [13].

4 Conclusion

The choice of operating frequencies in the design or modernization of the ICS should be performed considering the radio band load. It is shown that the choice of the operating frequency without taking into account the radio band load by radars can significantly worsen the characteristics of the system, up to its failure. The probability of this increases during operation in a complex electronic environment. Examples include communications in the sea, near airports, on the borders of states and in other cases of massive use of radar. Moreover, in the sea, the situation worsens multiple re-reflection of signals from the water surface. Combined with weather conditions, this can significantly impair the quality of local communication and radar detection.

The results obtained in the paper show that interference from stations in the millimeter range has the least impact on signals transmitted in ICS. Therefore, it is the most promising for the development of such systems. In particular, in the city it can be used to control unmanned objects, and in the long term by cars [14]. At the same time, the operation in the centimeter range will lead to failures in the operation of the control system and unsolvable problems with traffic. Therefore, data about frequency density in the information space will help developers reduce the likelihood of failures in communication systems.

References

1. Chenakin, A.: Frequency Synthesizers. Concept to Product. Artech House, Norwood (2011)
2. Ameri, H., Attaran, A., Moghavvemi, M.: Design an X-band frequency synthesizer. Microwaves RF **79**, 98–103 (2010)
3. Davydov, V.V., Ermak, S.V., Karseev, A.U., Nepomnyashchaya, E.K., Petrov, A.A., Velichko, E.N.: Fiber-optic super-high-frequency signal transmission system for sea-based radar station. In: Balandin, S., Andreev, S., Koucheryavy, Y. (eds.) NEW2AN 2014. LNCS, vol. 8638, pp. 694–702. Springer, Cham (2014). https://doi.org/10.1007/978-3-319-10353-2_65
4. Ermolaev, A.N., Krishpents, G.P., Davydov, V.V., Vysoczkiy, M.G.: Compensation of chromatic and polarization mode dispersion in fiber-optic communication lines in microwave signals transmittion. J. Phy. Conf. Ser. **741**(1), 012171 (2016). https://doi.org/10.1088/1742-6596/741/1/012171

5. Ryazantsev, L.B., Likhachev, V.P.: Assessment of range and radial velocity of objects of a broadband radar station under conditions of range cell migration. Meas. Tech. **60**(11), 1158–1162 (2018). https://doi.org/10.1007/s11018-018-1334-4

6. Tarasenko, M.Yu., Davydov, V.V., Lenets, V.A., Akulich, N.V., Yalunina, T.R.: Features of use direct and external modulation in fiber optical simulators of a false target for testing radar station. In: Galinina, O., Andreev, S., Balandin, S., Koucheryavy, Y. (eds.) NEW2AN/ruSMART/NsCC -2017. LNCS, vol. 10531, pp. 227–232. Springer, Cham (2017). https://doi.org/10.1007/978-3-319-67380-6_21

7. Streetly, M. (ed.): Jane's Radar and Electronic Warfare Systems, 22nd ed. IHS Jane's, London (2010)

8. Podstrigaev, A.S., Likhachev, V.P., Ryazantsev, L.B.: Technique for tuning microwave strip devices. Meas. Tech. **59**(5), 547–550 (2016). https://doi.org/10.1007/s11018-016-1005-2

9. Podstrigaev, A.S.: All-purpose adjuster for microwave microstrip devices. In: 24th International Crimean Conference on Microwave & Telecommunication Technology (CriMiCo), pp. 896–897 (2014). https://doi.org/10.1109/crmico.2014.6959682

10. Bystrov, V.V., Likhachev, V.P., Ryazantsev, L.B.: Experimental check of the coherence of radiolocation signals from objects with nonlinear electrical properties. Meas. Tech. **57**(9), 1073–1076 (2014). https://doi.org/10.1007/s11018-014-0582-1

11. Koo, V.C., et al.: A new unmanned aerial vehicle synthetic aperture radar for environmental monitoring. Prog. Electromagnet. Res. **122**, 245–268 (2012)

12. González-Partida, J.-T., Almorox-González, P., Burgos-Garcia, M., Dorta-Naranjo, B.P.: SAR system for UAV operation with motion error compensation beyond the resolution cell. Sensors **8**(5), 3384–3405 (2008)

13. Giancarlo, M., Broggi, A., Cerri, P.: Vehicle and guard rail detection using radar and vision data fusion. IEEE Trans. Intell. Transp. Syst. **8**(1), 95–105 (2007). https://doi.org/10.1109/TITS.2006.888597

14. Viikari, V.V., Varpula, T., Kantanen, M.: Road-condition recognition using 24-GHz automotive radar. IEEE Trans. Intell. Transp. Syst. **10**(4), 639–648 (2009). https://doi.org/10.1109/TITS.2009.2026307

EMC Provision Method of LTE-800 Networks and Air Traffic Control Radars Based on Mechanism of Cell Radius Management for LTE Base Stations

Valery Tikhvinskiy[1,2]([✉]), Victor Koval[3], Pavel Korchagin[3], and Sergey Terentyev[1]

[1] JSC National R&D Institute of Technologies and Communications, Moscow, Russian Federation
vtniir@mail.ru
[2] Moscow Technical University of Communications and Informatics, Moscow, Russian Federation
[3] Geyser-Telecom Ltd, Moscow, Russian Federation

Abstract. Geographical separation and frequency separation between a source and a victim of harmful interferences are two main methods for ensuring electromagnetic compatibility (EMC). Geographical separation was chosen as the main method of sharing in 800 MHz band between user terminal (UE) of LTE-800 and air traffic control radars (ATC radars) since frequency separation is not possible in case of co-channel LTE and ATC Radar operations.

The proposed interference protection method for ATC Radar receivers experiencing interference from UE transmitters of LTE-800 networks is based on providing separation distance between ATC Radar and UE. Ensuring of separation distance is implemented by the management and limiting of cell radius of LTE base stations located near ATC Radars. The method was proved by simulation tests and can provide a solution to the EMC issue by LTE base station parameter management which defines cell radius in LTE-800.

Keywords: Radar · Protection criteria · LTE-800 · Air-traffic control

1 Introduction

One of the issues of LTE-800 networks deployment in CIS and Eastern European countries (Region 1 in the Radio Regulations) is to provide EMC between LTE-800 networks and the Air Traffic Control Radars (ATC) (like DRL) in the airfield areas. These RLS systems are deployed in the vicinity of the airport and other building where LTE-800 user terminals can operate and cause harmful interference. The reason is that the UE transmitters of LTE-800 network channels can coincide with ATC radars operating frequencies and can cause significant interference to position indicators of the ATC radar, thus hiding the current location of aircrafts in the airfield area. Such cases can significantly affect the flight safety in the airfield area.

© Springer Nature Switzerland AG 2018
O. Galinina et al. (Eds.): NEW2AN 2018/ruSMART 2018, LNCS 11118, pp. 516–523, 2018.
https://doi.org/10.1007/978-3-030-01168-0_46

The traditional technical organizing method to provide EMC can be one of the methods to satisfy EMC issues between ATC radars and LTE-800 UEs operating co-frequency. This method is to ensure the separation distance, by management of LTE-800 base station parameters in the areas close to airfields with operating ATC radars.

2   Spectrum Sharing Features in 800 MHz Band

The demand for broadband mobile communication in rural and sparsely populated area requires spectrum identification in 800 MHz band (called "digital dividend" in ITU and Band 20 in 3GPP) for mobile wireless communication on a global basis. 3GPP and ITU agreed the LTE-800 frequency plan [2] (see Fig. 1) which includes the usage of 6 frequency channels with 5 MHz bandwidth and inverse usage of LTE UE frequency channels in the frequency band 832–862 MHz.

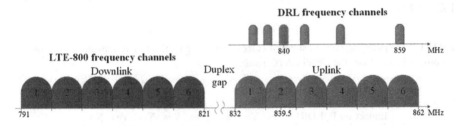

Fig. 1. Channeling arrangement of UE LTE-800 and air-traffic control radars [3].

The air traffic control radars named as DRL (Dispatch Radio Locator) also operate in the same frequency band. As Fig. 1 shows several frequency channels used by LTE-800 UEs are coinciding with the frequency carriers of ATC radars. Thus, the usage of geographical separation in the shared operation of LTE-800 UE and DRL is the only method to ensure EMC.

The possible interference impact scenario from LTE-800 transmitters to DRL receivers is shown in Fig. 2. Using this scenario the separation distance equation describing the EMC condition between LTE-800 UE transmitters and DRL receivers can be the following:

$$R_{UE-DRL} > R_{separation\ distance} = R_{BS-DRL} - R_{cell\_radius} \qquad (1)$$

The initial data in this equation are the following:

- base station cell radius R_{cell_radius};
- separation distance between AT LTE-800 transmitter and DRL receiver $R_{separation\_distance}$, ensuring EMC [3] (ERPI < 17 dBμV/m).

The experimental studies were performed to investigate the impact of LTE-800 UE on DRL receivers. The study results showed the impact of LTE-800 UE on DRL receivers with field strength of 17 dBμV/m at the DRL antenna input.

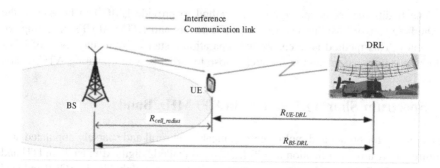

Fig. 2. Scenario of UE LTE-800 impact on Air-traffic control radar.

The interference field strength values created by LTE-800 UE transmitters deployed at various distances from the airfield ATC radar, obtained in the tests, are shown in Table 1 [3].

Table 1. The interference field strength values created by LTE-800 UE transmitters deployed at various distances from the airfield ATC radar.

$R_{UE\text{-}DRL}$, km	1.4	4.3	6.6	6.7	6.8	7.4	9.2	11.0
E_{DRL}, dBµV/m	44	42	30	24	>23	31	24	<17
Impact on PPI DRL	Yes	Yes	Yes	Yes	Yes	Yes	Yes	No

During the tests, a simulator of LTE-800 UE transmitters with power of 1 W was used as an interference source. The simulation scenario corresponds to the case when 5 user terminals are simultaneously communicating with the base station in co-channel uplink within one cell. The interference at the DRL receiver input was estimated by flashes at the plan position indicator (PPI) and by the interference level at the spectrum analyzer antenna simulating DRL receiver (see Fig. 3).

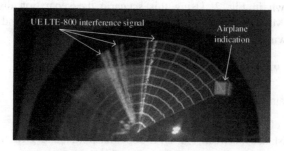

Fig. 3. LTE-800 UE interfering signal on the ATC Radar PPI.

The preliminary theoretical assessments of LTE-800 UE impact, given in the study [3], showed that separation distance of more than 13 km is required to provide EMC between LTE-800 UE at 1.5 m height and DRL. The tests for assessment of such separation distance showed that smaller separation distances $R_{separation\_distance} = 6, 8–11$ km are acceptable between the indicated systems.

As a conservative scenario the Government regulator defining the EMC conditions for such cases choses the separation distance of 10 km. The limits for LTE-800 base station cell radius are defined by the commercial operators of LTE-800 based on the marketing and commercial factors and they proposed to use the value of 1 km.

Thus from Eq. (1) with the minimal cell coverage radius of LTE-800 base station $R_{cell\_radius}= 1$ km we obtain the separation distance of 10 km from the DRL location prohibiting operation of LTE-800 UE frequency channel and the corresponding duplex frequency channel of the base station.

3 Providing Separation Distance by Base Station Cell Parameters Management, Other Than Power Control

Base station cell includes "downlink" cell and "uplink" cell. In ideal case two cells should coincide.

Since the UE but not base station is the interference source for DRL receiver then to provide EMC it is proposed not to limit base station EIRP level and "downlink" cell radius (DL) accordingly but to limit the "uplink" cell radius (UL) of base station by localization of Physical Random Access Channel (PRACH) cell. If UE is located in the "downlink" cell of base station and receiving data from the Physical Broadcast Channel (PBCH) and outside of PRACH cell in the uplink then the terminal will not have an success access to LTE-800 network and do not cause interference to DRL by channel PUSCH accordingly. Localization of Physical Random Access Channel (PRACH) cell is proposed to be made by choosing non-power parameters:

- PRACH preamble format;
- Zero Correlation Zone Config;
- Root Sequence Index.

The preamble format establishes the PRACH cell size with large resolution. The minimal PRACH cell radius up to 14 km is provided by preamble format number 0 (see Table 2) [4].

Table 2. PRACH cell radius according preamble format.

PRACH preamble format	0	1	2	3
Cell radius, km	14	75	28	108

If the cell size is smaller than the service area in accordance with PRACH preamble format then in case of high user density and to improve the main performance characteristics of KPI network operation ZeroCorrelationZoneConfig parameter is optimized. PRACH preamble format is Zadoff-Chu (ZC), an orthogonal pseudorandom sequence with 839 unit length. Zadoff-Chu (ZC) pseudorandom sequence is unique for each cell and determined by RootSequenceIndex with this all users within the cell use one and the same sequence but with various cyclic shifts. The resolution of cyclic shift is defined by cell size and by ZeroCorrelationZoneConfig parameter. The smaller the PRACH cell radius, the less resolution of cyclic shifts compensating preamble multipath propagation is required. The major manufactures establish on default ZeroCorrelationZoneConfig parameter corresponding to cell size of 10 km in base station configuration.

However, the limitation of PRACH cell radius by ZeroCorrelationZoneConfig parameter is not stable since under certain conditions (no multipath propagation, no other preamble from other UE at this time) the PRACH preamble can be received from UE deployed at the larger distance from the base station than it was defined in the calculation of ZeroCorrelationZoneConfig but at a smaller distance than it is established in the preamble format and decoded by the base station.

4 The Drive Test Results

The performed tests confirmed the possibility of ensuring separation distance between LTE-800 UE transmitters and DRL receivers by management of base station cell radius and providing their EMC in co-frequency operation.

TEMS, the mobile monitoring complex was used to define the experimental LTE-800 network cells. TEMS allows to assess distance between UE and BS during connection (BS coverage area). The experimental assessment results of BS coverage area during connection between UE and BS in geographical locations are given in Fig. 4 (Region 1) and in Fig. 5 (Region 2).

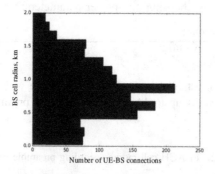

Fig. 4. Drive test estimated distances for connection with cell radius limit up to 1 km (Region 1).

The analysis of these results show that in spite of limiting the LTE-800 BS cell radius (Rcell_radius = 1 km), the coverage area radius of UE–BS will achieve 1.7–2.0 km.

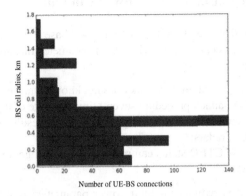

Fig. 5. Drive test estimated distances for connection with cell radius limit up to 1 km (Region 2).

Other option of limiting cell radius in the period after connection of UE with BS is to establish the threshold value of RSRP pilot signal on UE.

To limit the LTE-800 BS cell radius during connection with UE at small distances from BS and further in direction of DRL interference victim the threshold value of RSRP pilot signal is set as minus 92 dBm by Threshold Th4 parameter for RSRP and after it there is a disconnection.

The received experimental data corresponding to disconnection for this RSRP pilot signal level are given in Fig. 6. In Fig. 6 it is shown that disconnection appears while going outside of the cell radius of 1 km and can be at the distance of 6–7 km from the operating BS. In this case EMC between LTE-800 UE and DRL radar cannot be ensured. The distance between LTE-800 UE and DRL radar is 2–3 km and the interference level at the DRL receiver input is 65–70 dBμV/m.

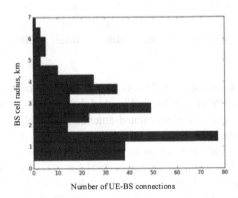

Fig. 6. Drive test estimated distances for disconnection between UE and BS based on RSRP pilot signal level.

5 EMC Method Based on BS Power Control

The method considered above of limiting the LTE-800 base station cell radius in uplink by Physical Random Access Channel (PRACH) parameter management has the following disadvantages:

- UE deployed in the BS cell in downlink does not know distance from BS and makes PRACH preamble in attach procedure several times despite of the distance from BS and each time it increases transmission power, and as a result it continues to cause interference to DRL radars;
- UE in RRC_CONNECTED status can move to the cell edge downlink in direction to DRL deployment.

Thus first of all it is required to limit power parameters in downlink to provide EMC of LTE-800 systems with DRL radars and regulation of the non-power parameters considered above should be made by operators in accordance with the specified requirements.

Equivalent isotropically radiated power (EIRP.) is a power parameter defining the BS cell radius. Equivalent isotropically radiated power is estimated as the sum of the transmitted signal power and the transmitter antenna gain minus feeder loss.

Increase/decrease of EIRP will lead to increase/decrease of BS cell radius accordingly.BS cell radius is an area covered by BS where the signal power at the UE receiver input is larger than its sensitivity.

In accordance with 3GPP TS 36.101 specification for the frequencies (see Fig. 1 band 20) the LTE-800 UE receiver sensitivity is −97 dBm with 5 MHz bandwidth and QPSK modulation (see Table 3).

Table 3. LTE-800 UE receiver sensitivity.

E-UTRA frequency band number	Bandwidth, MHz						Duplex
	1, 4	3	**5**	10	15	20	
20	Not applied	Not applied	**−97 dBm**	−94 dBm	−91.2 dBm	−90 dBm	FDD

UE estimates the receiving signal power from base station based on RSRP (Reference Signal Received Power). If MIMO is used then the RSRP pilot signal power is estimated for each logical port of a radiated antenna separately.

The reference signal (RS) bandwidth is 1/12 of Physical Resource Block (PRB) spectrum bandwidth and equals to 15 kHz. 25 Physical Resource Blocks are used in 5 MHz bandwidth as the maximum or 25*12 = 300 subcarrier frequencies. The required minimal average reference signal received power in 15 kHz bandwidth is estimated as −97 − log10(300) = −97 −24.77 = −121.77 dBm.

If the RSRP pilot signal level in 5 MHz bandwidth is less than −121.77 dBm then there is disconnection and it is considered that UE is out of the BS cell radius. Change of BS cell radius is achieved by EIRP changes and by antenna height, antenna angles.

BS EIRP is determined by the Reference Signal Power parameter which defines the LTE-800 BS transmitting pilot signal power and as a result defines UE cell radius. Therefore the UE cell radius can be limited up to the given radius, for example up to 1 km for this task by using the base station power control mechanism.

6   Conclusion

The studies and simulations performed by the authors of the article showed that ensuring EMC of LTE-800 mobile network with DRL radars operating co-frequency is a complicated scientific and engineering task. Unconventional approaches based on the usage of both separation distance and parameter control of LTE-800 base station coverage area were required to solved the task mentioned above.

The experimental test of control of LTE-800 base station cell parameters in order to provide a given separation distance between UE and DRL radars allowed to identify several disadvantages and the lack of reliability of such method of ensuring EMC. However, the usage of this method significantly reduces the interference level at the DRL receiver input.

To ensure limitation of LTE-800 cell coverage for connected UE, which is the interference source for DRL Radars, it is required to limit the base station transmit power by management of Reference Signal Power parameter in order to reduce the RSRP pilot signal level below -121.77 dBm outside of 1 km coverage area.

References

1. Report ITU-R M.2241. Compatibility studies in relation to Resolution 224 in the bands 698–806 MHz and 790–862 MHz, ITU (2012)
2. GPP TS 36.101. 3rd Generation Partnership Project, Technical Specification Group Radio Access Network, Evolved Universal Terrestrial Radio Access (E-UTRA), User Equipment (UE) radio transmission and reception
3. Tikhvinskiy, V., Bochechka, G., Korchagin, P., Seilov, S., Gryazev, A.: Protection criteria for sharing spectrum UE LTE-800 and air-traffic control radars based on experimental comparability results. In: Proceedings of the 2016 International Symposium on Electromagnetic Compatibility - EMC EUROPE 2016, Wroclaw, Poland, 5–9 September 2016
4. GPP TS 36.211. 3rd Generation Partnership Project, Technical Specification Group Radio Access Network, Evolved Universal Terrestrial Radio Access (E-UTRA), Physical Channels and Modulation
5. Tikhvinskiy, V.O., Terentyev, S.V., Vysochin, V.P.: LTE/LTE Advanced Mobile Communication Networks: 4G Technologies, Applications and Architecture. Media Publisher, Moscow (2014)

A Lower Bound on the Average Identification Time in a Passive RFID System

Nikita Stepanov[1]($\boxtimes$)(iD), Nikolay Matveev[1](iD), Olga Galinina[2,3](iD),
and Andrey Turlikov[1](iD)

[1] State University of Aerospace Instrumentation, Saint-Petersburg, Russia
{nstepanov,n.matveev,turlikov}@vu.spb.ru
[2] Tampere University of Technology, Tampere, Finland
olga.galinina@tut.fi
[3] Peoples Friendship University of Russia (RUDN University),
Moscow, Russian Federation

Abstract. One of the most well-known standards for radio frequency identification (RFID), the standard ISO 18000-6C, collects the requirements for RFID readers and tags and regulates respective communication protocols. In particular, the standard introduces the so-called Q-algorithm resolving conflicts in the channel (which occur when several RFID tags respond simultaneously). As of today, a vast amount of existing literature addresses various modifications of the Q-algorithm; however, none of them is known to significantly reduce the average identification time (i.e., the time to identify all proximate tags). In this work, we derive a lower bound for the average identification time in an RFID system. Furthermore, we demonstrate that in case of an error-free channel, the performance of the legacy Q-algorithm is reasonably close to the proposed lower bound; however, for the error-prone environment, this gap may substantially increase, thereby indicating the need for new identification algorithms.

1 Introduction

One of the most well-known standards for radio frequency identification (RFID), known as EPCglobal Class 1 Generation 2 (ISO 18000-6C) [2, 4], consolidates the requirements for RFID readers and tags as well as regulates respective communication protocols, operating at distances of 0.5–10 m and frequencies 860–960 MHz.

Inter alia, the standard ISO 18000-6C introduces a specific algorithm that allows an RF reader to poll and identify an unknown number of tags in its coverage area [6, 7]. This algorithm is often referred to as the Q-algorithm, owing to its core parameter typically denoted as Q.

As of today, a vast amount of existing literature addresses various modifications of the Q-algorithm; however, none of them is known to significantly reduce

© Springer Nature Switzerland AG 2018
O. Galinina et al. (Eds.): NEW2AN 2018/ruSMART 2018, LNCS 11118, pp. 524–534, 2018.
https://doi.org/10.1007/978-3-030-01168-0_47

the time required to identify all proximate tags (typically called the identification time). Most of the research attempts in this direction focus on defining parameters of the Q-algorithm, which would minimize the average identification time, and offer multiple heuristic methods as, e.g., proposed in [1,8,9]; however, the algorithm remains unchanged.

Viewed from another angle, the mentioned multiple available variations of the Q-algorithm could already be lying relatively close to a certain bound, thus, limiting the chances for further improvement. In the light of the above, we target our work towards developing a *hypothetical optimal algorithm* that ensures the minimum average time for identifying proximate tags by the reader. Having designed the hypothetical identification algorithm, we derive a recurrent expression for calculating its average identification time, which, in turn, represents *a lower bound* for the identification time within the class of similar algorithms, which also includes the standardized Q-algorithm. To illustrate the behavior of identification algorithms in an imperfect channel, we also extend our calculations by including the *probability of a channel error*.

This paper is organized as follows. In Sect. 2, we introduce the key assumptions of our system model, which is based on the standard ISO 18000-6C, and summarize the collision resolution algorithm (Q-algorithm). Further, in Sect. 3, we introduce a hypothetical algorithm and provide the derivation of our proposed lower bound for the average identification time. Section 4 illustrates the behavior of the obtained lower bound in comparison to the performance of the Q-algorithm (separately for the case of error-free and error-prone wireless channels) and demonstrates a substantial gap between the standard solution and our theoretical lower bound in case of channel errors.

2   System Model

We focus on a typical passive RFID scenario and study a wireless system that includes a *reader* and a set of RFID *tags* located in its coverage area, which communicate according to ISO/IEC 18000-6C RFID protocol. An example of the corresponding time diagram is illustrated in Fig. 1. Importantly, in our (passive RFID) scenario, the tags do not have an inbuilt source of power, but instead are able to harvest and use the energy received from the reader [3].

Below we introduce the key system assumptions, which on the one hand preserve the core features of algorithms advised by the standard ISO/IEC 18000-6C and, on the other hand, make the model analytically tractable for further evaluation.

2.1   Main System Assumptions

The system time is divided into *frames* so that each frame contains a variable number of intervals (termed *slots*), during which the reader may receive and decode the information from one transmitting tag. The number of slots within one frame is determined according to a specified rule and reliably broadcasted at the beginning of each frame by the reader.

After receiving a message from the reader, a tag may respond in one of the available slots of the subsequent frame; the slot is selected randomly according to the uniform distribution. As more than one device may independently decide to transmit during the same slot, we observe one out of the three following outcomes: no transmission, a conflict (collision) of two or more tags, and the successful transmission. We assume that the reader identifies the channel outcome instantaneously and correctly. Below we provide our assumptions on the above-listed three outcomes.

Assumption 1. If none of the tags responds, then no signal is detected by the reader (in our model, this outcome is further termed 'Empty'). The duration of an 'Empty' slot equals T_e time units.

Assumption 2. If two or more tags are transmitting simultaneously, the reader cannot identify any of them, and therefore, the corresponding slot is recognized by the reader as 'Conflict' and lasts for T_c time units.

Assumption 3. Finally, if exactly one tag transmits during the considered slot, then this tag is successfully identified with the probability $1 - p$, and its duration occupies precisely T_s time units (this outcome is termed 'Success'). With the complementary probability p (that is, the *probability of a channel error*), the tag will not be identified, and the slot length changes to T_c (equivalent to 'Conflict').

We assume that the probability p of the channel error is the same for any tag in our system, e.g., due to the fact that all tags are located at equal distances from the reader and in similar electromagnetic conditions.

If the tag is successfully identified at the end of its transmission (i.e., we observe 'Success'), the reader sends a corresponding command, which automatically excludes the successful tag from the further polling procedure. Otherwise, if the signal has been distorted and the transmission attempt fails, the tag continues contending for the channel in the next frame.

We refer to a sequence of frames, during which all proximate tags could be successfully identified, as the *identification time*. Below we aim at constructing an algorithm that defines a *dynamic sequence* of frame lengths and by that minimizes the average identification time.

2.2   Collision Resolution Algorithm (Q-algorithm)

In this subsection, we extend our assumptions introduced above and further elaborate on the random multiple access algorithm specified by the standard ISO 18000-6C (the Q-algorithm), which drew on a random selection of the frame length.

Typically, the length of the frame is represented by the number 2^Q, where Q is a variable parameter, which is initialized by default as $Q = 4$ and may vary from 0 to 15. We remind that the value of Q is broadcasted by the reader (as the

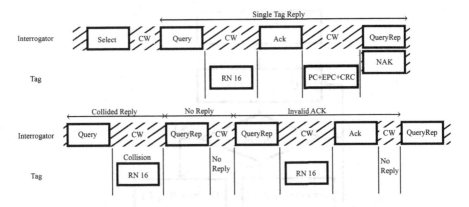

Fig. 1. Illustration of the RF identification process in the standard EPCglobal Class 1 Generation 2 [2]. The error in an RN16 sequence (uniquely identifying the tag) transmission results in the same processing time as the processing time in case of a collision. Contention window is denoted as CW.

Query command) at the beginning of each frame. All tags that have received the value of Q (that is, according to our assumptions, all proximate tags) generate a random slot number within the interval $[0, 2^Q - 1]$ and save it as a slot counter.

The tags, whose slot counters equal to 0, respond immediately (enter the 'Reply' stage) by sending a unique random sequence of 16 bits (called *RN16*), identifying an RFID tag.

Further, if exactly one tag replies, the value of the parameter Q for the next frame remains the same, and the reader transmits the QueryRep command, instructing the tags to decrement their slot counters by 1. When the tag counter reaches 0, it immediately starts transmitting its RN16 sequence to the reader. A block-scheme of the Q-algorithm is illustrated in Fig. 2.

If the tag is successfully identified after transmission of RN16, then the reader confirms by sending the ACK command to the tag and excludes the tag from further identification process. After the last tag is successfully identified, the reader stops the identification procedure.

If no tags or more than two tags access the channel, the reader sends the QueryAdjust command, which may increase/decrease the value of the parameter Q by the step C, where C is implementation-dependent and belongs to the interval $[0.1, 0.5]$. Usually, readers employ lower values of C, when Q is relatively large, and vice versa.

In case of a collision that is when more than two tags reply simultaneously, the reader cannot identify any of them; in this case, the tag counter values of all collided tags increase sharply (in particular, change to 32767, $2^{15}-1$, or 7FFFh) to prevent the tags from transmitting in the current frame.

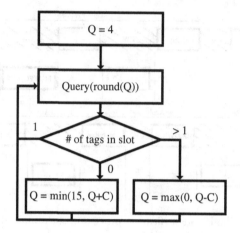

Fig. 2. A block-scheme for the Q-algorithm.

3   Derivation of the Lower Bound of the Average Identification Time

In this section, we introduce a class of RFID algorithms, which includes, among others, the Q-algorithm discussed above. Within the defined class of algorithms, we consider an optimal scheme, which provably delivers the least values of the average identification time and therefore, provides the lower bound for the entire class and, in particular, for the Q-algorithm.

We recall that for any identification procedure, the current frame length is broadcasted by the reader at the beginning of each frame so that every tag immediately selects a slot for the subsequent transmission of its RN16 sequence. Moreover, after any of the slots, the reader may interject the operation, that is, broadcast a new value of the frame length and restart activity of its tags. We further refer to this decision of the reader to change the operation to as an *interrupt*.

Here, we define *an RFID algorithm* as a set of two rules: (i) according to the first one, the reader selects the frame length, and (ii) the second rule drives decisions to interrupt. We further study a class of such algorithms, i.e., including all possible identification schemes that meet these two requirements. We also assume that the number of slots in the frame may be represented by an integer number from 1 to ∞ (it is unconstrained and not necessarily the power of 2 in contrast to the Q-algorithm).

Importantly, the standard Q-algorithm belongs to the defined class of RFID algorithms, and therefore, an algorithm that is optimal within this class will deliver equal or better performance in terms of the average identification time than any variation of the Q-algorithm.

Let us further assume that at any moment of time the exact number of tags is known (which is not possible in practice for the scenario in question) and consider a hypothetical algorithm that makes his decisions based on this knowledge. As any RFID algorithm targets selecting the frame length and making a decision on interruption optimally, we may formulate the following proposition.

Proposition 1. *The average identification time of the hypothetical RFID algorithm, aware of the exact current number of tags, is minimal within the considered class.*

Proof. The proof is trivial and left out of the scope of this paper.

Proposition 2. *If the exact current number of tags is known, an RFID algorithm that creates interruptions after the first slot delivers the average identification time not greater than that of an algorithm with any other interruption rule.*

Proof. The proof is left out of the scope of this paper.

We consider a particular state of the system with n unidentified tags. Let V_n represent *the optimal frame length* in case of n tags. In general, we may also interpret V_n as a rule, according to which the reader decides on the frame length to broadcast. Knowing this rule, we may calculate the average identification time for our optimal scheme, that as well constitutes *the lower bound* on for the Q-algorithm.

Theorem 1. *Given n currently unidentified tags and the corresponding optimal frame length V_n, the average identification time for the optimal algorithm may be obtained according to the following recurrent expression:*

$$l_n(V_n) = \frac{T_c P_c(V_n, n) + T_e P_e(V_n, n) + ((1 - p)(l_{n-1}(V_{n-1}) + T_s) + p T_c) P_s(V_n, n)}{1 - P_c(V_n, n) - P_e(V_n, n) - p P_s(V_n, n)},$$
(1)

where T_e, T_c, and T_s is the duration of the empty, conflict, and successful slots, correspondingly, while p is the probability of a channel error during the RN16 transmission, and $P_s(V_n, n)$, $P_c(V_n, n)$, $P_e(V_n, n)$ are the probabilities of the outcomes 'Success', 'Conflict', 'Empty' during the first slot of the current frame.

Proof. We consider our system in a state where $n > 0$ tags are to be identified and the optimal frame length is given by V_n. Naturally, we assume that $V_0 = 0$, thus, the average identification time is zero, $l_0(V_0) = 0$.

Let the random variable $[L_n|V_n]$ denote one realization of the identification time. For the sake of brevity, $[L_n|V_n]$ is further referred to as L_n. Then, for the system of n tags with frame interruptions, we may write down the following expression:

$$L_n = \text{I\{'Success'\}}\,(T_s + L_{n-1}) + \text{I\{'Failure'\}}\,(T_c + L_n) \\ + \text{I\{'Conflict'\}}\,(T_c + L_n) + \text{I\{'Empty'\}}\,(T_e + L_n),$$
(2)

where I{'Success'} is an indicator of that exactly one tag transmits and there are no errors during the subsequent transmission of the RN16 sequence, I{'Failure'} indicates that one tag accesses the channel but the transmission of RN16 fails, I{'Conflict'} and I{'Empty'} reflect a conflict and an empty slot, respectively. We remind that T_s, T_c, and T_e denote the duration of empty, conflict, and successful slots, correspondingly.

In order to obtain the average identification time $l_n = E[L_n]$, we find expected values of the left and right sides of the Eq. (2):

$$
\begin{aligned}
E[L_n] = E[\ I\{\text{'Success'}\}\ (T_s + L_{n-1})] + E[\ I\{\text{'Failure'}\}\ (T_c + L_n)] \\
+ E[I\{\text{'Conflict'}\}\ (T_c + L_n)] + E[\ I\{\text{'Empty'}\}\ (T_e + L_n)].
\end{aligned}
\tag{3}
$$

Taking into account the possibility of interruptions after the first slot of the current frame and the fact that:

$$
\begin{aligned}
E[\ I\{\text{'Success'}\}\ ] = P_s(V_n, n)(1 - p), \quad E[\ I\{\text{'Failure'}\}\ ] = P_s(V_n, n)p, \\
E[\ I\{\text{'Conflict'}\}\ ] = P_c(V_n, n), \quad E[\ I\{\text{'Empty'}\}\ ] = P_e(V_n, n),
\end{aligned}
\tag{4}
$$

we may arrive at the following expression for the average identification time:

$$
\begin{aligned}
l_n(V_n) = P_c(V_n, n)\,(l_n(V_n) + T_c) + P_e(V_n, n)\,(l_n(V_n) + T_e) \\
+ P_s(V_n, n)\,[(1 - p)\,(l_{n-1}(V_{n-1}) + T_s) + p\,(l_n(V_n) + T_c)],
\end{aligned}
\tag{5}
$$

where p is the probability of error in case of 'Success' (i.e., the reader identifies the outcome as 'Conflict'). The probabilities $P_s(V_n, n)$, $P_e(V_n, n)$, and $P_c(V_n, n)$ of 'Success', 'Empty', and 'Conflict', correspondingly, may be obtained as:

$$
P_s(V_n, n) = n\frac{1}{V_n}\left(1 - \frac{1}{V_n}\right)^{n-1},
\tag{6}
$$

since 'Success' corresponds to the case when one tag transmits in a particular slot with the probability $\frac{1}{V_n}$, and the remaining $n - 1$ tags are silent,

$$
P_e(V_n, n) = \left(1 - \frac{1}{V_n}\right)^{n},
\tag{7}
$$

where all n tags are not transmitting, and, finally,

$$
P_c(V_n, n) = 1 - P_s(V_n, n) - P_e(V_n, n).
\tag{8}
$$

Further, we simplify the expression (5), removing the parentheses and relocating the terms corresponding to $l_n(V_n)$ as follows:

$$
\begin{aligned}
l_n(V_n) - l_n(V_n) \cdot P_c(V_n, n) - l_n(V_n) \cdot P_e(V_n, n) - \\
p l_n(V_n) \cdot P_s(V_n, n) = T_c \cdot P_c(V_n, n) + T_e \cdot P_e(V_n, n) + \\
((1 - p)\,(T_s + l_{n-1}(V_{n-1})) + p T_c) \cdot P_s(V_n, n).
\end{aligned}
\tag{9}
$$

Here, we may divide both left and right sides of the equation by $(1 - P_c(V_n, n) - P_e(V_n, n) - pP_s(V_n, n))$ and, finally, obtain the following:

$$l_n(V_n) = \frac{T_c P_c(V_n, n) + T_e P_e(V_n, n) + [(1 - p)(l_{n-1}(V_{n-1}) + T_s) + pT_c] P_s(V_n, n)}{1 - P_c(V_n, n) - P_e(V_n, n) - pP_s(V_n, n)},$$

(10)

which was to be demonstrated.

4   Numerical Results

In this section, we compare the performance of our lower bound and the standard Q-algorithm described above. In our illustrative example below, we set realistic protocol timings borrowed from [5], in particular, we assume that $T_s = 8$ ms, $T_c = 0.9$ ms, $T_e = 0.61$ ms.

We begin with comparing our lower bound and the average identification time delivered by the standard Q-algorithm (for convenience the parameter C is set to 1) depending on the number of tags in the system (see Fig. 1). Here, we let the probability of error in the RN16 transmission be equal 0.5 as suggested in [5] for realistic conditions. In case of an ideal channel (the two lower curves), the average gain remains around 8%. Thus, we may conclude that *any* modification of the Q-algorithm cannot outperform the lower bound and, hence, the respective gain will not exceed this value.

If the RN16 transmission is not protected, that is when a potentially successful slot is treated as a conflict with the probability $p = 0.5$, and the tag remains unidentified (the upper set of curves), the Q-algorithm deviates from the lower by up to 15%. The latter indicates that the performance of the Q-algorithm may be further improved before it reaches its lower limit (Fig. 3).

Further, Fig. 4 shows the dependence of the average identification time on the value of the parameter $\tilde{V}_n$ if the total number of tags is ten and the probability p is varied. In particular, we assume that $V_i, i = 1, ..., 9$ are known and V_{10} is to be determined based on minimization over potential values of the optimal frame length denoted as $\tilde{V}_{10}$. We intentionally change our parameters to $T_s = 0.5$ ms, $T_c = 0.3$ ms, and $T_e = 1$ ms, as the previous set does not provide results illustrative enough for our purposes. As shown in Fig. 4, with the growth of p, the average identification time increases as more transmission attempts are required for the success. We may also observe a curious effect: all curves reach their minimums at the same value of $\tilde{V}_n$. We note that varying parameters T_s, T_c, T_e displays quite a similar behavior, however, the minimums might become less visible. As such, we conclude that V_n may not depend on p; however, strict proof of this hypothesis is out of the scope of this paper.

Finally, we increase the number of tags in the system up to 40 and vary the probability p (see Fig. 5). As the probability of incorrect decoding of the

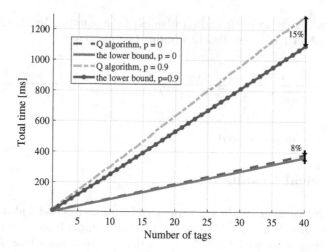

Fig. 3. Comparison of the average identification time of the standard Q-algorithm and the lower bound: (i) for an ideal channel and (ii) in case of RN16 transmission failure (error-prone channel).

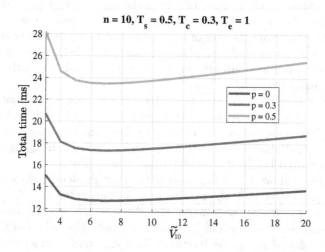

Fig. 4. Average identification time vs. the length of the frame.

RN16 sequence increases, the average identification time rises exponentially and naturally tends to infinity with $p \to 1$. We note that the difference between the lower bound and the standard algorithm will also widen with the growth of p. Hence, we may conclude that the presence of channel errors creates possibilities for improving the performance of the standard algorithm, and its modifications may bring significant benefits for the system in general.

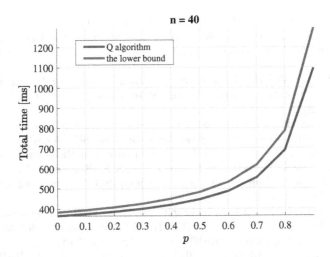

Fig. 5. Dependence of the total identification time on the probability p in case of 40 tags.

5   Conclusion

In this paper, we study an RFID system operation based on the standard Q-algorithm, for which there exist a variety of modifications that aim at minimizing the average identification time. Here, we derive a lower bound for the average identification time by introducing a hypothetical algorithm, which is aware of the current number of unidentified tags. We as well demonstrate that in case of an error-free channel, the performance of the legacy Q-algorithm is reasonably close to the proposed lower bound and any further modifications will not bring additional benefits. However, for the error-prone environment, where the RN16 sequence may be decoded incorrectly, the gap between the performance of the standard scheme and the lower bound substantially increases, thereby indicating the need for new identification algorithms in this area.

Acknowledgment. The work of N. Stepanov, N. Matveev, and A. Turlikov is supported by scientific project No. 8.8540.2017/8.9 "Development of data transmission algorithms in IoT systems with constraints on the devices complexity". The work of O. Galinina is supported by "RUDN University Program $5 - 100$".

References

1. Arjona, L., Landaluce, H., Perallos, A., Lopez-Garcia, P., Cmiljanic, N.: Analysis of RFID anti-collision protocols based on the standard EPCglobal Class-1 Generation-2. In: Proceedings of 21th European Wireless Conference on European Wireless 2015, pp. 1–6. VDE (2015)
2. Global, EPC.: EPC radio-frequency identity protocols class-1 generation-2 UHF RFID protocol for communications at 860 MHz–960 MHz. Version **1**, 23 (2008)

3. Instruments, Texas: TI UHF Gen2 Protocol Reference Guide
4. Kamrani, A.: Design and Development of a State Transition Table for the EPC global UHF Class 1 Gen2 RFID standard. Ph.D. thesis, University of Pittsburgh (2011)
5. Namboodiri, V., DeSilva, M., Deegala, K., Ramamoorthy, S.: An extensive study of slotted ALOHA-based RFID anti-collision protocols. Comput. Commun. **35**(16), 1955–1966 (2012)
6. Ometov, A., et al.: Secure and connected wearable intelligence for content delivery at a mass event: a case study. J. Sens. Actuator Netw. **6**(2), 5 (2017)
7. Prudanov, A., et al.: A trial of yoking-proof protocol in RFID-based smart-home environment. In: Vishnevskiy, V.M., Samouylov, K.E., Kozyrev, D.V. (eds.) DCCN 2016. CCIS, vol. 678, pp. 25–34. Springer, Cham (2016). https://doi.org/10.1007/978-3-319-51917-3_3
8. Uysal, I., Khanna, N.: Q-frame-collision-counter: a novel and dynamic approach to RFID Gen 2's Q algorithm. In: 2015 IEEE International Conference on RFID Technology and Applications (RFID-TA), pp. 120–125. IEEE (2015)
9. Zheng, F., Kaiser, T.: Adaptive ALOHA anti-collision algorithms for RFID systems. EURASIP J. Embed. Syst. **2016**(1), 7 (2016)

IoT Based Earthquake Prediction Technology

Rustam Pirmagomedov[1]([⊠]), Mikhail Blinnikov[2],
Alexey Amelyanovich[2], Ruslan Glushakov[3], Svyatoslav Loskutov[4],
Andrey Koucheryavy[2], Ruslan Kirichek[2], and Ekaterina Bobrikova[1]

[1] Peoples' Friendship University of Russia (RUDN University),
6 Miklukho-Maklaya Street, Moscow 117198, Russian Federation
prya.spb@gmail.com
[2] St. Petersburg State University of Telecommunication, Saint Petersburg
193232, Russian Federation
[3] Military Medical Academy, Saint Petersburg 194044, Russian Federation
[4] FSBSI ARRIAM, Saint Petersburg 196608, Russian Federation

Abstract. The article presents results of the first stage of the development of an animal monitoring system using IoT technology. Data on animal behavior collected using the system should confirm or refute the thesis about the ability of some animal species to respond to signs of an approaching earthquake. The article provides a review of the previous works on this topic and formulates new tasks, the solution of which will improve the previously available methodologies on the use of animals as a biosensor. The system exploits inertial sensors and computer vision to collect data on animal behavior. Data processing and analysis is carried out on the central server. It is expected that in the case of a systemic cause (for example, signs of an approaching earthquake) non-standard animal behavior should be massive.

Keywords: IoT · Earthquake prediction · Inertial sensors · Big data
Computer vision

1 Introduction

There are numerous earthquakes [1] around the world every year. Although the strength of most of them is insignificant (thus posing no threat to human society), those rare but powerful earthquakes are known for their devastating consequences and cannot be ignored. The number of victims from one such seismic event can number in the hundreds of thousands of people. To avoid these drastic consequences, researchers from different fields of science set a common goal to find an effective way to predict the earthquakes. Despite their considerable efforts, it is still impossible to predict with absolute accuracy, the date and time when a seismic phenomenon will begin.

The seriousness of this problem lies not only in the fact that the inaccurate detection of seismic hazard can lead to numerous human casualties and material losses, but also that actions taken due to a false alarm may unnecessarily damage the economy or create skepticism regarding warnings for individuals in the future.

A significant part of earthquake prediction methods is based on information about earthquakes that happened before. In this case, the more data, the higher the probability

© Springer Nature Switzerland AG 2018
O. Galinina et al. (Eds.): NEW2AN 2018/ruSMART 2018, LNCS 11118, pp. 535–546, 2018.
https://doi.org/10.1007/978-3-030-01168-0_48

of an accurate prediction. Such methods involve the processing of a massive flow of information. It is advisable to analyze not only the events that have already occurred but also process the reports obtained as a result of monitoring the earth's surface in real time [2–4]. New methods of monitoring are being developed to track the so-called earthquake precursors. Such methods of forecasting include the monitoring of changes in seismic wave velocities, the monitoring of groundwater levels, or radon and hydrogen content in rocks. In addition, studies aimed at finding the correlation of earthquakes with the motion of the earth's crust and a change in the concentration of ions in the ionosphere have been carried out, as well as the monitoring of changes in the behavior of biological objects (animals) as the seismic event approaches [5–7].

However, despite multiple types of research and the development of various systems of seismological monitoring using all-pervasive sensor networks, as well as artificial intelligence technologies and neural networks [8, 9], this issue remains open.

As it has been mentioned above, there is a theory that the behavior of animals can be a kind of indicator of natural disasters. When examining the results of numerous observations and experiments, researchers were able to determine that some animals are capable of predicting earthquakes more than a day in advance. A few of them might even be aware several weeks before the occurrence of a natural disaster (in particular, species of fish).

Considering the behavior of animals (as it relates to earthquakes) it is necessary to pay attention the sensitivity of the 'head compass,' that is, its ability to capture electromagnetic oscillations in response to the approach of an earthquake in different species of animals.

Thus, the mountain goats of the Sierra Nevada range descend from the highland pastures to the plains several days before the shocks. Other examples include lizards crawling out of their cracks into open spaces or foxes and wolves leaving the forests.

If we compare the response of the 'seismic sensor' to stimulation by electromagnetic or magnetic fields in humans and animals, we can see that there are differences between them, mainly due to the sensitivity under exogenous effects. In animals, the sensitivity threshold of the 'sensor' is much lower than that of a human.

It can be assumed that with a higher organization of the organism (human), the realization of its functions in ontogeny is provided not only by morphofunctional specialization and integration of regulatory systems but also due to the high sensitivity threshold of the 'seismic sensor.'

Many species of living organisms react to the upcoming seismic event with a genetically determined 'behavioral repertoire', despite the fact that quite often the dominant behavioral responses are unpredictable in view of the complex nature of their behavior. Therefore the problem of observing and deciphering the unusual behavior of animals remains relevant.

Many factors, even including ordinary ones (for example reaction to another animal), may cause non-standard behavior in animals. How then to distinguish the reaction of animals to local stimuli from a reaction to an impending earthquake? To answer this question, the project 'IoT Based Earthquake Prediction Technology.' was initated. The main goal of the project is to create a technological basis for monitoring a large number of animals simultaneously. Widespread data collection is a key aspect of the suggested approach and is supposed to reveal the presence of a systemic cause in 'not normal' behavior of animals, minimizing the impact on the result of commonplace or accidental causes.

This article is a description of the first stage of research and development conducted by the authors aimed at creating a client-server software and hardware system that allows to analyze automatically the behavior of biological objects. The task of the system is to monitor a large number of animals and perform online analysis of their behavior on the server. Among the parameters to be monitored are the motor activity, the level of noise created, the pulse, and other factors.

The information obtained as a result of monitoring of biological objects can be used in conjunction with existing methods for predicting earthquakes providing a more accurate result.

The article is organized in the following way: Sect. 2 gives a brief survey on the technologies of animal's behavior monitoring. Section 3 considers an architecture (software) of the system and integration with existing systems. Section 4 describes a subsystem for monitoring of the activity of animals using wearable inertial sensors. Section 5 describes a subsystem for birds and fishes monitoring based on the use of computer vision (OpenCV). In conclusion, the results of the work are analyzed, with ideas for future research being considered.

2 Related Works

Much development has been suggested to study the behavior of animals. Some of it involves the use of video cameras and algorithms for analyzing the images and recognizing various behavior patterns [10–12]. The main difference between these works is the type of animals on which the authors are focused, which gives specificity to the algorithms being developed. It is worth highlighting the work of Kriti Bhargava and others, which suggests using fog computation technology to perform all necessary calculations, pattern recognition, and behavior analysis of living organisms, among other studies. [13].

The use of various miniature devices (sensors placed on the body of the animal) has become a more economical solution. It is assumed that such sensors can measure parameters, the analysis of the changes of which can help to determine deviations from the normal behavior of animals. In [14], a device that is a wireless accelerometer measuring 3.2 × 2.5 cm, weighing 10.2 grams and a battery capacity of 150 mA, and an algorithm for recognizing behavior has been developed. Such an accelerometer turned out to be small enough to be used to monitor the activity of laboratory rats and was attached to them as a homemade vest. In the course of the experiment, the researchers obtained three models of the behavior of rats: Eat, Stand, and Groom. The authors note that grooming is of particular interest, because as a rule, rats cease to care for themselves when they are sick. Thus, with the help of the results of the work done, it is possible to monitor not only the behavior but also the health of the animal.

In addition, using the wireless sensor network (WSN), it is possible to track the behavior of not just one, but whole groups of animals and their social behavior. Handcock and others in [15] suggested putting on collars with built-in GPS-modules on cows for obtaining images from the satellite, providing the trajectory of the herd movement, the interaction between individuals, and formation of local groups.

In [16], like in [14], it is supposed to use wireless devices, which include an accelerometer and a gyroscope. A prototype of such a device is presented in the article, as well as an algorithm for calculating the level of activity of an animal. In accordance with this algorithm, the activity level is determined by the sum of the activity levels for all three axes. In turn, the activity along the axes is calculated as the average value of the deviations of the accelerometer values from the average value on this axis for the last 60 values.

Chandrakant and others put forward a new architecture for an earthquake prediction system. It is assumed that a system analyzing the unusual behavior of animals will be integrated with all existing systems, such as electromagnetic signals detectors, accelerometers detector for volcanism, air ionization detectors and enhanced infrared radiation detection images by satellite, a system analyzing the unusual behavior of animals will be integrated. In addition, a new algorithm is proposed, according to which danger can be determined and an alarm signal is given. The results of the experiment graphically represent the level of activity of dogs, cats, ducks, and hens, as well as readings of accelerometers detector for volcanism, air ionization detectors, and enhanced infrared radiation detection.

In [17, 18], a network architecture based on ZigBee is developed using a neural network for recognition of the patterns of behavior of animals derived from data from sensors. The neural network [18] is also described, and the simulation of energy consumption is carried out.

As can be seen from the analysis of previous works, the researchers have achieved considerable results in monitoring the behavior of biological objects, but not significant enough to be realized in the prediction of earthquakes. In this article, it is proposed to supplement and improve existing achievements through the use of technologies and methods of Big Data, which implies the ability to monitor a vast number of objects. This will allow avoiding erroneous conclusions when analyzing the causes of deviant behavior.

3   The Architecture of the System

The use of living organisms as earthquake sensors should complement the existing developments in the field of seismic activity prediction and the emerging developments. To achieve this goal, it is necessary to design the application architecture in a modular way (Fig. 1). Such architecture would allow easy integration of various elements using standard interfaces.

Data collection is carried out with use of IoT technologies. The data collection aspects are discussed in Sects. 4 and 5. For data collection, both existing and emerging IoT devices and technologies can be used. In order for applications to work successfully, gateways require both: converting data between protocols of data link and the network level, and providing compatibility at higher levels. This provides interoperability of devices using various protocols. Devices providing conversion at the semantic level (the application layer according to OSI model) are called semantic

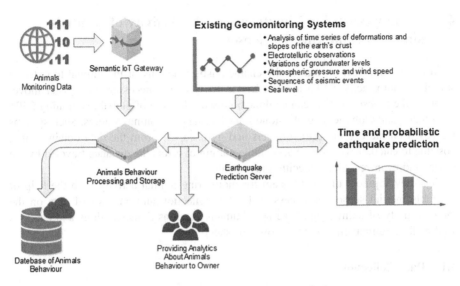

Fig. 1. The architecture of the earthquake prediction system

gateways (SG). They allow transformation between messaging protocols such as CoAP, MQTT, XMPP, HTTP v.1.1, HTTP/2, these protocols are widely used in IoT. The Gateway as Service Architecture for IoT Interoperability is considered in [19]. The alternative approach to already known principles suggested in [20], it is the employment of the SIP protocol as a container for M2M data. Using an SG in the system provides full connectivity with various types of IoT devices (animals monitoring devices).

The next element is the processing and storage unit. This device collects, analyzes and stores the data obtained from monitoring devices and provides access to the data for users (animal owners) and the Earthquake Prediction Server. At this stage of analysis, the non-typical and non-standard behavior of the monitored objects is revealed. The data is used by the Earthquake Prediction Server to compare with data from existing geological and environmental monitoring systems, presented in the form of a time series, to identify the correlation between geophysical factors and animal behavior. The time series trend selection is the first step in the analysis. Splitting the signal into "trend" and "remainder" is based on the assumption that smooth, low-frequency changes are due to the influence of some deterministic the reasons, while the relatively high-frequency oscillations reflect a random component of the process. The presence of a correlation between the identified trends in geo-monitoring and non-standard behavior of animals may be evidence of a pending earthquake.

Another function of the processing and storage unit is an analysis of data regarding animal health (a fitness tracker for animals) and providing the results of this analysis to animal owners in the form of a mobile application or website. This function is necessary to create a strong motivation for the owners of animals to use the proposed system. Such an approach may help to involve a more significant number of animals to the data collection process.

4   The Subsystem for Monitoring of the Activity of Animals Using Wearable Inertial Sensors

As follows from Sect. 2, one of the characteristic reactions of the animal to external stimuli is motor activity. To monitor motor activity, it is necessary to register movements in the timescale. This can be done both with the help of global positioning (GPS and others) and with the help of autonomous inertial positioning systems. Such systems are based on the use of accelerometers and gyroscopes. From the data obtained, it is possible to calculate (within an acceptable margin of error) the distance traveled by the object and even recognize specific movements.

The development of a subsystem for monitoring animal activity with the help of wearable inertial sensors is necessary for the collection and analysis of data on the motor activity of animals that live near humans (such as domestic dogs and cats), as well as farm animals like horses, cows, or sheep.

4.1   Data Collection

To carry out the experiment on data collection, a prototype of the data acquisition device was developed - a smart collar (Fig. 2). This device contains an accelerometer, a gyroscope, a massive battery pack and a NodeMCU microcontroller, which is an MQTT client. The server on which the data is processed was also an MQTT client and received it through the MQTT broker. The data received was subjected to low-frequency filtering, in order to separate the gravitational component of the signal.

Fig. 2. A dog with a prototype of the device for data collection during the experiment

During the experiment, activity data came to the server in CSV files containing a timestamp in the UNIX format and acceleration on three axes.

In connection with the fact that in future it is supposed to apply the methods of Machine Learning, the data was collected in three stages. The first stage, 'Walking,' implied walking down the street with the owner. The second stage, 'Playing,' was an artificially initiated (by the owner) intense action. The third stage, 'Rest', - the collection of data was carried out in conditions that excluded the presence of artificial factors that might stimulate activity. Each record has an 'Activity' parameter, which takes into account the stage at which data were collected (Table 1). This parameter is necessary for further training the neural network.

Table 1. The form for presenting activity data

Timestamp	x-axis	y-axis	z-axis	Activity
1525431335253	0.008776	0.00532	0.013132	Walking
1525431335307	−0.04818	−0.03654	0.025706	Walking
1525431335357	0.067857	−0.03084	0.046965	Walking

In total, about 262,420 lines with records of activity were collected in the resulting database. Figure 3 shows the visualization of activity data characterizing various types of activity. As seen in the graph, the dataset is quite different for various activity types.

4.2 Recognizing Activity Using a Recurrent Neural Network

To improve the quality of the animal activity analysis, it is necessary to exclude periods of activity caused artificially (walks, play) from the sample. To solve this problem, artificial neural networks have been used. The development and training of the neural network were carried out with the help of Python.

A recurrent neural network with Long Short-Term Memory (LSTM) was chosen as a neural network architecture. The LSTM network is universal in the sense that, with a sufficient number of network elements, it can perform any calculation that a conventional computer can do, which requires an appropriate matrix of weights that can be considered as a program.

The LSTM network receives a fixed length data entry, so the data is divided into 200-line segments or 10 s. Activity tags are converted to the unitary code. Data is divided into training and practice sets in the ratio of 8:2.

The network model contains 2 fully connected layers and 2 LSTM layers, each of which contains 64 hidden nodes.

The hyperparameters of learning:

- Optimizer: Adam
- Number of epochs: 50
- Number of samples per iteration: 1024
- Learning speed: 0.0025

During training, the parameter 'accuracy' was monitored - the function inverse to the error function in recognition of activity and the parameter 'loss' - cross entropy determining a slightly near-predictable distribution to the true one.

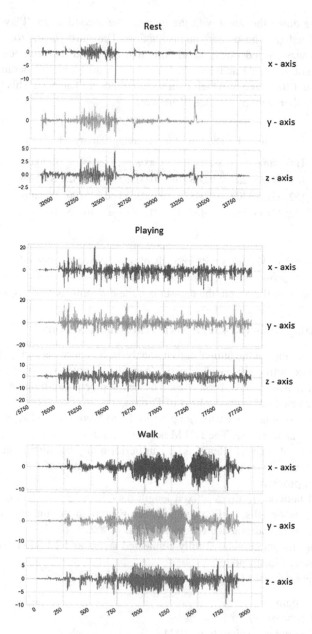

Fig. 3. Visualization of activity data

It is expressed as follows (Fig. 4):

As a result of learning, the neural network recognized the type of activity, with an accuracy of 97%. The codes developed as a result of the program and the source code are available for public access [21].

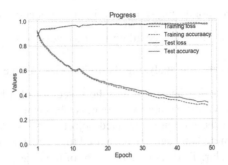

Fig. 4. Machine learning process

The considered method can be used for recognition of activity data, periods that are caused by artificial factors. Excluding these periods from the collected data, provides a data set that can be used to identify non-standard behavior.

4.3 Calculation of Motor Activity

To determine the deviations of the animal's behavior from the norm, the parameter 'level of motor activity' has been used in this project. For the case with a wearable inertial system, this parameter was calculated as the sum of the activity on all the three axes of the accelerometer. The activity on the axes is calculated as the average value of the deviation of the acceleration from the average value on this axis in the past minute.

$$A = \frac{1}{n}\sum\nolimits_{i=1}^{n} x_i - m(X)\vee + \frac{1}{n}\sum\nolimits_{i=1}^{n} y_i - m(Y)\vee + \frac{1}{n}\sum\nolimits_{i=1}^{n} z_i - m(Z)\vee \qquad (1)$$

Where n is the number of measurements made per minute, m is the average value of the accelerometer readings on the X, Y, and Z axes.

The values obtained are compared with the average activity at the same time in the previous days, as well as the average activity for the previous day, week, month, and year.

5 The Monitoring Subsystem Based on Use of Computer Vision Technologies

It is challenging to apply inertial methods of monitoring activity to some animal species. These animals include aquarium fish and animals in cages (birds, mice, rats, etc.). For these, it is necessary to develop a subsystem that allows monitoring activity without the use of wearable sensors. Taking into account that the area of movement of a biological object is limited to certain boundaries (an aquarium or a cage), computer vision systems can be used to solve the task of monitoring activity. In this project, the library of OpenCV computer vision algorithms has been used.

The OpenCV-based monitoring subsystem in this project consists of two modules:

1. The module for detecting and recording the trajectory of the movement of a biological object.
2. The module for storing and analyzing the trajectories of objects.

The module for detecting and recording the trajectory of a biological object is a device and software providing the ability to identify objects using the camera, and also record the trajectory of the motion of detected objects.

The Java programming language has been used to create a subsystem. The object is detected based on its color. To determine the color of the object, the HSV color model (Hue, Saturation, Value) has been used. Thus, by setting the boundaries of each parameter of the given color model, the sought object is detected.

Two cameras (Fig. 5) are used to record the trajectory of an object, which allows determining the trajectory in three-dimensional space. The trajectory is a collection of points (x, y, z), where a point of time is tied to each point. These trajectories are accumulated on the local computer center (in the experiment it has been implemented on Raspberry Pi 3) and sent to the server at the frequency of 1 time per minute. Each device that monitors a biological object has its own unique identifier and binding to a specific region.

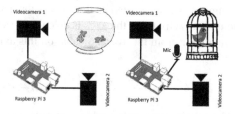

Fig. 5. Data collection schemes

The module for storing and analyzing trajectories of the objects has been implemented on the server. This module contains a database that stores a set of motion trajectories (Table 2).

Table 2. The format of the activity data stored in the database on the server

Location	Object type	Timestamp	Sound level	X	Y	Z
59.902938, 30.488745	Wavy parrot	1525431335253	44,48	0.87	5.32	3.01

In addition to the database, the module implements the functions of receiving and analyzing the obtained trajectories. It is possible to view the trajectory of movements from each device for a set period. The software of the module for detecting and recording the trajectory of a biological object developed during the experiment has been uploaded on GitLab [22].

6 Conclusion

The results presented in this article are the first stage of the project 'IoT Based Earthquake Prediction Technology.' A similar approach to data collection can be applied to other types of animals (like cats, horses, or cows). After the data collection and analysis system is done, another important task needs to be solved - replicating the system so that as many objects as possible can participate in the monitoring. Mass data collection is a key condition for the success of the project, according to the authors. To involve pet owners in the project, it is planned to implement a subsystem that will allow the owner to analyze the animal's behavior online, identify animal health problems promptly, and monitor the achievement of the required level of activity.

It should be noted that some of the proposed approaches, for example, recognition of the type of activity in dogs, can give inaccurate results in connection with the diversity of animal breeds and their essential differences. Perhaps, in this case, it will not be possible to exclude the activity caused by artificial factors from the data set. The calculation of the level of motor activity, in this case, will be performed on all the collected data.

One of the additional ways to use the developed system is to study the animals themselves. It can be assumed that the data collected allows us to study new aspects of animal physiology; for example, the influence of climatic conditions, magnetic storms, seasonal factors, noise, etc. on their behavior.

Acknowledgment. The publication has been prepared with the support of the "RUDN University Program 5-100" and funded by RFBR according to the research project No. No18-37-00084.

References

1. Spence, W., et al.: Measuring the Size of an Earthquake. https://earthquake.usgs.gov/learn/topics/measure.php. Accessed 31 May 2018
2. Wang, Q., et al.: Earthquake prediction based on spatio-temporal data mining: an LSTM network approach. IEEE Trans. Emerg. Top. Comput. (2017)
3. Saad, O.M., Shalaby, A., Sayed, M.S.: Automatic arrival time detection for earthquakes based on fuzzy possibilistic C-Means clustering algorithm. In: 2017 8th International Conference on Recent Advances in Space Technologies (RAST). IEEE (2017)
4. Rajabi, N., Rajabi, O.: Real time earthquake prediction using cross-correlation analysis & transfer function model. In: 2015 2nd International Conference on Knowledge-Based Engineering and Innovation (KBEI). IEEE (2015)
5. Bhargava, N., et al.: Earthquake prediction through animal behavior: a review. Indian J. Biomech. **7** (2009). 8
6. Chandrakant, N., et al.: Animals as wireless biomedical sensors in WSN for earthquake detection in advance. Int. J. Electro Comput. World Knowl. Interface **1**(2) (2011)
7. Kirschvink, J.L.: Earthquake prediction by animals: evolution and sensory perception. Bull. Seismol. Soc. Am. **90**(2), 312–323 (2000)
8. Asim, K.M., et al.: Seismic activity prediction using computational intelligence techniques in northern Pakistan. Acta Geophys. **65**(5), 919–930 (2017)

9. Ahmadi, M., Nasrollahnejad, A., Faraji, A.: Prediction of peak ground acceleration for earthquakes by using intelligent methods. In: 2017 5th Iranian Joint Congress on Fuzzy and Intelligent Systems (CFIS). IEEE (2017)

10. Zin, T.T., et al.: A general video surveillance framework for animal behavior analysis. In: 2016 Third International Conference on Computing Measurement Control and Sensor Network (CMCSN). IEEE (2016)

11. Noda, A., et al.: Behavior analysis of a small animal using IoT sensor system. In: 2017 International Conference on Intelligent Informatics and Biomedical Sciences (ICIIBMS). IEEE (2017)

12. Fan, J., Jiang, N., Wu, Y.: Automatic video-based analysis of animal behaviors. In: 2010 17th IEEE International Conference on Image Processing (ICIP). IEEE (2010)

13. Bhargava, K., et al.: Fog-enabled WSN system for animal behavior analysis in precision dairy. In: 2017 International Conference on Computing, Networking and Communications (ICNC). IEEE (2017)

14. Venkatraman, S., et al.: Wireless inertial sensors for monitoring animal behavior. In: 29th Annual International Conference of the IEEE Engineering in Medicine and Biology Society, EMBS 2007. IEEE (2007)

15. Handcock, R.N., et al.: Monitoring animal behaviour and environmental interactions using wireless sensor networks, GPS collars and satellite remote sensing. Sensors 9(5), 3586–3603 (2009)

16. Tan, S.-L., et al.: A wireless activity monitoring system for monkey behavioural study. In: 2011 IEEE 15th International Symposium on Consumer Electronics (ISCE). IEEE (2011)

17. Dominguez-Morales, J.P., et al.: Wireless sensor network for wildlife tracking and behavior classification of animals in Doñana. IEEE Commun. Lett. 20(12), 2534–2537 (2016)

18. Cerezuela-Escudero, E., et al.: Performance evaluation of neural networks for animal behaviors classification: horse gaits case study. Distributed Computing and Artificial Intelligence, 13th International Conference. AISC, vol. 474, pp. 377–385. Springer, Cham (2016). https://doi.org/10.1007/978-3-319-40162-1_41

19. Desai, P., Sheth, A., Anantharam, P.: Semantic gateway as a service architecture for IoT interoperability. In: 2015 IEEE International Conference on Mobile Services, New York, NY, pp. 313–319 (2015)

20. Masek, P., Hosek, J., Zeman, K., et al.: Implementation of true IoT vision: survey on enabling protocols and hands-on experience. Int. J. Distrib. Sens. Netw. 2016 (2016). Article ID 8160282

21. Pirmagomedov, R.: (2018) https://gitlab.com/prya.spb/IoT_Based_Earthquake_Prediction_Technology. Accessed 31 May 2018

22. Danilov, K.: IoT Based Earthquake Prediction Technology (2018). https://gitlab.com/danilovkn94/IoTBasedEarthquakePredictionTechnology. Accessed 31 May 2018

Interaction of AR and IoT Applications on the Basis of Hierarchical Cloud Services

Maria Makolkina[1,2(✉)], Van Dai Pham[1], Ruslan Kirichek[1,2],
Alexander Gogol[1], and Andrey Koucheryavy[1]

[1] The Bonch-Bruevich Saint-Petersburg State University of Telecommunications,
St.Petersburg, Russia
makolkina@list.ru, daipham93@gmail.com, kirichek@sut.ru,
al.gogol@mail.ru, akouch@mail.ru
[2] Peoples' Friendship University of Russia (RUDN University), Moscow, Russia

Abstract. Currently, the active development of all industries occurs within the framework of the "Industry 4.0" concept. One of the key applications of this concept is the implementation of the Industrial Internet of things (IIoT) and applications that will help to reduce the labor costs and costs of enterprises. Among such applications, undoubtedly, we can highlight Internet of things (IoT) and Augmented Reality (AR). The article discusses the interaction of applications of augmented reality with the Internet of things using the example of visualization of plant data collected by sensors and stored in a cloud service. A laboratory testbed is developed to measure the delay in the delivery of data from the cloud platform to the Augmented Reality application. The mechanism for uploading data to the local cloud is used to optimize the quality parameters. It helps to reduce the delay by introducing a new architecture for implementing IoT/AR applications.

Keywords: Internet of Things · Augmented reality · Cloud services

1 Introduction

Today, Internet of Things and its applications are found in almost all areas of modern society. In modern life is really difficult to imagine without a smartphone. Smartphone with several data transferring technologies and a lot of sensors that transmitting data from the real world to digital appear to be the Internet of Things. In the past years, with the advent of new models of smartphones, a new feature become very famous - the ability to display layers of information about the world around us based on augmented reality applications. The augmented reality has been known for a long time, but only in the last two years, video analytics, visualization, ultra-fast scanning and data transmission have made it possible to implement this application on the basis of a smartphone. The AR applications are distributed in education, mobile communications, medicine,

O. Galinina et al. (Eds.): NEW2AN 2018/ruSMART 2018, LNCS 11118, pp. 547–559, 2018.
https://doi.org/10.1007/978-3-030-01168-0_49

robotics, military science and e.t.c. [1]. At a glance, IoT and AR have different concepts, but they are having complementation to each other in order to, the conveying information to a person that he can perceive in an understandable form.

According to the forecast from the International Telecommunications Union (ITU), the number of devices connected to the Internet will reach 25 billion units by 2020 [2]. Such devices will be cars, engines, industrial equipment, wearable gadgets, all sensors and etc. Data will be collected from all devices for the purpose of subsequent analysis and the formation of control commands for executive devices. But if we think about what kind of data streams will be created, then it seems into a question the possibility of going deep to deeper, understanding, and comprehending data. Thanks to the use of IoT and AR, it will be possible not only to visualize the data for the people but also creating the opportunity to organize an individual interactive interaction with the interface of a device.

From the other hand, we can get the same IoT data without AR, but the combination of information received from the sensors (i.e., IoT) and the ability to infliction and display data directly next to the device itself make the commands context-sensitive. Augmented reality technologies also greatly help and increase different approaches to the perception of IoT objects, where information of devices and objects is accessible via the Internet. Human doesn't has ability to analyzing and processing huge amounts of data, and attention in the perception of more than 6 objects is generally dissipated. Adding augmented reality to them will help to visualize the information collected by sensors and to carry out high-quality human interaction with this information [1]. The article deals with the interaction models between IoT and AR. Based on these models, a mobile application has been developed that allows to extract real-time IoT data from the cloud service using a augmented reality. One of the key factors in augmented reality is the quality of experience, which depends on the delays in providing IoT data and the process of recognizing an object identifier. The article considers various architectural solutions that allow to reduce the delay when delivering data from the cloud service to the augmented reality application.

2   Related Works of IoT and AR

Issues of sharing and interaction IoT and AR have been prepared several scientific articles. The experimental scenarios are sufficiently different. Researchers from Auto-ID Labs have prepared a review about the augmented reality and the Internet of things technologies in [3]. In this article the question of why the augmented reality technology is necessary for development of gateways to the Internet of Things, was considered. Authors considered smartphones as a gateway and an intermediary between people, physical and digital things and the real environment.

In [4,5], the authors were offering a general augmented reality framework, supported by the extended infrastructure of the Internet of things. The implementation of the ARIoT application for visualizing the instructions of household appliances was demonstrated. In these articles, the ARIoT hardware-software complex is described, in which objects (or "things") are identified dynamically in the working space, important information about the purpose of one or another device is inferred and a quick recognition and spatial monitoring of objects inside the premises in order to output interactive information to provide a new service to the client was provided.

In [6] a survey on the question about the joint use of the Internet of things and augmented reality in the world was presented. This work describes architecture, technological aspects of implementation, applications for IoT and AR, which are already developed to date.

Another direction of the use of AR was presented in [7] for road traffic management if necessary for emergency services to driveway the crossroad (fire, police, and ambulance). This model includes the using of augmented reality to visualize the real information at the crossroad, which is displayed from the quadrocopter's camera implementing monitoring. The authors propose to inform beforehand the driver about possible difficulties in the passage of the intersection and an alternative route or maneuver.

The article [8] presented the results of the development and research of the AR web application and its interaction with mobile agents included in the resource-oriented architecture of the Internet of Things system. The AR mobile application allows to manage the coffee machine through the layers of augmented reality.

In [9] the authors investigated the interaction between augmented reality and flying ubiquitous sensor networks. For the subsequent implementation of such applications, the requirement to develop new traffic patterns was evaluated to further ensure the quality of experience. In this article, the authors considered the traffic structure for augmented reality applications. Traffic is created by a video camera, located on board a quadrocopter for control, surveillance, monitoring, and more.

In the context of smart cities in [10], the use of augmented reality and the Internet of things to observe people with physical disabilities were considered. The authors proposed a system that allows wheelchair users to interact with objects located in the surrounding world with the help of Augmented Reality and Radio Frequency Identification (RFID) technologies.

In the education, an investigation was conducted [11], which describes the system of augmented reality and the Internet of things used in laboratory exercises. This system helps students to study various experiments, as well as to reduce the factors associated with their extrinsic cognitive load, which they experience in laboratory classes constantly distracted.

The considered works have shown that augmented reality and the Internet of things are being studied for the implementation of various applications. Could be given to emphasize focus attention that articles only appear on this topic and in the next few years this topic will be very relevant. With the increase in the number of IoT devices antipersistent traffic flows will be created, it will be necessary to consider the quality of providing IoT/AR information in the development of such systems in order to ensure the acceptable quality of information to users.

3 Scheme of Interaction Between IoT and AR

We present the structural scheme of the interaction of IoT and AR applications, which is analyzed in this article.

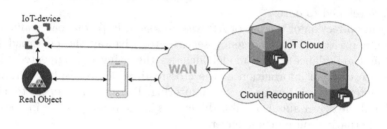

Fig. 1. The interaction between IoT and AR elements

Figure 1 shows the following elements:

- Real Object: it is existing things, about which users want to know information.
- IoT-Device: the micro-controller with a network interface and various sensors collects data of the real object and sends them to the IoT cloud in real time. If there are actuators on the IoT-Device, they can also receive control commands in real time.
- AR-Device: here it can be smartphones, smart glasses, smart lenses and another. As now the smartphones are the most popular for users, so development environments of AR applications for the smartphones are developing very quickly. In our case, the smartphone make the requests to the Cloud Recognition for recognizing the photo (image target) or object. After receiving the response of Cloud Recognition, the smartphone make the requests to the IoT-Cloud for receiving information about the state of the object. The received data will be visualized on phone screen.
- IoT-Cloud: places for processing and storing IoT data. And Cloud Recognition for recognizing the object identifier that needs to be supplemented with information. The clouds process the requests of IoT/AR devices and generate the responses with a minimum delay.

3.1 Structure of the Laboratory Stand

We are considering an example about the connection between the Internet of Things and the Augmented Reality applications. This application was developed in the IoT laboratory SPbSUT [12]. There were using the following components for the laboratory stand:

Hardware:

- Wi-Fi module NodeMCU.
- The DHT22 temperature and humidity sensor that can be connected to any digital pinout of the NodeMCU module (D0-D8) and the soil moisture sensor is connected to the analog pin A0.
- Smartphone (Android or iOS) with a high-resolution camera.

Software:

- Arduino IDE for uploading the firmware to the NodeMCU.
- Vuforia and Unity3D for creating the AR application.

IoT module NodeMCU connects to the Internet via Wi-Fi (connects to a Wi-Fi access point), reads sensor data and sends it to the IoT cloud via a secure HTTPS protocol with using RESTful requests. The connection scheme of the NodeMCU and the sensors is shown in Fig. 2. Basic information such as the identifier number, temperature, humidity, and soil is sent in JSON format to the cloud. Developed mobile AR-application visualizes information about temperature, humidity, and soil, when the phone camera recognized the target object, in our case, it is a flower. There was used the Vuforia framework in the development [14],which is one of the most popular for creating mobile Augmented Reality applications on iOS and Android operating systems and is supported in the Unity3D development system. Vuforia and Unity3D allow creating AR applications very quickly with a large set of functions. The process of data exchange between the AR application and the clouds is also performed by using a secure HTTPS protocol with using RESTful requests. Figure 3 shows a screenshot of the phone screen from the mobile application of augmented reality, on which the flower is supplemented with layers of received information from the sensors.

Fig. 2. The connection of WiFi module NodeMCU with sensors

Fig. 3. Display of information layers obtained from IoT sensors

3.2   Estimation of Service Time

To assess the performance of the system under consideration, we measure the response delay after a request from the AR application to the cloud service. For greater clarity, cloud services were considered, located in different countries, in order to simulate a different delay in the data delivery, *Thingspeak*, *thinger.io*, *Ubidots*. These cloud platforms are well known with the IoT developers. The experiment also uses its own cloud platform *AmazonWebService* (AWS), which is located in Singapore. The measurement results are presented in Table 1.

Table 1. Delays for receiving information after the request of AR client

IoT Cloud	Average Response Time (s)
Thingspeak	0,98
thinger.io	1,20
Ubidots	1,92
AWS-Sing	1,67

When using the framework Vuforia, image targets can be recognized on cloud service Vuforia. The free version for developers allows saving and recognizing up to 1000 target images in the cloud database. Estimates of the time of the object identifier recognition, in this case, this is target image, are presented in Table 2. The average response time is considered depending on the number of images that are stored in the database of the cloud service Vuforia.

Table 2. Delay for receiving response from cloud service Vuforia

Number of target in database	Average Response Time (s)
2	0,93
200	1,25
600	1,52

4   Tier Cloud Architecture for IoT/AR Services

As the results of investigation, IoT-Cloud can be located in different parts of the globe, which affects the data delivery time to augmented reality application. In view of the fact that data from sensors is updated 2 times for a minute and is not real-time data, it was decided to develop an architecture that will allow downloading data from the cloud server to a local server located in the immediate vicinity to the place of the AR application. Clouds can be distributed across districts and regions. IoT/AR devices can receive data from first or second tier cloud. The presented cloud levels are shown in Fig. 4.

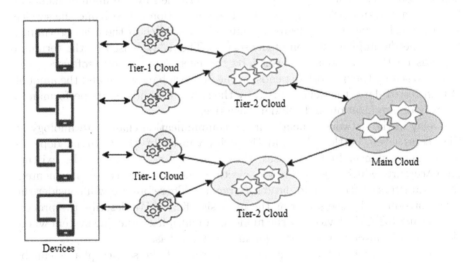

Fig. 4. The architecture of cloud tiers

Figure 4 shows that the IoT/AR data is stored at cloud levels. At first, device will request to a first tier cloud. If the answer is not received from it, then the request is directed to a second tier cloud, and etc. If the answer is received, the search stops.

In the proposed system, IoT/AR devices are connected to a nearest first tier cloud on the basis of geolocation. The first tier cloud can quickly process the request. Each group of the first tier clouds connects to a second tier cloud. The second tier cloud has a large storage space and is used to solve tasks that can not be processed by first tier cloud. Every second tier cloud plays the role of controller of the group of first tier clouds and is a gateway to the main cloud.

5   Service System Model

The idea of the multi-level service system, in this case, is to "approximate" the service system to the user, i.e. possibly, localize traffic and data. This is possible for the reason that much of the AR information is logically and geographically related to those objects that are within the reach of the user. This suggests that the overwhelming number of queries from users concentrated in some local area will refer to a well-defined data (information) cluster, i.e. there is a correlation between the geographic location and the data in the database. In this situation, the localization of the service will allow you to "enclose" a large percentage of traffic within the local zone. This solution will reduce the requirements for network bandwidth, server performance, and database size.

The time between the moment of requesting information and its receipt by the user (display) is one of the main indicators of the quality of service. It depends on several components, which are determined by the main elements of the system: the subscriber terminal, the communication network and the service server (cloud service). The degree of significance of each of them is determined by the specific implementation of the service. The delay caused by the terminal depends on the functions performed by the client application (software), the delay caused by the network depends on the network bandwidth and the amount of transferred data, the delay caused by the server depends on its performance, the way of organization and the database size.

Delay in the network depends on the communication channel technology at the access level (bottleneck), and the delay caused by the terminal depends on its performance. In most cases, these parameters are set by the existing infrastructure, which is available for selection when the service is implemented, the parameters of servers and databases remain. Therefore, we will consider the delay introduced by the server as the most significant. We describe the process of servicing IoT/AR devices by the model of a multiphase queuing system where clouds of each level represent the queuing system phase.

We will assume that the request flow at each of the service phases can be described by the model of the simplest flow. Such an assumption can be justified when the flow of primary arrivals, rather than packets or frames, is considered as a flow in the communication network. The properties of the flow generated by user actions, with a sufficiently large number of them, are usually close to those of the simplest flow. The characteristics of the service time in such a system are determined by the characteristics of the server and the database.

The queuing system model with three- service phase is shown in Fig. 5.

The incoming flow of requests comes from the user to the first level cloud service (first phase). With a probability of $p_1 = 1 - p_2$, the service is terminated in the first phase, and the request leaves the system, with a probability of p_2,

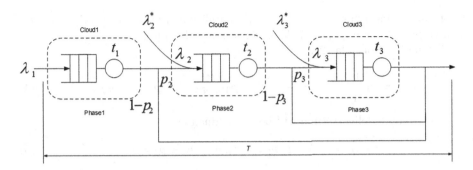

Fig. 5. Structure of the queuing system model

the request is sent to the second phase of the service system. With a probability of $1 - p_3$, the service is terminated in the second phase, and with a probability of p_3 the application is transferred to the third phase of service system. After the servicing in the third phase, the request leaves the system.

When implementing such a queuing system, the model is interested describing the quality of its functioning. As such a model, the probability can be chosen that the delivery time of the augmented reality data (latency) will not exceed some specified value T_0. This probability can be obtained based on service time distribution function:

$$p(t < T_0) = H(T_0) \tag{1}$$

where $H(t)$ - the distribution function of the data delivery time, and $H(T_0)$ the value of this function at the point T_0.

Obtaining an analytical distribution function for the multiphase queuing system, in general, is a non-trivial task. However, a solution is known for the particular case [13], for example, when the service times at each of the phase t_1, t_2 and t_3 are random and have an exponential probability distribution. In this case, the distribution function of the delivery time for the multiphase queuing system can be described by the Erlang distribution [13] with a probability density function:

$$f(t) = \frac{t^{k-1}}{\theta^k (k-1)!} e^{-\frac{t}{\theta}} \tag{2}$$

where θ and k are the distribution parameters, k is a number of service phases, a waiting delivery time is $\tau = \theta k$.

Assuming that the service completion events at each phase are independent, then the probabilities of p_1, p_2 and p_3 are also independent, then the probability density of the delivery time can be determined as:

$$h(t) = p_1 f_1(t) + p_2 f_2(t) + p_3 f_3(t), p_1 + p_2 + p_3 = 1 \tag{3}$$

where $f_1(t)$ is the probability density of the service time for a first phase, $f_2(t)$ is the probability density of the service time for a two-phase queuing system (first and second phases), $f_3(t)$ is the probability density of the service time for a three-phase queuing system (first, second and third phase).

Taking into formula (2):

$$h(t) = p_1 \frac{1}{\tau_1} e^{\frac{-t}{\tau_1}} + p_2 \frac{4t}{\tau_2^2} e^{-2\frac{t}{\tau_2}} + p_3 \frac{27t^2}{2\tau_3^3} e^{-3\frac{t}{\tau_3}}, p_1 + p_2 + p_3 = 1 \qquad (4)$$

The average delivery time will be determined as:

$$\tau = p_1\tau_1 + p_2\tau_2 + p_3\tau_3, p_1 + p_2 + p_3 = 1 \qquad (5)$$

where τ_1, τ_2 and τ_3 are average times for passing the request at the first, second and third phases of the service, respectively. When describing the model of each of the queuing system phase of the form M/M/1, these values can be calculated as:

$$\tau_i = \frac{t_i}{1 - \lambda_i t_i} \qquad (6)$$

where λ_i is a request rate at the input of phase i, i is a number of the service phase, t_i is a mean service time at i-phase.

The distribution function of delivery time for the system under consideration can be obtained from (4) and has the form:

$$H(t) = 1 - p_1 e^{\frac{-t}{\tau_1}} - p_2 \frac{\tau_2 + 2t}{\tau_2} e^{-2\frac{t}{\tau_2}} - p_3 \frac{1}{2}(\frac{6t}{\tau_3} + \frac{9t^2}{\tau_3^3} + 2)e^{-3\frac{t}{\tau_3}}, p_1 + p_2 + p_3 = 1 \quad (7)$$

The probability density function (4) and the distribution function (7) are shown in Fig. 6.

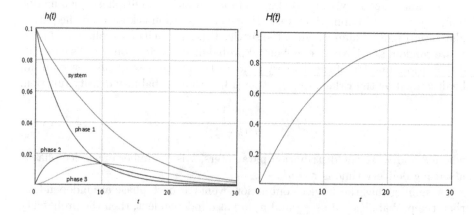

Fig. 6. The probability density and the distribution functions of delivery time

Figure 7 shows the empirical probability densities obtained by simulation for two variants of the service time distribution at each phase: a random (exponentially distributed) service time and a constant service time.

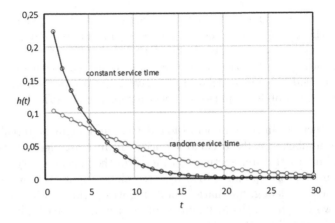

Fig. 7. Results of simulation modeling

The empirical probability density for the random service time, practically, coincides with the obtained above dependence (4) and Fig. 6. Simulation is performed in the AnyLogic system. The results also show that with a constant service time, the average data delivery time is almost half the time. From this, it can be concluded that the obtained distribution function (7) can be used as an estimate of the delivery time "above" while the accuracy of the evaluation is higher, the higher the coefficient of variation of service time to unity.

Using the condition (1) and the expression for the distribution function of delivery time (6), it is possible to control the parameters of the systems p_1, p_2, p_3, τ_1, τ_2 and τ_3 so as to provide the delivery time requirements. The most convenient for configuring the system is the service probabilities at each phase, then to find them need to solve (1) regarding these probabilities:

$$(p_1, p_2, p_3) = arg\{H(T_0) \leq \alpha\} \tag{8}$$

where α - the probability of meeting the delivery time requirements, T_0 - the standard (or norm) delivery time.

The values p_1, p_2, p_3, actually characterize the probability that the requested data will be found at the appropriate level of service. Managing their values can be the shares of subscriber traffic, closing at the appropriate level (the service phase).

The three-phase service system model considered above does not reduce the generality of this problem. The number of service levels (phases) can be arbitrary. For a different number of levels, the expressions (3) and (7) should be changed accordingly.

6   Conclusion

The combination of the Internet of Things and augmented reality technologies was implemented by using the WiFi NodeMCU module for collecting and sending data to IoT-Clouds. Framework Vuforia and Unity3D application were used for creating mobile augmented reality application. On the developed IoT/AR application, the evaluation of service requests time for receiving information from different clouds was estimated. The RESTFull HTTP requests were used to exchange data between IoT/AR devices and clouds. The results of the evaluations show considerable time delays for receiving the responses from the cloud. Consequently, the proposed multi-level service model of the IoT/AR application was considered.

With the increase in the number of IoT devices and augmented reality applications, the number of calls to cloud services will increase significantly, so the need for optimizing the architecture considered in the article is very relevant. A three-phase model corresponding to a three-tier cloud architecture was examined using analytical modeling. Simulation modeling was performed in the AnyLogic system for two variants of the distribution of service time in each phase: random (exponentially distributed) maintenance time and constant service time. The results show that with a constant service time, the average data of delivery time is almost half the time.

References

1. Yuen, S.C.Y., Yaoyuneyong, G., Johnson, E.: Augmented reality: an overview and five directions for AR in education. J. Educ. Technol. Dev. Exch. (JETDE) **4**(1), 119–140 (2011)
2. ITU. https://www.itu.int/net/pressoffice/press_releases/2015/31.aspx. Accessed 14 Feb 2018
3. Ilic, A., Fleisch, E.: Augmented Reality and the Internet of Things. ETH Zurich, p. 18 (2016)
4. Jo, D., Kim, G.J.: In-situ AR manuals for IoT appliances. In: 2016 IEEE International Conference on Consumer Electronics (ICCE), pp. 409–410 (2016)
5. Jo, D., Kim, G.J.: ARIoT: scalable augmented reality framework for interacting with Internet of Things appliances everywhere. IEEE Trans. Consum. Electron. **62**(3), 334–410 (2016)
6. Lanka, S., Ehsan, S., Ehsan, A.: A review of research on emerging technologies of the Internet of Things and augmented reality. In: 2017 International Conference on I-SMAC (IoT in Social, Mobile, Analytics and Cloud), pp. 770–774 (2017)
7. Makolkina, M., Paramonov, A., Vladyko, A., Dunaytsev, R., Kirichek, R., Koucheryavy, A.: The Use of UAVs, SDN, and augmented reality for VANET applications. DEStech Trans. Comput. Sci. Eng. (AIIE), 364–368 (2017)
8. Leppnen, T., Heikkinen, A., Karhu, A., Harjula, E., Riekki, J., Koskela, T.: Augmented reality web applications with mobile agents in the internet of things. In: Next Generation Mobile Apps, Services and Technologies (NGMAST), pp. 54–59 (2014)
9. Makolkina, M., Koucheryavy, A., Paramonov, A.: Investigation of traffic pattern for the augmented reality applications. In: Koucheryavy, Y., Mamatas, L., Matta, I., Ometov, A., Papadimitriou, P. (eds.) WWIC 2017. LNCS, vol. 10372, pp. 233–246. Springer, Cham (2017). https://doi.org/10.1007/978-3-319-61382-6_19
10. Rashid, Z., Meli-Segu, J., Pous, R., Peig, E.: Using augmented reality and internet of things to improve accessibility of people with motor disabilities in the context of smart cities. Futur. Gener. Comput. Syst. **76**, 248–261 (2017)

11. Srivastava, A., Yammiyavar, P.: Augmenting tutoring of students using tangible smart learning objects: an IOT based approach to assist student learning in laboratories. Internet Things Appl. (IOTA), 424–426 (2016)
12. Kirichek, R., Koucheryavy, A.: Internet of things laboratory test bed. In: Zeng, Q.-A. (ed.) Wireless Communications, Networking and Applications. LNEE, vol. 348, pp. 485–494. Springer, New Delhi (2016). https://doi.org/10.1007/978-81-322-2580-5_44
13. Kleinrock, L.: Queueing Systems, vol. 1, p. 448. Wiley, New York (1975)
14. Vuforia. https://developer.vuforia.com. Accessed 14 Feb 2018

AR Enabled System for Cultural Heritage Monitoring and Preservation

Ammar Muthanna[1,2], Abdelhamied A. Ateya[1,3(✉)], Aleksey Amelyanovich[1],
Mikhail Shpakov[1], Pyatkina Darya[2], and Maria Makolkina[1,2]

[1] St. Petersburg State University of Telecommunication, 22 Prospekt Bolshevikov,
St. Petersburg, Russia
ammarexpress@gmail.com, a_ashraf@zu.edu.eg,
lexler0596@gmail.com, sfairat31@gmail.com, makolkina@list.ru
[2] Peoples' Friendship University of Russia (RUDN University), 6 Miklukho-Maklaya St,
Moscow 117198, Russia
da@rudn.university
[3] Electronics and Communications Engineering, Zagazig University, Zagazig, Egypt

Abstract. Preventive conservation of cultural heritage is a serious matter for
people, especially for countries with a large patrimony. Enabling museums and
culture heritage sites with recent technologies is one way to save these sites and
enhance the visitor presence and experience. Augmented reality (AR) and internet
of things (IoT) are recent technologies that can be deployed for such purposes. In
this work, an end to end system structure is introduced for preventive conservation
of museums based on the IoT. The system enables to monitor and indicate any
climatic effects, and points for necessary maintenance. Moreover, the system
employs the AR to help museums visitors to get much experience. An application
interface is developed for both museum administrators and visitors to enable them
to extract the associated data. A prototype is developed for the experimental
evaluation and the results are indicated.

Keywords: IoT · AR · Cultural heritage · Monitoring · Museums · LoRa

1 Introduction

Deploying new technologies for preventive conservation of cultural heritage become a
demand. Recently, wireless sensor networks (WSN), internet of things (IoT), device to
device communication (D2D) and augmented reality (AR) are mentioned to be deployed
for monitoring, enhancing security and saving cultural heritage environments [1, 2].
Moreover, these technologies are used to increase visitors experience and provide them
with much knowledge of the distributed monuments.

Employing these technologies takes place recently in some big museums, as an
example, the louver museum starts deploying WSN since 2006, for monitoring the
museum field and increasing the place security [3]. WSN suffers from many issues
concerned with the transmission rate, which leads to the introduction of other technologies [4]. With the recent advances in sensory manufacturing and the innovation of IoT

© Springer Nature Switzerland AG 2018
O. Galinina et al. (Eds.): NEW2AN 2018/ruSMART 2018, LNCS 11118, pp. 560–571, 2018.
https://doi.org/10.1007/978-3-030-01168-0_50

technology, researches for deploying IoT for cultural heritage environments are developed [5].

The term culture defines many aspects includes; art, monuments and architectures [6]. Cultural heritage represents an important source of national income to some countries such as Egypt, Russia, France and Italy. Millions of tourists find their destinations to cultural places in these countries every year, providing an important source of national income to these countries. Thus, paying great attention to cultural heritage is a demand for all, especially, for these countries [7].

According to some statistical studies, for the Russian cultural sites, the visitors of Russian state museums and art objects in the period from 2012 to 2016 are increased by 41 percent [8]. For example, the Hermitage museum became the most visited museum in Russia and entered the top ten of most visited museums in the world; in 2017 its visitors are doubled [8]. For the mentioned statistics private museums didn't considered, which in some studies noted that their visitors doubled year by year.

From the previous studies, it can be inferred that the growth of public interest in cultural objects is highly increased. Thus, the recent innovation in technology and sensory should be deployed to serve for enhancing the visits of cultural sites and provide visitors with much experience. Moreover, these technologies and smart things should be used to conserve cultural heritage by providing smart systems for monitoring these domains and alarm for maintenance once it is needed [8].

Since artworks and monuments undergo deteriorations with the time, these environments and objects should be protected. The main factors affect cultural heritage and leads to these deteriorations are the weather conditions, lightings, material type of object and human effect. One way to save artworks and prevent them from damage is to keep them under controlled climatic conditions, periodically make maintenance and always monitor them [10]. Climatic conditions represent the main affecting factor and thus, temperature and humidity should be measured periodically to keep them at certain level that saves the cultural environments [11].

Introducing new technologies to provide a sustainable framework for cultural heritage conservation is a vital solution that reduces the risk of deterioration. The system should deploy a low cost reliable hardware, with an open software platforms [12].

The scope of this work is to deploy recent technologies (i.e. AR and IoT) for increasing the information accessibility and interactivity of art objects, which in turn helps in increasing number of visitors and preserving cultural heritage sites for future generations. Furthermore, the proposed framework monitors the cultural sites and alarms for any required maintenances. Thus, the proposed system reduces the risk of damage and save the cultural heritage. The system deploys an intelligent engine to achieve these benefits and provide a reliable system. Section 2 provides the background and related works, Sect. 3 introducing the proposed system for preventive conservation of cultural heritage and Sect. 4 provides the experimental work and testing results.

2   Background and Related Work

There are many systems and studies developed for cultural heritage; microclimate control is one of the most common systems for museums [13]. The system is mainly developed to ensure constant temperature and humidity in the closed rooms inside museums. The main disadvantage with the microclimate systems is the complexity.

In [14], authors introduced an IoT based system for remote monitoring of cultural heritage environments. The system deploys low cost hardware devises; and provides a prototype to measure temperature and humidity. The prototype uses the ESP8266 chip with the Wi-Fi. The gateway is programmed using Arduino IDE, and the system is tested for indoor measurements.

In [15], authors introduced IoT architecture for preventive conservation of cultural heritage environments. The work introduces the main requirements of the system from the artwork point of view. The system is based on LoRa and Sigfox technologies to provide a lifespan of more than ten years.

Moreover, many AR applications have been developed for cultural heritage. AR technology is a new paradigm aims to show virtual objects and embed multimedia to existing real environments [16]. Deploying AR for cultural sites introduces new terms such as Virtual archaeology, AR Heritage, Virtual Museums and Virtual History [16]. In [17], authors developed an AR game for introducing Philippine history. The main purpose for the application is the educational purposes.

The European Commission launched the "cultural heritage experiences through socio-personal interactions and storytelling" (CHESS) project, which aims to develop AR-based solutions for cultural heritage. The project consists of series of stages; include layering, scripting and production [18].

Another common AR heritage application is the Plaster ReCast developed for Carnegie Museum of Art's (CMOA) Hall of Architecture [19]. The hall represents the third largest architectural plaster cast collection in the world that contains monumental replicas of portions of buildings and fragments. The application provides 3D modules of historical and architectural information.

The novelty of our proposed work, come from the consideration of both cultural environment administrators and visitors. The proposed work merges both IoT and AR and integrates them in a single developed application.

3   AR - IoT Based Cultural Heritage Framework

The proposed structure is a modular design that increases the functionality and easily allows adding or removing elements for future purposes and updates. Figure 1 illustrates the proposed system structure. The system can be viewed as two main parts; the first part is the Preventive conservation block and the second part is the visitor assistance block. The Preventive conservation block is responsible for the preservation and management of art objects. This block is mainly handled by the manager of cultural sites (e.g. museums) and all collected data can only be shared with the managers and dedicated employees of museums. The visitor assistance block is responsible for increasing interactivity and information availability for site visitors. This block contains all public data and mainly deployed for interaction with the museum visitors. It provides visitors with services that increase their experience and help them to get much knowledge.

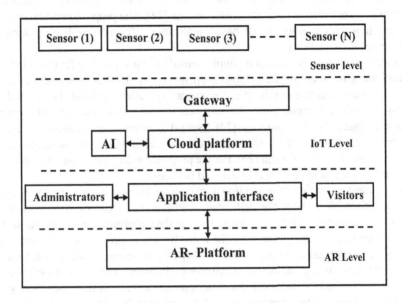

Fig. 1. End-to-end IoT/AR cultural heritage system.

3.1 The Preventive Conservation Part

The Preventive conservation part of the system is composed of radio module and special purpose sensors deployed for sensing certain parameters that are important for archaeologists to keep monuments. The min sensors deployed are illumination, temperature, humidity, ultraviolet radiation, air quality, pressure and shock load sensors. These sensors are distributed among monuments to cover all places in the museum, for which the proposed system is deployed. These sensors are controlled and managed through a base station centered at the museum. The communication standard for data transfer between sensors and the corresponding gateway is LPWAN-LoRa [20]. This interface technology is chosen because of the following:

1 The inability to create a wired infrastructure of the required scale in most existing museums,
2 The need to ensure long-term autonomy of each module,
3 The scalability and extensibility of the system in the future, and
4 The minimum cost of implementation.

The base station is connected with a cloud unit that is responsible for providing storing and processing capabilities of the sensed data. The cloud platform should deploy an intelligent system for processing the received sensor data and detect whether the monuments needs maintenance or not.

3.2 The Visitor Assistance Part

The visitor assistance block provides user assisted services by employing AR technology. At the moment, many museums equip their exhibits with QR-codes, which allow visitors to get more information about monuments [21]. The proposed system provides this interaction with visitors by AR based applications that achieve better interaction and information availability.

Visitor assistance services can be implemented in two ways. The first way involves the additional equipment of the exhibits with special marks that, when activated, can open additional interactive possibilities for interacting with the exhibit. The second way is the use of artificial intelligence (AI) (i.e. neural network) to analyze and recognize distributed objects in the museum [22]. Neural network can recognize things by comparing them with a record from the database and activates the same interactive possibilities, as in the first variant [23]. For the proposed work, the second way is chosen as the method for implementing user assisted services.

The simplest example of the additional interactive features is the output message of reference information about the dedicated object. A more complicated variant, for example, for paintings with battle scenes, may be the opportunity to see in real-life the equipment depicted in the picture. For the reproduction of projections, the most interesting is the use of glasses of augmented reality that museums could offer for visitors by analogy with existing audio guides. Moreover, in order to offer more facilitates for visitors, it is proposed to implement a mobile application with similar functionality. The functionality of the solution can be extended, for example, by adding an interactive map of the museum, which will help visitors better navigate in it, and guides will mark themselves with special tags that will help tourists find them, even if they fall behind the main group.

4   Experimental Evaluation

In this part, the previous system is implemented using real hardware and existing platforms. The two parts are considered, and the deployed hardware and software for each part is introduced. We develop a single application for the two system parts, while the application is able to distinguish between the two types of users (i.e. visitors and administrators). The associated data is displayed for each kind of users.

4.1 The Preventive Conservation Part

The main system components considered for experimental evaluation are:

1 IoT gateway:
   The IoT gateway is a NodeMCU V3 board with ESP8266 module, which was originally developed to create various IoT devices. The module is able to send and receive information to a local network or to the Internet using a built-in Wi-Fi. This module is considered to be the gateway, due to the cost efficiency, market availability and also convenience of work.

The IoT gateway NodeMCU is programmed using sketch in Arduino IDE, to initialize the wireless connection, and generate HTTP POST requests through the API to ThingSpeak cloud platform.

2 Cloud platform:

The considered cloud platform is the ThingSpeak, which is an open source application for the Internet of things and API for storing and retrieving data using the HTTP protocol. ThingSpeak is supported from the mathworks products and has built-in Matlab libraries for supporting IoT devices [24]. ThingSpeak enables users to analyze and visualize downloaded data using Matlab without having to purchase Matlab license from Mathworks.

3 Sensors:

For experimental evaluation we deploy a composite DHT11 sensor, which consists of two measuring sensors; thermometer and hygrometer. These two sensors are used to measure temperature and humidity.

Sensors measure and collect considered information and transfer them to the ThingSpeak cloud platform through the IoT gateway. The cloud platform analyzes and processes the received data and sends a report and alarms to the museum managers and employees through an application developed with the same platforms introduced in the next part. The structure of the proposed prototype is illustrated in Fig. 2.

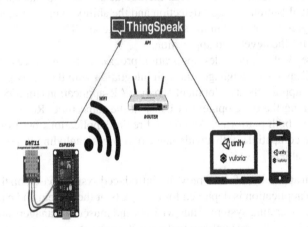

Fig. 2. Structure of the proposed prototype.

4.2 The Visitor Assistance Part

In order to provide visual information to visitors and employees, as mentioned earlier, augmented reality applications are developed. These applications are supported by augmented reality glasses and smart phones [25]. The following platforms are considered for building applications for assisting visitors:

1 Unity:

Unity is a cross-platform environment used to develop both two and three dimensional video games [26]. It enables the creation of heterogeneous applications for more than twenty different operating systems. The main advantages of Unity are the availability of a visual development environment, cross-platform support and a modular component system. In the other side, the main disadvantages of Unity include the emergence of complexities when working with multi-component schemes and difficulties in connecting external libraries.

2 Vuforia:

This is the Augmented Reality Platform and the Augmented Reality Software (SDK) toolkit for mobile devices developed by Qualcomm [27]. Vuforia uses technology of computer vision, as well as tracking flat images and simple bulk real objects (e.g. cubic ones) in real time. The ability to register images, made it possible to locate and orient virtual objects, such as 3D models and media content, in conjunction with real images when viewed through mobile device cameras. The virtual object is oriented on a real image so that the observer's point of view treats them in the same way to achieve the main effect (i.e. the feeling that the virtual object is part of the real world). Vuforia supports various 2D and 3D target types, including marker Image Target, 3D Multi-Target targets, and reference markers that select objects in the scene to recognize them. Additional functions include detection of obstacles using the so-called "Virtual Buttons", target detection and the ability to programmatically create and reconfigure targets within the self-modifying code. The Image Target is mainly considered for the developed application.

Furthermore, Vuforia provides application programming interfaces in C++, Java, Objective-C, and .Net languages through integration with the Unity gaming engine. Thus, SDK supports the development of native AR-applications for iOS and Android, while assuming the development in Unity at the same time. Results can easily be transferred to both platforms. Augmented reality applications built on the Vuforia platform are compatible with a wide range of devices, including iPhone, iPad, smart phones and tablets.

We build an application for the previous introduced system, based on the considered platforms. The application is deployed for both parts of the system and can be run over any appropriate operating system. Thus, visitors and museum managers and employees can use this application. The application defines the two categories of users; visitors and administrators, and offers the appropriate data and services for each category. Figure 3 illustrates the application interface.

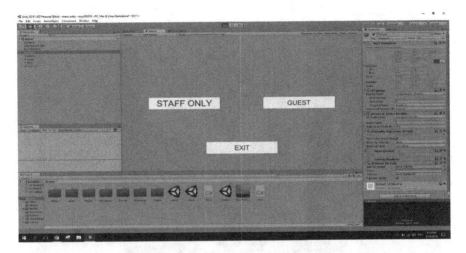

Fig. 3. General interface of the developed AR-application.

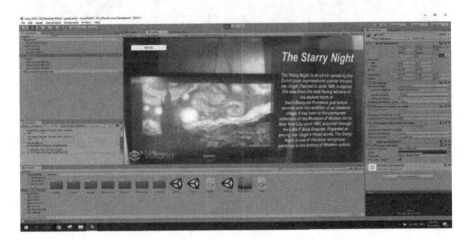

Fig. 4. Visitor interface.

We consider an art object as a prototype for testing the developed application. The art object considered is the image of the painting "Moonlight Night" by Vincent Van Gogh. The application provide two interfaces; one interface for the museum administrators and the other for visitors. Figure 5 illustrates the appropriate interface for museum administrators. Once an administrator hover over the picture's virtual blocks, the corresponding measured data is displayed. Figure 4 illustrates the appropriate interface for visitors, in which all public information about the object is displayed for the visitor.

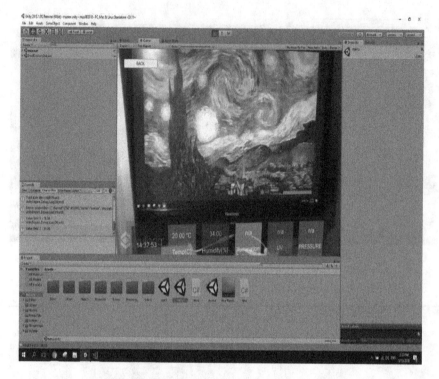

Fig. 5. Administrators interface.

Furthermore, readings for both considered sensors, for a day, are illustrated in Figs. 6 and 7.

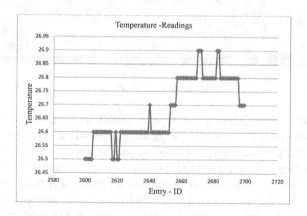

Fig. 6. Records for temperature sensor for a day.

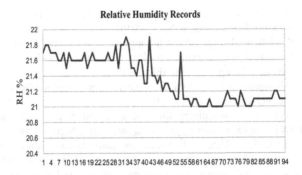

Fig. 7. Records for humidity sensor for a day.

5   Conclusion

This work provides a framework for preventive conservation of cultural heritage and helping visitors of cultural sites to get much experience. The system employs IoT and AR to achieve such purposes. Museums are fed with distributed sensors that support IoT and an IoT gateway is deployed for each museum and cultural site. The gateway is connected with a cloud platform that provides storage and processing capabilities for the sensed data. The cloud platform employs some means of neural network and supports the AR applications. An application is developed to support both museum administrators and visitors. The application toggles between the two types of users and displays the appropriate information for each user. Furthermore, the system alarms the administrators for any required maintenance.

Future Work

For future work, the prototype should be implemented for Hermitage museum and a single distributed database for storing information of all visitors and administrators will be introduced. Moreover, a system for connecting gateway for different museums should be developed.

Acknowledgement. The publication has been prepared with the support of the "RUDN University Program 5-100".

References

1. Piccialli, F., Chianese, A.: Editorial for FGCS special issue: the internet of cultural things: towards a smart cultural heritage. Futur. Gener. Comput. Syst. **81**, 514–515 (2018)
2. Olshannikova, E., Ometov, A., Koucheryavy, Y., Olsson, T.: Visualizing big data with augmented and virtual reality: challenges and research agenda. J. Big Data **2**(1) (2015)
3. Ouyang, J.S.: The key of sensor network applied in cultural heritage protection measurement and control technology. Sci. Cons. Archaeol **23**, 55–59 (2011)
4. Wang, Y., Dai, X., Jung, J.J., Choi, C.: Performance analysis of smart cultural heritage protection oriented wireless networks. Futur. Gener. Comput. Syst. **81**, 593–600 (2018)

5. De Carolis, B.N., Gena, C., Kuflik, T., Lanir, J.: Special issue on advanced interfaces for cultural heritage. Int. J. Hum.-Comput. Stud. (2018)
6. Chianese, A., Piccialli, F.: Designing a smart museum: When cultural heritage joins IoT. In: 2014 Eighth International Conference on Next Generation Mobile Apps, Services and Technologies (NGMAST), pp. 300–306. IEEE (2014)
7. Timothy, D.J., Nyaupane, G.P.: Cultural Heritage and Tourism in the Developing World: A Regional Perspective. Routledge, May 2009
8. Russian statistical analysis of cultural sites and museums. https://artinvestment.ru/invest/analytics/20170410_museum_visitors_in_russia_stats.html. Accessed May 2018
9. Chianese, A., Piccialli, F., Valente, I.: Smart environments and cultural heritage: a novel approach to create intelligent cultural spaces. J. Locat. Based Serv. **9**(3), 209–234 (2015)
10. Amato, F., et al.: Big data meets digital cultural heritage: design and implementation of SCRABS. A smart context-aware browsing assistant for cultural environments. J. Comput. Cult. Herit. (JOCCH) 10(1) (2017)
11. Ardito, C., et al.: Towards enabling cultural-heritage experts to create customizable visit experiences. In: Proceedings of the 2018 AVI-CH Workshop on Advanced Visual Interfaces for Cultural Heritage, vol. 2091 (2018)
12. Jara, A.J., Sun, Y., Song, H., Bie, R., Genooud, D., Bocchi, Y.: Internet of things for cultural heritage of smart cities and smart regions. In: 2015 IEEE 29th International Conference on Advanced Information Networking and Applications Workshops (WAINA), pp. 668–675. IEEE (2015)
13. Perles, A., et al.: An energy-efficient internet of things (IoT) architecture for preventive conservation of cultural heritage. Futur. Gener. Comput. Syst. **81**, 566–581 (2018)
14. Asinelli, M.G., Serra, M.S., Marimòn, J.M., Espaulella, J.S.: The smARTS_Museum_V1: an open hardware device for remote monitoring of cultural heritage indoor environments. HardwareX, e00028, May 2018
15. Shin, J.E., Park, H., Woo, W.: Connecting the dots: enhancing the usability of indexed multimedia data for AR cultural heritage applications through storytelling. In: Proceedings of the 15th International Workshop on Content-Based Multimedia Indexing, p. 11. ACM (2017)
16. Desai, N.: Recreation of history using augmented reality (2018)
17. Rodrigo, M.M., Caluya, N.R., Diy, W.D., Vidal, E.C.: Igpaw: intramuros-design of an augmented reality game for Philippine history. In: Proceedings of the International Conference on Computers in Education (2015)
18. The CHESS project. http://www.chessexperience.eu/. Accessed May 2018
19. Henry, M.: Plaster ReCast at the Carnegie Museum of Art. http://aiapgh.org/. Accessed May 2018
20. Sinha, R.S., Wei, Y., Hwang, S.H.: A survey on LPWA technology: LoRa and NB-IoT. ICT Express **3**(1), 14–21 (2017)
21. Wang, H.Y., Liu, G.Z., Hwang, G.J.: Integrating socio-cultural contexts and location-based systems for ubiquitous language learning in museums: A state of the art review of 2009–2014. Br. J. Edu. Technol. **48**(2), 653–671 (2017)
22. Liarokapis, F., Petridis, P., Andrews, D., de Freitas, S.: Multimodal serious games technologies for cultural heritage. Mixed Reality and Gamification for Cultural Heritage, pp. 371–392. Springer, Cham (2017). https://doi.org/10.1007/978-3-319-49607-8_15
23. Chen, C.: Research on the protection of cultural heritage in the background of global climate change based on the artificial neural network. Revista de la Facultad de Ingeniería **32**(11) (2017)

24. Pasha, S.: ThingSpeak based sensing and monitoring system for IoT with Matlab Analysis. Int. J. New Technol. Res. (IJNTR) **2**(6), 19–23 (2016)
25. Olshannikova, E., Ometov, A., Koucheryavy, Y.: Towards big data visualization for augmented reality. In: Proceedings of 16th Conference on Business Informatics (CBI), vol. 2, pp. 33–37
26. Unity. https://unity3d.com/. Accessed May 2018
27. Vuforia. https://www.vuforia.com/. Accessed May 2018

Distributed Streaming Data Processing in IoT Systems Using Multi-agent Software Architecture

Alexey Kovtunenko[1]([⊠]), Azat Bilyalov[2], and Sagit Valeev[1]

[1] Ufa State Aviation Technical University, Karl Marx st. 12, Ufa, Russia
askovtunenko@ugatu.su, vss2000@mail.ru
[2] Bashkir State Medical University, Teatralnaya st. 2a, Ufa, Russia
azat.bilyalov@gmail.com

Abstract. The problem of distributed storing and processing of streaming data in IoT systems is considered. A mathematical model and agent-based software architecture for distributed streaming data processing over heterogeneous computer network is offered. The software architecture determines the following features of IoT nodes software: structure of software components, models of interoperability, algorithms of resource management and also xml-based language which allows to descript distributed IoT applications. The offered architecture is implemented as a software framework ABSynth.

Keywords: Internet of Things · Streaming data · Software architecture
Distributed control system · Data stream management system

1 Introduction

Modern IoT systems include subsystems for processing and storing of streaming data (PSSD) as an integral part. Together they compose a large class of hardware-software complexes for the distributed collection, processing and storage of telemetry, monitoring of the state of technical and biological objects [1, 2]. Unlike traditional storage systems based on DBMS (database management systems), PSSD uses data stream management systems (DSMS) as a kernel and must provide a processing procedures priority higher than data storage procedures priority. They require faster access to data and less processing time. The specifics of the streaming data do not allow to use existing high-performance computing technologies in processing of streaming data [2].

In the article the agent-oriented approach to the organization of calculating process on the computer networks is discussed. It consists in establishment of special relations between program components and computing resources. Software complex components are presented by autonomous objects – software agents functioning in a certain environment – the agent platform. The agent platform is a software layer that provides a unified address space for software components, components interaction, and execution management.

Supported by Ufa State Aviation Technical University.

O. Galinina et al. (Eds.): NEW2AN 2018/ruSMART 2018, LNCS 11118, pp. 572–583, 2018.
https://doi.org/10.1007/978-3-030-01168-0_51

On the basis of the agent-oriented approach, a mathematical model of distributed data processing over on heterogeneous computer networks (HCN) with unpredictable loading (UL) is proposed. The problem of optimal resources allocation in heterogeneous computer network is formulated, and a multi-agent algorithm for dynamic control of computer network resources is discussed. The developed agent-oriented software architecture of distributed data processing systems is presented in the work.

The proposed approach and developed architectural and algorithmic solutions allow to design and create reliable distributed software for data processing in IoT systems based on HCN with UL.

2   State-of-the-Art Scientific Research in the Area of Streaming Data Processing

Streaming data processing is a modern approach, which regards the problem of data processing from a new point of view. Nowadays the efforts of researches are focused on the development of DSMS architectures in terms of providing the following functions: processing continuous queries (CQ) based on some query language or algebra for data streams [4] and providing a Quality of Service (QoS) based on some data stream model [5]. For example, Aurora [6] is a system for managing data streams for monitoring applications state. It uses simple dataflow diagrams to specify CQ and stream query algebra consisting of several primitive operators for expressing stream processing requirements [7].

Another aspect of scientific research in area of DSMS is complex event processing (CEP) systems: development of discrete-event models of different subject areas, formulating of logical rules for complex events generating. IBM's Active Middleware Technology (or AMiT) [8] is a tool that includes both a language and an efficient run-time execution engine, aimed at reducing the complexity of developing event-based applications. AMiT associates computations along with the definition of an event and uses event operators (following the ECA paradigm). AMiT is an example of a system that has tried to incorporate computations into an event specification.

Most of the considered systems represent a completed solution for event-based development, some of them even support distributed computations in common namespace, but they don't include a model of effectively computational resource allocation and don't allow dynamical management of computational resources.

3   Mathematical Models and Formulation of the Problem

The streaming data is a set of attributes, which has a different physical meaning (such as values of physical fields in different points, biometric indicators, simple and complex events et al.).

3.1 Model of Streaming Data Processing

The proposed mathematical model of PSSD is represented by the following tuple:

$$DPS = \langle A, P, C, I, M, V, T \rangle. \tag{1}$$

Here:

- A – set of attributes,
- P – set of computational procedures,
- C – a calculation relation that indicates which attributes are calculated by what procedure,

$$C = \{(p, a): p \text{ calculates } a; \ a \in A, \ p \in P\} \tag{2}$$

- I – a usage relation that indicates which attributes are used in calculations in each procedure, such that $C \cap I = \varnothing$,

$$I = \{(p, a): p \text{ uses } a; \ a \in A, \ p \in P\} \tag{3}$$

- $M{:}A \to \mathsf{N}\backslash\{0\}$ – a function that determines the memory consumption for storing attributes is such that $M(a)$ is the amount of memory required to store the attribute a in bytes,
- $V{:}P \to \mathsf{N}\backslash\{0\}$ – a function that determines the computational complexity of the data processing task is such that $V(p)$ is the computational volume of procedure p in the number of floating point operations,
- $T{:}P \to \mathsf{R}^+$ – the period of the procedure p in seconds.

Using the concept of the cut of the relation $Z{:}X \to Y$ with respect to the element x, as well as $Z(x) = \{y: (x, y) \in Z\}$ and the definition of the quotient set of Y determined by Z $Y/Z = \{Z(x)\}_{x \in X}$, we can define the following families of sets:

- partition A/C of a set of attributes over a set of computational procedures according to a calculation, where $C(p)$ is attributes whose values are calculating in procedure p,
- a family of subsets A/C, where $I(p)$ is attributes which procedure p uses for calculations and reflects the algorithmic relationship between attributes.

To formulate the problem of optimal computing resources allocation in the PSSD, it is necessary to formulate a mathematical model for the resource consumption in computer networks taking into account the specificity of the streaming data.

3.2 Model of the Resource Consumption in Computer Networks

The computing resource of the network is traditionally considered to consist of three components:

- performance (CPU resource),
- memory,

– speed of network interactions.

The following tuple corresponds to the computing network.

$$N = \langle W, R, S \rangle \tag{4}$$

Here:

– W – set of computing nodes,
– $R = \{cpu, mem, net^{in}, net^{out}\}$ – set of computing resources,
– $S_r(w)$ – nonnegative function that defines the following parameters:
  • $S_{cpu}(w)$ flops – the node w maximum achievable performance;
  • $S_{mem}(w)$ bytes– the node w available RAM;
  • $S_{net}^{in}(w)$, $S_{net}^{out}(w)$ bytes/second – the maximum achievable network rate for the incoming and outgoing traffic of the node w, respectively.

Various ways of specifying a function S lead to models of different network types.

3.3 Formulation of the Optimal Resource Allocation Problem

The model of the PSSD based on the computer network includes resource model of the computer network N, model of data processing DPS and the relation D describing system deployment in network.

$$IS = \langle N, DPS, D \rangle \tag{5}$$

Here:

– N – the computational network,
– DPS – the PSSD model,
– D – the deployment relation that shows $\forall p \in P$ on what node it physically deployed

$$D = \{(w, p): w \text{ executes } p;\ w \in W,\ p \in P\}. \tag{6}$$

Let $B: W \times R \rightarrow R^+$ – function that shows $\forall w \in W$ amount of the involved resource. The introduced notation makes it possible to formulate the problem of optimal allocation of computational resources in PSSD as the problem of combinatorial optimization. It is necessary to find such deployment relation D^*, that gives maximum to predefined efficiency function F.

$$D* = \arg \max_{D \subset W \times P} F(IS\langle D \rangle) \tag{7}$$

The following constraints apply.

$$B_r(w) \leq S_r(w) \tag{8}$$

4   Agent-Based Software Architecture for PSSD Systems

In line to proposed mathematical models the agent-based software architecture is developed [9]. It assumes software consisting of next structure components:

- the set of **target agents**, where each one is software entity that implements some data processing procedure;
- the set of processing **attributes** implemented as in-memory database (IMDB), that provides distributed storing of streaming data and a low-latency access to them.

To make a flexible, survivable and scalable software the agent approach is proposed, which means using a software agent as a basic software entity. It allows us hereinafter to talk about agent-based software architecture and implies the presence of some additional elements in the system, such as

- **agent platform** – a set of software components that supports the life cycle of target agents, interaction between them in the unified namespace, also the system event model
- **agent container** – a software environment that provides an access to the resources of each computing node and executes of software agents.

Thus, the streaming data processing model within the software architecture proposed above can be represented by the following structural diagram (Fig. 1).

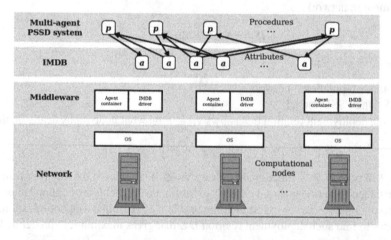

Fig. 1.   Software architecture for distributed streaming data processing

A unified storage space for streaming data is provided by a distributed cached in-memory storage that provides fast and transparent access to attribute values regardless of their physical location. The agent platform is a unified distributed execution space for data processing procedures.

4.1 The Target Agent's Architectural Features

The **target agent** is the basic software entity in the implementation of the data processing model within the proposed architecture [10]. It includes the following components:

- data processing procedure $p \in P$;
- local cache of used attributes $I(p) \subset A$;
- local cache of calculated attributes $C(p) \subset A$:
- mechanism for the management messages receiving and executing;
- mechanism for the agent initialization and self-destruction;
- agent's state.

The activity diagram of the **target agent** shows the process of the agent's functioning after creation (Fig. 2).

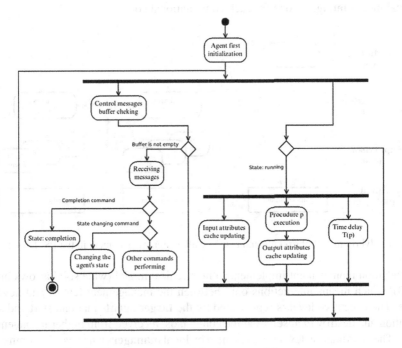

Fig. 2. The fragment of activity diagram of the target agent

In the process of distributed data processing, the **target agent** can be in the following states.

- "Created" – the initial state in the agent's lifecycle.
- "Initialized" – the calculated attributes are set to the initial values.
- "Running" – the agent's operating state when the attributes are recalculating in cyclic mode. "Holding" the agent's operating state when recalculating of attributes is temporary suspended.

- "Completion" is the final state in the agent's lifecycle (preparation for self-destruction, completion of all recording processes).

Changes of the **target agent's** state occurs either according to the stage of the lifecycle, or under the influence of management commands.

4.2   Management Within the Agent-Based Software Architecture

The management of the streaming data processing, as well as the allocation of resources within the agent-oriented software platform is carried out through special **service agents** (Fig. 3):

- **global system manager** – unique (if there is no global manager duplication mode) for each PSSD system;
- **local node manager** – one for each computational core.

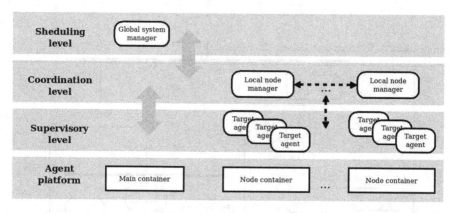

Fig. 3. Hybrid model of execute and resource allocation management in the PSSD system

Execution management implements a hierarchical model (wide solid arrows in the Fig. 3), which means interactions only between the management levels, and never – inside. The executive level is represented by the **target agents** that can start and stop execution and destroy themselves at the commands, received from higher management levels. The coordination level is represented by **local managers** who pass the commands of the **global manager** to the **target agents** of their node, within the management model. The planning level is represented by the **global manager**, as well as by the operator or supreme planning system [10].

Resource management is carried out within the hybrid model (thin dotted arrows in the Fig. 3), where there is a two-level hierarchical organization between **target agents** and **local managers**, as well as a multi-agent peer-to-peer model between **local managers**. The **service agents** interact with the **target agents** according to the specific protocol, proposed within the software architecture. It is based on the ACL [11] language of the FIPA standard and implies communication acts in the multi-agent software system within the following ontologies.

- **Management ontology** – managing commands within the hierarchical model (wide solid arrows at the Fig. 3).
- **Resource allocation ontology** – commands and data related to procedures of a computational resources allocation, such as: estimation of available resources for every node, heuristic algorithm for load balancing etc.

ABSynth allows to extend itself according to developer's needs by adding ontologies or supplementing of the existing ones.

4.3 Tasks Specification Description Language

For the formal description of multi-agent distributed PSSD systems the task specifications description language (TSDL) is offered. It is based on XML notation and consists of the following set of elements (in this section, the term "attribute" should be understood as an XML attribute – named string parameter of XML element which is written inside the angle brackets of the opening tag after its name):

- **<model > ... </model >** – root element containing all other elements inside. It can have attributes allowing to specify some default parameters of PSSD, for example, period and executing node (attributes "period" and "container"), name and developer of the system (attributes "name" and "author").
- **<agent > ... </agent >** – The element that defines the creation of the **target agent**. It contains all of the parameters and settings of the **target agent** as XML-attributes and subelements. Attributes of this element allow to set up for each target the following parameters: a unique name of the agent (attribute "name"), type of the agent (parameter "class"), period of recalculating values of PSSD attributes in milliseconds (attribute "period"), and placement (name of the container or computational node, where agent is physically deployed – attribute "container"). Subelements of this element allow to specify for each target agent a set of other target agents, whose PSSD attributes are used in its calculations (elements **< input name=""/>**), also a set of additional string parameters that need to be passed to it after creation, like initial values, settings, etc. (elements ****).

Also there are some elements that simplify writing of TSDL code and make it more readable.

- **<include > ... </include >** – for complex multi-file models delineates an optional section where files with other model fragments can be specified.
- **<constant > ... </constant >** – associates string content of an element to identifier defined as the attribute "name". Everywhere further identifier enclosed in braces will be interpreted as a string associated with it.

TSDL allows development of distributed PSSD systems specifications using text editors with syntax highlighting as simple as using special XML-based visual diagrams editors.

5   Software Implementation of Multi-agent Architecture for Distributed PSSD Systems and Examples of Using

Developed software architecture is implemented as a software framework **ABSynth**, based on java agent development environment (**JADE**) framework and inherits all its abilities and functionality. In addition to them **ABSynth** includes IMDB **Hazelcast** as a transport level for streaming data sharing.

Basic software entity of **ABSynth** in accordance with architecture is abstract java-class **BasicUnit** – implementation of **target agent**. **BasicUnit** extends the class **jade.Agent** from the JADE library and contains the abstract procedure **MainAction()** which is called with predefined period. Each data processing **target agent** should be implemented as a child class extending **BasicUnit**. The program code performing data processing should be placed into overridden **MainAction()** function. It can operate only with local cache of streaming data as with the class fields.

If one needs it is possible to create different agents (by extending the **BasicUnit** class) to perform one-type and diverse operations over the streaming data, such as, for example, assembling of raw data by sensors poll, signal filtration, indirect measuring (calculating not directly measureable parameters of business process), calculating of macroscopic parameters (some integral estimation, the sliding average), complex signals generation (events, automatic decisions) et cetera.

Let's consider some ways of possible using of the developed architectural and software solutions.

5.1   Networked Control Systems (NCS)

The most widespread scope of use of IoT systems is remote control of large-scale technical systems. The use of a heterogeneous computer network as part of automatic control systems has given rise to a new direction in control theory – networked control systems (NCS). The classical definition of NCS can be as follows: when a traditional feedback control system is closed via a communication channel, which may be shared with other nodes outside the control system, then the control system is called an NCS [12]. From the most general point of view the structure of NCS may be depicted as showed at the Fig. 4. A distinctive feature of the NCS as a software-hardware complex is that, the data, circulating between the network nodes, are streaming data. So, the implementation of control models and algorithms needs particular approach [13, 14].

The software layer of the NSC's technical equipment may be considered as a distributed PSSD, where the digital signals are streaming data, logical signals are complex events and control algorithms are streaming data processing procedures [15].

The example of the problem of semi-natural testing of the electronic control unit for gas-turbine engine was considered [16].

The software layer of the stand is based on ABSynth framework and contains the following types of target agents [17].

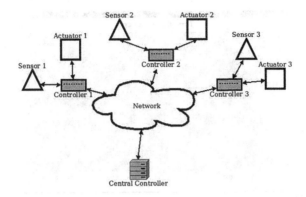

Fig. 4. The common schema of NCS

- Agent – element-wise computational model of the gas-turbin engine.
- Agents – drivers for all hardware modules (interfaces for the aircraft native communication buses, agents for simulation of the mechanical and electrical faults and others).
- Agents – simulation models for the other modules of aircraft automatics (for example the throttle).
- Agents – simulation models for flight dynamics of the aircraft (for example, geometrical position of the aircraft and its parts relative to the ground: the roll angles, pitch and yawing, the geostationary coordinates of the mass center).
- Agents – simulation models for the environment in flight conditions (the input air pressure of the engine, air resistance, the lift force depending on the geometrical position of the aircraft).
- Agents – elements of the graphical user interface (GUI), which allow the researchers to monitor the testing process (depicts all signals in the human-understandable form like numerical tables or time diagrams).

Using the ABSynth platform together with the classes presented above allows to perform the semi-natural testing as well as the full simulation (without natural parts) with involving of different types of computer networks. Different fragments of the model can be developed by separate research teams, copyrighted, and executed using computational equipment in different institutions.

For example, let's consider a simulation of starting and thrust-increment regime for the gas-turbine engine in conditions of different pressure amount of the input air. At the Fig. 5 time diagram of the rotation speed of the low-pressure turbine is represented.

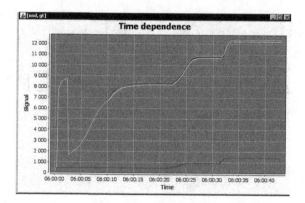

Fig. 5. Simulating of the low-pressure turbine rotation speed

Each instance of the engine executed autonomously at its own node of the computational network. Mechanisms of the ABSynth framework allowed to perform the simulation in real-time.

6   Conclusion

The main results of the work are presented below.

- The overview of state-of-the-art scientific research in the area of streaming data processing.
- The mathematical model of streaming data processing.
- The mathematical model of the resource consumption in computer networks.
- The formulation of the optimal resource allocation problem in heterogeneous networks
- Agent-based software architecture for distributed processing and storing of streaming data including the structure of components and TSDL, which is the language to describe multi-agent applications.
- The ABSynth platform is based on software agents technology, which implements all the developed models and architectural solutions.

References

1. Khakimov, A., Muthanna, A., Kirichek, R., Koucheryavy, A., Muthanna, M.S.A.: Investigation of methods for remote control IoT-devices based on cloud platforms and different interaction protocols. In: Proceedings of the 2017 IEEE Conference of Russian Young Researchers in Electrical and Electronic Engineering (EIConRus), Moscow, St. Petersburg (2017)
2. Ateya, A.A., Muthanna, A., Gudkova, I., Abuarqoub, A., Vybornova, A., Koucheryavy, A.: Development of intelligent core network for tactile internet and future smart systems. J. Sens. Actuator Netw. **7**(1) (2018)
3. Ramakrishnan, R.: Database Management Systems. WCB/McGraw-Hill (1998)

4. Soulé, R., et al.: A unified semantics for stream processing languages (extended). Technical report 2010-924, New York University (2010)
5. Chakravarthy, S., Jiang, Q.: Stream Data Processing: A Quality of Service Perspective Modeling, Scheduling, Load Shedding, and Complex Event Processing. Springer, Heidelberg (2009)
6. Abadi, D., et al.: Aurora: a new model and architecture for data stream management. VLDB J. **12**(2), 120–139 (2003)
7. Zdonik, S.B., Stonebraker, M., Cherniack, M., Çetintemel, U., Balazinska, M., Balakrishnan, H.: The aurora and medusa projects. IEEE Data Eng. Bull. **26**(1), 3–10 (2003)
8. Adi, A., Etzion, O.: AMiT – the situation manager. VLDB J. **13**(2), 177–203 (2004)
9. Valeev, S.S., Maslennikov, V.A., Kovtunenko, A.S.: Design of the middleware based on agent-oriented technologies for the automated control systems of complex technical objects. In: Proceedings of 6th International Conference "Parallel Computing and Control Problems" vol. 1, Moscow (2012)
10. Kovtunenko, A.S., Valeev, S.S., Maslennikov, V.A.: Agent-oriented software architecture for distributed processing and storing of streaming data. In: Proceedings of the 2nd International Conference on Intelligent Technologies for Information Processing and Management (ITIPM 2014), Ufa (2014)
11. FIPA ACL Message Structure Specification. Foundation for Intelligent Physical Agents (2002). http://www.fipa.org/specs/fipa00061/SC00061G.html
12. Zhao, Y.B., Sun, X.M., Zhang, J., Shi, P.: Networked control systems: the communication basics and control methodologies. Math. Prob. Eng. (2015)
13. Lin, H., et al.: Estimation and Control for Networked Systems with Packet Losses without Acknowledgement, Studies in Systems, Decision and Control, vol. 77. Springer (2017)
14. Volkov, A., Khakimov, A., Muthanna, A., Kirichek, R., Vladyko, A., Koucheryavy, A.: Interaction of the IoT traffic generated by a smart city segment with SDN core network. In: Koucheryavy, Y., Mamatas, L., Matta, I., Ometov, A., Papadimitriou, P. (eds.) WWIC 2017. LNCS, vol. 10372, pp. 115–126. Springer, Cham (2017). https://doi.org/10.1007/978-3-319-61382-6_10
15. Kovtunenko, A.S., Maslennikov, V.A.: Creation of distributed control information systems on the basis of agent-oriented approach. In: XII All-Russian Conference on Problems of Management, Moscow (2014)
16. Valeev, S.S., Zagitova, A.I., Kovtunenko, A.S.: Simulation of the gas-turbine aviation engine under flight conditions using the ABSynth multiagent platform. In: Proceedings of the 3rd International Conference on Information Technologies for Intelligent Decision Making Support ITIDS 2015, Ufa (2015)
17. Kovtunenko, A.S., Valeev, S.S., Maslennikov, V.A.: A multi-agent platform for distributed real-time processing. Nat. Tech. Sci. **2**(64) (2013)

Optimization of Routes in the Internet of Things

Omar Abdulkareem Mahmood[1,2(✉)] and Alexander Paramonov[1]

[1] Department of Communication Networks and Data Transmission,
The Bonch-Bruevich Saint-Petersburg State University of Telecommunications (SPbSUT),
Saint Petersburg, Russian Federation
mahmood_omar@list.ru, alex-in-spb@yandex.ru
[2] Department of Communications Engineering, College of Engineering,
University of Diyala, Diyala, Iraq

Abstract. This paper provides an analysis of approaches to choosing a route in the Internet of things networks. A method is proposed for choosing a route, taking into account the quality of its individual sections. The probability of collision is used as an indicator of the quality of the route segment. The proposed method is based on the use of the algorithm for finding the shortest routes in a weighted graph, the weights of which are the probability of collisions.

Keywords: Internet of Things · Collisions · Route searching · Optimization
IoT

1 Introduction

The development of the concept of the Internet of things was linked to enhance the growth in the number of different devices connected to communication networks. It aimed to further develop the information and communication system in terms of covering those areas of human activity that have not yet been involved in the information exchange process In other word, it involves the level of "smart" things, "smart" houses and the whole intellectual world [1–5]. This is made possible by the implementation of Internet technology of things.

At the same time, network access technologies can be very diverse. This can be both devices that support the standards of cellular networks and wireless broadband access, as well as specialized standards that focus on the construction of sensor wireless networks and the density of such devices can be very high. Around 32 billion connected devices are forecast by 2022, of which around 18 billion will be related to IoT. Potentially, such devices can be located both in the service area of mobile networks and outside them. The task of their operation is to deliver data to processing facilities or control commands in the opposite direction. Due to a wide variety of purposes and technologies, the structure of building networks can be different.

Specifically, for the organization of networks, various self-organization technologies can be used to allow building mesh and ad hoc networks. The use of these technologies makes it possible to significantly expand the capabilities of networks in terms of expanding the service area, ease of deployment, ensuring reliability and survivability.

© Springer Nature Switzerland AG 2018
O. Galinina et al. (Eds.): NEW2AN 2018/ruSMART 2018, LNCS 11118, pp. 584–593, 2018.
https://doi.org/10.1007/978-3-030-01168-0_52

The task of self-organization includes a subtask of the choice of routes for passing traffic between nodes of the network. The method of its implementation makes the network characteristics (bandwidth, delay in data delivery, packet loss, etc.) depend significantly on the network characteristics. Route selection work is associated with the transfer of additional traffic in the communication network, as well as the time consuming of network nodes. This reduces the quality of its functioning parameters. Therefore, it is expedient to minimize this work particularly because can be facilitated by the quality of the resulting routes.

This paper considers models and methods for selecting a route considering its quality. The quality of the route can be judged on the basis of various parameters: time of data delivery, loss factor, and throughput. Taking into account the peculiarities of wireless communication networks as a parameter characterizing the quality of the route, we will consider the probability of collisions (or potential collisions), which depends on both the parameters of the route and the traffic produced by the nodes of the network.

2 Overview and Related Work

The structure and choice of the structure of the network and the ways of passing traffic are the main problems of prospective communication networks, whose tasks include the maintenance of Internet of things traffic. There are many works devoted to the issues of the choice of traffic structure and routing.

In this works [1, 2], we propose a method for connectivity estimation in Wireless Sensor Networks (WSN). In [3], the results were about comparison of the routing protocols for the WSN and the choice of a specific protocol for constructing a sensor network depending on the density of nodes. In [4, 5], provides an analysis of the RPL protocol in terms of IoT, taking into account the following reference indicators: reliability and mobility, heterogeneity of resources and scalability. In work [6], results of researching on the field of optimization of routes in Internet of things networks. In [7–9], the methods of choosing a route by the criterion of minimum length were considered, and in this paper, the selection was made depending on the criterion minimal collisions.

Also, in a number of works, the connectivity for D2D communication was considered on long routes which were selected by the criterion of minimum attenuation.

In addition, we propose a method for selecting a route in the network according to its quality index, for which we accept a minimum of collisions (clear collisions). This indicator also allows us to indirectly take into account such indicators as the probability of loss of personnel and bandwidth.

3 Problem Statement

The task of choosing a route is one of the tasks of constructing a logical network structure. We will assume that there are n nodes (Internet traffic sources of things) and the network nodes are distributed in some invariable manner in the service area and form a homogeneous structure, the characteristics of which are unchanged for any fragment of the service area. In this case, the connection of the source with the traffic recipient can be organized either directly or through transit nodes, which can be any traffic source nodes.

The task of constructing a route between network nodes, as a rule, is optimization. Because the objective function depends on a certain metric and is subject to minimization or maximization. Such a metric can be conditional cost, distance and other parameters. The quality of the route in the communication network can be characterized by such quality indicators such as data delivery delay, bandwidth, and loss factor. Obviously, these indicators depend on the characteristics of the channels on all sections of the route. At each individual site, these parameters depend on the signal propagation and traffic conditions (use of the propagation medium). The traffic of neighboring (adjacent) nodes can lead to delays, frame loss and bandwidth reduction depending on the technology of channel realization. The growth of delay, loss and decrease in bandwidth is a consequence of the busy channel (propagation medium, radio frequency spectrum) transmitted by signals. If the standard used does not implement conflict prevention mechanisms, then channel employment leads to a collision.

Generally, a collision is understood as a situation in which two or more signals from different transmitters simultaneously arrive at the receiver's input. Because of signaling, none of the received messages is successfully received and all data elements (frames) transmitted on this interval are lost. It is theoretically possible that some data can be accepted, but the probability of this is small and this case is not considered further. Many current wireless standards implement collision avoidance mechanisms that reduce their probability by analyzing the state of the transmission medium and a certain timing algorithm for transmission. The use of such mechanisms increases the efficiency of using the transmission medium, but naturally leads to transmission delays. In this way, the probability of collision, even when using the mechanism for their prevention, characterizes the use (employment) of the transmission medium.

Accordingly, the probability of collisions directly or indirectly reflects the quality of the communication channel or the entire route. The parameters of bandwidth, packet loss and delay are directly related to this probability. Therefore, we decided to use it as a metric for selecting a route in a communication network.

4   Model and Method of Route Selection

We will assume that there are n nodes of the network that are also sources of traffic (Internet of things) and these network nodes are distributed in some invariable manner in the service area and form a homogeneous structure. The characteristics of which are unchanged for any fragment of the service area. In this case, the connection of the source with the traffic recipient can be organized either directly or through transit nodes, which can be any traffic source nodes.

The node may be equipped with an antenna, radio transmitter and/or radio receivers. All nodes, possibly except the gateway nodes, are equipped with standard equipment. The node communication zone is a circle of radius r centered at the node's location point. For most of the standards used for organizing IoT networks, the data transfer rate depends on the signal reception conditions. However, to simplify the task, we assume that the data rate within the communication zone is unchanged (independent of the distance to the node). This approach is often justified, because designing a network is trying to maximize its throughput and the node locations are selected in such a way that the communication conditions provide the maximum achievable speed. Each of the nodes, at the time of data transfer, occupies the transmission medium, whose dimensions are limited by the interference zone, and in this model also represents a circle of radius R and usually $R \geq r$. At the same time, transmission of data by two or more nodes results in data loss (collision), including when the receiving node is in the interference zone of at least one of the conflicting nodes.

If the average frame transmission time is τ, and the frame transmission rate λ in estimating the probability of collision, we will argue as follows: The collision occurs when the transmission of the frame, at least one more transmission is started or completed. In other words, when the transmission interval of the frame intersects with one or more transmission intervals, Fig. 1.

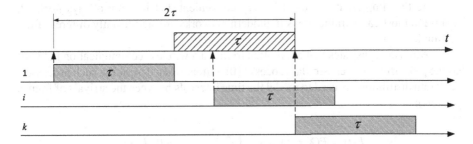

Fig. 1. Illustration of the process of collisions

It can be seen from the figure that in order for one or more transmissions to overlap on the frame in question, it is sufficient that one or more transmissions be started within a time interval equal to 2τ during the transmission of the frame under consideration, or no later than the time before the start of the reporting τ frame. Thus, the probability of collision is equal to the probability of the beginning of one or more frame transfers in the time interval 2τ.

$$p_c = p_{\geq 1}(2\tau) = 1 - p_0(2\tau) \tag{1}$$

If the traffic can be described as a model of the simplest flow, then the probability of collision can be calculated by substituting in (1) the formula for the Poisson distribution probability $p_k(x) = \dfrac{(\lambda x)^k}{k!} e^{-\lambda x}$, then

$$p_c = 1 - e^{-\lambda 2\tau} \tag{2}$$

A similar result could be obtained by considering the distribution function of time intervals between the moments of the beginning of frame transmission in the simplest flow.

All frames transmitted from the interval 2τ will be damaged. The average number of frames corrupted by the collision will be equal to:

$$m = 2\lambda\tau p_c \tag{3}$$

The intensity of collisions can be defined as

$$\eta = \lambda p_c \tag{4}$$

$$\lambda = \sum_{i=1}^{k} \lambda_i \tag{5}$$

λ_i - Transmission intensity (frames) by the i-m node in the communication area (frames/s),

k – number of nodes in the communication area.

Using the simplest flow model is very convenient, but it is not always justified, because in most cases traffic flows in modern networks are significantly different from the simplest flow.

Research demonstrates that most often traffic in a wireless communication network has the properties of a self-similar process [10]. Often, when describing such processes, the Pareto distribution is used to model the time intervals between the arrivals of frames (packets):

$$F(t) = P(X < x) = 1 - \left(\frac{t_m}{t}\right)^k, \quad x_m > 0, \ k > 0$$

$$M(x) = \frac{kt_m}{k-1}, \quad k > 1$$

Then the probability that the time interval will be less than 2τ is equal to:

$$p_c = p(X < 2\tau) = 1 - \left(\frac{t_m}{2\tau}\right)^k \tag{6}$$

The probability of collision characterizes the communication channel between nodes of the network. The smaller it is, the greater the resource of the channel. Hence, a large channel capacity is likely. Figure 2 shows the dependencies of the collision probability on the use of the channel for two flow models: the simplest flow and the flow, the intervals between the bids in which are subordinated to the Pareto distribution.

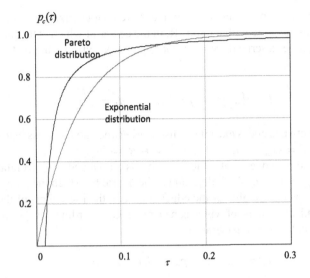

$p_c(\tau)$

Fig. 2. Dependence of the probability of collisions on the use of the channel

As can be seen from the graphs in the area of average loads the probability of collision is higher for a flow of the second type. But in the area of small and large load values, the collision probability is higher for the simple flow model. If traffic passes through a route from n independent sites, then the probability of losing data (frame) due to collisions for the entire route will be:

$$p_c = 1 - \prod_{i=1}^{m} \left(1 - p_{ci}\right) \qquad (7)$$

Where p_{ci} is the probability of collision on the i-m section of the route, as determined above. The task of finding a route with the minimal probability of collisions can be formulated as:

$$P = \arg\min_{r \in \Omega} \left(p_c\right) \qquad (8)$$

Where Ω is the set of possible routes.

As can be seen from (2) and (6), the more likely the collision is, the greater the traffic intensity (channel usage). So, when choosing a route, one can be guided both by the probability of collision (loss of a frame due to a collision), and by the amount of traffic intensity in the communication zone. However, the knowledge of the properties of traffic allows you to take into account the features that manifest themselves in the non-linear dependence of the channel properties on traffic (2) and (6). To search for a route according to the criterion (8), any algorithm for finding the shortest path in a weighted graph can be used. For this purpose, the network can be described by a weighted graph G (V, E) with a set of vertices $V = \{v_{ij}\}, \dots i,j = 1 \dots n$, that correspond to network nodes and a set of edges (arcs) E that correspond to potentially possible connections (channels) between network nodes. The nodes are characterized by a communication zone. In this case, a circle with a radius R, in the center of which is this node. Nodes of the network can be described by their coordinates (xi, yi) and the distances between them:

$$D = \{d_{ij}\}, \quad i,j = 1 \dots n, \quad d_{ij} = \sqrt{x_i^2 + x_j^2}$$

The edges between nodes exist if the distance between nodes is less than R, otherwise you can write: $E = \{e_{ij}\}, \quad i,j = 1 \dots n, \quad v_j \in R_i \Rightarrow \exists e_{ij}.$

Each of the edges of the graph is assigned a weighting factor: $C = \{c_{ij}\}, \dots i,j = 1 \dots n$, which determines the metric by which the path is optimized.

In this case, as the weights of the edges (or arcs), the logarithms of the probability of collisions and an estimate of which can be obtained according to (2) or (6) depending on the flow model under consideration:

$$c_{ij} = -\log(1 - p_{ij}), \quad i,j = 1 \dots n \tag{9}$$

Where *pij* is the probability of collisions in the route between the nodes i and j.

Figure 3 shows the results of simulation modeling. Model parameters: service area square with 200 m side, nodes are randomly distributed (x and y coordinates are random numbers subject to uniform distribution law), the number of nodes is 200, the coupling radius of the node is equal to the interference radius and is equal to 60 m. The result of choosing the "shortest" path (the path of the smallest collisions) using criterion (8) can be seen in Fig. 3b. Figure 3a for comparison shows the result of choosing the path by the criterion of least length.

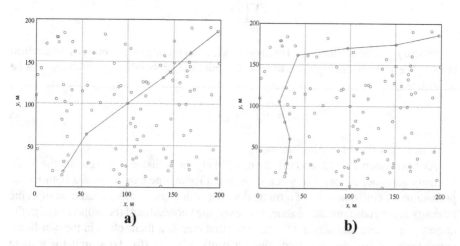

a)   b)

Fig. 3. The result of choosing the path with the least probability of collisions with a uniform distribution of network nodes

It can be seen from the figure that the route constructed on the basis of the criterion of the minimum of the collision probability differs significantly from the route constructed on the basis of the minimum length criterion. Figure 4 shows a similar example for the multimodal distribution of nodes in the service area (the distribution has five scattering centers). You can see that the route chosen by the minimum of collisions "tends to bypass" the region with a high node density. This is the expected result, because in a region with a high density of nodes, a greater number of them enters the interference zone (link), therefore, according to (2) or (b), there is a high probability of collisions.

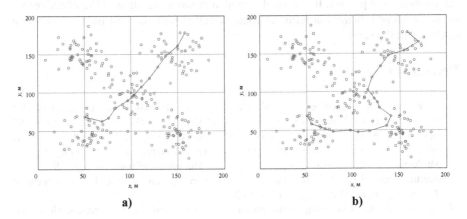

Fig. 4. Result of choosing the path with the least probability of collisions with multimodal distribution of network nodes

The results of simulation modeling, according to indirect estimates, have shown that this method allows finding routes with a larger bandwidth and a lower delay value due to the choice of the route with the minimum probability of collisions. On average, the search result yields a route with a collision probability of 20% less than the route found by the criterion of minimum length. The efficiency of the algorithm depends on the way the nodes of the network are located. The obtained results of modeling are received at casual placing of knots, namely, at uniform and multimodal distributions of knots of a network in service territory. This method is the most effective for different traffic intensities produced by network nodes.

5   Conclusions

The construction of Internet of things networks involves the use of various technologies and structures, many of which require solving the problem of selecting the route for passing traffic (message transmission). The choice of a route in the network can be made on the basis of various criteria and, in general, is an optimization problem, and its solution allows finding a route satisfying some quality criterion. Such a criterion may be one or more parameters used to describe the quality of the data transmission channel. An analysis of the main quality indicators of the wireless communication channel showed that most of them are directly or indirectly related to such a measure as the probability of collisions. Despite the fact that many protocols used use collision avoidance technologies, the probability of collision indirectly characterizes the use of the channel, the delay and its bandwidth (throughput) [11, 12]. The model proposed in this paper characterizes the channel quality on the basis of the collision probability and can be used both for the simplest flow model and for the self-similar traffic flow model. On the basis of the proposed model, a method of searching for a route with a minimum probability of collisions has been developed. Any algorithm for finding the shortest path in a graph can be used to implement the method. Simulation modeling showed the effectiveness of the developed method in terms of route selection with the best parameters of the quality of traffic servicing.

References

1. Paramonov, A., Nurilloev, I., Koucheryavy, A.: Provision of connectivity for (Heterogeneous) self-organizing network using UAVs. In: Galinina, O., Andreev, S., Balandin, S., Koucheryavy, Y. (eds.) NEW2AN/ruSMART/NsCC -2017. LNCS, vol. 10531, pp. 569–576. Springer, Cham (2017). https://doi.org/10.1007/978-3-319-67380-6_53
2. Muthanna, A., Koucheryavy, A., Prokopiev, A.: The mixed telemetry/image USN in the overload conditions. In: 16th International Conference on Advanced Communication Technology (ICACT) 2014, pp. 475–478 (2014)
3. Paramonov, A.: Clustering optimization for out-of-band D2D communications. In: Paramonov, A., Hussain, O., Samouylov, K., Koucheryavy, A., Kirichek, R., Koucheryavy, Y. (eds.) Wireless Communications and Mobile Computing, vol. 2017, p. 6747052 (2017)
4. Makolkina, M., Vikulov, A., Paramonov, A.: The augmented reality service provision in D2D network. In: Vishnevskiy, V.M., Samouylov, K.E., Kozyrev, D.V. (eds.) DCCN 2017. CCIS, vol. 700, pp. 281–290. Springer, Cham (2017). https://doi.org/10.1007/978-3-319-66836-9_24
5. Masek, P., Fujdiak, R., Zeman, K., Hosek, J., Muthanna, A.: Remote networking technology for IoT: cloud-based access for alljoyn-enabled devices. In: Proceedings of the 18th Conference of Open Innovations Association FRUCT and Seminar on Information Security and Protection of Information Technology, pp. 200–205 (2016)
6. Nurilloev, I., Paramonov, A., Koucheryavy, A.: Connectivity estimation in wireless sensor networks. In: Galinina, O., Balandin, S., Koucheryavy, Y. (eds.) NEW2AN/ruSMART -2016. LNCS, vol. 9870, pp. 269–277. Springer, Cham (2016). https://doi.org/10.1007/978-3-319-46301-8_22
7. Muthanna, A., et al.: Analytical evaluation of D2D connectivity potential in 5G wireless systems. In: Galinina, O., Balandin, S., Koucheryavy, Y. (eds.) NEW2AN/ruSMART -2016. LNCS, vol. 9870, pp. 395–403. Springer, Cham (2016). https://doi.org/10.1007/978-3-319-46301-8_33
8. Dao, N., Koucheryavy, A., Paramonov, A.: Analysis of routes in the network based on a swarm of UAVs. Information Science and Applications (ICISA) 2016. LNEE, vol. 376, pp. 1261–1271. Springer, Singapore (2016). https://doi.org/10.1007/978-981-10-0557-2_119
9. Kirichek, R., Paramonov, A., Vladyko, A., Borisov, E.: Implementation of the communication network for the multi-agent robotic systems. Int. J. Embed. Real-Time Commun. Syst. 7(1), 48–63 (2016)

10. Paramonov, A., Koucheryavy, A.: M2M traffic models and flow types in case of mass event detection. In: Balandin, S., Andreev, S., Koucheryavy, Y. (eds.) NEW2AN 2014. LNCS, vol. 8638, pp. 294–300. Springer, Cham (2014). https://doi.org/10.1007/978-3-319-10353-2_25

11. Hoang, T.: Adaptive routing in wireless sensor networks under electromagnetic interference. In: Hoang, T., Kirichek, R., Paramonov, A., Houndonougbo, F., Koucheryavy, A. (eds.) 31st International Conference on Information Networking (ICOIN) 2017, pp. 76–79 (2017)

12. Hoang, T., Kirichek, R., Paramonov, A., Houndonougbo, F., Koucheryavy, A.: Adaptive routing in wireless sensor networks under electromagnetic interference. In: 31st International Conference on Information Networking (ICOIN) 2017, pp. 76–79 (2017)

Chirped Fiber Grating Beamformer for Linear Phased Array Antenna

Sergey I. Ivanov$^{(\boxtimes)}$ ⓘ, Alexander P. Lavrov ⓘ, Igor I. Saenko ⓘ, and Daniil L. Filatov

Peter the Great St. Petersburg Polytechnic University, St-Petersburg, Russia
{ivanov_si,lavrov_ap,ig-i-saenko}@spbstu.ru

Abstract. Results of the development of a 6-channel chirped fiber grating beamformer (BF) for microwave linear phased array antenna (PAA) in receiving mode are presented. A BF incorporates DWDM technology based components of analogue fiber-optic transmission links with external modulation in microwave range 0.1–18 GHz and specially manufactured chirped fiber Bragg grating. In BF optical transmitters have optical carriers separated with 100 GHz step (ITU grid, DWDM standard in C-band). PAA beamforming is realized by introducing inter-channel time delays when microwave modulated optical carriers interact with chirped fiber Bragg grating in reflection mode, throughout the total operating spectral range occupied by 6-channel BF (6×100 GHz). The results of measuring the S-parameters of BF 6 analogue channels are given; these enable to synthesize far-field radiation patterns of the linear PAA with optical BF under investigation. Far-field patterns of 6 channel linear phased antenna array integrated with optical BF model have been measured and compared with calculated patterns taking into account amplitude and phase errors arising in beamformer channels.

Keywords: Phased array antenna · Optical beamformer · True-time delay
Chirped fiber bragg grating · Microwave fiber-optic link

1 Introduction

The optical (photonic) systems for microwave phased array antennas (PAA) beamforming is a subject of continuously growing attention due to their well known advantages compared to electronic beamforming systems [1, 2]. Especially ultrawideband large antenna arrays could benefit from development of optical beamforming system based on true-time-delay (TTD) technique instead of traditional phase-control technique. So, in the last two decades it has been shown that applying the microwave optical devices for PAA beamforming could afford the means to overcome conventional radio electronic beamformer limitations. Therefore, many beamformer architectures based on microwave optics for PAA have been proposed and evaluated, both in transmitting and receiving modes of operation. Some of the proposed beamformer architectures are based on fiber Bragg grating (FBG) usage in particular chirped FBG [3–5]. Chirped FBGs (CFBG) offer a simple arrangement for variable time delays and allow fast delays tuning. In addition they are more compact compared with, for example, time delay units exploited chromatic fiber dispersion. In this paper a true-time optical beamformer (BF)

© Springer Nature Switzerland AG 2018
O. Galinina et al. (Eds.): NEW2AN 2018/ruSMART 2018, LNCS 11118, pp. 594–604, 2018.
https://doi.org/10.1007/978-3-030-01168-0_53

based on a specially developed CFBG, components of fiber-optic analogue transmission lines with external modulation and DWDM technology is demonstrated. We consider the results of assembling and adjustment optical BF prototype using CFBG and its performance investigation including the microwave receiving linear PAA far-field pattern measurement in the 7–13 GHz frequency range.

2   Photonic Beamformer Model Design and Setup

In order to provide required time delays in optical BF based on dense wavelength division multiplexing (DWDM) technique one can use optical comb and DWDM multiplexing and realize wavelength depending interchannel time delays by using appropriate wavelength-dependent technique. Recently we have investigated an approach exploiting of chromatic dispersion in optical fiber segments [6, 7, 9, 10] but here we consider time delay unit (TDU), based on CFBG, in particular because of its compactness compared with TDU using chromatic fiber dispersion and possibility to provide variable time delays and allowing fast delays tuning. A flowchart in optical BF based on wavelength division multiplexing and chirped Bragg grating is shown in Fig. 1. The scheme in Fig. 1 is designed for the N-element's linear PAA and uses optical comb – a set of N lasers with different but uniformly spaced (step $\Delta\lambda$) wavelengths, the total wavelength band is $(N-1)\cdot\Delta\lambda$. The microwave (RF) signals from the antenna elements $A_1 .. A_N$ modulate a set of laser diodes optical carriers. An electrooptic conversion is achieved by using external modulators (Mach-Zehnder modulators) as represented in Fig. 1. Further, the intensity-modulated optical carriers are combined into a single fiber by a multiplexer (MUX N x 1 unit) and fed into TDU based on a wide bandwidth CFBG

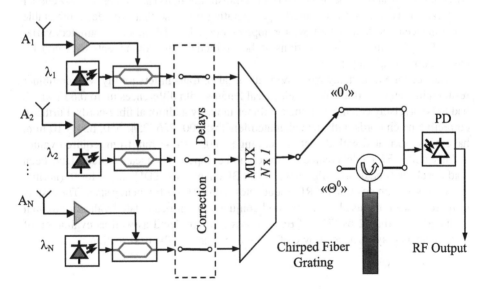

Fig. 1. A flowchart of optical beamformer based on wavelength multiplexing and chirped fiber Bragg grating

through an optical circulator. The reflection point inside the grating depends on the wavelength according to the specific chirping slope. The linear chirp slope results in linear relation between the reflection position and the wavelength. Time delay introduced between adjacent channels results in respective tilting of PAA beam. Photodiode (PD) at TDU output converts sum of delayed intensity-modulated optical carriers back to microwave domain. To align the electrical delays from RF sources (antennas $A_1 .. A_N$) to MUX output the "Correction Delays" unit comprises fiber-cords with strictly sized up lengths. Hence, it corresponds to a flat phase front tilted $0°$ to PAA base line – the PAA beam directed normal to PAA base. The RF signals modulating the optical comb of wavelengths are coherently summed at the photodetector output while optical carriers are summed incoherently. So due to chosen CFBG chirp slope and thus specified interchannel delays RF signals incoming from the corresponding direction (angle) are summed in phase.

The beamformer model developed for initial demonstration of linear PAA beam steering is based mainly on microwave photonics components commercially available on market of fiber optic telecom links (radio-over-fiber links). Among the optical key components depicted in Fig. 1 the units converting radio signal into optical domain (laser jointly with modulator) and backward (photodiode) are crucial for BF correct performance.

The designed optical BF model includes units of Optiva OTS-2 Microwave Band Fiber Optic Link from Emcore Corp. [8] namely 6 transmitter units OTS-2T with wavelengths from 1551.72 nm to 1555.75 nm, wavelength spacing of 0.8 nm (corresponding channels # 32 .. 27 of 100 GHz ITU grid), and one wideband receiver unit OTS-2R with in-built additional RF amplifier with 15 dB gain. The mentioned components provide optical link operation in 0.05..18 GHz frequency bandwidth. In-built microprocessor-based laser bias and temperature control (in transmitter units) as well as Mach-Zehnder modulator bias control for required $V_\pi/2$ operating point, so that provides more stable performance operation and allows for appreciably reduce BF model parameters variations. The performance specifications of the BF model main components were considered in more detail in [7].

The accurate time delay measurements of taken transmitters jointly with combiner resulted in a number of channel's electrical lengths with differences up to 0.6 m which had to be precisely equalized at the combiner input by additional fiber-cords. Further 6 corrective patch cords with the calculated lengths: 200, 226, 254, 550, 624, 810 mm, have been fabricated with deviations in range –0.02.. +2.43 mm from required values and inserted in BF model channels at the combiner inputs. The measurements have been made to check time delays alignment over 6 BF channels at TDU output for $0°$ (normal to base line) beam positions at RF frequencies with 10 GHz frequency span. The results of these measurements show their good conjunction with calculated values (maximal difference less than 2 ps [7]). CFBG features in more detail as well as evaluation of interchannel delays produced by CFBG are considered in the next section.

3   Chirped Fiber Bragg Grating Characterization

The chirped fiber Bragg grating used in developed optical BF was fabricated by the method of femtosecond point-by-point inscription. The advantage of the method is the ability to inscribe FBG with an arbitrary period and the refractive index modulation profile. Moreover, the method does not require removal of the fiber protective coating, and it can also be used for grating inscription in non-photosensitive optical fibers. A more detailed description of the experimental setup used for the inscription is given in details in [9]. The CFBG inscription was carried out through the protective polyimide coating of Fibercore SM1500(9/125)P fiber. To fabricate CFBG – to broad the FBG spectrum – a linear chip (1 nm/cm) is introduced into the FBG structure. For this, during inscription process, the fiber was moved at constantly accelerated speed from 0.535 µm/s to 0.537 µm/s, with 1 kHz repetition rate of laser pulses. The total length of the CFBG is $L = 20$ mm. The fabricated CFBG has 1-st order resonance near the wavelength of 1553 nm.

The CFBG was characterized in reflection mode with introducing optical radiation into each its side. We used tunable laser source 81606A and optical power meters N7745A (both from Keysight Technologies). The laser source radiation is routed to the CFBG through an optical three-port circulator CIR-3-15-L-1-2 (AFW Technologies). The CFBG reflection spectrum is shown in Fig. 2. One can see the CFBG working band covering the wavelengths from 1550.7 nm to 1556.5 nm, that is sufficient for work with multiplexed carriers of six ITU channels with 0.8 nm channel step (100 GHz grid), in particular Ch27 (1555.75 nm) – Ch32 (1551.72 nm).

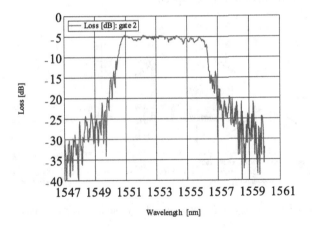

Fig. 2. The CFBG reflection spectrum

To evaluate time-delay properties of the CFBG we used a proper (suitable) microwave characterization with usage of amplitude modulated tunable optical carrier. Laser radiation from tunable laser source 81606A was amplitude modulated via an external LiNbO$_3$ Mach–Zehnder modulator IM-1550-20 (Optilab). This double-sideband microwave modulated optical signal was feed throw CIR-3-15-L-1-2 to the CFBG and back to photoreceiver module (Emcore) for demodulation. The microwave signal is generated and received by a vector network analyzer ZVA40 (Rohde & Schwartz). We measured S_{21} parameter, so we can see amplitude response of "CIR+CFBG" couple which is directly proportional to the amplitude of the received microwave signal. The measured phase response $\varphi(f)$ (as $\arg[S_{21}(f)]$) of "CIR+CFBG" couple after unwrapping gives an estimation of TDU time delay T_d (and its electrical length L_{el} also): $T_d = \Delta\varphi(f)/(2\pi\Delta f)$, where Δf – VNA sweep range (at least 10 GHz in our measurements). We measured time delay T_d for plurality of laser wavelengths with 80 pm step.

Figure 3a shows the measured dependences $T_{d1}(\lambda)$ and $T_{d2}(\lambda)$ when introducing optical signal into both CFBG sides. The total time-delay of CFBG reaches 160 ps. The CFBG dispersion is estimated by the slope of linear functions after a linear least square fit of the data. Slopes are $K_1 = 34.65$ ps/nm for short period side and $K_2 = -34.09$ ps/nm for long period side. For $T_{d2}(\lambda)$ measurements we used 6 transmitter modules OTS-2T (Emcore) with lasers on ITU channels #27...32 with step $\Delta\lambda = 0.8$ nm instead of the tunable laser 81606A, so we had only 6 measure data points in $T_{d2}(\lambda)$ function. 6 vertical color lines show wavelength of our 6 transmitters. When the CFBG is inserted in BF as time-delay unit it gives interchannel delay $\Delta\tau_{FBG} = K_1 \cdot \Delta\lambda = 27.7$ ps. Figure 3b shows the deviation of time-delays from a linear fits. The RMS deviation $\delta\tau$ of the time-delay $T_{d1}(\lambda)$ is approx. 1.65 ps. This value may be used for estimation of phase error $\delta\Phi$ along PAA at the upper frequency f up of its operation, e.g. 18 GHz: $\delta\Phi = 360\cdot\delta\tau\cdot f$ up, $\delta\Phi \approx 10.7°$. This is very small value. The interchannel delay $\Delta\tau_{FBG} = 27.7$ ps gives the change in PAA beam point angle θ according to $d\cdot\sin\theta = c\cdot\Delta\tau_{FBG}$, where d is PAA elements spacing, and c – EM-wave free space speed.

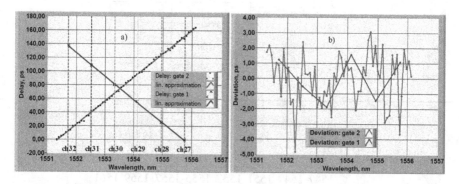

Fig. 3. The CFBG time-delay – (a) and time-delay deviation from linear fit – (b) versus wavelength

The selected method of CFBG microwave characterization with usage of amplitude modulated optical carrier was used also for characterization of BF separate channels – from RF input of each of 6 transmitters OTS-2T to BF common RF output (at the receiver module OTS-2R output). We found very large differences in electrical length of 6 microwave links each including OTS-2T, MUX, and common OTS-2R: up to 60 cm [7]. For initial adjustment of our optical BF channels we fabricated special correction fiber patch-cords for insertion in each BF channel [7]. Measurements of these special fiber patch-cords electrical lengths were done also by selected microwave characterization method with a vector network analyzer usage. The BF residual interchannel time-delay differences correspond to differences ΔL_{el} in electrical lengths approx. ± 0.5 mm. S_{21} measurements show that BF channels have high repeatability as in amplitude $|S_{21}(f)|$ so in phase $\arg[S_{21}(f)]$, see next section.

4 Far-Field Pattern Calculations and Measurements for PAA with Optical Time Delay Unit Based on Chirped FBG

4.1 6-Channel PAA Far Field Pattern Calculations

The results of chirped FBG and separate BF channels characteristics measurements presented in the previous section allow performing the synthesis (calculation) of linear broadband PAA with optical BF far field patterns (FFP). The calculations of the linear PAA characteristics are based on well known results [10, 11], taking into account the phase $\Phi_n(\omega)$ and amplitude $A_n(\omega)$ distributions of EM-wave field E along antenna aperture.

Let us consider the case of equidistant N-element array with a spatial period d. The PAA far field pattern as a function of angle θ measured from the PAA normal and signal frequency f is found by summing the fields of all emitters at the point of observation taking into account an inter-element phase shift:

$$\vec{E}(\theta, \omega) = \sum_{n=1}^{N} \vec{E}_n(\theta, \omega) = \sum_{n=1}^{N} \vec{e}_n(\omega, \theta) A_n(\omega) e^{-i\Phi_n(\omega)},$$

$$k = \omega/c = 2\pi/\lambda, \quad \omega = 2\pi f, \quad A_n(\omega) = \left[1 + a_n(\omega)\right] S_{21n}(\omega). \tag{1}$$

Here an $a_n(\omega)$ is centered relative amplitude error of the field distribution $E_n(\theta, \omega)$ pointed at the elements's locations over the PAA aperture, relative to the reference transfer characteristic $\langle S_{21}(\omega)\rangle$ averaged over all PAA channels. Vector $e_n(\omega, \theta)$ is a separate (partial) n-th element FFP related to the coordinate system with origin in the centre of corresponding element.

With a sufficient (large) number N of antenna elements the partial patterns $e_n(\theta, \omega)$ are almost identical for all elements, thus enabling them to be putting out of sum in (1). For a perfect linear PAA without amplitude and phase errors ($A_n(\omega) = A_0$, $\Phi_n = \Phi_0$) antenna FFP $E_0(\theta, \omega)$ can be represented as a product of array separate element vector complex FFP $e(\omega, \theta)$ to the array scalar complex multiplier FFP $F_0(\theta, \omega)$ as shown in (2) [10, 11].

$$\vec{E}_0(\theta, \omega) = \vec{e}(\omega, \theta) F_0(\theta, \omega) ,$$

where

$$F_0(\theta, \omega) = A_0(\omega) \sum_{n=1}^{N} e^{-inu},\tag{2}$$

$$u = kd \sin \theta.$$

For PAA in the receive mode with optical BF usage the phase distribution of EM-field on the antenna aperture $\Phi_n(\omega)$ (we determine it for case when a radiation source is in the PAA far-field zone with the angular position θ) can be calculated as

$$\Phi_n(\omega) = \omega \cdot \left[n(d \cdot \sin\theta/c - \Delta\tau_{FBG}) - \delta\tau 1_n + \delta\tau 2_n \right] + \Delta\Phi_n(\omega),\tag{3}$$

where $\Delta\tau_{FBG}$ is the average value of inter-channel time delay $\langle \tau_n - \tau_{n-1} \rangle$ implemented using chirped fiber Bragg grating; $\delta\tau 1_n$ is the deviation from the time delay linear dependence versus element number n; $\delta\tau 2_n$ is a residual uncompensated time delay due to the inaccuracy of the fiber-optic and RF cables lengths compared with the required ones; $\Delta\Phi_n(\omega)$ is a frequency-dependent phase deviation from linear approximation of the phase-frequency characteristics for n-th channel of BF.

The estimates of the parameters $\delta\tau 1_n$, $\delta\tau 2_n$ and deviations $a_n(\omega)$ allow calculate PAA statistical parameters – the isotropic background level and the FFP main lobe broadening.

Figure 4 shows the results of the power radiation pattern $|E_n(\theta, \omega)|^2$ calculations for linear 6-channel PAA with optical BF at the frequency of 10 GHz. The calculations were performed for equidistant array with $d = 19.2$ mm element spacing for an azimuth angle $\theta = 0$. Evaluation of amplitude and phase errors was performed according to the method given above with the values of the parameters and measured characteristics of the designed BF.

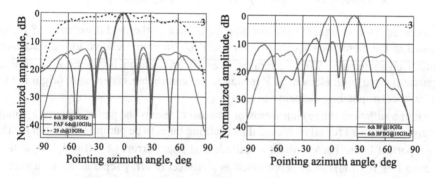

Fig. 4. PAA calculated far-field patterns

Dashed curve in Fig. 4a represents a typical FFP of ultra-wideband microstrip (patch) dipole element (Vivaldi type antenna) $e_0(\theta, \omega)$ which is an actual one for the linear PAA used in experiments [7, 12] at a frequency 10 GHz. PAA elements' partial patterns $e_n(\theta, \omega)$ are almost identical and coincide with $e_0(\theta, \omega)$, so it was used in the FFP calculation for the whole array according to (2). Figure 4a shows the array multiplier dependence in the form $|F(\theta, \omega)|^2$, which corresponds to the FFP of the PAA with isotropic emitters without any phase and amplitude errors. One can see that this FFP (PAF 6ch@10 GHz in Fig. 4a) has a lower level of the isotropic background than level of calculated FFP of 6-channel PAA (6ch BF@10 GHz) even considering the measured errors inserted by designed BFS. The 3 dB beam width of array FFP main lobe is about 13.3 degrees and coincides with the $|F(\theta, \omega)|^2$ main lobe width. The obtained data are consistent with data calculated in accordance with a statistical analysis of the array FFP features [10, 11]. Figure 4b presents the results of 6-channel PAA beam steering when BF time-delay unit (see Fig. 1) is represented by chirped FBG (6ch BFBG@10 GHz in Fig. 4b) at a frequency $f = 10$ GHz. Chirped FBG operation results in the main lobe axis shifting to the angle equal to 26°.

4.2 Measurements of Chirped FBG Beamformer Channels Characteristics

The above proposed method of PAA characteristics calculation (1–3) involves the use of a priori known frequency characteristics $A_n(\omega)$ and $\Delta\Phi_n(\omega)$ for each of 6 channels in developed BF with chirped FBG. To have this information we carried out measurements of the amplitude-frequency and phase-frequency characteristics for each of the BF 6 channels using vector network analyzer in the frequency range 0.1..20 GHz. Figure 5 illustrates the results of these measurements. 6-channel BFS operating range extends from 0.1 GHz to 18 GHz, that corresponds to datasheets on microwave optical transmission lines OTS-2 used in the developed BF. Note the high repeatability of $A_n(\omega)$ and $\Delta\Phi_n(\omega)$ (complex parameter $S_{21}(\omega)$ measured from RF inputs of BF to its common output) in different BF channels, which indicates (we believe) the decisive impact of broadband microwave amplifier in the photodetector module OTS-2R on amplitude-frequency and phase-frequency characteristics of BF. Consequently, it is possible to compensate the amplitude transmission coefficient $|S_{21}(\omega)|$ variations with an accuracy of 1 dB. A phase deviations in different channels at any fixed frequency is up to $\pm10°$ and the total phase change over the entire operating frequency range is about $\pm30°$.

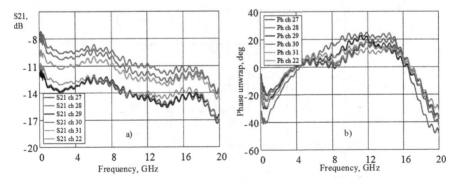

Fig. 5. Amplitude – (a) and phase – (b) characteristics versus frequency for 6 BF channels

The results of measuring of amplitude-frequency and phase-frequency characteristics of BF channels and these of chirped FBG will determine functions $A_n(\omega)$ and $\Delta\Phi_n(\omega)$ needed to calculate PAA FFP according to the model (1–3). Also calculated are the lengths of optical fiber cables required for the initial alignment of the electrical lengths of the channels, and the time delay errors $\delta\tau 2_n$. Note the high repeatability of dependencies $\Delta\Phi_n(\omega)$ in different BF channels (Fig. 5b).

4.3   Measurements of Far-Field Patterns of 6-Elements PAA with Chirped FBG Based BF

To verify the accuracy of PAA patterns calculation model based on measured BF with CFBG separate channels characteristics an experimental study of radiation pattern of a 6-elements PAA was carried out as according to the method described in [7, 12]. The frequency range 7–13 GHz for measurements was constrained by used PAA. The linear

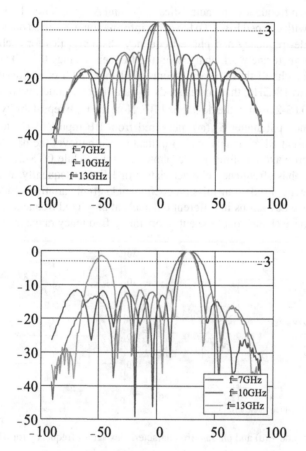

Fig. 6. The measured PAA far-field patterns at different frequencies: normal to PAA base – (a), at 25 angle – (b)

PAA consists of many ultra-wideband printed-circuit elements (Vivaldi type antenna) [7]. A typical radiation pattern of a single PAA element was shown above, see Fig. 4a.

Figure 6 shows the results of 6-elements PAA beam steering when using chirped FBG time-delay unit. More exactly: Fig. 6a shows PAA FFP without chirped FBG, one can see FFP patterns at frequencies 7, 10, and 13 GHz, and Fig. 6b shows PAA FFP with chirped FBG also at the same frequencies. One can clearly see an absence of frequency dependent beam position shift (in Fig. 6b), this is due to really true-time-delay approach in summing of PAA elements output signal. From measurements one can see the shift in PAA beam position $\theta \approx 25°$ (Fig. 6b). This is close to estimation value θ from relation $d \cdot \sin(\theta) = \Delta\tau_{FBG} \cdot c = 27.7$ ps$\cdot c = 8.31$ mm ($c = 3 \cdot 10^8$ m/s). In this experiment we used a PAA with element's spacing $d = 19.2$ mm. So one can expect PAA beam shift to angle $\theta \approx 25.6°$.

One can compare also calculated (Fig. 4) and experimental (Fig. 6) FFPs at 10 GHz, there is some discrepancy in the isotropic background level, the side lobes shape of the calculated and measured FFP is due to the presence of additional phase and amplitude errors produced by microwave amplifiers and RF matching schemes included in the experimental setup between PAA and optical BF.

The measured parameters of PAA with optical BF are in good agreement with the calculated ones.

5   Conclusion

We have investigated the basic features of optical chirped fiber Bragg grating beamformer (BF) to be applied to wideband linear PAA in receive mode. The paper contains the results of measurements the characteristics of chirped FBG with 2 cm length. Integration of this FBG with components of fiber-optic analogue transmission lines with external modulation and DWDM components has allowed to develop 6-channel optical BF prototype for steering a beam of linear PAA in the frequency range of 0.1..18 GHz. The results of measured antenna far-field patterns comparison with the calculated ones in view of channels amplitude and phase fluctuations show their good agreements. The results of BF model experimental testing with 6-elements PAA in frequency range 7..13 GHz show squint free array far-field pattern steering in wide frequency range and weak influence of practical amplitude and phase errors on its basic features.

Authors thank Alexander V. Dostovalov and Alexey A. Wolf from Institute of Automation and Electrometry of Russian Academy of Sciences (Novosibirsk, Russia) for producing chirped fiber Bragg grating with parameters required for our optical beamformer parameters.

References

1. Yao, J.P.: A tutorial on microwave photonics – part II. IEEE Photonics Soc. Newslett. **26**(3), 5–12 (2012)
2. Urick, V.J., McKinney, J.D., Williams, K.J.: Fundamentals of Microwave Photonics. Wiley, Hoboken (2015)

3. Liu, Y., Yao, J., Yang, J.: Wideband true-time-delay beam former that employs a tunable chirped fiber grating prism. Appl. Optics **42**(13), 2273–2277 (2003). https://doi.org/10.1364/ao.42.002273
4. Yang, J.L., Tjin, S.C., Ngo, N.Q.: All chirped fiber grating based true-time delay for phased-array antenna beam forming. Appl. Phys. B **80**(6), 703–706 (2005). https://doi.org/10.1007/s00340-005-1771-z
5. Hunter, D.B., Parker, M.E., Dexter, J.L.: Demonstration of a continuously variable true-time delay beamformer using a multichannel chirped fiber grating. IEEE Trans. Microw. Theor. Tech. **54**(2), 861–867 (2006). https://doi.org/10.1109/TMTT.2005.863056
6. Ivanov, S.I., Lavrov, A.P., Saenko, I.I.: Model of photonic beamformer for microwave phased array antenna. In: Galinina, O., Andreev, S., Balandin, S., Koucheryavy, Y. (eds.) NEW2AN/ruSMART/NsCC -2017. LNCS, vol. 10531, pp. 482–489. Springer, Cham (2017). https://doi.org/10.1007/978-3-319-67380-6_44
7. Volkov, V.A., Gordeev, D.A., Ivanov, S.I., Lavrov, A.P., Saenko, I.I.: Photonic beamformer model based on analog fiber-optic links' components. J. Phys. Conf. Ser. **737**(012002), 1–6 (2016). https://doi.org/10.1088/1742-6596/737/1/012002
8. Optiva OTS-2 18 GHz Amplified Microwave Band Fiber Optic Links. http://emcore.com/wp-content/uploads/2016/03/Optiva-OTS-2-18GHz-Unamplified.pdf
9. Dostovalov, A.V., Wolf, A.A., Parygin, A.V., Zyubin, V.E., Babin, S.A.: Femtosecond point-by-point inscription of Bragg gratings by drawing a coated fiber through ferrule. Opt. Express **24**(15), 16232–16237 (2016). https://doi.org/10.1364/OE.24.016232
10. Scolnik, M.I.: Introduction to Radar Systems, 3rd edn. Tata McGraw Hill Publishing Company Limited, New Delhi (2004). pp. 538–672
11. Shifrin, Y.S.: Statistical Antenna Theory. Golem Press, Boulder (1971). pp. 3–370
12. Ivanov, S.I., Lavrov, A.P., Saenko, I.I.: Application of microwave photonics components for ultrawideband antenna array beamforming. In: Galinina, O., Balandin, S., Koucheryavy, Y. (eds.) NEW2AN/ruSMART -2016. LNCS, vol. 9870, pp. 670–679. Springer, Cham (2016). https://doi.org/10.1007/978-3-319-46301-8_58

Design and Analysis of Circular-Polarized Patch Antenna at S-band for a Nanosatellite

Vasilii Semkin[1,2(✉)] [ORCID]

[1] Department of Electronics and Nanoengineering, Aalto University, Espoo, Finland
vasilii.semkin@gmail.com
[2] Peoples Friendship University of Russia (RUDN University),
6 Miklukho-Maklaya St, Moscow 117198, Russian Federation

Abstract. In this work, planar circular-polarized (CP) antenna operating at 2.40–2.45 GHz is presented. The developed antenna is suitable for the installation on a Nanosatellite or CubeSat, which are usually considered lightweight satellites (below 10 kg), and used for the different observation purposes. The following requirements for the antenna design are considered: operation in S-band (2–4 GHz), circular polarization, and 10×10 cm^2 ground platform for the antenna installation. Developed patch antenna is theoretically studied using high frequency electromagnetic field simulation software, then it is manufactured, and a set of validation measurements is performed. The radiation characteristics, matching and losses are presented. The simulated and measured results correspond well to each other and fit the requirements. The S_{11} of the antenna is below -6 dB in the desired frequency range and the simulated axial ratio values vary from 3 to 5.8 dB. The minimum measured axial ratio is 1.5 dB at 2.35 GHz. The results indicate that this antenna could be a good candidate for the installation on a nano-satellite chassis.

Keywords: Patch antenna · Circular polarized · Nanosatellite
S-band

1 Introduction

During the recent years the miniaturized satellites have become attractive due to their low manufacturing and launching costs, and relatively small time required for the development. Large number of small satellites has already been launched into the space. The small satellites can be classified by their weight, e.g., the weight of a nanosatellite is usually defined between 1 kg and 10 kg. Currently, the nanosatellites are developed by different countries and universities for various Earth observation missions [1,2]. For example, Finnish Aalto-1 nanosatellite [3]

The publication has been prepared with the support of the RUDN University Program 5-100.

© Springer Nature Switzerland AG 2018
O. Galinina et al. (Eds.): NEW2AN 2018/ruSMART 2018, LNCS 11118, pp. 605–612, 2018.
https://doi.org/10.1007/978-3-030-01168-0_54

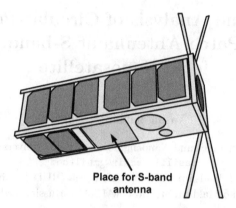

Place for S-band antenna

Fig. 1. Schematic view of a nanosatellite and highlighted place for the S-band antenna.

has been launched recently. The satellite weight is approximately 4 kg and the design is based on a CubeSat standard [4]. The communication with the satellite is critical for the data transmission, and the frequencies in S-band can be utilized for the downlink communication. The schematic view of a nanosatellite is shown in Fig. 1 and the place for the S-band antenna is highlighted, which size is usually limited by the chassis dimensions.

Antennas are one of the critical components for the satellite radio communication. The form factor has an important role and should be kept small so that the antennas fit the chassis of the satellite. It is commonly known that circularly polarized compact printed antennas are required in the satellite communications, Global Positioning System (GPS), etc. In satellite communication systems, circularly polarized antennas are preferred to ensure proper signal receiving because the transmitted signal can be received by arbitrary polarized antenna and the Faraday Rotation Effect is small [5]. A CP antenna with low profile, small size, and light weight is required in mobile satellite communications. Many types of microstrip antennas have been proposed and investigated by other researchers [6,7]. However, there is still a need for improving the antenna performance and developing new antenna structures. The purpose of this paper is to design CP patch antenna which can fit the required characteristics and dimensions for a nanosatellite [4], i.e. 10×10 cm^2. The important parameters for the implementation developed antenna on the satellite become the axial ratio (the ratio of orthogonal components of an E-field) and radiation patterns. These aspects are specifically addressed in this paper. As a result, it is shown that a good solution is found by properly modifying and optimizing existing designs of CP patch antennas. Usually, to generate circular polarization either single-fed or dual-fed type antennas are used. The feeding type defines the number of feeding points necessary for generating the CP [8]. The single-fed type has an advantage since an external polarizer, such as a 90° hybrid coupler, is not required. This type of feeding was used in the design of CP patch antenna presented in this work.

2 Antenna Design

Some technical specifications and the requirements for the antenna are set first. The main technical prerequisites are: achieve the matching properties in terms of S11 parameter of the order of -6 dB or lower; antenna should have circular-polarization in the required frequency range. The other requisitions include that size of the ground plane should be 10×10 cm^2 and the antenna height should be less than 6 mm. These dimensions represent the space available for the antenna element mounting on the satellite chassis (Fig. 1).

Considering the requirements mentioned above, circular shape patch antenna was considered as a good option. The actual antenna structure is illustrated in Fig. 2. The antenna consists of a ground plane and a round patch antenna element, placed above the ground plane. The properties of the proposed antenna structure are studied with the electromagnetic field simulations, where optimized antenna dimensions are obtained. There are two slots in the patch antenna and the feed point is moved 11 mm toward the edge of the antenna to provide circular polarization. Diameter of the circular patch element is 57 mm. Dimensions of the slots are 35×2 mm^2 and 24×2 mm^2 respectively. The distance between the ground plane and patch element is 5 mm. The manufactured antenna is installed on a metallic box, representing the satellite chassis, and the radiation characteristics are measured in the anechoic chamber.

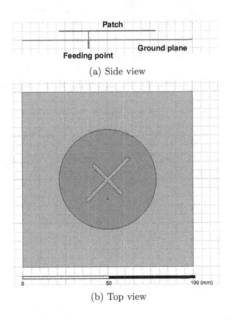

(a) Side view

(b) Top view

Fig. 2. Antenna model in HFSS.

3   Simulation and Measurement Results

Simulations are performed using electromagnetic simulators CST Studio and Ansoft HFSS^{TM}. The 3D realized gain and the gain dependence versus frequencies are shown in Figs. 3 and 4, respectively. As it can be observed, maximum value of the obtained realized gain is 9.35 dB at 2.5 GHz. In the frequency range from 2.4 up to 2.45 GHz gain value varies from 9.20 to 9.26 dB.

After reaching satisfactory antenna design by simulations the antenna was manufactured. The photograph of the manufactured antenna is presented in Fig. 5. Reflection coefficient is illustrated in Fig. 6. Results obtained from CST and HFSS correspond well to each other. There is small difference in the results in terms of resonance frequencies. This can appear due to different geometries of the port in the models. As can be seen in Fig. 6, the measured S11-parameter

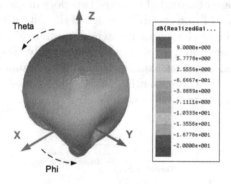

Fig. 3. Simulated 3D realized gain of the antenna.

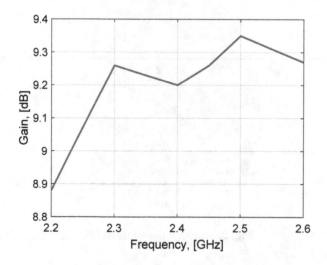

Fig. 4. Simulated antenna gain in HFSS versus frequency.

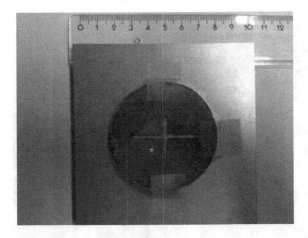

Fig. 5. Photograph of the manufactured antenna. Top view.

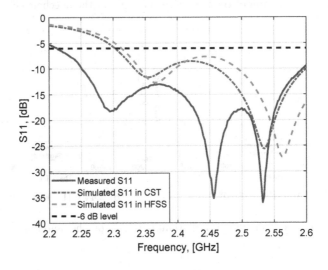

Fig. 6. Measured and simulated reflection coefficient of the antenna.

remains clearly below −6 dB for the range of frequencies from 2.2 up to 2.6 GHz. Comparing to simulated model, where 0.3 mm thick copper ground plane was used, in manufactured antenna double sided metallized FR-4 material was used. Relative permittivity of FR-4 material is 3.4. Styrofoam, which permittivity is near 1, was used to support patch element above the ground plane (in the simulations air gap was used). This can explain the difference in the S11 parameter level. The resonance frequency in the measurement results shifted to the lower frequencies. This can be explained by technological mistakes during manufacturing the antenna (errors in the slots dimensions). In order to perform realistic radiation pattern measurements, the antenna was mounted on a metal

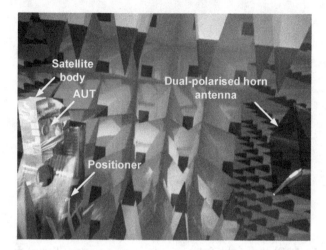

Fig. 7. A photograph of the measurement setup in the anechoic chamber.

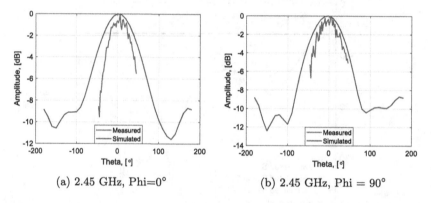

(a) 2.45 GHz, Phi=0°   (b) 2.45 GHz, Phi = 90°

Fig. 8. Measured and simulated radiation patterns of the antenna at frequency 2.45 GHz.

box, identical to the satellite body (Fig. 7). The measurements were performed in the anechoic chamber in the S-band. Dual-polarized horn antenna was used as a probe. The radiation patterns of the developed antenna or antenna under test (AUT) are presented at 2.45 GHz in Fig. 8. The measured main lobe radiation patterns are in a good agreement with the simulated ones. Since the main goal was to achieve good axial ratio values, the measurements of the AUT were done only for the ±50° range. Based on these measurements the axial ratio values can be calculated. In Fig. 9 axial ratio is presented and compared with the simulation results. In the desired range of frequencies minimum value of axial ratio is 3 dB and maximum value is 5.8 dB for simulations performed with CST. Simulations in HFSS provide minimum value of axial ratio is 4.6 dB and maximum value

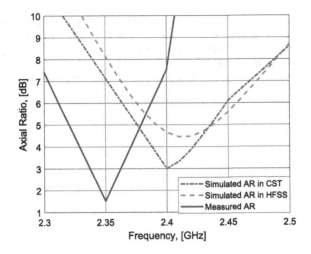

Fig. 9. Measured and simulated axial ratio of the antenna.

is 6 dB in the desired range of frequencies. The minimum measured axial ratio value is 1.5 dB at 2.35 GHz. The axial ratio value at 2.4 GHz is 7.4 dB.

4   Conclusions

In this paper we have shown by the electromagnetic field simulations that round-shape patch antenna is a potential candidate for the use on a nanosatellite. The antenna has almost circular polarization in the desired range of frequencies and good matching properties. Also the manufacturing costs per antenna are very low since the conventional material is used in the antenna manufacturing. Simulated results were confirmed by the measurements. As a part of future work an improvement of axial ratio in the operating frequency range can be performed.

References

1. Swartout, M.: The first one hundred cubesats - a statistical look. J. Small Satellites **2**(2), 213–233 (2013)
2. Nanosatellite database. http://www.nanosats.eu/
3. Leppinen, H., Kestil, A., Tikka, T., Praks, J.: The aalto-1 nanosatellite navigation subsystem: development results and planned operations. In: 2016 European Navigation Conference (ENC), pp. 1–8, May 2016. https://doi.org/10.1109/EURONAV.2016.7530545
4. California Polytechnic State University: CubeSat Design Specification. Revision 13 (2015)
5. Titheridge, J.E.: Faraday rotation of satellite signals across the transverse region. J. Geophys. Res. **76**(19), 4569–4577 (1971). https://doi.org/10.1029/JA076i019p04569
6. Gao, S., et al.: Antennas for modern small satellites. IEEE Antennas Propag. Mag. **51**(4), 40–56 (2009). https://doi.org/10.1109/MAP.2009.5338683

7. Rather, N.N., Suganthi, S.: Electrically small s-band antenna for cubesat applications. In: 2017 International Conference on Wireless Communications, Signal Processing and Networking (WiSPNET), pp. 1687–1691, March 2017. https://doi.org/10.1109/WiSPNET.2017.8300049
8. Balanis, C.A.: Antenna Theory: Analysis and Design. Wiley, Hoboken (2005)

The Casimir-Operated Microdevice for Application in Optical Networks

Galina L. Klimchitskaya[1,2] , Vladimir M. Mostepanenko[1,2] ,
and Viktor M. Petrov[3(✉)]

[1] Central Astronomical Observatory at Pulkovo of the Russian Academy of Sciences,
St. Petersburg 196140, Russia
[2] Institute of Physics, Nanotechnology and Telecommunications, Peter the Great
Saint Petersburg Polytechnic University, St. Petersburg 195251, Russia
[3] Institute of Advanced Manufacturing Technologies, Peter the Great Saint
Petersburg Polytechnic University, St. Petersburg 195251, Russia
vikpetroff@mail.ru

Abstract. The microdevice capable of producing light pulses from a continuous incident laser light do not using the mechanical wheels with a highly stable speed of rotation is suggested. This device, which is functioning like an optical chopper, is operated by a balance between the Casimir force and light pressure force acting in the Fabry-Pérot microresonator formed by the two atomically thin Ag mirrors deposited on the parallel sides of a Si microstructure. The separation distance between the mirrors only slightly exceeds the half wavelength of the laser light. It is shown that this condition leads to a cyclic process resulting in the pulses of transmitted light. The feasibility of the proposed device is confirmed with detailed computations of both the Casimir and light pressure forces taking into account the realistic properties of both the microstructure and mirror materials.

Keywords: Casimir force · Light pressure · Microdevices

1 Introduction

Currently various microdevices are finding increasing application ranging from computers and sensors to cellular and optical communications. It is common knowledge that the basic requirement to increase the functionality of microdevices while minimizing energy consumption inevitably leads to shrinking the device sizes. When device dimensions shrink to submicrometer level, the electromagnetic fluctuations of quantum origin come into play. They give rise to the so-called van der Waals and Casimir forces [1,2] which act even between electrically neutral constituent parts of microdevices. These forces increase much faster with decreasing separation distances than the characteristic electric forces and become dominant at separations of a few hundred nanometers. Because of this, the fluctuation-induced van der Waals and Casimir forces will play an important

© Springer Nature Switzerland AG 2018
O. Galinina et al. (Eds.): NEW2AN 2018/ruSMART 2018, LNCS 11118, pp. 613–623, 2018.
https://doi.org/10.1007/978-3-030-01168-0_55

role in the next generation microdevices intended for advanced optical networks and other prospective applications.

During the last few years, the Casimir force and its gradient have been measured in many high-precision experiments using metallic [3–8] and semiconductor [9–14] test bodies. In several experiments the test bodies with the structured (sinusoidally and rectangular corrugated) surfaces have also been used [15–19]. Simultaneously with fundamental investigations, the Casimir force was exploited as an operating force in various microdevices. Of the important steps forward that have been made, one should mention creation of the micromechanical oscillator driven by the Casimir force [20,21], investigation of the role of geometry and dielectric properties in the stability of Casimir-actuated microdevices [22,23], and the study of microdevice actuation by the Casimir force with account of surface roughness and amorphous to crystalline phase transition [24,25]. Special attention has been paid to the experimental study of the Casimir force acting on a micromechanical silicon chip [26,27].

In this paper, the new type of the Casimir-operated microdevice is proposed which interrupts a continuous laser beam and transforms it into the transmitted light pulses. In other words, the suggested microdevice is functioning like an optical chopper, but with no use of any mechanical wheels of complicated shape which should have a highly stable speed of rotation.

The general principle of operation of the proposed optical chopper driven by the fluctuation-induced force is the following. A continuous light beam of the wavelength λ is incident from the left on a Fabry-Pérot microresonator formed by two thin Ag mirrors deposited on two parallel walls of a Si microstructure. The left wall is sufficiently thick, but the right wall is by an order of magnitude thinner. It can be tilted under the influence of the Casimir and light pressure forces. The microresonator is fabricated in such a way that in the absence of light pressure and Casimir forces the separation distance between both mirrors exceeds $\lambda/2$ by some $\Delta\lambda$, which is large enough to violate the resonance condition in the microresonator. The Casimir force slightly tilts the right wall to the left one, so that the resonance condition is obeyed over the cross section of the light beam when the laser is switched on. As a result, the amplitude of a standing wave in the microresonator increases instantaneously leading to a high level of intensity of the transmitted light. At this point the repulsive force due to the light pressure compensates the Casimir force and the mirror deposited on the right wall takes the vertical position, i.e., the distance to the left mirror becomes equal to $\lambda/2 + \Delta\lambda$ in violation of the resonance condition. Hence, the wave amplitude and the light pressure drop leading to almost zero level of the transmitted light. Thus, the Casimir force returns the right mirror to its initial tilted position, where the resonance condition is obeyed, and the new cycle starts. One can say that the proposed chopper is operated using the balance between the Casimir and light pressure forces, whereas the elastic properties of a wall supporting the right mirror play only an auxiliary role.

The detailed quantitative analysis presented below demonstrates the feasibility of the proposed microdevice. Specifically, the Casimir force is calculated using the Lifshitz theory at the room temperature taking into account an anisotropy of the dielectric permittivity of atomically thin metallic films forming the microresonator mirrors [28]. The reflectance and transmittance of these mirrors are found with allowance made for an increased transparency of thin films [29]. The parameters of a microresonator which provide the desired balance between the Casimir and light pressure forces are determined.

2 Sketch of the Setup

The general scheme of the laboratory setup is shown in Fig. 1. As a source of the incident light of intensity I_{in} it is suggested to use c CW Nd-YAG laser with a wavelength of the second harmonic $\lambda = 532.0$ nm. The laser beam passes through the optical window of the vacuum chamber (with a pressure of about 10^{-6} – 10^{-7} Torr), the laser forming system and enters the Fabry-Pérot microresonator described below in more detail. This beam has a Gaussian profile and a diameter of approximately 40 μm. The transmitted light passes through another beam forming system, escapes from the vacuum chamber through the optical window and is registered using the photodetector and the two-channel oscilloscope (see Fig. 1).

The key element of the proposed setup is a SiO_2 microdevice consisting of a cube with the side $D_1 = 50$ μm and a square wall of the same side-length of thickness $D_2 = 5$ μm which are located at the joint base parallel to each other as shown in Fig. 2. The separation distance between the foot parts of the cube and of the wall is $a = \lambda/2 + \Delta\lambda$, where $\Delta\lambda$ is large enough to violate the resonance condition (see Sect. 4 for the specific value of $\Delta\lambda$). The left face of the wall and

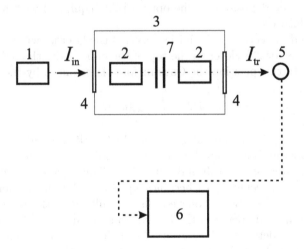

Fig. 1. Sketch of the optical chopper driven by the Casimir force: 1 — laser, 2 — beam-forming systems, 3 — vacuum chamber, 4 — optical windows, 5 — photodetector, 6 — two-channel oscilloscope, 7 — Fabry-Pérot microresonator.

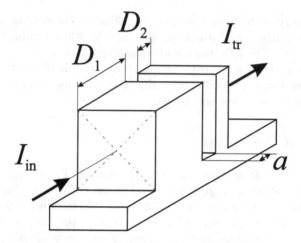

Fig. 2. Schematic of the SiO$_2$ Fabry-Pérot microresonator with the length of cavity a.

the right face of the cube in Fig. 2 are coated with thin Ag mirrors of thickness d ranging from 0.5 to 3 nm. The SiO$_2$ microdevice shown in Fig. 2 can be fabricated by means of the ion-beam etching. For this purpose, the technological processes have been elaborated allowing fabrication of large area microstructures with a uniform gap of several hundred nanometers width and vertical sidewalls [19, 26]. The Ag mirrors are deposited on Si surfaces by either sputtering or electroplating [19]. The high precision positioning technology elaborated for other purposes [30–32] allows to perform the assembly and alignment of the described device inside the vacuum chamber. To guarantee a stability of the setup, the vacuum chamber should be deposited on the optical table equipped with an active air-pumped stabilization.

As described in Sect. 1, in the initial position the laser is switched off and the top of the wall is for $\Delta\lambda$ closer to the top of the cube under the influence of the Casimir force. In this position the Casimir force is balanced by the elastic force

$$k\Delta\lambda = \frac{1}{2}P_{\text{Cas}}(a)S. \tag{1}$$

Here k is the spring constant of the 5 μm thick wall, $P_{\text{Cas}}(a) = F_{\text{Cas}}(a)/S$ is the Casimir pressure given by the Lifshitz formula [2], and $S = 50 \times 50\,\mu\text{m}^2$ is the common area of the wall and cube sides. Thus, the effective length of a microresonator in its central part becomes equal to $\lambda/2$ with sufficient precision, i.e., the resonance condition is obeyed. As a result, after the laser is switched on, the wave amplitude between the two mirrors instantaneously increases, and the photodetector detects high intensity I_{tr} of the transmitted light. Then, the increased repulsive force due to the light pressure in the microresonator compensates the Casimir force, and the wall becomes vertical violating the resonance condition. This position, where the elastic force is equal to zero, is, however, unstable. The point is that the wave amplitude in the microresonator instantaneously drops together with the light pressure, and a very low intensity of the transmitted light is detected. Thus, the Casimir force, which is not balanced by the light pressure any more, returns the wall in its initial, tilted, position, where the resonance condition is obeyed, and the new cycle starts. This cyclic process takes place under a condition that in the vertical position of the wall the attractive Casimir pressure is equal in magnitude to the repulsive light pressure. As shown below, this condition can be easily satisfied with quite reasonable values of all relevant parameters of the experimental setup.

3   The Casimir Pressure in the Microresonator

The Casimir pressure between two atomically thin Ag mirrors of thickness d at temperature T deposited on the Si cube of thickness D_1 and Si wall of thickness D_2, respectively (see Fig. 2), can be calculated using the Lifshitz formula adapted for the case of four material layers and a vacuum gap [2,33].

$$
P_{\text{Cas}}(a,T) = -\frac{k_B T}{\pi} \sum_{l=0}^{\infty}{}' \int_0^{\infty} k_\perp \, dk_\perp \, q_l \left\{ \left[\frac{e^{2aq_l}}{R_{\text{TM},l}^{(1)} R_{\text{TM},l}^{(2)}} - 1 \right]^{-1} \right.
$$
$$
\left. + \left[\frac{e^{2aq_l}}{R_{\text{TE},l}^{(1)} R_{\text{TE},l}^{(2)}} - 1 \right]^{-1} \right\}. \tag{2}
$$

Here, $k_\perp = |\mathbf{k}_\perp|$ is the magnitude of the projection of the wave vector on the plane of the wall, $\xi_l = 2\pi k_B T l/\hbar$ with integer l are the Matsubara frequencies, k_B is the Boltzmann constant, the prime divides the term with $l = 0$ by two, and

$$
q_l = \sqrt{k_\perp^2 + \frac{\xi_l^2}{c^2}}. \tag{3}
$$

The quantities $R_{\text{TM(TE)},l}^{(i)}$ are the reflection coefficients on the left ($i = 1$) and right ($i = 2$) mirrors for the transverse magnetic (TM) and transverse electric (TE) polarizations of the electromagnetic field. They are defined at the pure imaginary Matsubara frequencies

$$
R_{\text{TM(TE)},l}^{(i)} = \frac{r_{\text{TM(TE)},l}^{(v,m)} + R_{\text{TM(TE)},l}^{(g,i)} e^{-2dk_{\text{TM(TE)},l}^{(m)}}}{1 + r_{\text{TM(TE)},l}^{(v,m)} R_{\text{TM(TE)},l}^{(g,i)} e^{-2dk_{\text{TM(TE)},l}^{(m)}}}. \tag{4}
$$

The reflection coefficients at the boundary planes between vacuum and a semis-pace made of an anisotropic metal (Ag), entering Eq. (4), are given by [2,34,35]

$$
r_{\text{TM},l}^{(v,m)} = \frac{\varepsilon_{xx,l}^{(m)} q_l - k_{\text{TM},l}^{(m)}}{\varepsilon_{xx,l}^{(m)} q_l + k_{\text{TM},l}^{(m)}}, \qquad r_{\text{TE},l}^{(v,m)} = \frac{q_l - k_{\text{TE},l}^{(m)}}{q_l + k_{\text{TE},l}^{(m)}}, \tag{5}
$$

where $\varepsilon_{xx,l}^{(m)} \equiv \varepsilon_{xx}^{(m)}(i\xi_l) = \varepsilon_{yy}^{(m)}(i\xi_l)$, $\varepsilon_{zz,l}^{(m)} \equiv \varepsilon_{zz}^{(m)}(i\xi_l)$ are the frequency-dependent dielectric permittivities of a thin Ag film in the plane of the wall and perpendicular to it, respectively, and

$$k_{TM,l}^{(m)} = \sqrt{\frac{\varepsilon_{xx,l}^{(m)}}{\varepsilon_{zz,l}^{(m)}}k_\perp^2 + \varepsilon_{xx,l}^{(m)}\frac{\xi_l^2}{c^2}}, \qquad k_{TE,l}^{(m)} = \sqrt{k_\perp^2 + \varepsilon_{xx,l}^{(m)}\frac{\xi_l^2}{c^2}}. \tag{6}$$

The coefficients $R_{TM(TE),l}^{(g,i)}$, also entering Eq. (4), are somewhat more complicated. They are given by

$$R_{TM(TE),l}^{(g,i)} = \frac{r_{TM(TE),l}^{(m,g)} + r_{TM(TE),l}^{(g,v)}e^{-2D_i k_l^{(g)}}}{1 + r_{TM(TE),l}^{(m,g)}r_{TM(TE),l}^{(g,v)}e^{-2D_i k_l^{(g)}}}, \tag{7}$$

where $\varepsilon_l^{(g)} \equiv \varepsilon^{(g)}(i\xi_l)$ is the dielectric permittivity of quartz glass SiO_2 and

$$k_l^{(g)} = \sqrt{k_\perp^2 + \varepsilon_l^{(g)}\frac{\xi_l^2}{c^2}}. \tag{8}$$

The remaining reflection coefficients in Eq. (7) are at the boundary planes between the semispaces made of an anisotropic Ag and SiO_2 glass

$$r_{TM,l}^{(m,g)} = \frac{\varepsilon_l^{(g)}k_{TM,l}^{(m)} - \varepsilon_{xx,l}^{(m)}k_l^{(g)}}{\varepsilon_l^{(g)}k_{TM,l}^{(m)} + \varepsilon_{xx,l}^{(m)}k_l^{(g)}}, \qquad r_{TE,l}^{(m,g)} = \frac{k_{TE,l}^{(m)} - k_l^{(g)}}{k_{TE,l}^{(m)} + k_l^{(g)}} \tag{9}$$

and between a SiO_2 semispace and vacuum

$$r_{TM,l}^{(g,v)} = \frac{k_l^{(g)} - \varepsilon_l^{(g)}q_l}{k_l^{(g)} + \varepsilon_l^{(g)}q_l}, \qquad r_{TE,l}^{(g,v)} = \frac{k_l^{(g)} - q_l}{k_l^{(g)} + q_l}. \tag{10}$$

To compute the Casimir pressure using Eqs. (2)–(10), one needs to know the dielectric permittivities of SiO_2 and Ag at the pure imaginary Matsubara frequencies. For SiO_2 this information was taken from Ref. [36]. For Ag the dielectric permittivity along the imaginary frequency axis was obtained from the optical data for its complex index of refraction [37] extrapolated to zero frequency by means of the Drude model. The effects of anisotropy, which are important for films consisting of only a few atomic layers, were taken into account using the results of Ref. [28] where they have been found in the framework of the density functional theory.

The Casimir pressure was computed at separation distances a from 266 to 270 nm between the mirrors with a step of 0.1 nm for mirror thicknesses varying from 0.94 to 2.115 nm with a step of 0.235 nm. To take one example, for $a = 268.3$ nm ($\Delta\lambda = 2.3$ nm) and $d = 1.175$ nm (i.e., for the mirrors consisting of five atomic layers) the Casimir pressure is equal to $P_{Cas} = -180.4$ mPa and the respective attractive Casimir force $F_{Cas} = P_{Cas}D_1^2 = -0.45$ nN. These values of the microresonator parameters are used in the next section to demonstrate that the Casimir pressure can be precisely compensated by the pressure of light in the vertical position of the wall.

4   The Pressure of Light and the Cyclic Process

For the chosen laser, the power of light incident from the left on the cube is $N_{\text{in}} = 7$ mW. Taking into account high transparency of SiO_2 at $\lambda = 532$ nm ($\omega = 3.54 \times 10^{15}$ rad/s) one can assume that the same power falls on the outside of the left mirror. When considering the reflectivity properties of thin metallic mirror, the glass cube from the left and the vacuum gap from the right of it can be considered as semispaces. Because of this, the magnitude of the reflection coefficient on the left mirror at the normal incidence takes the form [38]

$$|R| = \left| \frac{\tilde{r}^{(v,m)} + \tilde{r}^{(m,g)} e^{-2i\frac{\omega}{c} d\sqrt{\varepsilon^{(m)}}}}{1 + \tilde{r}^{(v,m)} \tilde{r}^{(m,g)} e^{-2i\frac{\omega}{c} d\sqrt{\varepsilon^{(m)}}}} \right|, \tag{11}$$

where the complex dielectric permittivity of Ag at the frequency ω is

$$\sqrt{\varepsilon^{(m)}} = n^{(m)} + i\kappa^{(m)} \tag{12}$$

with $n^{(m)} = 0.129$ and $\kappa^{(m)} = 3.19$ [37]. The reflection coefficients $\tilde{r}^{(v,m)}$ and $\tilde{r}^{(m,g)}$ in Eq. (11) are given by Eqs. (5) and (9) where $i\xi_l$ is replaced with the real frequency ω and $\varepsilon_{xx}^{(m)} = \varepsilon_{zz}^{(m)}$ because the anisotropy of dielectric properties discussed in Sect. 3 does not play any role at rather high frequency ω of the laser light. Note also that $k_\perp = 0$ at the normal incidence and the TM reflection coefficient coincides with the TE one.

Strictly speaking, Eq. (11) can be used only for sufficiently thick films, whereas atomically thin films are characterized by the increased transparency [29,39]. It was shown phenomenologically [40–42] that for the ultrathin metallic films good agreement with the measurement data is reached if one puts $\kappa^{(m)} = 0$ in the coefficients $\tilde{r}^{(v,m)}$ and $\tilde{r}^{(m,g)}$ in Eq. (11), so that they become

$$\tilde{r}^{(v,m)} = \frac{n^{(m)} - 1}{n^{(m)} + 1}, \qquad \tilde{r}^{(m,g)} = \frac{n^{(g)} - n^{(m)}}{n^{(g)} + n^{(m)}}, \tag{13}$$

where $n^{(g)} = \sqrt{\varepsilon^{(g)}} \approx 1.46$ at the used frequency.

From Eqs. (11) and (13) for the reflectance of Ag mirror $\mathcal{R} = |R|^2$ at the given ω one obtains

$$\mathcal{R} = \frac{\tilde{r}^{(v,m)2} + \tilde{r}^{(m,g)2} e^{-2\alpha d} - 2\tilde{r}^{(v,m)} \tilde{r}^{(m,g)} e^{-\alpha d} \cos\psi}{1 + \tilde{r}^{(v,m)2} \tilde{r}^{(m,g)2} e^{-2\alpha d} - 2\tilde{r}^{(v,m)} \tilde{r}^{(m,g)} e^{-\alpha d} \cos\psi}, \tag{14}$$

where $\alpha = 4\pi\kappa^{(m)}/\lambda$ and $\psi = 4\pi n^{(m)} d/\lambda$. The transmittance of Ag mirror can be found similarly

$$\mathcal{T} = \frac{\tilde{t}^{(v,m)} \tilde{t}^{(m,g)} e^{-\alpha d}}{1 + \tilde{r}^{(v,m)2} \tilde{r}^{(m,g)2} e^{-2\alpha d} - 2\tilde{r}^{(v,m)} \tilde{r}^{(m,g)} e^{-\alpha d} \cos\psi}, \tag{15}$$

where the following notations are introduced:

$$\tilde{t}^{(v,m)} = \frac{4n^{(m)}}{(1 + n^{(m)})^2}, \qquad \tilde{t}^{(m,g)} = \frac{4n^{(m)} n^{(g)}}{(n^{(m)} + n^{(g)})^2}. \tag{16}$$

Then for the absorptance of light by the ultrathin Ag mirror one finds

$$\mathcal{A} = 1 - \mathcal{R} - \mathcal{T}. \tag{17}$$

Assuming that $\mathcal{R}$ and $\mathcal{T}$ for both mirrors are equal, the transmission coefficient and the amplification factor for the power entering the filter are given by [43]

$$\tau = \left(1 - \frac{\mathcal{A}}{1 - \mathcal{R}}\right)^2, \qquad q = \frac{1}{(1 - \mathcal{R})^2}. \tag{18}$$

In doing so, the power of light transmitted through the microresonator is $N_{\text{tr}} = \tau N_{\text{in}}$.

Now we consider the pressure of light amplified in the microresonator [43]

$$P_{\text{light}} = \frac{1}{c}(1 + \mathcal{R})I_{\text{f}}, \tag{19}$$

where $I_{\text{f}} = q I_{\text{in}} \mathcal{T}$ is the amplified intensity of light.

Integrating Eq. (19) over the section of the light beam, one finds for the light pressure force acting on the wall

$$F_{\text{light}} = (1 + \mathcal{R})\frac{N_{\text{f}}}{c}, \tag{20}$$

where $N_{\text{f}} = q N_{\text{in}} \mathcal{T}$ is the power of light amplified in the microresonator. It is easily seen that for the parameters of a microresonator considered in Sect. 3 ($\Delta\lambda = 2.3$ nm, $a = 268.3$ nm, $d = 1.175$ nm) one obtains $\mathcal{R} = 0.934$, $\mathcal{T} = 0.0436$, and $q = 230$. This results in $N_{\text{f}} = 70$ mW and in the light pressure force $F_{\text{light}} = 0.45$ nN, which is equal in magnitude but of the opposite direction to the Casimir force calculated in Sect. 3.

Thus, the microresonator with the considered above parameters ensures a balance between the Casimir and light pressure forces in the vertical position of the wall and, thus, a cyclic process of obtaining the discrete pulses of transmitted light from a continuous incident light. In doing so, the power of transmitted pulses leaving the microresonator is given by $N_{\text{tr}} = \mathcal{T} N_{\text{f}} = 3$ mW. Using various sources of light and parameters of a microresonator gives the possibility to obtain the transmitted pulses of varied duration and power.

5   Conclusions

This paper proposes the possibility of creating the novel microdevice operated by the Casimir force. This microdevice is capable to convert a continuous laser light into the light pulses, i.e., works as an optical chopper, do not using the mechanical wheels with a highly stable speed of rotation. Both the Casimir and light pressure forces, which are responsible for the cyclic process in the proposed microdevice, are calculated in detail and the possibility of a balance between them is demonstrated. The suggested optical chopper driven by the fluctuations of the electromagnetic field may find applications in the next generation of optical networks.

The experimental realization of this proposal would require measurements of the reflectance and transmittance of a unit SiO_2 wall with deposited on it atomically thin metallic films rather than using the tabulated optical data of respective materials. It is desirable also to measure the elastic properties of the SiO_2 wall for different wall thicknesses. These measurement results and their comparison with theory will allow to conclusively establish the feasibility of the proposed microdevice.

Acknowledgments. The authors are grateful to Prof. Dr. T. Tschudi for useful discussions.

References

1. Parsegian, V.A.: Van der Waals Forces: A Handbook for Biologists, Chemists, Engineers, and Physicists. Cambridge University Press, Cambridge (2005)
2. Bordag, M., Klimchitskaya, G.L., Mohideen, U., Mostepanenko, V.M.: Advances in the Casimir Effect. Oxford University Press, Oxford (2015)
3. Decca, R.S., López, D., Fischbach, E., Klimchitskaya, G.L., Krause, D.E., Mostepanenko, V.M.: Tests of new physics from precise measurements of the Casimir pressure between two gold-coated plates. Phys. Rev. D **75**, 077101-1-4 (2007)
4. Chang, C.-C., Banishev, A.A., Castillo-Garza, R., Klimchitskaya, G.L., Mostepanenko, V.M., Mohideen, U.: Gradient of the Casimir force between Au surfaces of a sphere and a plate measured using an atomic force microscope in a frequency-shift technique. Phys. Rev. B **85**, 165443-1-1–17 (2012)
5. Banishev, A.A., Chang, C.-C., Klimchitskaya, G.L., Mostepanenko, V.M., Mohideen, U.: Measurement of the gradient of the Casimir force between a nonmagnetic gold sphere and a magnetic nickel plate. Phys. Rev. B **85**, 195422-1-7 (2012)
6. Banishev, A.A., Klimchitskaya, G.L., Mostepanenko, V.M., Mohideen, U.: Demonstration of the Casimir force between ferromagnetic surfaces of a Ni-coated sphere and a Ni-coated plate. Phys. Rev. Lett. **110**, 137401-1-5 (2013)
7. Banishev, A.A., Klimchitskaya, G.L., Mostepanenko, V.M., Mohideen, U.: Casimir interaction between two magnetic metals in comparison with nonmagnetic test bodies. Phys. Rev. B **88**, 155410-1-20 (2013)
8. Bimonte, G., López, D., Decca, R.S.: Isoelectronic determination of the thermal Casimir force. Phys. Rev. B **93**, 184434-1-15 (2016)
9. Chen, F., Klimchitskaya, G.L., Mostepanenko, V.M., Mohideen, U.: Control of the Casimir force by the modification of dielectric properties with light. Phys. Rev. B **76**, 035338-1-15 (2007)
10. Torricelli, G., et al.: Switching Casimir force with phase-change materials. Phys. Rev. A **82**, 010101(R)-1-4 (2010)
11. de Man, S., Heeck, K., Wijngaarden, R.J., Iannuzzi, D.: Halving the Casimir force with conductive oxides. Phys. Rev. Lett. **103**, 040402-1-4 (2009)
12. Chang, C.-C., Banishev, A.A., Klimchitskaya, G.L., Mostepanenko, V.M., Mohideen, U.: Reduction of the Casimir force from indium tin oxide film by UV treatment. Phys. Rev. Lett. **107**, 090403-1-4 (2011)

13. Banishev, A.A., Chang, C.-C., Castillo-Garza, R., Klimchitskaya, G.L., Mostepanenko, V.M., Mohideen, U.: Modifying the Casimir force between indium tin oxide plate and Au sphere. Phys. Rev. B **85**, 045436-1–18 (2012)
14. Sedighi, M., Svetovoy, V.B., Palasantzas, G.: Casimir force measurements from carbide surfaces. Phys. Rev. B **93**, 085434-1–6 (2016)
15. Chan, H.B., et al.: Measurement of the Casimir force between a gold sphere and a silicon surface with nanoscale trench arrays. Phys. Rev. Lett. **101**, 030401-1–4 (2008)
16. Bao, Y., et al.: Casimir force on a surface with shallow nanoscale corrugations: Geometry and finite conductivity effects. Phys. Rev. Lett. **105**, 250402-1–4 (2010)
17. Chiu, H.-C., Klimchitskaya, G.L., Marachevsky, V.N., Mostepanenko, V.M., Mohideen, U.: Demonstration of the asymmetric lateral Casimir force between corrugated surfaces in nonadditive regime. Phys. Rev. B **80**, 121402(R)-1–4 (2009)
18. Chiu, H.-C., Klimchitskaya, G.L., Marachevsky, V.N., Mostepanenko, V.N., Mohideen, U.: Lateral Casimir force between sinusoidally corrugated surfaces: Asymmetric profiles, deviations from the proximity force approximation, and comparison with exact theory. Phys. Rev. B **81**, 115417-1–20 (2010)
19. Intravaia, F., et al.: Strong Casimir force reduction through metallic surface nanostructuring. Nat. Commun. **4**, 2515-1–8 (2013)
20. Chan, H.B., Aksyuk, V.A., Kleiman, R.N., Bishop, D.J., Capasso, F.: Quantum mechanical actuation of microelectromechanical system by the Casimir effect. Science **291**, 1941–1944 (2001)
21. Chan, H.B., Aksyuk, V.A., Kleiman, R.N., Bishop, D.J., Capasso, F.: Nonlinear micromechanical Casimir oscillator. Phys. Rev. Lett. **87**, 211801-1–4 (2001)
22. Barcenas, J., Reyes, L., Esquivel-Sirvent, R.: Scaling of micro- and nanodevices actuated by the Casimir force. Appl. Phys. Lett. **87**, 263106-1–3 (2005)
23. Esquivel-Sirvent, R., Pérez-Pascual, R.: Geometry and charge carrier induced stability in Casimir actuated nanodevices. Eur. Phys. J. B **86**, 467-1–6 (2013)
24. Broer, W., Waalkens, H., Svetovoy, V.B., Knoester, J., Palasantzas, J.: Nonlinear actuation dynamics of driven Casimir oscillators with rough surfaces. Phys. Rev. Appl. **4**, 054016-1–7 (2015)
25. Sedighi, M., Broer, W., Palasantzas, G., Kooi, B.J.: Sensitivity of micromechanical actuation on amorphous to crystalline phase transformations under the influence of Casimir forces. Phys. Rev. B **88**, 165423-1–5 (2013)
26. Zou, J., et al.: Casimir forces on a silicon micromechanical chip. Nat. Commun. **4**, 1845-1–6 (2013)
27. Tang, L., et al.: Measurement of non-monotonic Casimir forces between silicon nanostructures. Nat. Photonics **11**, 97–101 (2017)
28. Boström, M., Persson, C., Sernelius, B.E.: Casimir force between atomically thin gold films. Eur. Phys. J. B **86**, 43-1–4 (2013)
29. Ghosh, D.S.: Ultrathin Metal Transparent Electrodes for the Optoelectronics Industry. Springer, Cham (2013)
30. Petrov, V.M., Petrov, M.P., Bryksin, V.V., Petter, J., Tschudi, T.: Optical detection of the Casimir force between macroscopic objects. Opt. Lett. **32**, 3167–3169 (2006)
31. Petrov, V.M., Petrov, M.P., Bryksin, V.V., Petter, J., Tschudi, T.: Casimir force measurement using dynamic holography. JETP **104**, 696–703 (2007). [Zh. Eksp. Teor. Fiz. **131**, 798–807 (2007)]
32. Petrov, V., Hahn, J., Petter, J., Petrov, M., Tschudi, T.: Precise subnanometer control of the position of a macro object by light pressure. Opt. Lett. **30**, 3138–3140 (2005)

33. Tomaš, M.S.: Casimir force in absorbing multilayers. Phys. Rev. A **66**, 052103-1-7 (2002)
34. Klimchitskaya, G.L., Mostepanenko, V.M.: Casimir and van der Waals energy of anisotropic atomically thin metallic films. Phys. Rev. B **92**, 205410-1-7 (2015)
35. Mostepanenko, V.M.: Special features of the thermal Casimir effect across a uni-axial anisotropic film. Phys. Rev. A **92**, 012511-1-8 (2015)
36. Bergström, L.: Hamaker constants of inorganic materials Adv. Colloid Interface Sci. **70**, 125–169 (1997)
37. Palik, E.D. (ed.): Handbook of Optical Constants of Solids. Academic, New York (1985)
38. Landau, L.D., Lifshitz, E.M., Pitaevskii, L.P.: Electrodynamics of Continuous Media. Pergamon, Oxford (1984)
39. Nash, D.J., Sambles, J.R.: Surface plasmon study of the optical dielectric function of silver. J. Mod. Opt. **43**, 81–91 (1996)
40. Kudykina, T.A.: Boundary conditions in case of electromagnetic wave absorption. Phys. Stat. Sol. (b) **160**, 365–373 (1990)
41. Kudykina, T.A.: Dispersion and propagation of light in crystals in the exciton absorption region. Phys. Stat. Sol. (b) **165**, 591–598 (1991)
42. Kovalenko, S.A.: Optical properties of thin metal films. Semicond. Phys. Quantum Electr. Optoelectr. **2**, 13–20 (1999)
43. Born, M., Wolf, E.: Principles of Optics. Cambridge University Press, Cambridge (1999)

Features of Transmission of Intermediate Frequency Signals over Fiber-Optical Communication System in Radar Station

Alexey S. Podstrigaev[1,2], Roman V. Davydov[3($\boxtimes$)], Vasiliy Yu. Rud[5], and Vadim V. Davydov[3,4,5]

[1] Scientific-Research Institute Vector OJSC, St. Petersburg 196084, Russia
[2] Saint Petersburg Electrotechnical University "LETI", St. Petersburg, Russia
[3] Peter the Great Saint Petersburg Polytechnic University,
St. Petersburg 195251, Russia
davydovrv@spbstu.ru
[4] The Bonch-Bruevich Saint - Petersburg State University
of Telecommunications, St. Petersburg 193232, Russia
[5] Department of Ecology, All-Russian Research Institute of Phytopathology,
B. Vyazyomy, Odintsovo District, Moscow Region 143050, Russia

Abstract. Features of transmission at analog intermediate frequency signals over fiber-optical communication lines are considered. Considering these features the technique for checking of developed optical transmission systems at analog intermediate frequency signals is offered. The results of experimental investigations are presented.

Keywords: Radar · Fiber-optical communication
Automatic frequency control system · Amplitude-frequency characteristic
Mach-Zehnder modulator

1 Introduction

The fiber-optical transmission systems (FOTS) for information transmission have found a large number of applications [1–4]. The use of FOTS help to solve various actual tasks. In a number of cases, experiments have shown that only the use of FOTS allows to achieve the desired result [2–7]. Nowadays, one of these actual tasks is to ensure the reliable operation of the radar station in various conditions of its operation [3, 7–9].

In conditions of a high level disturbances of various kinds the effective protection of signal, coming to the antenna complex, especially the intermediate frequency signal (IF) that is the most sensitive to different disturbances [2, 8–10] appears to be a rather difficult problem. As the IF f_i in the given case is chosen frequency f in which the signal frequency transformed at the intermediate period of its processing in the radioelectron setup (superheterodyne receiver). Depending upon the radar exploitation condition f_i represent the sum or difference between the received signal frequency and frequency of heterodyne. As a result, all signal processing is carried out at the fixed frequency that is

O. Galinina et al. (Eds.): NEW2AN 2018/ruSMART 2018, LNCS 11118, pp. 624–630, 2018.
https://doi.org/10.1007/978-3-030-01168-0_56

more comfortable than to readjust the whole receive path in the order to find the necessary signal.

In transmission IF signal tract occurs the frequency f_i offset value on random law if there is interference in it. The automatic frequency control system (AFC) in most cases does not have the time that to follow up f_i, that leads to the appearance of various errors (for example, in determining the distance to the target, the size and speed of the target, etc.).

Besides that, in many cases it is necessary to reduce the weight loading on the scanning antenna platform, especially in the case when it is placed on flying device or on the top part of the ship mats. It is also important to ensure a free access to equipment for repairing even in unfavorable conditions [2, 3, 7, 8]. Moreover, in the case of radar placing on the non-stable object the electromagnetic disturbance level in the area of the communication network location is several times greater than that observed at the ground station. The modern methods of the coaxial connections screening almost exhausted their resources from the point of view of creating new methods of protecting against electromagnetic disturbances [7, 9–11]. The additional protection is realized in the most cases by increasing both the number of screening covers and its thickness. It leads to increased weight and volume of the cable and also decreased flexibility of it that creates difficult problems during radar operation [2, 3, 7–11].

One of the possible solution of this task can be received by our fiber-optical transmission system of analog IF signal from the master generator to heterodyne located at the antenna past.

It should be noted that for the microwave transmission from the receiving antenna to the processing devices located from it at various distances (from 30 to 500 m) a large number of FOPS was developed and manufactured, including by the authors [3, 5–7]. Also various FOPS have been developed for transmitting the microwave signal to the transmitting antenna from control devices [3, 5–7]. However, all these designs of the FOPS (as experiments have shown) due to the specifics of using the IF signal are not so suitable for its transmission. Therefore, in order to reduce the probability of errors in determining the position and parameters of the target, it is necessary to develop own design of FOTS for transmitting IF signals to the radar.

2 Features of Transmission of Intermediate Frequency Signals over Fiber-Optical Communication System

Our studies have shown that a number of features arise when FOTS is using for IF signal transmission and it is must be taken into account when FOTS is developing. These features can be divided into three groups. The first group is connected with energy and frequency characteristics: amplitude frequency characteristic (AFC), dynamic characteristic, the loss coefficient at the frequency of the transmitted analog signal, etc. It is related to the fact that the heterodyne is placed at the antenna in the area of a great number of different disturbances. That is why the signal from the optical receiver must be immediately sent to the heterodyne input. The using of noise amplifier in the case of low power of the optical receiver output signal, as was shown by the experiments, is unsuitable. The measurement error, for example, of the distance to the target during the received microwave signal, reflected from it, can be increased by the order and greater.

The second group is connected with the fact, that a part of FOTS is located in the area where the temperature change from 213 to 332 K is possible. The performed experiments showed that in transmitting and receiving optical modula (Dilaz) the supply driver with the thermostabilizer doesn't provide the necessary stability for the IF signal transmission. Therefore, it is necessary to find a new solution of this problem.

The third group relates to the high requirements which is applied to the IF signal itself. During transmission of this signal to the FOTS pulse shape distortion (even minor) should not happen, comparing to other analogue signals which are also transmitted in radar station by the FOTS [2, 3, 6, 7, 11, 12]. It is important especially in the case of impulses with high pulsing ratio.

It should be noted that the obtained result (the establishment of all the features and their systematization) appears for the first time in scientific publications and can be interesting for future developers of the FOTS. In Fig. 1 is a block diagram of the study FOTS for the IF signal transmission. In the developed design of the FOTS we took into account established by us features of the IF signal transmission.

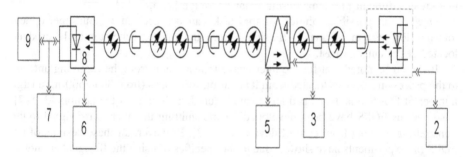

Fig. 1. The structural scheme of IF signal transmission: 1 – laser (λ = 1550 nm); 2 – power supply; 3 – generator IF; 4 – Mach - Zehnder modulator (Oplilab company); 5 – source DC bias; 6 – bias source (+12 V); 7 – heterodyne; 8 – optical detector (Miteq); 9 – oscilloscope.

In order to increase the reliability of the thermal stabilization circuit in our design of the FOTS it was decided to divide the optical transmitting module into two separate units (the transmitting laser module 1 and the modulator 4), in contrast to those FOTS which are developed for the transmission of microwave signals [2, 3, 7]. Such construction of FOTS also allow to increase the optical power of laser radiation to 12 dBm. The conducted studies showed that in the optical modulator (in the case of a temperature change) it is extremely difficult to adjust the minimum attenuation point of the IF signals for different values of f_i by the level of the output signal from the optical photodetector. In Fig. 2, as an example, the results of an investigation the changes in the transfer characteristic of the Mach-Zehnder optical modulator 4 at different temperatures. The optical power of laser radiation is 10 dBm at modulator input.

The analysis of obtained results shows that in the range of operating temperatures (from 223 to 333 K) of the radar it is difficult without additional devices to provide the

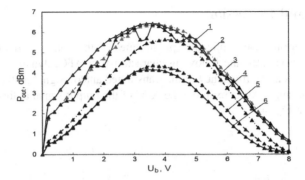

Fig. 2. The dependency of optical signal power change P_{out} on modulator input from bias voltage U_b. The graphs 1, 2, 3, 4, 5 and 6 correspond to the temperature in K: 273.1; 282.9; 293.2; 298.1; 315.3; 323.2.

location of the operating point of the modulator on the linear portion of its transfer characteristic. To solve this task, we proposed a new technique for automatically adjusting the position of the operating point in the linear section of the modulator transfer characteristic. On its basis, a new scheme for monitoring the operating point has been developed (Fig. 3).

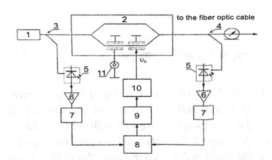

Fig. 3. The structural scheme of modulator operating point control: 1 – transmitting laser module; 2 – electro-optical modulator; 3, 4 – optical divider; 5 – photodetector; 6 – operational amplifier; 7 – low frequency filter; 8 – adder; 9 – proportional regulator; 10 – zone switch; 11 – input of high frequency signal.

Through the optical divider (points 3 and 4) a part of input and output radiation arrives at the control photodiodes. The output voltages from the photodiodes through amplifier 6 and low frequency filter 7 are transmitted to the adder inputs 8. At its output the error signal is formed (the operating point shift from the given value). The error signal is fed to the proportional regulator 9 and integral part of that allows to consider the history of changes output value. In the absence of temperature fluctuations the constant bias at the output of the switch 10 is stabilized. The integral part of the regulator 9 provides a constant bias voltage U_b.

3 Results and Discussion

In order to determine the application possibilities of the constructed FOTS in radar stations were investigated the following characteristics: AFC, dynamic characteristic, the loss coefficient at the frequency of the transmitted analog signal on the environment temperature T change. As an example the AFC of the output signal at the different T is presented in the Fig. 4.

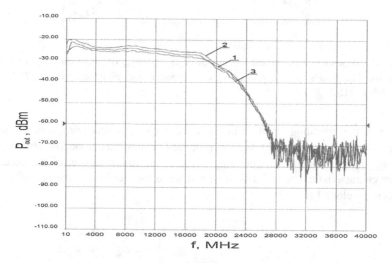

Fig. 4. AFC, graphs 1, 2 and 3 corresponds to T in K: 231.2; 293.1; 333.4.

The received results show that the developed by us device of electro-optical modulator 2 operating points and the scheme of the thermal stabilization for the transmitting laser module 1 allows to make insignificant the influence of temperature T change (over the whole range of the radar station operation) upon the value of optical signal power loss while the signal is moving over the FOTS. It gives a possibility to transmit IF signals along it without loss and distortions.

Moreover, the developed circuit for control the working point of the electrooptical modulator 4 and the thermostabilization scheme of the transmitting laser module 1 allow to make an impact of the changes T (over the entire range of radar operation) by the magnitude of the optical signal power loss insignificant when FOTS is using. Thus, transmission of IF signals can be done without losses and distortions.

As an example, in the Fig. 5 are presented the IF signals in the form of rectangular impulses transmitted to the heterodyne input from the master generator by developed FOTS and coaxial cable.

The received results show that there are no distortions in the forms of the impulses in the IF signal which is supplied to the heterodyne 7 input from the optical detector 9 output. The conducted experiments showed that distortions in the coaxial cable significantly change the shape of the pulses of the IF signal. This signal cannot be used in the heterodyne. For stable operation of the heterodyne, this signal must be further

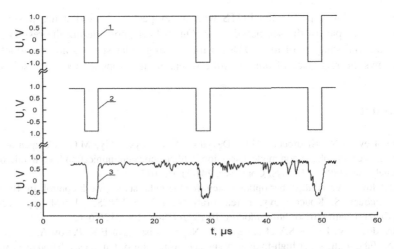

Fig. 5. The form of the pulses of the output signal: 1 – generator 3; 2 – at the exit FOTS (input oscilloscope 9); 3 – at the exit a coaxial cable.

processed in order to eliminate distortions. This creates additional problems, especially for radars located on moving objects. Elimination of these problems is an indisputable advantage of the use of FOTS in comparison with a coaxial cable.

4   Conclusion

The obtained results show that the use of developed FOTS allow to make insignificant the effect of the negative factors which are occurred when the IF signal is transmitted via a coaxial cable to the results of the work of the heterodyne.

The conducted researches allowed to establish new requirements for the parameters of FOTS:

1. Unevenness of amplitude frequency characteristic FOTS in the range of the transferred frequencies has to be less than ±3 dB.
2. Change of coefficient of losses in FOTS from the frequency of the transmitted microwave signal shouldn't exceed ±2 dB
3. The phase deviation of modulation at the distribution of a signal on FOTS from a temperature in the range from 238 to 323 K shouldn't exceed 3°.

If the requirements are not met, it will be difficult to ensure reliable transmission of IF signals in the radar. The presented results can be interesting for the developers of the FOTS, which provide the solution of this problem.

It should also be noted that the proposed FOPS construction allows to successfully use it in different types of radars in the entire frequency range of IF signals. This is, on the one hand, a new result. On the other hand, it is possible to widely apply the proposed design of the FOTS, since there are a lot of different radar station in operation.

Currently, the experimental FOTS with developed by us design has passed successful tests as part of the sea-based radar. On its basis, considering the results of the tests, industrial samples of the FOTS for the existing radar stations are manufactured. This shows the relevance of our scientific research and proposed technical solutions.

References

1. Ermolaev, A.N., Krishpents, G.P., Davydov, V.V., Vysoczkiy, M.G.: Compensation of chromatic and polarization mode dispersion in fiber-optic communication lines in microwave signals transmittion. J. Phys. Conf. Ser. **741**(1), 012071 (2016)
2. Davydov, V.V., et al.: Fiber-optics system for the radar station work control. In: Balandin, S., Andreev, S., Koucheryavy, Y. (eds.) ruSMART 2015. LNCS, vol. 9247, pp. 712–721. Springer, Cham (2015). https://doi.org/10.1007/978-3-319-23126-6_65
3. Davydov, V.V., Ermak, S.V., Karseev, A.U., Nepomnyashchaya, E.K., Petrov, A.A., Velichko, E.N.: Fiber-optic super-high-frequency signal transmission system for sea-based radar station. In: Balandin, S., Andreev, S., Koucheryavy, Y. (eds.) NEW2AN 2014. LNCS, vol. 8638, pp. 694–702. Springer, Cham (2014). https://doi.org/10.1007/978-3-319-10353-2_65
4. O'Mahony, M.J.: Future optical networks. IEEE OSA Journal of Light wave Technology. **24** (12), 4684–4696 (2006)
5. Agrawal, G.P.: Light Wave Technology: Telecommunication Systems. Wiley, Hoboken (2014). 480 p.
6. Tarasenko, M.Y., Davydov, V.V., Lenets, V.A., Akulich, N.V., Yalunina, T.R.: Features of use direct and external modulation in fiber optical simulators of a false target for testing radar station. In: Galinina, O., Andreev, S., Balandin, S., Koucheryavy, Y. (eds.) NEW2AN/ruSMART/NsCC -2017. LNCS, vol. 10531, pp. 227–232. Springer, Cham (2017). https://doi.org/10.1007/978-3-319-67380-6_21
7. Davydov, R.V., et al.: Fiber-optic transmission system for the testing of active phased antenna arrays in an anechoic chamber. In: Galinina, O., Andreev, S., Balandin, S., Koucheryavy, Y. (eds.) NEW2AN/ruSMART/NsCC -2017. LNCS, vol. 10531, pp. 177–183. Springer, Cham (2017). https://doi.org/10.1007/978-3-319-67380-6_16
8. Ameri, H., Attaran, A., Moghavvemi, M.: Design an X-band frequency synthesizer. Microwaves RF **79**, 98–103 (2010)
9. Bystrov, V.V., Likhachev, V.P., Ryazantsev, L.B.: Experimental check of the coherence of radiolocation signals from objects with nonlinear electrical properties. Meas. Tech. **57**(9), 1073–1076 (2014)
10. Petrov, A.A., Davydov, V.V.: Improvement frequency stability of caesium atomic clock for satellite communication system. In: Balandin, S., Andreev, S., Koucheryavy, Y. (eds.) ruSMART 2015. LNCS, vol. 9247, pp. 739–744. Springer, Cham (2015). https://doi.org/10.1007/978-3-319-23126-6_68
11. Friman, R.K.: Fiber-Optic Communication Systems. Wiley, Hoboken (2012). 496 p.
12. Marpaung, D., Roeloffzen, C., Heideman, R., Leinse, A., Sales, S., Capmany, J.: Experimental realization and application in a unidirectional ring mode-locked laser diode. Integr. Microw. Photonics (Laser Photonics Rev.) **4**(3), 506–511 (2013)

Fiber Optic Current Meter for IIoT in Power Grid

Valentina Temkina[1], Andrey Medvedev[1(✉)],
Alexey Mayzel[1], and Alexander Mokeev[2]

[1] Peter the Great St. Petersburg Polytechnic University (SPbPU),
Polytechnicheskaya, 29, St. Petersburg 195251, Russia
temkina.vs@edu.spbstu.ru, medvedev@rphf.spbstu.ru,
amayzel@gmail.com
[2] OOO 'Electro', Nauki pr. 17/6, St. Petersburg 195220, Russia
mokeeff@mail.ru

Abstract. The article is focused on the fiber optic current meter with FPGA-based data processing. The sensor element operation is based on the Faraday effect which occurs in the Spun optical fiber twisted around electric wire. The demodulation method used to process the sensor signal is the joint measuring of the first and second harmonics of the modulation signal and the processing of their ratio. It is widely known for fiber optic gyroscopes and traditionally is performed in the analog form, fully or partially. This leads to errors related to instability of the analog elements, limited operating range of the device etc. We managed to combine the optical sensor circuit with NI technologies to develop and implement digital data processing algorithms, exclude sources of errors from the optical device circuit. We used a scalable NI RIO architecture, LabVIEW FPGA and LabVIEW Real-Time to implement every step in optical sensing of electric current: modulation forming, response measuring, inline processing and analog/IEC 61850 output providing, comparing performance of high-end and low-cost RIO devices.

Keywords: Fiber optic current meter · Faraday effect · Fiber optic sensor
Spun fiber

1 Introduction

There is a century-old approach to measure high current values (from hundreds of amperes to hundreds of thousands of amperes) by transforming it to the appropriate for metrology equipment levels using inductive coupling. The devices traditionally used for this purpose are called measuring (or primary) transformers. This solution worked just fine for pre-digital era, but even then, it had some well-known significant limitations in measuring transient currents, especially in emergency shortage or significant overload, which caused protection malfunction or false triggering. After moving the power grid automation and protection systems to all-digital base, it appeared that a use of secondary transducers is needed to convert an output of measuring transformers to ADC input levels.

© Springer Nature Switzerland AG 2018
O. Galinina et al. (Eds.): NEW2AN 2018/ruSMART 2018, LNCS 11118, pp. 631–640, 2018.
https://doi.org/10.1007/978-3-030-01168-0_57

It was an industry demand to build a solution that performs electrical measurements using a physical process other than inductive transformation. In early 2000s, the pioneer in this field NXT Phase delivered the first commercial optical current transformer. Later there were several enhancements of this technology made by market players, bringing the optical current sensing to its modern look. Nowadays these are complex devices, ready to work in both analog environment and such-called digital substations, operating in IEC 61850 standard.

Despite a couple of decades of evolution, the technology is not flawless. One of the main challenges here is cost reduction for the product. One of the most expensive components of the optical current transformer is an optical delay line, which is a long (several hundreds of meters) polarization maintaining (PM) optic fiber. The exclusion of this delay line could make optical sensing far more competitive, but requires deep changes in physical approach, optical scheme and digital processing algorithms.

Another important challenge that we must solve is to improve the algorithm of data processing. Traditionally, signal processing is performed fully or partially in the analog form, but this leads to errors related to instability of analog elements, limitation of the operating range of the device etc. Therefore, in order to eliminate undesirable effects and improve the accuracy of the sensor, we realized the demodulation method completely in digital form.

We at Peter the Great Polytechnic University have made a theoretical study of the design of such solution and have used National Instruments (NI) technologies to implement it in the hardware.

2 Current Meter Optical Scheme

2.1 Description

The scheme of the current sensor is shown in Fig. 1.

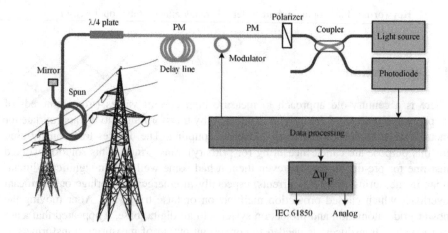

Fig. 1. Scheme of fiber-optic current meter.

The main components of the optical scheme of the meter are:

- light source;
- phase modulator made of a PM optical fiber coiled on a PZT cylinder;
- fiber delay line;
- fiber quarter wave plate;
- sensing element made of the Spun optical fiber;
- mirror;
- polarizer;
- photodiode;
- data processing unit.

The meter uses additional harmonic phase modulation:

$$\Delta \psi_M(t) = \psi_m \cdot \cos(2\pi f_M t) \tag{1}$$

where the modulation frequency $f_M = 40.1$ kHz corresponds to the piezoceramics resonant frequency. In addition, the meter uses digital phase detection, constructed according to the scheme of analysis of the ratio of harmonics of the modulation frequency [1].

2.2 Circuit Operation Principle

The optical wave passes through the fiber optic coupler, the polarizer and goes to the phase modulator. The polarizer axis is at 45° to the own polarization axes of the PM fiber of the modulator. Optical radiation propagates through the modulator in the form of two light waves with orthogonal linear polarizations. Then the light passes through the delay line whose length is determined by the frequency of the modulation signal. It provides such a lag between the light waves propagating in the forward and backward directions so that during the propagation of the wave from the phase modulator to the mirror and back, the phase of the modulating voltage changes by 180°. The phase plate, whose axes are oriented at the angle of 45° to the own axes of the supply path, converts linearly polarized modes to the circularly polarized modes. They propagate at different speeds in the sensitive Spun fiber, which is coiled around the wire with the current to be measured. In this case, a phase shift proportional to the current in the wire arises between them:

$$\Delta \psi_F = 2VNI \tag{2}$$

where V is the Verde constant for the fiber, N is the number of fiber turns around the conductor with current I [2].

The light wave, reflected from the mirror, propagates in the opposite direction. With a repeated pass of the phase plate, the circularly polarized modes are again transformed into linearly polarized modes. As a result, only the Faraday phase shift, connected with the measured electric current, is conserved. Linearly polarized modes, passing through the polarizer, interfere, and then the interference signal through the splitter goes to the photodiode.

3   Signal Demodulation Method

The optical power of the signal at the input of the photodiode is the sum of the harmonics of the modulation frequency:

$$
\begin{aligned}
P(\Delta\psi_F, t) &= P_0[1 + \cos(\Delta\psi_F + \psi_m \cos(2\pi f_M t))] \\
&= P_0[1 + \cos(\Delta\psi_F)\cos(\psi_m \cos(2\pi f_M t)) - \sin(\Delta\psi_F)\sin(\psi_m \cos(2\pi f_M t))]
\end{aligned}
\tag{3}
$$

Using expansion in terms of Bessel functions gives

$$
\begin{aligned}
P(\Delta\psi_F, t) &= P_0 + P_0\cos(\Delta\psi_F) \\
&\cdot [J_0(\psi_m) + 2J_2(\psi_m)\cos(4\pi f_M t) + 2J_4(\psi_m)\cos(8\pi f_M t) + \ldots] \\
&+ P_0\sin(\Delta\psi_F)[2J_1(\psi_m)\sin(2\pi f_M t) + 2J_3(\psi_m)\sin(6\pi f_M t) + \ldots]
\end{aligned}
\tag{4}
$$

wherein $J_k(\psi_m)$ are Bessel functions of the first kind of the k-th order.

It can be seen from Eq. (4) that the amplitudes of the harmonics depend on the optical power P_0 and on the values of the Bessel functions. Even harmonics are proportional to $\cos(\Delta\psi_F)$, and odd ones are proportional to $\sin(\Delta\psi_F)$. In the state of rest, the detected signal consists of even harmonics of the modulation frequency, and in the presence of a magnetic field created by an electric current, odd harmonics of the modulation frequency also appear.

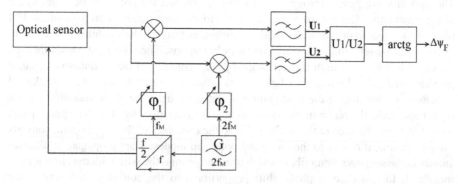

Fig. 2. Block diagram of signal demodulation method

The demodulation method used to process the sensor signal and is widely known in fiber optic gyroscopes is the joint recording of the first and second harmonics of the modulation frequency and the analysis of their ratio (see Fig. 2).

We used synchronous detection of the photodiode's output voltage at frequencies f_M and $2f_M$ to obtain two quadrature signals:

$$
U_1 \sim \sin(\Delta\psi_F) \cdot J_1(\psi_m) \cdot \cos(\Delta\varphi_1)
\tag{5}
$$

$$U_2 \sim \cos(\Delta\psi_F) \cdot J_2(\psi_m) \cdot \cos(\Delta\varphi_2) \tag{6}$$

where $\Delta\varphi_1$ and $\Delta\varphi_2$ – phase shifts between the corresponding reference signals and the detected frequency components. We achieved $\cos(\Delta\varphi_1) = \cos(\Delta\varphi_2) = 1$ by adjusting the phase shifters. The amplitude of the phase modulation should be chosen so that Bessel functions $J_1(\psi_m)$ and $J_2(\psi_m)$ are equal and maximum at the same time.

As a result, the output signal of the fiber optic current sensor is given by the following formula:

$$\Delta\psi_F = \text{arctg}\,\frac{U_1}{U_2} = \text{arctg}\left[\frac{J_1(\psi_m)}{J_2(\psi_m)}\,\text{tg}(\Delta\psi_F)\right] \tag{7}$$

4 Compensation of the Result Dependence on the Modulation Amplitude

It is clear from the Eq. (7) that the amplitude of the phase modulation affects the result of the demodulation. Therefore, it is necessary (a) to select optimal phase modulation amplitude and (b) to compensate demodulation errors occurring due uncontrolled amplitude variation.

Optimal amplitude was chosen based on the following two criteria:

- the first U_1 and second U_2 harmonics of the modulation frequency must be maximal and equal;
- optimal amplitude should correspond to minimum errors arising due amplitude variation.

We choose optimal amplitude $\psi_{m0} = 2.629874$ radians that corresponds to the minimum amplitude value giving $J_1(\psi_{m0}) = J_2(\psi_{m0})$.

If the modulation amplitude deviates from the optimal operating value, the ratio of the Bessel functions in Eq. (7) becomes different from 1. Hence, an error arises when calculating the arctangent. Therefore, we introduced compensation, thanks to which Bessel functions ratio in Eq. (7) is maintained equal to one in a certain range of amplitudes. In order to take into account the effect of the deviation of the modulation amplitude, it is convenient to use the ratio of the second and fourth harmonics of the photodiode's output signal

$$\frac{U_2}{U_4} = J_2(\psi_m)/J_4(\psi_m) \tag{8}$$

The Eq. (7) with compensation has the following form

$$\Delta\psi_F = arctg\left[\frac{U_1}{U_2}\cdot\left\{1+k_1\left(\frac{U_2}{U_4}-5.30176\right)+k_2\left(\frac{U_2}{U_4}-5.30176\right)^2+k_3\left(\frac{U_2}{U_4}-5.30176\right)^3\right\}\right]$$

$$= arctg\left[\frac{J_1(\psi_m)}{J_2(\psi_m)}\cdot g(\psi_m)\cdot tg(\Delta\psi_F)\right],$$

(9)

where

$$g(\psi_m) = 1+k_1\left(\frac{J_2(\psi_m)}{J_4(\psi_m)}-5.30176\right)+k_2\left(\frac{J_2(\psi_m)}{J_4(\psi_m)}-5.30176\right)^2+k_3\left(\frac{J_2(\psi_m)}{J_4(\psi_m)}-5.30176\right)^3$$

(10)

is the error correction function and $\frac{J_2(\psi_{m0})}{J_4(\psi_{m0})} = 5.30176$ for optimal phase modulation amplitude.

Modified block diagram of signal demodulation looks like in Fig. 3.

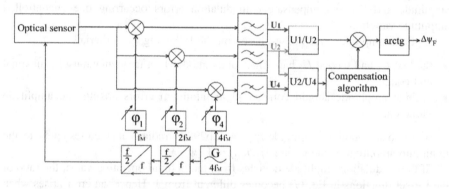

Fig. 3. Block diagram of signal demodulation with compensation

5 Implementation

There were three major cycles of development of our solution: PC-based modeling, first prototype and industrial prototype. From the very beginning our team started modeling the processing using LabVIEW graphical tools. First step model was developed in floating-point math with both calculated and measured signal waveforms. It showed the predicted results of basic demodulation and demodulation with compensation.

The modeling code was built with consideration of future conversion to FPGA-applicable integer and fixed-point math. Thus, there was no need to write completely new code for implementation in a prototype (as shown in Fig. 4).

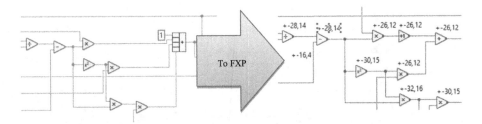

Fig. 4. An example of code transformation from host-based floating-point (left) to FPGA fixed-point (right). The code shown implements the compensation.

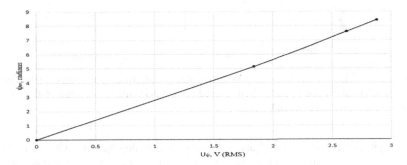

Fig. 5. Calibration of the phase modulator

When modeling in the LabVIEW program environment, the coefficients k_1, k_2 and k_3 were chosen so that the function $g(\psi_m)$ in Eq. (9) was as close as possible to one in the neighborhood of optimal amplitude ψ_{m0}. Figure 6 presents simulation results without compensation, floating-point compensation model, fixed-point FPGA implementation, as well as experimental data. Thus, as a result of compensation, within the range of amplitudes of $\pm6\%$ from the optimal amplitude ψ_{m0} the error was within the limits of $\pm0.1\%$ with the coefficients chosen in the program. Less variations of ψ_m gives smaller values of errors up to $\pm0.0001\%$.

Before starting the experiments, we calibrated phase modulator used in the sensor. Calibration was carried out in absence of any measured signal by measuring modulation amplitudes corresponding to zero values of Bessel functions of the first kind of even orders. As a result, a calibration curve was obtained for a modulator fed at resonance frequency (see Fig. 5). From this curve follows that the optimal modulation voltage corresponding to $\psi_{m0} = 2.629874$ is $U_{\psi 0} = 0.9163$ V.

The first prototype was implemented on entry-level hardware unit of National Instruments RIO platform named myRIO [3]. It's a low-cost device designed for educational purposes with limited IO lines and FPGA resources. With all these limitations this hardware allowed our team to build a functional prototype with fully synchronous inline FPGA processing, Real-Time OS supervision and control and desktop GUI. We managed to run verification tests on our theoretical research and selected system architecture.

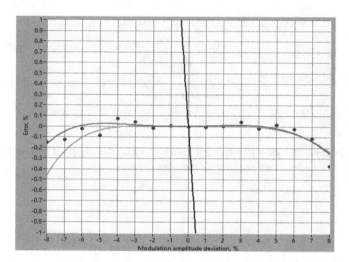

Fig. 6. Error RMS vs Modulation amplitude deviation graph. Black – uncompensated. Green – floating-point compensation model. Red – fixed-point FPGA implementation. Blue – experimental data. (Color figure online)

NI myRIO is equipped with 4 analog inputs and 2 analog outputs, which allowed us to generate a modulating signal, register the signal of the photodetector and generate an analog signal for feeding the comparison instrument. However, an inexpensive, compact controller designed primarily for educational purposes can generate an output signal with a sampling frequency of a maximum of 345 kHz and a resolution of 12 bits, that is, 8 points for a modulation cycle of 40 kHz and 4 points for a second harmonic of the modulation frequency. In this regard, the maximum accuracy that was achieved by the comparison instrument was about 1% in amplitude and half degree in phase. To increase the sampling frequency and bit capacity, external ADCs and DACs were used together with myRIO, which also increased the accuracy of the sensor analog output readings.

The front panel of the signal-processing program is shown in Fig. 7. The graphs at the upper part of the panel show the first (blue) and second (red) harmonics of the modulation frequency, as well as the photodiode's signal (green). On the bottom of the panel left graph indicates the error correction function $g(\psi_m) - 1$, zero value corresponding to phase modulation amplitude equal to ψ_{m0} and $g(\psi_m) = 1$. The analog signal used for measuring comparator instrument feeding is shown at the bottom right side of the panel.

It was an easiest transition between alpha prototype and industrial prototype because of keeping NI RIO platform. The NI System-On-Module (SOM) sbRIO-9651 was selected. NI SOM is an example of the same FPGA-based RIO architecture [4]. This allowed us to migrate code completely from alpha prototype, changing only IO numbers. The bigger FPGA and larger number of DIO lines opened the possibilities to achieve a faster timing, finer tuning and wider process control. The front panel of the signal-processing program using NI SOM are presented in Fig. 8.

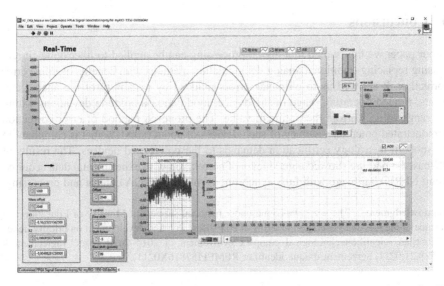

Fig. 7. The front panel of the signal-processing program for the current meter when using NI myRIO with external ADCs and DACs (Color figure online)

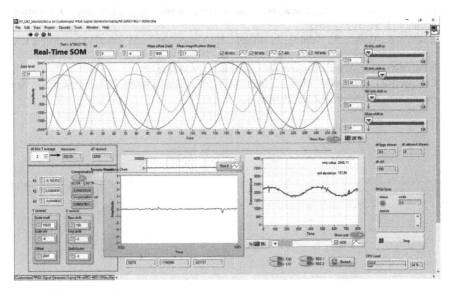

Fig. 8. The front panel of the signal-processing program for the current meter when using NI SOM.

In electrical metrology only specially designed high precision analog measuring comparator instruments are certified to calibrate sensors. So we used output digit to analog converter (DAC) to feed this comparator instrument and NI ready to use Toolkit to obtain standard IEC 61850 signal for future applications.

6 Conclusions

In our paper we had shown that digital processing can be applied to fiber optic current sensing replacing traditional analog demodulation methods with benefits of inline error correction. It allowed us to eliminate errors induced by instability of the analog elements, limited operating range of the device, modulation amplitude deviations within the range of $\pm 2\%$ from the optimum value. The proposed method of error correction demonstrated potential accuracy of such devices of about 0.0001% that is compliant to the modern metrology certification demands to the power grid devices. The results of PC-based modeling showed a good level of correlation with experimental data acquired using NI RIO hardware with fully synchronous onboard FPGA control and processing.

Acknowledgments. The work was done under financial support of Ministry of Education and Science of the Russian Federation in terms of FTP "Research and development on priority trends of Russian scientific-technological complex evolvement in 2014–2020 years" (agreement # 14.578.21.0211, agreement unique identifier RFMEFI57816X0211).

References

1. Berg, R.A., Lefevre, H.C., Shaw, H.J.: An overview of fiber-optic gyroscopes. J. Lightwave Technol. **2**, 91–107 (1984)
2. Bohnert, K., Gabus, P., Nehring, J., Brandle, H.: Temperature and vibration insensitive fiber-optic current sensor. J. Lightwave Technol. **20**, 267 (2002)
3. NI myRIO. http://www.ni.com/myrio/. Accessed 29 May 2018
4. CompactRIO System on Module. http://www.ni.com/en-us/shop/select/compactrio-system-on-module. Accessed 29 May 2018

Some Directions of Quantum Frequency Standard Modernization for Telecommunication Systems

Alexander A. Petrov[1,2(✉)], Vadim V. Davydov[1,3,4],
and Nadya M. Grebenikova[1]

[1] Peter the Great Saint-Petersburg Polytechnic University,
St. Petersburg, Russia
Alexandrpetrov.spb@yandex.ru
[2] Russian Institute of Radionavigation and Time, St. Petersburg, Russia
[3] The Bonch-Bruevich Saint - Petersburg State University
of Telecommunications, St. Petersburg, Russia
[4] Department of Ecology, All-Russian Research Institute of Phytopathology,
B. Vyazyomy, Odintsovo District, Moscow Region, Russia

Abstract. Quantum frequency standard is one of the essential elements of many telecommunication and satellite systems. In present work several directions of modernization of the quantum frequency standard on the atoms of cesium-133 are considered. Implementation a method of direct digital synthesis is allow to improve output signal characteristics of frequency synthesizer. Also it is allow developing a magnetic field stabilization system. Experimental research of the cesium atomic clock's metrological characteristics showed improvement Allan variance on 10%.

Keywords: Quantum frequency standard · Cesium atomic clock
Frequency synthesizer · Direct digital synthesis
Magnetic field stabilization system · Allan variance

1 Introduction

The accurate measurement of time and frequency is vital to the success of many fields of science and technology. For example atomic physics (atom-photon interactions, atomic collisions, and atomic interactions with static and dynamic electromagnetic fields), geodesy, radio-astronomy (very long baseline interferometry), pulsar astronomy, various communication equipment and metrological services rely on high frequency stability and uniform timescales [1–7].

Quantum frequency standard or atomic clock is a device that measures time by the frequency of radiation emitted by an atom or molecule when it makes a transition between two energy states. Atomic clocks are extremely precise and are used to keep universal time—the international basis for establishing legal and scientific times and for setting all public and private clocks worldwide [1–4, 7–9].

Atomic clock uses include measuring the rotation of the earth, which may vary by 4 to 5 ms per day, and aiding satellite navigational systems such as the global positioning

© Springer Nature Switzerland AG 2018
O. Galinina et al. (Eds.): NEW2AN 2018/ruSMART 2018, LNCS 11118, pp. 641–648, 2018.
https://doi.org/10.1007/978-3-030-01168-0_58

system in computing distances and synchronize the work of many sophisticated devices and instruments [6–12].

A necessary condition for reliable operation of communication systems is coordinated work of the primary generator and receiver, so that the receiving unit can correctly interpret the digital signal. The difference in the synchronization of various units in a network, may lead to missing or to reread the information by receiving unit.

In order to solve this problem is using quantum frequency standards.

But with the development of scientific - technical progress and large amount of data transmitted over networks, new requirements for measurement accuracy, reliability and frequency stability are produced to frequency standard [2, 4, 6, 10]. This leads constantly upgrade existing and develop new models.

The process of quantum frequency standards modernization includes various directions: reducing the weight and dimensions, reducing energy consumption, improvement metrological characteristics. And for frequency standards characterized by the fact that modernization may not be for the whole construction and may be for individual units or blocks. In present work development of new frequency synthesizer and magnetic field stabilization system are presented.

2   Principles of Cesium Atomic Clock Operation

The work of a cesium atomic clock is based on the principle of adjustment a highly stable voltage-controlled quartz crystal oscillator (VCXO) to quantum frequency transition of atoms of caesium-133 [1, 8, 10, 11]. The Fig. 1 shows a block diagram of a cesium atomic clock.

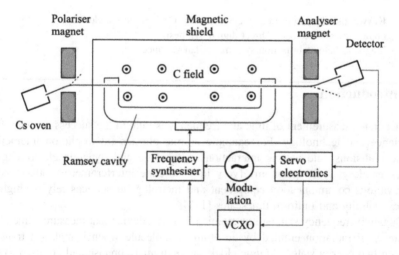

Fig. 1. Block diagram of a caesium atomic clock.

The output signal frequency 5 MHz of the VCXO is supplied to the frequency synthesiser. Frequency synthesiser consists of frequency converter, mixer signals and multiplier signals. In the frequency converter input signal frequency 5 MHz is converted to the signal frequency 12,631772 MHz and supplied to the input of mixer signals. In the multiplier signals input signal frequency 5 MHz is multiplied to the frequency 270 MHz and then to frequency 9180 MHz. This signal frequency 9180 MHz is also supplied to the input of mixer signals. As a result, the output signal of the frequency synthesiser is the signal of ultrahigh frequency 9192,631772 MHz. This signal is supplied in the Ramsey cavity.

In caesium atomic clock with the help of magnet polarizer 2 the atoms are prepared such that they are either in the $|F = 4, m_F = 0>$ or in $|F = 3, m_F = 0>$ state. Afterwards the atoms interact with an electromagnetic field that induces transitions into the former unoccupied state.

A magnetic field is used to separate energetically the otherwise degenerate magnetic sub-levels in order to allow the excitation of the clock transition $|F = 3, m_F = 0> \rightarrow |F = 4, m_F = 0>$ isolated from the other transitions. By convention such a field is referred to as the C-field as it is applied between the fields of the polarizer and the analyzer.

The magnitude of the C-field is chosen as a compromise between two conflicting requirements. First, it has to be large enough to separate the otherwise overlapping resonances. Second, the C-field shifts the resonance frequency quadratically which has to be corrected. However, in a larger field the frequency of the clock is influenced to a larger extent by fluctuations of the magnetic field. In the scheme of a commercial Cs clock the C-field is often generated by a coil with windings in the paper plane wound around the Ramsey resonator and hence, points perpendicularly to that plane. Owing to the dependence of the frequency of the clock transition from the magnetic field, efficient magnetic shielding has to be provided in order to attenuate the ambient magnetic field and the magnitude of the associated fluctuations.

The atoms in the former unoccupied state are detected and allow one to determine the frequency of the interrogating field where the transition probability has a maximum. The observed transition frequency is corrected for all known frequency offsets that would shift the transition frequency from the unperturbed transition and is used to produce a standard frequency or pulse per second every 9192631772 cycles [1, 8, 10].

Scanning the frequency v of the atomic resonance leads to a detector current like the one shown on the Fig. 2. The signal shows the Ramsey resonance structure on a broader, so-called, Rabi pedestal.

The central feature with the maximum at the transition frequency v_0 is used to stabilize the frequency of the crystal oscillator to the atomic transition frequency. To this end, the frequency from the synthesizer is modulated across the central peak. The signal from the detector is phase-sensitively detected in the servo electronics, integrated and this servo signal is used for stabilizing the frequency of the VCXO. From this suitable output frequencies are derived, such as 5 MHz or a 1 PPS signal.

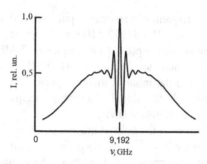

Fig. 2. Ramsey resonance structure on the Rabi pedestal.

3 Cesium Atomic Clock Frequency Synthesizer

Frequency synthesizer is one of the main blocks of quantum frequency standard. Frequency synthesizer takes a part of generating the microwave signal at ~ 9.2 GHz (used to interrogate the ^{133}Cs atoms hyperfine resonance transition) from the 5 MHz quartz oscillator frequency [1, 8, 10].

The main characteristic of the frequency synthesizer is ability to impact on the characteristic of frequency stability of the quantum frequency standard output signal. Frequency instability introduced by the synthesizer is determined by the lateral discrete spectrum components of the signal that occurs in dividing, multiplying, mixing frequency signals, the accuracy of the generated frequency, and the impact on the signal of natural and technical noise [5–7].

In order to provide the best possible frequency stability, it is crucial that the microwave signal which interrogates the ^{133}Cs atoms be as "clean" as possible; that is, free of unwanted sidebands and spurious signals which can cause Bloch-Siegert frequency shifts [1, 5].

Experimental study showed that the present method of generating the frequency synthesizer output signal needs to increase the accuracy. The large resolution of step frequency is necessary. New scheme of the frequency synthesizer is designed by using method of direct digital synthesis (DDS - Direct Digital Synthesis). This method allows to generate the output signal of the synthesizer with accuracy about 10^{-5} Hz. Step frequency tuning ΔF_{out} calculated by the formula below:

$$\Delta F_{out} = \frac{F_{clk}}{2^N}, \tag{1}$$

where F_{clk} is the clock frequency, N is the capacity of accumulator.

In our scheme the clock frequency is equal $F_{clk} = 15$ MHz, the capacity of accumulator is equal N = 40. Step frequency tuning is equal $\Delta F_{out} = 1,36 \cdot 10^{-5}$ Hz.

When we calculate step frequency tuning relatively resonant frequency of the microwave transitions of rubidium atoms, using parity (1), we get relative step frequency tuning ΔF_{outrel}:

$$\Delta F_{outrel} = \frac{\Delta F_{out}}{6.834\,\text{GHz}} = \frac{1.36^{-5}\,\text{Hz}}{9.192\,\text{GHz}} = 1.48 * 10^{-15} \tag{2}$$

The application of direct digital synthesis gave the possibility of obtaining the generated frequencies in a wide range (0–5 MHz), in contrast to previous schemes, where this feature was absent. This feature gets a possible to develop a magnetic field stabilization system. Range of generated output frequencies may be calculated by the formula below:

$$F_{out} = \frac{M * F_{clk}}{2^N}, \tag{3}$$

where M is the frequency code in decimal, F_{clk} is the clock frequency, N is the capacity of accumulator.

To meet the requirements for spectral purity of output signal 10 bit DAC was used. It is possible to obtain the suppression of lateral amplitude components in the spectrum of the output signal is not worse than −90 dB.

In Fig. 3 as an example, oscillograms measured in the band of 6 kHz of the output signal of a previously used design (a) and a new (b) of the frequency synthesizer are presented.

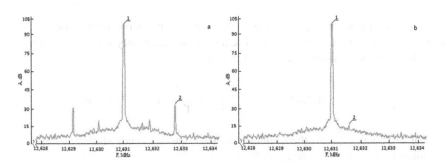

Fig. 3. Suppression of the lateral components in the band of 6 kHz.

The experimental results show that the suppression of lateral components in the spectrum of microwave-excitation signal in the band of 6 kHz is improved on 24 dB.

With a decrease of lateral components more fine-tuning on the center of the resonance line is occur. This leads to a more accurate determination of the value of the nominal output frequency of frequency standard and, consequently, improves the short-term frequency stability of a cesium atomic clock.

In addition new design of the frequency synthesizer allows eliminating one of the most important perturbing factors affecting on long-term frequency stability.

4 Magnetic Field Stabilization System

The stable isotope Cs-133 has a two hyperfine states F = 4 and F = 3 which are split in the magnetic field into 16 components. In accordance with the selection rules seven transitions between the components of hyperfine sublevels are possible [10, 11]. These are represented in Fig. 4.

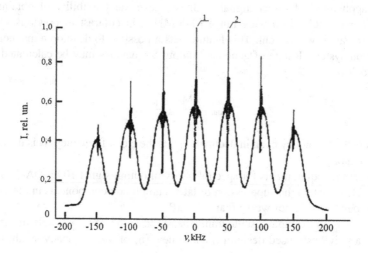

Fig. 4. Microwave resonances in the Ramsey cavity.

The central resonance $|F = 3, m_F = 0> \rightarrow |F = 4, m_F = 0>$ (marked on the Fig. 4 as a «1») due to the Zeeman effect expose a quadratic frequency shift. With the help of formula (1) we can calculate a frequency shift.

$$\Delta f_{B^2} \approx 4.2745 * 10^{-2} \, Hz * \left(\frac{6 * 10^{-6} T}{uT}\right)^2 = 1,5388 \, Hz \qquad (4)$$

For a typical value of the C field near 8 µT the frequency shift is 2.7 Hz corresponding to a relative frequency shift of $3 \cdot 10^{-10}$.

The accuracy of the output signal quantum frequency standard is dependent on the shift of the central resonance. It should be noted that not only the central resonance is exposed the frequency shift, but also all six transitions $(3, m_F) \leftrightarrow (4, m_F)$, which $\Delta m_F = 0$. To express these changes as a function of magnetic induction B and atomic constants use the equation Bright Rabi:

$$E(F, m_F) = -\frac{h\nu}{2(2I+1)} - g_I \mu_B B m_F + \varepsilon \frac{h\nu}{2} \left(1 + \frac{4m_F}{2I+1} x + x^2\right)^{\frac{1}{2}}, \qquad (5)$$

where $E(F, m_F)$ is the change energy of atoms in the ground state; I is the quantum number of nuclear spin; g_I is the factor Lande for electron; μ_B is the Bohr magneton; m_F is the magnetic quantum number; $x = \dfrac{(g_j + g_I)\mu_B B}{h\nu}$.

This formula can be used for calculation the frequency shift of any transition between two hyperfine sublevels, depending on the magnetic field. Revealing this expression, we find that the first member is proportional to the magnetic field B. For cesium beam primary frequency standards we must consider the quadratic member of this expression.

In theory, the frequency shifts can be taken into account in the calculation of the functional dependence on magnetic field values and the atomic constants using equation Bright-Rabi. But in practice, any changes of the magnetic field shift the resonance frequency. And values of these frequency shifts cannot be accounted for in advance.

Thanks to the development of a new frequency synthesizer [8, 10, 11] the range of generated output frequencies has been expanded. It allowed detuning output synthesizer's frequency to the neighboring resonance frequency of spectral line that makes it possible to adjust the C-field in quantum frequency standard.

Now in cesium atomic clock the magnetic field is maintained by the active stabilization system. For this purpose the neighboring transition $|F = 3, m_F = 1> \leftrightarrow |F = 4, m_F = 1>$ (marked on the Fig. 4 as a «2») is used. The method of C-field adjustment is similar to the method of frequency adjustment to the main maximum. For this purpose, the average value of the sampling frequency v_i is changed from the value of the v_{Cs} to $v_{Cs} + \Delta v$, where v_{Cs} is the frequency of the main transition of the cesium atom, Δv - difference between the transitions for a preset value of the magnetic field. Then the value of magnetic field is adjusted in a such way that the frequency of transition $|F = 3, m_F = 1> \leftrightarrow |F = 4, m_F = 1>$ match the preset frequency. This adjustment set up automatically several times per minute. The value of the applied field is automatically maintained at a predetermined level.

Alternately closing the ring-locked loop at the central and the neighboring transition we adjust the frequency of the VCXO to the frequency of the central atomic transition, and support the constant value of the magnetic field inside the Ramsey cavity.

In this case effects associated with any changes in the magnetic field (for example, long-term drift of the current source, temperature dependence, effect of external magnetic field, etc.) are excluded.

5   Conclusion

The use of new design of the frequency synthesizer and system for stabilizing magnetic field makes it possible to obtain a better frequency stability of quantum frequency standard. A pure spectrum of the frequency synthesizer output signal and best resolution of frequency step improved short-term frequency stability. System for stabilizing magnetic field eliminated one of the most important perturbing factors affecting on main metrological characteristic of cesium atomic clock.

Experimental research of the metrological characteristics of the modernizing quantum frequency standard on the atoms of cesium-133 showed improvement in frequency stability about 10–15%.

Obtained results allow extending the areas of applying the quantum frequency standard on atoms of cesium-133 in the international fiber-optic communication system and in the ground part of the GLONASS system [2, 3, 8].

References

1. Riechle, F.: Frequency Standards: Basics and Applications. Wiley-VCH, Weinheim (2004). 526 p.
2. Glazov, A.I., et al.: International comparisons of standards in the area of fiber-optic communication and information transmission systems. Meas. Tech. **60**(10), 1064–1070 (2018)
3. Kolmogorov, O.V., et al.: System for transmitting reference frequency and time signals to measurement resources of the Glonass ground complex by optical cable. Meas. Tech. **60**(9), 901–905 (2017)
4. Balaev, R.I., Malimon, A.N., Fedorova, D.M., Kurchanov, A.F., Troyan, V.I.: Estimation of the precision of transmission of the standard signals of a hydrogen oscillator analog a fiber-optic communication line with electronic compensation of disturbances. Meas. Tech. **60**(8), 806–812 (2017)
5. Semenov, V.V., Nikiforov, N.F., Ermak, S.V., Davydov, V.V.: Calculation of stationary magnetic resonance signal in optically oriented atoms induced by a sequence of radio pulses. Sov. J. Commun. Technol. Electron. **36**(4), 59–63 (1991)
6. Petrov, A.A., Davydov, V.V., Myazin, N.S., Kaganovskiy, V.E.: Rubidium atomic clock with improved metrological characteristics for satellite communication system. In: Galinina, O., Andreev, S., Balandin, S., Koucheryavy, Y. (eds.) NEW2AN/ruSMART/NsCC - 2017. LNCS, vol. 10531, pp. 561–568. Springer, Cham (2017). https://doi.org/10.1007/978-3-319-67380-6_52
7. Nazarov, L.E., Golovkin, I.V.: Symbol reception of signals corresponding to high-speed super-accurate codes and turbo codes based on them. J. Commun. Technol. Electron. **52**(10), 1125–1129 (2007)
8. Petrov, A.A., Davydov, V.V.: Improvement frequency stability of caesium atomic clock for satellite communication system. In: Balandin, S., Andreev, S., Koucheryavy, Y. (eds.) ruSMART 2015. LNCS, vol. 9247, pp. 739–744. Springer, Cham (2015). https://doi.org/10.1007/978-3-319-23126-6_68
9. Pakhomov, A.A.: Fast digital image processing of artificial Earth satellites. J. Commun. Technol. Electron. **52**(10), 1114–1118 (2007)
10. Petrov, A.A., Vologdin, V.A., Davydov, V.V., Zalyotov, D.V.: Dependence of microwave – excitation signal parameters on frequency stability caesium atomic clock. J. Phys. Conf. Ser. **643**(1), 012087 (2015)
11. Petrov, A.A., Davydov, V.V.: Digital frequency synthesizer for ^{133}Cs-Vapor atomic clock. J. Commun. Technol. Electron. **62**(3), 289–293 (2017)
12. Efimov, A.I., Lukanina, L.A., Samoznaev, L.N., Chashei, I.V., Bird, M.K.: Intensity of fluctuations in the frequency of radio signals of spacecraft in the near-solar plasma. J. Commun. Technol. Electron. **55**(11), 1253–1262 (2010)

Nanocommunication System with a Laser Activated Molecular Film

Elena Velichko$^{(\boxtimes)}$ ⓘ, Ekaterina Savchenko ⓘ,
Elina Nepomnyashchaya ⓘ, Dmitrii Dyubo ⓘ, and Oleg Tsybin ⓘ

Peter the Great Saint Petersburg Polytechnic University, Saint Petersburg, Russia
velichko-spbstu@yandex.ru

Abstract. Molecular communication systems became a promising paradigm of modern nanonetworks. Quantum carrier transfer creates communication channels between parts of a molecule and/or connects a few domains inside a molecular metamaterial network. This manuscript presents our studies of such channels activated by means of modulated laser beam irradiation resulting in molecular fluorescence, probably at a shifted frequency. The acquisition system for the electromagnetic field generated by a molecule based on the novel dynamic pin-photo diode is characterized by a high sensitivity and a low noise level due to the signal amplitude-time function integration.

Keywords: Molecular communication · Dynamic pin-photodiode
Registration of fluorescence · Nanonetworks · Nanocommunication
Nano electromagnetic field

1 Introduction

Nanoscale communication is one of frontiers in modern scientific research. The research activities focus on different types of nanomachines, nanoscale sensing, molecular communications, nanonetworks, communication with fluorescence resonance energy transfer, etc. [1–5]. The most promising nanomachines in the field of modern nanotechnology are presented by nanodevices with micrometer – sub-micrometer scale embedded molecular films which are installed, for instance, on an electro-conductive target. These are biosensors, medical biochips, molecular transmitters, molecular electron devices [6–9]. For example, light emitting diodes are nanoscale semiconductor structures which convert organic molecular signals into optical fluxes. One of the main consequences of their tiny dimensions is that their properties are between those of bulk semiconductors and discrete molecules [10]. An intense development of nanotechnologies mandates change of traditional communication architecture to nanoscales.

Molecular mono- or multi-layer ordered films on a solid surface typically consist of organic molecules or biomolecules, e.g., peptides and proteins. In the so-called "bottom-up" case, a metal or a semiconductor target has a smooth clean polycrystalline surface covered with a molecular layer. The lateral size of the surface operation zone can be as small as ~ 0.1–10 nm. In contrast, the "top-down" technology provides an

© Springer Nature Switzerland AG 2018
O. Galinina et al. (Eds.): NEW2AN 2018/ruSMART 2018, LNCS 11118, pp. 649–655, 2018.
https://doi.org/10.1007/978-3-030-01168-0_59

operation zone of "bio-hybrid" nanodevices of 1000 nm or greater which is larger than in the bottom-up case. Two types of a communication channel processing can be energy-activated in the form of

(i) intramolecular internal interaction,
(ii) interaction among separate nanomachines.

To activate the charge motion, one can use an electromagnetic field (EMF), laser beams, acoustic phonons, nano-mechanics, and molecular transport. At least two of them, nano electromagnetic fields and molecular transport, are appropriate for wireless nanocommunication [7]. However, a comprehensive description of energy phenomena for nanomachines is at its initial stage [8]. On the other hand, an effective micro- or nano-detector of a signal is required.

In this manuscript we analyze modulated laser beam irradiation of a molecular system resulting in molecular fluorescence, and detection of molecule-generated electromagnetic fields at shifted frequencies.

2 Experimental

2.1 Experimental Setup

The experimental setup (Fig. 1) consists of a laser source at wavelength 405 nm or 532 nm; attenuating filters and wavelength filter; a total internal reflection prism with the sample; a light-protected thermostat; and a dynamic-regime pin-photodiode (DPD) placed directly under the sample at a distance 10 cm. The DPD was operating in the dynamic regime recently described in [11, 12]. The dynamic-regime pin-photodiode is a new type of CMOS-compatible photodetector characterized by space charge trapping and signal integration inside the target volume. The novel pin diode provides a high visible light sensitivity and signal integration features. In our measurements we used a DPD detector for registration of fluorescence signal.

The triggering time of DPD was measured by a microelectronic circuit controller and simultaneously studied with a TDS-520 digital oscilloscope. The thermo-pair detector was installed in thermal contact with the DPD.

A laser beam (1) passed through a filter and a prism (2, 3) and was incident on a sample (4), thus activating a molecular layer. The reflected beam did not act on the DPD. Thus, only a molecular fluorescence signal was detected.

2.2 Samples

Light-transducing proteins are believed to be relevant examples of biomolecular systems with potential optoelectronic applications. The main features of light-transducing proteins which hold considerable promise as a photochromic biological material for a variety of technological applications are a light sensitivity, a remarkable stability toward chemical and thermal degradation, an enormous cyclicity, and non-linear optical properties [13].

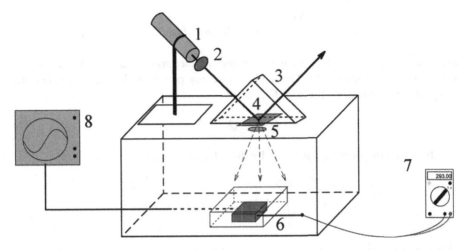

Fig. 1. Experimental setup: 1 – laser, 2 – attenuating filter, 3 – total internal reflection prism, 4 – sample, 5 – wavelength filter, 6 – pin-diode, 7 – thermo-detector, 8 – oscilloscope.

We measured the fluorescence of molecular samples in the visible light range. The first one was rhodamine 6G with the wavelength spectrum from 555 nm to 585 nm. The second one was the chlorophyll solution with the wavelength spectrum 600 nm–700 nm. Molecular samples of rhodamine 6G were dissolved in distilled water with concentration 2 vol. %, chlorophyll was dissolved in spirit in the ratio 1:4 (concentration was 25 vol. %).

To prepare the molecular layer, 0.1 ml of the sample solution was placed on the glass under the prism.

2.3   Detection System

The DPD forward current switched on after a triggering time delay caused by the space charge accumulation under the gate. The forward current magnitude is controlled by the applied forward voltage alone and is independent of light intensity. So, the photon flux signal is not presented by a time-dependent magnitude with additional noise. Instead, the time delay linearly depends on the absorbed light power.

A low noise level is explained by the fact that the time delay of a high forward current is measured instead of photocurrent amplitude. The photocurrent preamplifier typically demands a wide bandwidth, and, as a consequence, a high noise is generated. In our scheme, no preamplifier is required in the measuring circuit. No other tools, such as a comparator, integrator and analog signal digitizer are needed as well. As a result, a new device becomes a high-quality receiver for nanocommunication devices.

Immediately after the bias voltage switches from the initial negative into a working positive, the device stays closed because the total space charge Q_0 is trapped under the gate. At the same time, the neutralizing charge Q_n is generated. The generation rate is

$$\frac{dQ_n}{dt} = I_{photo} + I_{thermo} + I_{leak},\tag{1}$$

where I_{photo}, I_{thermo}, and I_{leak} are the photocurrent, thermo-generated dark current, and leakage current, respectively. The device switching occurs at the moment, when the charge difference reaches the specified value

$$Q_{min} = Q_0 - Q_n\tag{2}$$

The triggering time T_{trig} is given by the integral equation

$$Q_{min} = Q_0 - \int_0^{Ttrig} dt[I_{photo}(t) + I_{thermo}(t) + I_{leak}(t)],\tag{3}$$

where the total triggering time T_{trig} is considered to be inversely proportional to the integrated photocurrent. The error depends on the relative values of both cathode leakage current and thermo-generated dark current. The noise level is defined by the self-triggering time T_{trig} fluctuations mainly in the absence of incident photon irradiation.

The thermo-generated current is limited by the potential over-barrier drift transport of charge carriers. Absolute values of self-triggering time vary with temperature. Near 0 °C the signal-to-noise ratio reaches 50–60 dB, which allows detection of extremely low photon fluxes. Thus, a higher sensitivity may be achieved at temperatures below 10 °C. Temperature characteristics also depend on the ion film situated between the cathode and gate on the dielectric surface.

3   Results and Discussion

Fluorescence of molecular films were measured after excitation by two types of lasers – with wavelengths 405 nm and 532 nm. The triggering times when laser is switched on (t_{on}) and when laser is switched off (t_{off}) were measured. Relative values

$$T_{rel} = (t_{off} - t_{on})/t_{off},\tag{4}$$

were calculated for each measurement and averaged over 10 measurements.

To control the influence of solvent the same parameters were measured for distilled water and spirit applied on a prism. The averaged values of T_{rel} for distilled water and spirit were 0.28 ± 0.02 arb. units. The experimental results are presented in Table 1.

We analyzed the relative time T_{rel} to estimate the intensity of fluorescence of the samples. The use of dynamic pin-photodiode allowed us to increase the sensitivity in comparison with standard methods of registration of the fluorescence radiation and to record weak signals of laser-induced biomolecular fluorescence.

A schematic representation of the target excited via the laser EMF, the molecular layer, and detector are shown in Fig. 2.

Table 1. Experimental results.

Laser wavelength, nm	Object	T_{rel}, arb. units
405	Chlorophyll	0.98
405	Rhodamine 6G	0.63
532	Rhodamine 6G	0.97

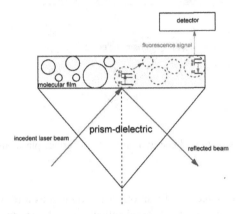

Fig. 2. Schematic representation of the target and the molecular layer, excited via laser radiation.

Light-transducing molecules are activated by laser light, then energy is transmitted from the activated molecules to others along the surface and the fluorescence of molecules at some distance is registered by the detector.

A theory of a high-frequency electromagnetic field-induced excitation of a surface covered with a molecular film is still missing. By using time-of-flight mass spectrometry, instantaneous desorption/ionization/fragmentation of molecular particles in vacuum was observed when a nanosecond time (1–10 ns) with an over-threshold current propagated through a metal or a semiconductor target irradiated by EMF [14, 15]. When the electron subsystem of the target is excited by an external short pulse EMF, hot electrons from the conduction band in the skin layer interact with the adsorbed molecules and thus stimulate: (a) hot electron emission from molecular orbitals; (b) the desorption induced by excitation of surface-molecular bonds; (c) tunnel neutralization of desorbed ions; and (d) electron capture/transfer dissociation-type reactions. Here we use laser beam irradiation at wavelength 405 nm which is beyond the diode working range (or in the low sensitivity range). So, only Raman shifted molecular fluorescence could be detected in the visible light band.

We may draw an analogy with principles of plasmon resonance [16]. The scheme of the process of excitation and subsequent energy transfer by molecular plasmons on a metal surface is presented in Fig. 3.

Laser radiation activates molecules, after which a travelling electromagnetic wave appears and propagates along the metal-dielectric boundary. The surface plasmons

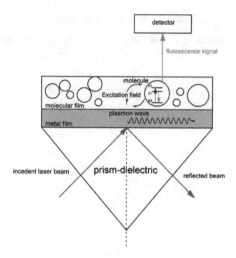

Fig. 3. Schematic representation of energy transfer from surface plasmons in excitation state to molecules.

cause luminescence of molecules. Plasmons are activated as a result of a high angular momentum of photons and a total internal reflection in the prism [17].

4   Conclusion

We have discussed two phenomena: (i) a direct excitation of (bio)molecules by an amplitude-time modulated laser beam which may initiate intramolecular charge and energy transfer, create fluorescence, and thus produce electromagnetic nano-communication channels; and (ii) detection of the fluorescence induced at a shifted frequency by means of a hybrid integrated photodiode. The tools employing these phenomena can be attractive for creation of molecular communication channels between remote nanomachines. The dynamic pin photodiode can be used effectively as a photodetector, namely, a weak signal linear integrator with noise suppression. The temperature range which proves a stable device operation with the highest sensitivity has been found. One can conclude that the dynamic pin photodiode has promising applications in low-power radiation detection systems in the visible light range, including luminometers, spectral analyzers and other analytical devices.

Acknowledgements. The authors are grateful to ActLight SA (Lausanne, Switzerland, [18]) for their donation of a pin photodiode for the research.

References

1. Akan, O.B., Ramezani, H., Khan, T., Abbasi, N.A., Kuscu, M.: Fundamentals of molecular information and communication science. J. Proc. IEEE **105**(2), 306–318 (2017)

2. Bush, S.F.: Nanoscale Communication Networks. Artech House, Norwood (2010)
3. Walsh, F., Balasubramaniam, S., Botvich, D., Donnelly, W.: Synthetic protocols for nano sensor transmitting platforms using enzyme and DNA based computing. Nano Commun. Netw. 1(1), 50–62 (2010)
4. Kuscu, M., Akan, O.B.: On the physical design of molecular communication receiver based on nanoscale biosensors. IEEE Sens. J. 16(8), 2228–2243 (2016)
5. Kuscu, M., Akan, O.B.: The Internet of molecular things based on FRET. IEEE Internet Things J. 3(1), 4–17 (2016)
6. Offenhäusser, A., Rinaldi, R.: Nanobioelectronics – for Electronics, Biology and Medicine. Springer, New York (2009). https://doi.org/10.1007/978-0-387-09459-5
7. Akuildiz, I.F., Jornet, J.M.: Electromagnetic wireless nanosensor networks. J. Nanocommun. Netw. 1, 3–19 (2010)
8. Heath, J.R., Ratner, M.A.: Molecular electronics. J. Phys. Today 56(5), 43–49 (2003)
9. Tsybin, O.: Nano-device with an embedded molecular film: mechanisms of excitation. In: Balandin, S., Andreev, S., Koucheryavy, Y. (eds.) ruSMART 2015. LNCS, vol. 9247, pp. 772–777. Springer, Cham (2015). https://doi.org/10.1007/978-3-319-23126-6_72
10. Giné, P.L., Akyildiz, I.F.: Molecular communication options for long range nanonetworks. J. Comput. Netw. 53, 2753–2766 (2009)
11. Okhonin, S., et al.: A dynamic operation of a PIN photodiode. J. Appl. Phys. Lett. 106, 031115 (2015)
12. Sallin, D.: A low-voltage CMOS-compatible time-domain photodetector, device & front end electronics. Ph.D. thesis 6869, EPFL, Lausanne, Switzerland (2005)
13. Thoma, R., Hampp, N., Bräuchle, C., Oesterhelt, D.: Bacteriorhodopsin films as spatial light modulators for nonlinear-optical filtering. Opt. Lett. 16, 651–653 (1991)
14. Tsybin, O.Y., Mishin, M.: Ion desorption from skin-current induced metal surface. ZTF Lett. 22(4), 21–24 (1996). (In Russian)
15. Zamiatin, A.V., Tsybin, O.Y.: Surface skin-current activated emission of electrons and ions. In: 20th International Workshop on Beam Dynamics and Optimization, BDO 2014, 6890100 (2014)
16. Merlo, J.M., et al.: Wireless communication system via nanoscale plasmonic antennas. Sci. Rep. 6, 31710 (2016)
17. Le Ru, E.C., Etchegoin, P.: Principles of Surface-Enhanced Raman Spectroscopy: and Related Plasmonic, 1st edn. Elsevier, Amsterdam (2009)
18. ActLight: The future of light based electronics. http://act-light.com/technology. Accessed 19 May 2018

Graphene-Coated Substrate as a Basis for Nano-Antennae

Vladimir S. Malyi$^{(\boxtimes)}$ ⓘ, Constantine C. Korikov ⓘ,
and Viktor M. Petrov ⓘ

Institute of Physics, Nanotechnology and Telecommunications, St. Petersburg
State Polytechnic University, St. Petersburg 195251, Russia
Voldemar778@gmail.com

Abstract. Nowadays, structures like "graphene-material" are in demand. To
design such structures, it is necessary to know how the layer of graphene
deposited to the substrate affects the reflection and absorption of electromagnetic
waves in a given range.

Detailed calculations of the reflective features of such structures are not well-
studied. In this paper theoretical description for the reflectivity properties of
dielectric, metal and semiconductor plates coated with graphene will be
demonstrated on an example of fused silica, Au, and Si substrates. The observed
results can be used in case of plates, which can be used for producing nano-
antennae.

Keywords: Nano-antenna · Graphene · Graphene-coated substrates

1 Introduction

Antennae are very important for many devices which are using electromagnetic radiation in the radiowave or microwave mode. Nevertheless, their optical analogue is not
widely used. The characteristics of the nano-antenna can be modified with geometrical
size and shape as well as material composites. However, the features cannot be
dynamically tunable once the nano-antenna has been already fabricated.

Graphene, which consists of a single atomic layer of carbon with a 2-dimensional
hexagonal lattice structure, has attracted intense scientific interests due to its unique
properties such as high electron mobility, high optical transparency, flexibility, and
tunable conductivity [1, 2]. One of the most important property is a capability of
dynamically modifying chemical potentials through tuning the gate voltage of the
graphene-metal or graphene-dielectric nano-antenna enables to fabricate controllable
devices by introducing the graphene sheet [2].

O. Galinina et al. (Eds.): NEW2AN 2018/ruSMART 2018, LNCS 11118, pp. 656–665, 2018.
https://doi.org/10.1007/978-3-030-01168-0_60

2 Structure

Analysis of the literature dedicated to application of graphene in the production of nano-antennae demonstrates that the following combinations of materials are actively developing nowadays:

1. CaF_2 + graphene + golden substrate. Such combination is using as a tunable ultra-narrowband mid-infrared TE-polarization absorber, which has potential applications in the tunable filtering and tunable coherent emission of thermal source [1].
2. For creating a graphene-based asymmetric nano-antenna microstructure that can be used to realize electrically controllable, unidirectionally propagating broadband surface plasmon polaritons authors in Ref. [3] offer a combination: SiO_2 + graphene + golden substrate;
3. To design a perfect metamaterial absorber authors in Ref. [4, 5] used combination: glass + graphene + golden substrate + dielectric. The absorption peaks of this device can be dynamically adjusted actively by tuning the Fermi energy and the refractive index of the dielectric with no degraded practically perfect peak absorption which provides a flexible method to tune the absorption channel without changing the geometric parameters of the structure. The number of absorption peaks can be increased with double-layer graphene arrays;
4. SiO_2 + Al_2O_3 + graphene + golden substrate. This architecture involves a conventional gold dipole antenna which consists of two Au arms separated by a feed gap, the graphene sheet, and aluminum oxide as an insulator inserted between the dipole antenna and graphene [2]. Such combination allows to control the characteristics of similar antennas;
5. SiO_2 + Si + graphene-hexagonal-boron-nitride-graphene layer + golden and silver substrate. Such structure can be used to create a high performance broadband optical modulator [6]. If golden and silver substrates were excluded and graphene-hexagonal-boron-nitride-graphene layer were replaced by graphene layer we can obtain chip-integrated nearly perfect graphene absorber [7].

It should be noted that there are a lot of papers devoted to designing of nano-antennae, but none provides information about the reflective properties of the substrates. So, in this paper we will introduce solution of the problem of determining the reflective properties of graphene deposited on different substrates.

3 Sketch of the Structure

To solve this problem it is necessary to develop a structure which is a combination of the following on Fig. 1 layers.

4 Theory

In the present paper, the complete theory allowing calculation of the reflection coefficients and reflectivities of graphene-coated material plates is applied. This theory exploits the polarization tensor of graphene derived in Ref. [8, 9] and, thus, is based on first principles of quantum electrodynamics. In addition to great fundamental interest, such kind theory is much needed in numerous technological applications of graphene. Graphene coatings are already used to increase the efficiency of light absorption on optical metal surfaces [10, 11]. This is important for the optical detectors. Graphene-coated substrates have potential applications as transparent electrodes (see Ref. [11] where graphene-silica thin films are used as transparent conductors). Deposition of graphene on silicon substrates provides an excellent anti-reflection, which is important for solar-cells [12]. One could also mention that graphene coating on metal surfaces is employed for corrosion protection [13]. In all these cases, the developed theory can be used to calculate the effect of graphene coating.

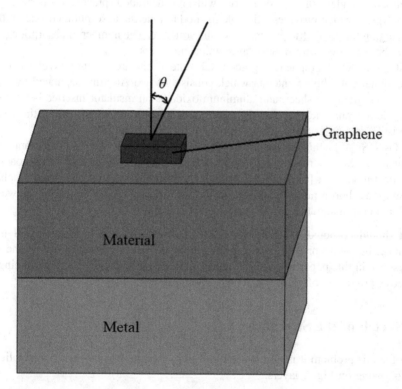

Fig. 1. Sketch of the substrate for graphene-based nano-antenna, where θ is an angle of the incidence of light.

In this section, we present general theoretical results for the reflectivity properties of thick material plate (semispace) coated with the layer of graphene. We assume that the material of the plate is nonmagnetic and is described by the frequency-dependent dielectric permittivity $\varepsilon(\omega)$. Graphene is considered as a pristine (gapless) one (the generalization for the case of a nonzero gap is straight forward). It is described by the polarization tensor $\Pi_{ik}(\omega, k_\perp)$ with $i, k = 0, 1, 2$ derived in Ref. [8], where $k_\perp$ is the magnitude of the wave vector projection on the plane of graphene.

The reflection coefficients on a graphene-coated plate, where graphene is described by the polarization tensor and plate material by the dielectric permittivity, were obtained in Ref. [14] at the imaginary Matsubara frequencies. Taking into account that the polarization tensor of Ref. [8] is valid also at the real frequencies, one can leave the same derivation unchanged. We only take into account that at real frequencies the mass-shell equation for photons is satisfied resulting in the equation

$$k_\perp = \frac{\omega}{c}\sin\theta_i, \tag{1}$$

where θ_i is the angle of incidence. As a result, the reflection coefficients of graphene-coated plates for TM and TE polarizations of graphene-coated plates for TM and TE polarizations of the electromagnetic field take the form

$$R_{TM}^{(g,p)}(\omega,\theta_i) = \frac{\varepsilon(\omega)\cos\theta_i - \sqrt{\varepsilon(\omega)-\sin^2\theta_i}}{\varepsilon(\omega)\cos\theta_i + \sqrt{\varepsilon(\omega)-\sin^2\theta_i}}\frac{\left[1-\Pi_{00}^*(\omega,\theta_i)\right]}{\left[1+\Pi_{00}^*(\omega,\theta_i)\right]}, \tag{2}$$

$$R_{TE}^{(g,p)}(\omega,\theta_i) = \frac{\cos\theta_i - \sqrt{\varepsilon(\omega)-\sin^2\theta_i} - \Pi^*(\omega,\theta_i)}{\cos\theta_i + \sqrt{\varepsilon(\omega)-\sin^2\theta_i} + \Pi^*(\omega,\theta_i)}. \tag{3}$$

Then using result from Ref. [10] we have the reflectivity of graphene-coated plate in the region optical and near-infrared spectral bands at the normal incidence

$$R^{(g,p)}(\omega,0) = \left|R_{TM}^{(g,p)}(\omega,0)\right|^2 = \left|R_{TE}^{(g,p)}(\omega,0)\right|^2 \approx \frac{[n_1(\omega)-1+\pi\alpha]^2 + n_2^2(\omega)}{[n_1(\omega)+1+\pi\alpha]^2 + n_2^2(\omega)}, \tag{4}$$

which should be compared with the reflectivity of an uncoated plate at the normal incidence

$$R^{(p)}(\omega,0) = \frac{[n_1(\omega)-1]^2 + n_2^2(\omega)}{[n_1(\omega)+1]^2 + n_2^2(\omega)}, \tag{5}$$

The computational results of the reflectivity of the graphene-coated and uncoated silica at the normal incidence as a function of frequency, which were introduced in Ref. [10], are presented in Fig. 2 by the line 2 and 1, respectively. As can be seen in Fig. 2, in the regions of optical and ultraviolet frequencies the graphene coating slightly (from 6% to 9%) increases the reflectivity of a silica plate at the normal incidence.

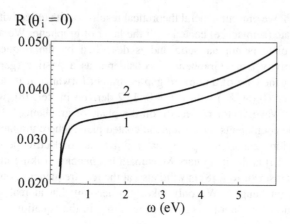

Fig. 2. The reflectivities of the graphene-coated and uncoated silica plates (the lines 2 and 1, respectively) at the normal incidence as functions of frequency. Calculations are performed for the high-frequency region.

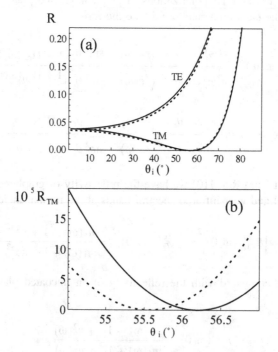

Fig. 3. (a, b) The TM and TE reflectivities of the graphene-coated silica plate $\omega = 3 \times 10^{15}$ rad/s as functions of the incidence angle. Solid lines – for graphene-coated silica plate, dashed lines – for uncoated silica plates.

Now we will describe the results of the case of amorphous SiO_2 plate coated with graphene film. The computational results at $\omega = 2$ eV $= 3 \times 10^{15}$ rad/s (visible light) are presented in Fig. 3(a) as a function of θ_i by the lower and upper solid lines (for the TM and TE polarizations, respectively). In the same figure, the dashed lines show the computational results in the absence of graphene coating.

As is seen in Fig. 3(a), the Brewster angle, at which the reflected light is fully (TE) polarized, is slightly different for the uncoated and graphene-coated silica plates. To make this effect more quantitative, in Fig. 3(b) we plot the reflectivities R_{TM} on a larger scale in the vicinity of the Brewster angle (the dashed and solid lines are for an uncoated and for a graphene-coated silica plates, respectively). As is seen in Fig. 3(b), graphene coating leads to the increase of the Brewster angle from $\theta_B = 55.5°$ to $\theta_B = 56.2°$.

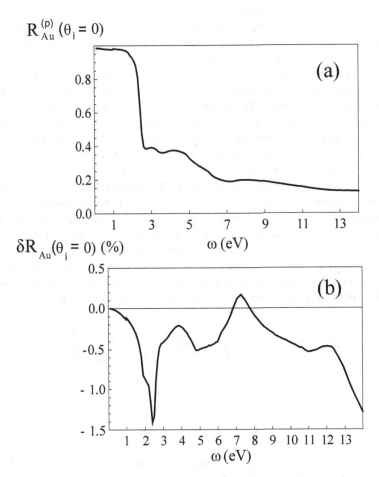

Fig. 4. (a) The reflectivities of the uncoated Au plate and (b) the relative change of reflectivity of the graphene-coated Au plate as functions of frequency. Normal incidence.

The case of Au plate. The complex index of refraction of Au is tabulated in Ref. [16] in the frequency region 0.125 eV < ω < 9919 eV. In the most part of this region (specifically, at ω > 0.26 eV) one can use the asymptotic expression at high frequencies to calculate the reflectivities of the graphene-coated Au plate. The reflectivity of an uncoated Au plate at the normal incidence is shown in Fig. 4(a) as a function of frequency. Note that the influence of graphene coating on metal surfaces is smaller than that on dielectric ones. Because of this it is not convenient to show the reflectivity of the graphene-coated Au plate in the same figure as the uncoated one. Instead, we calculate the relative quantity

$$\delta R_{Au}(\omega, \theta_i = 0) = \frac{R_{Au}^{(g,p)}(\omega, 0) - R_{Au}^{(p)}(\omega, 0)}{R_{Au}^{(p)}(\omega, 0)} \tag{6}$$

where $R_{Au}^{(g,p)}(\omega, 0)$ in the high-frequency region ω > 0.26 eV was computed in Ref. [10]. At ω < 0.125 eV the optical data were extrapolated by means of the Drude model [17]. The computational results for the quantity as a function of frequency are presented in Fig. 4(b) in the frequency range from 0.125 eV to 14 eV.

For an Au plate, as shown in Fig. 4(b), the reflectivity at the normal incidence becomes smaller due to the presence of graphene coating (this is opposite to the case of silica considered previously). The single exception from this observation is the frequency interval 6.7 eV < ω < 7.5 eV, where $\delta R_{Au}(\omega, \theta_i = 0) < 0$. The largest in magnitude influence of graphene coating $|\delta R_{Au}(\omega, \theta_i = 0)| = 1.4\%$ takes place at $\omega = 2.4$ eV = 3.65×10^{15} rad/s. It should be stressed also that in the whole region ω < 0.26 eV (the moderate and low frequencies) $\delta R_{Au}(\omega, \theta_i = 0) = 0$ with high accuracy, i.e., the graphene coating does not influence the reflectivity properties. As a result, the reflectivity of the graphene-coated Au plate does not depend on temperature. These observations are in agreement with the fact that graphene coatings do not influence the van der Waals and Casimir forces between metal plates [18].

Here the reflectivity of low-resistivity p-doped Si at the normal incidence in the region of moderate and low frequencies ω < 0.26 eV is presented. As above, for $\omega \leq 2.6$ meV the condition $\omega << \omega_T$ was used. In the frequency interval 2.6 meV < 0.26 eV computations were performed by the exact equations

$$\Delta_T \Pi_{00}(\omega, \theta_i) = \frac{8\alpha\hbar\omega}{\tilde{v}_F^2 c} \int_0^\infty \frac{dy}{\exp(\beta y) + 1} \left[1 + \sum_{\lambda = \pm 1} \frac{X_\lambda(\theta_i, y)}{2u(\theta_i)} \right], \tag{7}$$

$$\Delta_T \Pi(\omega, \theta_i) = \frac{8\alpha\hbar\omega^3}{\tilde{v}_F^2 c^3}$$
$$\int_0^\infty \frac{dy}{\exp(\beta y) + 1} \left\{ 1 + \frac{u(\theta_i)}{2} \sum_{\lambda = \pm 1} X_\lambda(\theta_i, y) + \tilde{v}_F^2 \frac{\sin^2 \theta_i}{X_\lambda(\theta_i, y)} \right\} \tag{8}$$

where $\beta \equiv \omega/(2\omega_T)$, and the thermal frequency is defined as $\omega_T = k_B T/\hbar$. The exact definitions of the X_λ are given in Ref. [10].

The computational results at T = 300 K as a function of frequency are shown in Fig. 5.

5 Discussion

Currently, graphene is considered as a suitable material for nano-antennae. Graphene coating of various materials significantly affects the reflective and absorbing properties of such structures. The ability to change the absorption and reflection properties of the antenna's surface is very important for operation in transmission and receive modes. That is why such antennae are called "smart".

We have proposed a theory that allows to design the reflective and absorbing properties of such antennae with high accuracy. Strictly speaking, the theory allows to do it in a wide spectral range. In this work we have focused our attention on the range most interesting for modern information nanocommunications, radiophotonics and terahertz bandwidth – from 4 to 300 THz.

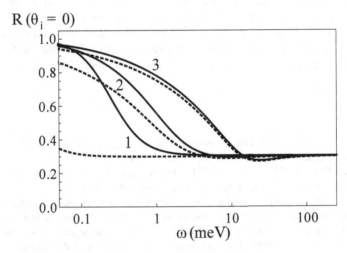

Fig. 5. The reflectivities of the low-resistivity graphene-coated Si plates at T = 300 K. The concentrations of free charge carriers $N_1 = 5 \times 10^{14}$ cm^{-3}, $N_2 = 5 \times 10^{16}$ cm^{-3}, and $N_3 = 5 \times 10^{17}$ cm^{-3} (curves 1, 2, 3, respectively). Dashed lines are the same plates without coating.

Our results allow us to calculate the reflection dependences with high accuracy both for TE and for TM modes, which is also important for practical applications.

The creation of various structures containing sets of layers of both graphene and other materials allow to create antennae, or as they are now called – "absorbers" with noticeable beam-forming properties [2].

Considered theoretical description for the reflective properties of the graphene-coated plates made of different materials (dielectric, metal and semiconductor). This theory is based on the Dirac model of graphene, which is described by the recently found polarization tensor in (2 + 1)-dimensional space-time, allowing the analytic continuation to the real frequency axis [10]. The materials of the plate are described by the frequency-dependent complex index of refraction. In the framework of this theory, the general formulas are obtained for the reflection coefficients and reflectivities of graphene-coated plates.

The demonstrated theory is applied to the cases of graphene-coated dielectric, metal, and semiconductor plates. It is shown that for a dielectric material in the region of optical and ultraviolet frequencies the reflectivity of the graphene-coated plate at the normal incidence is by several percent larger than of an uncoated one. In this frequency region, the graphene coating also results in some increase of the Brewster angle.

For metals, the developed theory is illustrated by the examples of Au plates. In the case of high and moderate frequencies, we have shown that the graphene coating decreases the plate reflectivity at the normal incidence.

The considered theory to graphene-coated semiconductor plates (Si with various concentrations of charge carriers) was applied. For the high-resistivity Si plate, the influence of graphene coating at high frequencies achieves several percent similar to the case of a silica plate. For the low-resistivity Si the effect of graphene coating was investigated at different concentrations of charge carriers. It is shown that at high frequencies the graphene coating does not influence the reflecting properties.

References

1. Liao, Y.-L., Zhao, Y.: Graphene-based tunable ultra-narrowband mid-infrared TE-polarization absorber. Opt. Express 25, 32080–32089 (2017)
2. Ren, X., Sha, W.E.I., Choy, W.C.H.: Tuning optical responses of metallic dipole nanoantenna using graphene. Opt. Express 21, 31824–31829 (2013)
3. Huang, L., et al.: Tunable unidirectional surface plasmon polariton launcher utilizing a graphene-based single asymmetric nanoantenna. Opt. Mater. Express 7, 569–576 (2017)
4. Meng, H., Wang, L., Liu, G., Xue, X., Lin, Q., Zhai, X.: Tunable graphene-based plasmonic multispectral and narrowband perfect metamaterial absorbers at the mid-infrared region. Appl. Opt. 56, 6022–6027 (2017)
5. Ning, Y., Dong, Z., Si, J., Deng, X.: Tunable polarization-independent coherent perfect absorber based on a metal-graphene nanostructure. Opt. Express 25, 32467–32474 (2017)
6. Chen, X., et al.: A broadband optical modulator based on a graphene hybrid plasmonic waveguide. J. Lightwave Technol. 34, N21 (2016)
7. Xu, W., et al.: Chip-integrated nearly perfect absorber at telecom wavelengths by graphene coupled with nanobeam cavity. Opt. Lett. 40(14), 3256–3259 (2015)
8. Bordag, M., Klimchitskaya, G.L., Mostepanenko, V.M., Petrov, V.M.: Quantum field theoretical description for the reflectivity of graphene. Phys. Rev. D 91, 045037 (2015)
9. Bordag, M., Klimchitskaya, G.L., Mostepanenko, V.M., Petrov, V.M., Erratum: Quantum field theoretical description for the reflectivity of graphene. Phys. Rev. D 93, 089907(E) (2016)
10. Klimchitskaya, G.L., Korikov, C.C., Petrov, V.M.: Theory of reflectivity of graphene-coated material plates. Phys. Rev. B 92, 125419 (2015)
11. Klimchitskaya, G.L., Korikov, C.C., Petrov, V.M.: Erratum: theory of reflectivity of graphene-coated material plates. Phys. Rev. B 93, 159906(E) (2016)
12. Vajtai, R. (ed.): Springer Handbook of Nanomaterials. Springer, Berlin (2013)
13. Dumée, L.F., He, L., Wang, Z., et al.: Carbon 87, 395 (2015)

14. Klimchitskaya, G.L., Mohideen, U., Mostepanenko, V.M.: Phys. Rev B **89**, 115419 (2014)
15. Klimchitskaya, G.L., Mostepanenko, V.M.: Phys. Rev B **91**, 174501 (2015)
16. Palik, E.D. (ed.): Handbook of Optical Constants of Solids. Academic, New York (1985)
17. Chang, C.-C., Banishev, A.A., Castillo-Garza, R., Klimchitskaya, G.L., Mostepanenko, V. M., Mohideen, U.: Phys. Rev. B **85**, 165443 (2012)
18. Klimchitskaya, G.L., Mostepanenko, V.M.: Phys. Rev. A **89**, 052512 (2014)

Synthesis of the Demodulation Algorithm for the Phase Modulated Signals in Presence of the Background Noise Using Complete Sufficient Statistics

Sergey I. Ivanov⬤, Leonid B. Liokumovich⬤,
and A. V. Medvedev$^{(\boxtimes)}$⬤

Peter the Great St. Petersburg Polytechnic University, St-Petersburg, Russia
medvedev@rphf.spbstu.ru

Abstract. A description is given of the algorithm for demodulating a linearly frequency-modulated signal against a background of white Gaussian noise with a constant component. The functions of complete sufficient statistics for estimating the phase and amplitude parameters of the signal, as well as the level of the constant component and the noise variance are found. The demodulation algorithm is implemented in the LabVIEW program environment. Computer modeling showed the asymptotic efficiency of the received estimates of the signal parameters.

Keywords: Estimation of parameters · Sufficient statistics
Phase modulated signal · Demodulation · A priori non-certainty
Gaussian noise

1 Introduction

Phase-modulated signals are very widely used in various areas of modern science and technology, for example, remote sensing of objects and media, navigation, communications. These signals have a large base - the product of the signal duration T by the occupied frequency band, which, under known signal parameters, allows processing with a small signal-to-noise ratio (SNR). However, in practice, signal processing has to be realized under conditions of a priori uncertainty with respect to the signal and noise parameters.

There are many methods to estimate the phase coefficients such us high-order ambiguity function [1], product high-order ambiguity function [2], least-square fitting [3], maximum likelihood [4], high-order phase function [5], Bayesian estimation [6], short-time Fourier transform [7], estimates based on integral (spectral) transformations [7] and others (see, for example, [8–12]).

The method of measurable functions of complete sufficient statistics was successfully used in [13, 14] for stable statistical estimation of the parameters of the phase-modulated signal under conditions of a priori uncertainty. This method is not accompanied by large computational costs and can be implemented in real time using compact digital processors, which distinguishes it from the above methods.

O. Galinina et al. (Eds.): NEW2AN 2018/ruSMART 2018, LNCS 11118, pp. 666–674, 2018.
https://doi.org/10.1007/978-3-030-01168-0_61

In [15], the method of measurable functions of complete sufficient statistics was used to obtain a statistical estimate of the parameters of a linearly frequency-modulated signal and, accordingly, its demodulation. The method of successive approximations was used to estimate the initial phase of the signal φ_0. In the first approximation, the initial phase φ_0 estimate was determined for the harmonic signal model. Such an approximation can limit the class of phase-modulated signals under consideration and increase the variance of the estimate of the initial phase. For some technical solutions, for example, for the analysis of interferograms, polarization measurements of parameters of the Stokes vector, the useful signal besides the periodic signal also contains a constant component [16].

In this paper, we describe an algorithm for a stable statistical evaluation of the parameters of a phase-modulated signal against a background of white Gaussian noise $N(0,\sigma^2)$ with a constant component U_0 under conditions of a priori uncertainty relative to the variance σ^2 and the level value U_0. The signal phase demodulation algorithm was implemented in the LabVIEW software environment, which allowed not only to calculate the statistical characteristics of the evaluation, but also to go over to the hardware implementation of the proposed algorithm.

2 Mathematical Model of a Signal. Synthesis of the Estimation Algorithm

Let the process $x(t)$ bean additive mixture of the useful phase-modulated signal $s(t)$ and stationary white Gaussian noise $n(t)$ with dispersion σ^2 and constant component U_0

$$
\begin{aligned}
x(t) &= s(t) + U_0 + n(t) = A_0 \cos(\omega_0 t + \varphi(t)) + U_0 + n(t) = \\
&= A_0 \cos(\omega_0 t + \varphi_0 + \Omega t + \beta t^2) + U_0 + n(t).
\end{aligned}
\tag{1}
$$

Here the carrier frequency of the signal $\omega_0 = 2\pi f_0$ is given by the user (known), the initial phase φ_0, the signal amplitude A_0, the constant component U_0, the dispersion of the noise samples σ^2, he linear frequency modulation parameters Ω and β are assumed to be a priori undefined. It is assumed that the following conditions are satisfied on the observation interval $[0, T]$:

$$
\omega_0 T > > 1; \ \Omega T < \pm \pi; \ \beta T^2 < \pm \pi.
\tag{2}
$$

Such a mathematical model of the observed process describes, for example, the signals of the interferometric system of fiber optic sensors and the radar signal when measuring the effective scattering area using the Doppler Effect [14, 17]. The result of the analog to digit transformation of the observable process $x(t)$ is the sampling vector $x = [x_0, \ldots x_{n-1}]^T$, for which $x_i = x(t_i)$ are the sampling times, $i = 0 \ldots, n - 1$.

To synthesize the algorithm for estimating the vector parameter Θ we represent the logarithm of the probability density $w(x, \Theta)$ of the vector sample x as a function Θ in the following form

$$\ln w(\textbf{\textit{x}}, \boldsymbol{\Theta}) = \ln L(\boldsymbol{\Theta}) = 1/2n \ln(-\theta_1/\pi) + \ln(-\theta_1) + n/8\theta_1(\theta_2^2 +$$
$$+ \theta_3^2 + 2\theta_8^2) + \theta_1 T_1 + \theta_2 T_2 + \theta_3 T_3 + \theta_4 T_4 + \theta_5 T_5 + \theta_6 T_6 + \theta_7 T_7 + \theta_8 T_8, \tag{3}$$

where the 8-dimensional (coordinate) vector-parameter $\boldsymbol{\Theta} = [\theta_1, \theta_2, \theta_3, \theta_4, \theta_5, \theta_6, _7, \theta_8]^T$ and the 8-dimensional vector statistic $\textbf{\textit{T}} = [T_1, T_2, T_3, T_4, T_5, T_6, T_7, T_8]^T$ are determined by the following relations:

$$\theta_1 = -1/2\sigma^2, \quad \theta_2 = -A_0 \sin \varphi_0/\sigma^2, \qquad \theta_3 = A_0 \cos \varphi_0/\sigma^2,$$
$$\theta_4 = -A_0\Omega_0 \sin \varphi_0/\sigma^2, \qquad \theta_5 = -A_0\beta_0 \sin \varphi_0/\sigma^2,$$
$$\theta_6 = -A_0\Omega_0 \cos \varphi_0/\sigma^2, \qquad \theta_7 = -A_0\beta_0 \cos \varphi_0/\sigma^2, \quad -\pi < \varphi_0 < \pi,$$
$$\theta_8 = U_0/\sigma^2,$$

$$T_1 = \sum_{i=0}^{n-1} x_i^2, \quad T_2 = \sum_{i=0}^{n-1} x_i \sin(i\,\omega_0), \quad T_3 = \sum_{i=0}^{n-1} x_i \cos(i\,\omega_0),$$

$$T_4 = \sum_{i=0}^{n-1} ix_i \cos(i\,\omega_0), \quad T_5 = \sum_{i=0}^{n-1} i^2 \cos(i\,\omega_0), \tag{4}$$

$$T_6 = \sum_{i=0}^{n-1} ix_i \sin(i\,\omega_0), \quad T_7 = \sum_{i=0}^{n-1} i^2 \sin(i\,\omega_0) \quad T_8 = \sum_{i=0}^{n-1} x_i.$$

Here ω_0, Ω_0 and β_0 are the signal phase parameters ω, Ω and β normalized to the sampling frequency.

For the probability density $w(\textbf{\textit{x}}, \boldsymbol{\Theta})$ considered for a fixed vector sample $\textbf{\textit{x}}$ as a function of $\boldsymbol{\Theta}$ (the likelihood function $L(\boldsymbol{\Theta})$) the factorization criterion is fulfilled, and therefore the vector-statistics $\textbf{\textit{T}}$ is sufficient [12, 13]. The distribution (3) belongs to an exponential family with sufficient statistics $\textbf{\textit{T}}$. For the unknown σ^2, A_0, φ, Ω and β the parameter $\boldsymbol{\Theta}$ takes values from the region $(-\infty,0) \times (-\infty,\infty) \times (-\infty,\infty)$ $(-\infty,\infty) \times (-\infty,\infty) \times (-\infty,\infty) \times (-\infty,\infty) \times (0, \infty)$ that is, it contains an 8-dimensional interval. Therefore, by the completeness theorem [10], a sufficient statistic $\textbf{\textit{T}}$ is complete.

The construction of estimates using the likelihood method in this approximation does not lead to the result: it is possible to determine only the energy parameters of the signal and noise. We propose the use of the Lehmann-Scheffe theorem [18] for constructing effective estimates of the parameters of the phase modulated signal. It follows from this theorem that an effective estimate of the parameter $\boldsymbol{\Theta}$ is a linear combination of moments of a complete sufficient statistic of an arbitrary order [13, 18].

It is easy to obtain an expression for the estimate of the components of the vector $\textbf{\textit{Y}} = [\theta_2/2\theta_1, -\theta_3/2\theta_1, \theta_4/2\theta_1, \theta_5/2\theta_1, \theta_6/2\theta_1, \theta_7/2\theta_1]^T$ as a linear combination of the elements of the vector of a complete sufficient statistic $\textbf{\textit{T}}_Y(x) = [T_2, T_3, T_4, T_5, T_6, T_7]^T$ of the first order and, accordingly, determine the estimate of the normalized signal parameters Ω_0 and β_0. The estimate of the vector $\textbf{\textit{Y}}$ is determined by solving a system of six linear algebraic equations

$$Y = \mathbf{A}^{-1} T_Y, \tag{5}$$

where $\mathbf{A}^{-1}$- square matrix 6×6, the elements of which depend on the sample x in accordance with the expression (4). Diagonal elements of the matrix $\mathbf{A}$

$$a_{11} = 0,5(n+1), \quad a_{22} = -a_{11}, \quad a_{33} = a_{55} = -\frac{1}{12}n(n-1)(2n-1),$$

$$a_{44} = a_{66} = -\frac{1}{60}n((n-1)(2n-1)[3n(n-1)-1].$$

The remaining elements of the matrix are defined in a similar way. Estimate of the components of a vector

$$\mathbf{Z} = \left[\theta_2^2 + \theta_3^2/4\theta_1^2 = A_0^2, \quad \theta_8^2/4\theta_1^2 = U_0^2, \quad 1/2\theta_1 = \sigma^2 \right]^{\mathrm{T}}$$

is determined by solving a system of three linear algebraic equations

$$Z = \mathbf{B}^{-1} T_Z \tag{6}$$

The matrix $\mathbf{B}$ is a square matrix of dimension 3×3, whose elements are equal to

$$\mathbf{B} = \begin{Bmatrix} n/2 & n & n \\ 0 & n^2 & n \\ n^2/4 & 0 & n \end{Bmatrix},$$

a linear combination of the elements of the vector of a complete sufficient statistic $T_Y(x)$ of the first and second degrees

$$T_Z(x) = \left[T_1, T_8^2, T_2^2 + T_3^2 \right]^{\mathrm{T}}.$$

Equation (6) gives the following expressions for estimating the energy parameters of the phase-modulated signal and the noise A_0^2, U_0^2, σ^2 in the form of measurable functions of complete sufficient statistics

$$A_{0e}^2 = \frac{4}{n^2(n-3)} \left[(n-1)T_2^2 + (n-1)T_3^2 - nT_1 + T_8^2 \right],$$
$$\sigma_e^2 = \frac{1}{n}(T_2^2 + T_3^2) - \frac{n}{4}A_{0e}^2, \tag{7}$$
$$U_{0e}^2 = \frac{T_8^2}{n^2} - \frac{\sigma_e^2}{n}.$$

For the estimates obtained, the following property is proved: if the estimate obtained is biased, then it provides a minimum of average losses in the class of all estimates having a displacement [10, 13, 18]. The shifted estimates of the signal and noise parameters (5), (7) can be regarded as asymptotically efficient, having an inessential for practice bias that decreases monotonically with increasing sample size n. In this case, for large n, the displacement becomes much less than the root-mean-square deviation from the true value. Synthesized estimates have the property of stability under conditions of a priori uncertainty in the sense that the signal processing algorithm does not depend on the unmeasured signal and interference parameters, and the average losses are thus minimally possible [15].

Estimates of the energy parameters of the phase-modulated signal and noise coincide with analogous estimates of the harmonic signal against the background of additive noise obtained in [15]. It is shown that for a large sample size n, the relative standard deviation of the estimate is inversely proportional to the square root of n and the signal-to-noise ratio.

3   The Simulation Results of the Phase Demodulation Algorithm

To test the efficiency of the developed algorithm for estimating the parameters of the phase-modulated signal against the noise background (1), computer simulation of signal processing was performed in the LabVIEW graphical software environment. The program code of the algorithm uses the built-in functions of LabVIEW - virtual instruments (VI) and works in quasi-real time.

In the modeling process, a comparison was made between the accuracy characteristics of the estimation of signal parameters using the algorithm considered above with the results of the authors of [15]. In Fig. 1 shows the dependence of the modulus of the difference $\Delta_\varphi(\varphi_0)$ for estimating the initial phase of the signal φ_0 and its true value. The simulation was performed at a ratio of the noise parameter σ to the signal amplitude A_0 equal to 0.0, the number of samples on the period τ of the carrier frequency equal to 27, the processing interval T equal to 16 periods τ and the observation interval of $2T$. Curve (1) is calculation by the above algorithm, curve (2) - calculation by the algorithm presented by the authors in [4]. It can be seen from the figure that the accuracy of the estimate depends weakly on the value of the initial phase of the signal φ_0. Estimates of the initial phase of the signal φ_0 using the algorithm considered above are, on average, more accurate by 0.06 radians (3.5°).

Figure 2 shows the dependence of the modulus of the difference $\Delta_\varphi(\sigma)$ of the estimate of the initial phase of the signal φ_0 and its true value as a function of the ratio of the noise parameter σ to the signal amplitude A_0. In the simulation, the value of the initial phase of the signal φ_0 was set equal to $\pi/8$. As expected, the accuracy of the estimate of φ_0 decreases with increasing noise level. As in the previous case, estimates of the initial phase of the signal φ_0 using the above algorithm are, on average, more accurate by 0.05–0.1 rad (3–6°).

During the simulation, the effect of the noise/signal ratio, sample size n on the displacement and variance of the signal phase parameters estimation was investigated. Also, the statistical characteristics of the parameter estimates of the given algorithm are compared with the statistical characteristics of parameter estimates obtained by spectral analysis [9].

In Fig. 3. the oscillogram of the observed signal $x(t)$ the informative phase-modulated signal $s(t)$ and its phase $\varphi(t)$ n accordance with the mathematical model (1) obtained with the help of VI LabVIEW is presented. The ratio of the noise parameter σ to the signal amplitude for this implementation is 1.0, the number of samples on

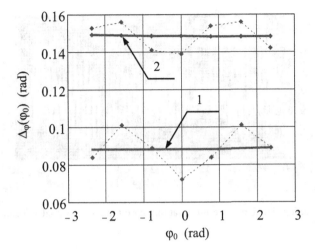

Fig. 1. The modulus of the difference $\Delta_\varphi(\varphi_0)$ vs the initial signal phase φ_0

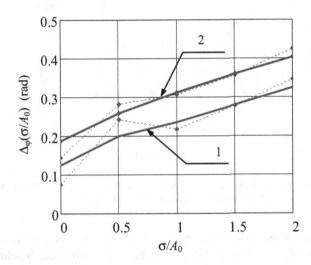

Fig. 2. The modulus of the difference $\Delta_\varphi(\sigma)$ vs the noise parameter σ to the signal amplitude A_0 ratio

periodτof the carrier frequency is 27, the processing interval T is 16 periods τ and the observation interval is $2T$.

In Fig. 4. the results of demodulation of the phase $\varphi(t)$ of the useful signal $s(t)$ are presented by two methods: the developed method of complete sufficient statistics and the method based on the Hilbert transform. The oscillogram in Fig. 4 corresponds to the case of the ratio σ/A equal to 1.0 and the processing interval T equal to 16 periods of the carrier frequency.

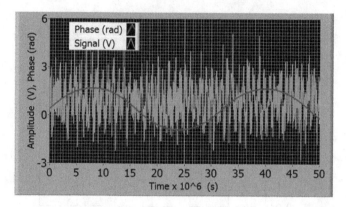

Fig. 3. The observed signal $x(t)$, the information phase modulated signal $s(t)$ and its phase $\varphi(t)$

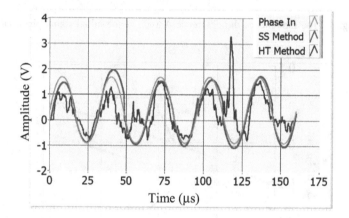

Fig. 4. The result of the phase $\varphi(t)$ demodulation (Phase In) by the methods of sufficient statistics (SS Method) and the Hilbert transform (HT Method)

A description of the algorithm for the phase demodulation process using the Hilbert transform is briefly described in [15]. From the results obtained, it follows that the algorithm developed by us for the model under consideration gives a better result than the phase demodulation algorithm with the use of the Hilbert transform.

The analysis of the dependence of the root-mean-square error (RMS) of the phase estimate on its true value as a function of the ratio of the noise parameter σ to the signal amplitude A_0 showed that the obtained estimates of the signal parameters are asymptotically unbiased and effective.

4 Conclusion

The developed demodulation algorithms for a phase-modulated signal based on complete sufficient statistics under conditions of a priori uncertainty relative to signal amplitude, constant component, and noise variance have advantages over other algorithms for large ratios σ/A, $\omega_0 T$ and large sample size n. The simulation results in the LabVIEW program environment showed the asymptotic unbiasedness and effectiveness of the estimates obtained.

Acknowledgment. The work was done under financial support of Ministry of Education and Science of the Russian Federation in terms of FTP "Research and development on priority trends of Russian scientific-technological complex evolvement in 2014–2020 years" (agreement# 14.578.21.0211, agreement unique identifier RFMEFI57816X0211).

References

1. Djurovic, I., Simeunovic, M., Djukanovic, S., et al.: A hybrid CPF-HAF estimation of polynomial-phase signals: detailed statistical analysis. IEEE Trans. Signal Process. **60**, 5010–5023 (2012)
2. Barbarossa, S., Scaglione, A., Giannakis, G.B.: Product high-order ambiguity function for multicomponent polynomial-phase signal modeling. IEEE Trans. Signal Process. **46**, 691–708 (1998)
3. McKilliam, R.G., Quinn, B.G., Clarkson, I.V.L., et al.: Polynomial phase estimation by least squares phase unwrapping. IEEE Trans. Signal Process. **62**, 1962–1975 (2014)
4. Ghogho, M., Nandi, A.K., Swami, A.: Cramér-Rao bound and maximum likelihood estimation for random amplitude phase modulated signals. IEEE Trans. Signal Process. **47**, 2905–2916 (1999)
5. Wang, P., Djurovíc, I., Yang, J.Y.: Generalized high-order phase function for parameter estimation of polynomial phase signal. IEEE Trans. Signal Process. **56**, 3023–3028 (2008)
6. Theys, C., Ferrari, A., Vieira, M.: Marginal Bayesian analysis of polynomial-phase signals. Signal Process. **81**, 69–82 (2001)
7. Djurovíc, I., Stankovíc, L.: Quasi-maximum-likelihood estimator of polynomial phase signals. IET Signal Process. **8**(4), 347–359 (2014)
8. Deng, Z., et al.: Compound time-frequency domain method for estimating parameters of uniform-sampling polynomial-phase signals on the entire identifiable region. IET Signal Proc. **10**(7), 743–751 (2016)
9. Kay, S.M.: Fundamentals of Statistical Signal Processing: Estimation Theory. Prentice Hall, New Jersey, 595 p. (1993)
10. Lehmann, E.L., Casella, G.: Theory of Point Estimation, 2nd edn. Springer, New York (1998). Chapter 4
11. Van Trees, H.L.: Optimum Array Processing. Part IV, 1443p. Wiley, Hoboken (2002)
12. Brynolfsson, S.J., Jakobsson, A., Hansson-Sandsten, M.: Sparse semi-parametric estimation of harmonic chirp signals. IEEE Trans. Signal Process. **64**(7), 1798–1807 (2016)
13. Vostretsov, A.G.: Efficient signal parameter estimation under the conditions of the a priori uncertainty using complete sufficient statistics. J. Communi. Technol. Electron. **44**(5), 512–517 (1999)

14. Ivanov, S.I., Kyrnyshev, A.M., Lavrov, A.P.: Measuring radar cross-section of complex-shaped objects using the Doppler shift. In: International Siberian Conference on Control and Communications, SIBCON 2015-Proceedings, pp. 1–4. https://doi.org/10.1109/SIBCON.2015.7147075

15. Ivanov, S.I., Liokumovich, L.B., Medvedev, A.V.: Estimation of the parameters of the phase modulated signal in presence of the background noise using complete sufficient statistics. In: 2017 XX IEEE International Conference on Soft Computing and Measurements (SCM), pp. 11–13 (2017). https://doi.org/10.1109/scm.2017.7970480

16. Molodjakov, S.A., Ivanov, S.I., Lavrov, A.P.: Optoelectronic pulsars' processor and its real-time software. In: 2017 IEEE II International Conference on Control in Technical Systems (CTS), pp. pp. 59–62 (2017). https://doi.org/10.1109/ctsys.2017.8109488

17. Kudryashov, A.V., Liokumovich, L.B., Medvedev, A.V.: Digital demodulation methods for fiber interferometers. Opt. Mem. Neural Netw. (Inf. Opt.) 22(4), 236–243 (2013)

18. Lehmann E.L., Scheffe H.: Completeness, similar regions, and unbiased estimation-Part I, Part II. In: Rojo J. (ed.) Selected Works of E. L. Lehmann. Selected Works in Probability and Statistics. Springer, Boston (2012). https://doi.org/10.1007/978-1-4614-1412-4_23

Dynamics of Polypeptide Cluster Dipole Moment for Nano Communication Applications

Elena Velichko$^{(\boxtimes)}$ (iD), Tatiana Zezina (iD), Maxim Baranov (iD),
Elina Nepomnyashchaya (iD), and Oleg Tsybin (iD)

Peter the Great St. Petersburg Polytechnic University, St. Petersburg, Russia
velichko-spbstu@yandex.ru

Abstract. Computer simulation of instantaneous time-dependent dipole moment and related nano-electromagnetic field of peptide network, or cluster, in vacuum and aquatic environment, reveal possibilities of molecular system control by means of external electric field. Protein water solution electrical conductivity measurements show some frequency resonances. Probably, revealed effects could be used in nano communication systems in RF-Microwave-THz frequency range.

Keywords: Biomolecular cluster · Dipole moment · Nano-electromagnetic field
Computer simulations

1 Introduction

Biomolecular nanomaterials are relevant to "bio" sciences like biomedical, biochemistry, biophysics, and else to some emerging applications in electronics, computers, and nano communication as well. A number of electron devices appear, based on biomolecular platform: transistors, switches, memory elements, sensors. Nano communication devices with embedded biomolecules appear as promising high efficiency tools [1–3]. Due to unique biomolecular properties, such innovative devices could possesses high performance, speed, stability, and low energy consumption. In order to create such devices, required are more data on charge transfer processes, dynamics of thin liquid films, influence of external conditions on biomolecular structure. Instantaneous time-dependent magnitude and orientation of a dipole moment is a convenient and almost universal characteristic, sensitive to any biomolecular system geometric changes [4], which arise under the influence of the external electrostatic field, surrounding molecules environment, temperature changes, etc. [5–8]. Computer simulation of biomolecular dipole moments dynamics is predictive theoretical platform for new devices development, and a self-reliant approach to complex biomolecular systems study. Computer modelling provides behavior of a single molecule placed in an external electric field, which stimulate effect of aggregate/self-assemble [6]. Further, time-dependent dipole moment of single peptides (polyalanin, 2–24 groups) at picosecond time scales in vacuum and aqueous solutions were determined with molecular dynamics method to evaluate effects of external environment (temperature, amplitude and the orientation vector of the electrostatic field) [8]. In [8], dynamic scenarios for single isolated

© Springer Nature Switzerland AG 2018
O. Galinina et al. (Eds.): NEW2AN 2018/ruSMART 2018, LNCS 11118, pp. 675–682, 2018.
https://doi.org/10.1007/978-3-030-01168-0_62

molecules were calculated with the time step of 1 fs, and the simulation time up to 100 ns under the external electrostatic field E up to 10^8 V/m. Such approach reveals required control of biomolecule dynamic parameters by means of an external electric field.

Instead of early single molecule investigation [8], here we study a biomolecular network, or cluster, research, namely the evolution of both instantaneous time-dependent dipole moment (theoretically, by means of computer modelling) and electro conductivity experimental measurements as well.

2   Methods

2.1   Computer Simulation

We realized dynamic simulation of molecular network, or cluster, consists of four diphenylalanine (FF) peptides. Software packages allow correctly assess internal physical processes and also to visualize biomolecular objects, analyze their dynamics, and restructuring under operation of external electric field. Initial data for computer simulation include some biomolecules in vacuum and in solution at various concentrations, temperatures, under various amplitudes and orientation of external electrostatic field. We used Avogadro program [9] to create the computational model of FF alpha-helix structure containing all coordinates of molecular atoms in one pdb-file. The VMD program [10] was utilized for molecular evolution analysis and for further system development. We placed four FF peptides in vertices of a square with sides distant for approximately 20 Å from each other. Each FF molecule had been turned around z-axis for 90 degrees. We studied the system evolution both in vacuum and in the aqueous environment built as a box distant from all peptide molecules for at least 5 Å and filled with water molecules with TIP3 [11] parameters. Positive Na^+ and negative Cl^- ions were added in solution for the concentration of 0.15 mol/L.

All MD simulations were performed with CHARMM27 [12] force field using the NAMD [13] package under periodic boundary conditions and a cut-off radius for non-bonded interactions with switching function starting at a distance of 12 Å and reaching zero at 14 Å for the aqueous environment and extended this value up to 65 Å for vacuum.

Total runtime was 5 ns for all dynamic scenarios with 1 fs–0.2 fs time-step for peptides in water and vacuum respectively. Initial thermalization in aqueous box was achieved by twice repeated cycle of minimizing energy of the system, heating the system up to 500 K, in increments of 20 K for time intervals of 1.5 ps and a few ps MD run at the highest temperature, followed by cooling of the system back to 300 K by reducing the temperature of the system at the same pace of 20 K decrements every 1.5 ps and a few ps MD run. The energy of the entire system was minimized in 5000 steps. During both cycles the simulation box was allowed to relax its size. First cycle was held with the peptide molecules fixed. Equilibrated system of approximately $44 \times 34 \times 26$ Å^3 in size was investigated with a Langevin thermostat set at 300 K and the Berendsen control of 1.01325 Bar target pressure. NAMD allows to run the molecular dynamics simulations of the biomolecular systems under constant external electrostatic field in units of 1 kcal/(mole·Å·e) $\approx 4.35 \cdot 10^8$ V/m. We used the field of 0.01 and 0.1 kcal/(mole· Å ·e) for our investigation of the protein system in aqueous medium. The system in vacuum was limited in space by adding building a cube ($80 \times 80 \times 80 \text{ Å}^3$) with periodic boundary conditions and was prepared only by minimizing the energy of the system for 500 steps and rescaling the velocities to the target temperature of 300 K.

Further, we calculated instantaneous time-dependent dipole vector moment $\vec{d_k}$ for each FF peptide in VMD program. Resulting instantaneous time-dependent dipole vector moment for whole cluster was estimated as a vector sum of each FF molecule's dipole,

$$\vec{D} = \sum_{k=1}^{4} \vec{d_k}.$$

Finally, module of vector $\vec{D}$ instantaneous value shows a level of mutual orientation of all molecules dipole vector moments $\vec{d_k}$, so demonstrates interaction and self-organisation processes inside the cluster.

2.2 Molecular Film Conductivity Measurements

It is known, that macromolecules in a solution are prone to self-assembly, and formation of some ordered structures. Ability of such molecules in solution to change certain properties and form self-assembled structures is discussed in a number of publications [14–17]. Moreover, the type of these structures, and consequently the parameters of conductivity of the solution, are influenced by some external factors such as electric and magnetic fields, frequency of the current flowing through the sample, the acidity of the solvent, the temperature and humidity of the environment. Solutions of albumin protein (molecular mass about 65 kDa) were investigated in the experimental part of this work. It is multifunctional, and therefore sensitive to many types of external influences. Electrical conductivity of biomolecular sample solutions were measured, which looks sensitive to an external electric field. So, such biomolecular network could be used as active media in biomolecular devices. The experimental setup is shown on Fig. 1. The gap width between the aluminum electrodes 1 was approx. 2 mm. A sample drop volume was around of 4 µl, placed in the gap. Alternative current electric voltage V_G (frequency f = 10–10000 kHz) was applied at the sample. The measurements were taken after time delay the source of the alternating current was turned on, during this time the molecules could undergo self-assembly in the solution.

To study electrical field effect on biomolecular solutions (with experimental setup shown in Fig. 1), one needs determine total amplitude of the current passing through the sample electrical circuit. Therefore, it is necessary to take into account the sample total impedance, or complex resistance, which include capacitive and inductive components. At that stage, the voltage drop $U_{out} = V_G - V$ at the sample of molecules was determined. Then sample impedance Rs value was calculated as follows:

$$R_s = \left(\frac{V_G}{V} - 1 \right) R_v.$$

$$R_s = \left(\frac{V_G}{V} - 1\right) R_v.$$

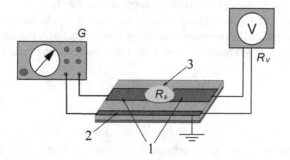

Fig. 1. Experimental setup for biomolecular solution sample electrical conductivity measurements. G—RF electrical generator with voltage Vg; 1—signal electrodes; 2—ground electrode; 3—sample on the dielectric target, situated in the gap between electrodes 1; Rs–electrical impedance of the sample; V—digital voltmeter; Rv–input impedance of voltmeter.

3   Results and Discussion

3.1   Computer Simulation

Received results are shown at figures below, showing both conformation and orientation of peptide molecules, and time-dependent instantaneous dipole moment realizations.

Figure 2 represents evolution of modules D of instantaneous time-dependent dipole moment for four diphenylalanine peptides P1-P4 and their vector sum in vacuum and zero external electric field at 5 ns period. Dipole moment modules dynamics reveal THz frequency band oscillations.

At initial stage, molecular cluster possesses excess internal energy and related high averaged dipole moment module as well. Stochastic transform occurs abruptly approximately at 1.3 ns, and new low dipole moment stage establishes, stable at as long period as 4 ns and more. One can see instantaneous vector dipole moments realizations at Fig. 2 before (a) and after (b) that transformation. Probably, differences of these values occur due to the variety directions of peptides' dipole moments.

In aqueous medium (Fig. 3) the dipole moment oscillations, for all four molecules have small amplitudes and remain steady around average values meanwhile the vector sum of these values has smaller frequencies and significantly higher magnitude. Apparently, one can see effects of induced dipole moments inside water environment (Fig. 4).

For a weak or zero electrostatic field, the probability of correlated molecular arrangement on average is almost equals to zero, as it can be seen on the graph (Fig. 3, a) for the weak electric field of 0.01 kcal/(mole·Å·e) $\approx 4.35 \cdot 10^6$ V/m. Under a higher external electric field of 0.1 kcal/(mole·Å·e) $\approx 4.35 \cdot 10^7$ V/m (Fig. 3, b) the correlated ordered state of the vectors of the electric dipole moments of peptides establishes, and as a result the summarized dipole moment on average corresponds to a higher value.

Application of an external electric field results in reorientation of the dipole moment of the molecules in the field and reaches a steady state stage. The investigation of average

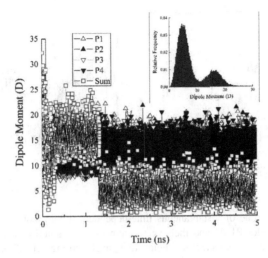

Fig. 2. Evolution in time of instantaneous time-dependent dipole moment modules for four diphenylalanine peptides P1-P4 and their **vector sum** in vacuum, zero external electric field.

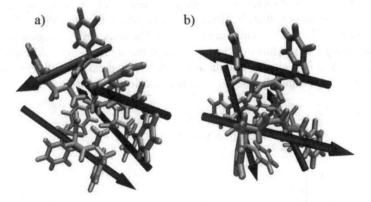

Fig. 3. Instantaneous vector dipole moments displacement before (a, disordered stage) and after (b, ordered stage) transition, occurred at 1.3 ns approximately.

dipole moment for different phenylalanine peptides in aqueous medium and in vacuum may provide a new data on the most probable conformational states of these molecules and give a more accurate knowledge on FF motif self-assembled structures formation under various conditions.

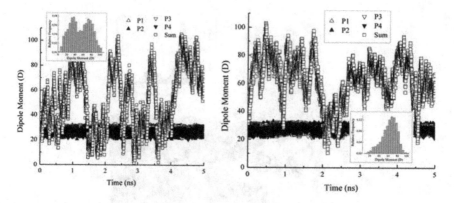

Fig. 4. Evolution in time of instantaneous time-dependent dipole moment modules for four diphenylalanine peptides P1-P4 and their vector sum in water solution and different external electric field. Left panel – weak electrical field, right panel – strong field 42 mV/Å.

3.2   Conductivity Measurements

Electrical conductivity of molecular sample dependence on time, temperature, concentration (data not shown), voltage V_G, frequency f of a current flowing through the sample, as well as on the pH of the solvent, were measured. The values of V voltage as a function of the frequency f of the flowing current for albumin solutions with different pH are shown at Fig. 5. It is evidently, that these factors would affect the self-organization of the protein and, consequently, the parameters of its electrical conductivity.

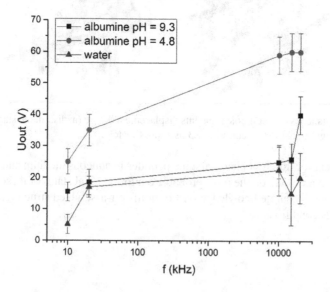

Fig. 5. Dependence of voltage on frequency in pure water and albumin molecular solutions with pH = 9.3 and pH = 4.8 (isoelectric point).

One can notice that the voltage on the sample is increasing with the frequency. The non-linear behavior of these plots can testify self-assembly process in the molecular solution. Probably, voltage difference observed with different pH could be explained by variations of ion concentration and self-organized structure of the sample in isoelectric point.

4 Conclusion

Ordered structures formation were investigated for a molecular networks, or clusters, in vacuum and water solution as well. For the first time, collective dipole moment oscillations in Microwave-THz band revealed with computer modelling. In water solution, these oscillations had two levels, one of them with a higher and other one with a lower value of average summed dipole magnitude. The lower level relates to self-ordered system with pairwise antiparallel arrangement of peptides' dipole moments, and so lowest total energy. In external electrostatic field, collective oscillations possesses excess energy, and so relate to high level. The corresponding nano-electromagnetic field is generated in two states with a possible controllable switching between them by means of an external signal. Protein water solution electrical conductivity measurements show frequency resonances in any parts of RF band.

These observations may facilitate future studies on the controlled formation of nanostructured aggregates of peptides and the understanding of their electro-mechanical properties. Actually, revealed effects could be used in a nano communication system in Microwave-THz frequency range.

References

1. Velichko, E., Zezina, T., Cheremiskina, A., Tsybin, O.: Nano communication device with embedded molecular films: effect of electromagnetic field and dipole moment dynamics. In: Balandin, S., Andreev, S., Koucheryavy, Y. (eds.) ruSMART 2015. LNCS, vol. 9247, pp. 765–771. Springer, Cham (2015). https://doi.org/10.1007/978-3-319-23126-6_71

2. Tsybin, O.: Nano-device with an embedded molecular film: mechanisms of excitation. In: Balandin, S., Andreev, S., Koucheryavy, Y. (eds.) ruSMART 2015. LNCS, vol. 9247, pp. 772–777. Springer, Cham (2015). https://doi.org/10.1007/978-3-319-23126-6_72

3. Dyubo, D., Tsybin, O.Y.: Nano communication device with an embedded molecular film: electromagnetic signals integration with dynamic operation photodetector. In: Galinina, O., Andreev, S., Balandin, S., Koucheryavy, Y. (eds.) NEW2AN/ruSMART/NsCC -2017. LNCS, vol. 10531, pp. 206–213. Springer, Cham (2017). https://doi.org/10.1007/978-3-319-67380-6_19

4. Vasanthi, H., Krishnaswamy, S.: Dipole moment in TIM alpha/beta fold proteins. Indian J. Biochem. Biophys. **40**(3), 194–202 (2003)

5. Miller, C.A., Hernández-Ortiz, J.P., Abbott, N.L., Gellman, S.H., Pablo, J.J.: Dipole-induced self-assembly of helical β-peptides. J. Chem. Phys. **129**, 015102 (2008)

6. Kelly, C.M., Northey, T., Ryan, K.: Conformational dynamics and aggregation behavior of piezoelectric diphenylalanine peptides in an external electric field. Biophys. Chem. **196**, 16–24 (2015)

7. Ripoll, D.R., Vila, J.A., Scheraga, H.A.: On the orientation of the backbone dipoles in native folds. Proc. Natl. Acad. Sci. U.S.A. **102**, 7559–7564 (2005)

8. Zezina, T.I., Tsybin, O.Y.: Subpicosecond dynamics of the molecular polyalanine dipole moment. Phys. Mathe. (2017). St. Petersburg Polytechnical University

9. Hanwell, M.D., Curtis, D.E., Lonie, D.C., Vandermeersch, T., Zurek, E., Hutchison, G.R.: Avogadro: an advanced semantic chemical editor, visualization, and analysis platform. J. Cheminf. **4**, 17 (2012). https://doi.org/10.1186/1758-2946-4-17

10. Humphrey, W., Dalke, A., Schulten, K.: VMD: visual molecular dynamics. J. Mol. Graph. Model. **14**(1), 33–38 (1996). https://doi.org/10.1016/0263-7855(96)00018-5

11. Jorgensen, W.L., Chandrasekhar, J., Madura, J.D., Impey, R.W., Klein, M.L.: Comparison of simple potential functions for simulating liquid water. J. Chem. Phys. **79**(2), 926–935 (1983). https://doi.org/10.1063/1.445869

12. Mackerell, A.D., Feig, M., Brooks, C.L.: Extending the treatment of backbone energetics in protein force fields: limitations of gas-phase quantum mechanics in reproducing protein conformational distributions in molecular dynamics simulations. J. Comput. Chem. **25**, 1400–1415 (2004). https://doi.org/10.1002/jcc.20065

13. Phillips, J.C., et al.: Scalable molecular dynamics with NAMD. J. Comput. Chem. **26**, 1781–1802 (2005). https://doi.org/10.1002/jcc.20289

14. Velichko, E., Baranov, M., Nepomnyashchaya, E., Cheremiskina, A., Aksenov, E.: Studies of biomolecular nanomaterials for application in electronics and communications. In: Balandin, S., Andreev, S., Koucheryavy, Y. (eds.) ruSMART 2015. LNCS, vol. 9247, pp. 786–792. Springer, Cham (2015). https://doi.org/10.1007/978-3-319-23126-6_74

15. Chen, Q., Liu, X., Chen, J., Zeng, J., Cheng, Z., Liu, Z.: A self-assembled albumin-based nanoprobe for in vivo ratiometric photoacoustic pH imaging. Adv. Mater. **27**(43), 6820–6827 (2015)

16. Chen, B., He, X.Y., Yi, X.Q., Zhuo, R.X., Cheng, S.X.: Dual-peptide-functionalized albumin-based nanoparticles with pH-dependent self-assembly behavior for drug delivery. ACS Appl. Mater. Interfaces. **7**(28), 15148–15153 (2015)

17. Choi, S.H., et al.: Inhalable self-assembled albumin nanoparticles for treating drug-resistant lung cancer. J. Controlled Release **197**, 199–207 (2015)

X-Ray Scattering by Antiphase Ferroelectric Domain Walls in the Antiferroelectric Phase of the $PbZr_{0.985}Ti_{0.015}O_3$

Sergej Vakhrushev[1,2]([✉]) [iD], Daria A. Andronikova[1,2] [iD],
Dmitry Y. Chernyshov[3] [iD], Alexey V. Filimonov[2] [iD],
Stanislav A. Udovenko[1,2] [iD], and N. V. Ravi Kumar[4] [iD]

[1] Ioffe Institute, Polytekhnicheskaya 26, St. Petersburg 194021, Russia
s.vakhrushev@mail.ioffe.ru
[2] Peter the Great St. Petersburg Polytechnic University, Polytechnicheskaya 29,
St.-Petersburg 195251, Russia
[3] Swiss-Norwegian Beamlines at ESRF, BP 220, 38043 Grenoble Cedex, France
[4] Indian Institute of Technology-Madras (IIT Madras), Chennai 600036, India

Abstract. The results of the X-ray diffuse scattering (DS) measurements of the Zr-rich $PbZrO_3$ - $PbTiO_3$ solid solution $PbZr_{0.985}Ti_{0.015}O_3$ (PZT1.5) are presented. Measurements were performed in zero electric field and in applied electric field E = 5 kV/cm. In the antiferroelectric phase diffuse scattering streaks around Σ superstructure peaks $(h + \frac{1}{4} \ k + \frac{1}{4} \ l)$ were found and interpreted as a scattering on ferroelectric antiphase domain walls. This conclusion is corroborated by the observation of a strong influence of the electric field on these streaks. Reported results are important for the prospective application of the antiferroelectrics as the basis for the high-density non-volatile memory devices.

Keywords: Antiferroelectrics · Domain walls · Diffuse scattering

1 Introduction

Antiferroelectricity was discovered nearly 70 years ago. First materials identified as antiferroelectrics (AF) were $PbZrO_3$ (PZ) [1,2] and WO_3 [3], both of these belongs to the group of the crystals with the displacive phase transitions. Later the order-disorder type antiferoelectrics like $(NH_4)_2H_2PO_4$ (ADP) were found [4]. Very soon after the discovery of the PZ the $(PbZrO_3)_x(PbTiO_3)_{1-x}$ (PZT) solid solutions were synthesized [5]. It was demonstrated that even rather small amount of the Ti doping results in the transformation of the antiferroelectric structure into the ferroelectric one [6]. For long time the main interest was concentrated on the ferroelectric compositions with x≈0.48 in the vicinity of the so-called morphotropic phase boundary (MPB), where PZT demonstrates

© Springer Nature Switzerland AG 2018
O. Galinina et al. (Eds.): NEW2AN 2018/ruSMART 2018, LNCS 11118, pp. 683–690, 2018.
https://doi.org/10.1007/978-3-030-01168-0_63

extremely large electromechanic response [7]. However in last decade a revival of the interest to the antiferroelectrics is observed. There are several reasons for this revival. First one is that despite the long history of the antiferroelectrics studies we do not have clear microscopic picture of the microscopic mechanism of the phase transitions. The other reason is practical significance of the antiferro-electrics. Recently it was demonstrated that AF can be considered as promising materials for the design of the capacitor-type fast energy storage devices [8] and as the basis of electrocaloric cooling systems [9,10]. We would like to emphasize the third practically important application of the antiferroelectrics based on the existence of the polar domain walls.

As in any other ferroic material phase transition from the high symmetry phase to the low symmetry one results in the appearance of the domain struc-ture in the way that the average symmetry of the macroscopic crystal is con-served. Ferroic domains recently attracted special attention. The details can be found in the book [11]. Along with domains of great interest are domain bound-aries and intergranular interfaces. The obtained results indicate the exceptional properties of ferroelectric interfaces, such as: the appearance of new phases in the broadened domain walls [12], ferroelectricity in ferroelastic domain walls [13], ferromagnetism in ferroelectric walls [14,15], ferroelectric properties of interfaces between two dielectric layers [16,17]. The mechanisms of most of these effects remain insufficiently studied.

Domain configuration of the antiferroelectrics deserves special consideration. Like in ferroelectrics the ferroelastic domains are created. Due to the absence of the polarisation we expect to see only 90 degrees ferroelastic domains. Instead of the 180 degrees ferroelectric domains translational antiphase boundaries (APB) are created. In the paper [18] the results of the atomic resolution TEM measure-ments were reported showing APB. Based on the atomic displacements pattern it was suggested that these APB are ferroelectric, i.e. have spontaneous polarisa-tion. The ferroelectric nature of these APB was confirmed using piezoresponse force microscopy technique [19]. However piezoresponse force microscopy mea-surements at variable temperatures, especially in the region of the temperatures corresponding to the phase transitions in the Zr-rich PZT, are practically impos-sible.

One of the most effective way of studying nanodomains and domain walls in ferroics is X-ray (or neutron) scattering [20]. In the case of the antiphase domains and APBs it is probably the only technique for such studies at the variable external conditions (temperature and electric field). To the best of our knowledge there are no papers where the domain configuration in the antiferroelectric single crystals was studied as a function of temperature. Such information is of crucial importance for the practical applications.

2   Experiment

Measurements were performed with the $PbZr_{0.985}Ti_{0.015}O_3$ (PZT1.5) single crys-tal grown in the Southern Federal University by the technique described in [21].

Sample in the shape of the rectangular rod 30 * 200 * 1000 μm³ was prepared by cutting, polishing, and subsequent etching in hydrochloric acid. Long axis of the rod was parallel to the [110] axis of the crystal. Silver electrodes were on the top and bottom of the rod providing the possibility of the high voltage application. Nominal electric field applied to the crystal during the measurements in the presence of the electric field was 5 kV/cm. The calculations of the field distribution inside the sample were performed using the COMSOL software [22]. The value of the field was shown to depend on the dielectric permittivity of the crystal and consequently on the sample temperature. The field at the sample center amounted to about 4.5 kV/cm at temperatures just above the transition temperature from cubic to the intermediate rhombohedral ferroelectric phase and to about 4 kV/cm above the transition from the ferroelectric to the antiferroelectric phase.

The sample was mounted on the special high voltage setup [23] that was placed on the goniometer of the PILATUS@SNBL diffractometer at the BM01 beamline of the European Synchrotron Radiation Facility (CRG Swiss-Norwegian beamlines). Diffraction data were recorded using PILATUS2M pixel area detector. The measurements were performed with the X-ray energy 13 keV just below the absorption edge for lead. Diffraction patterns from the PILATUS were preliminary treated using SNBL toolbox program [24] and final reconstruction of the scattering intensity in the reciprocal space was done using the π map software and by a custom 3D reconstruction program based on MATLAB.

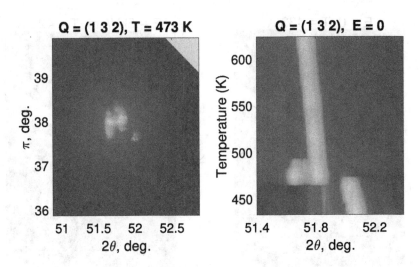

Fig. 1. (a) - $2\theta - \omega$ map of the X-ray diffraction at 330 K; (b) - temperature evolution of intensity distribution near [1 3 2] cubic node of reciprocal space.

In all cases measurements were performed on cooling. Sample was heated up to 620 K that is into the paraelectric phase and afterwards was slowly cooled to the desired temperature. Most measurements in the antiferroelectric phase were carried out at 430 K. Measurements in the applied electric field were performed in the field cooled (FC) regime. Voltage was applied to the sample at 630 K and following measurements were performed at the same conditions as without the field.

3   Results and Discussions

To control the transition temperature the series of fast measurements were performed on cooling. The results were presented in the form of the $2\theta - \omega$ maps for several Bragg peaks and temperature evolution of these maps was analysed. In the Figs. 1a and b $2\theta - \omega$ map for the (132) Bragg peak at 473 K is shown and the temperature evolution of this map is demonstrated. (132) reflection is of low symmetry and sensitive for all symmetry changes. The appearance of the splitting of the peak at around 490 K due to the transition from the cubic paraelectric phase to the rhombohedral ferroelectric one is clearly seen. Next transition from the ferroelectric to the orthorhombic antiferroelectric state is evidenced by the sharp change of the splitting pattern at around 463 K.

3.1   Measurements Without Electric Field

As it was mentioned above most of the results reported in the present paper were related to the study of the antiferroelectric state and were performed at

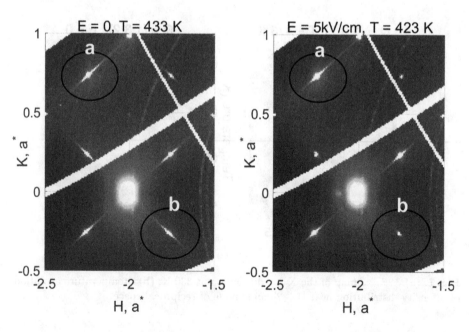

Fig. 2. PZT1.5 X-ray diffraction pattern at 430 K without electric field (left panel) and in the nominal field 5 kV/cm (right panel). Points a correspond to the (110)-type domains and points b to the $(1\bar{1}0)$-type domains.

430 K. In the Fig. 2a two-dimensional map of the scattering intensity around the $(\overline{2}00)$ reciprocal lattice point is shown. In addition to the typical single crystal diffraction spots some powder diffraction rings are observed probably due to some pieces of the thin gold wire connecting the electrodes on the sample with the high-voltage supply. In agreement with earlier published data no anisotropic diffuse scattering around the Bragg peak is observed. As it was demonstrated earlier this indicates strong suppression of the ferroelectric fluctuations in the antiferroelectric phase. Clear Σ peaks $\{h + \frac{1}{4} \ k + \frac{1}{4} \ l\}$ (points a and b in the Fig. 2a) are seen. Since the star of the $q_\Sigma = \{\frac{1}{4} \ \frac{1}{4} \ 0\}$ wavevector describing the antiferroelectric structure has 12 arms, we can expect up to 12 domains. However since the structures described by q_Σ and $-q_\Sigma$ are equivalent only 6 domains could be distinguished with diffraction. Wave vector q_Σ completely describes the antiferroelectric displacements in the domain. In the case of $q_\Sigma = (\frac{1}{4} \ \frac{1}{4} \ 0)$ (point a in the Fig. 2) the ionic displacements are in $[1\overline{1}0]$ direction, while in the case of $q_\Sigma = (\frac{1}{4} \ \frac{\overline{1}}{4} \ 0)$ (point b in the Fig. 2) the ionic displacements are in $[110]$ direction (in the Fig. 2 only (hk0) scattering plane is shown so the other domains are not seen). Without electric field the intensities of the different superstructure peaks are roughly the same indicating the uniform distribution of the domains in the sample.

Around all Σ peaks long and narrow diffuse streaks are seen. To our knowledge those streaks were never reported before. The shape of the scattering in the reciprocal space is the Fourier transform of the shape of the object in the real space. In the present case it means that the diffuse scattering (DS) streaks are related to some 2-d objects in the crystal. Taking into account that the streaks are around the Σ-superstructural peaks conclusion can be made that they are related to the APB. As it was mentioned before APB in antiferroelectrics are expected to be polar, with ionic displacements perpendicular to the q_Σ. To determine the APB size the quantitative model has to be developed (will be published later). Rough estimation can be made based on the one dimensional cuts of the scattering intensity map. In the Fig. 3 such 1-d plots for the $[110]$ direction (A) and the $[1\overline{1}0]$ direction (B) is shown. Without electric field the intensities of the DS in both directions are similar indicating similar density of the APBs. The width of the intensity distribution perpendicular to the streak (not shown in the plots) is related to the APB length. It was resolution limited that gives the lower bound for the length to be about 100 lattice parameters (400 Å). At the moment we can give only very rough estimations of the streaks "length" (APB width) presented in the insets ti the Fig. 3. This width seems to be 10 to 15 lattice parameters (40 Å– 60 Å). This value is twice as large as reported for the pure $PbZrO_3$ by Wei at all [18] based on the electron diffraction measurements. To find the reason for the discrepancy higher resolution X-ray measurements are needed.

3.2   Measurements on the Applied Electric Field

One of the main aims of our experiment was to trace the influence of the electric field on the DS in the PZT single crystal. As stated above the field was applied

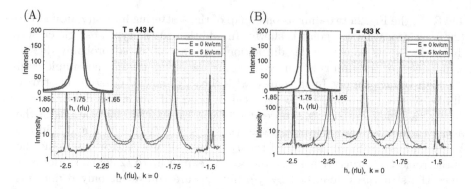

Fig. 3. Logarithmic plot of the one dimensional cuts of the scattering maps of the scattering intensity along the [110] direction (A) and the [1$\bar{1}$0] direction (B). In the insets small region around the superstructure peaks are shown with linear scale for the intensity

in the [110] direction (direction of the DS streaks inside the circles (a) in the Fig. 2). Both types of the 90-degrees domains are preserved in the presence of the electric field as it follows from the observation of (a) and (b) peaks in the right panel of the Fig. 2. On the other hand the diffuse streak inside the circle b is completely suppressed. This streak around $q_\Sigma = (\frac{1}{4}\,\frac{\bar{1}}{4}\,0)$ (point b in the Fig. 2) corresponds to the ionic displacements in the APB in [110] direction parallel to the applied electric field. The effect of the electric field is very clearly evidenced in the Fig. 3. In the left panel (panel A) the plot for the q_Σ arm along the field direction, i.e. ionic displacements perpendicular to the field, is shown, and in the right panel (panel B) - for the q_Σ arm perpendicular the field direction, i.e. ionic displacements along to the field. APBs with displacements perpendicular to the field are practically not affected by the electric field, while APBs with displacements parallel to the field disappear. There could be 2 explanations for this phenomena. The most likely one is that antiphase domains become energetically unfavourable and only 90-degrees domains survive. The other explanation is that the APBs get very thin (one lattice constant thick) and corresponding scattering becomes very broad and weak.

4   Summary

The influence of the electric field on the mesoscopic structure of the PZT1.5 in the antiferroelectric phase was studied using X-ray diffuse scattering technique. For the first time the existence of the narrow streaks of the diffuse scattering around the antiferroelectric superstructure peaks were reported. These streaks were attributed to the antiphase domain boundaries and the estimation of the APD width was obtained. We demonstrated that the application of the electric field in the FC regime does not affect the APBs polarized perpendicular to the field while the APBs polarized parallel to the field got extinct. Discovered effect demonstrates the principal possibility to control the APBs in antiferroelectric phase forming the basis for the use of such materials in high density memory devices.

Acknowledgement. We acknowledge N.G. Leontiev (AzovBlack Sea Engineering Institute, Don State Agrarian University) and I.N. Leontiev (Southern Federal University) for providing the single crystal of PZT. A.V. Filimonov and N.V. Ravi Kumar acknowledge the support of Russian Foundation for Basic Research (Grant No. 16-52-48016). S. Udovenko acknowledges the support of the Ministry of Science and Education of the Russian Federation, project no. 3.1150.2017/4.6. D. Andronikova acknowledges the support of Russian Foundation for Basic Research (Grant No. 16-29-14018) and the Russian President Scholarship No. SP-3762.2018.5

References

1. Roberts, S.: Dielectric properties of lead zirconate and barium-lead zirconate. J. Am. Ceram. Soc. **33**(1946), 63–66 (1950)
2. Shirane, G., Sawaguchi, E., Takagi, Y.: Dielectric properties of lead zirconate. Phys. Rev. **84**(3), 476–481 (1951)
3. Matthias, B.T., Wood, E.A.: Low temperature polymorphic transformation in WO_3. Phys. Rev. **84**(6), 1255–1255 (1951)
4. Mason, W.P.: The elastic, piezoelectric, and dielectric constants of potassium dihydrogen phosphate and ammonium dihydrogen phosphate. Phys. Rev. **69**(5–6), 173–194 (1946)
5. Shirane, G., Hoshino, S.: Crystal structure of the ferroelectric phase in $PbZrO_3$ containing Ba or Ti. Phys. Rev. **86**(2), 248–249 (1952)
6. Shirane, G., Suzuki, K., Takeda, A.: Phase transitions in solid solutions of $PbZrO_3$ and $PbTiO_3$ (II) X-ray study (1952)
7. Jaffe, B., Cook, W.J., Jaffe, J.: Piesoelectric Ceramics. Academic Press, London (1971)
8. Hao, X.: A review on the dielectric materials for high energy-storage application. J. Adv. Dielectr. **03**(01), 1330001 (2013)
9. Mischenko, A.S., Zhang, Q., Scott, J.F., Whatmore, R.W., Mathur, N.D.: Giant electrocaloric effect in PZT. Science **104**(1), 9–13 (2014)
10. Glazkova-Swedberg, E., Cuozzo, J., Lisenkov, S., Ponomareva, I.: Electrocaloric effect in $PbZrO_3$ thin films with antiferroelectric-ferroelectric phase competition. Comput. Mater. Sci. **129**, 44–48 (2017)
11. Tagantsev, A.K., Cross, L.E., Fousek, J.: Domains in Ferroic Crystals and Thin Films. Springer, New York (2010). https://doi.org/10.1007/978-1-4419-1417-0
12. Jia, C.L., et al.: Direct observation of continous electric dipole rotation in flux-closure domains in FE PZT. 1420–1424 (2011). 2101
13. Wada, S., Kakemoto, H., Tsurumi, T.: Enhanced piezoelectric properties of piezoelectric single crystals by domain engineering. Mater. Trans. **45**(2), 178–187 (2004)
14. Rao, W.F., Wang, Y.U.: Domain wall broadening mechanism for domain size effect of enhanced piezoelectricity in crystallographically engineered ferroelectric single crystals. Appl. Phys. Lett. **90**(4) (2007)
15. Tagantsev, A.K., Courtens, E., Arzel, L.: Prediction of a low-temperature ferroelectric instability in antiphase domain boundaries of strontium titanate. Phys. Rev. B **64**(22), 224107 (2001)

16. Goncalves-Ferreira, L., Redfern, S.A.T., Artacho, E., Salje, E.K.H.: Ferrielectric twin walls in CaTiO3. Phys. Rev. Lett. **101**(9), 1–4 (2008)
17. Bousquet, E., et al.: Improper ferroelectricity in perovskite oxide artificial superlattices. Nature **452**(7188), 732–736 (2008)
18. Wei, X.-K., Tagantsev, A.K., Kvasov, A., Roleder, K., Jia, C., Setter, N.: Ferroelectric translational antiphase boundaries in nonpolar materials. Nature Communi. **5**, 3031 (2014)
19. Andreeva, N.V.V., et al.: Domain structures and correlated out-of-plane and in-plane polarization reorientations in $Pb(Zr_{0.96}Ti_{0.04})O_1$ single crystal via piezoresponse force microscopy. AIP Adv. **6**(9) (2016)
20. Bruce, A.D., Cowley, R.A.: Structural Phase Transitions. Taylor and Francis, London (1981)
21. Leontiev, N.G., Smotrakov, V.G., Fesenko, O.E.: Phase diagram of $PbZr_{1-x}Ti_xO_3$ at x < 0,1. Izv. Akad. Nauk SSSR, Neorg. Mater. **18**(449) (1982)
22. COMSOL Multiphysics Reference Manual. COMSOL, Inc. www.comsol.com
23. Udovenko, S.A., Chernyshov, D.Y., Andronikova, D.A., Filimonov, A.V., Vakhrushev, S.B.: The technique of studying X-Ray scattering over wide temperature range in an electric field. Phys. Solid State **60**(5) (2018)
24. Dyadkin, V., Pattison, P., Dmitriev, V., Chernyshov, D.: A new multipurpose diffractometer PILATUS@SNBL. J. Synchrotron Radiat. **23**(3), 825–829 (2016)

Study of Self-assembled Molecular Films as a Method of Search for Promising Materials in Nanoelectronics and Nanocommunications

Elena Velichko⬤, Elina Nepomnyashchaya⬤,
and Maxim Baranov$^{(\boxtimes)}$⬤

Peter the Great Saint Petersburg Polytechnic University, Saint Petersburg, Russia
baranovmal993@gmail.com

Abstract. In the following article, we have considered problems and ways of transition from semiconductor to biomolecular electronics. We have used experimental studies of biomolecular materials to achieve goals for biomolecular electronics and communications. The paper studies a joint use of laser correlation spectroscopy and visual microscopic methods for investigation of self-assembling of proteins at transition from the liquid state to films. Furthermore, we have presented preliminary results of studies of molecular solutions. Finally, it is shown that structure evolution of drying out protein solution is a complex multi-stage process, which depends on external conditions.

Keywords: Biomolecules · Self-organization · Protein films · Nanoelectronics
Biomolecular electronics

1 Introduction

An urgent problem today is the need to develop a number of modern electronic components and network equipment to support a computer network. In general all components made of inorganic, semiconductor compounds. However, this undoubtedly limits a further development of communication technologies. As a result, to develop electronic basis and new materials, functional and methodical resources, a number of urgent research problems are set:

- search for new materials, processes and structures for changeover from silicon materials applied in communications;
- study of structure and properties of new materials, especially of perspective self-organized organic and biomolecular materials;
- investigation of electrical properties of semiconductor structures with built-in molecules to create new nanosized biomolecular electronic components;
- implementation of methods of diagnostics of molecular system self-organization processes; studies of geometrical, conformation and boundary non-uniformity during self-organization.

Consequently, nanocommunication is considered to become a major component for many novel applications. The recent developments are given in the scope of nano

O. Galinina et al. (Eds.): NEW2AN 2018/ruSMART 2018, LNCS 11118, pp. 691–701, 2018.
https://doi.org/10.1007/978-3-030-01168-0_64

machinery, coordination and control of these devices becomes the critical challenge to be solved [1].

Self-organizing molecular structures and the creation of functional devices are based on structured molecular films. This fact is an important part of the nano-communication study.

Nanocommunication devices can be based on molecular systems. What we mean is that these systems can be made of molecular self-assembled films. If we add external dynamic control to such films, it will allow us to control the parameters of the films.

As it can be seen, new materials, functional and methodical resources should be investigated [2]. It is important to emphasize, that biomolecules can become a new and perspective material for electronic components in networks and nanocommunications [3].

Self-organized materials will be needed in bottom-up nanofabrication of intelligent stimuli-driven 3D photonic materials and devices [4]. Therefore, creation of new biomolecular objects and study of their structural properties are important problems in modern physics. Especially, in such fields as: biotechnology, electronics and other scientific and technical fields [5–8]. Take for instance, the relevance of such research is confirmed by an increasing number of publications in the peer-reviewed journals devoted to this subject (Table 1).

Table 1. Number of publications in Scopus database [9]

Subject of publications	2014	2015	2016	2017
Biomolecular electronics	3 159	3 570	3 786	4 213
Molecular films	64 225	67 484	70 969	73 894

Expressively, interest in biomolecular technologies is explained by the fact that molecular systems are characterized by extremal miniaturization at which it is possible to increase spatial density of the device elements up to $10^{13} - 10^{14}$ elements per cm^2 in a nanometer layer. At the same time molecular elements have properties of self-organization that can make the nanodevice creation process quite simple. It is shown that biomolecules can effectively appear as conductors of an electric current, molecular switches, nanotransistors, nanodiodes, logical elements, nanobiochips, nanoengines, transformers of energy, biosensors, etc. There are laboratory and trial samples of devices based on the principles of biomolecular electronics (BME). One of the key questions for BME is studies of macromolecule self-organization processes. Self-organization of proteins in vitro can be considered as the simplest case.

It is impossible to deny, that self-organization can be studied at various levels, from single macromolecules of biopolymer to collective molecules. The hypothesis of protein folding or the "framework model" allows one to study this process at the level of separate molecules. That is why, this model postulated consecutive involvement of various interactions in formation of structure of protein. It also emphasized the importance of formation of α-spirals at early stages of self-organization [10].

In a living cell protein folding happens in macromolecular environment, therefore it is necessary to study self-organization of proteins in vitro in a protein – water system.

This process is considered within the theory of self-organization and is defined by the protein structure in nonequilibrium condition. In general the phenomenon of self-organization of protein depends on the physical and chemical properties of biomacromolecules.

In our previous work a number of preliminary experiments aimed at a study of electrical properties of protein solutions were carried out [11]. It was shown that proteins react differently to an electric field of different frequencies. But these experimental results require interpretation and theoretical background.

In the present work we have considered a possibility to create some models of the BME functional elements using textured self-organized protein films. In addition, we have also discussed development of the method of self-organized protein film formation at dehydration.

2 Experiments and Discussion

2.1 Experimental Techniques

The main stage of study of the processes of self-organization is the investigation of kinetics of formation of structures in the protein films obtained by dehydration of water solutions. Having considered several nondestructive methods of research of self-organization processes [12, 13], the method of visual control of dehydration process of protein solutions with the subsequent study of the obtained films was chosen. The experimental setup is presented in Fig. 1.

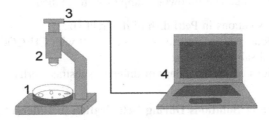

Fig. 1. Experimental setup. 1 — sample, 2 — microscope, 3 — CCD camera, 4 — computer.

The protein solutions were controlled by means of the laser correlation spectroscopic technique (Fig. 2). We have measured the sizes of nanoparticles and their agglomerates in the liquid by using this technique. The laser radiation scattered from nanoparticles was registered by photoelectron multiplier and was processed on computer with the help of the Tikhonov regularization method [14].

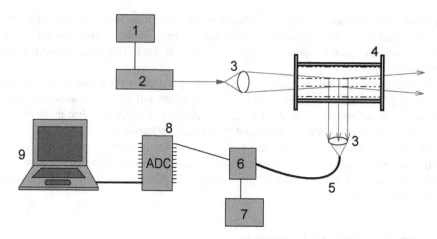

Fig. 2. Laser correlation spectrometer. 1 — laser power supply, 2 — junction laser λ = 665 nm, 3 — focusing lenses, 4 — cuvette with the sample, 5 — optical fiber, 6 — photoelectron multiplier, 7 — power supply of photoelectron multiplier, 8 — ADC, 9 — computer.

2.2   Solutions Under Study

We have studied a number of protein solutions that were dried on different substrate surfaces.

It is important to note, that we have chosen the albumin protein as the most explored protein, because it was suitable for us to work and analyze experimental results with this material. The following samples were studied:

– Albumin water solutions in Petri dish of different sizes;
– Albumin solutions near isoelectric point (pH = 3.9 with CH_3COOH acid);
– Albumin normal saline solutions;
– Albumin solutions on the boundary of different substrate surfaces.

2.3   Albumin Water Solutions During Dehydration in Different Petri Dishes

We have prepared the water solutions of albumin in a concentration of 100 g/l for our experiment. It is a well-known fact, that albumin protein is a globular protein. Therefore, under normal conditions it is a spheroid with a size of about 4.9 nm. Hence, it was confirmed by the laser correlation spectroscopic experiment. The size distribution of albumin molecules in water solution is shown in Fig. 3. Here N is the relative concentration and R is the radii of molecules.

Albumin solution was placed in two various Petri dishes to study the dehydration process. In the small dish we can see the solution contacts with the edges of the dish and in the big dish the drop of solution did not reach the dish's edges. The data were obtained after a one-day dehydration at room temperature (24 °C) (see Fig. 4).

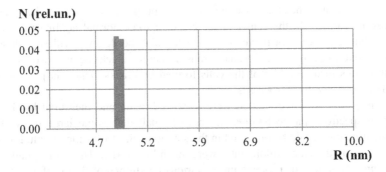

N (rel.un.)

Fig. 3. Size distribution of albumin molecules in water solution.

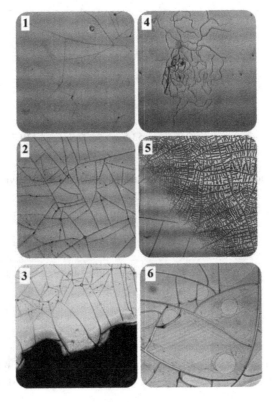

Fig. 4. Microscope photos of a protein film in various Petri dishes; 1, 2, 3 — photos of solution in the small dish taken from the center to edges; 4, 5, 6 — photos of solution in the big dish taken from the center to edges.

We can see that in the small dish the cracks are propagate from white circles and no structures are formed. In the center of the big dish some normal nonlinear cracks are observed, but if the camera moves closer to the dish border, we will see "dendrite" structures. Near the edge there is a sharp transition to the cracks again. Also we can see spiral structures in the middle of the cells formed by cracks. That can point to complicated self-organization processes.

The films that we have studied were dissolved again and measured by the laser correlation spectroscopic technique. In case of small dish some large parts of not-dissolved structures could be observed in the solution (Fig. 5). At the dissolution of the film from the big dish the agglomerates were much less in size. This corresponds to the other nature of binding of proteins under various conditions of drying (Fig. 6).

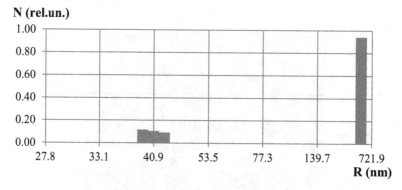

Fig. 5. Size distribution of albumin molecules in water solution: re-dissolving of film from the small Petri dish.

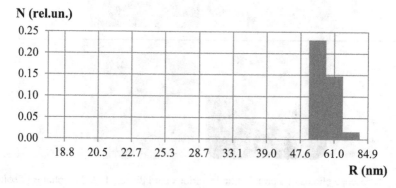

Fig. 6. Size distribution of albumin molecules in water solution: re-dissolving of film from the big Petri dish.

2.4   Albumin Solutions Near the Isoelectric Point

We have prepared the water solutions of albumin in a concentration of 100 g/l with addition of acetic acid for our experiment (for the solution pH = 3.9). This pH is close to the isoelectric point for albumin, in this regard albumin form agglomerates, which were detected by laser correlation spectroscopy (Fig. 7).

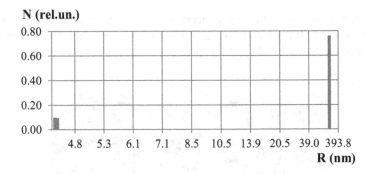

Fig. 7. Size distribution of albumin molecules in water solution near the isoelectric point.

The results of microscopy visualization of such protein film are presented in Fig. 8. Experiments were carried out in a big Petri dish.

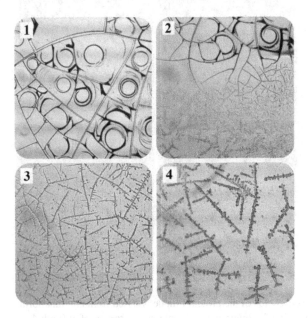

Fig. 8. Microscope photos of the protein film dehydrated from water solutions with pH = 3.9. 1 — spiral structures near the edge of film; 2 — the transition from spiral structures to "dendrites" while moving closer to the film center; 3, 4 — "dendrite" structures in the center of the dish.

The principal difference from solutions with normal pH could be noticed. In the center of the protein film the "dendrite" structures are formed. Their shapes quite differ from the previous experiment. It can be explained by presence of protein agglomerates in the initial solutions. It can also be noticed that spiral structures near the film edge are twirled counterclockwise.

2.5 Albumin Normal Saline Solutions

In these experiments, two different concentrations of the albumin in normal saline solution were used: 100 g/l and 30 g/l. The solutions were dehydrated in big Petri dishes at room temperature (24 °C) for three days. Results are presented in Fig. 9.

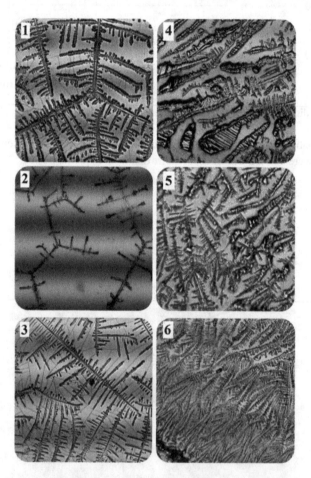

Fig. 9. Microscope photos of the protein film dehydrated from albumin normal saline solutions. 1, 2, 3 — dehydration of solution with albumin concentration of 100 g/l; 4, 5, 6 — dehydration of solution with albumin concentration of 100 g/l.

It can be seen that normal saline solution affects the formation of protein films absolutely differently. Concentration of albumin in normal saline solution considerably influences on the formation of protein film. It can be concluded that by dehydration from water and normal saline solutions self-organized films with different electrical properties can be produced.

2.6   Albumin Solutions on the Border of Different Substrate Surfaces

In these experiments we have used the albumin solutions at the isoelectric point (pH = 4.8). It is a well-known fact, that the surface charge of albumin can help it to form spiral structures during dehydration at the isoelectric point. The concentration of albumin was 100 g/l. The data was taken after one-day dehydration at room temperature (24 °C). The drop of this solution was placed on the border of glass-aluminium substrate surface. The results are presented in Fig. 10

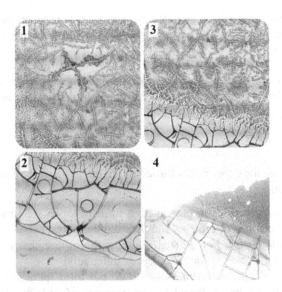

Fig. 10. Microscope photos of the protein film dehydrated on the border of glass-aluminium substrate surface. 1, 2, 3 — the solution drop on the glass part: photos are taken from the center to the border; 4 — the solution drop on the aluminium substrate surface.

Indeed, it can be seen that on the glass part of the substrate surface the characteristics of the dehydrated film are similar to Fig. 8. Unlike glass, there are no spiral structures between the cracks on the metal surface. Therefore, this leads to the conclusion that a metal surface redistributes the surface charges of proteins and no spiral structures can be formed.

2.7   Albumin Under the Influence of Various Magnetic Fields

In these experiments protein films were prepared from water solution and on the basis of a saline solution. In addition, we have conducted experiments on the influence of magnetic fields on the structure of the films. We have paid special attention to magnetic sensitivity of the protein, because this effect is not understood yet. However, number of structures formed under the influence of various magnetic fields is depicted in Fig. 11.

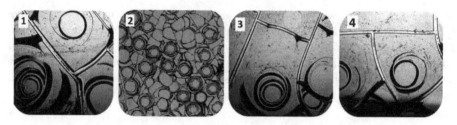

Fig. 11. Protein films: 1 – without external influence; 2 – with the application of a magnetic field 3 Oe; 3 - with the application of a magnetic field 10 Oe; 4 - with the application of a magnetic field 20 Oe.

As a result, it can be seen that weak external influences, such as a low magnetic field (3 Oe), renders a significant impact on formation of structures in albumin films.

3   Conclusions

In the final analysis, the experiments carried out in the framework of our investigation have revealed that:

- self-organization processes in protein films at dehydration are strongly influenced by the surface charges of proteins and their agglomerates;
- when the solution is dried on a metal substrate, the surface charges of proteins are redistributed and self-organization proceeds differently;
- albumin protein in normal saline solution do not form spiral structures in dehydrated films. It can be explained by the presence of sodium chloride ions that interfere in the self-organizing process and affects the albumin surface charge balance;
- near the isoelectric point albumin molecules form agglomerates. The agglomerates were measured by laser correlation spectroscopy. In dehydrated films a lot of spiral structures were detected. So it can be concluded that the surface charge strongly influence on self-organization process during dehydration;
- by dissolving the dried protein films we measured the stability of structures formed as a result of self-organization process. It was concluded that the protein films produced in the small Petri dish were more stable and protein agglomerates were bigger, while the film in the big dish was easier to dissolve and then dry again.
- weak external influences render a significant impact on proteins, in particular on albumin films.

Finally, these conclusions about self-organization processes during dehydration will be used in further experiments directed at creation of biomolecular nanoelectronic devices for communications.

References

1. Dressler, F., Fischer, S.: Connecting in-body nano communication with body area networks: challenges and opportunities of the internet of nano things. Nano Commun. Netw. **6**(2), 29–38 (2015)
2. Wang, L., et al.: Reversible near-infrared light directed reflection in a self-organized helical superstructure loaded with upconversion nanoparticles. J. Am. Chem. Soc. **136**(12), 4480–4483 (2014)
3. Velichko, E.N, Tsybin, O.Y.: Biomolecular electronics. Polytechnic University **260**(2012)
4. Wang, L., Li, Q.: Stimuli-Directing self-organized 3D liquid-crystalline nanostructures: from materials design to photonic applications. Adv. Funct. Mater. **26**(1), 10–28 (2016)
5. Yeltsina, B.N.: Microstructure and thermodynamic properties of systems of particles of various shapes - the final report on the research of M: Ural. Feder. Un-t them. The first President of Russia, 21 (2013)
6. Valueva, S.V., Borovikova, L.N.: Investigation of the process of self-organization and dependence of morphological characteristics of selenium-containing nanostructures on the basis of hydroxyethyl cellulose on the mass ratio of selenium: polymer in solution. J. Sci. Bull. Belgorod State University. Ser.: Nat. Sci. 2009
7. Gonchar, K.A., Osminkina, L.A., Sivakov, V., Lysenko, V., Timoshenko, V.Y.: Optical properties of filamentary nanostructures obtained by metal-stimulated chemical etching of plates of slightly doped crystalline silicon. Phys. Technol. Semicond. **48**(12), 1654–1659 (2014)
8. Doronin, I.S.: The device for measuring the dimensions of nanoparticles in liquid media. J. Polzunovsky Alm. **2**, 261–263 (2010)
9. Scopus data base Homepage. https://www.scopus.com. Accessed 1 June 2018
10. Deegan, R.D.: J. Phys. Rev. (61), 475 (2000)
11. Velichko, E., Baranov, M., Nepomnyashchaya, E., Cheremiskina, A., Aksenov, E.: Studies of biomolecular nanomaterials for application in electronics and communications. In: Balandin, S., Andreev, S., Koucheryavy, Y. (eds.) ruSMART 2015. LNCS, vol. 9247, pp. 786–792. Springer, Cham (2015). https://doi.org/10.1007/978-3-319-23126-6_74
12. Rapis, E.G.: Self-assembly of cluster protein films in the process of condensation (allotropic nonequilibrium non-crystalline form of it). Tech. Phys. **70**(1), 122–133 (2000)
13. Tarasevich, YuYu.: Mechanisms and models of the dehydration self-organization of biological fluids. J. Successes Phys. Sci. **174**(7), 780–784 (2004)
14. Nepomnyashchaya, E., Velichko, E., Aksenov, E.: Inverse problem of laser correlation spectroscopy for the analysis of polydisperse solutions of nanoparticles. IOP J. Phys. Conf. Ser. **769**(1), 012025 (2016)

Finally, these conclusions about self-organization processes during dehydration might be used in further experiments directed at creation of biomolecular nanoelectronic devices for communications [illegible].

References

1. [illegible]: [illegible] communication with body area [illegible] challenges and opportunities of the [illegible] nano things. Nano Commun. Netw. 6(2), 2[illegible]–28 (201[illegible])
2. Wang, L., et al.: [illegible] of light-driven [illegible] in a self-organized helical [illegible] with [illegible] upconversion nanoparticles. J. Am. Chem. Soc. 136(12), 4480–4483 (2014)
3. [illegible], M., [illegible], O.V.: [illegible]. Polytechnic University 260(2012)
4. Wang, L., Li, Q.: Stimuli-Directing self-organized 3D liquid-crystalline nanostructures: from materials design to photonic applications. Adv. Funct. Mater. 26(1), 10–28 (2016)
5. [illegible], B.[illegible]: [illegible] of thermodynamic properties of systems of particles [illegible] the doctor of [illegible]. [illegible] Press, [illegible] of Russia, 20[illegible]3
6. [illegible], S. [illegible]: [illegible] of the process of [illegible] and [illegible] dependence of [illegible] on the [illegible] of [illegible] for [illegible] of [illegible] polymer [illegible]. [illegible] Bull. [illegible] State University, Ser. Nat. [illegible] 2[illegible]
7. [illegible], K., [illegible], A., [illegible], V., [illegible], V.: [illegible] Optical properties of [illegible] nanostructures obtained by [illegible] chemical etching of [illegible] silicon. [illegible] 48(12), 1634–1639 (201[illegible])
8. [illegible]: The device for measuring the dimensions of [illegible] nanoparticles and [illegible]. [illegible] 2, [illegible]–26[illegible] (2010)
9. [illegible] [illegible] [illegible]. [illegible] 2015
10. [illegible], R.[illegible]: Phys. Rev. 76(1), 1[illegible] (20[illegible])
11. [illegible], [illegible], M., [illegible]: [illegible] studies of [illegible] for [illegible] applications in electronics and communications. In: [illegible], S., [illegible], S., [illegible], Y. (eds.) NEW2AN ruSMART 2015. LNCS, vol. 9247, pp. [illegible]. Springer, Cham (2015). https://doi.org/10.1007/978-3-319-23126-6_[illegible]
12. [illegible], [illegible]: [illegible] of [illegible] protein films in the process of condensation [illegible] nonequilibrium [illegible]. Tech. Phys. 79, [illegible] (20[illegible])
13. [illegible]: Mechanisms and models of the [illegible] self-organization of biological [illegible]. [illegible] Phys. Sci. 174(9), [illegible] (2004)
14. [illegible], E., [illegible], E.: [illegible] problem of [illegible] spectroscopy for the [illegible] of polydisperse solutions of nanoparticles. IOP J. Phys. Conf. Ser. 769(1), 012025 (2016)

Author Index